Discrete Mathematics

Discrete Mathematics

BABU RAM

Formerly Dean, Faculty of Physical Sciences,
Maharshi Dayanand University, Rohtak

Pearson

ISBN 9788131733103

First Impression, *2010*
Sixteenth Impression, 2023
Seventeenth Impression, 2024

Published by Pearson India Education Services Pvt. Ltd, CIN: U72200TN2005PTC057128.

Head Office: 1st Floor, Berger Tower, Plot No. C-001A/2, Sector 16B, Noida - 201 301, Uttar Pradesh, India.
Registered Office: Featherlite, 'The Address' 5th Floor, Survey No 203/10B, 200 Ft MMRD Road, Zamin Pallavaram, Chennai – 600 044.
Website: in.pearson.com; Email: companysecretary.india@pearson.com

Printed in India at Repro India Limited

In the Memory of

MY PARENTS

Smt. Manohari Devi and Sri. Makhan Lal

Contents

6 Lattices 281

7 Boolean Algebra 308

8 Graphs 351

List of Symbols

Symbol	Meaning
Chapter One	
$\in$	belongs to, is an element of
$\{a, b\}$	set consisting of a and b
$\notin$	does not belong to, is not an element of
R	the set of real numbers
Z	the set of integers
Z^+	the set of positive integers
N	the set of natural numbers
ϕ	empty set, null set, void set
$\subseteq$	set inclusion, is a subset of
$\not\subseteq$	is not a subset of
$\Leftrightarrow$	if and only if
P(A)	the power set of the set A, the set of all subsets of the set A
$\|A\|$	the number of elements in the set A, cardinality of a finite set A
U	universal set, universe of discourse
$\cup$	union of sets
$\cap$	intersection of sets
$B - A$	relative complement of the set A in the set B
A^c	complement of the set A
$A \Delta B$, $A \oplus B$	symmetric difference of sets A and B
$(x_1, x_2, \ldots, x_n)$	ordered n-tuple
D(R)	domain of the relation R
R(R)	range of the relation R
$a \equiv b \pmod m$	a and b are congruent modulo m, that is, m divides $(a - b)$
$\Rightarrow$	implies
a R b	a is related to b or $(a, b) \in R$
[a]	equivalence class of the element a
$A \mid R$	quotient of a set A by the equivalence relation R
g o f	composition of two functions/relations f and g
I	identity mapping

Symbol	Meaning
$\aleph_0$	aleph—naught, the cardinal number of N
$A \mid b$	a divides b
(A, R)	set A with partial order R, poset
$a \parallel b$	a and b are not comparable
R^∞	transitive closure of a relation R
D_n	the set of all divisors of a positive integer n
A*	set of all words on A
lub	least upper bound
glb	greatest lower bound
gcd	greatest common divisor
lcm	least common multiple

Chapter Two

Symbol	Meaning
$\|A\|$	the number of elements in the set A
$n\,P_m$	number of permutations of n objects taken m at a time
n!	n factorial which equals n (n – 1)…3.2.1
${}^nC_r$ or $\binom{n}{r}$	the number of r-combinations of an n element set which equals $\frac{n!}{r!(n-r)!}$
$\cap$	intersection
$\cup$	union
$\bar{A}$, A^c, A'	not A
$A \cap B = \phi$	mutually exclusive event A and B
$A \cap B'$	only A
$A' \cap B$	only B
$(A \cup B)'$ or $A' \cap B'$	neither A nor B
P(E)	probability of occurrence of event E
μ	mean of random variables
σ	variance of random variables

Chapter Three

Symbol	Meaning
n!	n-factorial
gcd (n, m)	greatest common divisor of n and m
A(z)	generating function of the numeric function (1, 1, …, 1, …)

Chapter Four

Symbol	Meaning
$p \wedge q$	conjunction of propositions p and q
$p \vee q$	disjunction of proposition p and q

$\sim p$	negation of proposition p
p XOR q, $p \oplus q$	p or q but not both (Exclusive OR)
τ	tautology
c	contradiction
$p \equiv q$	p logically equivalent to q
$p \rightarrow q$	if p then q S-4
$p \leftrightarrow q$	p if and only if q
$\forall x$	for all x
$\exists x$	there exists x

Chapter Five

F(A)	free semigroup on A
$l(w)$	length of the word w
Z_m	integer modulo m
[a]	equivalence class of a
a R b	a related to b
S/R	quotient semigroup of S by the relation R
a^{-1}	inverse of the element a
$A \cap B$	intersection of A and B
$A \cup B$	union of A and B
$H_1 H_2$	product of two subgroups H_1 and H_2
$xH, x \in G$	left coset of H in G
$Hx, x \in G$	right coset of H in G
[G : H]	number of left cosets of H in G
O(H)/O(G)	order of H divides order of G
$N \Delta G$	N is a normal subgroups of G
ker(f)	kernel (null space) of f
$G \simeq H$	G is isomorphic to H
gcd(m,n)	greatest common divisor of m and n
S_n	symmetric set of transformation of degree n
$\beta \circ \alpha$	composition of permutations α and β
(a b c)	$\begin{pmatrix} a & b & c \\ b & c & a \end{pmatrix}$
A_n	alternating group of degree n and order $\frac{n!}{2}$
$G_1 \times G_2 \times G_3$	direct product of groups G_1, G_2 and G_3

$\deg(f(x))$	degree of the polynomial f(x)
$R[x]$	polynomial ring over R

Chapter Six

$a \vee b$	LUB ({a, b}), join (sum) of a and b
$a \wedge b$	GLB ({a, b}), meet (product) of a and b
$\subseteq$	set inclusion
D_n	set of all positive divisors of n
$P(S)$	power set of the set S

Chapter Seven

B_n	switching algebra
$p(x_1, x_2, \ldots, x_n)$	Boolean polynomial
	OR gate with inputs x, y and output z=x+y
	AND gate with inputs x, y and output z=xy
	NOT gate with input x and output y=x′
	NOR gate with input x, y and output z=(x+y)′
	NAND gate with input x, y and output z=(xy)′
	Exclusive OR gate with input x, y and output z

Chapter Eight

	edge between the vertices v and w
	self-loop at the vertex v
	parallel edges between the vertices v and w
$\deg(v_i)$	degree of the vertex v_i
D_n	discrete graph with n vertices
K_n	complete graph with n vertices
$K_{m,n}$	complete bipartite graph
C_n	n-vertex cycle graph
CL_n	circular ladder graph
$G+v$	join of the graph G and a new vertex v
$\chi(G)$	chromatic number of a graph G

Symbol	Meaning
C(G)	number of components of the graph G
(T, v_0)	tree with root v_0
dist [x]	distance of x from a specified vertex
P(e)	priority value of an edge e
w(e)	weight of an edge e
diam (G)	diameter of a connected graph G
T_L	left subtree
T_R	right subtree

Chapter Nine

Symbol	Meaning
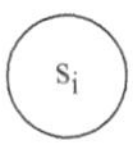	state s_i
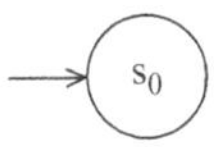	initial state
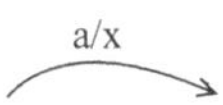	transition with input a and output x
FSM	finite state machine
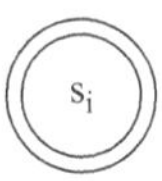	accepting state s_i
FSA	finite state automata
L(M)	language L of the machine M
NDFSA	non-deterministic finite state automaton
AC(M)	the set of strings accepted by the machine M

Chapter Ten

Symbol	Meaning
A*	the set of all strings over A
ε	empty string
L*	$\bigcup_{k=0}^{\infty} L^k$, Kleene closure of language L
N	finite set of non-terminal symbols
T	finite set of terminal symbols
$\|v\|$	length of word v

Preface

This book is intended to be a text book for a course in Discrete Mathematics at Bachelor of Technology, Master of Computer Applications and Master of Science (Mathematics) degree level. The contents of the book are, of course, mathematical but they have many applications in Computer Science and Electronics. The book is self contained and requires minimal mathematical computer science prerequisites. The concepts and basic theory presented in the text would be sufficient to understand advanced computer science applications. To make the text comprehend to readers, numerous examples, figures, tables, exercises and various C-programs have been included in the text.

The contents of the book have been divided into ten chapters. Chapter 1 consists of basic material required in the course. In this chapter, basic concepts related to sets, relations, mappings, countability of sets, partial order relation, Hashing function and principle of mathematical induction have been discussed. Chapter 2 deals with counting methods. The topics of Permutations, Combinations, Pigeonhole Principle and Probability Theory have been discussed in detail in this chapter. Recurrence relations and their solutions with various methods have been dealt in Chapter 3. Chapter 4 of the book deals with logic. The contents of the chapter increase the ability of the students to think abstractly. This topic increases the power of reasoning of the students. Chapter 5 is devoted to algebraic structures—semigroups, monoids, groups, rings, polynomials and group codes. These structures have wide applications in communication engineering. In Chapter 6, various characterizations of a lattice structure are discussed. Chapter 7 deals with Boolean algebra, the study of which is useful in switching algebra, circuit theory and digital computer designing. Graphs and trees have been studied in Chapter 8. Shortest path problems, matrix representation of graphs, coloring problem, tree-representation of algebraic expressions, transport networks and other important concepts have been dealt with in this chapter. In Chapter 9, topics on computability like finite state machines, finite state automata, Moore and Mealy machines and their equivalence have been discussed. In Chapter 10, various types of grammar and their corresponding languages have been dealt with.

Towards the end, answers and hints to various problems of exercises and appendix containing C-programming of some algorithms have been presented.

My thanks are due to all of my colleagues for extending co-operation and helpful discussions. This book would not have been possible without the active, moral and emotional support of my family. My daughter Indu, who is a hardware engineer, and son Aman Kumar, Software Engineer, offered their wise comments on the contents of the books. My wife, Meena, stood by me like a pillar of strength and courage at every level during this period.

I am thankful to Shri Jai Parkash Prajapati for typing the manuscript excellently. Special thanks are due to Thomas Mathew Rajesh, Anita Yadav, Anant Kumar and Vamanan Namboodiri of Pearson Education for their constructive support.

BABU RAM

1 Sets, Relations and Functions

In this chapter, we introduce the notions of sets, relations and functions which are the basic tools of discrete mathematics. The set theory was founded by Georg Cantor in late nineteenth century. The concept of a set appears in all mathematical structures.

1.1 SETS

A set is an undefined term of set theory just as sentence, true and false are undefined terms of logic. According to Georg Cantor, a **set** may be viewed as a well-defined collection of objects, called the **elements** or **members** of the set. The term "well defined" means that it is possible to decide if a given object belongs to the collection or not.

The sets are denoted by capital letters such as A, B, C, whereas the elements of a set are denoted by lowercase letters such as a, b, c. We indicate the fact that a is an element of a set A by writing $a \in A$ (to be read as "a belongs to A"). Similarly, we indicate the fact that a is not an element of a set A by writing $a \notin A$ (to be read as "a does not belong to A").

EXAMPLE 1.1

Let a, b and c be the members of a set X. Then it is described by listing the elements of the set between braces and separated by commas. Thus,

$$X = \{a, b, c\}.$$

It may be mentioned here that **the order in which the** elements of a set are listed is not important. Thus $\{a, b, c\}$, $\{b, a, c\}$, $\{c, a, b\}$, $\{b, c, a\}$ are the representations of the same set.

EXAMPLE 1.2

The set

$$\mathbf{R} = \{x : x \text{ is a real number}\}$$

is called a set of real numbers.
Similarly,

(i) $\mathbf{Z} = \{x : x \text{ is an integer}\}$ represents the set of all integers: … −4, −3, −2, −1, 0, 1, 2, 3, 4, …
(ii) $\mathbf{Z}^+ = \{x : x \text{ is a positive integer}\}$ represents the set of all positive integers 1, 2, 3, …
(iii) $\mathbf{N} = \{x : x \text{ is a positive integer or zero}\}$ represents the set consisting of 0, 1, 2, 3,…
(iv) The set having no element is represented by ϕ or $\{\ \}$ and is called the **empty** (or **null** or **void**) set.

EXAMPLE 1.3

Describe the set

$$\{x \in \mathbf{Z} : -3 < x < 8\}.$$

Solution.

$\{x \in \mathbf{Z} : -3<x<8\}$ is the set of all integers lying between -3 and 8. Thus, the given set is

$$\{-2, -1, 0, 1, 2, 3, 4, 5, 6, 7\}.$$

EXAMPLE 1.4

Describe the set

$$\{x : x \text{ is a real number satisfying } x^2=-1\}.$$

Solution.

We know that the square of a real number is always non-negative. Therefore, there is no element in the above set. Hence

$$\{x : x \text{ is a real number satisfying } x^2=-1\}=\phi \text{ (empty set)}.$$

Definition 1.1

Let A and B be sets. Then A is called a **subset** of B, written as $A \subseteq B$, if and only if every element of A is also an element of B.
Thus,

$$A \subseteq B \Leftrightarrow \forall a, \text{ if } a \in A, \text{ then } a \in B.$$

Clearly, a set A is not a subset of a set B, written as $A \not\subseteq B$, if and only if there is at least one element of A that is not an element of B.
Thus,

$$A \not\subseteq B \text{ if and only if there exist } a \in A \text{ such that } a \notin B.$$

For example, if

$$A=\{6, 5, 8) \text{ and } B=\{2, 3, 6, 7, 5, 8\},$$

then A is a subset of B.

Further, since every element in a set A is in A, it follows that any set A is a subset of itself.

Definition 1.2

Let A and B be sets. Then A is said to be a **proper subset** of B, written as $A \subset B$, if and only if, every element of A is in B but there is at least one element of B that is not in A.

For example, the set $\{6, 5, 8\}$ is a proper subset of the set $\{2, 3, 6, 7, 5, 8\}$.

Definition 1.3

Two sets A and B are called **equal** if every element of A is in B and every element of B is in A.

Thus, $A=B \Leftrightarrow A \subseteq B$ and $B \subseteq A$. Obviously, two sets are equal if they have the same elements.

EXAMPLE 1.5

Let

$$A=\{1, 2, 3, 4\}, B=\{x : x \text{ is a positive integer and } x^2<18\}.$$

Is $A=B$?

Solution.

We have

$$A=\{1, 2, 3, 4\} \text{ and } 1^2=1, 2^2=4, 3^2=9, 4^2=16,$$

but the square of any other positive integer is not less than 18.

Thus,

$$B = \{1, 2, 3, 4\} \text{ and so } A = B.$$

EXAMPLE 1.6

Let $A = \{3, -6\}$, $B = \{x : x^2 + 3x - 18 = 0\}$. Is $A = B$?

Solution.

We have $A = \{3, -6\}$ and note that 3 and -6 are roots of the polynomial $x^2 + 3x - 18$. Hence $B = \{3, -6\}$ and so $A = B$.

1.1.1 Venn Diagram

Let A and B be the sets represented as regions in the plane. Then the diagrams used to show relationship between A and B are called **Venn Diagram** after the British mathematician John Venn. For example, the relationship $A \subseteq B$ is shown in one of the following ways:

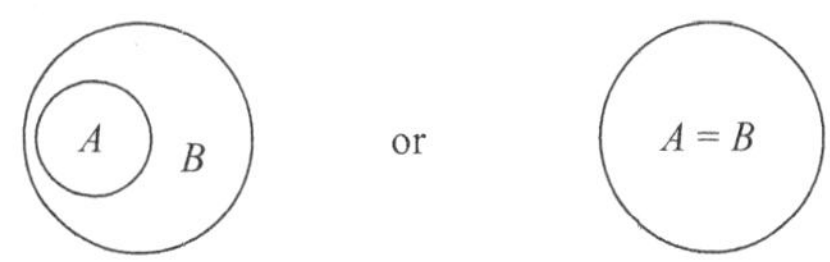

Figure 1.1 ($A \subseteq B$).

On the other hand, the relationship $A \not\subseteq B$ can be represented in the following three ways:

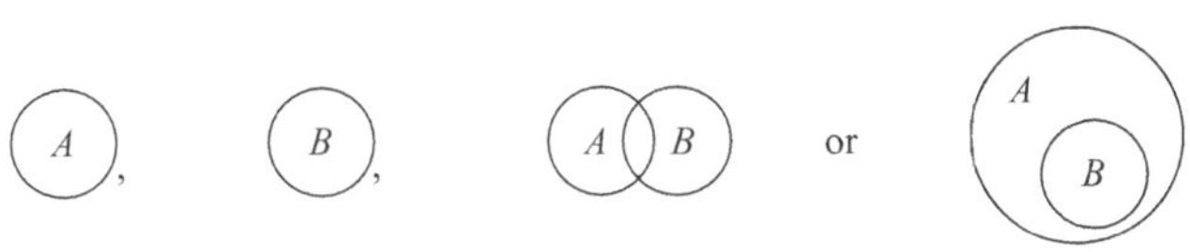

Figure 1.2 ($A \not\subseteq B$).

If A, B and C are sets and if $A \subseteq B$ and $B \subseteq C$, then $A \subseteq C$. This property is known as **Transitive Property of subsets**.

Definition 1.4

The set of all subsets (proper or not) of a set A, denoted by $P(A)$, is called the **power set** of A.

EXAMPLE 1.7

Let

$$A = \{a, b, c, d\},$$

then the members of $P(A)$ are

ϕ, $\{a\}$, $\{b\}$, $\{c\}$, $\{d\}$, $\{a, b\}$, $\{a, c\}$, $\{a, d\}$, $\{b, c\}$, $\{b, d\}$, $\{c, d\}$, $\{a, b, c\}$, $\{a, b, d\}$, $\{a, c, d\}$, $\{b, c, d\}$, $\{a, b, c, d\}$.

We note that except $\{a, b, c, d\}$, all other members of $P(A)$ are proper subsets of A.

Also, we note that

Number of elements in A, denoted by $|A|$ is 4

Number of member of $P(A)$, denoted by $|P(A)| = 2^4 = 16$

We prove this observation in the form of the following theorem.

Theorem 1.1

If $|A|=n$, then

$$|P(A)|=2^n.$$

Proof. We shall prove this theorem using induction on n. If $n=0$, then A is the empty set. The only subset of the empty set is the empty set itself. Thus,

$$|P(A)|=1=2^0=2^n.$$

Thus the result is true for $n=0$.

Assume that the result holds for n. We shall show that it holds for $n+1$. So, let A be a set with $n+1$ elements. Let $a \in A$. We claim that exactly half of the subsets of A contain a and exactly half of the subsets of A do not contain a. In fact, each subset S of A that contain a can be paired uniquely with the subset obtained by removing a from S. For example, for $A=\{a, b, c, d\}$, we have the following situation:

Subsets of A that contain a	*Subsets of A that do not contain a*
{a}	ϕ
{a,b}	{b}
{a,c}	{c}
{a,d}	{d}
{a,b,c}	{b,c}
{a,b,d}	{b,d}
{a,c,d}	{c,d}
{a,b,c,d}	{b,c,d}

Thus, exactly half of the subsets of A contain a and exactly half of the subsets of A do not contain a. Let B be the set obtained from A by removing a. Then B has n elements. Therefore, by induction hypothesis,

$$|P(B)|=2^n$$

But the subsets of B are the subsets of A that do not contain a. Therefore,

$$|P(B)|=\frac{|P(A)|}{2} \quad \text{(by the above argument)},$$

that is,

$$|P(A)|=2\,|P(B)|=2 \cdot 2^n=2^{n+1}.$$

Hence, the theorem holds for $n+1$. Therefore, by the Principle of Mathematical Induction, the theorem holds for all $n \geq 0$.

Second Proof. Let

$$A=\{a_1, a_2, \ldots, a_n\}.$$

A subset of A can be constructed in n successive steps:

Pick or do not pick a_1,
Pick or do not pick a_2,

____

____

Pick or do not pick a_n.

Thus, each step can be performed in two ways. Hence, the number of possible subsets of A is

$$\underbrace{2 \cdot 2 \cdot 2 \cdot \ldots \cdot 2}_{n \text{ factors}} = 2^n$$

that is, $|P(A)| = 2^n$.

Theorem 1.2

Let A and B be two sets. If $A \subseteq B$, then $P(A) \subseteq P(B)$.

Proof. Suppose $X \in P(A)$. Then $X \subseteq A$ by the definition of power set. But $A \subseteq B$. Hence, by transitive property of inclusion of sets, $X \subseteq B$. Hence, $X \in P(B)$, again by the definition of power set. It follows, therefore, that $P(A) \subseteq P(B)$.

Definition 1.5

Suppose we are dealing with sets all of which are subsets of a set U. Then this set U is called a **universal set** or a **universe of discourse** or a **universe.**

Definition 1.6

Let A and B be subsets of a universal set U. Then the **union** of A and B, denoted by $A \cup B$, is the set of all elements a in U such that a is in A or is in B.

Symbolically,

$$A \cup B = \{a \in U : a \in A \text{ or } a \in B\}$$

Definition 1.7

Let A and B be subsets of a universal set U. Then the **intersection** of A and B, denoted by $A \cap B$, is the set of all elements a of U such that $a \in A$ and $a \in B$.

Symbolically,

$$A \cap B = \{a \in U : a \in A \text{ and } a \in B\}$$

Definition 1.8

Let A and B be the subsets of a universal set U. Then the **difference** B minus A (or **relative complement** of A in B), denoted by B – A, is the set of all elements a in U such that $a \in B$ and $a \notin A$.

Symbolically,

$$B - A = \{a \in U : a \in B \text{ and } a \notin A\}.$$

Definition 1.9

Let A be a subset of a universal set U. Then **complement** of A, denoted by A^c, is the set of all elements a in U such that a is not in A.

Symbolically,

$$A^c = \{a \in U : a \notin A\}.$$

The union, intersection, difference and complement of sets can be represented by the following Venn diagrams:

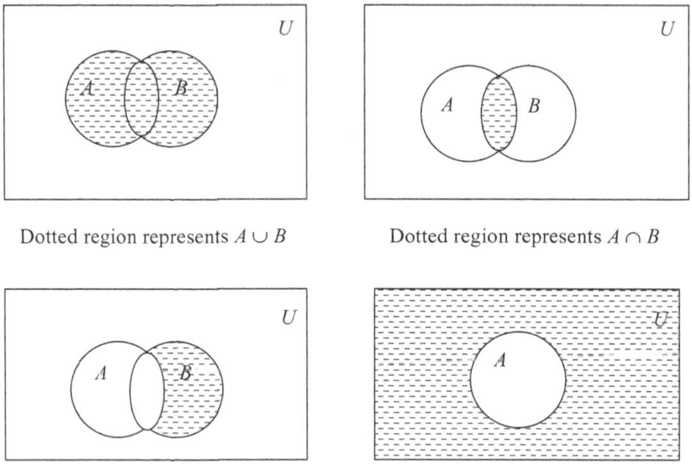

Dotted region represents $A \cup B$ Dotted region represents $A \cap B$

Dotted region represents $B - A$ Dotted region represents A^c

Figure 1.3

EXAMPLE 1.8

Let

$$A = \{1, 2, 4\}, B = \{4, 5, 6\}.$$

Find $A \cup B, A \cap B, A - B, B - A$.

Solution.

We have

$$A \cup B = \{1, 2, 4, 5, 6\}, A \cap B = \{4\}$$
$$A - B = \{1, 2\}, B - A = \{5, 6\}.$$

Also, we note that

$$A \cap B^c = \{1, 2, 4\} \cap \{4, 5, 6\}^c$$
$$= \{1, 2, 4\} \cap \{1, 2\} = \{1, 2\}.$$

Thus,

$$A - B = A \cap B^c.$$

It can be shown easily that for sets A and B,

(a) $A \cap B \subseteq A$ and $A \cap B \subseteq B$
(b) $A \subseteq A \cup B$ and $B \subseteq A \cup B$.

Definition 1.10

Let $A_1, A_2, \ldots, A_n$ be n distinct sets. A **fundamental product** of these sets is the set

$$A_1^* \cap A_2^* \cap \ldots \cap A_n^*,$$

where A_i^* is either A_i or A_i^c.

EXAMPLE 1.9

Let A, B and C be three sets. Determine all possible fundamental products of these sets.

Solution.

We note that the following eight fundamental products exist for these three sets—A, B and C:

$P_1 = A \cap B \cap C$, $P_2 = A \cap B \cap C^c$, $P_3 = A \cap B^c \cap C$
$P_4 = A \cap B^c \cap C^c$, $P_5 = A^c \cap B \cap C$, $P_6 = A^c \cap B \cap C^c$
$P_7 = A^c \cap B^c \cap C$, $P_8 = A^c \cap B^c \cap C^c$

We further note that

1. For n sets, there are 2^n fundamental products.
2. The fundamental products are mutually disjoint.
3. The universal set U is the union of all the fundamental products.

Definition 1.11

The **symmetric difference** between two sets A and B is the set containing all the elements that are in A or in B but not in both.

A symmetric difference of two sets A and B is denoted by $A \Delta B$ or $A \oplus B$.

Thus,

$$A \Delta B = (A \cup B) - (A \cap B).$$

EXAMPLE 1.10

Let $A = \{1, 2, 3\}$, $B = \{2, 3, 4, 5\}$. Find $A \Delta B$.

Solution.

We know that

$$A \Delta B = (A \cup B) - (A \cap B).$$

But,

$$A \cup B = \{1, 2, 3, 4, 5\}, A \cap B = \{2, 3\}.$$

Therefore,

$$A \Delta B = \{1, 2, 3, 4, 5\} - \{2, 3\} = \{1, 4, 5\}.$$

Also, we note that

$$A - B = \{1\} \text{ and } B - A = \{4, 5\}.$$

Therefore,

$$(A - B) \cup (B - A) = \{1\} \cup \{4, 5\} = \{1, 4, 5\}.$$

Thus,

$$A \Delta B = (A - B) \cup (B - A).$$

This result holds (to be proved later on in this chapter) for all sets A and B. Hence,

"The symmetric difference between two sets A and B is defined by

$$A \Delta B = (A - B) \cup (B - A)."$$

1.2 ALGEBRA OF SETS

The operations of union, intersection and complement on sets satisfy various laws or identities. We state these identities in the form of a theorem.

Theorem 1.3

Let A, B and C be the subsets of a universal set U. Then the following identities hold:

1. **Commutative Laws:**
   (a) $A \cup B = B \cup A$, (b) $A \cap B = B \cap A$
2. **Associative Laws:**
   (a) $A \cup (B \cup C) = (A \cup B) \cup C$ (b) $A \cap (B \cap C) = (A \cap B) \cap C$
3. **Distributive Laws:**
   (a) $A \cup (B \cap C) = (A \cup B) \cap (A \cup C)$
   (b) $A \cap (B \cup C) = (A \cap B) \cup (A \cap C)$
4. **Idempotent Laws:**
   (a) $A \cup A = A$, (b) $A \cap A = A$
5. **Properties of Universal set:**
   (a) $A \cup U = U$, (b) $A \cap U = A$
6. **Absorption Laws:**
   (a) $A \cup (A \cap B) = A$, (b) $A \cap (A \cup B) = A$
7. **Complement Law:**
   $A \cap A^c = \phi$
8. **Double Complement Law:**
   $(A^c)^c = A$
9. **De Morgan's Law:**
   (a) $(A \cup B)^c = A^c \cap B^c$
   (b) $(A \cap B)^c = A^c \cup B^c$
10. **Alternate representation for set difference:**

    $A - B = A \cap B^c$

Proof. We shall prove De Morgan Laws only. The other identities can be established similarly by the readers themselves. So let $x \in (A \cup B)^c$. By the definition of complement, $x \notin A \cup B$ and so $x \notin$ A and $x \notin$ B, that is, $x \in A^c$ and $x \in B^c$. This implies that $x \in A^c \cap B^c$. Thus,

$$(A \cup B)^c \subseteq A^c \cap B^c.$$

Conversely, suppose that $x \in A^c \cap B^c$. Thus, $x \in A^c$ and $x \in B^c$, that is, $x \notin A$ and $x \notin B$. This implies that $x \notin A \cup B$ and so $x \in (A \cup B)^c$. Thus,

$$A^c \cap B^c \subseteq (A \cup B)^c.$$

Hence,

$$(A \cup B)^c = A^c \cap B^c,$$

which proves 9(a). Further, 9(b) follows by the **principle of duality** which states that "if an equation E is an identity then its dual E^* is also an identity." Also, we know that dual of a statement E in set theory in obtained by replacing $\cup$, $\cap$, U and ϕ in E by $\cap$, $\cup$, ϕ and U respectively.

Definition 1.12

Two sets A and B are called **disjoint** if and only if they have no element in common.

Thus, symbolically,

$$A \text{ and } B \text{ are disjoint} \Leftrightarrow A \cap B = \phi.$$

EXAMPLE 1.11

Let A and B be two sets. Show that $A - B$ and B are disjoint.

Solution.

Suppose, on the contrary that $A-B$ and B are not disjoint. Thus $(A-B) \cap B \neq \phi$ and so there is an element $x \in (A-B) \cap B$. This implies,

$$x \in A-B \text{ and } x \in B$$
$$\Rightarrow x \in A, x \notin B \text{ and } x \in B.$$

Thus $x \in B$ and also $x \notin B$, which is a contradiction. Hence our supposition is wrong and $A-B$ and B are disjoint.

Definition 1.13

Sets $A_1, A_2, \ldots, A_n$ are said to be **mutually disjoint** (or **pairwise disjoint** or **non-overlapping**) if and only if for all $i, j=1, 2, \ldots, n$.

$$A_i \cap A_j = \phi,\ i \neq j.$$

Theorem 1.4

Let A and B be two sets. Then

$$A \,\Delta\, B = (A-B) \cup (B-A).$$

Proof. We know that

$$\begin{aligned} A \,\Delta\, B &= (A \cup B) - (A \cap B) \\ &= (A \cup B) \cap (A \cap B)^c \text{ (using alternate expression for set difference)} \\ &= (A \cup B) \cap (A^c \cup B^c) \text{ (using De-Morgan Law)} \\ &= (A \cap A^c) \cup (A \cap B^c) \cup (B \cap A^c) \cup (B \cap B^c) \text{ (using distributive law)} \\ &= \phi \cup (A \cap B^c) \cup (B \cap A^c) \cup \phi \text{ (using complement law)} \\ &= (A \cap B^c) \cup (B \cap A^c) \text{ (property of empty set)} \\ &= (A-B) \cup (B-A) \text{ (using alternate representation of set difference)} \end{aligned}$$

Definition 1.14

A set A is called **finite** if it has n distinct elements, where $n \in \mathbf{N}$. In this case, n is called the **cardinality** of A and is denoted by $|A|$.

Definition 1.15

A set that is not finite is called **infinite.**

For example, $\{1, 2, 4\}$ is a finite set whereas the set $\mathbf{N}$ of natural number is infinite.

Theorem 1.5

If A and B are disjoint finite sets, then $A \cup B$ is finite and

$$|A \cup B| = |A| + |B|.$$

Proof. To count the elements of $A \cup B$, we first count those element which are in A. There are $|A|$ of these elements. The remaining elements of $A \cup B$ are those that belong to B but not to A. Since A and B are disjoint, no element of B is in A. So the number of elements in B but not in A is $|B|$. Hence,

$$|A \cup B| = |A| + |B|.$$

Theorem 1.6 (Addition Principle)

If A and B are finite sets, then $A \cup B$ and $A \cap B$ are finite and

$$|A \cup B| = |A| + |B| - |A \cap B|.$$

Proof. If A and B are finite, then $A \cap B$ and $A \cup B$ are certainly finite. Further, if we count first the number of elements in A and then count the number of elements in B, then every element in $A \cap B$ would be counted twice, once in A and once in B. Hence,

$$|A \cup B| = |A| + |B| - |A \cap B|.$$

Geometrically, we would have a region as shown in Figure 1.4:

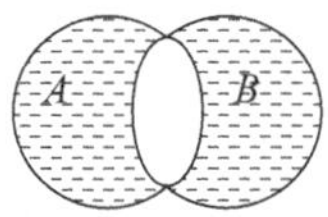

Dotted region represents $(A \cup B) - (A \cap B)$

Figure 1.4

Because of this figure, the addition principle is also called the **inclusion–exclusion principle.**

For three sets, the situation is somewhat complicated. The corresponding figure for three set addition principle becomes

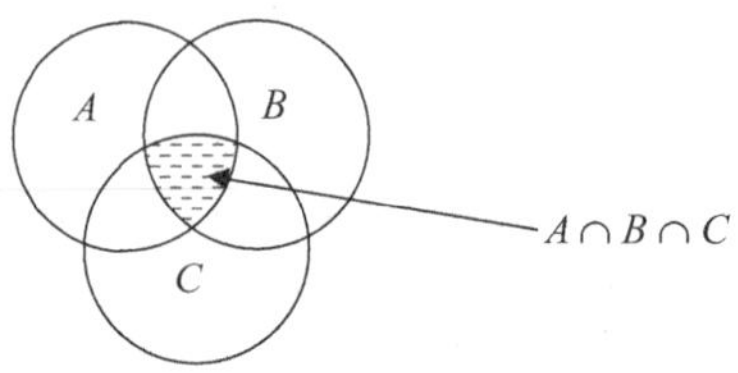

Figure 1.5

and we have

Theorem 1.7

Let A, B and C be finite sets. Then,

$$|A \cup B \cup C| = |A| + |B| + |C| - |A \cap B| - |B \cap C| - |A \cap C| + |A \cap B \cap C|$$

EXAMPLE 1.12

In a survey of 80 people, it was observed that 30 people read *Hindustan Times*, 25 read *Times of India*, 28 read *The Tribune*, 15 read both *Hindustan Times* and *The Tribune*, 18 read both *Times of India* and *The Tribune*, 20 read both *Hindustan Times* and *Times of India* and 5 read all three newspapers. Find

(a) The number of people who read at least one of the three newspapers
(b) The number of people who read no newspaper at all.

Solution.

(a) We have to find the number of people who read **at least one of the three newspapers.** In other words, we want to find $|A \cup B \cup C|$, where

A represents *Hindustan Times*
B represents *Times of India*
C represents *The Tribune*.

Therefore,

$$|A \cup B \cup C| = |A| + |B| + |C| - |A \cap B| - |B \cap C|$$
$$- |A \cap C| + |A \cap B \cap C|$$

Therefore, we have

$$|A \cup B \cup C| = 30 + 25 + 28 - 20 - 18 - 15 + 5 = 35.$$

(b) The number of people who read no newspaper at all $= 80 - 35 = 45$.

EXAMPLE 1.13

A software company requires 60 engineers to perform Java programming jobs and 35 engineers to perform c++ programming jobs. Out of this requirement, 15 are expected to perform both types of jobs. How many engineers have to be appointed for the purpose?

Solution.

Let A be the set of engineers to perform Java programming and B be the set of engineers to perform c++ programming. Then,

$$|A| = 60,\ |B| = 35 \text{ and } |A \cap B| = 15$$

so that,

$$|A \cup B| = |A| + |B| - |A \cap B|$$
$$= 60 + 35 - 15 = 80 \text{ engineers.}$$

Definition 1.16

Let $x_1, x_2, \ldots, x_n : n \in N$ be elements (not necessarily distinct). The **ordered n-tuple** $(x_1, x_2, \ldots, x_n)$ consists of $x_1, x_2, \ldots, x_n$ together with the ordering : first x_1, then x_2, and so on up to x_n.

An ordered 2-tuple is called **an ordered pair** and an ordered 3-tuple is called a **triple.**

Two ordered n-tuples $(x_1, x_2, \ldots, x_n)$ and $(y_1, y_2, \ldots, y_n)$ are called **equal** if and only if $x_1 = y_1$, $x_2 = y_2, \ldots, x_n = y_n$.

Definition 1.17

Let A and B be two sets. Then the **Cartesian product** of A and B, denoted by $A \times B$, is defined as the set of all ordered pairs (a, b), where $a \in A$ and $b \in B$.

Thus,

$$A \times B = \{(a, b) : a \in A, b \in B\}.$$

If $A = \phi$ or $B = \phi$, then clearly $A \times B = \phi$. Hence,

$$A \times \phi = \phi.$$

If we consider the Cartesian product of a set A with itself, then $A \times A$ is denoted by A^2. The set of elements (a, a) of $A \times A$ is called the **diagonal of** $A \times A$.

EXAMPLE 1.14

If $A = \{1, 2\}$, then

$$A \times A = A^2 = \{(1, 1), (1, 2), (2, 1), (2, 2)\}$$

EXAMPLE 1.15

Let

$$A\ \{1, 2, 3\} \text{ and } B = \{a, b\}.$$

Then,

$$A \times B = \{(1, a), (1, b), (2, a), (2, b), (3, a), (3, b)\}.$$

We can represent $A \times B$ graphically as

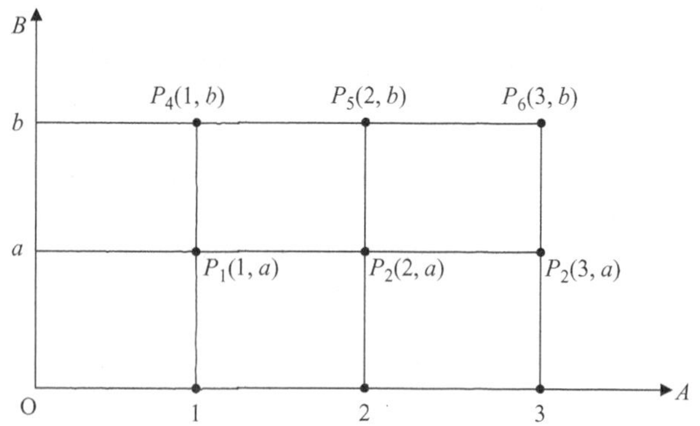

Figure 1.6

Similarly,

$$B \times A = \{(a, 1), (a, 2), (a, 3), (b, 1), (b, 2), (b, 3)\}.$$

Clearly,

$$A \times B \neq B \times A.$$

Also, we note that if A has m elements and B has n elements, then $A \times B$ has mn elements.

Definition 1.18

Let A and B be two sets. Then, a subset R of $A \times B$ is called a **relation in** A **and** B.

Given an ordered pair (a, b) in $A \times B$, a is related to b by R, written as $a\ R\ b$ if, and only if $(a, b) \in R$.

If $B = A$, then R is called a **relation on** A.

The set of first components of pairs in R is called **relation domain** of R.

Similarly, the set of last components of pairs in R is called **relation range** of R.

Thus,

$$\text{Relation domain of } R = \{a : (a, b) \in R\},$$
$$\text{Relation range of } R = \{b : (a, b) \in R\}.$$

If we denote the domain of R by $D\ (R)$ and the range by $R\ (R)$, then clearly

$$D\ (R) \subseteq A \text{ and } R\ (R) \subseteq B.$$

If R is a relation of A on B, then R^{-1}, the relation of B on A is defined by

$$R^{-1} = \{(b, a) : (a, b) \in R\}.$$

EXAMPLE 1.16

Let $A = \{1, 2, 3\}$ be a set. Then,

$$A \times A = \{(1, 1), (1, 2), (1, 3), (2, 1), (2, 2), (2, 3), (3, 1), (3, 2), (3, 3)\}.$$

and

$$R=\{(1, 1), (3, 2), (2, 3), (2, 1)\}$$

is a relation on A.

Moreover,

$$R^{-1}=\{(1, 1), (2, 3), (3, 2), (1, 2)\},$$
$$D(R)=\{1, 2, 3\},$$
$$R(R)=\{1, 2, 3\}.$$

Definition 1.19

A relation R on a set A is said to be an **equivalence relation** if it satisfies the following properties:

(i) Reflexivity:

$$(a, a) \in R \text{ for all } a \in A.$$

(ii) Symmetry:

$$(a, b) \in R \Rightarrow (b, a) \in R \text{ for all } a, b \in A$$

(iii) Transitivity:

$$(a, b) \in R \text{ and } (b, c) \in R \Rightarrow (a, c) \in R \text{ for all } a, b, c \in \text{A}.$$

Thus,

A relation R on a set A is called an equivalence relation if it is reflexive, symmetric and transitive.

EXAMPLE 1.17

If $A=\{1, 2, 3\}$, then

$$A\times A=\{(1, 1), (1, 2), (1, 3), (2, 1), (2, 2), (2, 3), (3, 1), (3, 2), (3, 3)\}.$$

We note that

$$R_1=\{(1, 1), (1, 2), (2, 1), (2, 2), (3, 3)\},$$

is an equivalence relation.

But, the relation defined by

$$R_2=\{(1, 2), (2, 1), (2, 2), (3, 1), (3, 3)\}$$

is **not** an equivalence relation of A since $1 \in A \not\Rightarrow (1, 1) \in R_2$. Also,

$$(3, 1) \in R_2 \not\Rightarrow (1, 3) \in R_2.$$

Similarly,

$$R_3=\{(1, 1), (2, 2), (3, 3), (1, 2)\}$$

is **not** an equivalence relation on A since $(1, 2) \in R_3 \not\Rightarrow (2, 1) \in R_3$.

EXAMPLE 1.18

Let X be a set and

$$S=\{A : A\subseteq X\}.$$

Then a subset

$$R=\{(A, B) : A\subseteq B; A, B \in S\}$$

of $S\times S$ is not an equivalence relation. Thus,

"Set inclusion is not an equivalence relation".

In fact, we observe that

(i) Every set is a subset of itself, that is $A \subseteq A$.

Therefore, $\subseteq$ is a reflexive relation.

(ii) If $A \subseteq B$, $B \subseteq C$, then $A \subseteq C$. Therefore, the relation $\subseteq$ is transitive.

(iii) If $A \subseteq B$, then it is not necessary that $B \subseteq A$. Thus,

$$(A, B) \in R \nRightarrow (B, A) \in R.$$

Hence, the relation $\subseteq$ is **not symmetric**.

It follows, therefore, that the relation $\subseteq$ is not an equivalence relation.

EXAMPLE 1.19

Let T be a set of triangles in a plane, and define R as the set

$$R = \{(a, b);\ a, b \in T, a \text{ is congruent to } b\}.$$

Is R is an equivalence relation?

Solution.

If a and b are triangles in a plane, then $(a, b) \in R$, if and only if a is congruent to b. Also we know that two triangles are congruent if the area of one triangle is equal to the area of the other, that is, $\Delta a = \Delta b$, if the triangles are congruent. We observe that

1. Reflexivity: Since every triangle is congruent to itself, therefore, $(a, a) \in R$ for all $a \in T$. Hence R is reflexive.
2. Symmetry

$$\begin{aligned}(a, b) \in R &\Rightarrow a \text{ is congruent to } b\\ &\Rightarrow \Delta a = \Delta b\\ &\Rightarrow \Delta b = \Delta a\\ &\Rightarrow b \text{ is congruent to } a\\ &\Rightarrow (b, a) \in R \text{ for all } a, b \in T.\end{aligned}$$

Hence, R is symmetric.

3. Transitivity: If

$$\begin{aligned}(a, b) \in R, (b, c) \in R &\Rightarrow a \text{ is congruent to } b \text{ and } b \text{ is congruent to } c.\\ &\Rightarrow \Delta a = \Delta b \text{ and } \Delta b = \Delta c\\ &\Rightarrow \Delta a = \Delta c\\ &\Rightarrow a \text{ is congruent to } c\\ &\Rightarrow (a, c) \in R \text{ for all } a, b \in T.\end{aligned}$$

Hence, R is transitive.

Thus, R is a reflexive, symmetric and transitive relation and so is an equivalence relation.

EXAMPLE 1.20

If R and S are equivalence relations in the set X, prove that $R \cap S$ is an equivalence relation.

Solution.

Here, we note that

1. Since R and S are equivalence relations, for all $a \in X$, $(a, a) \in R$ and $(a, a) \in S$. Thus,

$$\forall a \in X, (a, a) \in R \cap S.$$

Hence, $R \cap S$ is reflexive.

2. For $a, b \in X$,

$$\begin{aligned}(a, b) \in R \cap S &\Rightarrow (a, b) \in R \text{ and } (a, b) \in S\\ &\Rightarrow (b, a) \in R \text{ and } (b, a) \in S\\ &\quad \text{(because } R \text{ and } S \text{ are equivalence relations)}\\ &\Rightarrow (b, a) \in R \cap S.\end{aligned}$$

Hence, $R \cap S$ is symmetric.

3. For $a, b, c \in X$, let $(a, b) \in R \cap S$ and $(b, c) \in R \cap S$. Then $(a, b) \in R$ and $(a, b) \in S$; $(b, c) \in R$ and $(b, c) \in S$. Therefore,

$$\begin{aligned}&(a, b) \in R, (b, c) \in R \text{ and } (a, b) \in S, (b, c) \in S\\ &\Rightarrow (a, c) \in R \text{ and } (a, c) \in S \ (R \text{ and } S \text{ being equivalent relations})\\ &\Rightarrow (a, c) \in R \cap S.\end{aligned}$$

Hence, $R \cap S$ is an equivalence relation.

Definition 1.20

Let m be a fixed integer. Two integers a and b are said to be **congruent modulo** m, written as $a \equiv b \pmod m$, **if and only if** m **divides** $(a-b)$. Thus, $a \equiv b \pmod m$ if and only if $(a-b)=k\,m$ for some integer k.

EXAMPLE 1.21

Show that the relation of congruence modulo m, defined on the set **Z** of integers by $a \equiv b \pmod m$ is an equivalence relation.

Solution.

Consider the relation

$$R_m = \{ (a, b) : a \equiv b \pmod m,\ a, b \in \mathbf{Z}\}.$$

We note that

1. Reflexivity: Since $a-a=0$ is divisible by m, we have

$$a \equiv a \pmod m$$

that is,

$$(a, a) \in R_m.$$

Thus, R_m is Reflexive.

2. Symmetry: Let

$$a \equiv b \pmod m.$$

Therefore, $a-b$ is divisible by m. But this implies that $b-a=-(a-b)$ is also divisible by m. Hence,

$$b \equiv a \pmod m.$$

Thus,

$$(a, b) \in R_m \Rightarrow (b, a) \in R_m.$$

Hence R_m is Symmetric.

3. Transitivity: Let $(a, b) \in R_m$, $(b, c) \in R_m$. Then,

$$\begin{aligned}a \equiv b \pmod m \text{ or } a-b=k\,m\\ b \equiv c \pmod m \text{ or } b-c=l\,m,\end{aligned}$$

for some integers k and l.

This gives

$$\begin{aligned}(a-b)+(b-c)&=(k+l)m \qquad \text{for some integer } (k+l)\\ &\Rightarrow a-c=(k+l)\,m\\ &\Rightarrow m \text{ divides } (a-c)\\ &\Rightarrow a\equiv c \pmod m.\end{aligned}$$

Hence $(a, c) \in R_m$. Thus,

$$(a, b) \in R_m, (b, c) \in R_m \Rightarrow (a, c) \in R_m.$$

Hence R_m is transitive.

This proves that R_m is reflexive, symmetric and transitive and hence is an equivalence relation.

Definition 1.21

A relation R is called **circular** if $(a, b) \in R, (b, c) \in R \Rightarrow (c, a) \in R$.

EXAMPLE 1.22

Show that a relation is reflexive and circular if and only if it is reflexive, symmetric and transitive (equivalence relation).

Solution.

Suppose first that R is a relation which is reflexive and circular. Therefore $(a, a) \in R$ and $(a, b) \in R$, $(b, c) \in R \Rightarrow (c, a) \in R$. Since $(c, a) \in R$ and $(a, a) \in R$ and R is circular, we have $(a, c) \in R$. Thus,

$$(c, a) \in R \Rightarrow (a, c) \in R,$$

which shows that R is symmetric.

Further,

$$\begin{aligned}(a, b) \in R, (b, c) \in R &\Rightarrow (c, a) \in R, \text{ since } R \text{ is circular}\\ &\Rightarrow (a, c) \in R, \text{ since } R \text{ is symmetric (proved above).}\end{aligned}$$

Hence R is transitive.

Thus R is reflexive, symmetric and transitive and so is an equivalence relation.

Conversely, suppose R is reflexive, symmetric and transitive. We note that

$$\begin{aligned}(a, b) \in R, (b, c) \in R &\Rightarrow (a, c) \in R \text{ (by transitivity)}\\ &\Rightarrow (c, a) \in R \text{ (by symmetry).}\end{aligned}$$

Hence, R is circular and so R is reflexive and circular.

Definition 1.22

A collection of non-empty sets $\{A_1, A_2, \ldots, A_n\}$ is called a **partition of a set** A if, and only if, $A_1, A_2, \ldots, A_n$ are mutually disjoint and $A = A_1 \cup A_2 \cup \ldots \cup A_n$.

The subsets A_i are then called **Blocks of the partition.** Thus a partition of a set is a division of the elements in the set into disjoint subsets. Figure 1.7 given below illustrates the partition of a set A into five disjoint subsets.

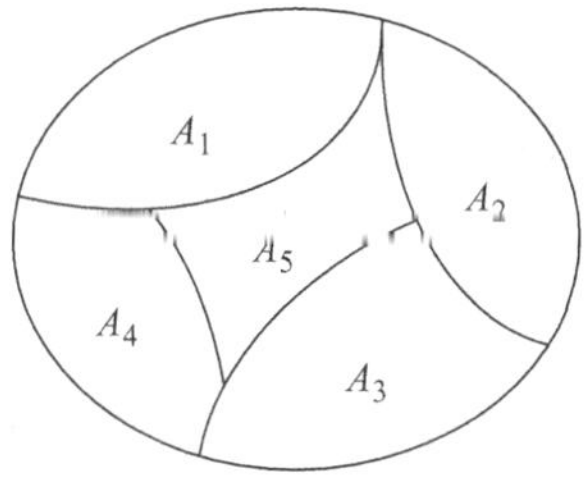

$A_i \cap A_j = \phi, i \neq j$

$A = A_1 \cup A_2 \cup \ldots \cup A_5$

Figure 1.7

Definition 1.23

Let $A = A_1 \cup A_2 \cup \ldots \cup A_n$ be a partition of a set A. A relation R on A is said to be **relation induced by the partition** if for all $a, b \in A$,

$$a\ R\ b \Leftrightarrow \text{there is a subset } A_i \text{ of the partition such that both } a \text{ and } b \text{ are in } A_i.$$

Theorem 1.8

Let A be a set with partition $\{A_1, A_2, \ldots, A_n\}$ and let R be the relation induced by the partition. Then R is an equivalence relation.

Proof.

(i) Suppose $a \in A$. Since $\{A_1, A_2, \ldots, A_n\}$ is a partition of A, $a \in A_i$ for some i. Thus, there is a set A_i of the partition such that $a \in A_i$ and $a \in A_i$. Hence $a\ R\ a$ proving that R is reflexive.

(ii) Let $a, b \in A$ such that $a\ R\ b$. Then there is a subset A_i of the partition such that both a and b are in A_i, that is both b and a belong to some A_i. Hence, $b\ R\ a$, proving that R is symmetric.

(iii) Let $a, b, c \in A$ such that $a\ R\ b$ and $b\ R\ c$. Thus, there are subsets A_i and A_j of the partition such that

$$a, b \in A_i \quad \text{and} \quad b, c \in A_j$$

We assert that $A_i = A_j$. If not, then $A_i \cap A_j = \phi$ since all sets of the partition are disjoint. But b is in both A_i and A_j. Hence $A_i \cap A_j \neq \phi$. Thus, we arrive at a contradiction. So $A_i = A_j$ showing that a, b and c are all in A_i. In particular a and c are both in A_i showing that $a\ R\ c$ and so R is transitive.

EXAMPLE 1.23

Let

$$A = \{a, b, c, d, e, f, g\}.$$

Then,

$$\{\{a\}, \{b, c, d\}, \{e, f\}, \{g\}\}$$

is a partition of A.

Definition 1.24

Let R be an equivalence relation on A and $a \in A$. Then the set

$$\{b \in A : (a, b) \in R\}$$

is called the **equivalence class** of a. It is denoted by $[a]$.

Thus,

$$[a]=\{b \in A : (a, b) \in R\}$$

EXAMPLE 1.24

Let **Z** be a set of all integers. Then

$$R=\{(a, b) : a \in \mathbf{Z}, b \in \mathbf{Z}, (a-b) \text{ is an even integer}\}$$

is an equivalence relation on **Z**. In fact,

(i) $a-a=0$ is an even integer. So $(a, a) \in R$. Therefore, R is reflexive.

(ii) Let $(a, b) \in R$. Then $a-b$ is an even integer. This implies that $b-a=-(a-b)$ is also an even integer which implies that $(b, a) \in R$.

Thus, $(a, b) \in R \Rightarrow (b, a) \in R$ and so R is symmetric.

(iii) Let $(a, b) \in R$, $(b, c) \in R$. Then $a-b$ is an even integer and $b-c$ is an even integer. We have

$$a-c=(a-b)+(b-c).$$

The sum of two even integers on the right is also even. Therefore $a-c$ is an even integer, which implies $(a, c) \in R$. Thus,

$$(a, b) \in R, (b, c) \in R \Rightarrow (a, c) \in R$$

and so R is transitive.

Hence R is reflexive, symmetric and transitive and so is an equivalence relation. The elements of the equivalence class of $a \in \mathbf{Z}$ are $a+2k$, $k=0, \pm 1, \pm 2, \pm 3, \ldots$.
Thus,

$$[0]=\{\ldots, -6, -4, -2, 0, 2, 4, 6, \ldots\},$$
$$[1]=\{\ldots, -5, -3, -1, 1, 3, 5, 7, \ldots\},$$
$$[2]=\{\ldots, -6, -4, -2, 0, 2, 4, 6, \ldots\},$$

which is same as [0].
Proceeding similarly, we note that there are **only two equivalence classes**

$$\{\ldots, -6, -4, -2, 0, 2, 4, 6, \ldots\}=E_0 \text{ (say)}$$

and

$$\{\ldots, -5, -3, -1, 1, 3, 5, 7, \ldots\}=E_1 \text{ (say).}$$

We also note that

$$\mathbf{Z}=E_0 \cup E_1 \text{ and } E_0 \cap E_1=\phi.$$

The equivalence classes E_0 and E_1 are blocks of the partition and so $\{E_0, E_1\}$ is a partition of the set of all integers.

EXAMPLE 1.25

Let A be a set of real numbers. Then

$$R=\{(a, b) : a \in A, b \in A, a=b\}.$$

is an equivalence relation.
The equivalence class of a has only one element and that is a itself.
In fact,

$$[a]=\{b \in A : (a, b) \in \mathbf{R}\}$$
$$=\{b \in A : a=b\}=\{a\}.$$

EXAMPLE 1.26

Let **Z** be the set of all integers, then consider the relation

$$R_m = \{(a, b) : a \equiv b \pmod{m},\ a, b \subset \mathbf{Z}\}.$$

We have already shown that R_m is an equivalence relation. The elements of the equivalence classes of $a \in \mathbf{Z}$ are $a+m K$, $K=0, \pm 1, \pm 2, \ldots$. The m different classes in this case are

$$[0]=E_0=\{\ldots, -3m, -2m, -m, 0, m, 2m, 3m, \ldots\},$$
$$[1]=E_1=\{\ldots, -3m+1, -2m+1, -m+1, 1, m+1, 2m+1, 3m+1, \ldots\},$$
$$[2]=E_2=\{\ldots, -3m+2, -2m+2, -m+2, 2, m+2, 2m+2, 3m+2, \ldots\},$$

$$[m-1]=E_{m-1}=\{\ldots, -2m-1, -m-1, -1, m-1, 3m-1, \ldots\},$$
$$[m]=E_m=\{\ldots, -3m, -2m, -m, 0, m, 2m, 3m, \ldots\}=E_0.$$

Similarly, $E_{m+1}=E_1=[1]$, and so on. Thus, there are m distinct equivalence classes $E_0, E_1, \ldots, E_{m-1}$. We note that

$$\mathbf{Z}=E_0 \cup E_1 \cup E_2 \cup \ldots \cup E_{m-1} \text{ and } E_i \cap E_j=\phi,\ i\neq j.$$

Also we note that $a \in [a]\ \forall\, a \in \mathbf{Z}$.

Theorem 1.9

Let R be an equivalence relation on A and $[a]$ be an equivalence class of $a \in A$. Then

(i) $a \in [a]\ \forall\, a \in A$,
(ii) $[a]=[b]$ and if and only if $(a, b) \in R$,
(iii) If $[a]\neq[b]$, then $[a] \cap [b]=\phi$,
(iv) A is the union of all disjoint equivalence classes.

Proof.

(i) Since R is an equivalence relation, reflexivity is there, that is, for every $a \in A$ we have $(a, a) \in R$. T his implies that $a \in [a]$.
(ii) Firstly, suppose that $(a, b) \in R$. We shall prove that $[a]=[b]$.

Let $x \in [b]$. Then, by definition of $[b]$, we have $(b, x) \in R$. Transitivity yields

$$(a, b) \in R \text{ and } (b, x) \in R \Rightarrow (a, x) \in R.$$

But,

$$(a, x) \in R \Rightarrow x \in [a].$$

Hence,

$$\begin{aligned} x \in [b] &\Rightarrow x \in [a] \\ &\Rightarrow [b] \subseteq [a]. \end{aligned} \tag{1}$$

On the other hand,

$$(a, b) \in R \Rightarrow (b, a) \in R \text{ (by symmetry)}.$$

Let $x \in [a]$. Then $(a, x) \in R$. Therefore, $(b, x) \in R$ by transitivity. Hence, $x \in [b]$. Thus,

$$[a] \subseteq [b] \tag{2}$$

By (1) and (2), we have

$$[a]=[b].$$

Conversely, if $[a]=[b]$, then $b \in [b]=[a]$, that is, $(a, b) \in R$.

(iii) We shall prove the result by contradiction. So, we assume that $[a] \cap [b] \neq \phi$. Therefore, there exists an element $c \in [a] \cap [b]$. Then,

$$\begin{aligned} c \in [a] \text{ and } c \in [b] &\Rightarrow a\,R\,c \text{ and } b\,R\,c \\ &\Rightarrow a\,R\,c \text{ and } c\,R\,b \text{ (by symmetry)} \\ &\Rightarrow a\,R\,b \text{ (by transitivity)} \\ &\Rightarrow (a, b) \in R. \end{aligned}$$

But, we have proved above that if $(a, b) \in R$, then $[a]=[b]$ which contradicts our hypothesis. Hence $[a] \cap [b]=\phi$.

(iv) Let

$$P=\cup\, [a],\ a \in A.$$

Then, we have

$$P \subseteq A.$$

On the other hand, for each $a \in A$, there exists an equivalence class $[a]$ containing a. Thus,

$$\begin{aligned} a \in A &\Rightarrow a \in [a], \\ &\Rightarrow a \in P, \\ &\Rightarrow A \subseteq P. \end{aligned}$$

Hence, $A=P$.

Definition 1.25

Let R be an equivalence relation on A. Then the collection of equivalence classes of the elements of A is called **Quotient of** A **by** R and is denoted by $A \mid R$.
Thus,

$$A \mid R=\{[a] : a \in A\}.$$

Theorem 1.10 (Fundamental Theorem on Relations)

If R is an equivalence relation on A, then $A \mid R$ is the partition of A.

Proof. We know that $A \mid R=\{[a] : a \in A\}$. Therefore it is sufficient to show that A is the union of all disjoint equivalence classes.
Let

$$P=\cup\, [a],\ a \in A.$$

Thus,

$$P \subseteq A.$$

On the other hand for each $a \in A$, there exists an equivalence class $[a]$ containing a.
Thus,

$$\begin{aligned} a \in A &\Rightarrow a \in [a], \\ &\Rightarrow a \in P, \\ &\Rightarrow A \subseteq P. \end{aligned}$$

Hence, $A=P$.

EXAMPLE 1.27

Let $A=\{0, 1, 2, 3, 4\}$. Find the equivalence classes of the equivalence relation $R=\{(0, 0), (0, 4), (1, 1), (1, 3), (2, 2), (3, 1), (3, 3), (4, 0), (4, 4)\}$ defined on A. Draw digraph of R and write down the partition of A induced by R.

Solution.

We note that

(i) The elements related to 0 are 0, 4. Therefore,

$$[0]=\{0, 4\}$$

(ii) The elements related to 1 are 1, 3. Therefore,

$$[1]=\{1, 3\}$$

(iii) The element related to 2 is 2 only. Hence,

$$[2]=\{2\}.$$

(iv) The elements related to 3 are 1, 3. Hence,

$$[3]=\{1, 3\}=[1]$$

(v) The elements related to 4 are 0 and 4. Hence,

$$[4]=\{0, 4\}=[0].$$

Thus, there are only three equivalence classes

$$\{0, 4\}, \{1, 3\} \text{ and } \{2\}.$$

The digraph of R is

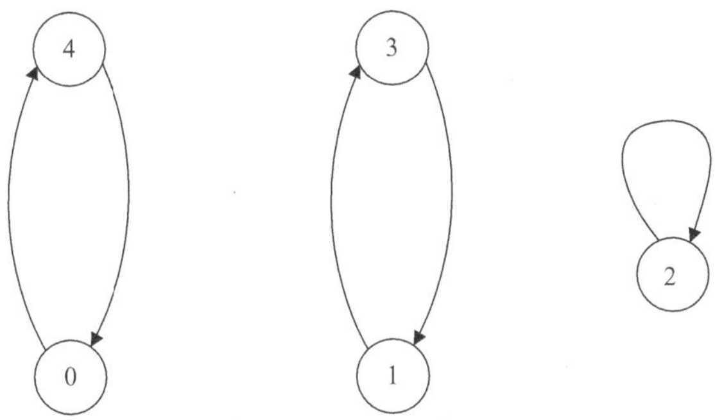

Figure 1.8

where three sub-graphs of the digraph represent the equivalence classes. We further observe that

$$A=\{0, 4\} \cup \{1, 3\} \cup \{2\}$$

and so $\{0, 4\}$, $\{1, 3\}$ and $\{2\}$ is the partition of A induced by R.

1.3 REPRESENTATION OF RELATIONS ON FINITE SET

A relation from a finite set A to a finite set B can be represented by the following three ways:

1. **Arrow diagram of a relation:** In this representation, the elements of the set A and the elements of the set B are written in two disjoint disks. If $a \in A$ is related to $b \in B$, then we draw an arrow from a to b. For example, let

$$A=\{1, 2\}, B=\{1, 2, 3\}.$$

Then the relation

$$R=\{(1, 2,), (1, 3), (2, 2), (2, 3)\}$$

is represented as shown in Figure 1.9.

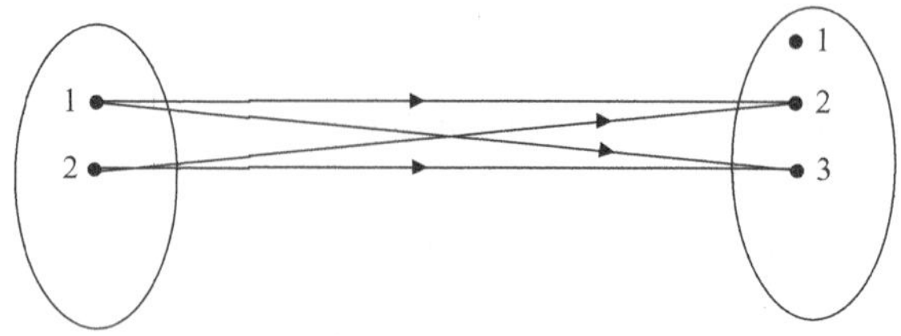

Figure 1.9 (Pictorial representation of R = {(1,2), (1, 3), (2, 2), (2, 3)}).

2. **Matrix representation of a relation:** It is a convenient way to represent a relation R from A to B. Matrix representation of a relation is used in computers. If $A=\{a_1, a_2, \ldots, a_m)$ and $B=(b_1, b_2, \ldots, b_n)$ are finite sets containing m and n elements, respectively, and R is a relation from A to B, then R is represented by the $m \times n$ matrix, denoted by $M_R=(c_{ij})$, defined by

$$c_{ij} = \begin{cases} 1 & \text{if } (a_i, b_j) \in R \\ 0 & \text{if } (a_i, b_j) \notin R. \end{cases}$$

The matrix M_R is called the **matrix of the relation** R.

For example, the matrix of the relation

$$R=\{(a_1, b_1), (a_1, b_4), (a_2, b_2), (a_2, b_3), (a_3, b_1), (a_3, b_3)\}$$

from $A=\{a_1, a_2, a_3\}$ to $B=\{b_1, b_2, b_3, b_4\}$ is

$$M_R = \begin{bmatrix} 1 & 0 & 0 & 1 \\ 0 & 1 & 1 & 0 \\ 1 & 0 & 1 & 0 \end{bmatrix}.$$

3. **Directed graph representation of relation from a finite set A to itself:** In this representation, we draw a small circle for each element of the set A and label the circle with the corresponding element of A. These circles are called **vertices.** We then draw arrow from vertex a_i to vertex a_j if and only if $(a_i, a_j) \in R$. These arrows are called **edges.** The pictorial representation of R so obtained is called **directed graph** or **digraph** of R.

EXAMPLE 1.28

Let $A=\{1, 2, 3\}$ and let

$$R=\{(1, 1), (1, 2), (1, 3), (3, 1), (2, 3), (2, 1)\}$$

be a relation on A. Draw the directed graph of R.

Solution.

We note that:

(i) Since 1 R 1, there is a loop from 1 to itself
(ii) Since 1 R 2, there is an edge from 1 to 2
(iii) Since 1 R 3, there is an edge from 1 to 3
(iv) Since 3 R 1, there is an edge from 3 to 1
(v) Since 2 R 3, there is an edge from 2 to 3
(vi) Since 2 R 1, there is an edge from 2 to 1

Thus, the directed graph of R is as shown in the Figure 1.10.

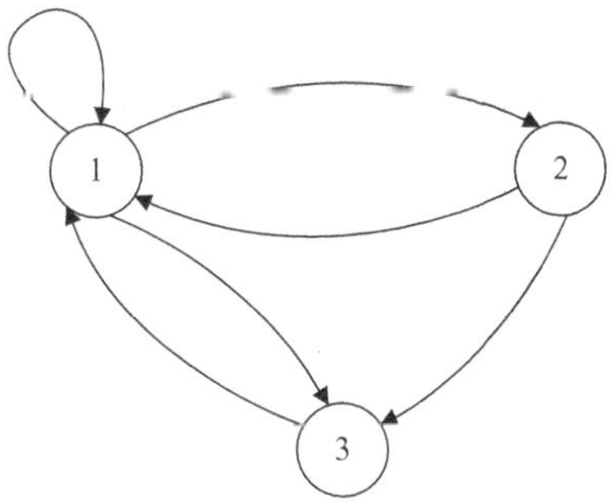

Figure 1.10

EXAMPLE 1.29

Find the relation R if

$$M_R = \begin{bmatrix} 1 & 0 & 0 \\ 0 & 1 & 1 \\ 1 & 1 & 0 \end{bmatrix}.$$

Solution.

Here, M_R is a 3×3 matrix. So, we take

$$A=\{a_1, a_2, a_3\},\ B=\{b_1, b_2, b_3\}.$$

By definition of M_R, $(a_i, b_j) \in R$ if $c_{ij}=1$. Thus, it follows that $a_1\ R\ b_1$, $a_2\ R\ b_2$, $a_2\ R\ b_3$, $a_3\ R\ b_1$, $a_3\ R\ b_2$. Hence

$$R=\{(a_1, b_1), (a_2, b_2), (a_2, b_3), (a_3, b_1), (a_3, b_2)\}.$$

EXAMPLE 1.30

Find the relation R whose digraph is shown in the Figure 1.11.

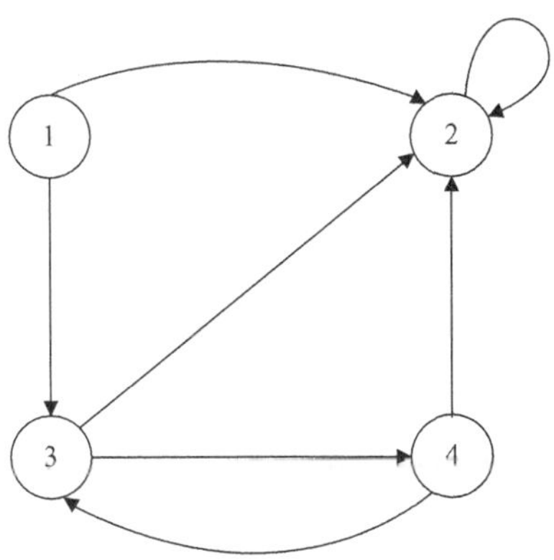

Figure 1.11

Solution.

Clearly, the relation R is defined on the set $A=\{1, 2, 3, 4\}$. Also, we know that $a_i \; R \; a_j$ if and only if there is an edge from a_i to a_j. Thus, we note that

$$1 \; R \; 2, \; 1 \; R \; 3, \; 2 \; R \; 2, \; 3 \; R \; 2, \; 3 \; R \; 4, \; 4 \; R \; 2, \; 4 \; R \; 3.$$

Hence,

$$R=\{(1, 2), (1, 3), (2, 2), (3, 2), (3, 4), (4, 2), (4, 3)\}.$$

1.4 MAPPINGS (FUNCTIONS)

Definition 1.26

Let A and B be two sets. Then a relation f from A into B is called a mapping if

(i) For every element $x \in A$, there is an element $y \in B$ such the $(x, y) \in f$.

(ii) $(x, y) \in f$; $(x, z) \in f \Rightarrow y=z$.

We denote this mapping by $f: A \rightarrow B$. The set A is called domain of f, the set B is called co-domain of f, and the range of f is a subset of B.

Thus, we may also define our function f as follows:

Let X and Y denote arbitrarily given sets. By a function

$$f: X \rightarrow Y$$

from X into Y, we mean a rule which assigns to each member x of X a unique member $f(x)$ of Y.

Let $f: A \rightarrow B$ be a mapping from a set A to another set B. If $(x, y) \in f$, then y is called the image of x and is denoted by $f(x)$. Also, $f(x)$ is called the value of f at x. Thus, a map $f: A \rightarrow B$ can be represented by

$$f=\{(x, f(x)) : x \in A\}.$$

EXAMPLE 1.31

Let $A=\{1, 2, 3, 4\}$, $B=\{1, 2, 3\}$. If f is a subset of $A \times B$ such that

$$f=\{(1, 1), (2, 1), (3, 2), (4, 2)\},$$

then $f: A \rightarrow B$ is a mapping whose domain and range are given by

$$D(f)=\{1, 2, 3, 4\}, R(f)=\{1, 2\} \subseteq B$$

EXAMPLE 1.32

If A is a set of real numbers and

$$f=\{(x, x^2) : x \in A\},$$

then f is clearly a subset of $A \times A$. The first components in any two ordered pair are not equal. Therefore f: $A \rightarrow A$ is a mapping such that

$$D(f)=\{x : (x, x^2) \in f\}=A.$$

We can write this function as $f(x)=x^2$

Moreover,

$$R(f)=\{x^2 : (x, x^2) \in f\}.$$

EXAMPLE 1.33

If

$$A=\{1, 2, 3\} \text{ and } B=\{1, 2, 3, 4\},$$

then the subset $\{(1, 2), (1, 3), (3, 4)\}$ of $A\times B$ is not mapping. In fact, $(1, 2) \in f$, $(1, 3) \in f \not\Rightarrow 2=3$.

Definition 1.27

If $f(x)$, $x \in X$, consists of a single point y of Y, then we say that

$$f: X\to Y$$

is a constant function from X into Y.

1.4.1 Types of Mappings

Definition 1.28

Let $f: A\to B$ be a mapping from A into B. If $f(x_1)=f(x_2)\Rightarrow x_1=x_2$, then f is said to be **one-one mapping** or **injective mapping.** Thus, a function $f: A\to B$ is injective if and only if the images of distinct points of A are distinct, that is, $a\neq b\Rightarrow f(a)\neq f(b)$.

EXAMPLE 1.34

Let **Z** be a set of positive integers and **Y** the set of even positive integers. Then the mapping

$$f=\{(x, y) : x \in \mathbf{Z}, y \in \mathbf{Y}, y=2x\},$$

that is, when $f(x)=2x$ is injective. In fact, if $(x_1, y), (x_2, y) \in f$, then,

$$x_1 = \frac{y}{2}, \tag{1}$$

$$x_2 = \frac{y}{2}. \tag{2}$$

From (1) and (2), we have $x_1=x_2$.

EXAMPLE 1.35

Let $a\neq 0$ and let b denote arbitrary given real number. Consider the function

$$f: \mathbf{R}\to\mathbf{R}$$

defined by

$$f(x)=ax+b \text{ for every real number } x.$$

We note that

$$f(x_1)=a x_1+b, f(x_2)=a x_2+b$$

and therefore

$$\begin{aligned} f(x_1)=f(x_2) &\Rightarrow a x_1+b=a x_2+b \\ &\Rightarrow x_1=x_2. \end{aligned}$$

Hence, this function is injective.

EXAMPLE 1.36

Consider the case $X\subset Y$. The function $i : X\to Y$ defined by $i(x)=x \in Y$ for every $x \in X$ will be called the **inclusion function** of X into Y. Obviously, every inclusion function is injective.

Definition 1.29

If a function f is not one-one mapping, then it is called a **many-one mapping.**

For example,

$$f=\{(x, y) : y=x^2, -\infty<x<\infty$$

is not a one-to-one mapping since 4 is the image of both –2 and 2. Hence f is a many to one mapping.

Definition 1.30

Let $f: X \to Y$ be a map. If $f(X)=Y$, that is $R(f)=Y$, then we say that $f: X \to Y$ is a function from X **onto** (or **surjective**) Y. **Therefore,** $f: X \to Y$ **is surjective if and only if for every point** y **in** Y **there exists at least one point** x **in** X **such that** $f(x)=y$.

For example, the linear function $f: R \to R$, defined by $f(x)=a\,x+b$ is surjective. In fact, for every $y \in R$, we have an x in R such that

$$x = \frac{f(x)-b}{a}.$$

On the other hand, if **R** is a set of real numbers and $f: \mathbf{R} \to \mathbf{R}$ is function defined by $f(x)=\sin x$. Then, it is not surjective. In fact, by trigonometry, there is no element x in **R** for which $\sin x=2$. Thus, $R(f) \neq \mathbf{R}$. Hence f is not surjective.

EXAMPLE 1.37

Let $f: \mathbf{R} \to \mathbf{R}$ be defined by

$$f=\{(x, y) : x, y \in \mathbf{R} \text{ (set of real numbers)}, y=x^3\}$$

is a **surjective** (onto) mapping.

EXAMPLE 1.38

Show that the mapping $f: \mathbf{Z}^+ \to \mathbf{Z}^+$ defined by $f(x)=x^2$, $x \in \mathbf{Z}^+$, where $\mathbf{Z}^+$ is set of +ve integers, is **one-one** and **into** mapping.

Solution.

We observe that, if $x, y \in \mathbf{Z}^+$, then,

$$\begin{aligned} f(x)=f(y) &\Rightarrow x^2=y^2, \\ &\Rightarrow (\pm x)^2=(\pm y)^2, \\ &\Rightarrow \pm x=\pm y, \\ &\Rightarrow x=y \ (\because x, y \in \mathbf{Z}^+, \text{ so we consider only +ve values}). \end{aligned}$$

Hence f is one-to-one mapping. Further,

$$R(f)=\{1, 4, 9, \ldots\},$$

which is a proper subset of $\mathbf{Z}^+$. Hence f is **not surjective** and so f is **into** mapping.

Definition 1.31

A mapping $f: A \to B$ from a set A into the set B is called **bijective** if it is both injective and surjective.

EXAMPLE 1.39

Let $X=\{x \in \mathbf{R}, x \neq 0\}$. Show that $f: X \to X$ defined by $f(x)=1/x$ is one-one and onto (and hence bijective).

Solution.

Here X is the set of **non-zero real numbers.** Let $x, y \in X$. Then,

$$f(x) = f(y) \Rightarrow \frac{1}{x} = \frac{1}{y},$$
$$\Rightarrow y = x.$$

Hence, f is **injective.**

On the other hand, for every non-zero real number $x \in X$, there exists a non-zero real number $1/x \in X$ such that

$$f\left(\frac{1}{x}\right) = \frac{1}{1/x} = x$$

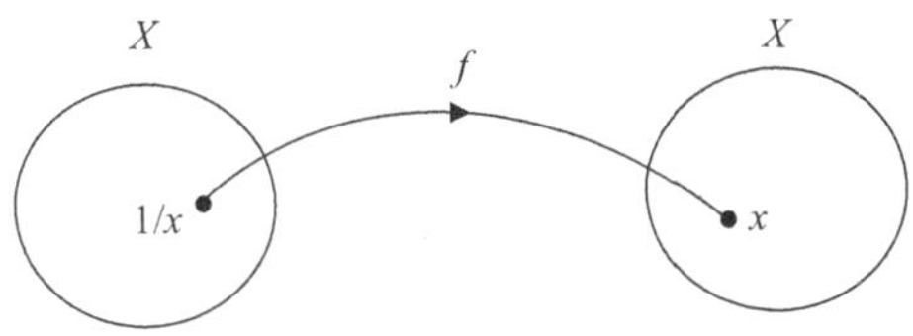

Figure 1.12

Hence f is onto (surjective) mapping. Thus, f is bijective.

EXAMPLE 1.40

Is an injective map from a set to itself a surjective map? Give reasons.

Solution.

Let A be a set and let $f: A \to A$ be an injective map. Therefore,

$$x \neq y \Rightarrow f(x) \neq f(y).$$

Thus, if A is **finite** with n distinct elements, then images of n distinct elements are n distinct elements of A under injective map f. Thus to each element y in finite set A(range), there is an element x in the finite set A(domain) such that $f(x)=y$. Hence f is **surjective.**

If A is infinite, then the injective map f may or may not be surjective depending upon the definition of f. For example, the map $f: A \to A$ defined by $f(x)=2x \; \forall x \in A$, where A is the set of all integers is one-one because for $x, y \in A$

$$f(x)=f(y) \Rightarrow 2x=2y,$$
$$\Rightarrow x=y.$$

But range of f is

$$R_f = \{0, \pm 2, \pm 4, \ldots\},$$

which is a proper subset of A. Hence f is not onto.

Note: From this example we conclude that "A set is infinite if and only if there exists a bijective mapping from the set onto a proper subset of itself."

A set is said to be finite if it is not infinite.

Let X be a non-empty set. Then the function f defined by

$$f = \{(x, x) : x \in X\},$$

that is, $f(x)=x$ is called **identity mapping.**

The identity mapping is clearly injective and surjective and hence bijective.

Definition 1.32

Let f and g be functions from X to Y. Then f **equals** g, written as $f=g$, if and only if $f(x)=g(x)$ for all $x \in X$.

For example, let **R** be the set of real number and let $f: \mathbf{R} \to \mathbf{R}$, $g: \mathbf{R} \to \mathbf{R}$ be defined by $f(x)=|x|$ and $g(x)=\sqrt{x^2}$ for all $x \in \mathbf{R}$. Since $|x|=\sqrt{x^2}$ for all $x \in \mathbf{R}$, it follows that $f=g$.

1.4.2 Inverse Mapping

Let $f: A \to B$ be a mapping. Since a mapping is a relation, therefore it induces inverse relation f^{-1}. **But it is not necessary that $f^{-1}: B \to A$ is a mapping.** The reasons are:

(i) When $D(f^{-1}) \neq B$ and this happens when $R(f) \neq B$, that is, when f is not onto mapping.

(ii) When $(y, x), (y, z) \in f^{-1} \nRightarrow x=z$, that is, when $(x, y), (z, y) \in f \nRightarrow x=z$, i.e., when $f(x)=f(z) \nRightarrow x=z$, i.e., when f **is not one-to-one mapping**.

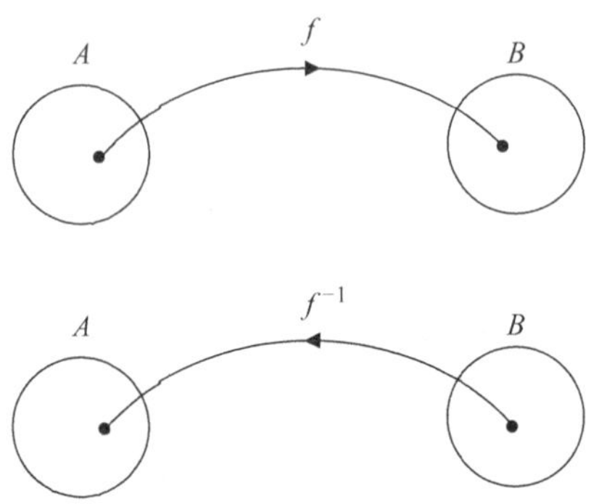

Figure 1.13

Thus f^{-1} exists if and only if f is bijective.

Definition 1.33

If $f: A \to B$ is a bijective mapping, then the relation f^{-1} of B on A is called **inverse mapping of f**. Thus, if

$$f=\{(x, y) : x \in A, y \in B\}$$

is bijective, then

$$f^{-1}=\{(y, x) : (x, y) \in f\}$$

is called **inverse mapping of f**.

Theorem 1.11

If $f: A \to B$ is a bijective mapping, then $f^{-1}: B \to A$ is also bijective.

Proof. Suppose that

$$f=\{(x, y) : x \in A, y \in B\},$$

then

$$f^{-1}=\{(y, x) : x \in A, y \in B, (x, y) \in f\}.$$

Let $(y_1, x), (y_2, x) \in f^{-1}$. Then, $(x, y_1), (x, y_2) \in f$.

But f is a mapping. Therefore, $y_1 = y_2$.
Thus,

$$(y_1, x), (y_2, x) \in f^{-1} \Rightarrow y_1 = y_2,$$
$$\Rightarrow f^{-1} \text{ is injective.}$$

Moreover,

$$R(f^{-1}) = \{x : (y, x) \in f^{-1}\}$$
$$= \{x : (x, y) \in f\} = D(f) = A,$$

which implies that f^{-1} is onto mapping. Hence, f^{-1} is bijective.

1.5 COMPOSITION OF MAPPINGS

Definition 1.34

Any two functions f and g are said to **composable** if and only if the range of f is equal to the domain of g.

Definition 1.35

Let $f: A \rightarrow B$ and $g: B \rightarrow C$ be two mappings. Then the set

$$\{x, g(f(x)) : x \in A\}$$

is called the **composition** of the two functions f and g and it is denoted by $g \circ f : A \rightarrow C$. Thus,

$$g \circ f = \{(x, g(f(x)) : x \in A\}$$

or

$$(g \circ f)(x) = g(f(x)).$$

EXAMPLE 1.41

Let $f: \mathbf{R} \rightarrow \mathbf{R}$ and $g : \mathbf{R} \rightarrow \mathbf{R}$ be functions defined by

$$f = \{(x, \sin x) : x \in \mathbf{R}\}$$

and

$$g = \{(x, x^2) : x \in \mathbf{R}\}$$

Then the composition $g \circ f$ is determined as follows:

$$g(f(x) = g(\sin x) = (\sin x)^2.$$

Hence,

$$g \circ f = \{(x, \sin^2 x) : x \in \mathbf{R}\}.$$

EXAMPLE 1.42

Let $f, g: \mathbf{R} \rightarrow \mathbf{R}$ denote the functions defined by

$$f(x) = 2x + 3,\ g(x) = x^2 \text{ for every } x \in \mathbf{R}.$$

Then, the composition $f \circ g$ and $g \circ f$ are both defined , and are given by

$$(g \circ f)(x) = g[f(x)] = g(2x + 3) = (2x + 3)^2$$
$$(f \circ g)(x) = f[g(x)] = f(x^2) = 2x^2 + 3.$$

This example shows that $(f \circ g)(x) \neq (g \circ f)(x)$.

EXAMPLE 1.43

Let

$$f=\{(x, y) : x \in A, y \in B\}$$

be a bijective mapping from A onto B. Then,

$$f^{-1}=\{(y, x) : x \in A, y \in B\}$$

and

$$\begin{aligned} f^{-1} \circ f &= \{(x, f^{-1}(f(x))) : x \in A\} \\ &= \{(x, f^{-1}(y) : x \in A\}= \{(x, x) : x \in A\} \\ &= I_A \text{ (identity mapping).} \end{aligned}$$

Similarly, it can be shown that $f \circ f^{-1}=I_B$.

Theorem 1.12

Let $f: A\to B$ and $g: B\to C$ be two functions. Then,

(i) $g \circ f: A\to C$ is onto if f and g are onto.

(ii) $g \circ f: A\to C$ is one-to-one if f and g are one-to-one.

Proof.

(i) Since g is surjective, therefore for $c \in C$, there exists an element $b \in B$ such that $g(b)=c$. Now, since f is surjective and $b \in B$, there is an element $a \in A$ such that $f(a)=b$. Hence, $(g \circ f)(a)= g(f(a))=g(b)=c$, that is, to every $c \in C$, there exists at least one point a in A such that $(g \circ f)(a)=c$. Therefore, $g \circ f$ is surjective.

(ii) Suppose that

$$(g \circ f)(a)=(g \circ f)(a_1), a, a_1 \in A.$$

Then,

$$\begin{aligned} & g(f(a))=g(f(a_1)) \\ & \Rightarrow f(a)=f(a_1) \quad (\because g \text{ is injective}) \\ & \Rightarrow a=a_1 \quad (\because f \text{ is injective}) \end{aligned}$$

Thus $(g \circ f)(a)=(g \circ f)(a_1)\Rightarrow a=a_1$. Hence, $g \circ f$ is injective.

Theorem 1.13

If $f: A\to B$ and $g : B\to C$ are bijective mapping, then

$$(g \circ f)^{-1}=f^{-1} \circ g^{-1}.$$

Proof. The function f and g being bijective, the function $g \circ f$ is also bijective. Hence, the existence of $(g \circ f)^{-1}$ is valid. We have,

$$\begin{aligned} (f^{-1} \circ g^{-1}) \circ (g \circ f) &= f^{-1} \circ [g^{-1} \circ (g \circ f)] \\ &= f^{-1} \circ [(g^{-1} \circ g) \circ f] \text{ (associativity of composition)} \\ &= f^{-1} \circ (I \circ f) \text{ since } g^{-1} \circ g=I \text{ (identity mapping)} \\ &= f^{-1} \circ f \text{ property of identity mapping)} \\ &= I \end{aligned}$$

and

$$\begin{aligned} (g \circ f) \circ (f^{-1} \circ g^{-1}) &= g \circ [f \circ (f^{-1} \circ g^{-1})] \\ &= g \circ [(f \circ f^{-1}) \circ g^{-1}] \\ &= g \circ [I \circ g^{-1}] \\ &= g \circ g^{-1}=I. \end{aligned}$$

Hence,

$$(g \circ f)^{-1} = f^{-1} \circ g^{-1}.$$

1.6 COUNTABILITY OF SETS

Definition 1.36

By a **sequence** of real numbers, we mean a function $f: \mathbf{N} \to \mathbf{R}$ from the set of natural number **N** into the real line.

Thus, for each $n \in \mathbf{N}$, we have

$$f(n) = x_n, \; x_n \in \mathbf{R}.$$

The term $x_n = f(n)$ is called the **nth term** of the sequence f. The sequence f is generally denoted by

$$f = \{x_1, x_2, \ldots, x_n, \ldots\}$$

For example,

(i) $\left\{\frac{1}{n}\right\}$ is the sequence $\left\{1, \frac{1}{2}, \frac{1}{3}, \ldots\right\}$ with $x_n = \frac{1}{n}$.

(ii) $\{(-1)^n\}$ is the sequence $\{-1, 1, -1, \ldots\}$ with $x_n = (-1)^n$.

Definition 1.37

A sequence $\{x_n\}$ is called **increasing** if $x_n \leq x_{n+1}$ for all n and decreasing if $x_n \geq x_{n+1}$ for all n.

For example, the sequence $\{x_n\}$, where $x_n = 2n$ is increasing because

$$x_n = 2n \leq 2(n+1) = x_{n+1} \text{ for all } n.$$

Definition 1.38

Two sets A and B are said to be **Equivalent**, **Equipotent** or to **have the same cardinality** (in symbol $A \simeq B$) if and only if there exists a bijective mapping $f: A \to B$ from A onto B.

Thus,

Two sets A and B are said to be **equivalent** if there exists a one-to-one correspondence between them.

Cardinal number is a symbol assigned to sets in such a way that two sets are assigned the same symbol if and only if they have the same cardinality. The cardinal number of infinite set N of positive integer is $\aleph_0$(aleph-naught). In general, cardinal number of set A is denoted by $|A|$.

Theorem 1.14

The relation $\simeq$ of sets being equivalent is an equivalence relation.

Proof. We note that

(i) Every set A is equivalent to itself because the identity map $I : A \to A$ defined by $I(x) = x \, \forall x \in A$ is bijective. Thus, $\simeq$ is reflexive.

(ii) If $A \simeq B$, then there exists a one-to-one correspondence f between them. Since if f is bijective, then f^{-1} exists and is also bijective, it follows that $f^{-1} : B \to A$ is bijective. Hence, $B \simeq A$. Thus, the relation $\simeq$ is symmetric.

(iii) If $A \simeq B$, there exists $f: A \to B$ which is bijective. Similarly if $B \simeq C$, there exists $g: B \to C$ which is bijective. Then $h = g \circ f$ is mapping from A to C which is bijective (because composition of bijective function is bijective). Hence $A \simeq C$. Thus the relation $\simeq$ is transitive.

Hence $\simeq$ is an equivalence relation. In other words, **cardinality is an equivalence relation.**

Definition 1.39

A set A is called **finite** having n elements if A is equipotent (equivalent) to $\{1, 2, \ldots, n\}$. Then n is called **cardinality** of A.

The empty set is also considered finite having cardinality 0.

Sets which are not finite are called **infinite sets.**

The chief difference between a finite set and an infinite set is that an infinite set must be equivalent to some proper subset of itself, whereas a finite set cannot be equivalent to any proper subset of itself.

For example, the infinite set

$$A = \{1, 2, 3, \ldots\}$$

of all positive integers is equivalent to its proper subset (2, 4, 8, 16, ...} consisting of powers of 2. The bijective function between these two set is

$$f(x) = 2^x \, \forall x \in A.$$

Definition 1.40

A set is said to be **countable** or **(countably infinite)** if and only if there exists an injective mapping $f: A \to \mathbf{N}$ from A into the set **N** of all natural numbers.

In particular, **N** is countable.

Equivalently, we may define the countability of a set as following:

Definition 1.41

A set A is said to be countable if and only if there exists a surjective mapping g: $\mathbf{N} \to A$ from the set **N** of all natural number onto A.

Theorem 1.15

Every set which is equipotent with a countable set is countable.

Proof. Let A be a set which is equipotent with a countable set C. Therefore, by definition, there exists a bijective function

$$g : A \to C.$$

Now since C is countable, there exists an injective mapping

$$h : C \to \mathbf{N}.$$

Then the mapping

$$h \circ g : A \to \mathbf{N},$$

being composition of two injective functions, is injective. Hence, A is countable.

Thus, we are now in a position to define countable set as

Definition 1.42

A set A is said to be **countable** (or **countably infinite**) if it is equipotent to **N**, the set of all natural numbers.

The following are the other equivalent definitions:

1. A set A is said to be countable (or countably infinite) if there exists a bijective mapping $f: A \to \text{N}$ from A onto the set N of natural numbers.
2. A set A is said to be countable (or countably infinite) if there exists a one-to-one correspondence between the set A and the set **N** of natural numbers.
3. A set A with cardinality $\aleph_0$ is said to be **denumerable** or **countably infinite**.

Theorem 1.16

Every subset of a countable set is countable.

Proof. Let A be any subset of a countable set C. Since C is countable, there exists an injective function

$$h : C \to \mathbf{N}.$$

On the other hand, the inclusion mapping

$$i : A \to C$$

is also bijective. Therefore composition of these two functions

$$h \circ i : A \to \mathbf{N}$$

is also injective. Hence, A is countable.

We note that **the set of terms in any infinite sequence** $\{a_1, a_2, \ldots\}$ **of distinct elements is countable.** Infact, a sequence is a function $f : \mathbf{N} \to \mathbf{R}$ defined by $f(n) = a_n$, $n \in \mathbf{N}$. So, if the a_n are distinct, the function f is one-one and onto.

Accordingly, each of the following set is countable

$$A = \left\{1, \frac{1}{2}, \frac{1}{3}, \ldots\right\}, \quad B = \{1, -2, 3, -4, \ldots\} \quad C = \{(1, 1), (4, 8), (9, 27), \ldots, (n^2, n^3), \ldots\}$$

Thus, we can define a countable set as

Definition 1.43

A set is said to be **countable** if it can be written as a sequence of distinct elements.

Theorem 1.17

The Cartesian square $\mathbf{N}^2 = \mathbf{N} \times \mathbf{N}$ of the set of natural numbers is countable.

Proof. The elements of $\mathbf{N} \times \mathbf{N}$ are

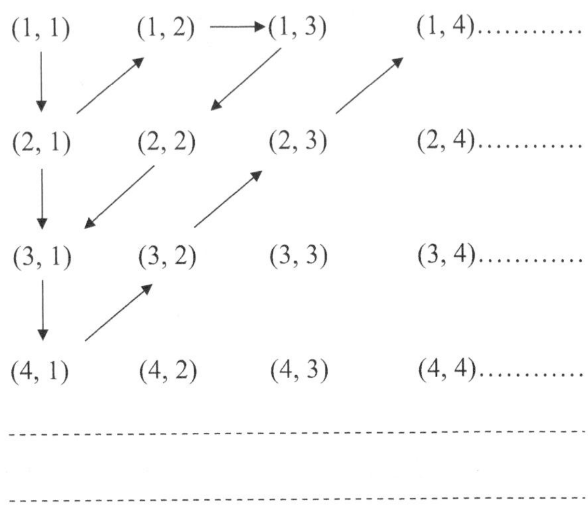

The set $\mathbf{N} \times \mathbf{N}$ can be written in the form of an infinite sequence of distinct elements as follows (by following the arrows in the above diagram):

$$\{(1, 1), (2, 1), (1, 2), (1, 3), (2, 2), (3, 1), (4, 1), \ldots\}.$$

Hence $\mathbf{N} \times \mathbf{N}$ is countable.

Theorem 1.18

Let $\{A_1, A_2, A_3, \ldots\}$ be a countable disjoint class of countable sets. Then $\bigcup_{i=1}^{\infty} A_n$ is countable.

Proof. Since the sets A_i are countable, they can be arranged in the form of sequences with distinct elements.

Let,

$$A_1 = \{a_{11}, a_{12}, a_{13}, \ldots\}$$
$$A_2 = \{a_{21}, a_{22}, a_{23}, \ldots\}$$

―――――――――――――

―――――――――――――

Then,

$$\bigcup_{i=1}^{\infty} A_i = \{a_{ij} : (i, j) \in \mathbf{N}\times\mathbf{N}\}.$$

The function

$$f : \bigcup_{i=1}^{\infty} A_i \to \mathbf{N}\times\mathbf{N}$$

defined by

$$f(a_{ij}) = (i, j)$$

is one-one and onto. Thus $\bigcup_{i=1}^{\infty} A_i$ is **equipotent** to $\mathbf{N}\times\mathbf{N}$. But $\mathbf{N}\times\mathbf{N}$ is countable. Hence $\bigcup_{i=1}^{\infty} A_i$ is countable.

Theorem 1.19

The set **Z** of all integers is countable.

Proof. The set **Z** of all integers is certainly not finite. To show that it is countably infinite, consider the following one-to-one correspondence between **N** and **Z**:

N:	1	2	3	4	5	6	7	8	9 …
	↓	↓	↓	↓	↓	↓	↓	↓	↓
Z:	0	1	−1	2	−2	3	−3	4	−4 …

This correspondence is nothing but the function $f: \mathbf{N} \to \mathbf{Z}$ defined by

$$f(n) = \begin{cases} \dfrac{n}{2} & \text{if } n \text{ is even} \\ \dfrac{1-n}{2} & \text{if } n \text{ is odd} \end{cases}$$

which is one-one and onto. Hence cardinality of **Z** is $\aleph_0$ and so it is countable.

Theorem 1.20

The set of all rational numbers is **countable (denumerable).**

Proof. We know that the set of integers

$$\mathbf{Z} = \{0, 1, -1, 2, -2, \ldots\}$$

is countable.

The set

$$A_n = \left\{\frac{0}{n}, \frac{1}{n}, \frac{-1}{n}, \frac{2}{n}, \frac{-2}{n}, \ldots\right\}$$

is a set of rational numbers with denominator n.

For $n=1, 2, 3, \ldots$, we have

$$A_1 = 0, 1, -1, 2, -2, \ldots\}$$
$$A_2 = \left\{\frac{0}{2}, \frac{1}{2}, \frac{-1}{2}, \frac{2}{2}, \frac{-2}{2}, \ldots\right\}$$

Then each A_1, A_2, … is countable. Further, the set of rational number Q is countable union of these sets, i.e.,

$$Q = \bigcup_{i=1}^{\infty} A_i.$$

But countable union of countable set is countable. Hence the set Q of rational numbers is countable.

Corollary 1.1

The set of rational numbers in [0, 1] is countable.

Proof. The set of rational numbers in [0, 1] is infinite subset of Q, the set of rational number. But Q is countable. Also we know that infinite subset of a countable set is countable. Hence the set of rational numbers in [0, 1] is also countable.

Theorem 1.21

The unit interval $I=\{x : 0 \leq x \leq 1\} = [0, 1]$ of real numbers is uncountable.

Proof. Suppose on the contrary that [0, 1] is countable. Then each element of

$$I = \{x_1, x_2, x_3, \ldots\}$$

can be written in the form of sequence (having distinct elements).

So, let each elements in I be written in the form

$$x_1 = 0 \cdot a_{11}\, a_{12}\, a_{13} \ldots a_{1n} \ldots$$
$$x_2 = 0 \cdot a_{21}\, a_{22}\, a_{23} \ldots a_{2n} \ldots$$
$$x_3 = 0 \cdot a_{31}\, a_{32}\, a_{33} \ldots a_{3n} \ldots$$

$$x_n = 0 \cdot a_{n1}\, a_{n2}\, a_{n3} \ldots a_{nn} \ldots$$

---,

where $a_{ij} \in \{0, 1, 2, \ldots, 9\}$ and where each decimal expression contains an infinite number of non-zero elements.

Now construct the real number.

$$y = 0.\, b_1\, b_2\, b_3 \ldots b_n \ldots, \tag{1}$$

where b_1 is any digit such that $b_1 \neq 0$ and $b_1 \neq a_{11}$, $b_2 \neq 0$ and $b_2 \neq a_{22}$ and so on. Observe that

$$y \neq x_1 \text{ because } b_1 \neq a_{11}$$
$$y \neq x_2 \text{ because } b_2 \neq a_{22}$$

Hence, $y \notin I$. But clearly (1) shows that $y \in I$. Thus we arrive at a contradiction. Hence $I=[0, 1]$ is not countable.

Corollary 1.2

Set of all real numbers is uncountable (Non-denumerable).

Proof. It is sufficient to show that [0, 1] is uncountable.

(Now proceed as in Theorem 1.21).

Theorem 1.22

Let A be the set of all sequence whose element are the digit 0 and 1. Then A is uncountable.

Proof. Let E be a countable subset of A and let E consists of the sequences S_1, S_2, S_3, ... We construct a sequence S as follows:

If the nth digit in S_n is 1, we let nth digit in S be 0 and vice versa. Then the sequence S differs from every member of E in at least one place and hence $S \notin E$. But clearly $S \in A$. So, E is a proper subset of A.

We have shown that countable subset of A is a proper subset of A. It follows therefore that A is **uncountable** (if not it would be a proper subset of A, which is absurd)

Corollary 1.3

Since we can represent any real number in binary system (base 2), its elements will be the digits 0 and 1. It follows therefore that **the set of real numbers is uncountable.**

Theorem 1.23

Every finite set is countable.

Proof. Let A be a finite set. Then there exists a natural number n such that A is equipotent with $\{1, 2, 3, \ldots, n\}$. Now $\{1, 2, 3, \ldots, n\}$, being a subset of the countable set N of natural numbers, is countable. Thus, A is equipotent to a countable set. Hence A is countable.

Theorem 1.24

Set of irrational number is uncountable,

Proof. Suppose on the contrary that the set of irrational numbers is countable. Also we know that the set of rational number is countable. Further the set of real numbers is the union of set of all rational numbers and set of all irrational numbers. Since the union of two countable set is countable, it follows that the set of real numbers is countable which is absurd. Thus we arrive at a contradiction. Hence the set of **irrational number is not countable.**

Definition 1.44

A real number is called an **algebraic number** if it is a root of a polynomial with integral coefficients.

Definition 1.45

A real number which is not algebraic is called a **transcendental number.**

We now state two theorems without proof.

Theorem 1.25

The collection of all polynomial is countable.

Theorem 1.26

The set of all algebraic number is countable.

EXAMPLE 1.44

Show that the set of transcendental number is not countable.

Solution.

We know that

$$\mathbf{R} = \text{set of algebraic number} \cup \text{set of transcendental number.}$$

Also we know that union of two countable sets is countable. Therefore, if we assume contrary that the set of transcendental number is countable, then the set of algebraic number being countable, it follows that **R** is countable, which is absurd. Hence the set of transcendental number is not countable.

1.7 PARTIALLY ORDERED SETS

Definition 1.46

A relation R on a set X is said to be **antisymmetric** if $a\ R\ b$ and $b\ R\ a$ imply $a=b$.

Definition 1.47

A relation R on a set X is called a **partial order relation** if it is reflexive, anti-symmetric and transitive.

A set X with the partial order R is called a **partially ordered set** or **poset** and is denoted by (X, R)

EXAMPLE 1.45

Let $\tilde{A}$ be a collection of subsets of a set S. The relation $\subseteq$ of **set inclusion** is a partial order relation on $\tilde{A}$. In fact, if $A, B, C \in \tilde{A}$, then,

(i) $A \subseteq A$, that is, A is a subset of itself which is true.
(ii) If $A \subseteq B$, $B \subseteq A$, then $A = B$
(iii) If $A \subseteq B$, $B \subseteq C$, then A is a subset of C, that is, $A \subseteq C$.

EXAMPLE 1.46

Let **N** be the set of positive integers. Then the usual relation $\leq$ (read "less than or equal to") is a partial order on **N**.

Similarly, $\geq$ (read "greater than or equal to") is a partial order on **N**.

But the relation $<$ (read "less than") is not a partial order on **N**. In fact, this relation is **not reflexive**.

EXAMPLE 1.47

Let **N** be the set of positive integers. Then the **relation of divisibility** is a partial order on **N**. We say that "a divides b" written as a | b, if there exists an integer c such that $a\,c=b$. We note that for $a, b, c \in \mathbf{N}$.

(i) $a \mid a$
(ii) $a \mid b,\ b \mid a \Rightarrow a = b$
(iii) $a \mid b,\ b \mid c \Rightarrow a \mid c$.

Thus, relation of divisibility is a partial order on **N**.

EXAMPLE 1.48

The relation of divisibility is not a partial order on the set of integers. For example, $3 \mid -3$, $-3 \mid 3$ but $3 \neq -3$, that is, the relation is not anti-symmetric and so cannot be partial order.

EXAMPLE 1.49

If R is a partial order on A, then R^{-1} (inverse relation) is also a partial order.

We know that if R is a relation on A, then

$$R^{-1} = \{(b, a) : (a, b) \in R\},\ a, b \in A.$$

Since R is a partial order relation,

(i) $a\,R\,a\ \forall\, a \in A$

(ii) $a\,R\,b,\ b\,R\,a \Rightarrow a = b$

(iii) $a\,R\,b,\ b\,R\,c \Rightarrow a\,R\,c.$

We observe that

(i) Since R is a relation, $(a, a) \in R\ \forall\, a \in A$

$\Rightarrow (a, a) \in R^{-1}$

$\Rightarrow a\,R^{-1}\,a.$

Thus the relation R^{-1} is reflexive.

(ii) If $(b, a) \in R^{-1}$ and $(a, b) \in R^{-1}$, then

$(a, b) \in R$ and $(b, a) \in R$,

$\Rightarrow a\,R\,b$ and $b\,R\,a$,

$\Rightarrow a = b$, since R is anti-symmetric.

Hence, R^{-1} is anti-symmetric.

(iii) If $(b, a) \in R^{-1}$ and $(c, b) \in R^{-1}$, then

$(a, b) \in R$ and $(b, c) \in R$,

$\Rightarrow (a, c) \in R$, since R is transitive

$\Rightarrow (c, a) \in R^{-1}$.

Thus $(c, b) \in R^{-1}$ and $(b, a) \in R^{-1} \Rightarrow (c, a) \in R^{-1}$ and so R^{-1} is transitive.
Hence R^{-1} is a partial order.

The poset (A, R^{-1}) is called the **dual of the poset** (A, R) and the partial order R^{-1} is called the **dual of the partial order** R.

Definition 1.48

A relation R on a set A is said to be **quasi order** if

(i) R is **irreflexive**, that is, $(a, a) \notin R$ for any $a \in A$

(ii) R is **transitive**, that is, $a\,R\,b,\ b\,R\,c \Rightarrow a\,R\,c$ for $a, b, c \in A$.

Definition 1.49

Let (A, R) be a poset. The elements a and b of A are said to be **comparable** if

$$a\,R\,b \text{ or } b\,R\,a.$$

EXAMPLE 1.50

We know that the relation of divisibility is a partial order on the set of natural numbers. But we see that $3 \nmid 7$ and $7 \nmid 3$.

Thus, 3 and 7 are positive integers in **N** which are **not comparable** (In such a case we write $3 \parallel 7$).

Definition 1.50

If every pair of elements in a poset (A, R) is comparable, we say that A is **linearly ordered (totally ordered** or **a chain**). The partial order is then called **linear order** or **total ordering relation**. The number of elements in a chain is called the **length of the chain**.

EXAMPLE 1.51

The set **N** of positive integers with relation $\leq$ (to read as "less than or equal to)" is linearly ordered. In fact, every ordered subset of **N** is also linearly ordered.

EXAMPLE 1.52

Let A be a set with two or more elements and let $\subseteq$ (set inclusion) be taken as the relation on the **subsets of** A. If a and b are two elements of A, then $\{a\}$ and $\{b\}$ are subsets of A but they are not comparable. Hence $P(A)$ is **not a chain**. A subset of A is called **Antichain** if no two distinct elements in the subsets are related.

But, if we consider the subsets ϕ, $\{a\}$ and A of A, then this collection (subsets $\{\phi, \{a\}, A\}$ of $P(A)$) is a chain because $\phi \subseteq \{a\} \subseteq A$. Similarly, ϕ, $\{b\}$ and A form a chain.

EXAMPLE 1.53

The set **N** of positive integers along with the relation of divisibility is not a chain. For example, the pair 5, 7 is not comparable.

But the subset $\{3, 6, 12, 24\}$ is a chain under the relation of divisibility.

EXAMPLE 1.54

Let (A, R) and (B, R') be posets. Then $(A\times B, R'')$ is a poset with partial order R'' defined by

$$(a, b)\ R''\ (a', b') \text{ if } a\ R\ a' \text{ in } A \text{ and } b\ R'\ b' \text{ in } B$$

Solution.

We note that

(i) Since R is a relation on A, $a\ R\ a$ by reflexivity. Similarly since R' is a relation on B, $b\ R'\ b$ by reflexivity. Let $(a, b) \in A\times B$. Since $a\ R\ a$ and $b\ R'\ b$,

$$(a, b)\ R''\ (a, b).$$

Thus R'' is reflexive.

(ii) Let $(a, b)\ R''\ (a', b')$ and $(a', b')\ R''\ (a, b)$. Then, by definition,

$$a\ R\ a', \quad a'\ R\ a \text{ in } A \qquad \text{(i)}$$
$$b\ R'\ b', \quad b'\ R'\ b \text{ in } B. \qquad \text{(ii)}$$

Since (A, R) and (B, R') are posets, (i) and (ii) respectively imply

$$a=a'$$

and

$$b=b'.$$

Thus, $(a, b)\ R''\ (a', b')$ and $(a', b')\ R''\ (a, b)$ imply

$$(a, b)=(a', b').$$

Hence R'' is anti-symmetric.

(iii) Let $(a, b)\ R''\ (a', b')$ and $(a', b')\ R''\ (a'', b'')$, where $a, a', a'' \in A$ and $b, b', b'' \in B$. Then

$$a\,R\,a' \text{ and } a'\,R\,a'' \tag{iii}$$
$$b\,R'\,b' \text{ and } b'\,R'\,b''. \tag{iv}$$

By transitivity of R, (iii) gives

$$a\,R\,a'',$$

while (iv) yields

$$b\,R'\,b''.$$

Hence,

$$(a, b)\,R''\,(a'', b'').$$

Hence R'' is transitive and so $(A \times B, R'')$ is a poset.

The partial order R'' defined on the Cartesian product $A \times B$, as above, is called the **Product Partial Order**.

Definition 1.51

A relation R on a set A is called **asymmetric** if $a\,R\,b$ and $b\,R\,a$ do not both hold for any $a, b \in A$.

Definition 1.52

A transitive, asymmetric relation R is called a **Strict Partial Ordering**.

Theorem 1.27

If $\leq$ is a partial order of the set A, then a relation $<$ defined by

$$a < b \text{ if } a \leq b \text{ and } a \neq b$$

is a **strict partial order of** A.

Proof. We shall show that $<$ is transitive and asymmetric.

(i) Transitivity: Suppose that $a<b$ and $b<c$. Then, by definition,

$$a \leq b \text{ and } b \leq c,\ a \neq b,\ b \neq c. \tag{1}$$

Since $\leq$ is partial order, it is transitive and so $a \leq c$. It remains to show that $a \neq c$. Suppose on the contrary that $a=c$. Then,

$$c \leq b \text{ (using } a \leq b \text{ from (1)).} \tag{2}$$

From (1) and (2), we have $b \leq c$ and $c \leq b$ and so $b=c$ which is contradiction. Hence,

$$a<b \text{ and } b<c \text{ implies } a<c.$$

Proving that $<$ is transitive.

(ii) Asymmetry: Suppose that $x<y$ and $y<x$ both holds. Therefore,

$$x \leq y \text{ and } y \leq x.$$

Since $\leq$ is **anti-symmetric**, it follows that $x=y$, which contradicts $x<y$. Hence $x<y$ and $y<x$ cannot both hold. Thus $<$ is asymmetric.

Hence $<$ is **strict partial order of** A.

Remark 1.1 If $<$ is a strict partial order of A, then the relation $\leq$ defined by $x \leq y$ if $x<y$ or $x=y$ is a partial order of A (can be proved using the definitions).

Definition 1.53

A sequence of letters or other symbols, written without commas is called a **string.** Further,

(i) A string of length p may be considered as an ordered p-tuple.

(ii) An infinite string such as *abababab*... may be regarded as the infinite sequence $a, b, a, a, b, ab, \ldots$

(iii) If S is any set with a partial order relation, then the set of strings over S is denoted by S^*.

Definition 1.54

Let $(A, \leq)$ and $(B, \leq)$ be chains (linearly ordered sets). Then the order relation (which is in fact totally ordered) $<$ on the Cartesian product $A \times B$ defined by

$$(a, b) < (a', b') \text{ if } a < a' \text{ or if } a = a' \text{ and } b \leq b'$$

is called **Lexicographic order** or **Dictionary order.**

EXAMPLE 1.55

Consider the plane $\mathbf{R}^2 = \mathbf{R} \times \mathbf{R}$. It is linearly ordered by lexicographic order. In fact, each vertical line has usual order (less than or equal to) and points on a line (e.g., $x = a_1$ in Fig. 1.14) are less than any point on a line farther to the right (e.g. $x = a_2$ in Fig. 1.14). Thus the point $p_1\,(a_1, b_1) < p_2\,(a_2, b_2)$ because $a_1 < a_2$. Further, $p_2(a_2, b_2) < p_3(a_2, b_3)$ because in this case $a_2 = a_2$ and $b_2 \leq b_3$.

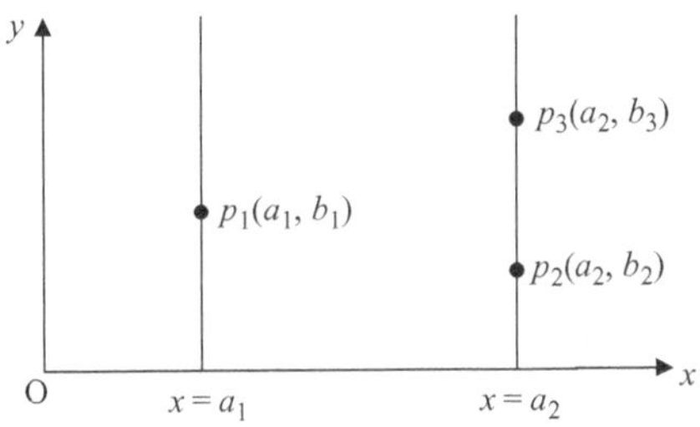

Figure 1.14

Theorem 1.28

The digraph (directed graph) of a partial order has no cycle of length greater than 1.

Proof. Suppose on the contrary that the digraph of the partial order $\leq$ on the set A contains a cycle of length $n \geq 2$. Then there exist distinct elements $a_1, a_2, \ldots, a_n$ such that

$$a_1 \leq a_2,\ a_2 \leq a_3,\ \ldots,\ a_{n-1} \leq a_n,\ a_n \leq a_1.$$

By the transitivity of partial order, used $n-1$ times, $a_1 \leq a_n$. Thus we have

$$a_1 \leq a_n \text{ and } a_n \leq a_1.$$

Since $\leq$ is partial order, anti-symmetry implies $a_1 = a_n$, which is a contradiction to the assumption that $a_1, a_2, \ldots, a_n$ are distinct. Hence the result.

Definition 1.55

The **Transitive closure** of a relation R is the smallest transitive relation containing R. It is denoted by R^∞.

EXAMPLE 1.56

Let $A=\{1, 2, 3, 4\}$ and $R=[(1, 2), (2, 3), (3, 4), (2, 1)]$. Find the transitive closure of R.

Solution.

The digraph of R is

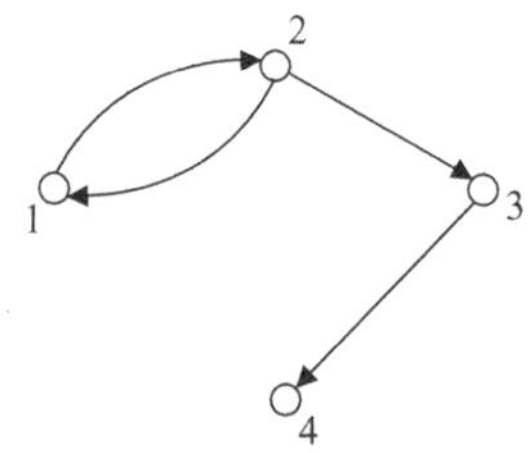

Figure 1.15

We note that from vertex 1, we have paths to the vertices 2, 3, 4 and 1. Note that path from 1 to 1 proceeds from 1 to 2 to 1. Thus we see that the ordered pairs (1, 1), (1, 2), (1, 3) and (1, 4) are in R^{∞}. Starting from vertex 2, we have paths to vertices 2, 1, 3 and 4 so the ordered pairs (2, 1), (2, 2), (2, 3) and (2, 4) are in R^{∞}. The only other path is from vertex 3 to 4, so we have

$$R^{\infty}=\{(1, 1), (1, 2), (1, 3), (1, 4), (2, 1), (2, 2), (2, 3), (2, 4), (3, 4)\}$$

1.8 HASSE DIAGRAM

Let A be a finite set. By the theorem proved above, the digraph of a partial order on A has only cycles of length 1. In fact, since a partial order is reflexive, every vertex in the digraph of the partial order is contained in a cycle of length 1. To simplify the matter, we shall delete all such cycles from the digraph. Thus, the digraph shown in the Figure 1.16(a) would be drawn as shown in Figure 1.16(b).

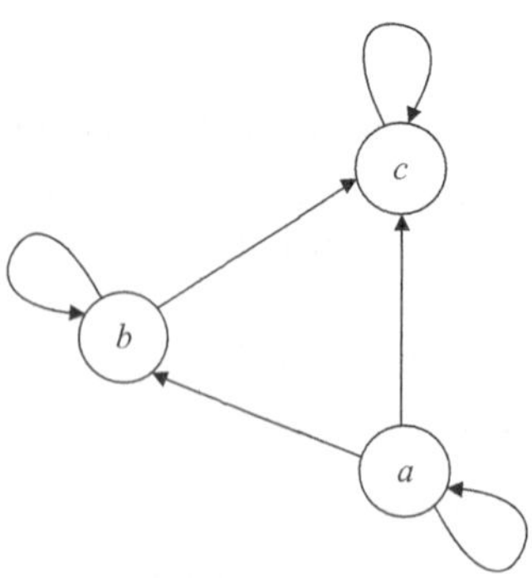

Figure 1.16(a)

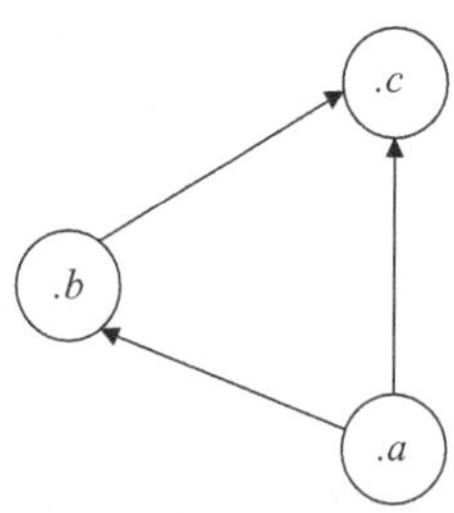

Figure 1.16(b)

We also eliminate all edges that are implied by transitivity property. Thus, if $a \leq b$, $b \leq c$, it follows that $a \leq c$. In this case, we omit the edge form a to c. Thus we have the digraph as shown in Figure 1.16(c).

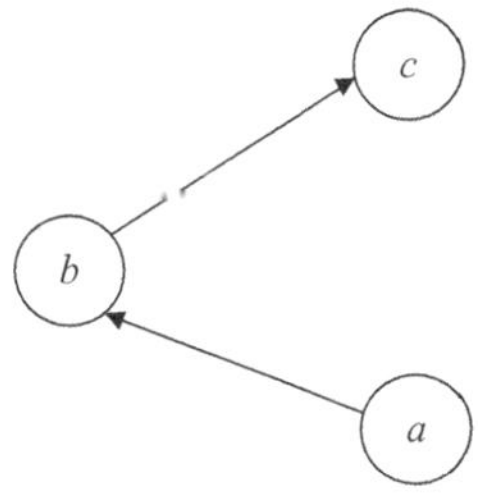

Figure 1.16(c)

We also agree to draw the digraph of partial order with all edges pointing upward, omit the arrows and to replace then the circles by dots. Hence, the final form of the digraph (Fig. 1.16(d)) becomes

Figure 1.16(d)

Thus,

"The diagram of a partial order obtained from its digraph by omitting cycles of length 1, the edges implied by transitivity and the arrows (after arranging them pointing upward) is called **Hasse diagram** of the partial order of the poset".

EXAMPLE 1.57

Let $A = \{1, 2, 3, 4, 12\}$. Consider the partial order of divisibility on A. That is, if a and b are in A, $a \leq b$ if and only if $a \mid b$.

The digraph of the poset is shown in the Figure 1.17.

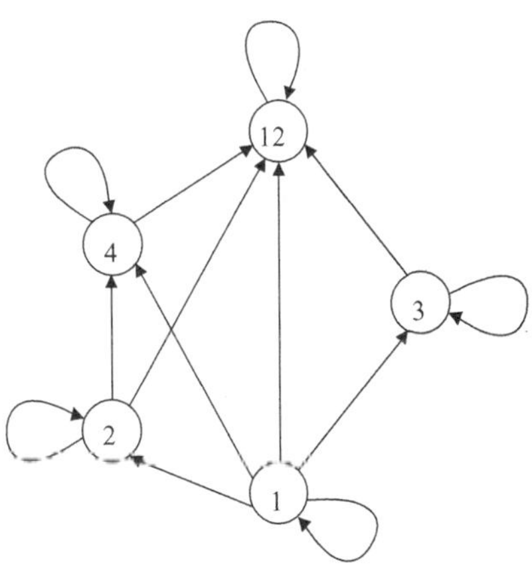

Figure 1.17

Therefore, the **Hasse diagram** of the poset $(A, \leq)$ is as shown in Figure 1.18.

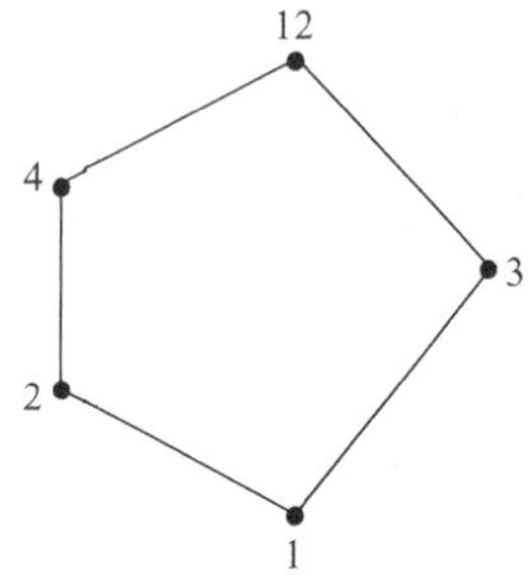

Figure 1.18

EXAMPLE 1.58

Let $S=\{a, b, c\}$ and $\tilde{A}=P(S)$ (power set of S).
Consider the partial order of set inclusion ($\subseteq$). We note that

$$\tilde{A}=P(S)=\{\phi, \{a\}, \{b\}, \{c\}, \{a, b\}, \{a, c\}, \{b, c\}, \{a, b, c\}\}$$

Then the Hasse diagram of the poset $(\tilde{A}, \subseteq)$ is as shown in Figure 1.19.

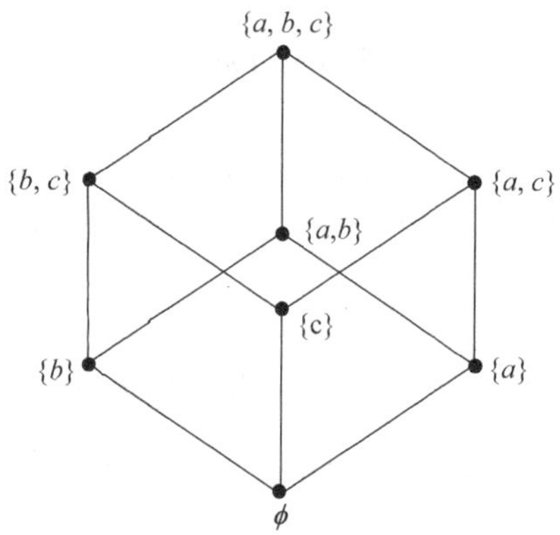

Figure 1.19

Remark 1.2 Hasse diagram of a finite linearly ordered set is always of the form as shown in the Figure 1.20 and thus consists of simply one path. Hence diagram of a totally order set is a chain.

Figure 1.20

1.8.1 Hasse Diagram of Dual Poset

If $(A, \leq)$ is a poset and $(A, \geq)$ is the dual poset, then the Hasse diagram of $(A, \geq)$ is just the Hasse diagram of $(A, \leq)$ turned upside down.

For example, let $A = \{a, b, c, d, e, f\}$ and let

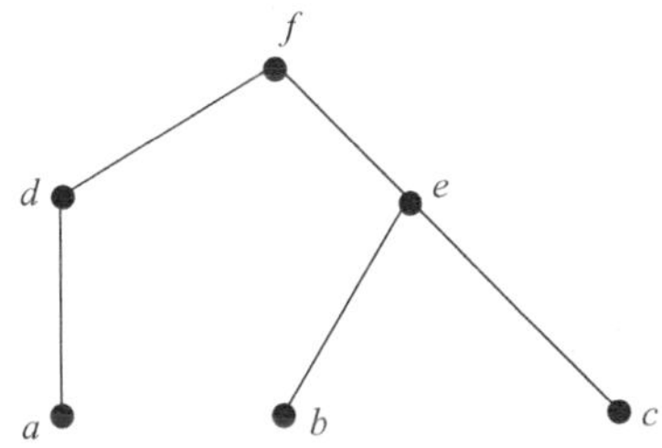

Figure 1.21

be the Hasse diagram of poset $(A, \leq)$. Then the Hasse diagram of dual poset $(A, \geq)$ is

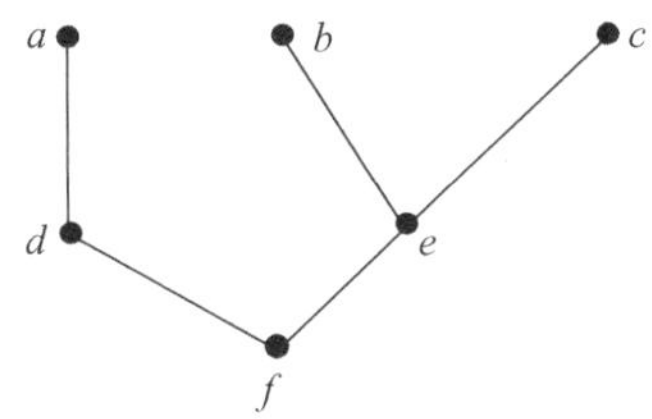

Figure 1.22

which can be constructed by turning the Hasse diagram of $(A, \leq)$ upside down.

EXAMPLE 1.59

Let $A = \{a, b, c, d, e\}$. Then the Hasse diagram

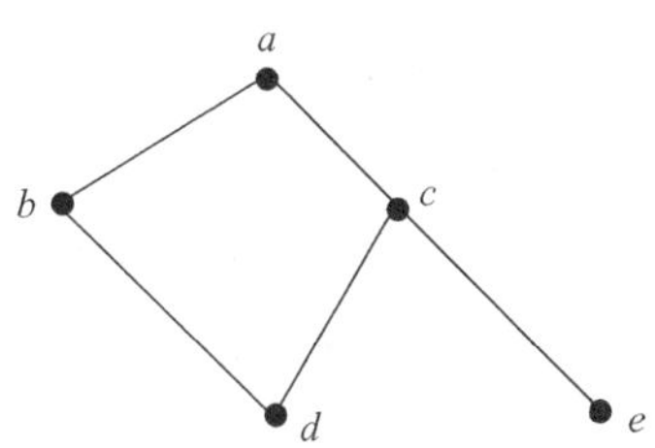

Figure 1.23

defines a partial order on B in the natural way. That is, $d \leq b$, $d \leq a$, $e \leq c$ and so on.

EXAMPLE 1.60

Let n be a positive integer and D_n denote the set of all divisors of n. Considering the partial order of divisibility in D_n, draw Hasse diagram D_{24}, D_{30} and D_{36}.

Solution.

We know that

$$D_{24} = \{1, 2, 3, 4, 6, 8, 12, 24\},$$
$$D_{30} = \{1, 2, 3, 5, 6, 10, 15, 30\},$$
$$D_{36} = \{1, 2, 3, 4, 6, 9, 12, 18, 36\}.$$

Therefore, the Hasse diagram of D_{24}, D_{30} and D_{36} are as shown in the Figure 1.24.

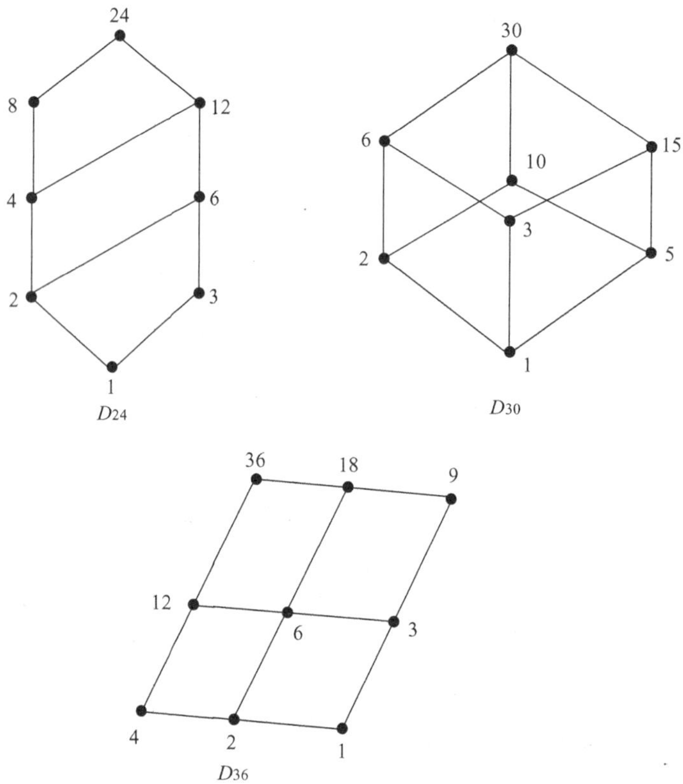

Figure 1.24

EXAMPLE 1.61

Find the directed graph of the partial order relation having the Hasse diagram given in the Figure 1.25.

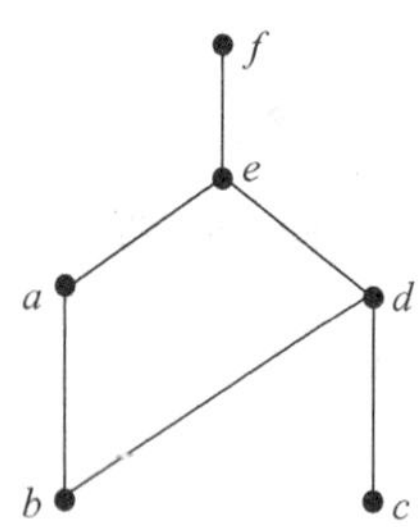

Figure 1.25

Solution.

First of all introduce arrow heads pointing upward and get Figure 1.26(a).

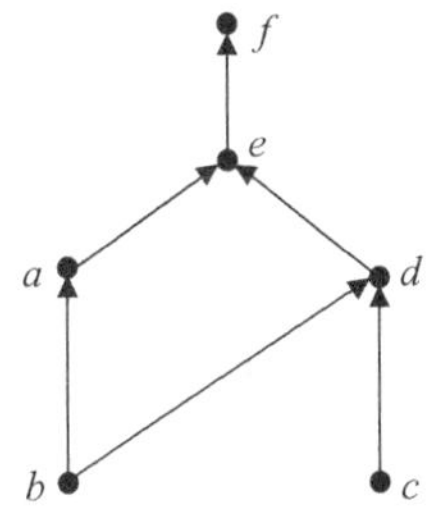

Figure 1.26(a)

Now introduce the edges which were deleted on account of transitivity and then introduce cycle of length 1. We get the directed graph as shown in the Figure 1.26(b).

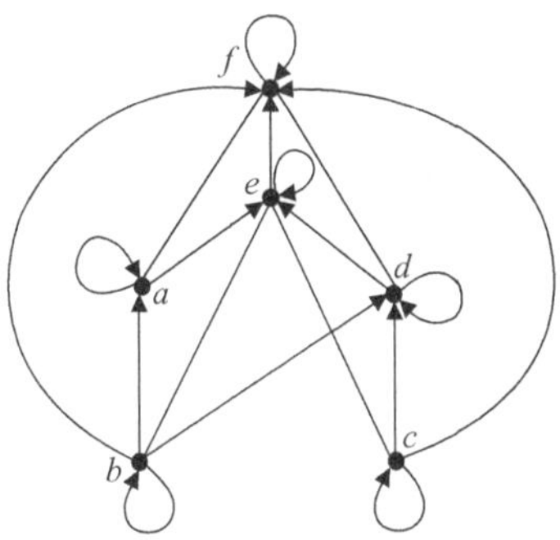

Figure 1.26(b)

Definition 1.56

A partition of a positive integer m is a set of positive integers whose sum is m.

EXAMPLE 1.62

There are seven partitions of $m = 5$ as follows:

$$\{5\}, \{3, 2\}, \{2, 2, 1\}, \{1, 1, 1, 1, 1\}, \{4, 1\}, \{3, 1, 1\}, \{2, 1, 1, 1\}.$$

We order the partitions of an integer m as follows:

A partition P_1 precedes a partition P_2 if the integers in P_1 can be added to obtain integers in P_2 or we can say that if the integers in P_2 can be further subdivided to obtain the integers in P_1. For example, $\{1, 1, 1, 1, 1\}$ precedes $\{2, 1, 1, 1\}$. On the other hand, $\{3, 1, 1\}$ and $\{2, 2, 1\}$ are non-comparable.

The Hasse diagram of the partition of $m = 5$ is

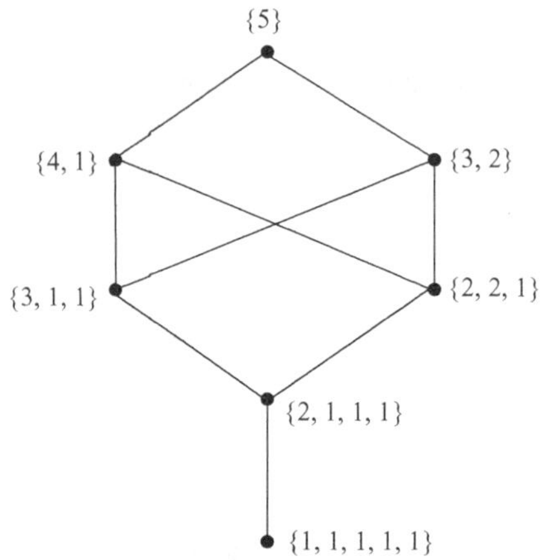

Figure 1.27

Definition 1.57

Let A be a (non-empty) linearly ordered alphabet. Then **Kleene closure** of A consists of all words w on A and is denoted by A^*.

Also then $|w|$ denotes the length of w.

We have following two order relations on A^*.

(A) **Alphabetical (Lexicographical) order:** In this order we have

(i) $\lambda < w$, where λ is empty word and w is any non-empty word.

(ii) Suppose $u = a\ u'$ and $v = b\ v'$ are distinct non-empty words where $a,\ b \in A$ and $u', v' \in A^*$. Then,

$$u < v \text{ if } a < b \text{ or if } a = b \text{ but } u' < v'$$

(B) **Short-lex order:** Here A^* is ordered first by length and then alphabetically, that is, for any distinct words u, v, in A^*,

$$u < v \text{ if } |u| < |v| \text{ or if } |u| = |v| \text{ but } u \text{ precedes } v \text{ alphabetically.}$$

For example, "to" precedes "and" since $|\text{to}| = 2$ but $|\text{and}| = 3$.

Similarly, "an" precedes "to" since they have the same length but "an" precedes "to" alphabetically.

This order is also called **free semi-group order**.

Definition 1.58

Let A be a partially ordered set with respect to a relation $\leq$. An element a in A is called a **maximal element** of A if and only if for all b in A, either $b \leq a$ or b **and** a **are not comparable**.

An element a in A is called **greatest element** of A if and only if for all b in A, $b \leq a$.

An element a in A is called **minimal element** of A if and only if for all b in A, **either** $a \leq b$ or b **and** a **are not comparable.**

An element a in A is called a **least element** of A if and only if for all b in A, $a \leq b$.

A greatest element is maximal but a maximal element need not be greatest element.

Similarly, a least element is minimal but a minimal element need not be a least element.

Remark 1.3 A partially ordered set with respect to a relation can have at most one greatest element and one least element but **it may have more than one maximal or minimal element.**

EXAMPLE 1.63

Consider the poset A whose Hasse diagram is given in the Figure 1.28.

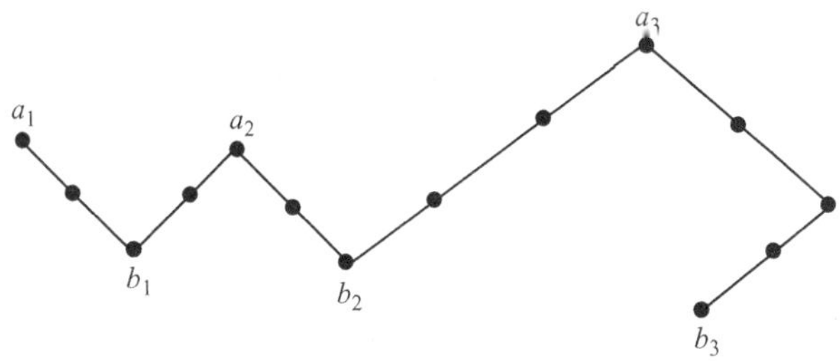

Figure 1.28

The elements a_1, a_2 and a_3 are **maximal elements** of A, and the elements b_1, b_2 and b_3 are **minimal elements**. Observe that since there is no line between b_2 and b_3 we can conclude that neither $b_3 \leq b_2$ nor $b_2 \leq b_3$ showing that b_2 and b_3 are not comparable.

EXAMPLE 1.64

Let $A = \{a, b, c, d, e, f, g, h, i\}$ have the partial ordering $\leq$ defined by the Hasse diagram (Figure 1.29).

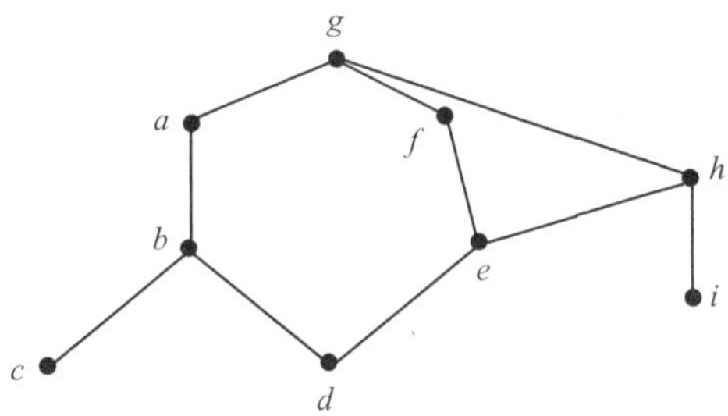

Figure 1.29

Find the maximal, minimal, greatest and least elements of A.

Solution.

Here g is the only maximal element which is also the greatest element. The minimal elements are c, d and i. There is no least element.

EXAMPLE 1.65

Let A be the poset of non-negative real numbers with usual partial order $\leq$ (read as "less than or equal to"). Then 0 is the minimal element of A. There is no maximal element of A.

EXAMPLE 1.66

The poset **Z** with the usual partial order $\leq$ has no maximal element and has no minimal element.

EXAMPLE 1.67

Let $S = \{a, b, c\}$ and consider the poset

$$\tilde{A} = P(S) = \{\phi, \{a\}, \{b\}, \{c\}, \{a, b\}, \{b, c\}, \{c, a\}, \{a, b, c\}\}.$$

The empty set is the least element and the set S is a greatest element of $\tilde{A}$.

Theorem 1.29

Let A be a finite non-empty poset with partial order $\leq$. Then A has at least one maximal element and at least one minimal element.

Proof. Let a be any element of A. If a is not maximal, we can find an element $a_1 \in A$ such that $a<a_1$. If a_1 is not maximal, we can find an element $a_2 \in A$ such that $a_1<a_2$. This argument cannot be continued indefinitely since A is finite. Consequently, we obtain finite chain

$$a<a_1<a_2<\ldots<a_{k-1}<a_k,$$

which cannot be extended. Thus, we **cannot have** $a_k<b$ for any $b \in$ A. Hence a_k is a maximal element of $(A, \leq)$.

The same argument implies that the dual poset $(A, \geq)$ has maximal element and so $(A, \leq)$ has a minimal element.

Theorem 1.30

A poset $(A, \leq)$ has at most one greatest element and at most one least element.

Proof. Suppose on the contrary that a and b are greatest elements of a poset $(A, \leq)$. Since b is greatest element, $a\leq b$. Similarly, since a is greatest element, we have $b\leq a$. Hence, by anti-symmetry of $\leq$, we have $a=b$. Hence, the greatest element, if exists, **is unique.**

By the same argument, the dual poset $(A, \geq)$ has at most one greatest element and so $(A, \leq)$ has at most one least element.

The greatest element of a poset, if exists, is often called the **unit element** (dented) by I). Similarly, the least element of a poset, if it exists, is called the **zero element** (denoted by 0).

Definition 1.59

Let $(A, \leq)$ be a poset and let B be a subset of A. An element $a \in A$ is called an **upper bound of** B if $b\leq a$ for all $b \in B$.

Similarly, an element $a \in A$ is called a **lower bound of** B if $a\leq b$ for all $b \in B$.

EXAMPLE 1.68

Consider the poset $A=\{a, b, c, d\}$ having Hasse diagram given in the Figure 1.30.

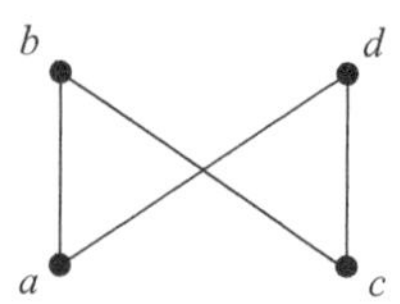

Figure 1.30

Find the lower bounds of $\{b\}$ and $\{d\}$.

Solution.

The lower bounds of $\{b\}$ are a and c whereas the lower bounds of $\{d\}$ are a and c.

EXAMPLE 1.69

Consider the poset $A=\{a, b, c, d, e, f, g, h\}$ having Hasse diagram shown in the Figure 1.31.

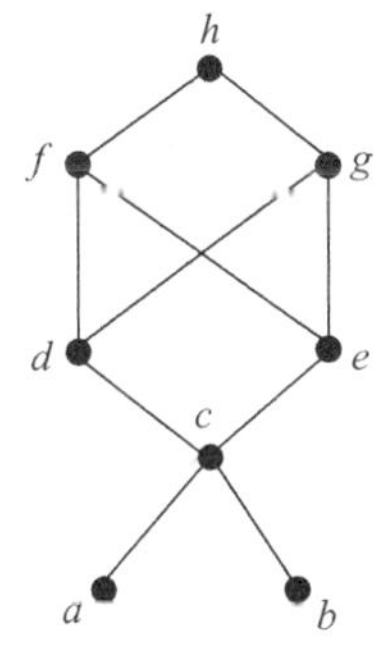

Figure 1.31

Find all upper and lower bounds of

(i) The subset $\{a, b\} = B_1$ (say)
(ii) The subset $\{c, d, e\} = B_2$ (say)

Solution.

(i) B_1 has no lower bound. The upper bounds of B_1 are c, d, e, f, g and h.
(ii) The upper bounds of B_2 are f, g and h. The lower bounds of B_2 are c, a and b.

It follows from the above example that

(i) A subset B of a poset may or may not have upper or lower bounds in A.
(ii) An upper or lower bound of B may or may not belong to B.

Definition 1.60

Let $(A, \leq)$ be a poset and B a subset of A. An element $a \in A$ is called a **least upper bound** (supremum) of B if

(i) a is an upper bound of B, that is, $b \leq a \; \forall \, b \in B$
(ii) $a \leq a'$ whenever a' is an upper bound of B.

An element $a \in A$ is called a **greatest lower bound** (infimum) of B if

(i) a is a lower bound of B, that is, $a \leq b \; \forall \, b \in B$
(ii) $a' \leq a$ whenever a' is a lower bound of B.

Further, upper bounds in the poset $(A, \leq)$ correspond to lower bounds in the dual poset $(A, \geq)$ and the lower bounds in $(A, \leq)$ correspond to upper bound in $(A, \geq)$.

Similar statements also hold for greatest lower bounds and least upper bounds.

Consider Example 1.69 above:

(i) Since $B_1 = \{a, b\}$ has no lower bound, it has no greatest lower bound. However,

$$\text{lub}\,(B_1) = c$$

(ii) Since the lower bounds of $B_2 = \{c, d, e\}$ are c, a and b, we have

$$\text{glb}\,(B_2) = c$$

The upper bounds of B_2 are f, g, h. Since f **and** g **are not comparable**, we conclude that B_2 has no least upper bound.

(Here d and e are not upper bounds of $\{c, d, e\}$ because $d \not\leq e \in B_2$ and $e \not\leq d \in B_2$)

EXAMPLE 1.70

Let $A = \{1, 2, 3, 4, 5, \ldots, 11\}$be the poset whose Hasse diagram is given (Figure 1.32).

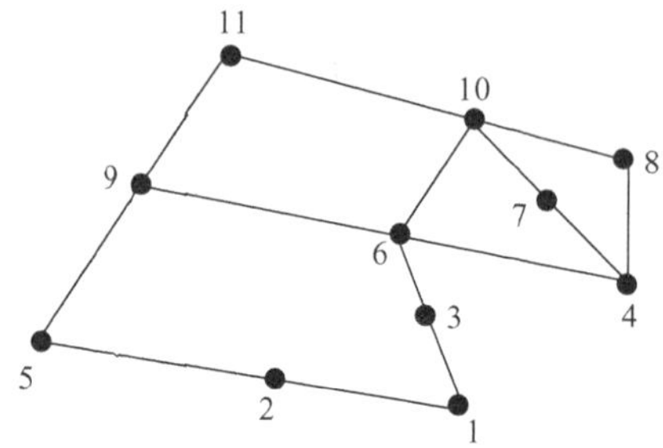

Figure 1.32

Find lub and glb of $B=\{6, 7, 10\}$, if they exist.

Solution.

Exploring all upward paths from 6, 7 and 10 we find that lub $(B)=10$. Similarly, by examining all downward paths from 6, 7 and 10, we find that glb $(B)=4$.

EXAMPLE 1.71

Let D_n denote the set of factors of a positive integer n partially ordered by divisibility. Then,

$$\left.\begin{array}{l}\operatorname{glb}(x, y)=\operatorname{gcd}(x, y) \\ \operatorname{lub}(x, y)=\operatorname{lcm}(x, y)\end{array}\right\} \text{ for all } x, y \in D_n.$$

Definition 1.61

Let $\leq$ and $\leq'$ be two partial order relations on a set A. Then $\leq'$ is said to be **compatible with** $\leq$ if and only if

$$a \leq b \Rightarrow a \leq' b \text{ for all } a, b \in A.$$

Definition 1.62

The process of constructing a linear order (total order) which is compatible to a given partial order on a given set is called **topological sorting**.

The construction of a topological sorting for a general finite partially order set is based on the fact that any partially ordered set that is finite and non-empty has a minimal element.

To create a total order for a partially ordered set $(A, \leq)$, we proceed as follows:

(i) Pick any minimal element and make it number one. Let this element be a.

(ii) Consider $A - \{a\}$. It is a subset of A and so is partially ordered. If it is empty, stop the process. If not, pick a minimal element from it and call it element number 2. Let it be b.

(iii) Consider $A - \{\{a\}, \{b\}\}$. If this set is empty, stop the process. If not, pick a minimal element and call it number 3. Continue in this way until all the elements of the set have been used up.

We now give algorithm to construct a topological sorting for a relation $\leq$ defined on a non-empty finite set A.

1.8.2 Algorithm for Topological Sorting

1. Pick any minimal element a in A (minimal element exists since A is non-empty)
2. Set $A := A - \{a\}$

3. Repeat steps x to z while $A \neq \phi$

x : Pick any minimal element b in A.
y : Define $a \leq' b$
z : Set $A : A - \{a\}$ and $a := b$.

EXAMPLE 1.72

Let $A = \{2, 3, 4, 6, 18, 24\}$ be partially ordered with relation of divisibility. The Hasse diagram of this relation is given below. Find topological sorting of this poset.

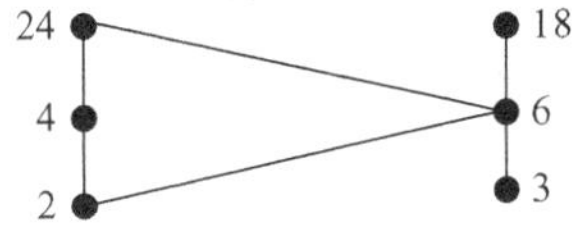

Figure 1.33

Solution.

The ordinary "less than or equal to" relation on this set is a topological sorting for it since for positive integers a and b, if $a \mid b$, then a is less than or equal to b.

Another topological sorting for this set. This set has two minimal elements 2 and 3. Let us pick up 3. The beginning of the total order is

total order: 3

Set $A = A - \{3\}$. We can indicate it by removing 3 from the Hasse diagram as shown below:

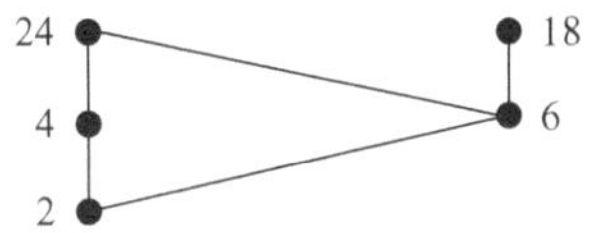

Figure 1.34

Next choose a minimal element from $A - \{3\}$. Only 2 is minimal and so pick it. The total order thus for is

total order: $3 \leq' 2$

Set $A = (A - \{3\}) - \{2\} = A - \{3, 2\}$. We indicate it by removing 2 from the Hasse diagram as shown below:

Figure 1.35

Choose a minimal element from $A - \{3, 2\}$. We have two minimal elements 4 and 6. We pick up 6. The total order for the elements chosen thus for it

total order: $3 \leq' 2 \leq' 6$

Set $A = (A - \{3, 2\}) - \{6\}$. The corresponding diagram becomes

Figure 1.36

Then, remove one of 4 and 18. If we remove 18, we get

Figure 1.37

total order : $3 \preceq' 2 \preceq' 6^* \preceq' 18$

Then,

$$A = (A - \{3, 2, 6, 18\})$$

Now minimal element is 4. We remove it and we get

$$A = A - \{3, 2, 6, 18, 4\} = \{24\}$$

total order : $3 \preceq' 2 \preceq' 6 \preceq' 18 \preceq' 4 \preceq' 24$

Figure 1.38

Remark 1.4 There would have been another total order if we had selected the other minimal points. Thus, "**There are many ways of topologically sorting of a given poset**."

EXAMPLE 1.73

Let $A = \{a, b, c, d, e\}$ and let the Hasse diagram of a partial order $\leq$ on A be as given below

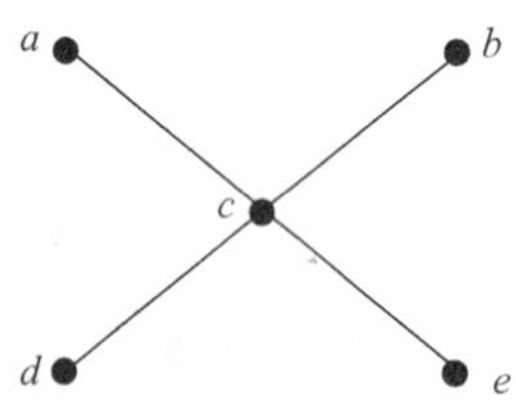

Figure 1.39

The minimum elements of this poset are d and e. We pick up d and put it in sort [1]. The Hasse diagram of $A - \{d\}$ is then as shown below:

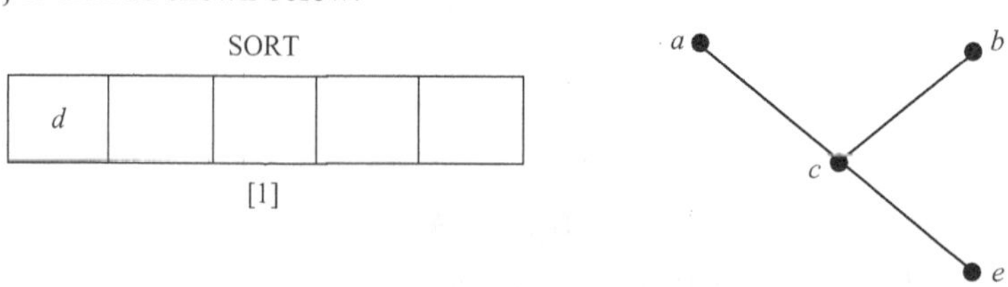

Figure 1.40 (Hasse diagram of $A - \{d\}$).

The minimal element of $A - \{d\}$ is e and we put e in sort [2]. The Hasse diagram of $A - \{d, e\}$ is then as shown below (Figure 1.41).

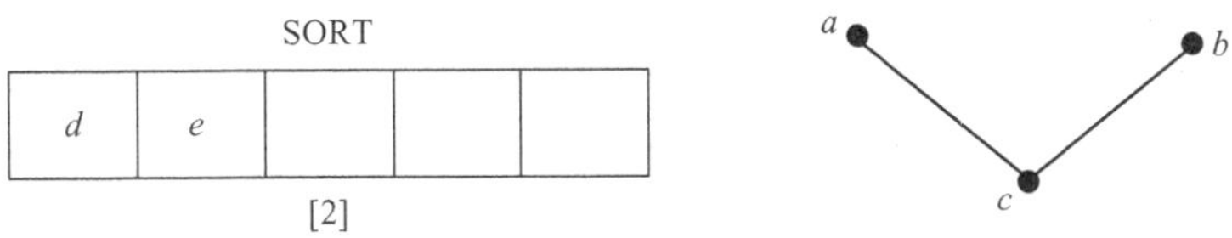

Figure 1.41 (Hasse diagram of $A - \{d, c\}$).

The minimal element of $A - \{d, e\}$ is c and we put e in sort [3]. The Hasse diagram of $A - \{d, e, c\}$ is as shown below (Figure 1.42(a)).

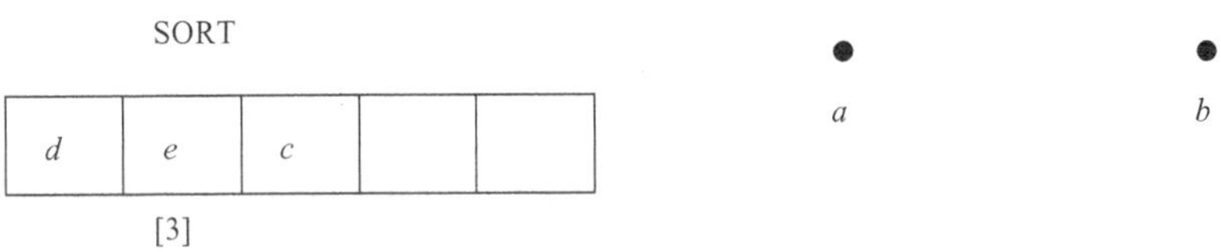

Figure 1.42(a) (Hasse diagram of $A - \{d, e, c\}$).

The minimal element of $A - \{d, e, c\}$ are a and b. We pick b and put it in sort [4]. The Hasse diagram of $A - \{d, e, c, b\}$ is shown below:

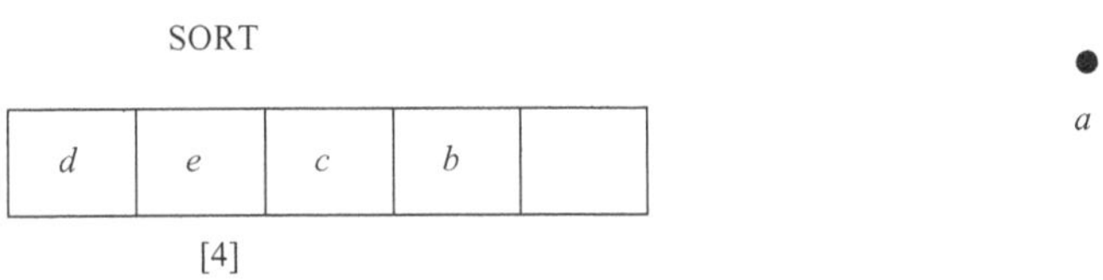

Figure 1.42(b) (Hasse diagram of $A - \{d, e, c, b\}$).

The minimal element of $A - \{d, e, c, b\}$ is a and we put it in sort [5].

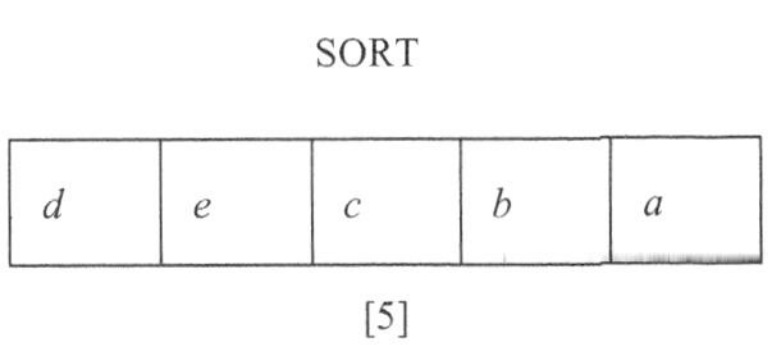

Figure 1.42(c)

The topological sorting of $(A, \leq)$ is therefore $(A, <)$, where

total order : $d<e<c<b<a$

and the Hasse diagram of $(A, <)$ is as shown in the Figure 1.42(d).

Figure 1.42(d)

1.9 ISOMORPHIC (SIMILAR) ORDERED SETS

Definition 1.63

Let $(A, \leq)$ and $(A', \leq')$ be posets. An one-to-one (injective) function f: $A \to A'$ is called an **isomorphism (similarity mapping)** from $(A, \leq)$ to $(A', \leq')$ if for any a and b in A

(i) $a \leq b$ if and only if $f(a) \leq' f(b)$
(ii) $a \parallel b \Rightarrow f(a) \parallel f(b)$, that is, if a and b are non-comparable then $f(a)$ and $f(b)$ are non-comparable.

Obviously, if A and A' are linearly ordered, then only (i) is needed for isomorphism.

Definition 1.64

Two ordered sets $(A, \leq)$ and $(A', \leq')$ are said to be **isomorphic** or **similar** (written as $A \cong A'$) if there exists a one-to-one correspondence (bijective mapping) $f: A \to A'$ which is isomorphism.

EXAMPLE 1.74

Let $(A, \leq)$ be poset of positive integers ordered with usual order (less than or equal to) and let $(A', \leq')$ be poset of positive even integer with usual order $\leq'$. Then the function $f: A \to A'$ given by

$$f(a) = 2a \; \forall \, a \in A$$

is an isomorphism from $(A, \leq)$ to $(A', \leq')$. In fact,

(i) We note that

$$f(\text{a}) = f(b) \Rightarrow 2a = 2b \Rightarrow a = b$$

and so f is one-one. Also domain of f is whole of the set of positive integers and range of f is equal to whole set of even positive integers. Hence, f is onto proving that f is bijective.

(ii) We note that

$$a \leq b \Leftrightarrow 2a \leq' 2b.$$

Also, if $a \parallel b$, then $2a \parallel 2b$.
Hence, f is an isomorphism.

EXAMPLE 1.75

Suppose $A = \{1, 2, 6, 8, 12\}$ is ordered by divisibility relation and suppose that $B = \{a, b, c, d, e\}$ is isomorphic to A. Let $f: A \to B$ be isomorphism:

$$f = \{(1, e), (2, d), (6, b), (8, c), (12, a)\}$$

The Hasse diagram of A is

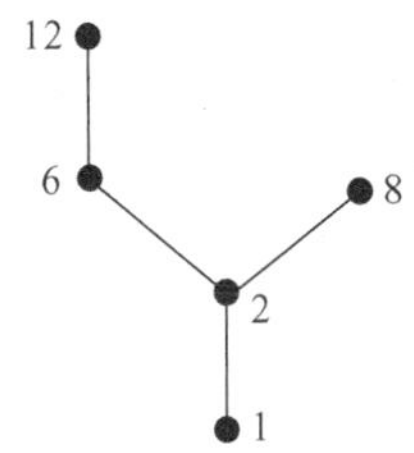

Figure 1.43

The isomorphism preserves the order of A and is bijective. Thus, f is simply a relabelling of the vertices in the above Hasse diagram. Hence the Hasse diagram for B is

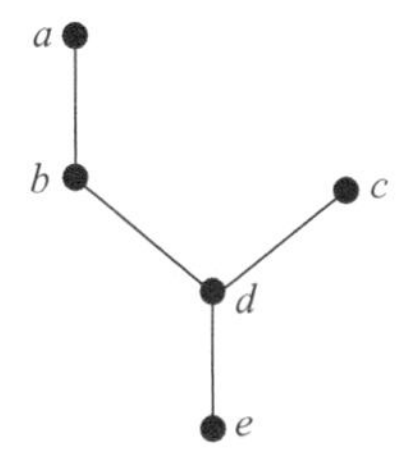

Figure 1.44

EXAMPLE 1.76

Let $A=\{1, 2, 3, 6\}$ and $\leq$ be the divisibility relation. Then, Hasse diagram of $(A, \leq)$ is

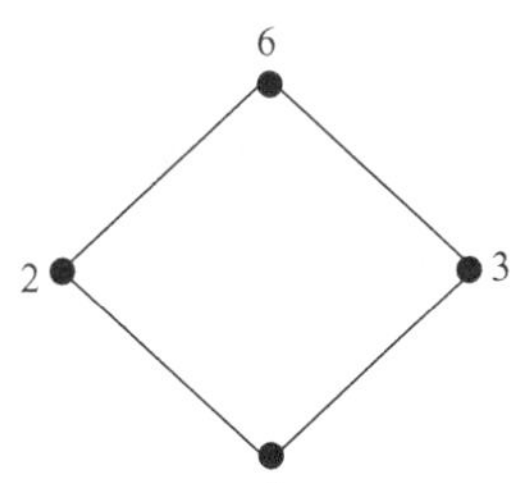

Figure 1.45

Let $B=P(\{a, b\})=\{\phi, \{a\}, \{b\}, \{a, b\}\}$ be ordered by set inclusion. If $f: A \rightarrow B$ is defined by

$$f(1)=\phi, f(2)=\{a\}, f(3)=\{b\}, f(6)=\{a, b\}.$$

Thus, f is one-to-one and onto. Also, if $a \mid b$ for $a, b \in A$, we have

$$f(a) \subseteq f(b).$$

Hence f is isomorphism and Hasse diagram of $(B, \subseteq)$ is

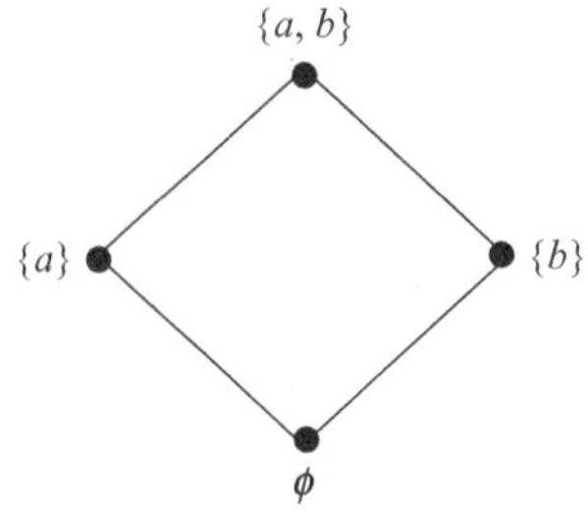

Figure 1.46

EXAMPLE 1.77

Show that the posets $(A, \leq)$ and $(A', \leq')$ whose Hasse diagrams are, respectively,

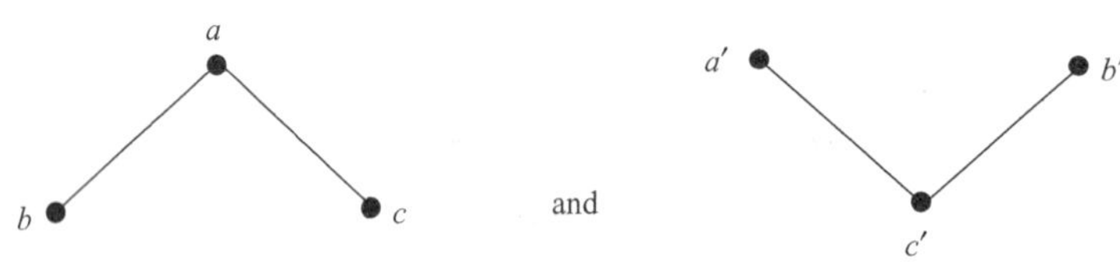

Figure 1.47

are not isomorphic.

Solution.

The two posets are not isomorphic because $(A, \leq)$ has a greatest element a while $(A', \leq')$ does not have a greatest element. Similarly, we can say that these posets are not isomorphic because $(A, \leq)$ does not have a least element, while $(A', \leq')$ does have a least element.

Definition 1.65

A partially ordered set is called **well-ordered** if **every** non-empty subset of it has a **least element (first element)**

EXAMPLE 1.78

A simple example of a well ordered set is $I_n = \{1, 2, \ldots, n\}$ or the set $I = \{1, 2, 3, \ldots\}$ with usual order. Similarly the sets $I_n \times I_n$ or $I \times I$ are well ordered under the natural ordering of "less than or equal to".

We further note that

(i) A well-ordered set A is linearly ordered. For if $a, b \in A$, then $\{a, b\}$ has a least element and hence a and b are comparable.

(ii) Every subset of a well-ordered set is well ordered.

(iii) If A is well ordered and B is isomorphic to A, then B is well ordered.

(iv) All finite linearly ordered sets with the same number of elements, n, are well ordered and are all isomorphic to each other. In fact, they are all isomorphic to $\{1, 2, \ldots, n\}$ with usual order.

(v) Every element $a \in A$ (well ordered), other than greatest (last) element, has an immediate successor. For, let $M(a)$ denote the set of element which strictly succeeds a. Then the first element of $M(a)$ is the immediate successor of a.

(vi) Every totally ordered (linearly ordered) set need not be well ordered. For example, the set **Z** of integers with usual order $\le$ (less than or equal to) is linearly ordered. Every element in **Z** has an immediate successor and an immediate predecessor. But **Z** is not well ordered, because **Z** itself has no least (first) element. However, any subset of **Z** which is bounded below is well ordered.

EXAMPLE 1.79

The set Q of rational numbers with the usual order $\le$ is linearly ordered but no element in Q has immediate successor or immediate predecessor. For if $a, b \in Q$, say $a<b$, then $(a+b)/2 \in Q$ and

$$a < \frac{a+b}{2} < b.$$

EXAMPLE 1.80

Consider the disjoint well-ordered sets

$$A=\{1, 3, 5, \ldots\} \text{ and } B=\{2, 4, 6, \ldots\}.$$

Then the following ordered set

$$S=\{A; B\}=\{1, 3, 5, \ldots; 2, 4, 6, \ldots\}$$

is well ordered. Note that, besides the first element 1, the element 2 does not have an immediate predecessor.

Remark 1.5 If A and B are disjoint ordered sets, then $\{A; B, \ldots\}$means the set $A \cup B \cup \ldots$ ordered point wise from left to right, that is, the elements in the same set keep their order, and any element on the left precedes any element in a set on its right. Thus every element in A precedes every element in B.

1.10 HASHING FUNCTION

To save space and time, each record stored in a computer is assigned an address (memory location) in the computer's memory. The task of assigning the address is performed by the Hashing function (or Hash function) $H : K \rightarrow L$, which maps the set K of keys to the set L of memory addresses. Thus a Hashing function provides means of identifying records stored in a table. The function H should be one-to-one. In fact, if $k_1 \neq k_2$ implies $H(k_1) = H(k_2)$, then two keys will have same address and we say that *collision* occurs. To resolve collisions, the following methods are used to define the hash function.

1. **Division Method.** In this method, we restrict the number of addresses to a fixed number (generally a prime) say m and define the hash function $H : K \rightarrow L$ by
$$H(k) = k \pmod{m}, \; k \in K,$$
where k (mod m) denotes the remainder when k is divided by m.
2. **Midsquare Method.** As the name suggest, we square the key in this method and define hash function $H : K \rightarrow L$ by $H(k) = l$, where l is obtained by deleting digits from both ends of k^2.
3. **Folding Method.** In this method the key k is partitioned into a number of parts, where each part, except possibly the last, has the same numbers of digits as the required memory address. Thus, if $k = k_1 + k_2 + \cdots + k_n$, then the hash function $H : K \rightarrow L$ is defined by
$$H(k) = k_1 + k_2 + \cdots + k_n,$$
where the last carry has been ignored.

For example, if we want to find a three digit address for 817324, then using folding we have

$$k_1 + k_2 = 817 + 324 = 1141,$$

and so

$$H(k) = H(817324) = 141, \textit{ ignoring the carry digit } 1 \textit{ in } k_1 + k_2.$$

Similarly, for a two—digit address, we have

$$k_1 + k_2 = 81 + 73 + 24 = 178,$$

$$H(817324) = 78, \textit{ ignoring carry digit } 1.$$

EXAMPLE 1.81

Let $H : K \to L$ be a hash function, where L consists of two digits addresses 00,01,02,…,49. Find $H(12304)$ using (i) Division method (ii) Midsquare method and (iii) Folding method.

Solution.

(i) **Division method.** The prime number close to 49 is 47. So, we take m = 47. Then

$$H(12304) = 12304 \pmod{47} = 37$$

(ii) **Midsquare method.** Let $k = 12304$. Then $k^2 = 151388416$. Therefore deleting four digits from both the ends of k^2, we have

$$H(12304) = 08.$$

(iii) **Folding Method.** Partitioning k, we have

$$12304 = 12 + 30 + 4 = 46.$$

Therefore

$$H(12304) = 46.$$

1.11 PRINCIPLE OF MATHEMATICAL INDUCTION

Mathematical induction is a fully developed technique of proof used to check conjectures about the outcomes of processes that occur repeatedly and in some definite patterns. The principle of mathematical induction can be stated in the following forms:

First form: Let A be a subset of the set **N** of positive integers with the following properties:

(i) $1 \in A$
(ii) If $n \in A$, then $n+1 \in A$.

Then, $A = \mathbf{N}$.

(this form of mathematical induction is one of Peano's axiom's for natural numbers)

Second form: Let A be a subset of the set **N** of positive integers with the following two properties:

(i) $1 \in A$
(ii) If $k \in$ A for $1 \le k < n$, then $n \in A$.

Then, $A = \mathbf{N}$

(this form is equivalent to the fact that **N** is well ordered).

Third form: Let P be a proposition defined on the positive integer **N**. If

(i) $P(1)$ is true
(ii) $P(n)$ is true implies $P(n+1)$ is true,

then P is true for every positive integer.

EXAMPLE 1.82

Show, by mathematical induction, that the sum of first n odd numbers is n^2.

Solution.
We have

$$P(n)=1+3+5+\cdots+(2n-1)=n^2.$$

Therefore,

$$P(1)=1=1^2.$$

and so $P(n)$ is true for $n=1$.
Suppose now that $k \geq 1$ and

$$P(k)=1+3+5+\cdots+(2k-1)=k^2. \tag{1}$$

We want to show that

$$P(k+1)=1+3+5+\cdots+(2(k+1)-1)=(k+1)^2$$

We have

$$\begin{aligned}1+3+\cdots+(2k-1)+(2k+1)&=k^2+(2k+1) \text{ using (1)}\\&=(k+1)^2.\end{aligned}$$

Thus, $P(k)$ is true implies $P(k+1)$ is true. Hence,

$$1+3+5+7+\cdots+(2n-1)=n^2.$$

EXAMPLE 1.83

Show, by mathematical induction, that for $n \geq 1$,

$$1+2+3+\cdots+n=\frac{n(n+1)}{2}.$$

Solution.
$P(1)$ is the statement

$$1=\frac{1(1+1)}{2},$$

which is clearly true.

Let $k \geq 1$. If $P(k)$ is true, we must show that $P(k+1)$ is true. Since $P(k)$ is true, we have

$$1+2+\cdots k=\frac{k(k+1)}{2}. \tag{1}$$

We shall show that

$$1+2+\cdots+(k+1)=\frac{(k+1)((k+1)+1)}{2}.$$

In fact, we have

$$\begin{aligned}1+2+\cdots+(k+1)&=(1+2+\cdots k)+(k+1)\\&=\frac{k(k+1)}{2}+(k+1)\quad \text{using (1)}\\&=(k+1)\left(\frac{k}{2}+1\right)\\&=\frac{(k+1)(k+2)}{2}=\frac{(k+1)((k+1)+1)}{2}\end{aligned}$$

showing that $P(k+1)$ is true.

EXAMPLE 1.84

Use mathematical induction to show that

$$n!\geq 2^{n-1},\ n=1,2,3,\ldots$$

Solution.

We know that n factorial, denoted by n!, is defined as

$$n!=n\cdot(n-1)\cdot(n-2)\cdot\ldots\cdot 3\cdot 2\cdot 1$$

and that

$$\begin{aligned}&0!=1,\\&1!=1,\\&n!=n\cdot(n-1)!.\end{aligned}$$

If $n=1$, then $n!=1\geq 1=2^{1-1}$. Hence $P(1)$ is true.

Suppose that $P(n)$ is true, that is,

$$n!\geq 2^{n-1}. \tag{1}$$

We want to show that

$$(n+1)!\geq 2^{(n+1)-1}=2^n.$$

We note that

$$\begin{aligned}(n+1)!&=(n+1)\cdot n!.\\&\geq (n+1)\,2^{n-1},\ \text{using (1)}\\&\geq \mathbf{2}\cdot\mathbf{2}^{n-1}\text{since } n+1\geq 2\\&=2^n.\end{aligned}$$

Hence $P(n+1)$ is true whenever $P(n)$ is true. Hence the result holds by the principle of mathematical induction.

EXAMPLE 1.85

Using mathematical induction, show that $5^n - 1$ is divisible by 4 for $n=1, 2, \ldots$.

Solution.

We note that $P(1)=5-1=4$ is divisible by 4, which is true. Hence $P(1)$ holds.

Suppose $P(n)$ is true, that is,

$$5^n - 1 \text{ is divisible by } 4,\ n=1, 2, \ldots$$

We want to show that $5^{(n+1)} - 1$ is divisible by 4.

We have

$$\begin{aligned} 5^{(n+1)} - 1 &= 5 \cdot 5^n - 1 \\ &= (5^n - 1) + 4 \cdot 5^n. \end{aligned}$$

Now $5^n - 1$ is divisible by 4 because of (1) and also $4 \cdot 5^n$ is divisible by 4. Hence the sum of $5^n - 1$ and $4 \cdot 5^n$ is divisible by 4. This proves that $P(n+1)$ is true.

EXAMPLE 1.86

Let $A_1, A_2, \ldots, A_n$ be any n sets. Show by mathematical induction that

$$\overline{\left(\bigcup_{i=1}^{n} A_i\right)} = \bigcap_{i=1}^{n} \overline{A_i}.$$

Solution.

We note that

$$P(1) = \overline{A_1} = \overline{A_1}$$

and so $P(1)$ is true. Suppose that $P(n)$ is true, that is,

$$P(n) = \overline{\left(\bigcup_{i=1}^{n} A_i\right)} = \bigcap_{i=1}^{n} \overline{A_i}.$$

We want to show that $P(n+1)$ is true. Toward this end, we note that

$$\begin{aligned} \overline{\left(\bigcup_{i=1}^{n+1} A_i\right)} &= \overline{(A_1 \cup A_2 \cup \ldots \cup A_n \cup A_{n+1})} \\ &= \overline{(A_1 \cup A_2 \cup \ldots \cup A_n) \cup A_{n+1}} \\ &= \overline{(A_1 \cup A_2 \cup \ldots \cup A_n)} \cap \overline{A_{n+1}} \quad \text{(using De Morgan Law)} \\ &= \left(\bigcap_{i=1}^{n} \overline{A_i}\right) \bigcap \overline{A_{n+1}} \quad \text{(since } P(n) \text{ is true)} \\ &= \bigcap_{i=1}^{n+1} \overline{A_i}. \end{aligned}$$

Hence, $P(n)$ is true implies $P(n+1)$ is true and so, by mathematical induction, the result holds.

EXAMPLE 1.87

Using mathematical induction, show that

$$1+2+2^2+2^3+\cdots+2^n=2^{n+1}-1$$

Solution.

We note that

$$P(1)=1+2=2^2-1=3.$$

Hence $P(1)$ is true.
Suppose that $P(n)$ is true, that is,

$$1+2+\cdots+2^n=2^{n+1}-1. \tag{1}$$

We shall show that $P(n+1)$ is true. We note that

$$\begin{aligned}1+2+2^2+\cdots+2^n+2^{n+1}&=2^{n+1}-1+2^{n+1} \qquad \text{(using (1))}\\ &=2\cdot 2^{n+1}-1=2^{n+2}-1.\end{aligned}$$

Thus,

$$P(n+1)=2^{n+2}-1$$

is true.
This proves the result.

EXAMPLE 1.88

If $m_1, m_2, m_3, \ldots$ is a sequence defined by the recurrence relation

$$m_k=2m_{k-1}+1, \qquad k\geq 2$$
$$m_1=1,$$

show, by using mathematical induction, that

$$m_n=2^n-1, \qquad n\geq 1.$$

Solution.

We note that $m_1=1$, and $2^1-1=1$. Hence $m_1=2^1-1$. Thus the formula holds for $n=1$. Suppose that the formula holds for $n=k$, that is,

$$m_k=2^k-1, k\geq 1. \tag{1}$$

Then,

$$\begin{aligned}m_{k+1}&=2m_{(k+1)-1}+1=2m_k+1\\ &=2(2^k-1)+1 \qquad \text{(using (1))}\\ &=2^{k+1}-2+1=2^{k+1}-1.\end{aligned}$$

Thus, by mathematical induction, the formula $m_n=2^n-1$ holds for all integers $n\geq 1$.

Remark 1.6 The sequence defined by the above recurrence relation is known as **Tower of Hanoi sequence.**

EXERCISES

1. If A and B are sets, show that $A-B=B^c-A^c$.
2. In a class of 100 students, 39 play Tennis, 58 play Cricket, 32 play Hockey, 10 play Cricket and Hockey, 11 play Hockey and Tennis and 13 play Tennis and Cricket. Find number of students who play all the three games.
3. Let $X=\{x \in \mathbf{Z} : 100 \le x \le 999\}$ and let A_i be the set of numbers in X whose ith digit is i. Find the cardinality of the set $A_1 \cup A_2 \cup A_3$.
4. Let $\mathbf{R}$ be the set of real numbers. Consider the relation C on $\mathbf{R}$:

$$C=\{(x, y) : x^2+y^2=1\}$$

Then

(i) Is $1\ C\ 0$?
(ii) Is $-2\ C\ 0$?
(iii) Is $0\ C\ (-1)$?
(iv) Draw the graph for C.

5. Let X be a set with six elements. How many relations can be there on A? How many relations on X are reflexive?
6. Let $A=\mathbf{Z}^+\times\mathbf{Z}^+$ and let a relation R on A be defined by :
$(a, b)\ R\ (c, d)$ if and only if $a+d=b+c$.
Show that R is reflexive and symmetric.
7. Let $A=\{1, 2, 3, 4\}$ and let $R=\{(1, 2), (1, 4), (2, 1), (2, 2), (3, 1)\}$ be a relation. Draw directed graph for R. Is it anti-symmetric relation?
8. Let $A=\{1, 2, 3, 4\}$. Give an example of a relation on A which is both an equivalence relation and partial order relation.
9. Find the matrix of the relation $R=\{(1, 2), (2, 3), (3, 4), (4, 5)\}$ on the set $A=\{1, 2, 3, 4, 5\}$.
10. Let $A=\{1, 2, 3, 4\}$ and $R=\{(x, y) : x>y\}$ be a relation on A. Draw the digraph of R.
11. If $A=\{2, 3, 6, 12, 24, 36\}$ and R is the relation such that $x\ R\ y$ if x divides y, draw the Hasse diagram of (A, R).
12. Let $A=\{a, b, c\}$. Which of the following digraphs represent a partial order relation?

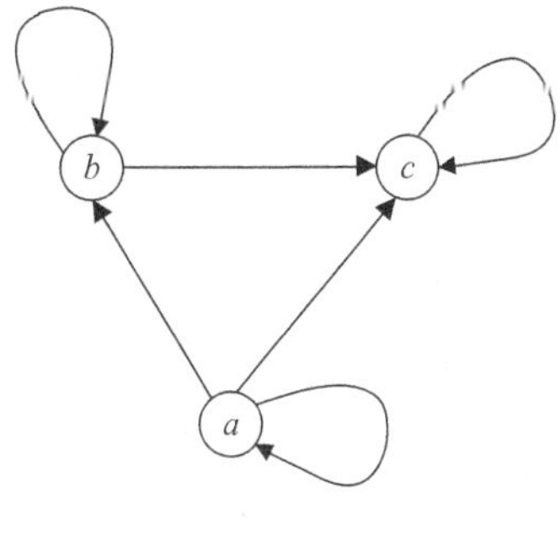

(a)

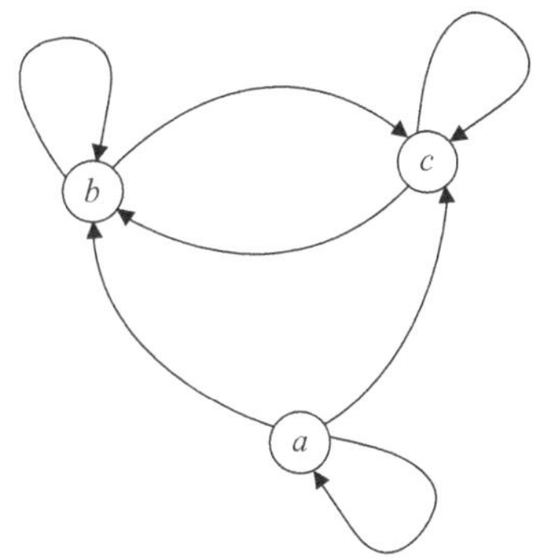

(b)

13. Let $f: \mathbf{R}\to\mathbf{R}$ and $g : \mathbf{R}\to\mathbf{R}$ be a real-valued functions defined by

$$f(x)=2x^3-1, x \in \mathbf{R}$$

and

$$g(x)=\left[\frac{1}{2}(x+1)\right]^{1/3}, x \in \mathbf{R}$$

Show that f and g are bijective and that each is the inverse of the other.
14. Let f: $\mathbf{R}\to\mathbf{R}$ be a real-valued function defined by $f(x)=x^2$, $x \in \mathbf{R}$. Is f invertible? Give reasons.
15. Let $R=\{(1, 2), (2, 3), (3, 4), (2, 1)\}$ be a relation on a set $A=\{1, 2, 3, 4\}$. Find the transitive closure of R.
16. Let $A=\{1, 2, 3, 4, 5, 6\}$. Define a relation R on A as follows:
$x\ R\ y$ if x divides y
Draw the Hasse diagram of the poset (A, R).

17. $R_1 = \{(1, 1), (1, 2), (3, 4), (4, 2)\}$ and $R_2 = \{(1, 1), (2, 1), (3, 1), (4, 4), (2, 2)\}$ be relations on the set $A = \{1, 2, 3, 4\}$. Find R_1 o R_2 and R_2 o R_1.
18. If an equivalence relation has only one equivalence class, how must the relation look like?
19. Let $A = \{1, 2, 3, 4, 5, 6\}$ and let
$R = \{(1, 1), (1, 5), (2, 2), (2, 3), (2, 6), (3, 2), (3, 3), (3, 6), (4, 4), (5, 1), (5, 5), (6, 2), (6, 3), (6, 6)\}$
be an equivalence relation on A. Find the equivalence classes of R and draw digraph of R.
20. Use mathematical induction to show that for any real r except 1, and any integer $n \geq 0$,

$$\sum_{m=0}^{n} r^m = \frac{r^{n+1} - 1}{r - 1}.$$

21. Use mathematical induction to prove that

$$1 + 2 + 2^2 + \cdots + 2^n = 2^{n+1} - 1,$$

where $n \geq 0$ is integer.

2 Counting

In our everyday life, we come across many different counting problems. There are two basic counting principles—Addition Rule and Multiplication Rule.

2.1 ADDITION RULE

If a set A is the union of n distinct mutually disjoint subsets $A_1, A_2, \ldots, A_n$, then

$$|A| = |A_1| + |A_2| + \ldots + |A_n|,$$

where $|A_i|$ denotes the number of elements in the set A_i.
The addition rule can also be stated as
If an event E_1 can occur in m ways and an event E_2 can occur in n ways, then E_1 or E_2 can occur in $m+n$ ways.

EXAMPLE 2.1

Let E_1 be the event of choosing an odd number between 10 and 20 and E_2 be the event of choosing an even number between 10 and 20. Since odd numbers between 10 and 20 are $\{11, 13, 15, 17, 19\}$, E_1 can occur in five ways. Similarly even numbers between 10 and 20 are $\{12, 14, 16, 18\}$ and so E_2 can occur in four ways. Thus, E_1 or E_2 can occur in $5+4=9$ ways.

2.2 MULTIPLICATION RULE

If an operation is performed in k steps and the first step can be performed in n_1 ways, the second step can be performed in n_2 ways (regardless of how the first step was performed) and similarly if the kth step can be performed in n_k ways (regardless of how the preceding steps were performed), then the entire operation can be performed in

$$n_1 \, n_2 \ldots n_k \text{ ways.}$$

EXAMPLE 2.2

Suppose a license plate contains three letters followed by four digits with first digit not zero. How many license plates can be manufactured?

Solution.

We note that

(i) Each letter can be chosen in 26 different ways.

(ii) The first digit can be selected in 9 ways, whereas each of second, third and fourth digits can be chosen in 10 different ways. Thus, the number of different plates manufactured shall be

$$26 \cdot 26 \cdot 26 \cdot 9 \cdot 10 \cdot 10 \cdot 10 = 26^3 \cdot 10^3 \cdot 9.$$

EXAMPLE 2.3

How many six-bit strings begin with 01 or 11?

Solution.

A six-bit string that begins with 01 can be constructed in four successive steps:

(i) Select the third bit
(ii) Select the fourth bit
(iii) Select the fifth bit
(iv) Select the sixth bit

Since each of the four bits can be selected in two ways, the multiplication rule implies that there are $2\times2\times2\times2=16$ six-bit stings that begin with 01.

Similarly, the number of six-bit strings that begin with 11 is 16.

Now addition rule implies that there are $16+16=32$ six- bit strings that begin with 01 or 11.

EXAMPLE 2.4

A virus was spread to 100 computers though e-mail. Each of recipients in turn sent e-mail to 100 computers. How many copies of the virus have been sent?

Solution.

We note that

(i) In the beginning, there is only 1 copy of the virus
(ii) In the second stage, by multiplication rule, 1×100 copies were sent
(iii) In the third stage, by multiplication rule,

$$100\times100=10^4 \text{ copies were sent}$$

Then, by addition rule, the total number of copies of the virus sent is

$$1+100+10^4=10{,}000+100+1=10{,}101.$$

EXAMPLE 2.5

Show that a set $\{x_1, x_2, \ldots, x_n\}$containing n elements has 2^n subsets.

Solution.

A subset of the given set can be constituted in n successive steps: pick or do not pick x_1, pick or do not pick x_2, …, pick or do not pick x_n. Further, each step can be done in two ways. Hence, by multiplication rule, the number of possible subset is

$$\underbrace{2\cdot2\cdot\ldots\cdot2}_{n \text{ times}}=2^n.$$

EXAMPLE 2.6

How many three-digit integers are divisible by 5?

Solution.

The three digit integers are 100, 101, 102, …, 999. We know that the integers that are divisible by 5 end in either 5 or 0. Thus, the set of all three-digit integers partitions into two sets A and B, where A consists of those three-digit integers which end in 0 and B consists of the three-digit integers ending in 5. We consider the set A first. The first digit can be any one out of 1, 2, …, 9 and so there are nine ways to write the first digit. The second digit can be chosen from 0, 1, 2, …, 9 and so there are 10 ways to write to second digit. Thus, by multiplication rule, there are $9\times10=90$

ways to form three-digit integers ending in 0. Similarly, there are 90 ways to form the three-digit integers ending in 5. Hence, by addition law, the number of three digit integers divisible by 5 is $90+90=180$.

EXAMPLE 2.7

If the license plates of a certain state require three English letters followed by 4 digits, then

(i) How many plates can be manufactured if only the letters can be repeated?
(ii) How many plates can be manufactured if only the digits can be repeated?
(iii) How many plates can be manufactured if no repetitions are allowed at all?

Solution.

(i) Since the letters can be repeated, the sequence of three letters can be formed in 26^3 ways. The digits are not allowed to repeat. Therefore, the first digit can be taken in 10 ways, the second digit in 9 ways, the third in 8 ways and the fourth digit can be selected in 7 ways. Hence, by the multiplication rule, the number of possible plates is

$$26^3 \cdot 10 \cdot 9 \cdot 8 \cdot 7.$$

(ii) In this case, only digits are allowed to repeat, therefore by multiplication rule, the number of possible plates is

$$26 \cdot 25 \cdot 24 \cdot 10^4.$$

(iii) In this case, neither the letters nor digits are allowed to repeat and so the number of possible plates is

$$26 \cdot 25 \cdot 24 \cdot 10 \cdot 9 \cdot 8 \cdot 7.$$

Theorem 2.1

Let A be a finite set with n elements and $1 \leq m \leq n$, then the number of sequences of length m that can be formed from elements of A, **allowing repetition**, is n^m.

Proof. Let A be the set consisting of n elements. To form a sequence, where the elements can be repeated, the first element of the sequence can be chosen in n ways. Since repetition is allowed, each of the second, third, …, mth element can be chosen in n ways, therefore, by multiplication rule, the number of sequence that can be formed is

$$\underbrace{n \cdot n \cdot \ldots \cdot n}_{m \text{ factors}} = n^m.$$

Theorem 2.2

Let A be a finite set with n elements and $1 \leq m \leq n$. Then the number of sequence of m **distinct elements** that can be formed from elements of A is $n(n-1)(n-2) \ldots (n-m+1)$.

Proof. Let A be a set with n elements and $1 \leq m \leq n$. The first element of the sequence can be chosen in n ways since any of the n elements can be chosen for the first position. Since repetition is not allowed, after the selection of first element of the sequence, only $(n-1)$ elements remain. So, the second element of the sequence can be chosen in $(n-1)$ ways. Similarly, the third element can be chosen in $(n-2)$ ways and so on. Thus the mth element of the sequence can be chosen in $(n-[m-1]) = n-m+1$ ways. Hence, by multiplication rule, the number of sequences of m distinct elements formed from the elements of A is

$$n(n-1)(n-2) \ldots (n-m+1).$$

2.3 PERMUTATIONS

Definition 2.1

A sequence of m **distinct elements** of a finite set A is called a **permutation on the finite set A taken m elements at a time.**

In Theorem 2.2, we have shown that "Number of permutations of n objects taken m at a time is equal to n $(n-1)$ $(n-2)$ ... $(n-m+1)$". This number of permutation is denoted by nP_m. Therefore, Theorem 2.2 may be stated as:

Theorem 2.3

If $1 \leq m \leq n$, then the number of permutations of n objects taken m at a time in $n\,P_m = n\,(n-1)\,(n-2)\ldots(n-m+1)$.

If $m=n$, then the definition of permutation becomes as stated below:

Definition 2.2

A sequence of n distinct elements of a finite set with n elements is called a **permutation** of A.

The number of permutation of a finite set with n elements is $n\,(n-1)\,(n-2)\ldots 3\cdot 2\cdot 1$ if $n \geq 1$. This number, called n **factorial**, is written as n!

EXAMPLE 2.8

Let $A=\{a, b, c\}$ be a finite set. Find all permutations on A.

Solution.

The set A has three elements and so number of permutations on A is

$$3 \cdot 2 \cdot 1 = 6.$$

The permutations are

$$abc,\ acb,\ cba,\ bac,\ bca,\ cab.$$

EXAMPLE 2.9

How many words of three distinct letters can be formed from the letters of the word COMPUTER?

Solution.

There are eight letters in the word COMPUTER. The number of permutations of eight letters taken three at a time is

$$8\,P_3 = 8 \cdot 7 \cdot 6 = 336.$$

Hence 336 words of three letters can be formed from the letters of the word COMPUTER.

EXAMPLE 2.10

In how many ways can the letters in the word COMPUTER be arranged if the letters CO must remain next to each other (in order) as a unit?

Solution.

We treat the letter group CO as a unit. Then there are effectively seven letters

CO, M, P, U, T, E, R,

which are to be arranged in a row. Hence the number of such arrangement is $7! = 5040$.

EXAMPLE 2.11

If $1 \leq m \leq n$, show that

$$nP_m = \frac{n!}{(n-m)!}.$$

Solution.

We note that

$$\begin{aligned} nP_m &= n(n-1)(n-2)\ldots(n-m+1), n \geq 1 \\ &= \frac{n(n-1)(n-2)\ldots(n-m+1)(n-m)\ldots 2\times 1}{(n-m)(n-m+1)\ldots 3\cdot 2\cdot 1} \\ &= \frac{n!}{(n-m)!}. \end{aligned}$$

EXAMPLE 2.12

Show that

$$n\,P_n = n\,(n-1\,P_{n-1}).$$

Solution.

We have

$$\begin{aligned} n\,(n-1\,P_{n-1}) &= n\,(n-1)! \\ &= n! = n\,P_n. \end{aligned}$$

EXAMPLE 2.13

Show that for all integers $n \geq 2$,

$$n\,P_2 + n\,P_1 = n^2.$$

Solution.

Let n be an integer that is greater than or equal to 2, then

$$nP_1 = \frac{n!}{(n-1)!} = \frac{n(n-1)!}{(n-1)!} = n,$$

$$nP_2 = \frac{n!}{(n-2)!} = \frac{n(n-1)(n-2)!}{(n-2)!} = n(n-1).$$

Hence,

$$\begin{aligned} n\,P_1 + n\,P_2 &= n + n\,(n-1) \\ &= n + n^2 - n = n^2. \end{aligned}$$

Theorem 2.4

The number of distinguishable permutations that can be formed from a collection of n objects in which the first object appears k_1 times, the second object appears k_2 times and so on, is

$$\frac{n!}{k_1!k_2!\ldots k_r!}.$$

Proof. Suppose that x denotes the number of distinguishable permutations formed from the collection of n objects with the given repetition of the object. If the first object were all different, there would be $x\ k_1!$ permutations since each old permutation would give rise to $k_1!$ new permutations.

Similarly, if second object were all different, there would be $x\ k_1!\ k_2!$ new permutations and so on. Thus the permutations with all objects as distinct would be

$$x\ k_1!\ k_2!\ k_3!\ \ldots\ k_r!.$$

But we know that the number of permutations on a set of n distinct element is $n!$ Therefore

$$x\ k_1!\ k_2!\ \ldots\ k_r! = n!$$

and so

$$x = \frac{n!}{k_1!k_2!\ldots k_r!}.$$

EXAMPLE 2.14

Find the number of distinguishable permutations of the letters in MATHEMATICS.

Solution.

The total number of letters in the word MATHEMATICS is 11. We note further that M occurs twice, A twice and T twice; and so the number of distinguishable permutations is

$$\frac{11!}{2!2!2!} = \frac{11\cdot 10\cdot 9\cdot 8\cdot 7\cdot 6\cdot 5\cdot 4\cdot 3\cdot 2\cdot 1}{2\cdot 2\cdot 2}$$

$$= 11\cdot 10\cdot 9\cdot 8\cdot 7\cdot 6\cdot 5\cdot 3 = 4{,}989{,}600.$$

EXAMPLE 2.15

Find the number of distinct permutations that can be formed from the letters of the word RADAR.

Solution.

In the word RADAR, there are five letters out of which two are R and two are A. Hence the number of distinguishable permutations is

$$\frac{5!}{2!2!} = \frac{5\cdot 4\cdot 3\cdot 2\cdot 1}{2\cdot 2} = 30.$$

EXAMPLE 2.16

Find the number of different message that can be represented by sequences of 4 dots and 6 dashes.

Solution.

Here the total number of symbols is 10. Further, dot occurs 4 times, whereas dash occurs 6 times. Therefore, the number of distinct messages is

$$\frac{10!}{4!6!} = \frac{10\cdot 9\cdot 8\cdot 7\cdot 6\cdot 5\cdot 4\cdot 3\cdot 2\cdot 1}{6\cdot 5\cdot 4\cdot 3\cdot 2\cdot 1\cdot 4\cdot 3\cdot 2\cdot 1}$$

$$= \frac{10\cdot 9\cdot 8\cdot 7}{4\cdot 3\cdot 2} = 210.$$

EXAMPLE 2.17

Determine the number of four-digit decimal numbers if

(i) Repetition of digits is not allowed

(ii) 0 is not the leading digit in any of the four-digit number and repetition of digits is not allowed

Solution.

(i) The number of four-digit numbers formed out of 0, 1, 2, … , 9 is

$$10P_4 = \frac{10!}{(10-4)!} = \frac{10!}{6!} = 5040.$$

(ii) Since 0 is not the leading digit, the first digit can be any one of 1, 2, 3, …, 9. Thus the first digit can be selected in 9 ways. The second digit can be any one of the remaining 9 digits. The third digit can be any one out of the remaining 8 digits and the fourth digit can be any one of the remaining 7 digits. Here, by multiplication rule, the number of four digit decimal number in this case is

$$9\times 9\times 8\times 7=4536.$$

EXAMPLE 2.18

In how many ways can five physics books, four mathematics books, 3 history books and two computer science books be arranged on a shelf so that all books of the same subject are together?

Solution.

The books shall be arranged in four units according to subject classifications. The first unit can be filled up by any of the four subjects, the second unit can be filled up by any of the remaining three subjects, the third unit by any of the remaining two subjects and the last unit by the last subject. Thus, there are

$$4\cdot 3\cdot 2\cdot 1=24$$

ways to arrange the books on the shelf according to subject classification.

Further,

(i) The books on physics can be arranged in 5! ways

(ii) The books on mathematics can be arranged in 4! ways

(iii) The books on history can be arranged in 3! ways

(iv) The books on computer science can be arranged in 2! ways

Hence, by multiplication rule, the total number of arrangements is

$$24\cdot 5!\ 4!\ 3!\ 2!=24\cdot 120\cdot 24\cdot 6\cdot 2=829{,}440.$$

2.4 COMBINATIONS

Definition 2.3

An r-element subset of an n-element set A is called a **r-combination** of A. The number of r-combinations of an n element set is denoted by ${}^nC_r$ or $\binom{n}{r}$.

Theorem 2.5

The number of r-combinations that can be chosen from a set of n elements is

$$= \binom{n}{r} = \frac{nP_r}{r!} = \frac{n!}{r!(n-r)!}.$$

Proof. The total number of r-combinations of A is $\binom{n}{r}$. Take any one of these combinations. The r objects in the combination can be arranged in $r!$ ways. Hence the total number of arrangements (permutation) is $n\,C_r\,r!$ But we know that total number of permutation is $n\,P_r$.

Hence,

$$\binom{n}{r} r! = nP_r$$

or

$$= \binom{n}{r} = \frac{nP_r}{r!} = \frac{n!}{r!(n-r)!}.$$

EXAMPLE 2.19

In how many ways can we constitute a committee of three professors and two readers from a group of five distinct professors and eight readers?

Solution.

We note that three professors can be selected in

$$\binom{5}{3} = \frac{5!}{3!2!}$$

$$= \frac{5\cdot 4\cdot 3\cdot 2\cdot 1}{3\cdot 2\cdot 1\cdot 2\cdot 1} = 10$$

ways and that two readers can be selected in ways.

$$\binom{8}{2} = \frac{8!}{2!6!} = \frac{8\cdot 7}{2} = 28$$

Since the committee can be constituted in two successive steps: select the professors, select the readers, by multiplication rule, the committee can be constituted in

$$10 \cdot 28 = 280 \text{ ways.}$$

EXAMPLE 2.20

How many five-person committees constituted from a group of six men and five women consists of (i) at least one man (ii) at most one man.

Solution.

(i) In all, there are 11 persons. Therefore the total number of five-person committees is

$$\binom{11}{5} = \frac{11!}{5!6!} = \frac{11\cdot 10\cdot 9\cdot 8\cdot 7}{5\cdot 4\cdot 3\cdot 2}$$

$$= 11 \cdot 3 \cdot 2 \cdot 7 = 462.$$

Further, the number of five-person committees consisting entirely of women is

$$\binom{5}{5}=1.$$

But,
Number of committees with at least one man=total number of five-person committees minus the number of five-person committees not containing any man=462−1=461.

(ii) Let
A=number of committees with at most one man
B=number of committees without any man
C=number of committees with one man

Then,

$$A=B+C.$$

But,

$$B=\binom{5}{5}=1,$$

$$C=\binom{6}{1}\cdot\binom{5}{4}=6\cdot 5=30.$$

Hence,

$$A=1+30=31.$$

EXAMPLE 2.21

How many committees of four persons with a given chairperson can be selected from 10 persons?

Solution.

The chairman can be selected in 10 ways since he can be any of the 10 persons. The remaining 3 persons are to be selected from the remaining 9 persons. This can be done in $\binom{9}{3}$ ways. So, by multiplication rule, the number of committees constituted is

$$10\cdot\binom{9}{3}=10\cdot\frac{9!}{3!6!}$$
$$=10\cdot 84=840.$$

EXAMPLE 2.22

A woman has 11 close friends and she wants to invite 5 of them to dinner. In how many ways can she invite them if

(i) There is no restriction on the choice
(ii) Two particular persons will not attend separately
(iii) Two particular persons will not attend together

Solution.

(i) Out of the 11 persons, 5 can be invited in

$$\binom{11}{5}=\frac{11!}{5!6!}=462 \text{ ways.}$$

(ii) As per condition, two particular persons will not attend separately. So, either they both should be invited or they both should not be invited.

If they both are invited, then three more persons should be invited out of nine persons. This task can be done in

$$\binom{9}{3} = \frac{9!}{3!6!} = 84 \text{ ways.}$$

If they both are not invited, the five persons should be invited out of nine remaining friends. This task can be done in

$$\binom{9}{5} = \frac{9!}{5!4!} = 126 \text{ ways.}$$

By addition rule, the total number of ways in which invitation can be sent is $84 + 126 = 210$ ways.

(iii) Since, two particular persons cannot be invited together, either one of them should be invited or none of them should be invited.

If one of them is invited, then four more friends are to be invited out of nine friends. This task of inviting can be performed in

$$\binom{9}{4} = \frac{9!}{4!5!} = 126 \text{ ways.}$$

Similarly, if the other is selected, then there are 126 ways.

If both of them are not invited, then all the five friends to be invited are to be selected from the remaining friends. This task of inviting can be performed in

$$\binom{9}{5} = \frac{9!}{5!4!} = 126 \text{ ways.}$$

Thus, by addition rule, the total number of ways in which she can invite her friends is $126 + 126 + 126 = 378$ ways.

EXAMPLE 2.23

A box contains six white balls and five red balls. In how many ways can four balls be drawn from the box if

(i) They can be of any colour
(ii) Two balls should be white and two red
(iii) All the balls should be of same colour

Solution.

(i) Since there is no restriction on the selection, the number of ways in which 11 balls can be drawn out of 11 balls is

$$\binom{11}{4} = \frac{11!}{4!7!} = 330.$$

(ii) Two white balls are to be drawn out of six balls. This task can be performed in

$$\binom{6}{2} = \frac{6!}{2!4!} = 15 \text{ ways.}$$

Similarly, two red balls are to be drawn out of five red balls. This task can be performed in

$$\binom{5}{2}=\frac{5!}{2!3!}=10 \text{ ways.}$$

By multiplication rule, the task of drawing two white balls and two red balls can be performed in $15\times 10=150$ ways.

(iii) Since the balls should be of the same colour, they all can be of white colour or they all can be red.

If all the four balls drawn are to be white, we have to choose four balls out of six white balls. This task can be performed in

$$\binom{6}{4}=\frac{6!}{4!2!}=15 \text{ ways.}$$

If all the four balls drawn are to be red, then we have to draw four balls out of five red balls. This task of selection can be performed in

$$\binom{5}{4}=\frac{5!}{4!}=5 \text{ ways.}$$

Therefore, by addition rule, the number of ways in which all the four balls are of same colour is $15+5=20$.

EXAMPLE 2.24

Show that

(i) $\binom{n+1}{r}=\binom{n}{r}+\binom{n}{r-1}$,

(ii) $r\cdot n\,C_r=n\cdot(n-1)\,C_{r-1}$.

Solution.

(i) We have

$$\binom{n}{r}+\binom{n}{r-1}=\frac{n!}{r!(n-r)!}+\frac{n!}{(r-1)!(n-r+1)!}$$

$$=\frac{n!}{r!(n-r)!}\cdot\frac{n-r+1}{n-r+1}+\frac{n!}{(r-1)!(n-r+1)!}\cdot\frac{r}{r}$$

$$=\frac{n!(n-r+1)}{r!(n-r+1)!}+\frac{n!r}{r!(n-r+1)!}$$

$$=\frac{n!(r+n-r+1)}{r!(n-r+1)!}=\frac{n!(n+1)}{r!(n-r+1)!}$$

$$=\frac{(n+1)!}{r!(n-r+1)!}=\binom{n+1}{r}.$$

(ii) We have

$$r\cdot nC_r = \frac{r\cdot n!}{r!(n-r)!}$$
$$= \frac{r\cdot n!}{r(r-1)!(n-r)!} = \frac{n!}{(r-1)!(n-r)!}$$
$$= \frac{n\cdot(n-1)!}{(r-1)!(n-r)!} = n\cdot\binom{n-1}{r-1}.$$

EXAMPLE 2.25

Show that

$$\binom{18}{5} = \binom{17}{5} + \binom{17}{4}.$$

Solution.

We know that

$$\binom{n+1}{r} = \binom{n}{r} + \binom{n}{r-1}.$$

Taking $n=17$, $r=5$, we see that

$$\binom{18}{5} = \binom{17}{5} + \binom{17}{4}.$$

2.5 COMBINATIONS WHERE REPETITIONS ARE ALLOWED

We now consider the case for combinations where repetitions are allowed. In this regard we have:

Theorem 2.6

The number of r-combinations out of n elements, with repetition allowed, is $\binom{n+r-1}{r}$.

Proof. Consider a sequence $\{s_i\} = \{s_1, s_2, \ldots, s_r\}$ which may be regarded combination of r elements. Adding $i-1$ to s_i, we get the new unique sequence

$$\{t_i\} = \{s_i + (i-1)\} = \{s_1, s_2+1, s_3+2, \ldots, s_r+r-1\}.$$

The sequence $\{t_i\}$ is a selection of r numbers without replacement from the set $\{1, 2, 3, \ldots, n+r-1\}$. Also, the sequence $\{s_i\}$ can be deduced from the sequence $\{t_i\}$ by subtracting $i-1$ from t_i and this sequence may be regarded as the one selected with replacement from the set $\{1, 2, 3, \ldots., n\}$. Since there are $\binom{n+r-1}{r}$ ways of choosing the sequence $\{t_i\}$, the number of ways of choosing the sequence $\{s_i\}$ is $\binom{n+r-1}{r}$.

EXAMPLE 2.26

Find the number of 3-combinations with repetition allowed, that can be formed from the set $\{1, 2, 3, 4, 5\}$.

Solution.

Using Theorem 2.6 with $n=5$, $r=3$, we have the number of 3-combinations equal to

$$=\binom{5+3-1}{3}=\binom{7}{3}=\frac{7!}{3!4!}=35.$$

EXAMPLE 2.27

A woman hosting a birthday party wants to purchase 16 cans of soft drinks for her invited guests. The shop she visited for the purpose has four different types of soft drinks. Determine

(i) How many different selections of 16 cans of soft drinks can she make?

(ii) If Coca-Cola is one of the soft drink available and she purchases at least five cans of Coca-Cola, how many different selections can she make?

Solution.

(i) Here $n=16$, $r=4$. So, the number of possible selection is

$$\binom{16+4-1}{4}=\binom{19}{4}=\frac{19!}{4!15!}=3876.$$

(ii) Since 5 of the 16 cans should be of Coca-Cola, we have for the rest

$$\binom{11+4-1}{4}=\binom{14}{4}=\frac{14!}{4!10!}=1001$$

ways of selection.

Remark 2.1 The following table is useful to select formula for choosing r elements from n elements:

	Order matters	*Order does not matter*
Repetition is allowed	n^r	$\binom{n+r-1}{r}$
Repetition is not allowed	$n P_r$	$\binom{n}{r}$

2.6 COUNTING THE ELEMENTS OF A LIST

To count the elements of a given list, the following theorem is helpful.

Theorem 2.7

If m and n are integers and $m \le n$, then there are $n-m+1$ integers from m to n inclusive.

Proof. We shall prove the result by induction on n. If $n=m$, then there is only one integer from n to n. Also, $m-n+1=1$ in that case. Thus, the result is true for $n=m$. We want to show that it is also true for $n=m+1$. If $n=m+1$, then the elements in the list are m, $m+1$. Thus, the number of elements in the list is 2. Also,

$$(n-m+1)=m+1-m+1=2.$$

Hence the theorem is true by mathematical induction.

EXAMPLE 2.28

How many three-digit integers are divisible by 5?

Solution.

We have solved this example already using multiplication rule. We now solve it using Theorem 2.7. We write the three-digit numbers in a row as shown below:

$$100, 101, 102, 103, 104, 105, 106, \ldots, 110, 111, \ldots, 995, \ldots, 999.$$

We note that the integers divisible by 5 are

$$100, 105, 110, 115, \ldots, 995.$$

On division by 5, we get

$$20, 21, 22, \ldots, 199.$$

Thus, the number of three-digit integers divisible by 5 is equal to the number of elements in the list

$$20, 21, 22, \ldots, 198, 199.$$

Hence, Theorem 2.7 implies that the number of three-digit integers divisible by 5 is

$$199-20+1=180.$$

EXAMPLE 2.29

If the largest of 87 consecutive integers in 326, what is the smallest?

Solution.

The total number of integers is 87. The largest is 326. Let

$$n=326$$
$$m=?$$

Then number of integers from m to n (both inclusive) is

$$n-m+1=87 \text{ (given)}$$
$$\Rightarrow 326-m+1=87$$
$$\Rightarrow m=326-87+1=240.$$

EXAMPLE 2.30

Suppose $A[1], A[2], \ldots, A[n]$ is a one-dimensional array and $n \geq 30$. How many elements are in the sub-array

$$A[6], A[7], \ldots, A[23]?$$

Solution.

Using Theorem 2.7, the number of elements in the given sub-array is

$$23-6+1=18.$$

2.7 PIGEONHOLE PRINCIPLE

This principle is also known as **Dirichlet Drawer Principle** or the **Shoe Box Principle.** Like principle of mathematical induction, this principle is also a proof technique. The three versions of the pigeonhole principle are **discussed below:**

First version. If n pigeons fly into m pigeonholes and $m < n$, then at least one pigeonhole contains two or more pigeons.

The pigeonhole principle tells us nothing about how to locate the pigeonhole that contains two or more pigeons. It simply asserts the existence of a pigeonhole containing two or more pigeons.

Proof. Label the m pigeonholes with the numbers 1 through m and the n pigeons with the numbers 1 through n. Now starting from the pigeon with label 1, we assign each pigeon in order to the pigeonhole with the same number. Thus, m pigeons are assigned m holds. Since $m<n$, there are $n-m$ pigeons that have not been assigned any pigeonhole. Thus at least one pigeonhole will be assigned a second pigeon. This completes the proof.

Second version. A function from one finite set to a smaller finite set cannot be one-to-one. There must be at least two elements in the domain that have the same image in the co-domain.

Proof. Let X and Y be finite sets such that $|X|>|Y|$ and let $f: X \to Y$ be a function. Suppose $|Y|=m$ and let $y_1, y_2, \ldots, y_m$ be elements of the finite set Y. For each $y_i \in Y$, the inverse image set is defined by

$$f^{-1}(y_i)=\{x \in X; f(x)=y_i\}.$$

Then the inverse image set for the elements of Y are

$$f^{-1}(y_1), f^{-1}(y_2), \ldots, f^{-1}(y_m).$$

Since f is a function, each element of X is mapped by f into an element of Y. Hence each element of X is in one of the inverse image sets.
Hence,

$$X=\bigcup_{i=1}^{m} f^{-1}(y_i).$$

Also, by the definition of function, no element of X is mapped by f to more than one element of Y. Thus each element of X is in only one of $f^{-1}(y_i)$ and thus the sets $f^{-1}(y_i)$ are disjoint. Therefore, addition rule implies

$$|X|=|f^{-1}(y_1)|+|f^{-1}(y_2)|+\ldots+|f^{-1}(y_m)|.$$

Suppose, on the contrary, that f is one-to-one. Then each set $f^{-1}(y_i)$ has at most one element and so

$$|f^{-1}(y_1)|+|f^{-1}(y_2)|+\ldots+|f^{-1}(y_m)| \leq 1+1+\ldots+1=m.$$

Hence,

$$|X| \leq m=|Y|,$$

which contradicts the fact that $|X|>|Y|$. Hence f is not one-to-one.

EXAMPLE 2.31

Among 13 people, there are at least 2 of them who were born in the same month.

EXAMPLE 2.32

If 14 shoes are selected from 13 pairs of shoes, there must be a pair of matched shoes among the selection.

EXAMPLE 2.33

Let $A=\{1, 2, 3, 4, 5, 6, 7, 8, 9\}$. If six integers are selected from A, there must be at least one pair of integers that have a sum of 10. In fact, the set A can be partitioned into the four subsets:

$$\{1, 9\}, \{2, 8\}, \{3, 7\}, \{4, 6\}, \{5\}.$$

Each of the six integers chosen must belong to one of the sets. Since there are only five sets, the pigeonhole principle tells us that two of the integers must belong to the same set and so in that case their sum will be 10.

Theorem 2.8

(Third Form of Pigeonhole Principle)

Generalized pigeonhole principle. If n pigeons fly into m pigeonholes and $m<n$, then one of the pigeonhole will contain at least $\left[\frac{n-1}{m}\right]+1$ pigeons.

Proof. Assume on the contrary that every pigeonhole will contain no more than $\left[\frac{n-1}{m}\right]$ pigeons. Therefore, the total number of pigeons in the m pigeonholes will be at most equal to $m\left[\frac{n-1}{m}\right]$ which is less than $m\cdot\frac{n-1}{m}=n-1$. This contradicts the fact that number of pigeons is n. This means that at least one pigeonhole will contain more than $\left[\frac{n-1}{m}\right]$ pigeons, that is, at least one pigeonhole will contain at least $\left[\frac{n-1}{m}\right]+1$ pigeons.

Another form of generalized pigeonhole principle. If m pigeonholes are occupied by $k\ m+1$ or more pigeons, where k is a positive integer, then at least one pigeonhole is occupied by $k+1$ or more pigeons.

EXAMPLE 2.34

How many people at least in a group of 85 people have the same last initials?

Solution.

Here 85 people are the pigeons and 26 letters are the pigeonholes. Thus, by generalized pigeonhole principle, the minimum number of people having same last initial will be

$$\left[\frac{n-1}{m}\right]+1=\left[\frac{85-1}{26}\right]+1$$

$$\left[\frac{84}{26}\right]+1=3+1=4.$$

Note: Using second form of generalized pigeonhole principle, we may solve this problem as follows: We note that $n=85$, $m=26$ and that $85>3.\ 26=78$. Thus, $k=3$ and so the minimum number of people having same last initials is $k+1=4$.

EXAMPLE 2.35

Show that if 25 journals in a library contain a total of 52,042 pages, then one of the journal must contain at least 2,082 pages.

Solution.

In this example, the pages are pigeons and journals are pigeonholes. Thus,

$$n=52{,}042 \text{ and } m=25.$$

By generalized pigeonhole principle, at least one pigeonhole must contain

$$\left[\frac{n-1}{m}\right]+1=\left[\frac{52,042-1}{25}\right]+1=2,082 \text{ pigeons.}$$

Hence, one of the journals must contain at least 2,082 pages.

EXAMPLE 2.36

Five friends on a tour of Mumbai found that they have a total of Rs 5,263 with them. Show that at least one of them has Rs 1,053 with him.

Solution.

Here, number of rupees is the number of pigeons and number of friends is the number of pigeonholes. Thus,

$$n=5263 \quad \text{and} \quad m=5.$$

By generalized pigeonhole principle, at least one of the friend will have

$$\left[\frac{n-1}{m}\right]+1=\left[\frac{5,262}{5}\right]+1=1,053 \text{ rupees with him.}$$

EXAMPLE 2.37

Consider an equilateral triangle whose sides are of length 1 unit. Suppose there are five points lying on or inside the triangle. Show that at least two these points must be not more than $\frac{1}{2}$ unit apart.

Solution.

Let A B C be the given equilateral triangle and let D, E and F be the middle points of BC, CA and AB, respectively. Joining D and E, E and F, and F and D, the triangle ABC is divided into four equilateral triangles T_1, T_2, T_3 and T_4 as shown in the Figure 2.1. The sides of these triangles are of length $\frac{1}{2}$ unit.

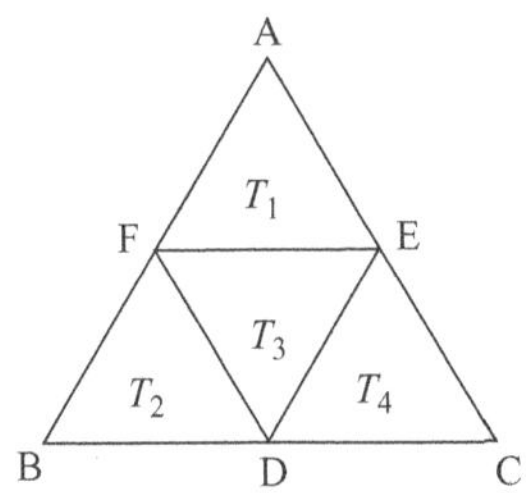

Figure 2.1

Suppose that these four triangles T_1, T_2, T_3 and T_4 be pigeonholes and the five points be the pigeons. By pigeonhole principle, at least one triangle out of the four triangles T_1, T_2, T_3 and T_4, contains two points and as per our subdivision, their distance is not greater than $\frac{1}{2}$ unit.

EXAMPLE 2.38

How many people among 200,000 people are born at the same time (hour, minute, seconds)?

Solution.

There are 24 hours in a day. Thus, in the language of pigeonhole principle, we have

$$\text{Number of pigeons } (n)=200,000,$$
$$\text{Number of pigeonhole } (m)=24.$$

Therefore, one of the pigeonhole must contain at least

$$\left\lceil \frac{n-1}{m} \right\rceil + 1 = \left\lceil \frac{199,999}{24} \right\rceil + 1 = 8,334 \text{ pigeons.}$$

This shows that 8,334 people are born during the same hour.

Now, there are 60 minutes in an hour. So we have

$$n=8334 \text{ and } m=60.$$

Therefore, one of the pigeonhole must contain at least

$$\left\lceil \frac{n-1}{m} \right\rceil + 1 = \left\lceil \frac{8,333}{60} \right\rceil + 1 = 139 \text{ pigeons.}$$

Thus, 139 people are born during the same hour and at the same minute.

Now, there are 60 seconds in a minute. So, we have $n=139$ and $m=60$.

Therefore, one of the pigeonholes must contain at least

$$\left\lceil \frac{n-1}{m} \right\rceil + 1 = \left\lceil \frac{138}{60} \right\rceil + 1 = 3 \text{ pigeons.}$$

Hence, by generalized pigeonhole principle, the number of people born in same hour, minute and second is 3.

EXAMPLE 2.39

A publication list of a professor consists of 45 research papers each marked "Foreign Journal" or "Indian Journal". The number of papers in Foreign Journals is 26. Show that there are at least two research papers marked "Foreign Journal" in the list exactly six research papers apart.

Solution.

Let a_i denote the position of *i*th Foreign Journal research paper. Consider the numbers

$$a_1, a_2, a_3, \ldots, a_{26} \leq 45$$

and

$$a_1+6, a_2+6, \ldots, a_{26}+6 \leq 45+6=51.$$

We thus have 52 positive integers

$$a_1, a_2, \ldots, a_{26}, a_1+6, a_2+6, \ldots, a_{26}+6,$$

which have possible values between 1 and 51 (both inclusive). So, we have 52 pigeons and 51 pigeonholes. By pigeonhole principle, two of the numbers must coincide. Since $a_1, a_2, \ldots, a_{26}$ are all different and $a_1+6, a_2+6, \ldots, a_{26}+6$ are also all different, it follows that one of $a_1, a_2, \ldots, a_{26}$ is equal to one of $a_1+6, a_2+6, \ldots, a_{26}+6$. Hence, there are *i* and *j* such that

$$a_i=a_j+6 \text{ or } a_i-a_j=6.$$

Thus, research papers published in Foreign Journals in the list appears exactly six papers apart.

EXAMPLE 2.40

Find the minimum number of persons selected so that at least eight of them will have birthdays on the same day of week.

Solution.

Let n be the minimum number of persons. The number of days in week is 7. Using generalized pigeonhole principle, we have

$$\frac{n-1}{7}+1=8 \text{ or } n=50.$$

Hence the required number of persons is 50.

EXAMPLE 2.41

A doctor gives a prescription of 20 tablets to a patient with the instructions to take at least one tablet per day for 15 days. Show that there is a period of consecutive days during which the patient takes a total of 9 tablets.

Solution.

Suppose a_i denotes the number of tablets taken by the patient till the ith day. The patient takes at least one tablet everyday. So we have

$$1 \le a_1 < a_2 < \ldots < a_{15} \le 20$$

and so

$$a_1+9 < a_2+9 < \ldots < a_{15}+9 \le 20+9=29.$$

We thus have 30 positive integers

$$a_1, a_2, \ldots, a_{15}, a_1+9, a_2+9, \ldots, a_{15}+9$$

all of which lie between 1 and 29. In the language of pigeonhole principle, we have 30 pigeons and 29 pigeonholes. So, two of these numbers must be equal. But $a_1, a_2, \ldots, a_{15}$ are all distinct and a_1+9, $a_2+9, \ldots, a_{15}+9$ are all distinct. So, it follows that one of $a_1, a_2, \ldots, a_{15}$ is equal to one of a_1+9, a_2+9, $\ldots, a_{15}+9$. Hence there are i and j such that $a_i=a_j+9$ or $a_i-a_j=9$. Therefore, between days i and j, the patient takes exactly 9 tablets.

2.8 PROBABILITY

Probability theory was developed in the seventeenth century to analyse games and so directly involved counting. It is mathematical modelling of the phenomenon of chance or randomness.

Definition 2.4

The measure of "Chance" or "likelihood" for a statement to be true is called the **probability** of the statement.

Thus, probability is an expression of an outcome of which we are not certain.

For example, if we toss a coin, we cannot predict in advance whether a head or tail will show up. Similarly, if a dice (die) is thrown, then any one of the six faces can turn up. We cannot predict in advance which number (face) is going to turn up.

Consider a pack of 52 playing cads. There are two colours, black and red, and four suits namely spades, hearts, diamonds and clubs. Each suit has 13 cards. If we shuffle the pack of cards and draw a card, we are not sure to get a desired card.

Definition 2.5

An **experiment** is a process that yields an outcome.

Definition 2.6

A **random experiment** or **experiment of chance** is an experiment in which

(i) All the outcomes of the experiment are known in advance.
(ii) The exact outcome of any specific performance of the experiment is not known in advance.

For example, tossing of a fair coin is a random experiment. The possible outcomes of the experiment are head and tail. But we do not know in advance what the outcome will be on any performance of experiment.

Definition 2.7

The set of all the possible outcomes of a random experiment is called the **sample space** of that random experiment. It is denoted by S.

An element of a sample space is called a **sample point.**

Definition 2.8

An **event** is a subset of a sample space.

An event may not contain any element. Such event is represented by ϕ and is called **impossible event.**
An event may include the whole sample space S. Such event is called **sure event.**
An event containing exactly one element is called a **simple event.**
For example, if we toss a fair coin, the sample space is

$$S_1 = \{T, H\},$$

where T stands for tail and H stands for head. Thus, S_1 consists of $2^1 = 2$ sample points.
If the same coin is tossed twice, then

$$S_2 = \{TT, TH, HT, HH\}$$

consists of $2^2 = 4$ sample points.
Thus, in case of n toss, the sample space S_n shall have 2^n sample points.

2.8.1 Tree Diagram to Find Sample Space

The sample space of a random experiment can also be determined with the help of a tree diagram. For example, if a fair coin is tossed thrice, then the tree diagram for the sample space is as given below:

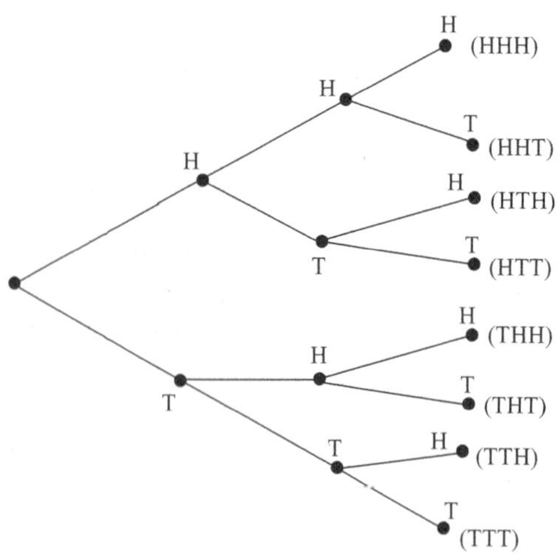

Figure 2.2

Thus,

$$S_3 = \{\text{HHH, HHT, HTH, HTT, THH, THT, TTH, TTT}\}.$$

Similarly, if an unbiased cubical dice is thrown, then

$$S_1 = \{1, 2, 3, 4, 5, 6\}.$$

If it is thrown again, then S_2 shall consist of $6^2 = 36$ sample points. These points can be determined in the following way:

	⚀	⚁	⚂	⚃	⚄	⚅
⚀	11	12	13	14	15	16
⚁	21	22	23	24	25	26
⚂	31	32	33	34	35	36
⚃	41	42	43	44	45	46
⚄	51	52	53	54	55	56
⚅	61	62	63	64	65	66

If two coins are tossed simultaneously, then the first coin may show up either H or T and the second coin may also show up either H or T. Therefore the outcomes of the experiment are

$$S = \{\text{HH, HT, TH, TT}\}.$$

In general, when two random experiment having m outcomes $e_1, e_2, \ldots, e_m$ and n outcomes $p_1, p_2, \ldots, p_n$, respectively are performed **simultaneously**, the sample space consists of $m\,n$ sample points and so

$$S = \{(e_1, p_1), (e_1, p_2), \ldots, (e_1, p_n), \ldots, (e_m, p_1), \ldots, (e_m, p_n)\}.$$

Definition 2.9

The **complement of an event** A with respect to the sample space S is the set of all elements of S which are not in A. It is denoted by $\bar{A}$ or **not by** A.

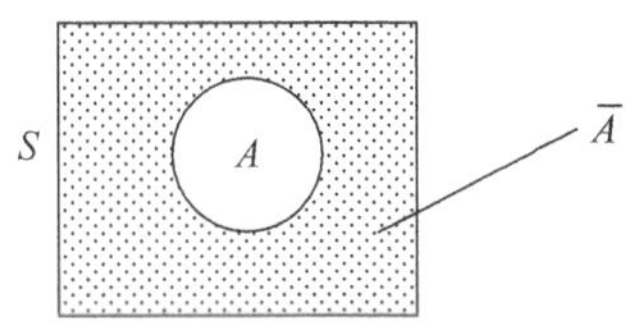

Figure 2.3

Definition 2.10

The **intersection** of two events A and B, denoted by $A \cap B$, consists of all points that are common to A and B.

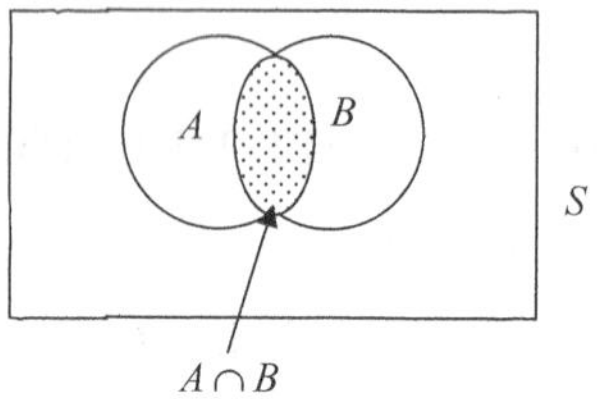

Figure 2.4

Thus $A \cap B$ denotes **simultaneous occurrence** of A and B.

Definition 2.11

Two events A and B are called **mutually exclusive** or **disjoint** if $A \cap B = \phi$.

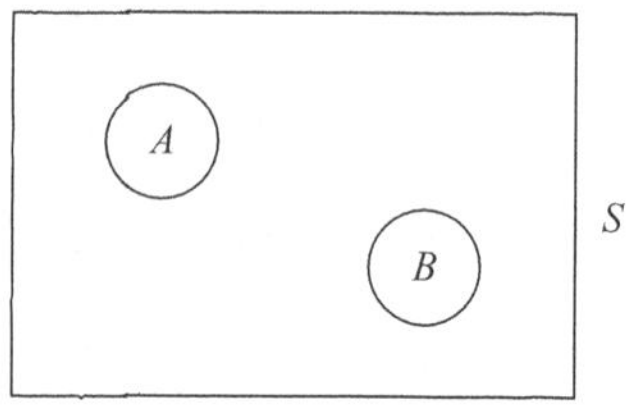

Figure 2.5 $(A \cap B = \phi)$

Definition 2.12

The **union** of the two events A and B, denoted by $A \cup B$, is the event containing all the elements that belong to A or to B or to both.

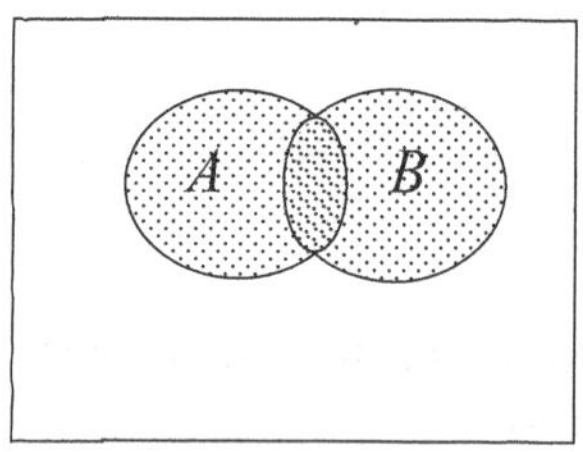

Figure 2.6 $(A \cup B)$

EXAMPLE 2.42

Let A be the event that a "sum of 6" appears on the dice when it is rolled **twice** and B denote the event that a "sum of 8" appears on the dice when rolled twice. Then,

$$A = \{15, 24, 33, 42, 51\},$$
$$B = \{26, 35, 44, 53, 62\}.$$

We observe that $A \cap B = \phi$. Therefore, A and B cannot occur simultaneously and are mutually exclusive (disjoint).

The following combinations of events are usually needed in probability theory:

Combination	*Meaning*
$A \cup B$	Either A or B or both
$A \cap B$	Both A and B
$\overline{A}$ or A^c or A'	not A
$A \cap B = \phi$	Mutually exclusive events A and B
$A' \cap B'$ or $(A \cup B)'$	Neither A nor B
$A \cap B'$	only A
$A' \cap B$	only B
$(A \cap B') \cup (A' \cap B)$	Exactly one of A and B
$A \cup B \cup C$	Exactly one of A, B and C
$A \cap B \cap C$	All the three A, B and C

Definition 2.13

A collection of events $E_1, E_2, \ldots, E_n$ of a given sample space S is said to be **mutually exclusive** and **exhaustive system** of events if

(i) $E_i \cap E_j = \phi,\ i \neq j;\ i, j = 1, 2, \ldots, n$
(ii) $E_1 \cup E_2 \cup \ldots \cup E_n = S.$

Definition 2.14

A collection of events is said to be **equally likely** if all the outcomes of the sample space have the same chance of occurring.

EXAMPLE 2.43

A coin is tossed thrice. If the event E denotes the "number of heads is odd" and event F denotes the "number of tails is odd", determine the cases favourable to $E \cap F$.

Solution.

The coin is tossed thrice, therefore the sample space is

$$S = \{\text{HHH, HHT, HTH, HTT, THT, THH, TTH, TTT}\}$$

The events E and F are

$$E = \{\text{HHH, HTT, THT, TTH}\},$$
$$F = \{\text{HHT, HTH, THH, TTT}\}.$$

We note that $E \cap F = \phi$.

EXAMPLE 2.44

From a group of two men and three women, two persons are to be selected. Describe the sample space of the experiment. If E is the event in which a man and a woman are selected, determine the favourable cases to E.

Solution.

Let M_1, M_2 and W_1, W_2, W_3 be the men and women in the group. Then number of ways selecting two persons is equal to

$$\binom{5}{2} = \frac{5!}{3!2!} = 10.$$

The sample space is

$$S = \{M_1 M_2, W_1 W_2, W_2 W_3, W_1 W_3, M_1 W_1, M_1 W_2, M_1 W_3, M_2 W_1, M_2 W_2, M_2 W_3\}.$$

If E is the event where one man and one woman is selected, then

$$E = \{M_1 W_1, M_1 W_2, M_1 W_3, M_2 W_1, M_2 W_2, M_2 W_3\}.$$

Thus, there are six favourable cases to the event E.

Definition 2.15

If S is a finite sample space having n mutually exclusive, equally likely and exhaustive outcomes out of which m are favourable to the occurrence of an event E, then the **probability** of occurrence of E, denoted by $P(E)$, is defined by

$$P(E) = \frac{\text{The number of favourable outcomes in } E}{\text{The total number of outcomes in } S}$$

$$= \frac{|E|}{|S|} = \frac{m}{n}.$$

It follows from the definition that

1. The probability of the sure event is 1, that is, $P(S) = 1$
2. The probability of the impossible event is 0, that is, $P(\phi) = 0$
3. Since $0 \leq m \leq n$, we have

$$0 \leq \frac{m}{n} \leq 1 \text{ or } 0 \leq P(E) \leq 1.$$

This relation is called the **axiom of calculus of probability.**

4. The cases favourable to non-occurrence of event E is $n - m$.

Therefore,

$$P(\text{not E}) = \frac{n-m}{n} = 1 - \frac{m}{n} = 1 - P(E),$$

that is,

$$P(\bar{E}) = 1 - P(E)$$

or

$$P(E) + P(\bar{E}) = 1.$$

EXAMPLE 2.45

Three coins are tossed simultaneously. What is the probability that at least two tails are obtained?

Solution.

The sample space consists of $2^3 = 8$ outcomes and

$$S = \{\text{HHH, HHT, HTH, THH, HTT, THT, TTH, TTT}\}.$$

Let E be the event obtaining at least 2 tails. Then,

$$E = \{\text{HTT, THT, TTH, TTT}\}.$$

Thus, there are four favourable cases to the event E. Hence

$$P(E)=\frac{4}{8}=\frac{1}{2}.$$

EXAMPLE 2.46

In a single throw of two distinct dice, what is the probability of obtaining

(i) A total of 7
(ii) A total of 13
(iii) A total as even number

Solution.

The sample space shall consist of $6^2=36$ points. We list the total number of outcomes as given below:

11	12	13	14	15	16
21	22	23	24	25	26
31	32	33	34	35	36
41	42	43	44	45	46
51	52	53	54	55	56
61	62	63	64	65	66

(i) Let E_1 be the event in which a total of seven is obtained. Then

$$E_1=\{61, 52, 43, 34, 25, 16\}$$

and so number of favourable outcomes to the event E_1 is 6.
Hence,

$$P(E_1)=\frac{6}{36}=\frac{1}{6}.$$

(ii) Since the sum of outcomes on the two dices cannot exceed $6+6=12$, there is no favourable outcome to an event E_2 having sum 13. Hence,

$$P(E_2)=\frac{0}{36}=0.$$

(iii) Let E_3 be the event in which we get even number as the sum. Then,

$$E_3=\{11, 13, 15, 22, 24, 26, 31, 33, 35, 42, 44, 46, 51, 53, 55, 62, 64, 66\}$$

Thus, number of favourable outcomes to the event E_3 is 18. Hence,

$$P(E_3)=\frac{18}{36}=\frac{1}{2}.$$

EXAMPLE 2.47

What is the probability that

(i) A non-leap year will have 53 Sundays
(ii) A leap year will have 53 Sundays

Solution.

(i) A non-leap year contains 365 days. So it has $\frac{365}{7}=52$ complete weeks and 1 extra day. The extra day can be any one of seven days—Sunday, Monday, Tuesday, Wednesday, Thursday,

Friday and Saturday. Out of these seven possibilities, the first one is only favourable to the event "53 Sundays". Therefore,

$$P(53 \text{ Sundays}) = \frac{1}{7}.$$

(ii) A leap year contains 366 days. So, it has 52 complete weeks and two extra days. These days can be any one of the following seven combinations

Sunday and Monday (boxed)	Monday and Tuesday
Tuesday and Wednesday	Wednesday and Thursday
Thursday and Friday	Friday and Saturday
Saturday and Sunday (boxed)	

Out of these seven possibilities, only two possibilities (enclosed in boxes) are favourable to the event "53 Sundays". Hence,

$$P(53 \text{ Sundays in a leap year}) = \frac{2}{7}.$$

EXAMPLE 2.48

Ten persons, among whom are A and B, sit down at random at a round table. Find the probability that there are three persons between A and B.

Solution.

Let A occupy any seat at the round table. Then there are nine seats available to B. If there are three persons between A and B, then B has only **two ways** to sit as shown in the diagram below:

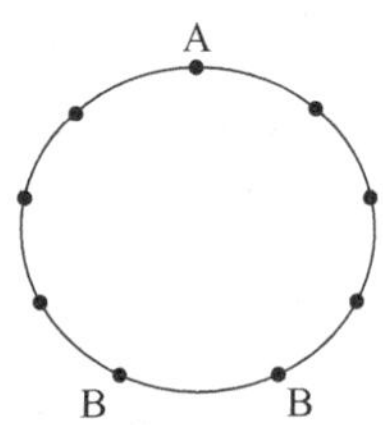

Figure 2.7

Thus, the probability of the required event is $\frac{2}{9}$.

EXAMPLE 2.49

From a lot of 20 microprocessors among which 5 are defective, 4 microprocessors are randomly selected. Find the probability of obtaining no defective microprocessor.

Solution.

The sample space will consist of $\binom{20}{4}$ sample points since there are $20\,C_4$ ways to select 4 microprocessors out of 20 microprocessors. Further, since 5 microprocessors are defective, the number of favourable outcomes to the event "no defective microprocessor is obtained" is $\binom{15}{4}$.

Hence,

$$= P(\text{no defective microprocessor}) = \frac{\binom{15}{4}}{\binom{20}{4}} = \frac{15\cdot 14\cdot 13\cdot 12}{20\cdot 19\cdot 18\cdot 17}$$

$$= \frac{32,760}{116,280} = 0.2817337.$$

EXAMPLE 2.50

A bag contains five distinct white and ten distinct black balls. Random samples of **three** balls are taken out **without replacement**. Find the probability that the sample contains

(i) Exactly one white ball
(ii) No white ball

Solution.

The total number of ways of choosing 3 balls out of 15 balls is $15\ C_3$. Thus the sample space consists of $15\ C_3$ points. Now,

(i) The number of ways of choosing one white ball out of five white balls is $5\ C_1$. Similarly the number of ways of choosing 2 black balls out of 10 is $10\ C_2$. Therefore, by multiplication rule, the total number of outcomes for the event "sample consists exactly one white ball" is $5\ C_1 \cdot 10\ C_2$. Hence,

$$= P(\text{exactly one white ball}) = \frac{5C_1 \cdot 10C_2}{15C_3}$$

$$= \frac{5\cdot 10\cdot 9\cdot 3\cdot 2}{2\cdot 15\cdot 14\cdot 13} = \frac{45}{91}.$$

(ii) The event "no white ball" means that all balls selected should be black. So we have to choose 3 balls out of 10 black balls. Hence the number of favourable outcomes to the event is $10\ C_3$. Therefore,

$$P(\text{no white ball}) = \frac{10C_3}{15C_3} = \frac{24}{91}.$$

EXAMPLE 2.51

Given a group of four persons, find the probability that

(i) No two of them have their birthday on the same day
(ii) All of them have the same birthday

Solution.

Each of the four persons can have his birthday on any of 365 days. Thus the sample space consists of $(365)^4$ points. Now

(i) Since no two persons have their birthday on the same day, the number of favourable outcomes to this event is

$$365 \cdot 364 \cdot 363 \cdot 362.$$

Hence,

$$P(\text{distinct birthday}) = \frac{365 \cdot 364 \cdot 363 \cdot 362}{(365)^4}$$
$$= \frac{364 \cdot 336 \cdot 362}{(365)^3} = \frac{364P_3}{(365)^3}.$$

(ii) If all the four persons have their birthday on the same day, then we have to choose just one day out of 365. Thus, the number of favourable outcomes to the event is 365.

Hence,

$$= P(\text{birthday on the same day}) = \frac{365}{(365)^4} = \frac{1}{(365)^3}.$$

EXAMPLE 2.52

A bag contains n distinct white and n distinct red balls. Pair of balls are drawn **without replacement** until the bag is empty. Show that the probability that each pair consists of one white and one red ball is $\frac{2^n}{2nC_n}$.

Solution.

The bag contains $2n$ distinct balls. Since the pairs are drawn without replacement, the total number of outcomes in the sample space is

$$\binom{2n}{2} \cdot \binom{2n-2}{2} \cdots \binom{4}{2} \cdot \binom{2}{2} = \frac{(2n)!}{2!(2n-2)!} \cdot \frac{(2n-2)!}{2!(2n-4)!} \cdots \frac{4!}{2!2!}$$
$$= \frac{(2n)!}{2^n}$$

Now, suppose that E is the event in which a pair of balls drawn consists of one white ball and one red ball. Then the first pair can be chosen in $n \cdot n$ ways. Since there is no replacement, the second pair can be selected in $(n-1) \cdot (n-1)$ ways and so on. Therefore, the number of favourable outcomes to the event is

$$n^2 (n-1)^2 (n-2)^2 \ldots 2^2 \cdot 1^2 = [n (n-1) (n-2) \ldots 2 \cdot 1]^2 = (n!)^2$$

Hence,

$$P(E) = \frac{(n!)^2}{(2n)!} \cdot 2^n = \frac{2^n}{(2n)!/(n!n!)} = \frac{2^n}{\binom{2n}{n}}.$$

Theorem 2.9

If E and F are two mutually exclusive events of a random experiment, then

$$P(E \text{ or } F) = P(E \cup F) = P(E) + P(F).$$

Thus, the probability that at least one of the mutually exclusive event E or F occurs is the sum of their individual probabilities.

Proof. Suppose that a random experiment results in n mutually exclusive, equally likely and exhaustive outcomes of which m_1 are favourable to the occurrence of the event E and m_2 are favourable to the occurrence of the event F. Then,

$$P(E)=\frac{m_1}{n} \quad \text{and} \quad P(F)=\frac{m_2}{n}.$$

Since E and F are mutually exclusive, by addition rule, the number of favourable outcomes to the occurrence of E or F is m_1+m_2. Hence,

$$= P(E \text{ or } F) = P(E \cup F) = \frac{m_1+m_2}{n}$$

$$= \frac{m_1}{n} + \frac{m_2}{n} = P(E)+P(F).$$

Corollary 2.1

If $E_1, E_2, \ldots, E_n$ are n mutually exclusive events, then

$$P(E_1 \cup E_2 \cup \ldots \cup E_n) = P(E_1)+P(E_2)+\cdots+P(E_n).$$

Proof. We shall prove the result by mathematical induction on n. By Theorem 2.9,

$$P(E_1 \cup E_2) = P(E_1)+P(E_2).$$

Let the result be true for $n=k$, that is,

$$P(E_1 \cup E_2 \cup \ldots \cup E_k) = P(E_1)+P(E_2)+\cdots+P(E_k). \tag{1}$$

We put

$$E = E_1 \cup \ldots \cup E_k.$$

Then,

$$P(E_1 \cup E_2 \cup \ldots \cup E_{k+1}) = P(E \cup E_{k+1})$$
$$= P(E_1)+P(E_2)+\cdots+P(E_k)+P(E_{k+1}) \text{ using (1)}.$$

Hence, the result holds by mathematical induction.

Corollary 2.2

If $E_1, E_2, \ldots, E_n$ are n mutually exclusive and exhaustive events, then

$$P(E_1)+P(E_2)+\ldots+P(E_n)=1.$$

Proof. Since $E_1, E_2, \ldots, E_n$ are mutually exclusive and exhaustive,

$$E_1 \cup E_2 \cup \ldots \cup E_n = S, \text{ sample space.}$$

Since $P(S)=1$, we have

$$1 = P(S) = P(E_1 \cup E_2 \cup \ldots \cup E_n) = P(E_1)+P(E_2)+\cdots+P(E_n).$$

Corollary 2.3

If E and F are two events, then

$$P(E \cap \bar{F}) = P(E) - P(E \cap F).$$

Proof. The Venn diagram shown below shows that the events $E \cap \overline{F}$ and $E \cap F$ are mutually exclusive.

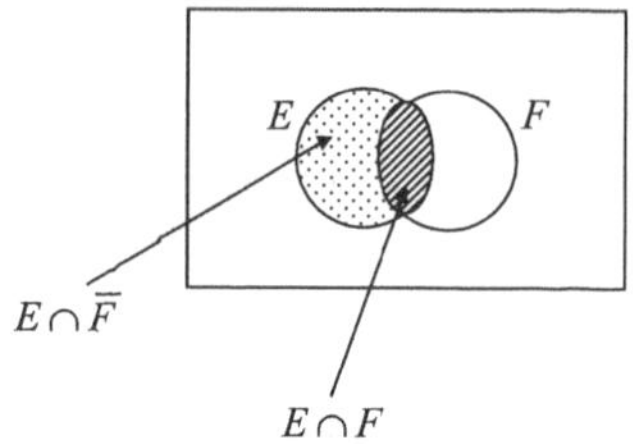

Figure 2.8

Also $(E \cap \overline{F}) \cup (E \cap F)=E$. Hence, by Theorem 2.9

$$P(E)=P(E \cap \overline{F})+P(E \cap F),$$

that is,

$$P(E \cap \overline{F})=P(E)-P(E \cap F).$$

Corollary 2.4

If E and F are two events such that $E \subsetneq F$, then $P(E) \le P(F)$.

Proof. Since $E \subseteq F$, we have $F=E \cup (F-E)$.

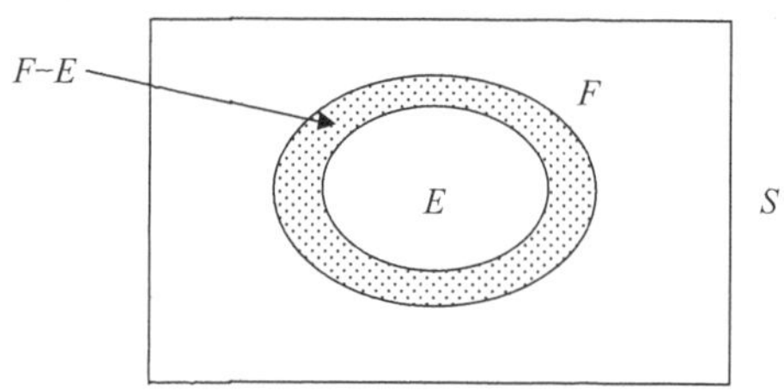

Figure 2.9

Also,

$$E \cap (F-E)=\phi,$$

Hence, by Theorem 2.9

$$P(F)=P(E)+P(F-E). \tag{1}$$

Since $P(F \backslash E)$ or $P(F-E) \ge 0$, it follows from (1) that $P(F) \ge P(E)$.

Theorem 2.10 (Addition Rule or Law of Addition of Probability)

If E and F are any arbitrary events associated with a random experiment, then

$$P(E \text{ or } F)=P(E \cup F)=P(E)+P(F)-P(A \cap B).$$

Proof. From the Venn diagram of the events and the corresponding sample space S, it follows that $E \cap \overline{F}$ and F are two mutually exclusive events and

$$(E \cap \overline{F}) \cup F=E \cup F.$$

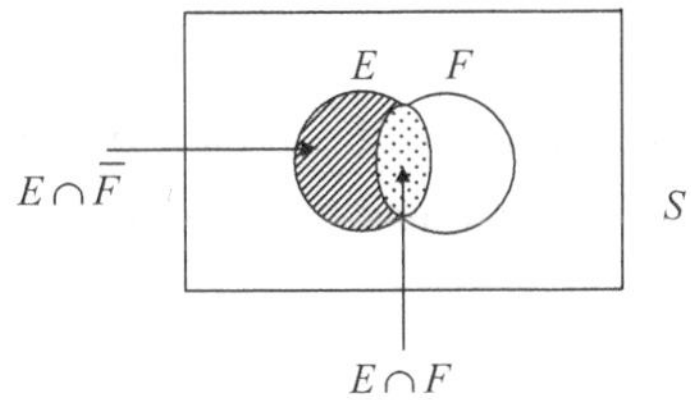

Figure 2.10

Hence,

$$P(E \cap \overline{F}) + P(F) = P(E \cup F). \quad (1)$$

But $E \cap F$ and $E \cap \overline{F}$ are mutually exclusive, that is,

$$(E \cap \overline{F}) \cup (E \cap F) = E$$

and so

$$P(E \cap \overline{F}) + P(E \cap F) = P(E).$$

Thus,

$$P(E \cap \overline{F}) = P(E) - P(E \cap F). \quad (2)$$

From (1) and (2) it follows that

$$P(E \cup F) = P(E) + P(F) - P(E \cap F).$$

Remark 2.2 If E and F are mutually exclusive, then $E \cap F = \phi$ and $P(\phi) = 0$, and so the above result reduces to

$$P(E \cup F) = P(E) + P(F),$$

which has been proved already.

EXAMPLE 2.53

Two fair dice are rolled. Find the probability of getting doubles (two dices showing the same numbers) or the sum of 7?

Proof. The sample space S is given by

11	12	13	14	15	16
21	22	23	24	25	26
31	32	33	34	35	36
41	42	43	44	45	46
51	52	53	54	55	56
61	62	63	64	65	66.

The total number of outcomes in S is 36.

Let E_1 be the event "get doubles" and E_2 is the event "sum of 7". Then

$$E_1 = \{11, 22, 33, 44, 55, 66\}, E_2 = \{16, 25, 34, 43, 52, 61\}.$$

We notice that E_1 and E_2 are mutually exclusive. Therefore,

$$P(E_1 \text{ or } E_2) = P(E_1 \cup E_2) = P(E_1) + P(E_2).$$

But

$$P(E_1) = \frac{\text{The number of favourable outcomes in } E_1}{\text{Number of outcomes in } S}$$

$$= \frac{6}{36} = \frac{1}{6}$$

Similarly,

$$P(E_2) = \frac{6}{36} = \frac{1}{6}.$$

Hence,

$$P(E_1 \text{ or } E_2) = \frac{1}{6} + \frac{1}{6} = \frac{1}{3}.$$

EXAMPLE 2.54

Two fair dices are thrown simultaneously. Find the probability of getting doubles or a multiple of 3 as the sum.

Proof. The sample space S consists of the points:

11	12	13	14	15	16
21	22	23	24	25	26
—	—	—	—	—	—
—	—	—	—	—	—
61	62	63	64	65	66.

Thus, S consists of 36 outcomes.

Let E_1 be the event of getting doubles. Then

$$E_1 = \{11, 22, 33, 44, 55, 66\}$$

and so number of favourable outcomes to the event E_1 is 6. So,

$$P(E_1) = \frac{6}{36} = \frac{1}{6}.$$

Let E_2 be the event of getting a multiple of 3 as the sum. Then,

$$E_2 = \{12, 15, 21, 24, 33, 36, 42, 45, 51, 54, 63, 66\}$$

and so number of favourable outcomes to the event E_2 is 12. Thus,

$$P(E_2) = \frac{12}{36} = \frac{1}{3}.$$

Further,

$$E_1 \cap E_2 = \{33, 66\}.$$

Thus,

$$P(E_1 \cap E_2) = \frac{2}{36} = \frac{1}{18}.$$

But

$$P(E_1 \cup E_2) = P(E_1) + P(E_2) - P(E_1 \cap E_2)$$

$$= \frac{1}{6} + \frac{1}{3} - \frac{1}{18} = \frac{4}{9}.$$

EXAMPLE 2.55

A bag contains 5 white, 7 black and 8 red balls. A ball is drawn at random. What is the probability that it is a red ball or a white ball?

Proof. The number of outcomes in the sample space is

$$5\,C_1 + 7\,C_1 + 8\,C_1 = 5 + 7 + 8 = 20.$$

Let E_1 be the event where red ball is obtained and E_2 be the event where white ball is obtained. Then,

$$P(E_1) = \frac{8C_1}{20} = \frac{8}{20} = \frac{2}{5}, \qquad P(E_2) = \frac{5C_1}{20} = \frac{5}{20} = \frac{1}{4}.$$

Also the events are mutually exclusive. Therefore,

$$\begin{aligned} P(E_1 \text{ or } E_2) &= P(E_1) + P(E_2) - P(E_1 \cap E_2) \\ &= \frac{2}{5} + \frac{1}{4} - P(\phi) \\ &= \frac{2}{5} + \frac{1}{4} - 0 = \frac{13}{20}. \end{aligned}$$

EXAMPLE 2.56

Let A and B be two mutually exclusive events of an experiment. If P (not A)=0.65, $P(A \cup B)=0.65$ and $P(B)=p$, find p.

Solution.

We have

$$P(\text{not } A) = P(\bar{A}) = 0.65.$$

But,

$$P(A) + P(\bar{A}) = 1$$

and so

$$P(A) = 1 - P(\bar{A}) = 1 - 0.65 = 0.35.$$

Further, since A and B are mutually exclusive,

$$P(A \cup B) = P(A) + P(B) = P(A) + p$$

and so

$$p = P(A \cup B) - P(A) = 0.65 - 0.35 = 0.30.$$

2.9 CONDITIONAL PROBABILITY

Definition 2.16

Let E and F be events and let $P(F) > 0$. Then the conditional probability of E given F is defined as

$$P(E \setminus F) = \frac{P(E \cap F)}{P(F)}.$$

EXAMPLE 2.57

Let two fair dice be rolled. If the sum of 7 is obtained, find the probability that at least one of the dice shows 2.

Solution.

Let E be the event "sum of 7 is obtained". Thus,

$$E=\{16, 25, 34, 43, 52, 61\}.$$

Let F be the event "at least one dice shows 2". Then

$$F=\{12, 22, 32, 42, 52, 62, 21, 23, 24, 25, 26\}.$$

We have

$$E \cap F=\{25, 52\}.$$

We want to find $P(F\backslash E)$. We have, by the definition of conditional probability,

$$P(F\backslash E)=\frac{P(E\cap F)}{P(E)}=\frac{2/36}{6/36}=\frac{1}{3}.$$

EXAMPLE 2.58

Weather records show that the probability of high barometric pressure is 0.82 and the probability of rain and high barometric pressure is 0.20. Find the probability of rain given high barometric pressure?

Solution.

Let E denote the event "rain" and F denote the event "high barometric pressure". We know that

$$P(E\backslash F)=\frac{P(E\cap F)}{P(F)}.$$

Therefore,

$$P(E\backslash F)=\frac{0.20}{0.82}=0.2446.$$

Theorem 2.11 (Multiplication Law of Probability)

Let $P(A\backslash B)$ denote the conditional probability of A when B has occurred. Then

$$P(A\cap B)=P(B)\,P(A\backslash B)=P(A)\,P(B\backslash A).$$

Proof. We know that

$$P(A\backslash B)=\frac{P(A\cap B)}{P(B)} \qquad (1)$$

and

$$P(B\backslash A)=\frac{P(A\cap B)}{P(A)}. \qquad (2)$$

From (1) and (2), we have

$$P(A\cap B)=P(\mathrm{B})\,P(A\backslash B)=P(A)\,P(B\backslash A).$$

EXAMPLE 2.59

A fair coin is tossed four times. Find the probability they are all heads if the first two tosses results in head.

Solution.

The sample space consists of $2^4 = 16$ outcomes. Let A be the event "all heads". Then

$$A = \{HHHH\}.$$

Let B be the event "first two heads". Then,

$$B = \{HHHH, HHHT, HHTH, HHTT\}.$$

We notice that

$$A \cap B = \{HHHH\}.$$

Therefore,

$$= P(B) = \frac{4}{16} = \frac{1}{4}, \qquad P(A \cap B) = \frac{1}{16}$$

and so

$$P(A \backslash B) = \frac{P(A \cap B)}{P(B)} = \frac{1/16}{1/4} = \frac{1}{4}.$$

2.10 INDEPENDENT EVENTS

Definition 2.17

Two events A and B are said to be **independent** if the occurrence or non-occurrence of one event does not affect the probability of the occurrence or non-occurrence of the other event. Mathematically, A and B are independent if

$$P(A \backslash B) = P(A).$$

Thus, if A and B are independent events, then

$$P(A) = P(A \backslash B) = \frac{P(A \cap B)}{P(B)}$$

or

$$P(A \cap B) = P(A)\,P(B).$$

This relation is called **multiplication rule for independent events.**

Hence, we can also define independence of events as follows:

Definition 2.18

Events A and B are called **independent** if

$$P(A \cap B) = P(A)\,P(B).$$

EXAMPLE 2.60

A married couple (husband and wife) appear for an interview for two vacancies against the same post. The probability of husband's selection is $\frac{1}{6}$ and the probability of wife's selection is $\frac{2}{5}$. What is the probability that

(i) Both of them will be selected
(ii) Only one of them will be selected
(iii) None of them will be selected
(iv) At least one of them will be selected

Solution.

Let E be the event "husband is selected" and F denote the event "wife is selected". We are given that

$$P(E)=\frac{1}{6} \quad \text{and} \quad P(F)=\frac{2}{5}.$$

Since there are two vacancies, selection of one does not affect the other. Hence E and F are independent events. Then,

(i) P (both of them are selected) $=P(E \cap F)$

$=P(E)\,P(F)$ since E and F are independent

$$=\frac{1}{6}\cdot\frac{2}{5}=\frac{1}{15}.$$

(ii) The Venn diagram shown below shows that $E \cap \overline{F}$ and $\overline{E} \cap F$ are exclusive. Hence,

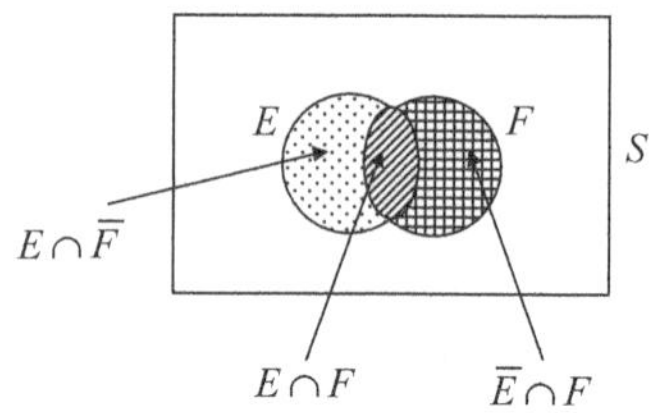

Figure 2.11

P (only one of them is selected) $=P[(E \cap \overline{F}) \cup (\overline{E} \cap F)]$

$=P(E \cap \overline{F})+P(\overline{E} \cap F)$ (exclusive events)

$=P(E)\,P(\overline{F})+P(\overline{E})\,P(F)$ (since E and F are independent)

$=P(E)\,(1-P(F))+(1-P(E))\,P(F)$

$$=\frac{1}{6}\left(1-\frac{2}{5}\right)+\left(1-\frac{1}{6}\right)\frac{2}{5}$$

$$=\frac{1}{10}+\frac{1}{3}=\frac{13}{30}.$$

(iii) We have

P (none of them is selected) $=P$ (not E and not F)

$=P(\overline{E} \cap \overline{F})$

$=P(\overline{E})\,P(\overline{F})$ (since E and F are independent)

$=(1-P(E))\,(1-P(F))$

$$=\left(1-\frac{1}{6}\right)\left(1-\frac{2}{5}\right)=\frac{5}{6}\cdot\frac{3}{5}=\frac{1}{2}.$$

(iv) We have

P (at least one of them gets selected) $=P$ (E or F)

$=P(E \cup F)$

$=P(E)+P(F)-P(E \cap F)$

$$= \frac{1}{6} + \frac{2}{5} - \frac{1}{15}, \text{ using (i)}$$

$$= \frac{1}{2}.$$

Remark 2.3 Using Venn diagram, we can also argue as follows:

$$P(E \cup F) = 1 - P\left(\overline{E \cup F}\right) = 1 - P(\overline{E} \cap \overline{F}) = 1 - \frac{1}{2} = \frac{1}{2}.$$

EXAMPLE 2.61

If $P(B) \neq 1$, show that

$$P\left(\overline{A} \backslash \overline{B}\right) = \frac{1 - P(A \cup B)}{P(\overline{B})}.$$

Solution.

We have

$$P\left(\overline{A} \backslash \overline{B}\right) = \frac{P(\overline{A} \cap \overline{B})}{P(\overline{B})} = \frac{P(\overline{A \cup B})}{P(\overline{B})} = \frac{1 - P(A \cup B)}{P(\overline{B})}.$$

EXAMPLE 2.62

A problem in mathematics is given to three students whose chances of solving the problem are $\frac{1}{2}, \frac{1}{3}, \frac{1}{4}$. What is the probability that the problem is solved?

Solution.

Let

A be the event "first student solves the problem"
B be the event "second student solves the problem"
C be the event "third student solves the problem"

It is given that

$$P(A) = \frac{1}{2}, \qquad P(B) = \frac{1}{3}, \qquad P(C) = \frac{1}{4}$$

and so

$$P\left(\overline{A}\right) = 1 - \frac{1}{2} = \frac{1}{2}, \qquad P\left(\overline{B}\right) = 1 - \frac{1}{3} = \frac{2}{3},$$

$$P\left(\overline{C}\right) = 1 - \frac{1}{4} = \frac{3}{4}.$$

Hence,

$$\begin{aligned} P\,(\text{the problem is solved}) &= P\,(A \text{ or } B \text{ or } C) \\ &= P\,(A \cup B \cup C) \end{aligned}$$

$$
\begin{aligned}
&= 1 - P\left[\overline{(A \cup B \cup C)}\right] \\
&= 1 - P\left(\overline{A} \cap \overline{B} \cap \overline{C}\right) \\
&= 1 - P\left(\overline{A}\right)P\left(\overline{B}\right)P\left(\overline{C}\right) \text{ since } A, B, C \text{ are independent} \\
&= 1 - \frac{1}{2}\cdot\frac{2}{3}\cdot\frac{3}{4} = \frac{3}{4}.
\end{aligned}
$$

Theorem 2.12 (Baye's Theorem)

Let $A_1, A_2, \ldots, A_m$ be pairwise mutually exclusive and exhaustive random events, where $P(A_i) \geq 0$, $i = 1, 2, \ldots, m$. Then for any arbitrary event B of the random experiment,

$$
P(A_i \backslash B) = \frac{P(A_i)P(B \backslash A_i)}{\sum_{i=1}^{m} P(A_i)P(B \backslash A_i)}.
$$

Proof. Let S be the sample space of the random experiment. Since the events $A_1, A_2, \ldots, A_m$ are pairwise exclusive and exhaustive, we have

$$
S = A_1 \cup A_2 \cup \ldots \cup A_m.
$$

Therefore we have,

$$
\begin{aligned}
B = S \cap B &= (A_1 \cup A_2 \cup \ldots \cup A_m) \cap B \\
&= (A_1 \cap B) \cup (A_2 \cap B) \cup \ldots \cup (A_m \cap B).
\end{aligned}
$$

Since $A_1 \cap B, A_2 \cap B, \ldots, A_m \cap B$ are mutually exclusive, it follows by addition law that

$$
\begin{aligned}
P(B) &= P(A_1 \cap B) + P(A_2 \cap B) + \ldots + P(A_m \cap B) \\
&= P(B \backslash A_1)\, P(A_1) + P(B \backslash A_2)\, P(A_2) + \ldots + P(B \backslash A_m)\, P(A_m).
\end{aligned}
$$

This relation is called the "**Theorem on Total Probability**".

Using this relation, we have

$$
\begin{aligned}
P(A_i \backslash B) &= \frac{P(A_i \cap B)}{P(B)} = \frac{P(B \backslash A_i)P(A_i)}{P(B)} \\
&= \frac{P(A_i)P(B \backslash A_i)}{\sum_{i=1}^{m} P(A_i)P(B \backslash A_i)}.
\end{aligned}
$$

EXAMPLE 2.63

A university purchased computers from three firms. The percentage of computer purchased and percentage of defective computers is shown in the table below:

	Firm		
	HCL	WIPRO	IBM
Percent purchase	45	25	30
Percent defective	2	3	1

Let

A be the event "Computer purchased from HCL"
B be the event "Computer purchased from WIPRO"
C be the event "Computer purchased from IBM"
D be the event "Computer was defective"
Find $P(A)$, $P(B)$, $P(C)$, $P(D\backslash A)$, $P(D\backslash B)$, $P(D\backslash C)$ and $P(D)$.

Solution.
We note that

$$P(A)=\frac{45}{45+25+30}=0.45,$$
$$P(B)=\frac{25}{100}=0.25,$$
$$P(C)=\frac{30}{100}=0.30,$$
$$P(D\backslash A)=\frac{2}{100}=0.02,$$
$$P(D\backslash B)=\frac{3}{100}=0.03,$$
$$P(D\backslash C)=\frac{1}{100}=0.01.$$

Then,

$$\begin{aligned}P(D)&=P(D\backslash A)\,P(A)+P(D\backslash B)\,P(B)+P(D\backslash C)\,P(C)\\&=(0.02)\,(0.45)+(0.03)\,(0.25)+(0.01)\,(0.30)\\&=0.0090+0.0075+0.0030=0.0195.\end{aligned}$$

EXAMPLE 2.64

In a test, an examinee either guesses, or copies or knows the answer to multiple choice questions with four choices. The probability that he makes a guess is $\frac{1}{3}$ and the probability that he copies the answer is $\frac{1}{6}$. The probability that his answer is correct, given that he copied it, is $\frac{1}{8}$. Find the probability that he knew the answer to the question given that he correctly answered it.

Solution.
Let us consider the following events:

A: the examinee guesses the answer
B: the examinee copies the answer
C: the examinee knows the answer
D: the examinee answers correctly.

It is given that

$$P(A)=\frac{1}{3},\qquad P(B)=\frac{1}{6},\qquad P(D\backslash B)=\frac{1}{8}.$$

Also, the hypothesis that examinee either guesses or copies or knows the answer implies that

$$P(C)=1-P(A)-P(B)=1-\frac{1}{3}-\frac{1}{6}=\frac{1}{2}.$$

Further,

$P(D\backslash C)=1$ since he knows the answer correctly.

$P(D\backslash A)=\frac{1}{4}$ (since if he guesses, he can tick any one of the four choices).

Then, by Baye's law,

$$P(C\backslash D)=\frac{P(D\backslash C)P(C)}{P(D\backslash A)P(A)+P(D\backslash B)P(B)+P(D\backslash C)P(C)}$$

$$=\frac{1\cdot(1/2)}{(1/4)\cdot(1/3)+(1/8)\cdot(1/6)+1\cdot(1/2)}=\frac{24}{29}$$

EXAMPLE 2.65

The following observations were made at a clinic where HIV test is performed.

(i) 15% of the patients at the clinic have HIV

(ii) Among those who have HIV, 95% test positive on the ELISA test

(iii) Among those who do not have HIV, 2% test positive on the ELISA test.

Find the probability that a patient has HIV if the ELISA test is positive.

Solution.

We consider the following events:

A: "has the HIV"

B: "does not have the HIV"

C: "test positive"

We are given that

$$P(A)=\frac{15}{100}=0.15.$$

Therefore,

$$P(B)=P(\bar{A})=1-P(A)=1-0.15=0.85,$$

$$P(C\backslash A)=\frac{95}{100}=0.95,$$

$$P(C\backslash B)=\frac{2}{100}=0.02.$$

We want to find $P(A\backslash C)$. By Baye's theorem, we have

$$P(A\backslash C)=\frac{P(C\backslash A)P(A)}{P(C\backslash A)P(A)+P(C\backslash B)P(B)}$$

$$=\frac{(0.95)(0.15)}{(0.95)(0.15)+(0.02)(0.085)}=0.89.$$

EXAMPLE 2.66

An item is manufactured by there factories F_1, F_2 and F_3. The number of units of the item produced by F_1, F_2 and F_3 are $2x$, x and x, respectively. It is known that 2% of the items produced by F_1 and F_2 are defective and 4% of the items produced by F_3 are defective. All units produced by these factories are put together in one stockpile and one unit is chosen at random. It is found that this item is defective. What is the probability that this defective unit came from (i) Factory F_1, (ii) Factory F_2 or (iii) Factory F_3?

Solution.

Consider the events:

A: "the unit is defective"
B: "the defective unit came from F_1"
C: "the defective unit came from F_2"
D: "the defective unit came from F_3"

We have then, as per given hypothesis,

$$P(B)=\frac{2x}{4x}=\frac{1}{2},$$

$$P(C)=\frac{x}{4x}=\frac{1}{4},$$

$$P(D)=\frac{x}{4x}=\frac{1}{4},$$

$$P(A\backslash B)=\frac{2}{100}=0.02,$$

$$P(A\backslash C)=\frac{2}{100}=0.02,$$

$$P(A\backslash D)=\frac{4}{100}=0.04.$$

Then Theorem on Total Probability implies that

$$P(A)=P(A\backslash B)\,P(B)+P(A\backslash C)\,P(C)+P(A\backslash D)\,P(D)$$

$$=0.02\left(\frac{1}{2}\right)+0.02\left(\frac{1}{4}\right)+0.04\left(\frac{1}{4}\right)$$

$$=0.025.$$

We then have, by Baye's theorem,

$$P(B\backslash A)=\frac{P(A\backslash B)P(B)}{P(A)}=\frac{0.02(1/2)}{0.025}=0.4,$$

$$P(C\backslash A)=\frac{P(A\backslash C)P(C)}{P(A)}=\frac{0.02(1/4)}{0.025}=0.2,$$

$$P(D\backslash A)=\frac{P(A\backslash D)P(D)}{P(A)}=\frac{0.04(1/4)}{0.025}=0.4.$$

2.11 PROBABILITY DISTRIBUTION

Definition 2.19

Let S be a sample space of an experiment. A **random variable** X is a function of the possible events of S which assigns a numerical value of each outcome in S.

Definition 2.20

Let a random variable X assumes the values $x_1, x_2, \ldots, x_m$ corresponding to various outcomes of a random experiment. If the probability of x_i is $P(x_i)=p_i$, $1 \le i \le n$ such that $p_1+p_2+\ldots+p_n=1$, then the function $P(X)$ is called the **probability function** of the random variable X and the set $\{P(x_i)\}$ is called the **probability distribution** of X. Since random variable X takes a finite set of values, it is called ***discrete variate*** and $\{P(x_i)\}$ is called the discrete ***probability distribution***.

The probability distribution of X is denoted by the table:

X	x_1	x_2	x_3	...	x_n
$P(X)$	p_1	p_2	p_3	...	p_n.

2.11.1 Mean and Variance of a Random Variable

Let X be a random variable which takes the values $x_1, x_2, \ldots, x_m$ with corresponding probabilities $p_1, p_2, \ldots, p_m$. Then the **mean** (also called ***expectation***) and **variance** of the random variables are defined by

Mean:
$$\mu = \frac{\sum_{i=1}^{m} p_i x_i}{\sum_{i=1}^{m} p_i} = \sum_{i=1}^{m} p_i x_i \quad \text{since } \sum_{i=1}^{m} p_i = 1.$$

Variance:
$$\begin{aligned}\sigma^2 &= \sum_{i=1}^{m} (x_i - \mu)^2 p_i \\ &= \sum_{i=1}^{m} \left(x_i^2 - 2\mu x_i + \mu^2\right) p_i \\ &= \sum_{i=1}^{m} p_i x_i^2 - 2\mu \sum_{i=1}^{m} p_i x_i + \mu^2 \sum_{i=1}^{m} p_i \\ &= \sum_{i=1}^{m} p_i x_i^2 - 2\mu^2 + \mu^2 \quad \text{since } \sum_{i=1}^{m} p_i x_i = \mu \text{ and } \sum_{i=1}^{m} p_i = 1 \\ &= \sum_{i=1}^{m} p_i x^2 - \mu^2,\end{aligned}$$

where σ is called the ***standard deviation of the distribution***.

EXAMPLE 2.67

Two cards are drawn successively **with replacement** from a well-shuffled pack of 52 playing cards. Find the probability distribution of the number of aces.

Solution.

Let X be the random variable that is the number of aces obtained in the draw of two cards. There are three possibilities: (i) there is no ace, (ii) there is one ace and (iii) there are two aces. Thus, the random variable takes the value 0, 1, 2. Then,

$$P\,(\text{no ace is drawn}) = P\,(X=0) = \frac{48}{52} \cdot \frac{48}{52} = \frac{144}{169}.$$

$$P\text{ (one ace is drawn)} = (X=1)$$
$$= P\text{ (one ace is drawn in the first draw and no ace is drawn in the second draw)}$$
$$+ P\text{ (no ace is drawn in the first draw and one ace is drawn in the second draw)}$$
$$= \frac{4}{52}\cdot\frac{48}{52}+\frac{48}{52}\cdot\frac{4}{52}=\frac{24}{169}.$$

$$P\text{ (two aces are drawn)} = P(X=2) = \frac{4}{52}\cdot\frac{4}{52}=\frac{1}{169}.$$

Hence the probability distribution is

X	0	1	2
$P(X)$	$\frac{144}{169}$	$\frac{24}{169}$	$\frac{1}{169}$.

EXAMPLE 2.68

Find the probability distribution of the number of green balls drawn when three balls are drawn one by one **without replacement** from a bag containing three green and five white balls.

Solution.

Let X be the random variable which is the number of green balls drawn when three balls are drawn without replacement. The random variable takes the values 0, 1, 2, 3.

We represent green ball by G and white ball by W. Then we have

$$P\text{ (no green ball is drawn)} = P(X=0)$$
$$= P(\text{WWW}) = \frac{5}{8}\cdot\frac{4}{7}\cdot\frac{3}{6}=\frac{5}{28}.$$

$$P\text{ (one green ball is drawn)} = P(X=1)$$
$$= P(\text{GWW})+P(\text{WGW})+P(\text{WWG})$$
$$= \frac{3}{8}\cdot\frac{5}{7}\cdot\frac{4}{6}+\frac{5}{8}\cdot\frac{3}{7}\cdot\frac{4}{6}+\frac{5}{8}\cdot\frac{4}{7}\cdot\frac{3}{6}=\frac{15}{28}.$$

$$P\text{ (two green balls are drawn)} = P(X=2)$$
$$= P(\text{GGW})+P(\text{GWG})+P(\text{WGG})$$
$$= \frac{3}{8}\cdot\frac{2}{7}\cdot\frac{5}{6}+\frac{3}{8}\cdot\frac{5}{7}\cdot\frac{2}{6}+\frac{5}{8}\cdot\frac{3}{7}\cdot\frac{2}{6}=\frac{15}{56}.$$

$$P\text{ (three green balls are drawn)} = P(X=3) = P(\text{GGG}) = \frac{3}{8}\cdot\frac{2}{7}\cdot\frac{1}{6}=\frac{1}{56}.$$

Therefore, the probability distribution is

X	0	1	2	3
$P(X)$	$\frac{5}{28}$	$\frac{15}{28}$	$\frac{15}{56}$	$\frac{1}{56}$.

2.11.2 Binomial Distribution

Let S be a sample space for a random experiment. Let A be an event associated with a subset of S and let $P(A)=p$, then we know that $P(\overline{A})=1-p$. If we denote $P(\overline{A})=q$, then $p+q=1$.

If we call the occurrence of the event A as "success" and non-occurrence of the event A as a "failure", then

$$P(\text{failure})=1-P(\text{success})$$

or

$$P(\text{failure})+P(\text{success})=1.$$

Suppose that X is a random variable on the sample space as the "number of success". Then the probability distribution associated with the above random experiment is

X	0	1
$P(X)$	q	p.

If the experiment is conducted two times, then the possible outcomes are success success, success failure, failure success, failure failure. Since the trails are independent, we have

$$\begin{aligned}P(\text{success success})&=P(\text{both success})\\&=P(\text{success})\,P(\text{success})\\&=p\,.\,p=p^2\end{aligned}$$

$$P(\text{success failure})=P(\text{success})\,P(\text{failure})=p\,.\,q$$

$$P(\text{failure success})=P(\text{failure})\,P(\text{success})=q\,.\,p$$

$$P(\text{failure failure})=P(\text{failure})\,P(\text{failure})=q^2.$$

Thus, in terms of random variable,

$$P(X=0)=P(\text{failure failure})=q^2,$$

$$P(X=1)=p\,q+q\,p=2\,p\,q,$$

$$P(X=2)=p\,(\text{success success})=p^2.$$

Also we note that

$$P(X=0)+P(X=1)+P(X=2)=p^2+q^2+2\,p\,q=(p+q)^2=(1)^2=1.$$

Thus, the probability distribution associated with the two experiments is

X	0	1	2
$P(X)$	q^2	$2\,p\,q$	p^2.

The term of $P(X)$ are the terms in the binomial expansion of $(q+p)^2$.

Similarly, the probability distribution associated with the three experiments is

X	0	1	2	3
$P(X)$	q^3	$3\,q^2\,p$	$3\,q\,p^2$	p^3.

Thus probabilities are the terms in the binomial expansions of $(q+p)^3$.

If the experiment is repeated n times, then the probability distribution is

X	0	1	2	...	r	... n
$P(X)$	q^n	${}^nC_1q^{n-1}p$	${}^nC_2q^{n-2}p^2$	...	${}^nC_rq^{n-r}p^r$	... p^n.

Clearly the probabilities are terms in the binomial expansion of $(q+p)^n$.

This probability distribution is called the **binomial distribution** and X is called a **binomial random variable**.

Further, **mean of the binomial distribution** is given by

Mean: $$\mu = \sum_{r=0}^{n} rP(r)$$
$$= P(1) + 2P(2) + \ldots + nP(n)$$
$$= {}^nC_1\, q^{n-1}\, p + 2\,{}^nC_2\, q^{n-2}\, p^2 + \ldots + {}^nC_n\, p^n$$
$$= npq^{n-1} + \frac{2n(n-1)}{2!}p^2q^{n-2} + \cdots + np^n$$
$$= n\,p[q^{n-1} + (n-1)p\, q^{n-2} + \cdots + p^{n-1}]$$
$$= n\,p[(q+p)^{n-1}]$$
$$= n\,p, \text{ since } q+p=1.$$

The **variance of the binomial distribution** is

Variance: $$\sigma^2 = \sum_{r=0}^{n} r^2 P(r) - \mu^2. \tag{1}$$

Now

$$\sum_{r=0}^{n} r^2 P(r) = \sum_{r=0}^{n} [r + r(r-1)]\, p(r)$$
$$= \sum_{r=0}^{n} rP(r) + \sum_{r=0}^{n} r(r-1)P(r)$$
$$= \mu + \sum_{r=0}^{n} r(r-1)P(r)$$
$$= np + \sum_{r=2}^{n} r(r-1)P(r) \quad \text{since } \mu = np$$
$$= np + \sum_{r=2}^{n} r(r-1)\ {}^nC_r\, p^r q^{n-r}$$
$$= np + n(n-1)\,p^2\,(q+p)^{n-2}$$
$$= np + n(n-1)\,p^2 \quad \text{since } (q+p) = 1.$$

Hence,

$$\sigma^2 = n\,p + n\,(n-1)p^2 - \mu^2$$
$$= n\,p + n\,(n-1)p^2 - n^2\,p^2 \text{ since } \mu = n\,p$$
$$= n\,p + n^2\,p^2 - n\,p^2 - n^2\,p^2$$
$$= n\,p(1-p) = n\,p\,q.$$

EXAMPLE 2.69

Find the probability of 4 turning up at least once in **two** tosses of a fair dice.

Solution.

Let X denote the number of times the number 4 turns up. We note that

$$P\,(4 \text{ turns up}) = p = \frac{1}{6}$$

and so

$$q = 1 - p = 1 - \frac{1}{6} = \frac{5}{6}.$$

Thus the probability distribution is

X	0	1	2
$P\,(X)$	q^2	$2p\,q$	p^2.

Hence,

$$\begin{aligned} P\,(4 \text{ turns up at least once}) &= P\,(X=1) + P\,(X=2) \\ &= 2p\,q + p^2 \\ &= 2 \cdot \frac{1}{6} \cdot \frac{5}{6} + \left(\frac{1}{6}\right)^2 = \frac{11}{36}. \end{aligned}$$

EXAMPLE 2.70

A coin is tossed five times. What is the probability of getting at least three heads?

Solution.

Let X denote the "number of heads obtained". We know that

$$p = P\,(\text{head obtained}) = \frac{1}{2}.$$

Therefore,

$$q = 1 - p = 1 - \frac{1}{2} = \frac{1}{2}.$$

The random variable X takes the values 0, 1, 2, 3, 4, 5 and $n=5$. Hence

$$\begin{aligned} P\,(\text{at least three heads}) &= P\,(X \geq 3) \\ &= P\,(X=3) + P\,(X=4) + P\,(X=5) \\ &= {}^5C_3 p^3 q^2 + {}^5C_4 p^4 q + {}^5C_5 p^5 \\ &= 10\left(\frac{1}{2}\right)^3 \left(\frac{1}{2}\right)^2 + 5\left(\frac{1}{2}\right)^4 \left(\frac{1}{2}\right) + \left(\frac{1}{2}\right)^5 \\ &= \frac{10}{32} + \frac{5}{32} + \frac{1}{32} = \frac{1}{2}. \end{aligned}$$

EXAMPLE 2.71

The mean and variance of a binomial variable X are 2 and 1, respectively. Find the probability that X takes a value greater than 1.

Solution.

Suppose n is number of independent trails. Since x is a binomial variate, we have

$$\text{Mean} = n\,p = 2 \qquad \text{(given)}, \tag{1}$$

$$\text{Variance} = n\,p\,q = 1 \qquad \text{(given)}. \tag{2}$$

Dividing (2) by (1), we get $q = \frac{1}{2}$, which yields $p = 1 - q = \frac{1}{2}$. Also then (1) gives $n=4$. Hence,

$$
\begin{aligned}
P(X>1) &= 1 - [P(X=0) + P(X=1)] \\
&= 1 - [{}^1C_0 q^1 + {}^1C_1 q^3 p] \\
&= 1 - [q^4 - 4pq^3] \\
&= 1 - \left[\left(\frac{1}{2}\right)^4 - 4\left(\frac{1}{2}\right)\left(\frac{1}{2}\right)^3\right] \\
&= 1 - \frac{5}{16} = \frac{11}{16}.
\end{aligned}
$$

EXERCISES

1. If seven colours are used to paint 50 desks, show that at least 8 desks will be painted by the same colour.
2. An inventory consists of a list of 80 items, each marked "available" or "not available". There are 50 available items. Show that there are at least two unavailable items in the list either 3 or 6 items apart.
3. Find the minimum number of integers to be selected from $\{1, 2, \ldots, 9\}$ so that sum of two of them is even.
4. How many cards must be picked from a pack of 52 playing cards to get at least one black car?
5. How many integers from 1 through 99 do not have any repeated digits?
6. How many integers are there between 5 and 1,004 that are multiples of 3?
7. Show that

$$98\, P_2 + 98\, P_1 = 9{,}604.$$

8. Show that

$$\binom{n}{0} - \binom{n}{1} + \binom{n}{2} - \cdots + \binom{n}{n} = 0.$$

9. In a single throw of two distinct dice, what is the probability of getting a total of 11?
10. Find the probability that a randomly chosen three-digit integer is divisible by 5.
11. Show that the number of distinguishable words that can be formed from the letters of MISSISSIPPI is 34,650.
12. A certain defective dice is tossed. The probabilities of getting the faces 1 to 6 are respectively

$$p_1 = \frac{2}{18},\quad p_2 = \frac{3}{18},\quad p_3 = \frac{4}{18},\quad p_4 = \frac{3}{18},$$

$$p_5 = \frac{4}{18},\quad p_6 = \frac{2}{18}.$$

13. What is the probability that a prime number is on the top?
14. Let A and B be two events such that $P(A)=0.4$, $P(B)=p$ and $P(A \cup B)=6$. Find p so that A and B are independent.
15. A bag contains three red and five black balls and a second bag contains six red and four black balls. A ball is drawn from each bag. Find the probability that one ball is red and the other is black.
16. The probability of a man hitting a target is $\frac{1}{3}$. If he fires six times, what is the probability that he hits the target
    (i) At least twice
    (ii) At most twice
17. A candidate takes on 20 questions with four multiple choice examination. One of the choices in every question is incorrect. The candidate makes guess of the remaining choices. Find the expected number of correct answers and the standard deviation.

3 Recurrence Relations

A recurrence relation relates the nth term of a sequence to its predecessors. These relations are related to recursive algorithms.

3.1 RECURRENCE RELATIONS

Definition 3.1

A **recurrence relation** for a sequence $a_0, a_1, a_2, \dots$ is a formula (equation) that relates each term a_n to certain of its predecessors $a_0, a_1, \dots, a_{n-1}$.

The **initial conditions** for such a recurrence relation specify the values of $a_0, a_1, a_2, \dots, a_{n-1}$.

For example, recursive formula for the sequence

$$3, 8, 13, 18, 23$$

is

$$a_1 = 3,\ a_n = a_{n-1} + 5,\ 2 \le n < \infty.$$

Here, $a_1 = 3$ is the initial condition.

Similarly, the infinite sequence

$$3, 7, 11, 15, 19, 23, \dots$$

can be defined by the recursive formula

$$a_1 = 3,\ a_n = a_{n-1} + 4,\ 2 \le n < \infty.$$

EXAMPLE 3.1

Find the sequence represented by the recursive formula

$$a_1 = 5,\ a_n = 2a_{n-1},\ 2 \le n \le 6.$$

Solution.

The initial condition is $a_1 = 5$ and n satisfies the condition $2 \le n \le 6$. Thus,

$$\begin{aligned} a_2 &= 2a_1 = 10, \\ a_3 &= 2a_2 = 20, \\ a_4 &= 2a_3 = 40, \\ a_5 &= 2a_4 = 80, \\ a_6 &= 2a_5 = 160. \end{aligned}$$

Hence the given recurrence formula defines the finite sequence

$$5, 10, 20, 40, 80, 160.$$

3.2 EXPLICIT FORMULA FOR A SEQUENCE

Consider the sequence

$$1, 4, 9, 16, 25, 36, 49, \ldots,$$

which is a sequence of the squares of all positive integers.

This sequence is described by the formula

$$a_n = n^2, \quad 1 \leq n < \infty.$$

Thus, the terms of the sequence have been described using only its positive number. This type of formula is called **Explicit formula**.

We note that the explicit formula

$$a_n = (-4)^n, \quad 1 \leq n < \infty$$

describes the infinite sequence

$$-4, 16, -64, 256, \ldots$$

EXAMPLE 3.2

Find the explicit formula for the finite sequence

$$87, 82, 77, 72, 67.$$

Can this sequence be described by a recursive relation?

Solution.

The explicit formula for the given finite sequence is

$$a_n = 92 - 5\,n, \quad n = 1, 2, \ldots.$$

Also, it can be described by the recursive formula

$$a_1 = 87, \qquad a_n = a_{n-1} - 5, \qquad 2 \leq n \leq 5.$$

Definition 3.2

The product of all the integers from 1 to n is called n **factorial.** It is denoted by $n!$.

Thus,

$$n! = 1 \cdot 2 \cdot 3 \ldots (n-1) \cdot n.$$

Further, **zero factorial** is defined to be 1, that is, $0! = 1$.

EXAMPLE 3.3

Find recursive formula for the factorial function.

Solution.

We note that

$$\begin{aligned}
0! &= 1,\\
1! &= 1 = 1 \cdot 1 = 1 \cdot 0!,\\
2! &= 1 \cdot 2 = 2 \cdot 1!,\\
3! &= 1 \cdot 2 \cdot 3 = 3 \cdot 2!,\\
4! &= 1 \cdot 2 \cdot 3 \cdot 4 = 4 \cdot 3!,\\
&\text{-----------------}\\
&\text{-----------------}
\end{aligned}$$

$$n! = 1 \cdot 2 \cdot 3 \ldots (n-1) \cdot n = n \cdot (n-1)!.$$

It follows therefore that

(i) If $n=0$, then $n!=1$,
(ii) If $n>0$, then $n!=n \cdot (n-1)!$.

Hence, the factorial is described by the recursive formula

$$0!=1,\ n!=n \cdot (n-1)!,\ n>0.$$

EXAMPLE 3.4

Fibonacci (1202) posed the following problem:

A single pair of rabbits (male and female) is born at the beginning of a year. Find the number of rabbits at the end of the year subject to the following conditions:

(a) Rabbit pairs are not fertile during their first month of life but there after give birth to one new male/female pair at the end of every month,
(b) No rabbit die.

Solution.

We observe that the number of rabbit pairs alive at the end of month n is equal to the number of rabbit pairs alive at the end of month $n-1$ plus the number of rabbit pairs born at the end of month n. This sum is equal to the number of rabbit pairs alive at the end of month $n-1$ plus the number of rabbit pairs alive at the end of month $n-2$. Thus if F_n represents the number of rabbit pairs alive at the end of month n and if F_0 be the initial number of rabbit pair, then, from the above discussion, we have

$$F_0=1,\ F_1=1 \text{ (since pairs is not fertile untill the second month)}$$
$$F_n=F_{n-1}+F_{n-2}.$$

If we compute $F_2, F_3, \ldots, F_{12}$, then find

$$F_2=2,\ F_3=3,\ F_4=5,\ F_5=8,\ F_6=13,\ F_7=21,$$
$$F_8=34,\ F_9=55,\ F_{10}=89,\ F_{11}=144,\ F_{12}=233.$$

Thus at the end of 12th month there are 233 pairs of rabbits, that is, 466 rabbits in all.

Definition 3.3

The sequence

$$1, 1, 2, 3, 5, 8, 13, 21, 34,\ldots$$

defined by the recurrence relation

$$f_0=1, \qquad f_1=1, \qquad f_n=f_{n-1}+f_{n-2}$$

is called **Fibonacci sequence.**

EXAMPLE 3.5

Derive recurrence relation for obtaining the amount A_n at the end of n years on the investment of Rs 10,000 at 5% interest compounded annually.

Solution.

Suppose

$$A_n = \text{amount at the end of } n \text{ years.}$$

Then,

$$A_n = A_{n-1} + \text{interest during } (n-1)\text{th year on } A_{n-1}$$

$$= A_{n-1} + \frac{5}{100} A_{n-1} = A_{n-1}(1+0.05) = 1.05 A_{n-1}.$$

Thus, the recurrence relation for calculating amount becomes

$$A_0 = \text{Rs } 10{,}000,$$
$$A_n = 1.05\ A_{n-1},\ n>0.$$

Using this recurrence relation, we can compute value of A_n for any n. For example,

$$A_1 = 1.05\ (10{,}000) = (1.05)^1\ (10{,}000) = \text{Rs } 10{,}500,$$
$$A_2 = (1.05)\ (1.05)\ (10{,}000) = (1.05)^2 (10{,}000),$$

$$A_n = (1.05)^n\ (10{,}000).$$

Remark 3.1 If the annual interest rate i is compounded m times per annum, then the interest rate per period is $\frac{i}{m}$ and so the recurrence formula becomes

$$A_0 = \text{initial amount},$$
$$A_n = A_{n-1} + A_{n-1}\left(\frac{i}{m}\right) = A_{n-1}\left(1+\frac{i}{m}\right).$$

EXAMPLE 3.6

Developing a recurrence formula, find the number of bit strings of length 4 that do not contain the pattern 111.

Solution.

Let S_n denote the number of bit strings n that do not contain the pattern 111. Any string in S_n begins with either a 0 or 10 or 11.

If an n-bit string begins with 0 and it does not contain the pattern 111, then the $(n-1)$ bit string following the initial 0 does not contain the pattern 111. Since any $(n-1)$-bit string not containing 111 can follow the initial 0, there are S_{n-1} strings of the type that begin with 0.

If an n-bit string begins with 10 and does not contain pattern 111, then the $(n-2)$-bit string following the initial 10 cannot contain the pattern 111. Therefore there are S_{n-2} string that begin with 10.

If an n-bit string begins with 11 and does not contain the pattern 111, then the third bit must be 0, otherwise the string would contain the pattern 111. The $(n-3)$-bit string following the initial 110 cannot contain the pattern 111. Therefore, there are S_{n-3} strings that begin with 11.

Hence, by the addition rule,

$$S_n = S_{n-1} + S_{n-2} + S_{n-3},\quad n \geq 4$$

and the initial conditions are

$$S_0 = 1 \text{ (empty string)},\ S_1 = 1,\ S_2 = 4,\ S_3 = 7.$$

Infact,

Number of string of length 0 = ε,
Number of strings of length 1 = 0, 1,
Number of strings of length 2 = 00, 01, 10, 11,
Number of strings of length of 3 = 000, 001, 010, 100, 101, 110.

From the formula, derived above, we have

$$S_4=S_3+S_2+S_1=7+4+1=12.$$

3.3 SOLUTIONS OF RECURRENCE RELATIONS

To study general properties of sequences, the recurrence relation with initial conditions are solved to get explicit formula. Such an explicit formula is called a **solution** of the given recurrence relation.

Definition 3.4

A technique for finding an explicit formula for the sequence defined by a recurrence relation is **called backtracking.** In this technique, the values of a_n are back tracked, substituting the values of a_{n-1}, a_{n-2} and so on, till a pattern is clear.

EXAMPLE 3.7

Find an explicit formula for the recurrence relation

$$a_0=1,\ a_n=a_{n-1}+2.$$

Solution.

The recurrence relation

$$a_0=1,\ a_n=a_{n-1}+2$$

defines the sequence

$$1, 3, 5, 7, \ldots.$$

We backtrack the value of a_n by the substituting the definition of a_{n-1}, a_{n-2} and so on untill there is a pattern. We have

$$\begin{aligned} a_n &= a_{n-1}+2 \\ &= a_{n-2}+2+2=a_{n-2}+2\cdot 2 \\ &= a_{n-3}+2+2+2=a_{n-3}+2\cdot 3 \\ &= a_{n-4}+2+2+2+2=a_{n-4}+2\cdot 4 \text{ and so on.} \end{aligned}$$

Thus, backtracking yields

$$a_n=a_{n-k}+2k.$$

If we set $k=n$, then

$$a_n=a_{n-n}+2\,n=a_0+2\,n=1+2\,n,$$

which is the required **explicit formula**.

EXAMPLE 3.8

Backtrack to find explicit formula for the sequence defined by the recurrence relation

$$a_1=1,\ a_n=3\,a_{n-1}+1,\ n\geq 2.$$

Solution.

The recurrence relation defines the sequence

$$1, 4, 13, 40, \ldots.$$

Backtracking yields

$$\begin{aligned}a_n &= 3\,a_{n-1}+1\\ &= 3(3\,a_{n-2}+1)+1 = 3^2 a_{n-2}+3^1+1\\ &= 3\{3(3\,a_{n-3}+1)+1\}+1 = 3^3\,a_{n-3}+3^2+3^1+1\\ &= 3[3\{3(3a_{n-4}+1)+1\}+1]+1\\ &= 3^4\,a_{n-4}+3^3+3^2+3^1+1 \text{ and so on.}\end{aligned}$$

The backtracking will end at

$$a_n = 3^k\,a_{n-k}+3^{k-1}+3^{k-2}+\ldots+3^2+3^1+1.$$

If we set $k=n-1$, then we have

$$\begin{aligned}a_n &= 3^{n-1}\,a_{n-(n-1)}+3^{n-2}+\ldots+3^3+3^2+3^1+1\\ &= 3^{n-1}\,a_1+3^{n-2}+\ldots+3^3+3^2+3^1+1\\ &= 3^{n-1}+3^{n-2}+\ldots+3^3+3^2+3^1+1\\ &= \frac{3^n-1}{3-1} = \frac{3^n-1}{2}.\end{aligned}$$

Hence

$$a_n = \frac{3^n-1}{2}$$

is the required explicit formula.

Definition 3.5

A sequence a_0, a_1, a_2, ... is called an **arithmetic sequence** if and only if there is a constant d such that

$$a_n = a_{n-1}+d \text{ for all integers } n \geq 1.$$

For example, the recurrence relation

$$a_n = a_{n-1}+3,\ a_1 = 2$$

defines an arithmatic sequence

2, 5, 8, 11, 14, ...

with constant difference 3.

The explicit formula for the arithmatic sequence $a_n = a_{n-1}+d$ is

$$a_n = a_0+d\,n, \qquad n \geq 0.$$

EXAMPLE 3.9

Let K_n be the picture obtained by drawing n dots (vertices) and joining each pair of vertices by an edge. Develop a recurrence relation for the number of edges of K_n and find explicit formula for it.

Solution.

The pictures of complete graphs K_1, K_2, K_3 and K_4 are respectively shown in the Figure 3.1.

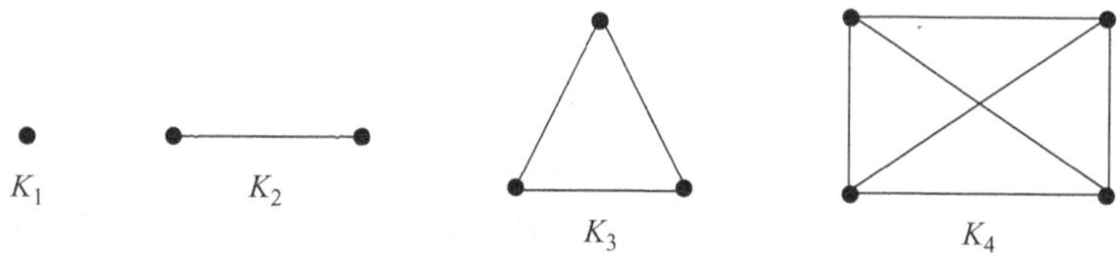

Figure 3.1

Now the complete graph K_5 may be obtained from K_4 by adding one more vertex and drawing edges between this new vertex and all the vertices of K_4. Thus, the picture of K_5 is as shown in the Figure 3.2.

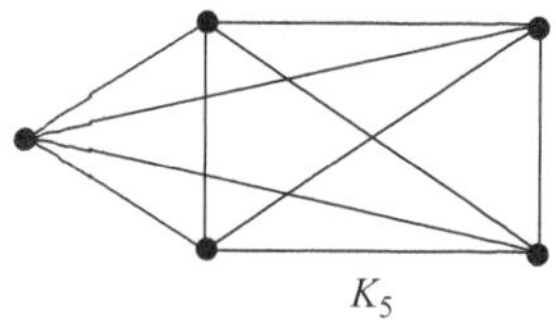

Figure 3.2

We note that

The number of edges of $K_5 = 4 +$ the number of edges of K_4,
The number of edges of $K_4 = 3 +$ the number of edges of K_3,
The number of edges of $K_3 = 2 +$ the number of edges of K_2,
The number of edges of $K_2 = 1 +$ the number of edges of K_1.

Thus, if

$$S_n = \text{the number of edges of } K_n,$$

then

$$S_n = S_{n-1} + (n-1), \text{ for } n \geq 2$$

and

$$S_1 = 0$$

is the recurrence relation for the number of edges of K_n.

To obtain the explicit formula, we use the technique of backtracking:

$$\begin{aligned} S_n &= S_{n-1} + (n-1) \\ &= S_{n-2} + (n-2) + (n-1) \\ &= S_{n-3} + (n-3) + (n-2) + (n-1) \\ &= S_{n-4} + (n-4) + (n-3) + (n-2) + (n-1) \end{aligned}$$

Thus, in general, we have
Thus,

$$S_n = S_{n-k} + (n-k) + (n-(k-1)) + \ldots + (n-2) + (n-1)$$

If we set $k = n-1$, we get

$$\begin{aligned} S_n &= S_{n-(n-1)} + (n-(n-1)) + \ldots + (n-4) + (n-3) + (n-2) + (n-1) \\ &= S_1 + 1 + 2 + 3 + \ldots + (n-1) \\ &= 0 + 1 + 2 + 3 + \cdots + (n-1) = \frac{(n-1)n}{2}, \end{aligned}$$

which is the required explicit formula.

EXAMPLE 3.10

Solve the recurrence relation

$$p_n = a - \frac{b}{k} \cdot p_{n-1}$$

for the price in the economics model, where a, b, k are positive parameters and p_0 is the initial price.

Solution.

To obtain the solution (explicit formula), we use the technique of backtracking. We put $-\frac{b}{k} = c$ and have

$$\begin{aligned}
p_n &= a + c\,p_{n-1} \\
&= a + c\,(a + c\,p_{n-2}) = a + a\,c + c^2\,p_{n-2} \\
&= a + a\,c + c^2(a + c\,p_{n-3}) = a + a\,c + a\,c^2 + c^3\,p_{n-3} \text{ and so on.}
\end{aligned}$$

In general, we have

$$p_n = a + a\,c + a\,c^2 + a\,c^{k-1} + c^k\,p_{n-k}.$$

If we set $k=n$, then

$$\begin{aligned}
p_n &= a + a\,c + a\,c^2 + \ldots + a\,c^{n-1} + c^n\,p_0 \\
&= a(1 + c + c^2 + \ldots + c^{n-1}) + c^n\,p_0 \\
&= \frac{a(1-c^n)}{1-c} + c^n p_0 = \frac{a - ac^n}{1-c} + c^n p_0 \\
&= c^n\left(\frac{-a}{1-c} + p_0\right) + \frac{a}{1-c} \\
&= \left(-\frac{b}{k}\right)^n \left(\frac{-ak}{k+b} + p_0\right) + \frac{ak}{k+b}.
\end{aligned}$$

If $\frac{b}{k} < 1$, the term

$$\left(-\frac{b}{k}\right)^n \left(\frac{-ak}{k+b} + p_0\right)$$

becomes very small for large n and thus the price p_n tends to stabilize at approximately $\frac{ak}{(k+b)}$. If $\frac{b}{k} = 1$, then p_n oscillates between p_0 and p_1. If $\frac{b}{k} > 1$, then the difference between the successive prices increase.

3.4 HOMOGENEOUS RECURRENCE RELATIONS WITH CONSTANT COEFFICIENTS

Sometimes during backtracking, a pattern is not obtained easily. So we have to use another technique. We consider another technique in this section.

Definition 3.6

A **linear recurrence relation of order** k with constant coefficient is a recurrence relation of the form

$$a_n = c_1 a_{n-1} + c_2 a_{n-2} + \dots + c_k a_{n-k}, \qquad c_k \neq 0.$$

For example,

(i) The relation

$$a_n = (-2)\, a_{n-1}$$

is a linear homogeneous recurrence relation of order 1.

(ii) the recurrence relation

$$a_n = a_{n-1} + a_{n-2}$$

is a linear recurrence relation of order 2.

(iii) The recurrence relation

$$a_n = 2n\, a_{n-1}$$

is **not** a linear recurrence relation with constant coefficients because the coefficient $2n$ is not constant. It is a linear homogeneous recurrence relation with **non-constant coefficients**.

(iv) The recurrence relation

$$a_n = a_{n-1} + 2$$

is **not** a linear homogeneous recurrence relation because $a_n - a_{n-1} \neq 0$. It is an inhomogeneous recurrence relation.

(v) The recurrence relation

$$a_n + 7a_{n-2} = 0$$

is a **second order linear recurrence relation** with constant coefficients.

(vi) The recurrence relation

$$f_n = f_{n-1}^2 + f_{n-2}$$

is **not** a linear homogeneous relation.

Definition 3.7

The equation

$$x^k = r_1 x^{k-1} + r_2 x^{k-2} + \dots + r_k$$

of degree k is called the **characteristic equation** of the linear homogeneous recurrence relation

$$a_n = r_1 a_{n-1} + r_2 a_{n-2} + \dots + r_k a_{n-k}$$

of order k.

Theorem 3.1

If the characteristic equation $x^2 - r_1 x - r_2 = 0$ of the homogeneous recurrence relation

$$a_n = r_1 a_{n-1} + r_2 a_{n-2}$$

has two distinct roots s_1 and s_2, then

$$a_n = u\, s_1^{\,n} + v\, s_2^{\,n},$$

where u and v depend on the initial conditions, is the explicit formula for the sequence.

(To say "u and v depend on the initial conditions" means that u and v are the solutions of the system of simultaneous equation $a_1=us_1+vs_2$ and $a_2=us_1^2+vs_2^2$)

Proof. Since s_1 and s_2 are roots of the characteristic equation $x^2-r_1x-r_2=0$, we have

$$s_1^2-r_1s_1-r_2=0, \quad (1)$$
$$s_2^2-r_1s_2-r_2=0. \quad (2)$$

Let

$$a_n=u\,s_1^n+v\,s_2^n \quad \text{for} \quad n\geq 1. \quad (3)$$

It is sufficient to show that (3) defines the same sequence as $a_n=r_1a_{n-1}+r_2\,a_{n-2}$. We have

$$a_1=u\,s_1+v\,s_2,$$
$$a_2=u\,s_1^2+v\,s_2^2$$

and the initial conditions are satisfied. Further,

$$\begin{aligned}a_n&=us_1^n+v\,s_2^n\\&=us_1^{n-2}\cdot s_1^2+v\,s_2^{n-2}\cdot s_2^2\\&=us_1^{n-2}(r_1\,s_1+r_2)+v\,s_2^{n-2}\,(r_1s_2+s_2)\ \text{(using (1) and (2))}\\&=r_1u\,s_1^{n-1}+r_2us_1^{n-2}+r_1vs_2^{n-1}+r_2vs_2^{n-2}\\&=r_1(us_1^{n-1}+vs_2^{n-1})+r_2(us_1^{n-2}+vs_2^{n-2})\\&=r_1a_{n-1}+r_2a_{n-2}\ \text{(using expressions of } a_{n-1} \text{ and } a_{n-2} \text{ from (3)).}\end{aligned}$$

Hence (3) defines the same sequence as $a_n=r_1a_{n-1}+r_2a_{n-2}$. Hence $a_n=us_1^n+vs_2^n$ is the solution to the given linear homogeneous recurrence relation.

Theorem 3.2

If the characteristic equation $x^2-r_1\,x-r_2=0$ of the linear homogeneous recurrence relation $a_n=r_1\,a_{n-1}+r_2\,a_{n-2}$ has a single root s, then the **explicit formula** (solution) for the recurrence relation is $a_n=us^n+vns^n$, where u and v depend on the initial conditions.

Proof. Since s is the root of the characteristic equation, we have

$$s^2-r_1\,s-r_2=0. \quad (1)$$

Let

$$a_n=u\,s^n+v\,n\,s^n,\ n\geq 1. \quad (2)$$

It suffices to show that (2) defines the same sequence as $a_n=r_1a_{n-1}+r_2a_{n-2}$. We have

$$a_1=u\,s+v\,s,$$
$$a_2=u\,s^2+2v\,s^2$$

and the initial conditions are satisfied. Also

$$\begin{aligned}a_n&=u\,s^n+v\,n\,s^n\\&=u\,s^{n-2}\cdot s^2+v\,n\,s^{n-2}\cdot s^2\\&=u\,s^{n-2}\,(r_1s+r_2)+vns^{n-2}\,(r_1s+r_2)\ \text{(using (1))}\\&=r_1us^{n-1}+r_2us^{n-2}+r_1vns^{n-1}+r_2vns^{n-2}\\&=r_1(us^{n-1}+vns^{n-1})+r_2(us^{n-2}+vns^{n-2})\\&=r_1a_{n-1}+r_2a_{n-2}\ \text{(using the expression for } a_{n-1} \text{ and } a_{n-2} \text{ from (2)).}\end{aligned}$$

Thus (2) defines the same sequence as $a_n=r_1a_{n-1}+r_2a_{n-2}$ and so is the explicit formula for the recurrence relation.

EXAMPLE 3.11

Find an explicit formula for the sequence defined by the recurrence relation

$$a_n = a_{n-1} + 2a_{n-2},\ n \geq 2,$$

with the initial conditions

$$a_0 = 1 \text{ and } a_1 = 8.$$

Solution.

The recurrence relation

$$a_n = a_{n-1} + 2a_{n-2}$$

is a linear homogeneous relation of order 2. Its characteristic equation is

$$x^2 - x - 2 = 0$$

which yields

$$x = \frac{1 \pm \sqrt{1+8}}{2} = \frac{1 \pm 3}{2} = 2, -1.$$

Hence

$$a_n = u(2)^n + v(-1)^n \tag{1}$$

and we have

$$a_0 = u + v = 1 \text{ (given)},$$
$$a_1 = 2u - v = 8 \text{ (given)}.$$

Solving for u and v, we have

$$u = 3,\ v = -2.$$

Hence,

$$a_n = 3(2)^n - 2(-1)^n, \quad n \geq 0$$

is the explicit formula for the sequence.

Remark 3.2 In Example 3.11, if we use the recurrence relation

$$a_n = a_{n-1} + 2a_{n-2},\ a_0 = 1,\ a_1 = 8,$$

we obtain the sequence

$$1, 8, 10, 26, \ldots.$$

Also, we note that the explicit formula

$$a_n = 3 \cdot 2^n - 2\,(-1)^n, \quad n \geq 0$$

yields

$$a_0 = 3 - 2 = 1,$$
$$a_1 = 6 + 2 = 8,$$
$$a_2 = 12 - 2 = 10,$$
$$a_3 = 24 + 2 = 26$$

and so it yields the sequence

$$1, 8, 10, 26, \ldots.$$

EXAMPLE 3.12

Solve the recurrence relation

$$d_n = 2d_{n-1} - d_{n-2}$$

with initial conditions $d_1 = 1.5$ and $d_2 = 3$.

Solution.

The relation $d_n = 2d_{n-1} - d_{n-2}$ is a linear homogeneous recurrence relation of order 2. The characteristic equation (or associated equation) for this recurrence relation is

$$x^2 - 2x + 1 = 0$$

which yields

$$x = \frac{2 \pm \sqrt{4-4}}{2} = 1, 1.$$

Thus, the characteristic equation has a multiple root 1.

Hence,

$$d_n = u \cdot 1^n + nv \cdot 1^n = (u + nv) \cdot 1^n$$

and so

$$d_1 = u + v = 1.5 \text{ (given)},$$
$$d_2 = u + 2v = 3 \text{ (given)}.$$

Solving for u and v, we get

$$u = 0, \quad v = 1.5$$

and so

$$d_n = 1.5n$$

is the explicit formula (homogeneous solution) for the given recurrence relation.

EXAMPLE 3.13

Find explicit formula for **Fibonacci sequence**.

Solution.

We know that the Fibonacci sequence is

$$1, 1, 2, 3, 5, 8, 13, 21, \ldots.$$

The recurrence relation for the Fibonacci sequence is

$$f_n = f_{n-1} + f_{n-2}, \quad n \geq 3$$

with the initial conditions $f_1 = f_2 = 1$.

It is a linear homogeneous recurrence relation of order 2. The characteristic equation for this recurrence relation is

$$x^2 - x - 1 = 0,$$

whose roots are

$$x = \frac{1 \pm \sqrt{1+4}}{2} = \frac{1+\sqrt{5}}{2}, \frac{1-\sqrt{5}}{2}.$$

Thus, the Fibonacci sequence satisfies explicit formula

$$f_n = u\left(\frac{1+\sqrt{5}}{2}\right)^n + v\left(\frac{1-\sqrt{5}}{2}\right)^n,$$

where u and v are the numbers whose values are determined by the fact that $f_1 = f_2 = 1$. We have

$$f_1 = u\left(\frac{1+\sqrt{5}}{2}\right)^1 + v\left(\frac{1-\sqrt{5}}{2}\right)^1 = 1,$$

$$f_2 = u\left(\frac{1+\sqrt{5}}{2}\right)^2 + v\left(\frac{1-\sqrt{5}}{2}\right)^2 = 1.$$

Solving the above equations for u and v, we have

$$u = \frac{1}{\sqrt{5}}\left(\frac{1+\sqrt{5}}{2}\right), \qquad v = -\frac{1}{\sqrt{5}}\left(\frac{1-\sqrt{5}}{2}\right).$$

Hence, the explicit formula for Fibonacci sequence is

$$f_n = \frac{1}{\sqrt{5}}\left(\frac{1+\sqrt{5}}{2}\right)^{n+1} - \frac{1}{\sqrt{5}}\left(\frac{1-\sqrt{5}}{2}\right)^{n+1}, \quad \text{for all } n \geq 0.$$

Surprisingly, even though the formula for f_n involves the irrational number $\sqrt{5}$, all the values of the Fibonacci sequence are integers. Also, the numbers $\frac{1+\sqrt{5}}{2}, \frac{1-\sqrt{5}}{2}$ are related to the **golden ratio of Greek mathematics**.

It may be mentioned here that **it took over two hundred years to find explicit formula for the Fibonacci sequence.**

EXAMPLE 3.14

Find the explicit formula for the Fibonacci sequence in term of the roots of golden ratio in Greek mathematics.

Solution.

Consider a rectangle of length ϕ units and breadth 1, where $\phi > 1$. Divide the rectangle into a rectangle and a square as shown below:

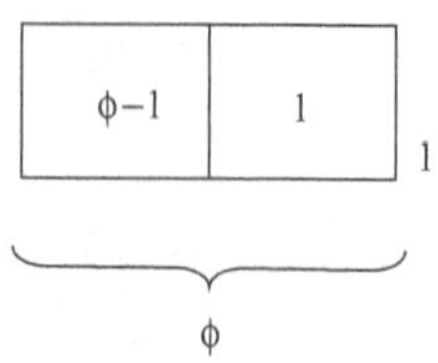

Greeks considered the outer rectangle to be perfectly proportioned, that is, its sides are in golden ratio if the ratio of the length to the width of outer rectangle is equal to the ratio of the length to width of the inner rectangle. Thus,

$$\frac{\phi}{1} = \frac{1}{\phi - 1} \text{ or } \phi^2 - \phi - 1 = 0,$$

which is same as characteristic equation of recurrence relation for Fibonacci sequence. The roots are $\frac{1 \pm \sqrt{5}}{2}$. Thus,

$$f_n = \frac{1}{\sqrt{5}} \phi_1^{n+1} - \frac{1}{\sqrt{5}} \phi_2^{n+1}.$$

EXAMPLE 3.15

Suppose that the population of a village is 100 at time $n=0$ and 110 at time $n=1$. The population increases from time $n-1$ to time n is twice the increase from time $n-2$ to time $n-1$. Find a recurrence relation and the initial conditions for the population at time n and then find the explicit formula for it.

Solution.

Let a_n denote the population at time n. We have the initial conditions as

$$a_0 = 100,\ a_1 = 110.$$

The increase from time $n-1$ to time n is $a_n - a_{n-1}$, whereas the increase from time $n-2$ to time $n-1$ is $a_{n-1} - a_{n-2}$. Thus, as per given hypotheses, we have the recurrence relation as

$$a_n - a_{n-1} = 2(a_{n-1} - a_{n-2}) \quad \text{or} \quad a_n = 3a_{n-1} - 2a_{n-2},$$

which is a linear homogeneous recurrence relation of order 2. The characteristic equation for this recurrence relation is

$$x^2 - 3x + 2 = 0,$$

whose roots are

$$x = \frac{3 \pm \sqrt{9-8}}{2} = \frac{3 \pm 1}{2} = 2, 1.$$

Hence,

$$a_n = u \cdot 2^n + v \cdot 1^n$$

Now

$$a_0 = u + v = 100 \text{ (given)},$$
$$a_1 = 2u + v = 110 \text{ (given)}.$$

Solving for u and v, we get $u=10$, $v=90$.

Hence, the explicit formula is

$$a_n = 10 \cdot 2^n + 90.$$

Obviously, the growth is exponential.

Remark 3.3 For a recurrence relation of order k with constant coefficients, if x is a root of

$$x^k - c_1 x^{k-1} - c_2 x^{k-2} - \dots - c_k = 0$$

of multiplicity m, then the explicit formula for the recurrence relation will be

$$a_n = (a_1 + n\, a_2 + n^2\, a_3 + \ldots + n^{m-1}\, a_m)\, x^n,$$

where $a_1, a_2, \ldots, a_m$ are determined from the initial conditions.

EXAMPLE 3.16

Find the explicit formula for the difference equation (recurrence relation)

$$a_n + 6\, a_{n-1} + 12\, a_{n-2} + 8\, a_{n-3} = 0.$$

Solution.

The given recurrence relation is a homogeneous recurrence relation of order 3. Its characteristic equation is

$$x^3 + 6x^2 + 12x + 8 = 0 \quad \text{or} \quad (x+2)^3 = 0,$$

whose roots are –2, –2, –2. Hence –2 is root of multiplicity 3. Hence the explicit formula (homogeneous solution) for this recurrence relation is

$$a_n = (u + n\, v + n^2\, w)\, (-2)^n.$$

EXAMPLE 3.17

Find the homogeneous solution of the difference equation

$$4\, a_n - 20\, a_{n-1} + 17 a_{n-2} - 4\, a_{n-3} = 0.$$

Solution.

The given difference equation is a linear homogeneous recurrence relation of order 3. Its characteristic equation is

$$4\, x^3 - 20\, x^2 + 17\, x - 4 = 0.$$

By inspection, we notice that 4 is a root of this equation and the reduced equation is

$$4\, x^2 - 4\, x + 1 = 0,$$

whose roots are

$$x = \frac{4 \pm \sqrt{16-16}}{2} = \frac{1}{2}, \frac{1}{2}.$$

Thus the characteristic roots are $\frac{1}{2}, \frac{1}{2}$, 4. Hence the explicit formula is

$$a_n = (u + nv)\left(\frac{1}{2}\right)^n + w \cdot 4^n.$$

Definition 3.8

The (total) solution of a linear difference equation (linear recurrence relation)

$$a_n = r_1\, a_{n-1} + r_2\, a_{n-2} + \ldots + r_k\, a_{n-k} = f(n),$$

where $f(n)$ is constant or a function of n, with constant coefficients is the sum of two parts, the homogeneous solution satisfying the difference equation when the right-hand side of the equation is set to be 0, and the particular solution, which satisfies the difference equation with $f(n)$ on the right-hand side.

3.5 PARTICULAR SOLUTION OF A DIFFERENCE EQUATION

We have so far discussed the linear homogeneous recurrence relation with constant coefficients and obtained their homogeneous solutions. We now turn our attention to find particular solution of a difference equation. We do not have a general procedure to find particular solution of a given difference equation. So, the particular solution is obtained by the method of inspection. A general form of particular integral is set up according to the form of $f(n)$. We discuss the following cases.

Case I: If $f(n)$ is a polynomial in n of degree m, then we take

$$P_1 n^m + P_2 n^{m-1} + \dots + P_{m+1}$$

as the particular solution of the difference equation. Putting this solution in the given difference equation, the values of $P_1, P_2, \dots, P_{m+1}$ are determined.

EXAMPLE 3.18

Find the particular solution of the difference equation

$$a_n - a_{n-1} - 2a_{n-2} = 2n^2.$$

Also write down the total solution.

Solution.

Suppose that the particular solution is of the form

$$P_1 n^2 + P_2 n + P_3, \tag{1}$$

where P_1, P_2 and P_3 are constants to be determined. Substituting (1) in the given difference equation, we obtain

$$(P_1 n^2 + P_2 n + P_3) - [P_1 (n-1)^2 + P_2 (n-1) + P_3] \\ - 2[P_1 (n-2)^2 + P_2 (n-2) + P_3] = 2n^2$$

or

$$P_1 n^2 + P_2 n + P_3 - [P_1 (n^2 + 1 - 2n) + P_2 (n-1) + P_3] \\ - 2[P_1 (n^2 + 4 - 4n) + P_2 (n-2) + P_3] = 2n^2$$

or

$$-2P_1 n^2 + n(10P_1 - 2P_2) + (-9P_1 + 5P_2 - 2P_3) = 2n^2.$$

Comparing coefficients of the powers of n, we have

$$-2P_1 = 2,$$
$$10P_1 - 2P_2 = 0,$$
$$9P_1 - 5P_2 + 2P_3 = 0,$$

which yield

$$P_1 = -1,\ P_2 = -5,\ P_3 = -8.$$

Therefore, the particular solution is

$$-n^2 - 5n - 8.$$

The homogeneous solution of this recurrence relation (cf. Example 3.11) is

$$3(2)^n - 2(-1)^n.$$

Hence the total solution is

$$3.2^n - 2(-1)^n - n^2 - 5n - 8.$$

EXAMPLE 3.19

Find the particular solution of the difference equation

$$a_n + 5a_{n-1} + 6a_{n-2} = 3n^2 - 2n + 1.$$

Solution.

Here the right-hand side is a function of n and it is a polynomial of degree 2. So suppose that particular solution is of the form

$$P_1 n^2 + P_2 n + P_3, \qquad (1)$$

where P_1, P_2 and P_3 are to be determined. Substituting (1) in the given difference equation, we get

$$P_1 n^2 + P_2 n + P_3 + 5[P_1 (n-1)^2 + P_2 (n-1) + P_3] + 6[P_1 (n-2)^2 + P_2 (n-2) + P_3] = 3n^2 - 2n + 1,$$

which yields

$$12P_1 n^2 - n(34P_1 - 12P_2) + (29P_1 - 17P_2 + 12P_3) = 3n^2 - 2n + 1.$$

Comparing the coefficients of the powers of n, we have

$$12P_1 = 3,$$
$$34P_1 - 12P_2 = 2,$$
$$29P_1 - 17P_2 + 12P_3 = 1$$

and so

$$P_1 = \frac{1}{4}, \quad P_2 = \frac{13}{24} \quad \text{and} \quad P_3 = \frac{71}{288}.$$

Hence, the particular solution is

$$\frac{1}{4}n^2 + \frac{13}{24}n + \frac{71}{288}.$$

Case II: If $f(n)$ is a constant, then the particular solution of the difference equation will also be a constant P, **provided that 1 is not a characteristic root of the difference equation**.

EXAMPLE 3.20

Find the particular solution of the difference equation $a_n - 4a_{n-1} + 5a_{n-2} = 2$. Hence find the total solution of this recurrence relation.

Solution.

Here $f(n) = 2$ (constant) and 1 is not characteristic root. So the particular solution will also be a constant P. Putting in the given recurrence relation of order 2, we have

$$P - 4P + 5P = 2,$$
$$\Rightarrow 2P = 2,$$
$$\Rightarrow P = 1.$$

Also, the characteristic equation of the given difference equation is

$$x^2 - 4x + 5 = 0,$$

whose roots are

$$x = \frac{4 \pm \sqrt{16-20}}{2} = \frac{4 \pm 2i}{2} = 2 \pm i.$$

Thus the homogeneous solution is

$$u \cdot (2+i)^n + v \cdot (2-i)^n, \quad n \geq 0.$$

Hence the total solution of the given difference equation is

$$a_n = u \cdot (2+i)^n + v \cdot (2-i)^n + 1.$$

Case III: If $f(n)$ is of the form α^n, the corresponding particular solution is of the form $P\,\alpha^n$ provided that α is not a characteristic root of the difference equation of order n.

EXAMPLE 3.21

Find the particular solution of the difference equation $a_n + 5\,a_{n-1} + 4a_{n-2} = 56 \cdot 3^n$. Hence, find the total solution of this difference equation?

Solution.

The characteristic equation of the difference equation is

$$x^2 + 5x + 4 = 0,$$

whose roots are -4 and -1. Thus the homogeneous solution is

$$u\,(-4)^n + v\,(-1).$$

Since $f(n) = 56 \cdot 3^n$ and 3 is not a characteristic root, the particular solution is of the form $P\,3^n$. Substituting this in the difference equation, we get

$$P \cdot 3^n + 5P\,(3)^{n-1} + 4P\,(3)^{n-2} = 56 \cdot 3^n$$

$$\Rightarrow \quad P(3^n + 5 \cdot 3^{n-1} + 4 \cdot 3^{n-2}) = 56 \cdot 3^n$$

$$\Rightarrow \quad P(3^n + \frac{5}{3} \cdot 3^n + \frac{4}{9} \cdot 3^n) = 56 \cdot 3^n$$

$$\Rightarrow \quad P\left(1 + \frac{5}{3} + \frac{4}{9}\right) = 56$$

$$\Rightarrow \quad P\left(\frac{9+15+4}{9}\right) = 56$$

$$\Rightarrow \quad P = 18.$$

Hence the particular solution is $18 \cdot 3^n$.

Case IV: If α is not a characteristic root of the difference equation and $f(n)$ is of the form

$$(c_1\,n^m + c_2\,n^{m-1} + \ldots + c_{n+1})\,\alpha^n,$$

then the particular solution is of the form

$$(P_1\,n^m + P_2\,n^{m-1} + \ldots + P_{n+1})\,\alpha^n.$$

EXAMPLE 3.22

Find the total solution of the difference equation

$$a_n - a_{n+1} - 2a_{n-2} = 3n \cdot 4^n.$$

Solution.

The characteristic equation of the given difference equation is

$$x^2 - x - 2 = 0,$$

whose roots are 2 and –1. Hence its homogeneous solution is

$$u \cdot 2^n + v \cdot (-1)^n.$$

Further, $f(n) = 3n \cdot 4^n$ and 4 is not a characteristic root of the difference equation. Hence particular solution is of the form

$$(nP_1 + P_2)\, 4^n. \tag{1}$$

Substituting (1) in the given difference equation, we get

$$(nP_1 + P_2)\, 4^n - ((n-1)\, P_1 + P_2)\, 4^{n-1} - 2((n-2)\, P_1 + P_2)\, 4^{n-2} = 3n \cdot 4^n$$

$$\Rightarrow \quad (nP_1 + P_2)\, 4^n - \frac{1}{4}((n-1)\, P_1 + P_2)\, 4^n - \frac{2}{16}((n-2)\, P_1 + P_2)\, 4^n = 3n \cdot 4^n$$

$$\Rightarrow \quad n \cdot 4^n \left(P_1 - \frac{1}{4} P_1 - \frac{1}{8} P_1 \right) + 4^n \left(\frac{P_1}{2} + P_2 - \frac{P_2}{4} - \frac{P_2}{8} \right) = 3n \cdot 4^n$$

$$\Rightarrow \quad n4^n \left(\frac{5P_1}{8} \right) + 4^n \left(\frac{P_1}{2} + \frac{5P_2}{8} \right) = 3n \cdot 4^n$$

Comparing coefficients of both sides, we have

$$P_1 = \frac{24}{5} \quad \text{and} \quad \frac{1}{2} P_1 + \frac{5}{8} P_2 = 0 \quad \text{or} \quad P_2 = -\frac{96}{25}.$$

Hence the particular solution is

$$\left(\frac{24}{5} n - \frac{96}{25} \right) 4^n.$$

Therefore, the total solution is

$$a_n = u \cdot 2^n + v(-1)^n + \left(\frac{24}{5} n - \frac{96}{25} \right) 4^n.$$

Case V: If α is a characteristic root of multiplicity m–1 and $f(n)$ is of the form

$$(c_1\, n^p + c_2\, n^{p-1} + \ldots + c_{p+1})\, \alpha^n,$$

the corresponding particular solution of the recurrence relation will be of the form

$$n^{m-1}(P_1\, n^p + P_2\, n^{p-1} + \ldots + P_{p+1})\, \alpha^n.$$

EXAMPLE 3.23

Find the particular solution of the difference equation $a_n - 4a_{n-1} = 6 \cdot 4^n$.

Solution.

The characteristic equation of the given difference equation is

$$x - 4 = 0$$

and so 4 is a root of multiplicity 1. Therefore, the particular solution is of the form

$$n\,P \cdot 4^n. \tag{1}$$

Substituting (1) in the given difference equation, we get

$$\begin{aligned} & nP \cdot 4^n - 4(n-1)P \cdot 4^{n-1} = 6 \cdot 4^n \\ \Rightarrow\ & nP4^n - (n-1)P \cdot 4^n = 6 \cdot 4^n \\ \Rightarrow\ & nP - nP + P = 6 \\ \Rightarrow\ & P = 6. \end{aligned}$$

Hence the **Particular solution is**

$$6n \cdot 4^n,$$

whereas the **total solution** of the given difference equation is

$$4^n\,(u + 6n).$$

EXAMPLE 3.24

Find the total solution of the difference equation

$$a_n - 6a_{n-1} + 9a_{n-2} = \mathrm{n} \cdot 3^n.$$

Solution.

In this difference equation, $f(n)$ is $n \cdot 3^n$. Also, the characteristic equation is

$$x^2 - 6x + 9 = 0,$$

which has $x=3$ as characteristic root of multiplicity 2. The homogeneous solution of the given difference equation is therefore

$$(u + nv) \cdot 3^n.$$

Further, the particular solution is of the form

$$n^2(P_1 n + P_2)\,3^n. \tag{1}$$

Substituting (1) in the given difference equation, we obtain

$$\begin{aligned} & n^2(P_1 n + P_2) \cdot 3^n - 6\,(n-1)^2\,[P_1(n-1) + P_2] \cdot 3^{n-1} \\ & + 9(n-2)^2\,[P_1(n-2) + P_2] \cdot 3^{n-2} = \mathrm{n} \cdot 3^n \\ \Rightarrow\ & n^2(P_1 n + P_2) \cdot 3^n - 2(n-1)^2\,[P_1(n-1) + P_2] \cdot 3^n \\ & + (n-2)^2\,[P_1(n-2) + P_2] \cdot 3^n = \mathrm{n} \cdot 3^n \\ \Rightarrow\ & n^2(P_1 n + P_2) - 2(n-1)^2\,[P_1(n-1) + P_2] \\ & + (n-2)^2\,[P_1(n-2) + P_2] = n \\ \Rightarrow\ & P_1 n^3 + P_2 n^2 - 2P_1(n-1)^3 - 2(n-1)^2 P_2 + P_1(n-2)^3 + P_2(n-2)^2 = n \\ \Rightarrow\ & P_1 n^3 + P_2 n^2 - 2P_1 n^3 + 6n^2 P_1 - 6nP_1 + 2P_1 - 2P_2 n^2 + 4nP_2 \\ & -2P_2 + P_1 n^3 - 6n^2 P_1 + 12nP_1 - 8P_1 + P_2 n^2 - 4nP_2 + 4P_2 = n \\ \Rightarrow\ & n^3(P_1 - 2P_1 + P_1) + n^2(P_2 + 6P_1 - 2P_2 - 6P_1 + P_2) \\ & + n\,(-6P_1 + 4P_2 + 12P_1 - 4P_2) + (2P_1 - 2P_2 - 8P_1 + 4P_2) = n \\ \Rightarrow\ & n(6P_1) + (-6P_1 + 2P_2) = n. \end{aligned}$$

Comparing coefficients on both sides, we get

$$6P_1=1,$$
$$3P_1-P_2=0.$$

Hence $P_1=\frac{1}{6}, P_2=\frac{1}{2}$ and so the particular solution is

$$n^2\left(\frac{1}{6}n+\frac{1}{2}\right)\cdot 3^n.$$

Therefore the total solution is

$$a_n=(u+nv)\cdot 3^n+n^2\left(\frac{1}{6}n+\frac{1}{2}\right)\cdot 3^n.$$

EXAMPLE 3.25

Find the total solution of the difference equation

$$a_n-a_{n-1}=4.$$

Solution.

The characteristic equation of the given difference is

$$x-1=0.$$

Hence, 1 is the characteristic root. Further the difference equation can be expressed as

$$a_n-a_{n-1}=4\cdot 1^n.$$

Therefore, the particular solution will be of the form

$$n\,(P)\cdot 1^n. \qquad (1)$$

Substituting (1) in the given difference equation, we get

$$nP-(n-1)\,P=4 \text{ or } P=4.$$

Therefore, the particular solution is $4n$. Hence the total solution is

$$u\cdot 1^n+4n\cdot 1^n \text{ or } u+4n.$$

EXAMPLE 3.26

Find the particular solution of the difference equation

$$a_n-2a_{n-1}+a_{n-2}=3.$$

Solution.

The given difference equation can be written as

$$a_n-2a_{n-1}+a_{n-2}=3\cdot 1^n.$$

The characteristic equation of this difference equation is

$$x^2-2x+1=0,$$

and so the characteristic roots are 1, 1, that is, 1 is characteristic root of multiplicity 2. Hence the particular solution is of the type

$$n^2\,(P)\cdot 1^n. \qquad (1)$$

Putting (1) in the given difference equation, we have

$$n^2P - 2(n-1)^2 P + (n-2)^2 P = 3 \cdot 1^n$$

$$\Rightarrow \quad n^2P - 2P(n^2 + 1 - 2n) + (n^2 + 4 - 4n)P = 3 \cdot 1^n$$

$$\Rightarrow \quad -2P + 4P = 3$$

$$\Rightarrow \quad 2P = 3$$

$$\Rightarrow \quad P = \frac{3}{2}.$$

Hence the particular solution is $\frac{3}{2}n^2$.

EXAMPLE 3.27

Find the particular solution of the difference equation $a_n - 5a_{n-1} + 6a_{n-2} = 3^n + n$.

Solution.

Here $f(n) = n + 3^n$. We also note that the characteristic equation of the difference equation is

$$x^2 - 5x + 6 = 0$$

and so 3 and 2 are the characteristic roots. We note that particular solution for n is of the form $nP_1 + P_2$. Further, particular solution for 3^n is of the form $nP_3 \cdot 3^n$. Thus the particular solution for the difference equation will be of the form

$$nP_1 + P_2 + nP_3 \cdot 3^n. \tag{1}$$

Substituting (1) in the given difference equation, we get

$$(nP_1 + P_2 + nP_3 \cdot 3^n) - 5[(n-1)P_1 + P_2 + (n-1)P_3 \cdot 3^{n-1}]$$
$$+ 6[(n-2)P_1 + P_2 + (n-2)P_3 \cdot 3^{n-2}] = 3^n + n$$

or

$$3^n\left[nP_3 - \frac{5}{3}nP_3 + \frac{2}{3}nP_3 + \frac{5}{3}P_3 - \frac{4}{3}P_3\right] + n(P_1 - 5P_1 + 6P_1)$$
$$+ (P_2 + 5P_1 - 5P_2 - 12P_1 + 6P_2) = n + 3^n.$$

Comparing coefficients, we get

$$\frac{1}{3}P_3 = 1 \quad \Rightarrow \quad P_3 = 3$$

$$2P_1 = 1 \quad \Rightarrow \quad P_1 = \frac{1}{2}$$

$$2P_2 - 7P_1 = 0 \quad \Rightarrow \quad P_2 = \frac{7}{4}$$

Hence the particular solution of the given difference equation is

$$\frac{1}{2}n + \frac{7}{4} + n \cdot 3^{n+1}.$$

3.6 RECURSIVE FUNCTIONS

Definition 3.9

A function is said to be a **recursive function** if its rule of definition refers to itself. Such functions are used in the theory of computation in computer science.

3.6.1 Ackermann Function

Defined by Wilhelm Ackermann, the Ackerman function answers the question of what can and what cannot be computed on a computer.

Ackermann function is defined on the set of all pairs of non-negative integers by the recurrence relations

$$A(m, 0) = A(m-1, 1), \; m = 1, 2, \ldots \tag{1}$$

$$A(m, n) = A(m-1, A(m, n-1)), \qquad m, n = 1, 2, \ldots \tag{2}$$

and the initial conditions

$$A(0, n) = n+1, \; n = 0, 1, 2, \ldots. \tag{3}$$

The rate of growth of an Ackermann function is rapid. This function appears in the **time complexity** of certain algorithms such as the time to execute union/find algorithm.

We note that

(i) $A(0, 0) = 0+1 = 1,$ by (3)

(ii) $A(1, 1) = A(0, A(1, 0))$ by (2)
$= A(0, A(0, 1))$ by (1)
$= A(0, 2)$ by (3)
$= 3,$ by (3)

(iii) $A(1, 2) = A(0, A(1, 1))$ by (2)
$= A(0, 3)$ by (ii)
$= 4,$ by (3)

(iv) $A(1, 3) = A(0, A(1, 2))$ by (2)
$= A(0, 4)$ by (iii)
$= 5,$ by (3)

(v) $A(2, 2) = A(1, A(2, 1))$ by (2)
$= A(1, A(1, A(2, 0))$ by (2)
$= A(1, A(1, A(1, 1)))$ by (1)
$= A(1, A(1, 3))$ by (ii)
$= A(1, A(0, A(1, 2)))$ by (2)
$= A(1, A(0, 4))$ by (iii)
$= A(1, 5)$ by (3)
$= A(0, A(1, 4))$ by (2)
$= A(0, A(0, A(1, 3)))$ by (3)
$= A(0, A(0, 5))$ by (iv)
$= A(0, 6)$ by (3)
$= 7.$ by (3)

Similarly,

$$A(3, 3) = 61,$$

while

$$A(4,4) = 2^{2^{65,536}}$$

and the value of A (n, n) further increases extraordinarily rapidly.

Remark 3.4 It can be shown by Mathematical Induction that A (1, n)=n+2, A (2, n)=3+2n, A (3, n)=8 · 2^n – 3 for all non-negative integers n.

3.6.2 McCarthy's 91 Function

The function M: $\mathbf{Z}^+ \rightarrow \mathbf{Z}$ defined by

$$M(n) = \begin{cases} n-10 & \text{if } n > 100 \\ M(M(n+11)) & \text{if } n \leq 100 \end{cases}$$

for all positive integers n is called **McCarthy's 91 Function**.

We observe that

$$\begin{aligned}
M(21) &= M(M(21+11)) && \text{since } 21 \leq 100 \\
&= M(M(32)) && \\
&= M(M(32+11)) && \text{since } 32 \leq 100 \\
&= M(M(43) && \\
&= M(M(54)) && \text{since } 43 \leq 100 \\
&= M(M(65)) && \text{since } 54 \leq 100 \\
&= M(M(76) && \text{since } 65 \leq 100 \\
&= M(M(87) && \text{since } 76 \leq 100 \\
&= M(M(98)) && \text{since } 87 \leq 100 \\
&= M(M(109) && \text{since } 98 \leq 100 \\
&= M(99) && \text{since } 109 > 100 \\
&= M(M(110)) && \text{since } 99 \leq 100 \\
&= M(100) && \text{since } 110 > 100 \\
&= M(M(111)) && \text{since } 100 \leq 100 \\
&= M(101) && \text{since } 111 > 100 \\
&= 91 && \text{since } 101 > 100.
\end{aligned}$$

From this calculation, it is clear that

$$M(21) = M(99) = M(100) = M(101) = 91.$$

Interestingly, the value of this function comes out to be 91 for all positive integers less than or equal to 101. Also, M (n) is well defined for n> 101 because then it is equal to n – 10. Thus, McCarthy 91 function is well defined.

For example,

$$M(102) = 102 - 10 = 92,$$
$$M(106) = 106 - 10 = 96$$

and so on.

3.6.3 The Collatz Function

The function F: $\mathbf{Z}^+ \to \mathbf{Z}$ defined by

$$\text{F(n)=}\ F(n) = \begin{cases} 1 & \text{if } n = 1 \\ 1 + F\left(\dfrac{n}{2}\right) & \text{if } n \text{ is even} \\ F(3n+1) & \text{if } n \text{ is odd and } n > 1, \end{cases}$$

is called the **Collatz Function.**

Collatz has conjectured that the function is well defined on the set of all positive integers. At present, $F(n)$ is computable for all integers n with $1 \leq n < 10^9$.

For example,

$$\begin{aligned} F(1) &= 1, \\ F(2) &= 1 + F(1) = 1 + 1 = 2, \\ F(3) &= F(9+1) = F(10) = 1 + F(5) = 1 + (1 + F(16)) \\ &= (1+(1+(1+(1+F(8))))) \\ &= (1+(1+(1+(1+F(4))))) \\ &= (1+(1+(1+(1+(1+F(2)))))) \\ &= (1+(1+(1+(1+(1+2))))) \\ &= (1+1+1+1+3) = 7 \end{aligned}$$

and so on.

EXAMPLE 3.28

Let G: $\mathbf{Z}^+ \to \mathbf{Z}$ be defined by

$$G(n) = \begin{cases} 1 & \text{if } n = 1 \\ 1 + G\left(\dfrac{n}{2}\right) & \text{if } n \text{ is even} \\ G(3n-1) & \text{if } n \text{ is odd and } n > 1. \end{cases}$$

Show that G is not well defined.

Solution.

Using the definition of G, we have

$$\begin{aligned} G(1) &= 1, \\ G(2) &= 1 + G(1) = 2, \\ G(3) &= G(8) = 1 + G(4) = 1 + (1 + G(2)) = 1 + 3 = 4, \\ G(4) &= 1 + G(2) = 1 + 2 = 3. \end{aligned}$$

But,

$$\begin{aligned} G(5) &= G(14) = 1 + G(7) = 1 + G(20) \\ &= 1 + (1 + G(10)) = 1 + (1 + (1 + G(5))) \\ &= 3 + G(5)), \end{aligned}$$

which yields $0 = 3$, that is absurd. Hence G is not well defined.

3.6.4 Euclidean Algorithm

A well-known and oldest recorded non-trivial algorithm for finding the **greatest common divisor** of two non-negative integers is known as **Euclidean Algorithm**. If m and n are non-negative integers, then the greatest common divisor function is denoted by

$$\gcd(m,n)=\begin{cases}\gcd(n,m) & \text{if } n>m \quad (1)\\ m & \text{if } n=0 \quad (2)\\ \gcd(n,m(\text{mod}\, n)) & \text{otherwise,} \quad (3)\end{cases}$$

where m (mod n) is the remainder on dividing m by n.

For example,

$$\begin{aligned}\gcd(22, 9) &= \gcd(9, 4) && \text{using} \quad (3)\\ &= \gcd(4, 1) && \text{using} \quad (3)\\ &= \gcd(1, 0) && \text{using} \quad (3)\\ &= 1 && \text{using} \quad (2).\end{aligned}$$

Similarly,

$$\begin{aligned}\gcd(81, 36) &= \gcd(36, 9) && \text{using} \quad (3)\\ &= \gcd(9, 0) && \text{using} \quad (3)\\ &= 9. && \text{using} \quad (2).\end{aligned}$$

3.7 GENERATING FUNCTIONS

Definition 3.10

A function whose domain of definition is the set of natural numbers and whose range is the set of real numbers is called a **discrete numeric function** or simply, numeric functions.

Thus, a sequence is a numeric function. For a numeric function a we use $a_0, a_1, a_2, \ldots, a_n, \ldots$ to denote the value of the function at 0, 1, 2, …, n, …. Instead of representing a numeric function by listing its terms, we use short representation for its terms. For example, the sequence $\{1, 2, 4, 8, 16, \ldots, 2^r, \ldots\}$ is denoted by

$$a_r = 2^r,\ r \geq 0.$$

On the other hand, the representation

$$a_r = 1{,}000\ (1 \cdot 10)^r,\ r \geq 0$$

represents the sequence

$$[1{,}000,\ 1{,}100,\ 1{,}210,\ 1{,}331,\ \ldots].$$

Definition 3.11

An infinite series

$$a_0 + a_1 z + a_2 z^2 + \ldots + a_n z^n + \ldots$$

is called the **generating function** of the numeric function $(a_0, a_1, a_2, \ldots, a_n, \ldots)$.

We note that the coefficient of z^n in the generating function is the value of the numeric function at n. For example,

(i) The generating function of the numeric function $(1, 1, 1, \ldots, 1, \ldots)$ is

$$1 + z + z^2 + \ldots + z^n + \ldots$$

which can be written in compact form as

$$A(z)=\frac{1}{1-z}.$$

(ii) The generating function of the numeric function $(1, 3^1, 3^2, \ldots, 3^n, \ldots)$ is

$$3^0+3^1z+3^2z^2+\ldots+3^nz^n+\ldots$$

which can be written in compact form as

$$A(z)=\frac{1}{1-3z}.$$

We note that

(i) Let a and b be numeric functions and α be a scalar such that $b=\alpha a$. Then,

$$B(z)=\alpha A(z).$$

(ii) Let a, b and c be the numeric functions such that $c=a+b$. Then,

$$C(z)=A(z)+B(z).$$

(iii) Let a be a numeric function and $A(z)$ its generating function. If b is a numeric function such that

$$b_n=\alpha^n a_n \quad \text{for some scalar } \alpha,$$

then

$$B(z)=A(\alpha z).$$

In fact, the generating function of b is

$$\begin{aligned} B(z)&=a_0+\alpha a_1z+\alpha^2a_2z^2+\ldots+\alpha^na_nz^n+\ldots \\ &=a_0+a_1(\alpha z)+a_2(\alpha z)^2+\ldots+a_n(\alpha z)^n+\ldots \\ &=A(\alpha z). \end{aligned}$$

EXAMPLE 3.29

Find the generating function of the numeric function

$$a_n=5\cdot 2^n,\ n\geq 0.$$

Solution.

The generating function for the numeric function $(1, 2, 4, \ldots)$ is $\frac{1}{1-2z}$. Hence, the generating function for $a_n=5\cdot 2^n$ is

$$A(z)=5\left(\frac{1}{1-2z}\right)=\frac{5}{1-2z}.$$

EXAMPLE 3.30

Find the generating function of the numeric function

$$a_n=1^n+2^n+3^n, \quad n\geq 0$$

Solution.

The generating function of $<1^n>$ is $\frac{1}{(1-z)}$, the generating function of $<2^n>$ is $\frac{1}{(1-2z)}$ and the generating function of $<3^n>$ is $\frac{1}{(1-3z)}$. Hence the generating function for $a_n = 1^n + 2^n + 3^n$, $n \geq 0$ is

$$A(z) = \frac{1}{1-z} + \frac{1}{1-2z} + \frac{1}{1-3z}$$

EXAMPLE 3.31

Find the generating function for the numeric function $a_n = 2^{n+3}$, $n \geq 0$.

Solution.

We have

$$\begin{aligned} a_n &= 2^{n+3}, n \geq 0 \\ &= 2^3 \cdot 2^n = 8 \cdot 2^n. \end{aligned}$$

Hence,

$$A(z) = 8\left(\frac{1}{1-2z}\right) = \frac{8}{1-2z}.$$

EXAMPLE 3.32

Find the numeric function corresponding to the generating function

$$A(z) = \frac{3+2z-6z^2}{1-3z}.$$

Solution.

We note that

$$A(z) = 2z + A(z) = 2z + \frac{3}{1-3z} = 3 + 11z + 27z^2 + 81z^3 + \cdots.$$

Thus,

$$\begin{aligned} a_0 &= 3(3^0) = 3, \\ a_1 &= 2 + 3(3^1) = 11, \\ &\cdots\cdots \\ a_r &= 3 \cdot 3^r = 3^{r+1}, \\ &\cdots\cdots \\ &\cdots\cdots \end{aligned}$$

Hence,

$$a_n = \begin{cases} 3 & n = 0 \\ 11 & n = 1 \\ 3^{n+1} & n \geq 2. \end{cases}$$

EXAMPLE 3.33

Find the numeric function corresponding to generating function

$$A(z) = \frac{3z}{(1-z)(1+2z)}.$$

Solution.

We have

$$A(z)=\frac{3z}{(1-z)(1+2z)}=\frac{1}{1-z}-\frac{1}{1+2z}.$$

Therefore,

$$a_n=1^n+(-2)^n,\ n\geq 0.$$

Remark 3.5 If a and b are numeric function and $c=a\ b$, we do not have any simple form to express $C\,(z)$ in terms of $A\,(z)$ and $B\,(z)$.

3.8 CONVOLUTION OF NUMERIC FUNCTIONS

Let $a=(a_0, a_1, \ldots, a_n, \ldots)$ and $b=(b_0, b_1, \ldots, b_n, \ldots)$ be numeric functions. Then, $A\,(z)=a_0+a_1\,z+\ldots+a_nz^n+\ldots$ and $B\,(z)=b_0+b_1z+\ldots+b_nz^n+\ldots$ are their generating functions.

Let $c=a * b$. Then

$$c_n=a_0\,b_n+a_1b_{n-1}+a_2\,b_{n-2}+\ldots+a_{n-1}b_1+a_nb_0=\sum_{i=0}^{n}a_ib_{n-i}$$

is the coefficient of z^n in the Cauchy product

$$(a_0+a_1z+\ldots+a_nz^n+\ldots)\,(b_0+b_1z+\ldots+a_nz^n+\ldots).$$

Hence,

$$C\,(z)=A\,(z)\,B\,(z).$$

EXAMPLE 3.34

Let $(a_0, a_1, \ldots, a_n, \ldots)$ be an arbitrary numeric function and $(1, 1, 1, \ldots, 1, \ldots)$ be numeric function. Suppose c be the convolution of these two numeric functions. Find generating function $C\,(z)$.

Solution.

We have

$$c=a * b,$$

where,

$$a=(a_0, a_1, \ldots, a_n, \ldots),$$
$$b=(b_0, b_1, \ldots, b_n, \ldots)=(1, 1, \ldots, 1, \ldots)$$

so that

$$\begin{aligned}c_n&=a_0b_n+a_1b_{n-1}+\ldots+a_nb_0\\&=a_0+a_1+\ldots+a_n+\ldots\ \text{since each } b_i=1\end{aligned}$$

and the generating function of c is

$$C(z)=A(z)B(z)=A(z)\frac{1}{1-z}.$$

In particular, if we take $A(z)=\frac{1}{(1-z)}$, then

$$C(z)=\frac{1}{(1-z)^2}$$

is the generating function of the numeric function (1, 2, 3, …, n, …) because (using Cauchy product)

$$c_0=a_0\,b_0=1\cdot 1=1,$$
$$c_1=a_0\,b_1+a_1b_0=1+1=2,$$
$$c_2=a_0\,b_2+a_1\,b_1+a_2\,b_0=1+1+1=3=2+1,$$
$$c_3=a_0\,b_3+a_1\,b_2+a_2\,b_1+a_3\,b_0=1+1+1+1=4=3+1,$$

$$c_n=1+1+1+\ldots+(n+1\text{ times})=n+1.$$

Thus the generating function of the sequence $a_n=n+1$ is $\frac{1}{(1-z)^2}$.

EXAMPLE 3.35

Let $c=a * b$, where

$$a_n=2^n,\quad n\geq 0,$$
$$b_n=4^n,\quad n\geq 0.$$

Determine the generating function $C(z)$.

Solution.

We have

$$A(z)=\frac{1}{1-2z}\quad\text{and}\quad B(z)=\frac{1}{1-4z}.$$

Since $c=a * b$, we have

$$C(z)=A(z)B(z)=\frac{1}{1-2z}\cdot\frac{1}{1-4z}=\frac{2}{1-4z}-\frac{1}{1-2z}.$$

or

$$a_n=2(4)^n-1(2)^n=2^{2n+1}-2^n.$$

EXAMPLE 3.36

Show that the generating function $\frac{1}{(1-4z)^2}$ can be expressed as

$$a_n=(n+1)4^n.$$

Solution.

We know that

$$\frac{1}{1-4z}=1+4z+4^2z^2+\cdots+4^nz^n+\cdots.$$

can be written as $a_n=4^n$. Its numeric function is

$$(1, 4, 4^2, \ldots, 4^n, \ldots).$$

Taking the convolution, we observe that

$$(1, 4, 4^2, \ldots, 4^n, \ldots) * (1, 4, 4^2, \ldots, 4^n, \ldots)$$
$$=(1, 8, 48, 256, \ldots).$$

Hence

$$\frac{1}{(1-4z)^2} = 1+8z+144z^2+\cdots$$

or

$$a_n=(n+1)\,4^n.$$

Thus generating function of (1, 8, 48, 256, …) or that of $a_n=(n+1)4^n$ is $\frac{1}{(1-4z)^2}$.

EXAMPLE 3.37

Find the corresponding generic function for the generating function $\frac{z}{(1-z)^2}$.

Solution.

We have

$$\frac{z}{(1-z)^2} = \frac{A}{1-z} + \frac{B}{(1-z)^2}$$

$$z=A\,(1-z)+B$$

or

$$z=A+B-Az$$

Comparing coefficients, we get $A=-1$ and $B=1$. Therefore,

$$\frac{z}{(1-z)^2} = \frac{-1}{1-z} + \frac{1}{(1-z)^2}.$$

Hence the sequence is

$$a_n=-1+(n+1)=n.$$

Note: Some of the sequences and their generating functions are given in the table below:

Sequence	*Generating function*
$a_n=1$	$\frac{1}{1-z}$
$a_n=m^n$	$\frac{1}{1-mz}$
$a_n=(-1)^n$	$\frac{1}{1+z}$

(Continued)

$a_n=(-m)^n$	$\frac{1}{1+mz}$
$a_n=n+1$	$\frac{1}{(1-z)^2}$
$a_n=n$	$\frac{z}{(1-z)^2}$
$a_n=\frac{1}{n!}$	e^z
$a_n=\binom{m}{n}$	$(1+z)^m$
$a_n=(n+1)(n+2)$	$\frac{2}{(1-z)^3}$
$a_n=n(n+1)$	$\frac{2z}{(1-z)^3}$
$a_n=n^2$	$\frac{z(1+z)}{(1-z)^3}$.

3.9 SOLUTION OF RECURRENCE RELATIONS BY THE METHOD OF GENERATING FUNCTION

In this method, the given recurrence relations are first converted in the form of a generating function and then solved.

EXAMPLE 3.38

Find explicit formula for the recurrence relation

$$a_n=3a_{n-1}+1, \quad n\geq 2.$$

with the initial condition $a_0=0$, $a_1=1$.

Solution.

We are given that

$$a_n=3a_{n-1}+1. \tag{1}$$

Multiplying both sides of (1) by z^n, we have

$$a_n z^n=3a_{n-1}z^n+z^n, \quad n\geq 2. \tag{2}$$

Summing (2) for all $n\geq 2$, we obtain

$$\sum_{n=2}^{\infty} a_n z^n = 3\sum_{n=2}^{\infty} a_{n-1} z^n + \sum_{n=2}^{\infty} z^n. \tag{3}$$

But,

$$\sum_{n=2}^{\infty} a_n z^n = a_2 z^2 + a_3 z^3 + \cdots a_n z^n$$

$$= A(z) - a_1 z - a_0 = A(z) - z, \text{ since } a_0 = 0,\ a_1 = 1.$$

$$\sum_{n=2}^{\infty} a_{n-1} z^n = z \sum_{n=2}^{\infty} a_{n-1} z^{n-1}$$

$$= z\,(a_1 z^1 + a_2 z^2 + \ldots + a_n z^n + \ldots)$$

$$= z\,(A(z) - a_0) = zA(z)$$

$$\sum_{n=2}^{\infty} z^n = z^2 \sum_{n=2}^{\infty} z^{n-2} = \frac{z^2}{1-z}.$$

Thus, (3) reduces to

$$A(z) - z = 3zA(z) + \frac{z^2}{1-z}$$

$$\Rightarrow (1-3z)A(z) = z + \frac{z^2}{1-z}$$

$$= \frac{z - z^2 + z^2}{1-z} = \frac{z}{1-z}$$

$$\Rightarrow A(z) = \frac{z}{(1-z)(1-3z)} = \frac{1/2}{1-3z} - \frac{1/2}{1-z}.$$

Hence,

$$a_n = \frac{1}{2}(3^n) - \left(\frac{1}{2}\right)1, \quad n \geq 0$$

$$= \frac{3^n - 1}{2}, \quad n \geq 0.$$

EXAMPLE 3.39

Using generating function methods, find the explicit formula for Fibonacci sequence.

Solution.

We know that the Fibonacci sequence is

1, 1, 2, 3, 5, 8, 13, 21, …

and that its recurrence relation is

$$f_n = f_{n-1} + f_{n-2}, \quad n \geq 2, \tag{1}$$

with the initial conditions $f_0 = 1, f_1 = 1$.

Multiplying both sides of (1) by z^n, we obtain

$$f_n z^n = f_{n-1} z^n + f_{n-2} z^n. \tag{2}$$

Summing (2) for all $n \geq 2$, we have

$$\sum_{n=2}^{\infty} f_n z^n = \sum_{n=2}^{\infty} f_{n-1} z^n + \sum_{n=2}^{\infty} f_{n-2} z^n. \tag{3}$$

But,

$$\begin{aligned}\sum_{n=2}^{\infty} f_n z^n &= f_2 z^2 + f_3 z^3 + \cdots + f_n z^n + \cdots \\ &= A(z) - f_1 z - f_0 \\ &= A(z) - z - 1, \text{ using initial conditions } f_0 = 1 \text{ and } f_1 = 1,\end{aligned}$$

$$\begin{aligned}\sum_{n=2}^{\infty} f_{n-1} z^n &= z \sum_{n=2}^{\infty} f_{n-1} z^{n-1} \\ &= z\left(f_1 z + f_2 z^2 + \cdots + f_n z^n + \cdots\right) \\ &= z\left(A(z) - f_0\right) = z\,A(z) - z\end{aligned}$$

$$\begin{aligned}\sum_{n=2}^{\infty} f_{n-2} z^n &= z^2 \sum_{n=2}^{\infty} f_{n-2} z^{n-2} \\ &= z^2\left(f_0 + f_1 z + f_2 z^2 + \cdots + f_n z^n + \cdots\right) = z^2 A(z).\end{aligned}$$

Thus, (3) reduces to

$$\begin{aligned}& A(z) - z - 1 = zA(z) - z + z^2 A(z) \\ \Rightarrow\quad & (1 - z - z^2)\, A(z) = 1 + z - z \\ \Rightarrow\quad & A(z) = \frac{1}{1 - z - z^2} = \frac{1}{\left(1 - \frac{1+\sqrt{5}}{2} z\right)\left(1 - \frac{1-\sqrt{5}}{2} z\right)},\end{aligned}$$

which on using partial fraction reduces to

$$A(z) = \frac{\frac{1}{\sqrt{5}}\left(\frac{1+\sqrt{5}}{2}\right)}{\left(1 - \frac{1+\sqrt{5}}{2} z\right)} - \frac{\frac{1}{\sqrt{5}}\left(\frac{1-\sqrt{5}}{2}\right)}{\left(1 - \frac{1-\sqrt{5}}{2} z\right)}.$$

Hence

$$f_n = \frac{1}{\sqrt{5}}\left(\frac{1+\sqrt{5}}{2}\right)^{n+1} - \frac{1}{\sqrt{5}}\left(\frac{1-\sqrt{5}}{2}\right)^{n+1}.$$

EXAMPLE 3.40

Solve the recurrence relation

$$a_n - 4a_{n-1} = 6 \cdot 4^n, \quad a_0 = 1.$$

Solution.

We are given that

$$a_n = 4a_{n-1} + 6 \cdot 4^n. \tag{1}$$

Multiplying both sides of (1) by z^n, we have

$$a_n z^n = 4a_{n-1} z^n + 6 \cdot 4^n z^n, \quad n > 0. \tag{2}$$

Summing (2) for all $n \geq 1$, we have

$$\sum_{n=1}^{\infty} a_n z^n = 4\sum_{n=1}^{\infty} a_{n-1} z^n + 6\sum_{n=1}^{\infty} 4^n z^n. \tag{3}$$

But,

$$\sum_{n=1}^{\infty} a_n z^n = a_1 z + a_2 z^2 + \cdots + a_n z^n + \cdots$$
$$= A(z) - a_0 = A(z) - 1,$$
$$\sum_{n=1}^{\infty} a_{n-1} z^n = z\sum_{n=1}^{\infty} a_{n-1} z^{n-1}$$
$$= z(a_0 + a_1 z + \ldots + a_n z^n + \ldots)$$
$$= zA(z),$$
$$\sum_{n=1}^{\infty} 4^n z^n = 4z + 4^2 z^2 + \cdots$$
$$= 4z(1 + 4z + \ldots) = \frac{4z}{1-4z}.$$

Hence, (3) reduces to

$$A(z) - 1 = 4zA(z) + \frac{24z}{1-4z}$$
$$\Rightarrow (1-4z)A(z) = 1 + \frac{24z}{1-4z} = \frac{20z+1}{1-4z}$$
$$\Rightarrow A(z) = \frac{20z+1}{(1-4z)^2},$$

which on partial fraction gives

$$A(z) = -\frac{5}{1-4z} + \frac{6}{(1-4z)^2}.$$

Hence,

$$a_n = -5(4^n) + 6(n+1) \cdot 4^n = 4^n(1+6n).$$

EXERCISES

1. Using technique of backtracking, find the explicit formula for the recurrence relation
$$S_n = 2S_{n-1},\quad S_0 = 1.$$
2. Using technique of backtracking, find explicit formula for the recurrence relation
$$a_n = a_{n-1} + n,\quad a_1 = 4.$$
3. Solve the recurrence relation
$$a_n = 2a_{n-1} - a_{n-2},\quad n \geq 2$$
with the initial conditions: $a_0 = 1$, $a_1 = 4$.
4. Solve the recurrence relation
$$a_n = 4a_{n-1} - 4a_{n-2}$$
with the initial conditions: $a_0 = 1$, $a_1 = 1$.
5. Find the total solution of the difference equation
$$a_n + 5a_{n-1} + 6a_{n-2} = 3n^2 - 2n + 1.$$
6. Find the total solution of the difference equation
$$a_n - 5a_{n-1} + 6a_{n-2} = 1.$$
7. Find the particular solution of the difference equation
$$a_n + a_{n-1} = 3n \cdot 2^n.$$
8. For McCarthy's 91 function 91, show that $M(86) = 91$.
9. Using the definition of Ackermann function, find the value of $A(2, 1)$ and $A(2, 3)$.
10. Find the numeric function corresponding to the generating function
$$A(z) = \frac{-4z}{(1+z)(1-3z)}.$$
11. Show that the generating function $\frac{1}{(1-2z)^2}$ can be expressed as $a_n = (n+1)\,2^n$.
12. Using the generating function method, solve the recurrence relation $a_n - 7a_{n-1} + 10a_{n-2} = 0$ for $n \geq 2$ with the boundary conditions $a_0 = 10$ and $a_1 = 41$.

4 Logic

Logic is the study of reasoning and is specifically concerned with whether a particular reasoning is valid. It is a science of the necessary laws of thought, without which no employment of the understanding and the reason takes place.

4.1 PROPOSITIONS

Definition 4.1

A declarative sentence that is either true or false, but not both is called a **Proposition** or a **Statement.**

EXAMPLE 4.1

Which of the following are propositions?

(i) London is the capital of France
(ii) Open the door
(iii) Take two tablets of medicine
(iv) $x+y>0$, $x, y \in \mathbf{Z}$
(v) The only positive integers that divide a prime number are 1 and the number itself.
(vi) The sun is hot

Solution.

(i) The sentence is declarative and false. Hence, it is a proposition.
(ii) The sentence "Open the door" is not declarative. It is rather a command. Therefore, it is not a proposition.
(iii) The given sentence is not declarative and so it is not a proposition.
(iv) The sentence is not a statement because it is true for some values of x and y whereas for other values of x and y it is false. For example, if $x=2$, $y=1$, then it is true but if $x=-2$, $y=1$, then it is false.
(v) The given sentence is declarative as well as true. Hence it is a statement.
(vi) The sentence "The sun is hot" is both a declarative sentence and true. Hence it is a proposition.

Notations. The propositions are represented by lower case letters such as p, q and r. Thus, the notation p: $3+7=10$ means that p is a proposition $3+7=10$.

Many propositions are composite, that is, composed of sub-propositions and various connectives. Thus, we have

Definition 4.2

Composite propositions are called **compound propositions.**

A proposition which is not compound is said to be **primitive.**

Thus, a primitive proposition cannot be broken into simpler propositions.

EXAMPLE 4.2

Consider the sentence: The sun is shining and it is cold. This is a compound proposition composed of two propositions:

The sun is shining

and

It is cold

connected by the connective "and".

On the other hand, the proposition

London is in Denmark

is a **primitive statement**.

Definition 4.3

The truth values of a compound statement in terms of its component parts is called a **truth table.**

4.2 BASIC LOGICAL OPERATIONS

The three basic logical operations are

1. **Conjunction**
2. **Disjunction**
3. **Negation**

which correspond, respectively, to "and", "or" and "not".

Definition 4.4

The **conjunction** of two propositions p and q is the proposition p and q.
It is denoted by $p \wedge q$.

EXAMPLE 4.3

Let

p: This child is a boy
q: This child is intelligent.

Then

$p \wedge q$: This child is a boy and intelligent.

Thus, $p \wedge q$ is true, if the child is a boy and intelligent both.
Even if one of the component is false, $p \wedge q$ is false.
Thus,

"the proposition $p \wedge q$ is true if and only if the propositions p and q are both true".

The truth value of the compound proposition $p \wedge q$ is defined by the truth table given below:

p	q	$p \wedge q$
T	T	T
T	F	F
F	T	F
F	F	F

EXAMPLE 4.4

If

p: London is capital of India
q: A decade is 10 years

then p is false, q is true and the conjunction

$p \wedge q$: London is capital of India and a decade is 10 years

is **false**.

Definition 4.5

The **disjunction** of two proposition p and q is the proposition p or q and is denoted by $p \vee q$.

The compound statement $p \vee q$ is true if at least one of p or q is true. **It is false when both p and q are false.**

The truth value of the compound proposition $p \vee q$ is defined by the following truth table:

p	q	$p \vee q$
T	T	T
T	F	T
F	T	T
F	F	F

For example, if

p: London is capital of India
q: A decade is 10 years,

then p is false, q is true. The disjunction

$p \vee q$: London is capital of India or a decade is 10 years

is **true**.

EXAMPLE 4.5

Form the disjunction of p and q for each of the following:

(a) p: 2 is a positive integer, q: $\sqrt{2}$ is a rational number
(b) p: $2+3 \neq 5$, q: London is the capital of France.

Solution.

(a) $p \vee q$: 2 is a positive integer or $\sqrt{2}$ is a rational number. Since p is true, the disjunction $p \vee q$ is true, even though q is false.
(b) $p \vee q$: $2+3 \neq 5$ or London is the capital of France. Since both p and q are false, $p \vee q$ is false.

Remark 4.1 It is clear from the above example that in logic, unlike in ordinary English, **we may join totally unrelated statements by the connective "or".**

EXAMPLE 4.6

Consider the following four statements:

(i) Taj is in Agra and 7 is a prime number
(ii) Paris is in France and $\sqrt{2}$ is a rational number

(iii) Paris is in England and $2+2=4$
(iv) Paris is in England and $2+2=5$

Solution.

Here the propositions are combined by connective "and". Therefore, the compound proposition shall be conjunction. We observe that in (i) both sub-propositions are true. In the rest three statements, at least one sub-proposition is false. Hence only (i) is true. The truth table is

p	q	$p \wedge q$
T	T	T
T	F	F
F	T	F
F	F	F

Definition 4.6

If p is a statement, the **negation** of p is the statement "not p", denoted by $\sim p$.

Thus, $\sim p$ is the statement "it is not the case that p".

Hence, if p is true, then $\sim p$ is false and if p is false, then $\sim p$ is true.

The truth table for negation is

p	$\sim p$
T	F
F	T

EXAMPLE 4.7

Give the negation of the following statements:

(a) p: $2+3>1$
(b) q: It is cold.

Solution.

(a) $\sim p$: 2+3 is not greater than 1, that is, $\sim p$: $2+3 \leq 1$.

Since p is true in this case, $\sim p$ is false.

(b) $\sim q$: It is not the case that it is cold. More simply, $\sim q$: It is not cold.

Remark 4.2

(i) In expressions that include the symbol $\sim$ as well as $\wedge$ or $\vee$, the order of operation is that $\sim$ **is performed first**. For example, $\sim p \wedge q = (\sim p) \wedge q$.
(ii) An expression such as $p \wedge q \vee r$ is considered ambiguous. This expression must be written as either $(p \wedge q) \vee r$ or $p \wedge (q \vee r)$ to have meaning.

4.2.1 Translating from English to Symbols

We consider

EXAMPLE 4.8

Write each of the following sentences symbolically, letting p: "It is hot" and q: "It is sunny":

(a) It is not hot but it is sunny
(b) It is neither hot nor sunny

Solution.

(a) In logic, the words "but" and "and" mean the same thing. Generally, "but" is used in place of and when the part of the sentence that follows is in some way unexpected.

The given sentence is equivalent to " It is not hot and it is sunny" which can be written symbolically as $\sim p \wedge q$.

(b) The phrase neither A nor B means the same as not A and not B. Thus, to say "It is neither hot nor sunny" means that it is not hot and it is not sunny. Therefore, the given sentence can be written symbolically as $\sim p \wedge \sim q$.

Remark 4.3 The notation for inequalities involves "and" and "or" statements. For example, if x, a and b are particular real numbers, then

$$x \leq a \quad \text{means} \quad x < a \quad \text{or} \quad x = a,$$
$$a \leq x \leq b \quad \text{means} \quad a \leq x \quad \text{and} \quad x \leq b.$$

We note that $2 \leq x \leq 1$ means $2 \leq x$ and $x \leq 1$, which is false, no matter what x happens to be. We have taken x, a and b as particular real numbers to ensure that sentences such as $x < a$ and $x \geq b$ are either true or false and hence that such sentences are statements.

Definition 4.7

A "Statement form" or "Propositional form" is an expression made up of **statement variables** (such as $\sim$, $\wedge$, $\vee$) that becomes a statement when actual statements are substituted for the component statement variable. The **truth table** for a given statement form displays the truth values that correspond to the different combinations of truth values for the variables.

4.2.2 Truth Table for Exclusive OR

When OR is used in its exclusive sense, the statement "p or q" means "p or q but not both" or "p or q and not both p and q" which translates into symbols as

$$(p \vee q) \wedge \sim(p \wedge q).$$

This is sometimes abbreviated as

$$p \oplus q \quad \text{or} \quad p \text{ XOR } q.$$

The truth table for $p \oplus q$ is

p	q	$p \vee q$	$p \wedge q$	$\sim(p \wedge q)$	$(p \vee q) \wedge \sim(p \wedge q)$
T	T	T	T	F	F
T	F	T	F	T	T
F	T	T	F	T	T
F	F	F	F	T	F

It can be shown that $(p \wedge \sim q) \vee (q \wedge \sim p)$ have same truth table as $(p \vee q) \wedge \sim(p \wedge q)$.. Hence these are logically equivalent (see Definition 4.8).

EXAMPLE 4.9

Construct a truth table for the statement form:

$$(p \wedge q) \vee \sim r.$$

Solution.

The truth table for the given statement form is

p	q	r	$p \wedge q$	$\sim r$	$(p \wedge q) \vee \sim r$
T	T	T	T	F	T
T	T	F	T	T	T
T	F	T	F	F	F
T	F	F	F	T	T
F	T	T	F	F	F
F	T	F	F	T	T
F	F	T	F	F	F
F	F	F	F	T	T

Definition 4.8

Two different compound propositions (or statement forms) are said to be **logically equivalent** if they have the same truth value no matter what truth values their constituent propositions have.
Thus,

"Two different compound propositions (or statement forms) are said to be **logically equivalent** if they have **identical truth tables**".

We use the symbol $\equiv$ for logical equivalence.

EXAMPLE 4.10

Negation of the negation of a statement is equal to the statement. Symbolically, $\sim(\sim p) \equiv p$.

Solution.

The truth table of $\sim(\sim p)$ is

p	$\sim p$	$\sim(\sim p)$
T	F	T
F	T	F

Thus, truth values for p and $\sim(\sim p)$ are same and hence p and $\sim(\sim p)$ are logically equivalent. The logical equivalence $\sim(\sim p) \equiv p$ is called **Involution Law.**

EXAMPLE 4.11

Show that the statement forms $\sim(p \wedge q)$ and $\sim p \wedge \sim q$ are not logically equivalent.

Solution.

Construct the truth table for both statement forms:

p	q	$\sim p$	$\sim q$	$p \wedge q$	$\sim(p \wedge q)$		$\sim p \wedge \sim q$
T	T	F	F	T	F		F
T	F	F	T	F	T	$\neq$	F
F	T	T	F	F	T	$\neq$	F
F	F	T	T	F	T		T

Thus we have different truth values in rows 2 and 3 and so $\sim(p \wedge q)$ and $\sim p \wedge \sim q$ are not logically equivalent.

Remark 4.4 If we consider $\sim p \vee \sim q$, then its truth values shall be

F
T
T
T

and hence $\sim(p \wedge q)$ and $\sim p \vee \sim q$ are logically equivalent. Symbolically,

$$\sim(p \wedge q) \equiv \sim p \vee \sim q.$$

Analogously,

$$\sim(p \vee q) \equiv \sim p \wedge \sim q.$$

The above two logical equivalence are known as **De Morgan's Laws of Logic**.

EXAMPLE 4.12

Use De Morgan's Law to write the negation of $-4 < x \leq 2$.

Solution.

The given statement is equivalent to

$$-4 < x \quad \text{and} \quad x \leq 2.$$

By De Morgan's Law, the negation is

$$-4 \nless x \quad \text{or} \quad x \nleq 2,$$

which is equivalent to

$$-4 \geq x \quad \text{or} \quad x > 2.$$

EXAMPLE 4.13

Use De Morgan's Laws to write the negation of

p: John is a boy and John is handsome

Solution.

The negation of p is

$\sim p$: John is not a boy or John is not handsome

Remark 4.5 In logic, the connectives "and" and "or" are used between complete statements. Thus, we cannot write "John is not a boy and handsome."

Definition 4.9

A compound proposition which is **always true** regardless of truth values assigned to its component propositions is called a **Tautology**.

Definition 4.10

A compound proposition which is **always false** regardless of truth values assigned to its component propositions is called a **Contradiction**.

Definition 4.11

A compound proposition which can be either true or false depending on the truth values of its component propositions is called a **Contingency**.

EXAMPLE 4.14

Consider the statement form

$$p \vee \sim p.$$

The truth table for this statement form is

p	$\sim p$	$p \vee \sim p$
T	F	T
F	T	T

↑
all T's

Hence $p \vee \sim p$ is a tautology.

EXAMPLE 4.15

Consider the statement form

$$p \wedge \sim p.$$

The truth table for this statement form is

p	$\sim p$	$p \wedge \sim p$
T	F	F
F	T	F

↑
all F's

Hence the statement form $p \wedge \sim p$ is a **Contradiction**.

Remark 4.6 If τ and c denote tautology and contradictions, respectively, then we notice that

$$\sim \tau \equiv c \tag{1}$$

and

$$\sim c \equiv \tau. \tag{2}$$

Also from the above two examples

$$p \vee \sim p \equiv \tau \tag{3}$$

and

$$p \wedge \sim p \equiv c. \tag{4}$$

The logical equivalence (1), (2), (3) and (4) are known as **Complement Laws**.

4.3 LOGICAL EQUIVALENCE INVOLVING TAUTOLOGIES AND CONTRADICTIONS

If τ is a tautology and c is a contradiction, then the truth tables for $p \wedge \tau$ and $p \wedge c$ are:

p	τ	$p \wedge \tau$
T	T	T
F	F	F

↑ p and p ∧ τ: Same truth values

p	c	$p \wedge c$
T	F	F
F	T	F

↑ c and p ∧ c: Same truth values

Hence, we have

$$p \wedge \tau \equiv p \quad \text{and} \quad p \wedge c \equiv c.$$

Similarly, the truth tables for $p \vee \tau$ and $p \vee c$ are

p	τ	$p \vee \tau$
T	T	T
F	T	T

↑ τ and p ∨ τ: Same truth values

p	c	$p \vee c$
T	F	T
F	F	F

↑ p and p ∨ c: Same truth values

Hence, we have

$$p \vee \tau \equiv \tau \quad \text{and} \quad p \vee c \equiv p.$$

Thus we have the following logical equivalence:

$$p \wedge \tau \equiv p, \; p \wedge c \equiv c$$

$p \vee \tau \equiv \tau, \; p \vee c \equiv p$ (**universal bound laws**).

These four logical equivalence are known as **Identity Laws**.

EXAMPLE 4.16 (Idempotent Laws)

Consider the truth tables for $p \wedge p$ and $p \vee p$ given below:

p	p	$p \wedge p$	$p \vee p$
T	T	T	T
F	F	F	F

We note that

(i) $p \wedge p$ and p have same truth values
(ii) $p \vee p$ and p have same truth values

Hence,

$$p \wedge p \equiv p \quad \text{and} \quad p \vee p \equiv p.$$

These two logical equivalence are known as **Idempotent Laws**.

EXAMPLE 4.17 (Commutative Laws)

Consider the truth tables of $p \wedge q$, $q \wedge p$; $p \vee q$ and $q \vee p$. We have

p	q	$p \wedge q$	$q \wedge p$
T	T	T	T
T	F	F	F
F	T	F	F
F	F	F	F

(Same truth values and so $p \wedge q \equiv q \wedge p$)

and

p	q	$p \vee q$	$q \vee p$
T	T	T	T
T	F	T	T
F	T	T	T
F	F	F	F

(Same truth values and so $p \vee q \equiv q \vee p$).

It follows therefore that $p \wedge q \equiv q \wedge p$ and $p \vee q \equiv q \vee p$.
These logical equivalences are known as **Commutative Laws**.

EXAMPLE 4.18 (Absorption Laws)

Consider the truth tables of $p \wedge (p \vee q)$ and $p \vee (p \wedge q)$. We have

p	q	$p \wedge q$	$p \vee (p \wedge q)$
T	T	T	T
T	F	F	T
F	T	F	F
F	F	F	F

(Same truth values and so $p \vee (p \wedge q) \equiv p$)

and

p	q	$p \vee q$	$p \wedge (q \vee p)$
T	T	T	T
T	F	T	T
F	T	T	F
F	F	F	F

(Same truth values and so $p \wedge (p \vee q) \equiv p$)

Hence,

$$p \wedge (p \vee q) \equiv p$$

and

$$p \vee (p \wedge q) \equiv p.$$

These logical equivalence are known as **Absorption Laws**.

EXAMPLE 4.19 (Associative Laws and Distributive Laws)

If p, q and r are propositions, then:

$$(p \wedge q) \wedge r \equiv p \wedge (q \wedge r), \qquad (p \vee q) \vee r \equiv p \vee (q \vee r) \text{ (Associative Laws)}$$

and

$$p \wedge (q \vee r) = (p \wedge q) \vee (p \wedge r), \qquad p \vee (q \wedge r) = (p \vee q) \wedge (p \vee r) \text{ (Distributive Laws).}$$

These laws can be established by the readers themselves.

EXAMPLE 4.20

Simplify the logical statement

$$\sim(\sim p \wedge q) \wedge (p \vee q).$$

Solution.

We have

$$\begin{aligned}
\sim(\sim p \wedge q) \wedge (p \vee q) &\equiv (\sim(\sim p) \vee \sim q) \wedge (p \vee q) && \text{(De Morgan Law)}\\
&\equiv (p \vee \sim q) \wedge (p \vee q) && \text{(Involution Law)}\\
&\equiv p \vee (\sim q \wedge q) && \text{(Distributive Law)}\\
&\equiv p \vee (q \wedge \sim q) && \text{(Commutative Law)}\\
&\equiv p \vee c && \text{(Complement laws or Negation Law)}\\
&\equiv p && \text{(Identity law).}
\end{aligned}$$

4.4 CONDITIONAL PROPOSITIONS

Definition 4.12

If p and q are propositions, the compound proposition

if p then q or p implies q

is called a **conditional proposition** or **implication** and is denoted by

$$p \rightarrow q.$$

The proposition p is called the **hypothesis** or **antecedent** whereas the proposition q is called the **conclusion** or **consequent.**

The connective if……then is denoted by the symbol $\rightarrow$.

It is false when p is true and q is false, otherwise it is true. In particular, if p is false, then $p \rightarrow q$ is true for any q.

Definition 4.13

A conditional statement that is true by virtue of the fact that its hypothesis is false is called **true by default** or **vacuously true**.

For example, the conditional statement

"If 4 is a prime number, then I am President of America" is **true** simply because p: 4 is a prime number is false. So it is not the case that p is true and q is false simultaneously.

Thus the truth values of the conditional proposition $p \to q$ are defined by the truth table:

p	q	$p \to q$
T	T	T
T	F	F
F	T	T
F	F	T

Each of the following expressions is an equivalent form of the conditional statement $p \to q$:

p implies q
q if p
p only if q
p is sufficient condition for q
q is necessary condition for p.

EXAMPLE 4.21

Restate each proposition in the form of a conditional proposition:

(a) I will eat if I am hungry.
(b) $3+5=8$ if it is snowing.
(c) When you sing, my ears hurt.
(d) Ram will be a good teacher if he teaches well.
(e) A sufficient condition for Sohan to visit Calcutta is that he goes to Disney land.

Solution.

(a) If I am hungry, then I will eat
(b) If it is snowing, then $3+5=8$.
(c) If you sing, then my ears hurt.
(d) If Ram teaches well, then he will be a good teacher.
(e) If Sohan visits Calcutta, then he goes to Disney land.

4.4.1 Representation of "if ... then" as "or"

Lemma 4.1

Show that for propositions p and q,

$$p \to q \equiv \sim p \vee q.$$

Proof. The truth values for $p \to q$ and $\sim p \vee q$ are given below:

p	q	$p \to q$	$\sim p$	$\sim p \vee q$
T	T	T	F	T
T	F	F	F	F
F	T	T	T	T
F	F	T	T	T

Same truth values (of $p \to q$ and $\sim p \vee q$)

Hence,

$$p \to q \equiv \sim p \vee q.$$

EXAMPLE 4.22

Rewrite the statement in "If … then" form:

Either you teach well or you are terminated.

Solution.

Let

$$\sim p\text{: you teach well}$$

and

$$q\text{: you are terminated.}$$

Then the given statement is $\sim p \vee q$. But,

$$p\text{: you do not teach well.}$$

Hence, the equivalent "If….then" version of the given statement is

If you do not teach well, then you are terminated.

Negation of a conditional statement. We know that $p \to q$ is false if and only if p is true and its conclusion q is false. Also, we have shown above that

$$p \to q \equiv \sim p \vee q.$$

Taking negation of both sides, we have

$$\begin{aligned}\sim(p \to q) &\equiv \sim(\sim p \vee q)\\ &\equiv \sim(\sim p) \wedge (\sim q) \text{ (De Morgan's Law)}\\ &\equiv p \wedge \sim q \text{ (Double negative Law or Involution Law).}\end{aligned}$$

This can also be obtained by constructing the truth tables for $\sim(p \to q)$ and $p \wedge \sim q$; the truth tables would have the same truth values proving the logical equivalence.
Thus,

The negation of "If p then q" is **logically equivalent to** "p and not q".

EXAMPLE 4.23

Write negations for each of the following statements:

(a) If I am ill, then I cannot go to university.
(b) If my car is in the repair shop, then I cannot attend the class.

Solution.

We know that negation of "If p then q" is logically equivalent to "p and not q". Using this fact, the negations of (a) and (b) are respectively

1. I am ill and I can go to university.
2. My car is in the repair shop and I can attend the class.

Remark 4.7 The negation of (a) "if … then" proposition does not start with the word "if".

Definition 4.14

If $p \to q$ is an implication, then the **converse** of $p \to q$ is the implication $q \to p$.

Definition 4.15

The **contrapositive** of a conditional statement "If p then q" is "If $\sim q$ then $\sim p$".
In symbols,

The contrapositive of $p \to q$ is $\sim q \to \sim p$.

Lemma 4.2

A conditional statement is logically equivalent to its contrapositive.

Proof. The truth tables of $p \to q$ and $\sim q \to \sim p$ are:

$p \to q$

p	q	$p \to q$
T	T	T
T	F	F
F	T	T
F	F	T

$\sim q \to \sim p$

p	q	$\sim p$	$\sim q$	$\sim q \to \sim p$
T	T	F	F	T
T	F	F	T	F
F	T	T	F	T
F	F	T	T	T

Same truth values

Hence,

$$p \to q \equiv \sim q \to \sim p.$$

EXAMPLE 4.24

Give the converse and contrapositive of the implications:

(a) If it is hot, then I take cold drinks.
(b) If today is Monday, then tomorrow is Tuesday.

Solution.

(a) We have

p: It is hot
q: I take cold drinks

The converse is $q \to p$: If I take cold drinks, then it is hot.
The contrapositive is $\sim q \to \sim p$: If I do not take cold drinks, then it is not hot.

(b) We have

p: Today is Monday
q: Tomorrow is Tuesday

The converse is $q \to p$: If Tomorrow is Tuesday, then today is Monday.
The contrapositive is $\sim q \to \sim p$: If tomorrow is not Tuesday, then today is not Monday.

Definition 4.16

The **inverse** of the conditional statement $p \to q$ is $\sim p \to \sim q$.

For example, the inverse of "If today is Sunday, then tomorrow is Monday" is

"If today is not Sunday, then tomorrow is not Monday".

Remark 4.8 If a conditional statement is true, then its converse and inverse may or may not be true.

Only if: "p only if q" means that p can take place only if q also takes place. That is, if q does not take place, then p cannot take place, i.e., $\sim q \to \sim p$. Therefore, equivalence between a statement and its contrapositive imply that " if p occurs, then q must also occur". Hence,

If p and q are statements, "p only if q" means "if not q, then not p" or equivalently "if p then q".

Remark 4.9 "p only if q" does not mean "p if q".

EXAMPLE 4.25

Use contrapositive to rewrite the following statement in "if then" form:

"Ram will stand first in the class only if he works 12 hours a day."

Solution.

Version 1: We have

p: Ram will stand first in the class,
q: He works 12 hours a day.

The contrapositive is $\sim q \to \sim p$: If Ram does not works 12 hours a day, then he will not stand first in the class.

Version 2: If Ram stands first in the class, then he will work 12 hours a day.

Definition 4.17

If p and q are statements, the compound statement "p if and only if q" is called a **Biconditional statement** or an **equivalence**. It is denoted by $p \leftrightarrow q$.

Observe that $p \leftrightarrow q$ is true only when both p and q are true or when both p and q are false (i.e., if both p and q have same truth values) and is false if p and q have opposite truth values. Thus, the biconditional statement has the following truth table:

$p \leftrightarrow q$

p	q	$p \leftrightarrow q$
T	T	T
T	F	F
F	T	F
F	F	T

Lemma 4.3

If p and q are propositions, then

$$p \leftrightarrow q \equiv (p \to q) \wedge (q \to p).$$

Hence, $p \leftrightarrow q \equiv (\sim p \vee q) \wedge (\sim q \vee p)$.

Proof. We know that "p if and only if q" means that both "p if q" and "p only if q" hold. This means $p \leftrightarrow q$ should be logically equivalent to $(p \to q) \wedge (q \to p)$. We verify it using the truth table:

p	q	$p \to q$	$q \to p$	$p \leftrightarrow q$	$(p \to q) \wedge (q \to p)$
T	T	T	T	T	T
T	F	F	T	F	F
F	T	T	F	F	F
F	F	T	T	T	T

Same truth values.

Hence,

$$P \leftrightarrow q \equiv (p \to q) \wedge (q \to p).$$

It follows therefore that biconditional statement can be written as the **conjunction** of two "if then" statements namely $p \to q$ and $q \to p$. Also we know that

$$p \to q \equiv {\sim}p \vee q$$

and so

$$q \to p \equiv {\sim}q \vee p.$$

Hence,

$$\begin{aligned} p \leftrightarrow q &\equiv (p \to q) \wedge (q \to p) \\ &\equiv ({\sim}p \vee q) \wedge ({\sim}q \vee p). \end{aligned}$$

Thus, the statements having $\to$ or $\leftrightarrow$ symbol are logically equivalent to statement having ~, $\wedge$ and $\vee$.

Definition 4.18

Let p and q be statements. Then, p is a sufficient condition for q means "if p then q" whereas p is a necessary condition for q means "if not p then not q".

4.4.2 Hierarchy of Operations of Logical Connectives

The order of operations of connectives are

$$\sim, \quad \wedge, \quad \vee, \quad \to, \quad \leftrightarrow.$$

Definition 4.19

An **argument** is a sequence of statements. **All statements but the final one** are called **premises** (or **assumptions** or **hypothesis**). The **final statement** is called the **conclusion**.

The symbol $\therefore$, read "therefore", is generally placed just before the conclusion.

4.4.3 Logical Form of an Argument

The logical form of an argument can be obtained from the contents of the given argument. For example, consider the argument:

If a man is a bachelor, he is unhappy.
If a man is unhappy, he dies young.
$\therefore$ Bachelors die young.

The logical form of the argument is:

If p then q.
If q then r.
$\therefore p \to r$,

where

p: He is bachelor
q: He is unhappy
r: He dies young

Definition 4.20

An argument is said to be valid if the conclusion is true whenever all the premises are true.

Definition 4.21

An **argument** which is **not true** is called a **fallacy**.

4.4.4 Method to Test Validity of an Argument

1. Identify the premises and conclusion of the argument.
2. Construct a truth table showing the truth values of all the premises and conclusion.
3. Find the rows (called **critical rows**) in which all the **premises are true.**
4. In each critical row, determine whether the conclusion of the argument is also true.
   (a) If in each critical row the conclusion is also true, then the **argument form is valid**.
   (b) If **there is at least one critical row** in which conclusion is **false**, the argument form is **fallacy** (invalid).

EXAMPLE 4.26

Show that the argument

$$\begin{array}{c} p \\ p \to q \\ \therefore q \end{array}$$

is valid.

Solution.

The premises are p and $p \to q$. The conclusion is q. The truth table is

p	q	p (Premises)	$p \to q$ (Premises)	q (Conclusion)	
T	T	T	T	T	→ Critical row
T	F	T	F	F	
F	T	F	T	T	
F	F	F	T	F	

In the first row, all the premises are true. Therefore, the first row is critical row. The conclusion in this critical row is also true. Hence the argument is valid.

The argument (discussed above)

$$\begin{array}{c} p \\ p \to q \\ \therefore q \end{array}$$

is known as **Law of Detachment.** The validity of this law may also be established by showing that $p \wedge (p \to q) \to q$ is a tautology.

EXAMPLE 4.27

Verify the validity of the following argument form:

$$p \to q$$
$$p$$
$$\therefore q.$$

Solution.

The truth table for the premises and conclusion is

		Premises		Conclusion	
p	q	$p \to q$	p	q	
T	T	T	T	T	→ Critical row
T	F	F	T	F	
F	T	T	F	T	
F	F	T	F	F	

The first row is critical row and the conclusion in the critical row is true. Hence, the given argument form is **Valid**. Thus $(p \to q) \wedge p \to q$ is a tautology.

The fact that this argument form is valid is called **Modus ponens**. This Latin term means "**Method of affirming**" (since the conclusion is an affirmation).

EXAMPLE 4.28

Verify the validity of the argument form

$$p \to q$$
$$\sim q$$
$$\therefore \sim p.$$

Solution.

The truth table for the premises and conclusion is

		Premises		Conclusion	
p	q	$p \to q$	$\sim q$	$\sim p$	
T	T	T	F	F	
T	F	F	T	F	
F	T	T	F	T	
F	F	T	T	T	→ Critical row

The last row is critical row and conclusion in this row is also true. Hence, the argument form is valid.

The fact that this argument is valid is called **Modus Tollens** which means **Method of denying**, since the conclusion is a denial.

Theorem 4.1

(Rule of Inference, Law of Syllogism or Hypothetical Syllogism)
The argument

$$\begin{array}{r} p \to q \\ q \to r \\ \therefore p \to r \end{array}$$

is universally valid and so is a **rule of inference**.

In other words,

$$(p \to q) \wedge (q \to r) \to (p \to r)$$

is a Tautology.

Proof. The truth table for the argument and the conclusion is given below:

p	q	r	$p \to q$	$q \to r$	$p \to r$	
T	T	T	T	T	T	→ Critical row
T	T	F	T	F	F	
T	F	T	F	T	T	
T	F	F	F	T	F	
F	T	T	T	T	T	→ Critical row
F	T	F	T	F	T	
F	F	T	T	T	T	→ Critical row
F	F	F	T	T	T	→ Critical row

We observe that the critical rows for the premises $p \to q$, $q \to r$ are first row, fifth row, seventh row and eighth row. The conclusion $(p \to r)$ in these rows is also true. Hence the argument is valid. Also we note that $(p \to q) \wedge (q \to r) \to (p \to r)$ is a tautology because all the truth values in this case are true.

EXAMPLE 4.29

Consider the argument

If you invest in the stock market, then you will get rich.
If you get rich, then you will be happy.
∴ If you invest in the stock market, then you will be happy.

Solution.

By rule of inference, the argument is valid.

EXAMPLE 4.30

Verify the validity of the argument:

Drinking is healthy.
If drinking is healthy, then wine is prescribed by the physicians.
∴ Wine is prescribed by the physicians.

Solution.

In symbols, the argument is

$$\begin{array}{c} p \\ p \rightarrow q \\ \therefore q. \end{array}$$

The argument is of the form **Modus Ponens** (or Law of Detachment) and so it is valid. However, the **conclusion is false**. Observe that the first premises, p: "Drinking is healthy", is **false**. The second premises, $p \rightarrow q$ is then true and conjunction of the two premises $(p \wedge (p \rightarrow q))$ is false.

EXAMPLE 4.31

The following arguments are valid:

$$\begin{array}{cc} \text{(a)} & \text{(b)} \\ p & q \\ \therefore p \vee q & \therefore p \vee q. \end{array}$$

These arguments are called "**Disjunctive Addition**" and are used for **making generalizations**.

Solution.

(a) The truth table is

Premise		Conclusion	
p	q	$p \vee q$	
T	T	T	→ Critical row
T	F	T	→ Critical row
F	T	T	
F	F	F	

The conclusion in the critical rows is also true. Hence, the argument is valid.

(b) The truth table is

	Premise	Conclusion	
p	q	$p \vee q$	
T	T	T	→ Critical row
T	F	T	
F	T	T	→ Critical row
F	F	F	

The conclusion in the critical rows is true. Hence the argument is valid.

EXAMPLE 4.32

The following arguments are valid:

$$\begin{array}{cc} \text{(a)} & \text{(b)} \\ p \wedge q & p \wedge q \\ \therefore \quad p & \therefore \quad q. \end{array}$$

(These arguments are called **Conjunctive Simplification** and are used for particularizing.)

For example (a) says that if both p and q are true, then in particular, p is true. The validity of these arguments can be proved using truth tables.

EXAMPLE 4.33

The arguments

(a)		(b)
$p \vee q$	and	$p \vee q$
$\sim q$		$\sim p$
$\therefore\ p$		$\therefore\ q$

are valid

For example, (a) says that there are two possibilities p or q but q is not there. Hence p is there. So (a) is valid. Similarly, (b) is valid. These arguments are called "**Disjunctive Syllogism**".

EXAMPLE 4.34 Dilemma: (Proof by Division into Cases)

The argument

$$\begin{array}{l} p \vee q \\ p \rightarrow r \\ q \rightarrow r \\ \therefore\ r \end{array}$$

is valid because $(p \vee q) \wedge (p \rightarrow r) \wedge (q \rightarrow r) \rightarrow r$ is a tautology.

EXAMPLE 4.35

Using the following true statements, locate the treasure hidden in the estate:

(i) If the house is next to a lake, then the treasure is not in the kitchen.
(ii) If the tree in front yard is a mango tree, then the treasure is in the kitchen.
(iii) The house is next to a lake.
(iv) The tree in the front yard is a mango tree or the treasure is placed behind the mirror in the wall of the bed room.
(v) If the tree in the backyard is an oak tree, then the treasure is in the drawing room.

Solution.

Let

p: the house is next to a lake
q: the treasure is in the kitchen
r: the tree in the front yard is a mango tree
s: the treasure is placed behind the mirror in the wall of the bed room
t: the tree in the backyard is an oak tree
u: the treasure is in the drawing room

Then the logical forms of the clues are:

(i) $p \rightarrow \sim q$
(ii) $r \rightarrow q$
(iii) p
(iv) $r \vee s$
(v) $t \rightarrow u$

The following deductions can be made:

(1)	$p \to \sim q$	by (i)
	p	by (iii)
$\therefore$	$\sim q$	by Modus Ponens
(2)	$r \to q$	by (ii)
	$\sim q$	by the conclusion of (1)
$\therefore$	$\sim r$	by Modus Tollen
(3)	$r \vee s$	by (iv)
	$\sim r$	by conclusion of (2)
$\therefore$	s	by Disjunctive Syllogism

Hence, the treasure is placed behind the mirror in the wall of the bed room. The fifth clue has not been used to get the conclusion.

EXAMPLE 4.36

If there is a gas in the car, then John will go to a store. If John goes to the store, then he will get a soda. There is gas in the car. Will John get a soda?

Solution.

Let

p: There is a gas in the car

q: John will go to a store

r: John will get a soda.

Then, logical forms of the given statements are

(a) $p \to q$ (b) $q \to r$ (c) p

The following deductions can be made:

(1)	$p \to q$	by (a)
	p	by (c)
	$\therefore q$	by Modus Ponens
(2)	$q \to r$	by (b)
	q	by the conclusion of (1)
	$\therefore r$	by Modus Ponens

Hence, John will get a soda.

EXAMPLE 4.37

A software engineer makes the following observations in a computer programming:

(i) There is an undeclared variable or there is a syntax error in the first five lines.

(ii) If there is a syntax error in the first five lines, then there is a missing semicolon or a variable name is misspelled.

(iii) There is not a missing semicolon.

(iv) There is not a misspelled variable name.

Find the mistake in the program.

Solution.

Let

p: There is undeclared variable

q: There is a syntax error in the first five lines

r: There is a missing semicolon

s: Variable name is misspelled.

The logical forms of the given statements are

(i) $p \vee q$
(ii) $q \rightarrow r \vee s$
(iii) $\sim r$
(iv) $\sim s$.

Deductions:

(1)	$\sim r$	by (iii)
	$\sim s$	by (iv)
	$\therefore \sim r \wedge \sim s$	by conjunction

But $\sim r \wedge \sim s = \sim(r \vee s)$. Thus the first conclusion is $\sim(r \vee s)$.

(2)	$q \rightarrow r \vee s$	by (ii)
	$\sim(r \vee s)$	by the conclusion of (1)
	$\therefore \sim q$	by Modus Tollen
(3)	$p \vee q$	by (i)
	$\sim q$	by the conclusion of (2)
	$\therefore p$	by Disjunctive Syllogism

Hence, there is an undeclared variable.

EXAMPLE 4.38

Show that the following argument is invalid.

If taxes are lowered, then income rises
Income rises
∴ Taxes are lowered.

Solution.
Let

p: Taxes are lowered, q: Income rises

Then the argument form is

$$p \rightarrow q$$
$$q$$
$$\therefore p.$$

The truth table for the premises and conclusion is

	Premises		Conclusion	
p	q	$p \rightarrow q$	p	
T	T	T	T	← Critical row
T	F	F	T	
F	T	T	F	← Critical row
F	F	T	F	

In the third critical row, the conclusion is not true. Hence, the argument is **not valid**.

EXAMPLE 4.39

Is the following argument valid?

If two sides of a triangle are equal, then the opposite angles are equal
Two sides of a triangle are **not** equal
∴ The opposite angles are not equal.

Solution.

In symbolic form the argument is

$$p \to q$$
$$\sim p$$
$$\therefore \sim q.$$

The truth table is

		Premises		Conclusion	
p	q	$p \to q$	$\sim p$	$\sim q$	
T	T	T	F	F	
T	F	F	F	T	
F	T	T	T	F	← Critical row
F	F	T	T	T	← Critical row

In this case, the third and fourth rows are critical rows. But in the third row, the conclusion is false. Hence the argument is a **fallacy**.

EXAMPLE 4.40

Consider the following argument for validity:

If I study, then I will not fail in mathematics.
If I do not play basketball, then I will study.
But I failed in mathematics.
∴ I must have played basketball.

Solution.

The symbolic form of the argument is

(a) $p \to q$
(b) $r \to p$
(c) $\sim q$
$\therefore \sim r.$

The validity can be verified by the following two methods:

First Method: The following deductions can be made:

(1) $p \to q$ by (a)
$\sim q$ by (c)
$\therefore \sim p$ by Modus Tollen
(2) $r \to p$ by (b)
$\sim p$ by (1)
$\therefore \sim r$ by Modus Tollen

which is true (conclusion). Hence the argument is **valid**.

Second Method: The truth table for the argument form is:

			Premises			Conclusion	
p	q	r	$p \to q$	$r \to p$	$\sim q$	$\sim r$	
T	T	T	T	T	F	F	
T	T	F	T	T	F	T	
T	F	T	F	T	T	F	
T	F	F	F	T	T	T	
F	T	T	T	F	F	F	
F	T	F	T	T	F	T	
F	F	T	T	F	T	F	
F	F	F	T	T	T	T	← Critical row

We see that there is only one critical row (eighth row) in the truth table. The conclusion is also true in the critical row. Hence the argument is valid.

Remark 4.10 It is possible for a valid argument to have a false conclusion and for an invalid argument to have a true conclusion. For example:

(1) If Sohan was an artist, then Sohan had green hair
Sohan was an artist.
∴ Sohan had green hair.

In symbol, we have

$$p \to q$$
$$p$$
$$\therefore q.$$

The argument is valid by Modus Ponens. But its major premises is false and so is its conclusion.

(2) If Delhi is a big city, then Delhi has tall buildings.
Delhi has tall buildings.
∴ Delhi is a big city.

The symbolic form of the argument is

$$p \to q$$
$$q$$
$$\therefore p.$$

Using truth table, it can be shown that the **argument is fallacy**. But the conclusion is true.

If we replace first premise by $q \to p$, then argument becomes

$$q \to p$$
$$q$$
$$\therefore p \text{ which is valid.}$$

That is why, such type of fallacy is called **Converse Error**.

Remark 4.11 Consider the argument form

$$p \to q$$
$$\sim p$$
$$\therefore \sim q.$$

We can see that the argument form is **invalid**. But if we replace the premises by $\sim p \to \sim q$, then it becomes valid.

Such type of fallacy is called **Inverse Error**.

4.4.5 Contradiction Rule

If the supposition that the statement p is false leads logically to a contradiction, then you can conclude that p is true.

In symbols,

$$\sim p \to c, \text{ where } c \text{ is a contradiction}$$
$$\therefore p.$$

The truth table for the premise and the conclusion of this argument are given below:

p	$\sim p$	c	$\sim p \to c$	p	
T	F	F	T	T	→ Critical row
F	T	F	F	F	

The premises and conclusion are both true in the critical row and hence the argument is valid.

EXAMPLE 4.41

Knights and Knaves (Raymond Smullyan's description of an island containing two types of people):

This island contains two types of people: knights who always tell the truth and Knaves who always lie. A visitor visits the island and approached two natives who spoke to the visitor as follows:

A says: B is a knight
B says: A and I are of opposite type.

What are A and B?

Solution.

Suppose A is a knight. Because A always tells the truth, it follows that B is a knight. Therefore what B says is true (by the definition of Knight). Therefore A and B are of opposite type. Thus we arrive at a contradiction: A and B are both Knights and A and B are of opposite type. Therefore supposition is wrong. Hence A is not a Knight. So A is a Knave. Therefore what A says is false. Hence B is not a Knight and so is a Knave. Hence A and B **are both Knaves**.

4.4.5.1 Summary of Rules of Inference

The following are the valid argument forms:

1. **Modus Ponens:**

$$p \to q$$
$$p$$
$$\therefore q.$$

2. Modus Tollens:

$$p \to q$$
$$\sim q$$
$$\therefore \sim p.$$

3. Disjunctive Addition:

(a) p
$\therefore p \vee q$ for any q

(b) q
$\therefore p \vee q$ for any p.

4. Conjunctive Simplification:

(a) $p \wedge q$
$\therefore p.$

(b) $p \wedge q$
$\therefore q.$

5. Conjunctive Addition:

$$p$$
$$q$$
$$\therefore p \wedge q.$$

6. Disjunctive Syllogism:

(a) $p \vee q$
$\sim q$
$\therefore p.$

(b) $p \vee q$
$\sim p$
$\therefore q.$

7. Hypothetical Syllogism:

$$p \to q$$
$$q \to r$$
$$\therefore p \to r.$$

8. Dilemma: proof by division into cases:

$$p \vee q$$
$$p \to r$$
$$q \to r$$
$$\therefore r.$$

9. Rule of contradiction:

$\sim p \to c$, where c is a contradiction
$\therefore p.$

4.5 QUANTIFIERS

So far we have studied the compound statements which were made of simple statements joined by the connectives $\sim$, $\wedge$, $\vee$, $\to$ and $\leftrightarrow$. That study cannot be used to determine validity in the majority of everyday and mathematical situations. For example, the argument

All men are mortal,
All teachers are men,
$\therefore$ All teachers are mortal,

is intuitively correct. Yet its validity cannot be derived using the methods studied so far. To check the validity of such argument it is necessary to separate the statements into parts, subjects and predicates.

Also we must analyse and understand the special role played by words denoting quantities such as All or Some.

Definition 4.22

The symbolic analysis of predicates and quantified statements is called the **predicate calculus** whereas the symbolic analysis of ordinary compound statements is called the **statement calculus** or **propositional calculus**.

In English grammar, the predicate is the part of a sentence that gives information about the subject. For example, in the sentence "Ram is a resident of Karnal", the word Ram is the subject and the phrase "is a resident of Karnal" is the predicate. Thus, **predicate is the part of the sentence from which the subject has been removed**.

In logic, predicates can be obtained by removing any nouns from a statement. For example, if P stands for "is a resident of Karnal" and Q stands for "is a resident of", then both P and Q are predicate symbols. The sentences "x is a resident of Karnal" and "x is a resident of y" are denoted as $P(x)$ and $Q(x, y)$, respectively, where x and y are predicate variables that take values in appropriate sets.

Definition 4.23

A **predicate** is a sentence that contains a finite number of variables and becomes a statement when specific values are substituted for the variables.

The domain of a predicate variable is the set of all values that may be substituted in place of the variables. The predicates are also known as "**propositional functions** or **open sentences**".

Definition 4.24

Let $P(x)$ be a predicate and x has domain D. Then the set

$$\{x \in D : P(x) \text{ is true}\}$$

is called the **truth set** of $P(x)$.

For example, let $P(x)$ be "x is an integer less than 8" and suppose the domain of x is the set of all positive integers. Then the truth set of $P(x)$ is $\{1, 2, 3, 4, 5, 6, 7\}$,

Let $P(x)$ and $Q(x)$ be predicates with common domain D of x. The notation $P(x) \Rightarrow Q(x)$ means that every element in the truth set of $P(x)$ is in the truth set of $Q(x)$.

Similarly, $P(x) \Leftrightarrow Q(x)$ means that $P(x)$ and $Q(x)$ have **Identical truth sets**.

For example, let the domain of x be the set of positive integers and let

$$P(x)\text{: "}x \text{ is an integer less than 8"},$$
$$Q(x)\text{: "}x \text{ is a factor of 4"}.$$

Then,

$$\text{Truth set of } P(x) \text{ is } \{1, 2, 3, 4, 5, 6, 7\},$$
$$\text{Truth set of } Q(x) \text{ is } \{1, 2, 4\}.$$

Since every element in the truth set of $Q(x)$ is in the truth set of $P(x)$, $Q(x) \Rightarrow P(x)$.

Definition 4.25

The words that refer to quantities such as "All", or "some" and tell for how many elements a given predicate is true are called **quantifiers**.

By adding quantifier, we can obtain statements from a predicate.

Definition 4.26

The symbol $\forall$ denotes "for all" and is called the **Universal quantifier**.

Thus the sentence

All men are strong,

can be written as

$\forall x \in S$, x is strong,

where S denotes the set of all men.

Definition 4.27

Let $P(x)$ be a predicate and D the domain of x. A statement of the form "$\forall x \in D, P(x)$" is called a **universal statement.**

A universal statement $P(x)$ is true if and only if $P(x)$ is true **for every** x in D and a universal statement $P(x)$ is false if and only if $P(x)$ is false **for at least one** $x \in D$.

A value for x for which $P(x)$ is **false** is called a **Counterexample** to the universal statement.

EXAMPLE 4.42

Let $D=\{1, 2, 3, 4\}$ and consider the universal statement

$$P(x): \forall x \in D, x^3 \geq x.$$

This is true for all values of $x \in D$ since $1^3 \geq 1$, $2^3 \geq 2$ and so on.

But the universal statement

$$Q(x): \forall n \in N, n+2>8$$

is not true because if we take $n=6$, then $8>8$ which is absurd.

Definition 4.28

The symbol $\exists$ denotes "there exists" and is called the **existential quantifier**.
For example, the sentence "There is a University in Kurukshetra" can be expressed as

$\exists$ a university u such that u is in Kurukshetra.

or, we can write

$\exists u \in U$ such that u is in Kurekshetra, where U is the set of universities.

The words **such that** are inserted just before the predicate.

Definition 4.29

Let $P(x)$ be a predicate and D be the domain of x. A statement of the form "$\exists x \in D$ such that $P(x)$" is called an **Existential Statement**. It is defined to be true if and only if $P(x)$ **is true for at least one** $\boldsymbol{x}$ **in** $\boldsymbol{D}$. It is false if and only if $P(x)$ is false for all x in D.

For example, the existential statement

$$\exists n \in N : n+3<9$$

is **true** since the set

$$\{n : n+3<9\}=\{1, 2, 3, 4, 5\}$$

is not empty.

EXAMPLE 4.43

Let $A=\{2, 3, 4, 5\}$, then the existence statement

$$\exists n \in A : n^2=n$$

is false because there is no element in A whose square is equal to itself.

Definition 4.30

A statement of the form

$$\forall x, \text{ if } P(x) \text{ then } Q(x)$$

is called **universal conditional statement**.

For example, consider the statement

$$\forall x \in \mathbf{R}, \text{ if } x>2 \text{ then } x^3>8$$

is a universal conditional statement.

Definition 4.31

The negation of a universal statement

$$\forall x \text{ in } D, P(x)$$

is logically equivalent to a statement of the form

$$\exists x \text{ in } D \text{ such that } \sim P(x).$$

Thus,

$$\sim(\forall x \in D, P(x)) \equiv \exists x \in D, \sim P(x).$$

Hence,

The negation of a universal statement "all are" is logically equivalent to an existential statement "some are not".

For example, the negation of

(i) " For all positive integers n, we have $n+2>9$"

is

" There exists a positive integer n such that $n+2 \not> 9$".

(ii) The negation of

"All students are intelligent"

is

"Some students are **not** intelligent"

or

"$\exists$ a student who is **not** intelligent".

(iii) the negation of

" No politicians are honest"

is

"$\exists$ a politician x such that x is honest."

or

"Some politicians are honest".

Definition 4.32

The **negation of a universal conditional statement** is defined by

$$\sim(\forall x, P(x) \rightarrow Q(x)) \equiv \exists x \text{ such that } \sim(P(x) \rightarrow Q(x)).$$

Also we know that the negation of if-then statement is

$$\sim(P(x) \rightarrow Q(x)) \equiv P(x) \wedge \sim Q(x).$$

Hence,

$$\sim(\forall x, P(x) \rightarrow Q(x)) \equiv \exists x \text{ such that } P(x) \wedge \sim Q(x),$$

that is,

$$\sim(\forall x, P(x) \rightarrow Q(x)) \equiv \exists x \text{ such that } P(x) \text{ and } \sim Q(x).$$

EXAMPLE 4.44

The negation of universal conditional statement

"$\forall$ human beings x, if x is a man, then x is strong"

is

"$\exists$ a human being x such that x is a man and x is not strong"

If $P(x)$ is a predicate and the domain of x is $D=\{x_1, x_2, \ldots, x_n)$, then the statement

$$\forall x \in D, P(x) \text{ and } P(x_1) \wedge P(x_2) \wedge \ldots \wedge P(x_n)$$

are logically equivalent.
For example, let $P(x)$ be

"$x \cdot x = x$"

and let $D=\{0, 1\}$. Then

$$\forall x \in D, P(x),$$

can be written as

$\forall$ binary digits x, $x \cdot x = x$.

This is equivalent to

$$0 \cdot 0 = 0 \text{ and } 1 \cdot 1 = 1,$$

which can be written as

$$P(0) \wedge P(1).$$

Similarly, if $P(x)$ is a predicate and $D=(x_1, x_2, \ldots, x_n\}$ then the statements

$$\exists x \in D, P(x) \text{ and } P(x_1) \vee P(x_2) \vee \ldots \vee P(x_n)$$

are logically equivalent.

Definition 4.33

Let

$$\forall x \in D, \text{ if } P(x) \text{ then } Q(x)$$

be a statement. Then

(i) **Contrapositive** of this statement is

$$\forall x \in D, \text{ if } {\sim}Q(x) \text{ then } {\sim}P(x).$$

(ii) **Converse** of this statement is

$$\forall x \in D, \text{ if } Q(x) \text{ then } P(x).$$

(iii) **Inverse** of this statement is

$$\forall x \in D, \text{ if } {\sim}P(x) \text{ then } {\sim}Q(x).$$

4.6 UNIVERSAL MODUS PONENS

The following argument form is valid

Formal Version	*Informal Version*
$\forall$ x if P (x) then Q (x)	If x makes P (x) true, then x makes Q (x) true
P (a) for a particular a	a makes P (x) true
$\therefore$ Q (a).	$\therefore$ a makes Q (x) true.

An argument of this form is called a **Syllogism**. The first and second premises are called its **major premises** and **minor premises**, respectively.

EXAMPLE 4.45

Consider the argument

All men are good.
Ramesh is a man.
∴ Ramesh is good.

The major premise of this argument is

$\forall$ x, if x is a man, then x is good.

Let,

P (x): x is a man,
Q (x): x is good,
P (a): Ramesh is a man.

Therefore, by Modus Ponens, the argument is valid.

4.7 UNIVERSAL MODUS TOLLENS

The following argument form is valid

Formal Version	*Informal Version*
$\forall$ x if P (x) then Q (x)	If x makes P (x) true, then x makes Q (x) true
~Q (a) for a particular a	a does not makes Q (x) true
$\therefore$ ~P (a).	$\therefore$ a does not makes P (x) true.

EXAMPLE 4.46

Consider the argument form:

All intelligent persons are engineers.
John is not an engineer.
∴ John is not intelligent.

The major premise of this argument can be rewritten as

$\forall$ x, if x is intelligent, then x is an engineer.

Let

P (x): x is intelligent.
Q (x): x is an engineer.
Z=John.

Then we have,

$$\forall x, \text{if } P(x) \text{ then } Q(x),$$
$$\sim Q(Z),$$
$$\therefore \sim P(Z),$$

which is valid by Universal Modus Tollens.

EXAMPLE 4.47

Consider the argument:

All men are strong.
Ritu is not strong.
∴ Ritu is not a man.

The major premise can be written as

$$\forall x, \text{if } x \text{ is a man, then } x \text{ is strong.}$$

Let

$$P(x): x \text{ is a man},$$
$$Q(x): x \text{ is strong},$$
$$Z = \text{Ritu}.$$

Then we have

$$\forall x, \text{if } P(x) \text{ then } Q(x)$$
$$\sim Q(Z)$$
$$\therefore \sim P(Z).$$

Hence, by Universal Modus Tollens, Ritu is not a man.

4.8 USE OF DIAGRAMS FOR VALIDITY OF ARGUMENTS

Consider the argument:

All human beings are mortal
Zeus is not mortal
∴ Zeus is not a human being.

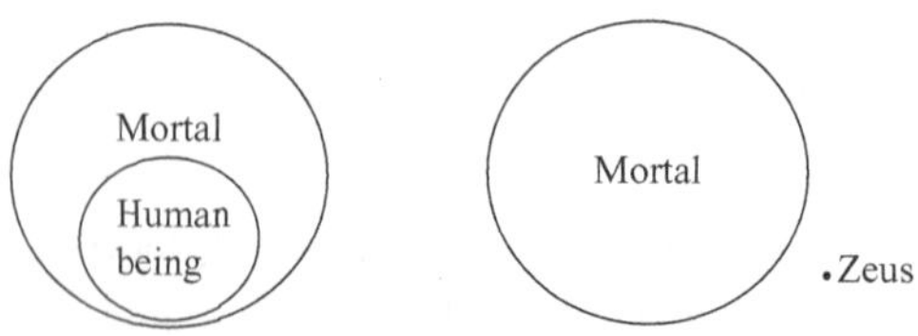

Major premise Minor premise

The two diagrams fit together in only one way as shown below:

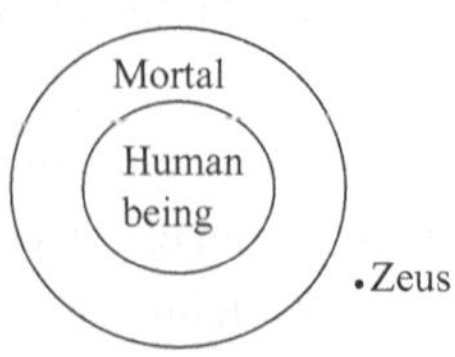

Since Zeus is outside the mortal disc it is necessarily outside the human beings disc. Hence the Conclusion is true.

EXAMPLE 4.48

Use a diagram to show the invalidity of the arguments

All men are strong.
Ritu is strong.
∴ Ritu is a man.

Solution.

The major premise and a minor premise of the arguments are shown in the diagrams below:

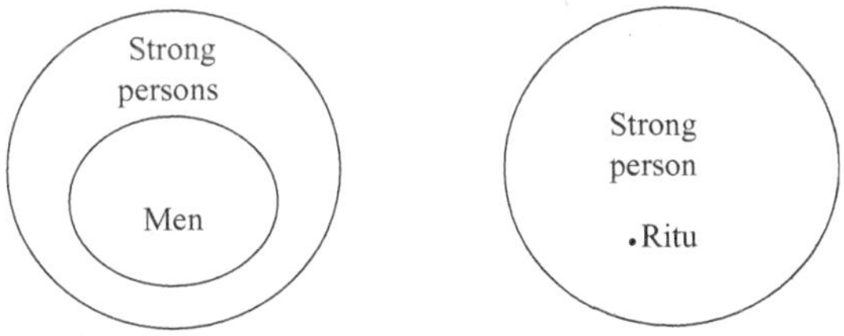

There are two possibilities to fit these two diagrams into a single one.

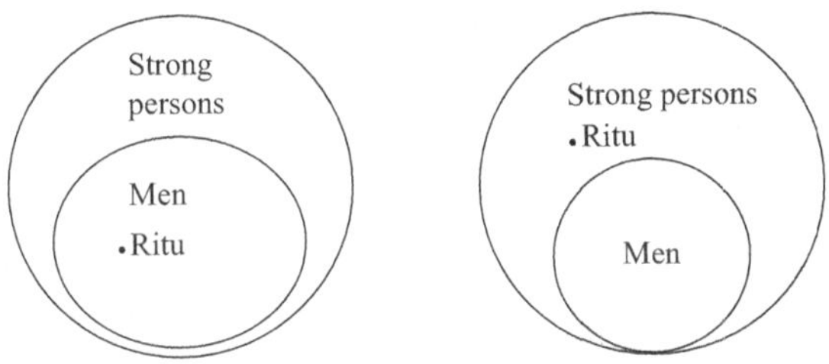

(1) (2)

The conclusion "Ritu is a man" is true in the first case but not in the second. Hence the argument is invalid.

EXERCISES

1. Show that the logical expression

$$\{[p \to (q \vee r)] \wedge (\sim q)\} \to (p \to r)$$

is a tautology.

2. Restate the following as implications "If..., then ...":
   (i) A sufficient condition that a figure be rectangle is that it be a square.
   (ii) I am citizen of India if I am living in Bangalore.
   (iii) A necessary condition for Pakistanis to win a cricket match is that they have at least two left-handed batsmen.

3. Write negations of the following propositions:
   (i) No rectangle is a square.
   (ii) He takes his bath if and only if the water is cold.
   (iii) If it rains, then they do not go for a walk.

4. Verify the validity of the following arguments:
   (i) (a) If there are more pigeons than there are pigeonholes, then at least two pigeons fly into the same pigeonhole.

(b) There are more pigeons than there are pigeonholes.

(c) Therefore, at least two pigeons fly into the same pigeonhole.

(ii) (a) If this number is divisible by 9, then it is divisible by 3.

(b) This number is not divisible by 3.

(c) Therefore, this number is not divisible by 9.

5. State the converse, inverse and contrapositive to the following statement:

   "If triangle ABC is a right triangle, then $|AB|^2+|BC|^2=|AC|^2$".

6. Show that the following argument is a fallacy:

   If today is Smith's birthday, then today is May 18.
   Today is May 18
   Therefore, today is Smith's birthday.

7. Let x be a real number. Is the following argument valid?

   x is positive or x is negative
   If x is positive, then $x^2>0$
   If x is negative, then $x^2>0$
   $\therefore x^2>0$.

8. Verify the validity of the following argument by using rules of inference:

   If Julia does not live in Italy, then she does not speak Italian.
   Julia does not drive a car
   If Julia lives in Italy, then she travels by train.
   Either Julia speaks Italian or she drives a car.
   Therefore, Julia travels by train.

9. Verify the validity of the following argument:

$$\begin{array}{l} \sim r \\ p \to q \\ q \to r \\ \therefore \sim p. \end{array}$$

10. Fill in the blanks:

    (i) If the graphs are isomorphic, then their degree spectrum will be the same
    Their degree spectrums are different
    Hence ________________

    (ii) If it rains, Bill will be happy
    Bill is not happy
    Therefore ________________

11. Write negation of the following sentence by changing quantifiers:

    "Every complete bipartite graph is not planar".

12. Express the following statement in terms of existential quantification:

13. "The number 24 can be written as a sum of two even integers".

5 Algebraic Structures

The algebraic structures like semigroup, monoid, group, ring and field have wide applications in many disciplines and in particular to binary coding.

5.1 BINARY OPERATIONS

Definition 5.1

A collection of objects with operations defined on them and the accompanying properties form a **Mathematical Structure or System**.

For example, the collection of sets with the operations of union, intersection, and complement and their accompanying properties is a discrete mathematical structure. We denote this structure by [sets, $\cup$, $\cap$, $-$].

Similarly, the collection of 3×3 matrices with the operations of addition, multiplication and transpose and their accompanying properties is a mathematical structure. We denote this structure by [3×3 matrices, $+$, $*$, T].

Definition 5.2

A structure is said to be **closed with respect to an operation** if that operation always produces another member (element) of the collection of objects.

For example, let A be the set of even integers. Then the structure [A, +] is **closed with respect to addition** operation because the sum of two even integers is always even.

On the other hand, the structure [A, +], where A is the set of **odd integers**, is **not closed** with respect to addition because the sum of two odd integers is an even integer.

Definition 5.3

Let A be a non-empty set. Then a function $f: A\to A$ from A into A is called a **unary operation**.

For example, $f: \mathbf{Z}\to\mathbf{Z}$ defined by

$$f(n)=|n|$$

is a unary operation on **Z**.

Similarly, taking complement of a set A is a unary operation of the power set $P(S)$ of a set S.

Definition 5.4

Let A be a non-empty set. Then a mapping $f: A\times A\to A$ is called a **binary operation**. Thus, a binary operation is a rule that assigns to each ordered pair $(a, b)\in A\times A$ an element of A.

For the sake of simplicity, we write $a * b$ in place of $f(a, b)$.

EXAMPLE 5.1

Let **Z** be the set of integers. Then $f: \mathbf{Z}\times\mathbf{Z}\to\mathbf{Z}$ defined by $f(a, b)=a * b=a+b$, $a, b\in\mathbf{Z}$ is a binary operation on **Z** because the sum of two integers a and b is again an integer.

Thus, **addition of integers** is a binary operation.

EXAMPLE 5.2

Let **N** be the set of positive integers. Then f: $\mathbf{N}\times\mathbf{N}\rightarrow\mathbf{N}$ defined by $f(a, b)=a * b=a-b$ is **not a binary operation** because difference of two positive integers need not be positive integer. For example, $2-5$ is not a positive integer.

EXAMPLE 5.3

For the set **N** of positive integers, let f: $\mathbf{N}\times\mathbf{N}\rightarrow\mathbf{N}$ be defined by $f(a, b)=\frac{a}{b}$. Then, f is not a binary operation. For example, if $a=2$, $b=7$, then $\frac{a}{b}=\frac{2}{7}$ is not a positive integer.

EXAMPLE 5.4

Let **Z** be the set of all integers. Then f: $\mathbf{Z}\times\mathbf{Z}\rightarrow\mathbf{Z}$ defined by

$$f(a, b)=\max(a, b)$$

is a binary operation. For example,

$$f(2, 4)=2 * 4=\max(2,4)=4 \in \mathbf{Z}.$$

EXAMPLE 5.5

Let $A=\{a, b, c\}$. Define $*$ by

$$x * y=x, \quad x, \quad y \in A.$$

Then the table given below defines the operation $*$

$*$	a	b	c
a	a	a	a
b	b	b	b
c	c	c	c

Further, if we define "•" by

$$x \cdot y=y, \quad x, \quad y \in A,$$

then the table given below defines the operation.

$\cdot$	a	b	c
a	a	b	c
b	a	b	c
c	a	b	c

EXAMPLE 5.6

If $A=\{0, 1\}$. Then the binary operations $\wedge$ and $\vee$ are defined by the following tables:

$\wedge$	0	1
0	0	0
1	0	1

and

∨	0	1
0	0	1
1	1	1

5.2 PROPERTIES OF BINARY OPERATION

Definition 5.5

A binary operation $*$ on a set A is said to be **commutative if**

$$a * b = b * a$$

for any elements a and b in A.

For example, consider the set **Z** of integers. Since,

$$a+b=b+a \quad \text{and} \quad a \cdot b = b \cdot a,$$

for $a, b \in \mathbf{Z}$, the addition and multiplication operations on **Z** are commutative. But, on the other hand, subtraction in **Z** is not commutative since, for example,

$$2-3 \neq 3-2.$$

EXAMPLE 5.7

Fill in the following table so that the binary operation $*$ is commutative.

$*$	a	b	c
a	b	–	–
b	c	b	a
c	a	–	c

Solution.

We note that $b * a = c$, therefore, for commutativity we must have $a * b = c$.
Also, $c * a = a$, hence $a * c$ should also be a.
Further, for commutativity we should have

$$c * b = b * c = a.$$

Thus $c * b$ should be a.

Note that for commutativity of $*$, the entries in the table are symmetric with respect to the main diagonal.

Definition 5.6

A binary operation $*$ on a set A is said to be **associative** if for any elements a, b, c in A, we have

$$a * (b * c) = (a * b) * c.$$

For example, addition and multiplication of integers are associative. But subtraction of integers is **not** associative. For example,

$$(2-4)-5=-7,$$

but

$$2-(4-5)=3.$$

Theorem 5.1

Let $*$ be an associative binary operation on a set A. Then, any product $a_1 * a_2 * \ldots * a_n$ requires no parenthesis, that is, all possible products are equal.

Proof. We shall prove this result by induction on n. Since $*$ is associative, the theorem holds for $n=1, 2$ and 3. Suppose $[a_1\ a_2 \ldots a_n]$ denote any product and

$$(a_1\ a_2 \ldots a_n) = (\ldots (a_1\ a_2)\ a_3 \ldots)a_n.$$

It is sufficient then to show that

$$[a_1\ a_2 \ldots a_n] = (a_1\ a_2 \ldots a_n).$$

Since $[a_1\ a_2 \ldots a_n]$ denote arbitrary product, there is an $m<n$ such that the induction yields

$$\begin{aligned}
[a_1\ a_2 \ldots a_n] &= [a_1\ a_2 \ldots a_m]\ [a_{m+1} \ldots a_n] \\
&= [a_1\ a_2 \ldots a_m]\ (a_{m+1} \ldots a_n) \\
&= [a_1\ a_2 \ldots a_m]\ ((a_{m+1} \ldots a_{n-1})a_n) \\
&= ([a_1\ a_2 \ldots a_m]\ (a_{m+1} \ldots a_{n-1}))a_n \\
&= [a_1 \ldots a_{n-1}]\ a_n = (a_1 \ldots a_{n-1})a_n = (a_1\ a_2 \ldots a_n),
\end{aligned}$$

which proves the result.

Definition 5.7

Let $*$ be a binary operation on a set A. An element e in A is called an **identity** element for $*$ if for any element $a \in A$,

$$a * e = e * a = a.$$

Further, e is called **right identity** if $a * e = a$ and **left identity** if $e * a = a$ for any $a \in A$.

Let e_1 the left identity and e_2 be the right identity for a binary operation $*$. Then,

$$e_1\ e_2 = e_2 \quad \text{since } e_1 \text{ is left identity}$$

and

$$e_1\ e_2 = e_1 \quad \text{since } e_2 \text{ is right identity.}$$

Hence, $e_1 = e_2$ and so **identity element for a binary operation is unique**.

Definition 5.8

Let $*$ be a binary operation on a set A and let A has identity element e. Then **inverse** of an element a in A is an element b such that

$$a * b = b * a = e.$$

We shall see later on that if $*$ is associative, then the inverse of an element, if it exits, is unique.

Definition 5.9

A binary operation $*$ on a set A is said to satisfy the **left cancellation law** if

$$a * b = a * c \Rightarrow b = c.$$

A binary operation $*$ on a set A is said to obey **right cancellation law** if

$$b * a = c * a \Rightarrow b = c.$$

Let **Z** be the set of integers. Since

$$a+b=a+c \Rightarrow b=c$$

and

$$b+a=c+a \Rightarrow b=c \text{ for } a, b, c \in \mathbf{Z},$$

it follows that addition of integers in **Z** obeys both cancellation laws.

Similarly, multiplication of integers also obey cancellation laws.

On the other hand, matrix multiplication does not obey cancellation laws. To see it, let

$$A=\begin{bmatrix} 1 & 1 \\ 0 & 0 \end{bmatrix}, \quad B=\begin{bmatrix} 1 & 1 \\ 0 & 1 \end{bmatrix}, \quad C=\begin{bmatrix} 0 & -3 \\ 1 & 5 \end{bmatrix}.$$

Then,

$$AB=AC=\begin{bmatrix} 1 & 2 \\ 0 & 0 \end{bmatrix},$$

but $B \neq C$.

Definition 5.10

A non-empty set together with a number of binary operations on it is called an **algebraic system**.

In what follows, we shall develop some algebraic systems:

5.3 SEMIGROUPS AND MONOIDS

Definition 5.11

A non-empty set S is said to be a **semigroup** if in S there is defined a binary operation $*$ satisfying the following property:

If $a, b, c \in S$, then

$$a * (b * c)=(a * b) * c \quad \text{(associative law).}$$

Thus

A non-empty set S together with an associative binary operation $*$ defined on S is called a **semigroup**.

We denote the semigroup by $(S, *)$.

Definition 5.12

A semigroup $(S, *)$ is called **commutative** if the binary operation $*$ is a commutative operation, i.e., if

$$a * b=b * a \text{ for } a, b \in S.$$

EXAMPLE 5.8

Let **Z** be the set of all integers. Then $(\mathbf{Z}, +)$ is a commutative semigroup. In fact, if $a, b, c \in \mathbf{Z}$, then

(i) $a * b=a+b$ is an integer. Therefore, the operation $+$ on **Z** is a binary operation.
(ii) $a+(b+c)=(a+b)+c$, because associative law holds in the set of integers.
(iii) $a+b=b+a$, because addition in **Z** is commutative.

EXAMPLE 5.9

The set **Z** of integers with the binary operation of subtraction is not a semigroup since subtraction is not associative in **Z**.

EXAMPLE 5.10

Let S be a finite set and let $F(S)$ be the collection of all functions $f: S \to S$ under the operation of **composition of functions**. We know that composition of functions is associative, i.e.,

$$f \circ (g \circ h) = (f \circ g) \circ h, \quad f, g, h \in F(S).$$

Hence F (S) is a semigroup.

EXAMPLE 5.11

The set $P(S)$, where S is a set, together with the operation of union is a commutative semigroup.

EXAMPLE 5.12

The integers modulo m, denoted by $\mathbf{Z}_m$, refer to the set

$$\mathbf{Z}_m = \{0, 1, 2, \ldots, m-1\}.$$

The addition in $\mathbf{Z}_m$ is defined as

$$a+b=r,$$

where r is the remainder when $a+b$ is divided by m. The multiplication in $\mathbf{Z}_m$ is defined by

$$a \cdot b=r,$$

where r is the remainder when $a \cdot b$ is divided by m.

For example, consider

$$\mathbf{Z}_4 = \{0, 1, 2, 3\}.$$

The addition table is

+	0	1	2	3
0	0	1	2	3
1	1	2	3	0
2	2	3	0	1
3	3	0	1	2

We note that

$$(1+2)+3=3+3=2 \text{ and } 1+(2+3)=1+1=2.$$

Hence,

$$(1+2)+3=1+(2+3).$$

In gencral,

$$(a+b)+c=a+(b+c),\ a, b, c \in \mathbf{Z}_4.$$

Hence $\mathbf{Z}_4$ is a semigroup.

Definition 5.13

A non-empty set S is said to be a **monoid** if in S there is defined a binary operation $*$ satisfying the following properties.

(i) If $a, b, c \in S$, then

$$a * (b * c) = (a * b) * c \quad \text{(associative law).}$$

(ii) There exists an element $e \in S$ such that

$$e * a = a * e = a \text{ for all } a \in S \quad \text{(existence of identity element).}$$

Thus, an algebraic system $(S, *)$ is said to be a **monoid** if

(i) $*$ is a binary operation on non-empty set S
(ii) $*$ is an associative binary operation on S
(iii) There exists an identity element e in S

It, therefore, follows that

A monoid is a semigroup $(S, *)$ that has an identity element.

EXAMPLE 5.13

In Example 5.10 above, identity function is an identity element for $F(S)$. Hence, $F(S)$ is a monoid.

EXAMPLE 5.14

Let **M** be the set of all $n\times n$ matrices and let the binary operation $*$ of **M** be taken as addition of matrices. Then $(\mathbf{M}, *)$ is a monoid. In fact,

(i) The sum of two $n\times n$ matrices is again a matrix of order $n\times n$. Thus, the operation of matrix addition is a binary operation.
(ii) If $A, B, C \in \mathbf{M}$, then

$$A+(B+C)=(A+B)+C \quad \text{(associative law).}$$

(iii) The zero matrix acts as additive identity of this monoid because

$$A+\mathbf{0}=\mathbf{0}+A=A \text{ for } A \in \mathbf{M}.$$

Definition 5.14

Let A be a non-empty set. **A word** w on A is a finite sequence of its elements.

For example,

$$w = ab\ ab\ bb = ab\ ab^3$$

is a word on $A=\{a, b\}$.

Definition 5.15

The number of elements in a word w is called **its length** and is denoted by $l(w)$.

For example, length of w in the above example is $l(w)=6$.

Definition 5.16

Let u and v be two words on a set A. Then the word obtained by writing down the elements of u followed by the elements of v is called the **concatenation** of the words u and v on A.

For example, if $A=\{a, b, c\}$, $u=ab\ a\ bbb$ and $v=a\ c\ b\ a\ b$,

then

$$w = ab\ abbb\ ac\ bab = abab^3acbab$$

is the concatenation of u and v.

Let $F(A)$ denote the collection of all words on A under the operation of concatenation. We note that

$$(u\ v)\ w = u\ (v\ w)$$

for $u, v, w \in F(A)$. Hence, $F(A)$ is a **semigroup** known as **free semigroup on** A. The elements of A are called the **generator** of $F(A)$.

Also, we note that if u, v are two words, then

$$l(uv) = l(u) + l(v).$$

Further, the empty sequence, denoted by λ, is also considered as a word on A. However, we do not assume that λ belongs to the free semigroup $F = F(A)$. The set of all words on A including λ is usually denoted by $A*$. Thus, $A*$ **is a monoid under concatenation**. It is called the **free monoid** on A.

Definition 5.17

Let $(S, *)$ be a semigroup and T be a subset of S. If T is closed under the operation $*$ that is, $a * b \in T$ whenever $a, b \in T$, then $(T, *)$ is called a **sub-semigroup** of $(S, *)$.

Definition 5.18

Let $(S, *)$ be a monoid with identity e, and let T be a non-empty subset of S. If T is closed under the operation $*$ and $e \in T$, then $(T, *)$ is called a **submonoid** of $(S, *)$.

Clearly, the associative property holds in any subset of a semigroup and so **a sub-semigroup $(T, *)$ of a semigroup $(S, *)$ is itself a semigroup.**

Similarly, a submonoid of a monoid is itself a monoid.

EXAMPLE 5.15

Let A be the set of even positive integers. Then $(A, \cdot)$, where "$\cdot$" denotes ordinary multiplication is a sub-semigroup of $(N, \cdot)$ since A is closed under multiplication.

Similarly, the set B of odd positive integers form a subsemigroup $(B, \cdot)$ of $(N, \cdot)$.

Also $(A, +)$ is a sub-semigroup of $(N, +)$. But $(B, +)$ is not a subsemigroup of $(N, +)$ because B is not closed under addition. For example, $1+3=4$ which is not odd.

EXAMPLE 5.16

Let $(S, *)$ be a semigroup and $a \in S$. If $T = \{a^i : i \in N\}$, then $(T, *)$ is a sub-semigroup of $(S, *)$.

EXAMPLE 5.17

Let $F(A)$ be a free semigroup on the set $A = \{a, b\}$. Let G consists of all even words, that is, words with even length. The concatenation of two such words is also even. Thus, G is a sub-semigroup of $F(A)$.

Theorem 5.2

The inverse of every element in a **semigroup with identity** e is unique.

Proof. We shall use associativity of the binary operation $*$ to prove the uniqueness of the inverse element. So, suppose that b and c are two inverses of an element a in a monoid $(S, *)$. Therefore, we have

$$a * b = b * a = e, \qquad \text{(i)}$$

$$a * c = c * a = e. \qquad \text{(ii)}$$

We note that

$$b * (a * c) = b * e, \quad \text{by (ii)}$$
$$= b, \text{ because } e \text{ is identity} \qquad \text{(iii)}$$

and

$$(b * a) * c = e * c, \quad \text{by (i)}$$
$$= c, \text{ because } e \text{ is identity.} \qquad \text{(iv)}$$

But associativity of binary operation $*$ implies

$$b * (a * c) = (b * a) * c.$$

Hence, from (iii) and (iv) it follows that $b=c$, proving that the inverse, if exists, of every element in a monoid is unique.

5.4 HOMOMORPHISM OF SEMIGROUPS

We discuss now a method for comparing the algebraic structures of the two semigroups.

Definition 5.19

Let $(S, *)$ and $(T, *')$ be two semigroups. A function $f: S \to T$ is called a **semigroup homomorphism** if

$$f(a * b) = f(a) *' f(b)$$

for all $a, b \in S$.

If, in addition, f is also onto, we say that T is a **homomorphic image** of S.

Definition 5.20

Let $(S, *)$ and $(T, *')$ be two semigroups. If $f: S \to T$ is both one-one and onto in addition to being a homomorphism, then f is called an **isomorphism** from $(S, *)$ onto $(T, *')$.

Definition 5.21

A homomorphism f from $(S, *)$ to $(T, *')$ is called a **monomorphism** if f as a map is injective (one-one).

Definition 5.22

A homomorphism f from $(S, *)$ to $(T, *')$ is called an **epimorphism** if f as a map is surjective (onto).

Thus, we may define isomorophism between two semigroups $(S, *)$ and $(T, *')$ in the following way.

Definition 5.23

Let $(S, *)$ and $(T, *')$ be two semigroups. Then a homomorphism f: $(S, *) \to (T, *')$ is called an **isomorphism** if it is both monomorphism and epimorphism.

Thus, we have,

Definition 5.24

Let $(S, *)$ and $(T, *')$ be two semigroups. Then a mapping $f: S \to T$ is called an **isomorphism** if

(i) $f(a * b) = f(a) *' f(b)$ for all $a, b \in S$ (**semigroup homomorphism**)

(ii) f as a map is bijective.

Definition 5.25

Let $(S, *)$ and $(T, *')$ be two semigroups. If $f: S \to T$ is an isomorphism, then the semigroups $(S, *)$ and $(T, *)$ are called **isomorphic**. In such a case $(T, *')$ is called isomorphic image of $(S, *)$.

EXAMPLE 5.18

Let $F(A)$ be the free semigroup of a set A, and let $\mathbf{Z}$ be the semigroup of integers under addition. Let

$$f: F(A) \to \mathbf{Z}$$

be defined by

$$f(w) = l(w),\ w \in F(A).$$

We note that, if $u, v \in F(A)$, then

$$f(uv) = l(uv) = l(u) + l(v) = f(u) + f(v).$$

Hence f is a **homomorphism**. Here, the operation in $F(A)$ is written multiplicatively, whereas the operation in $\mathbf{Z}$ is addition.

EXAMPLE 5.19

Let $\mathbf{Z}$ be the set of integers and T be the set of all even integers. Then $(\mathbf{Z}, +)$ and $(T, +)$ are semigroups. Let

$$f: \mathbf{Z} \to T$$

be defined by

$$f(a) = 2a, \quad a \in \mathbf{Z}.$$

We note that

(i) $f(a+b) = 2(a+b) = 2a + 2b = f(a) + f(b)$.

Thus, f is a homomorphism.

(ii) $f(a) = f(b) \Rightarrow 2a = 2b$
$\Rightarrow a = b$.

Hence f is one-one, that is, f is monomorphism.

(iii) Let b be an even integer. Then $a = \frac{b}{2} \in \mathbf{Z}$ and

$$f(a) = f\left(\frac{b}{2}\right) = 2\left(\frac{b}{2}\right) = b.$$

Thus to every $b \in T$, there is an $a \in \mathbf{Z}$ such that $f(a) = b$. Therefore, f is onto, i.e., f is epimorphism. Hence f is an isomorphism.

EXAMPLE 5.20

Let $A = \{0, 1\}$ and let $F(A)$ be the free semigroup on A. Let the binary operation $+$ on A be defined by the table

+	0	1
0	0	1
1	1	0

Then $(F(A), \cdot)$ and $(A, +)$ are semigroups, where $\cdot$ denotes catenation operation.

Define

$$f: F(A) \to A$$

by

$$f(u) = \begin{cases} 1 & \text{if } u \text{ has an odd number of 1's} \\ 0 & \text{if } u \text{ has an even number of 1's} \end{cases}$$

Then, if $u, v \in F(A)$, we have

$$f(u \cdot v) = f(u) + f(v).$$

Thus, f is a homomorphism. Further, we note that

$$f(0)=0 \text{ and } f(1)=1,$$

that is, to each element a of A, there is an element u in $F(A)$ such that $f(u)=a$. Hence, f is onto. But the condition $f(u)=f(v) \Rightarrow u=v$ is not satisfied showing that f is not one-one. Hence f is **epimorphism**.

EXAMPLE 5.21

Let $\mathbf{Z}_4$ and $\mathbf{Z}_{10}$ denote, respectively, the integers modulo 4 under addition and integer modulo 10 under multiplication. Their addition and multiplication tables are then

+	0	1	2	3
0	0	1	2	3
1	1	2	3	0
2	2	3	0	1
3	3	0	1	2

$(\mathbf{Z}_4, +)$

·	1	3	7	9
1	1	3	7	9
3	3	9	1	7
7	7	1	9	3
9	9	7	3	1

$(S, \cdot)$ in $\mathbf{Z}_{10}$

where $S=\{1, 3, 7, 9\}$. Let

$$f: \mathbf{Z}_4 \to S$$

be defined by

$$f(0)=1, f(1)=3, f(2)=9, f(3)=7.$$

Then for any $a, b \in \mathbf{Z}_4$, we have

$$f(a+b)=f(a) \cdot f(b).$$

For example, let $a=1$, $b=3$, then

$$f(a+b)=f(1+3)=f(0)=1=3 \cdot 7 \text{ in } \mathbf{Z}_{10}=f(1) \cdot f(3).$$

Also f is bijective. Hence $\mathbf{Z}_4$ is isomorphic to S.

Theorem 5.3

Let $(S, *)$ and $(T, *')$ be monoids with identities e and e', respectively. Let $f: S \to T$ be a homomorphism from $(S\ *)$ onto $(T, *')$. Then $f(e)=e'$.

Proof. Let b be any element of T. Since f is surjective, there is an element $a \in S$ such that $f(a)=b$. Since e is identity of S, we have

$$a * e = a = e * a \tag{i}$$

and so

$$\begin{aligned} b &= f(a) = f(a * e), \quad \text{by (i)} \\ &= f(a) *' f(e), \text{ because } f \text{ is homomorphism} \\ &= b *' f(e). \end{aligned}$$

Also,

$$\begin{aligned} b &= f(a) = f(e * a) \\ &= f(e) *' f(a) = f(e) *' b. \end{aligned}$$

Hence,

$$b *' f(e) = f(e) *' b = b$$

and so $f(e)$ is identity for T. Thus, $f(e)=e'$.

Theorem 5.4

If f is a homomorphism from a commutative semigroup $(S, *)$ onto a semigroup $(T, *')$, then $(T, *')$ is also commutative, **that is, homomorphic image of an abelian (commutative) semigroup is abelian.**

Proof. Let $t_1, t_2 \in T$. Since f is onto, there exist $s_1, s_2 \in S$ such that

$$f(s_1) = t_1 \text{ and } f(s_2) = t_2.$$

Then,

$$\begin{aligned} t_1 *' t_2 &= f(s_1) *' f(s_2) \\ &= f(s_1 * s_2), \text{ since } f \text{ is homomorphism} \\ &= f(s_2 * s_1), \text{ since } S \text{ is abelian} \\ &= f(s_2) *' f(s_1), \text{ since } f \text{ is homomorphism} \\ &= t_2 *' t_1. \end{aligned}$$

Hence $(T, *')$ is abelian.

Remark 5.1 The converse of the above theorem is not true.

Theorem 5.5

Let $f: (S, *) \to (T, *')$ be semigroup homomorphism. If S' is a sub-semigroup of $(S, *)$, then the image of S' under f is a subsemigroup of $(T, *')$.

Proof. Let $f(S')$ be the image of S' under f and let t_1, t_2 be in $f(S')$. Then there are s_1 and s_2 in S' such that

$$t_1 = f(s_1) \text{ and } t_2 = f(s_2).$$

We claim that $f(S')$ is closed under the binary operation $*'$. It is sufficient to show that $t_1 *' t_2 \in f(S')$. We have, in this direction,

$$\begin{aligned} t_1 *' t_2 &= f(s_1) *' f(s_2) \\ &= f(s_1 * s_2), \text{ because } f \text{ is homomorphism.} \end{aligned}$$

Now since S' is a semigroup and $s_1, s_2 \in S'$, we have $s_1 * s_2 \in S'$(due to closeness of the operation $*$). Hence $f(s_1 * s_2) \in f(S')$. It follows, therefore, that $t_1 *' t_2 \in f(S')$. Further, since the associativity hold in T, it also holds in $f(S')$. Hence $f(S')$ is a sub-semigroup of $(T, *')$.

Theorem 5.6

The intersection of two sub-semigroups of a semigroup $(S, *)$ is subsemigroup of $(S, *)$.

Proof. Let $(S_1, *)$ and $(S_2, *)$ be two sub-semigroups of the semigroup $(S, *)$. Let $a \in S_1 \cap S_2$ and $b \in S_1 \cap S_2$. Then,

$$a \in S_1 \cap S_2 \Rightarrow a \in S_1 \text{ and } a \in S_2,$$
$$b \in S_1 \cap S_2 \Rightarrow b \in S_1 \text{ and } b \in S_2.$$

Since S_1 is a sub-semigroup, therefore, $a, b \in S_1$ implies $a * b \in S_1$. Similarly, since S_2 is a sub-semigroup, $a, b \in S_2$ implies $a * b \in S_2$. Hence,

$$a * b \in S_1 \cap S_2.$$

Hence, $S_1 \cap S_2$ is closed under the operation $*$. Further associativity in S_1 and S_2 implies the associativity of $S_1 \cap S_2$ since $S_1 \cap S_2 \subseteq S_1$ and $S_1 \cap S_2 \subseteq S_2$. Hence, $S_1 \cap S_2$ is a sub-semigroup of $(S, *)$.

Corollary 5.1

Intersection of two submonoids of a monoid $(S, *)$ is a semimonoid of $(S, *)$.

(Proof follows the same line as that in Theorem 5.6).

Remark 5.2 Union of two sub-semigroups of a semigroup $(S, *)$ need not be a sub-semigroup of $(S, *)$.

For example,

$$(S_1, *) = \{0, \pm 2, \pm 4, \pm 6, \pm, \ldots\}$$

and

$$(S_2, *) = \{0, \pm 3, \pm 6, \pm 9, \pm, \ldots\}$$

are sub-semigroups of the semigroup $(\mathbf{Z}, +)$ of integers. But,

$$S_1 \cup S_2 = \{0, \pm 2, \pm 3, \pm 4, \pm 6, \pm \ldots\}$$

is not a sub-semigroup of $(\mathbf{Z}, +)$, because

$$2 \in S_1 \cup S_2,\ 3 \in S_1 \cup S_2,$$

but, $2+3=5 \notin S_1 \cup S_2$ showing that $S_1 \cup S_2$ is not closed under addition.

5.5 QUOTIENT STRUCTURES

Definition 5.26

An equivalence relation R on a semigroup $(S, *)$ is called a **congruence relation** if a R a' and b R b' imply $(a * b)$ R $(a' * b')$.

EXAMPLE 5.22

Let (**Z**, +) be the semigroup of integers. Consider the relation R defined on **Z** by

$$a \text{ R } b \text{ if and only if } a \equiv b \pmod{m}.$$

We know that $a \equiv b \pmod{m}$ if m divides $a-b$. We note that

(i) For any integer a, we have $a \equiv a \pmod{m}$, i.e., a R a.
(ii) If a R b, then $a \equiv b \pmod{m} \Rightarrow m \mid (a-b) \Rightarrow m \mid (b-a)$ and so $b \equiv a \pmod{m}$ which means b R a.
(iii) If a R b and b R c, then

$$\begin{aligned} & a \equiv b \pmod{m} \text{ and } b \equiv c \pmod{m} \\ & \Rightarrow m \mid (a-b) \text{ and } m \mid (b-c) \\ & \Rightarrow m\ [(a-b)+(b-c)] \\ & \Rightarrow m \mid (a-c) \\ & \Rightarrow a \equiv c \pmod{m}, \text{ which means that } a \text{ R } c. \end{aligned}$$

Thus R is reflexive, symmetric and transitive and so is an **equivalence relation**. Further, if

$$a \equiv c \pmod{m} \text{ and } b \equiv d \pmod{m},$$

then

$$\begin{aligned} & m \mid (a-c) \text{ and } m \mid (b-d) \\ & \Rightarrow m \mid [(a-c)+(b-d)] \\ & \Rightarrow m \mid [(a+b)-(c+d)] \\ & \Rightarrow (a+b) \equiv (c+d) \pmod{m} \\ & \Rightarrow (a+b) \text{ R } (c+d). \end{aligned}$$

Hence R is a congruence relation.

EXAMPLE 5.23

Consider the semigroup (**Z**, ·), where · denotes ordinary multiplication. Let us again consider the relation R on **Z** defined by

$$a \text{ R } b \text{ if and only if } a \equiv b \pmod{m}.$$

This relation is an equivalence relation. Further, if $a \equiv c \pmod{m}$ and $b \equiv d \pmod{m}$, then

$$\begin{aligned} & m \mid (a-\text{c}) \quad \text{and} \quad m \mid (b-d) \\ & \Rightarrow m \mid b\,(a-c) \quad \text{and} \quad m \mid c\,(b-d) \\ & \Rightarrow n \mid (ab-bc) \quad \text{and} \quad m \mid (bc-cd) \\ & \Rightarrow m \mid [(ab-bc+bc-cd)] \\ & \Rightarrow m \mid (ab-cd) \\ & \Rightarrow ab \equiv cd \pmod{m}. \end{aligned}$$

Hence the relation is a congruence relation on (**Z**, ·).

EXAMPLE 5.24

Let $F(A)$ be the free semigroup on a set A. Define u R v if u and v have the same length. We note that

(i) u R u because u has same length as u.
(ii) If u R v, then u and v have same length $\Rightarrow v$ and u have same length $\Rightarrow v$ R u.

(iii) If u R v and v R w, then u and v have same length and also v and w have same length and so u and w have same length, that is, u R w.

Hence R is an equivalence relation. Further, let u R v and u' R v'. Then,

$$l(u)=l(v) \text{ and } l(u')=l(v').$$

Then

$$l(uu')=l(vv')=m+n,$$

that is

$$l(uu')=l(vv')$$
$$\Rightarrow uu' \text{ R } vv'.$$

Hence, R is a congruence relation on $F=F(A)$.

EXAMPLE 5.25

Let (**Z**, +) be the semigroup of integers and let $f(x)=x^2-x-2$. Let R be a relation defined on **Z** by

$$a \text{ R } b \text{ if and only if } f(a)=f(b).$$

It can be shown that R is an equivalence relation. Further we note that

$$f(-1)=f(2)=0 \text{ and so } -1 \text{ R } 2,$$
$$f(-2)=f(3)=4 \text{ and so } -2 \text{ R } 3.$$

But

$$f(-3)=10 \text{ and } f(5)=18,$$

and so

$$-3 \not{R} \; 5.$$

Hence R is not a congruence relation.

5.6 EQUIVALENCE CLASSES

If R is an equivalence relation on the semigroup $(S, *)$, it will partition S into equivalence classes. Let $[a]$ be the equivalence class containing a in S and let S/R denote the set of all equivalence classes, where R is congruence relation.

We define an operation ✡ on the equivalence classes S/R by

$$[a] ✡ [b]=[a * b], \qquad a, b \in S$$

that is, ✡: $S/\text{R} \times S/\text{R} \rightarrow S/\text{R}$ is defined by

$$✡([a], [b])=[a] ✡ [b]=[a * b].$$

Then we have the following theorem.

Theorem 5.7

Let R be a congruence relation on the semigroup (S, $*$). Then ✡: $S/\text{R} \times S/\text{R} \rightarrow S/\text{R}$ defined by

$$✡([a], [b])=[a] ✡ [b]=[a * b], \; a, b \in S$$

is a binary operation on S/R and (S/R, ✡) is a semigroup.

Proof. Suppose that ([a], [b])=([a'], [b']). Then a R a' and b R b'. Since R is congruence relation, this implies $a * b$ R $a' * b'$. Thus. [$a * b$]=[$a' * b'$], that is, ✡ is a well-defined function. Hence ✡ is a **binary operation** on S/R.

Further, we note that

$$\begin{aligned}[a] \text{✡} ([b] \text{✡} [c]) &= [a] \text{✡} [b * c] \text{ (by definition of ✡)}\\ &= [a * (b * c)] \text{ (by definition of ✡)}\\ &= [(a * b) * c] \text{ (associativity of } * \text{ in } S)\\ &= [a * b] \text{✡} [c] \text{ (by definition of ✡)}\\ &= ([a] \text{✡} [b]) \text{✡} [c] \text{ (by definition of ✡).}\end{aligned}$$

Hence ✡ is an associative operation. This implies that (S/R, ✡) is a semigroup.

The operation ✡ is called **quotient binary relation** on S/R constructed from the given binary relation $*$ on S by the congruence relation R.

The semigroup (S/R, ✡) **is called Quotient Semigroup** or **Factor Semigroup** or **the Quotient of** S **by** R.

Theorem 5.8

Let R be the congruence relation on the monoid (S, $*$), then (S/R, ✡) is a monoid.

Proof. We have shown above that (S/R, ✡) is a semigroup. Further, if e is an identity element in (S, $*$), then [e] is the identity in (S/R, ✡). Thus, (S/R, ✡) is semigroup having identity element [e] and so is a monoid.

EXAMPLE 5.26

Let (**Z**, +) be semigroup of integers. We have already shown that the relation R defined on **Z** by

$$a \text{ R } b \text{ if and only if } a \equiv b \pmod{m}$$

is a congruence relation.

Let us take, for example, $m=5$. Then we note that

$$\begin{aligned}[0]&=\{\ldots, -10, -5, 0, 5, 10, 15, \ldots\}=[5]=[10]=\ldots\\ [1]&=\{\ldots, -14, -9, -4, 1, 6, 11, 16, \ldots\}=[6]=[11]=\ldots\\ [2]&=\{\ldots, -13, -8, -3, 2, 7, 12, 17, \ldots\}=[7]=[12]=\ldots\\ [3]&=\{\ldots, -12, -7, -2, 3, 8, 13, 18, \ldots\}=[8]=[13]=\ldots\\ [4]&=\{\ldots, -11, -6, -1, 4, 9, 14, 19, \ldots\}=[9]=[14]=\ldots\end{aligned}$$

are the equivalence classes that form the quotient set **Z**/≡(mod 5).

We denote the quotient set Z/ ≡ (mod m) by $\mathbf{Z}_m$ or by Z/mZ. Also, by the above theorem **Z**/≡(mod m) or $\mathbf{Z}_m$ is a monoid under operation ⊕ with identity [0].

We note that for addition operation ⊕, we have

$$[a] \oplus [b] = [a+b] = [r],$$

where r is the remainder when $a+b$ is divided by m.

Thus, for example, in $\mathbf{Z}_5$, we have

$$\begin{aligned}&[0] \oplus [1] = [0+1] = [1],\\ &[1] \oplus [2] = [1+2] = [3],\\ &[1] \oplus [3] = [1+3] = [4],\end{aligned}$$

$$[1] \oplus [4] = [1+4] = [5] = [0],$$
$$[2] \oplus [3] = [2+3] = [5] = [0],$$
$$[2] \oplus [4] = [2+4] = [6] = [1].$$

Thus, the addition table for the semigroup $\mathbf{Z}_5$ with operation $\oplus$ becomes

$\oplus$	[0]	[1]	[2]	[3]	[4]
[0]	[0]	[1]	[2]	[3]	[4]
[1]	[1]	[2]	[3]	[4]	[0]
[2]	[2]	[3]	[4]	[0]	[1]
[3]	[3]	[4]	[0]	[1]	[2]
[4]	[4]	[0]	[1]	[2]	[3]

Remark 5.3 It is clear that $\mathbf{Z}_m$ has m equivalence classes

$$[0], [1], [2], \ldots, [m-1].$$

Further, $\mathbf{Z}/m\mathbf{Z}$ is represented by

$$\begin{aligned} \mathbf{Z}/m\mathbf{Z} &= \{n + m\mathbf{Z}, n \in \mathbf{Z}\} \\ &= \{\ 0 + m\mathbf{Z},\ 1 + m\mathbf{Z},\ 2 + m\mathbf{Z},\ \ldots\ (m-1) + m\mathbf{Z}\ \}. \end{aligned}$$

It has m elements which correspond to ([0], [1], …, [m–1]).

Thus, it follows that there is no essential difference between $\mathbf{Z}_m$ and $\mathbf{Z}/m\mathbf{Z}$ and so, they can be used interchangeably.

The next theorem shows that to each element a in a semigroup, we can assign its equivalence class $[a]$.

Theorem 5.9

Let R be a congruence relation on a semigroup $(S, *)$ and let $(S/\text{R}, ✡)$ be the corresponding quotient semigroup. Then the mapping ϕ: $S \to S/\text{R}$ (called the **natural mapping**) defined by $\phi(a) = [a]$ is an **onto homomorphism**, known as **Natural homomorphism**.

Proof. According to definition of ϕ, to each $[a]$ in S/R, there is $a \in S$ such that $\phi(a) = [a]$. Hence ϕ is surjective. Now let $a, b \in S$. Then,

$$\phi(a * b) = [a * b] = [a] ✡ [b] = \phi(a) ✡ \phi(b).$$

Hence ϕ is homomorphism onto.

Theorem 5.10 (Fundamental Theorem of Semigroup Homomorphism)

Let $f: S \to T$ be a homomorphism of the semigroup $(S, *)$ onto the semigroup $(T, *')$. Let R be the relation on S defined by

$$a \text{ R } b \text{ if } f(a) = f(b) \text{ for } a, b \in S.$$

Then,

(i) R is a congruence relation on S,
(ii) $(S/\text{R}, ✡)$ is isomorphic to $(T, *')$.

(If f is not onto, then (ii) shall be "S/R is isomorphic to $f(S)$".

Proof. First we show that R is an equivalence relation. We note that

(i) Since $f(a)=f(a)$, we have a R a.
(ii) If a R b, then $f(a)=f(b)$ or $f(b)=f(a)$ and hence b R a.
(iii) If a R b and b R c, then

$$f(a)=f(b) \quad \text{and} \quad f(b)=f(c)$$

and hence

$$f(a)=f(c)$$

and so a R c. Thus, the relation R is reflexive, symmetric and transitive and so an equivalence relation. Suppose now that

$$a \mathrm{~R~} a' \quad \text{and} \quad b \mathrm{~R~} b'.$$

Then,

$$f(a)=f(a') \quad \text{and} \quad f(b)=f(b').$$

Since f is homomorphism,

$$f(a * b)=f(a) *' f(b)=f(a') *' f(b')=f(a' * b').$$

Hence,

$$(a * b) \mathrm{~R~} (a' * b')$$

and so R is a congruence relation.

Define ψ: S/R → T by ψ ([a])=f (a). We claim that ψ is well defined. Suppose [a]=[b]. The mapping ψ will be well defined if $f(a)=f(b)$. Now [a]=[b] implies a R b, that is, $f(a)=f(b)$. Hence ψ is a function (well defined).

Further, if [a], [b] ∈ S/R, then

$$\begin{aligned} \psi([a] \,✡\, [b]) &= \psi([a * b]),\ a, b \in S \\ &= f(a * b)=f(a) *' f(b), \text{ because } f \text{ is homomorphism} \\ &= \psi[a] *' \psi[b]. \end{aligned}$$

So ψ is semigroup homomorphism. Also,

$$\begin{aligned} \psi([a]=\psi([b]) &\Rightarrow f(a)=f(b) \\ &\Rightarrow a \mathrm{~R~} b \\ &\Rightarrow [a]=[b] \end{aligned}$$

and so ψ is one-one.

Thus, ψ, as a map, is bijective and homomorphism. Hence ψ is an isomorphism and S/R ≅ **T**

Remark 5.4 We have proved that the mapping ϕ: S → S/R is natural homomorphism. Also, we proved that the mapping ψ: S/R → T is an isomorphism. Thus, diagram of the situation becomes

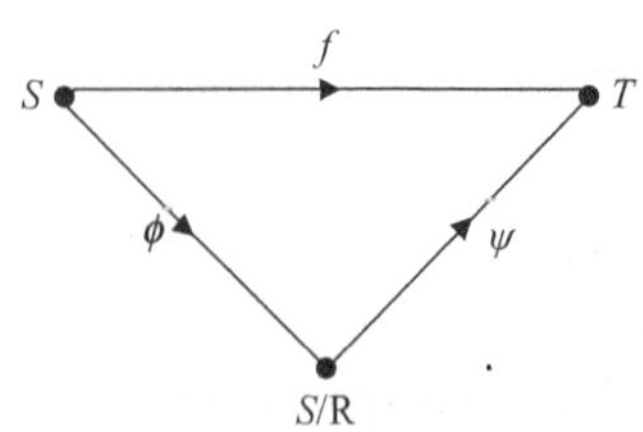

Also, we note that

$$(\psi \circ \phi)(a)=\psi(\phi(a))=\psi([a])=f(a) \text{ for all } a \in S.$$

Hence,

$$\psi \circ \phi=f.$$

5.7 DIRECT PRODUCT OF SEMIGROUPS

Let $(S, *)$ and $(T, *')$ be two semigroups. Consider the Cartesian product $S\times T$. Define a binary operation $*''$ on $S\times T$ by

$$(s_1, t_1) *'' (s_2, t_2)=(s_1* s_2, t_1*' t_2).$$

In what follows, we prove that $(S\times T, *'')$ is a semigroup.

Theorem 5.11

Let $(S, *)$ and $(T, *')$ be semigroups. Then $(S\times T, *'')$ is a semigroup under the binary operation $*''$ defined by

$$(s_1, t_1) *'' (s_2, t_2)=(s_1* s_2, t_1*' t_2).$$

Proof. If (s_1, t_1), (s_2, t_2) and $(s_3, t_3) \in S\times T$, then

$$\begin{aligned}[(s_1, t_1) *'' (s_2, t_2)] *'' (s_3, t_3) &=(s_1* s_2, t_1*' t_2) *'' (s_3, t_3)\\ &= ((s_1* s_2) * s_3, (t_1*' t_2) *' t_3)\\ &= (s_1* (s_2* s_3), t_1*' (t_2*' t_3))\\ &= (s_1, t_1) *'' (s_2* s_3, t_2*' t_3)\\ &= (s_1, t_1) *'' [(s_2, t_2) *'' (s_3, t_3)].\end{aligned}$$

Hence $*''$ is associative and so $(S\times T, *'')$ is a semigroup.

Corollary 5.2

If $(S, *)$ and $(T, *')$ are monoids, then $(S\times T, *'')$ is also a monoid.

Proof. We have proved above that $(S\times T, *'')$ is a semigroup. We further note that if e_S is identity of $(S, *)$ and e_T is identity of $(T, *')$, then for $(s_1, t_1) \in S\times T$, we have

$$(e_S, e_T) *'' (s_1, t_1)=(e_S * s_1, e_T *' t_1)=(s_1, t_1)$$

and

$$(s_1, t_1) *'' (e_S, e_T)=(s_1* e_S, t_1*' e_T)=(s_1, t_1).$$

Thus,

$$(s_1, t_1) *'' (e_S, e_T)=(e_S, e_T) *'' (s_1, t_1)=(s_1, t_1)$$

showing that (e_S, e_T) is identity element of $(S\times T, *'')$, that is, $(S\times T, *'')$ is a semigroup with identity (e_S, e_T) and hence is a monoid.

5.8 GROUPS

Definition 5.27

A non-empty set G is said to be a **group**, if in G, there is defined a binary operation $f: G \times G \to G$, denoted by $*$, satisfying the following properties:

(i) If $a, b, c, \in G$, then

$$a * (b * c) = (a * b) * c \text{ (Associative Law).}$$

(ii) There exists an $e \in G$, called identity element, such that

$$a * e = e * a = a \text{ for all } a \in G \text{ (existence of identity element).}$$

(iii) For every element $a \in G$, there exists an element $b \in G$, called inverse of a, such that

$$a * b = b * a = e \text{ (existence of inverse element).}$$

Thus, an algebraic system $(G, *)$, where $*$ is a binary operation, is called a group if

(i) $*$ is an associative binary operation,
(ii) There exists an identity element e in G,
(iii) Every element in G has an inverse.

In term of a monoid, we can define a group G as follows:

Definition 5.28

A monoid $(G, *)$, in which every element has an inverse, is called a **group**.
Since a group is a monoid, the following results hold good.

Theorem 5.12

The identity element of a group is unique.

Theorem 5.13

Inverse of every element in a group is unique.

Definition 5.29

Let G be a group and let $a \in G$. Then an element $b \in G$ is called **left inverse** of a if $b * a = e$. Similarly, b will be called a **right inverse** of a if $a * b = e$.

Theorem 5.14

In a group G, a left inverse of an element of G is also a right inverse of that element.

Proof. Let b be a left inverse of $a \in G$ and let c be a left inverse of b. Then

$$b * a = e \quad (1)$$

and

$$c * b = e. \quad (2)$$

We, therefore, have

$$(b * a) * b = e * b \text{ using (1)}$$
$$= b \quad (3)$$

and so

$$c * (b * a) * b = c * b \quad \text{using (3)}$$
$$= e \quad \text{using (2).} \tag{4}$$

Also,

$$c * (b * a) * b = ((c * b) * a) * b \quad \text{(by associativity in } G)$$
$$= (e * a) * b \quad \text{using (2)}$$
$$= a * b. \tag{5}$$

It follows from (4) and (5) that $a * b = e$, showing that b is the right inverse of a.

Notation. In what follows, the inverse of an element a will be denoted by a^{-1}. Also we shall write $a * b$ as $a\,b$.

Theorem 5.15

Let G be a group. If $a, b \in G$, then $(a\,b)^{-1} = b^{-1}\,a^{-1}$.

(Thus, inverse of the product of two elements in a group is the product of the inverses of the elements taken in the reverse order.)

Proof. We note that,

$$(a\,b)\,(b^{-1}\,a^{-1}) = a\,(b\,b^{-1})\,a^{-1} \quad \text{by associativity}$$
$$= a\,e\,a^{-1} \quad \text{since } b\,b^{-1} = e$$
$$= a\,a^{-1} \quad \text{since } a\,e = a$$
$$= e \quad \text{since } a\,a^{-1} = e$$

and therefore by definition of inverse of an element, it follows that

$$(a\,b)^{-1} = b^{-1}\,a^{-1}.$$

Definition 5.30

Let $f: E \times E \to E$ defined by $f(m, n) = m * n$ be a binary operation in the set E. Then an element $a \in E$ is said to be **left cancellative** if

$$a\,x = a\,y \Rightarrow x = y \quad \forall\, x, y \in E.$$

Similarly, if

$$x\,a = y\,a \Rightarrow x = y \quad \forall\, x, y \in E,$$

then $a \in E$ is called **right cancellative**.

If an element is left as well as right cancellative, then it is called **cancellative** or **regular element**. If every element in a set E is regular, then E is said to obey **cancellation laws**.

Theorem 5.16

Let G be a group and $a, b, c \in G$. Then,

$$a\,b = a\,c \Rightarrow b = c,$$
$$b\,a = c\,a \Rightarrow b = c.$$

Proof. Since G is group and $a \in G$, there exists an element a' such that $a\,a' = a'\,a = e$. Pre-multiplication of $a\,b = a\,c$ by a' gives

$$\begin{aligned} & a'\,(a\,b)=a'\,(a\,c) \\ \Rightarrow\quad & (a'\,a)\,b=(a'\,a)\,c \quad \text{(by associativity)} \\ \Rightarrow\quad & e\,b=e\,c \quad \text{since } a'\,a=e=a'a \\ \Rightarrow\quad & b=c \quad \text{since } e\,b=b \text{ and } e\,c=c. \end{aligned}$$

Similarly, post-multiplication of $b\,a=c\,a$ by a' yields $b=c$.

Definition 5.31

The number of elements in a group G is called the **order of the group** G and is denoted by $O(G)$. If this number is finite, then the group is called a **finite group**.

Definition 5.32

A group G is said to be **Abelian** or **commutative** if for every $a, b \in G$, we have $a * b=b * a$.

EXAMPLE 5.27

Let **Z** be the set of integers and we define a binary operation in **Z** by

$$f: \mathbf{Z}\times\mathbf{Z}\to\mathbf{Z},$$
$$f(a, b)=a+b,\ a, b \in \mathbf{Z}.$$

Then **Z** is an additive infinite abelian group. Infact,

(i) $a+(b+c)=(a+b)+c \quad \forall\ a, b, c \in \mathbf{Z}$ (associativity).

(ii) There exists an element 0 in **Z** such that

$$a+0=0+a=a \quad \forall\ a \in \mathbf{Z} \quad \text{(existence of addition identity).}$$

(iii) For every $a \in \mathbf{Z}$, there exists an element $-a \in \mathbf{Z}$ such that

$$a+(-a)=(-a)+a=0 \quad \text{(existence of additive inverse in } \mathbf{Z}\text{).}$$

(iv) For every $a, b \in \mathbf{Z}$, we have

$$a+b=b+a.$$

EXAMPLE 5.28

If in **Z**, we take binary operation f: $\mathbf{Z}\to\mathbf{Z}$ defined by $f(a, b)=a\,b,\ a, b \in \mathbf{Z}$, then **Z cannot be a multiplicative group** since only elements which have inverses in **Z** are 1 and −1. The multiplicative inverses of all other integers are not in **Z**. But, as we have seen, it is a monoid.

EXAMPLE 5.29

The set of rational numbers **Q** is an additive abelian group.

EXAMPLE 5.30

The set of rationals **Q** excluding zero is an abelian multiplicative group. The identity element being 1 and the inverse of $a \in \mathbf{Q}$ being $\frac{1}{a}$.

EXAMPLE 5.31

If we define binary composition f: $V\times V\to V$ by $f\left(\vec{a},\vec{b}\right)=\vec{a}+\vec{b},\ \vec{a},\ \vec{b}\in V$ in the set of vectors V, then V is an additive abelian group. In fact,

(i) $\vec{a}+(\vec{b}+\vec{c})=(\vec{a}+\vec{b})+\vec{c} \quad \forall\ \vec{a},\ \vec{b},\ \vec{c} \in V$ (associativity).

(ii) There is zero vector $\vec{0}$ in V such that

$$\vec{a}+\vec{0}=\vec{0}+\vec{a}=\vec{a} \quad \forall\ \vec{a} \in V$$

(iii) For every $\vec{a} \in V$ there is an vector $-\vec{a} \in V$ such that

$$\vec{a}+(-\vec{a})=(-\vec{a})+\vec{a}=\vec{0}.$$

(iv) For every pair $\vec{a}$, $\vec{b} \in V$

$$\vec{a}+\vec{b}=\vec{b}+\vec{a}.$$

EXAMPLE 5.32

The set $\{-1, 1\}$ is a multiplicative abelian group.

EXAMPLE 5.33

The set M of $m \times n$ matrices is an additive abelian group. The set of all non-singular $n \times n$ matrices form a group under multiplication with unit matrix as the identity element and A^{-1} as the inverse of a matrix A. This group is not abelian.

EXAMPLE 5.34

If ω is the cube root of unity, then the set $\{1, \omega, \omega^2\}$ is a multiplicative group.

EXAMPLE 5.35

Set $\{0\}$ is an additive group.

EXAMPLE 5.36

Set of integers module n,

$$\mathbf{Z}/n\mathbf{Z}=\{m+n\mathbf{Z}: m \in \mathbf{Z}\}$$

is an additive abelian group.

If $m_1+n\mathbf{Z}$, $m_2+n\mathbf{Z} \in \mathbf{Z}/n\mathbf{Z}$, then define addition as

$$(m_1+n\mathbf{Z})+(m_2+n\mathbf{Z})=(m_1+m_2)+n\mathbf{Z}.$$

We observe that if $m_1+n\mathbf{Z}$, $m_2+n\mathbf{Z}$, $m_3+n\mathbf{Z} \in \mathbf{Z}/n\mathbf{Z}$, then

(i) **Associativity:**

$$\begin{aligned}
&[(m_1+n\mathbf{Z})+(m_2+n\mathbf{Z})]+(m_3+n\mathbf{Z})\\
&\quad=[(m_1+m_2)+n\mathbf{Z})]+(m_3+n\mathbf{Z})\\
&\quad=[(m_1+m_2)+m_3]+n\mathbf{Z}\\
&\quad=[(m_1+(m_2+m_3)]+n\mathbf{Z} \qquad \text{[by associative law in integers]}\\
&\quad=(m_1+n\mathbf{Z})+[(m_2+n\mathbf{Z})+(m_3+n\mathbf{Z})].
\end{aligned}$$

(ii) **Existence of identity element:**

$$\begin{aligned}
(m_1+n\mathbf{Z})+(0+n\mathbf{Z})&=(m_1+0)+n\mathbf{Z}\\
&=m_1+n\mathbf{Z}\\
&=(0+m_1)+n\mathbf{Z}\\
&=(0+n\mathbf{Z})+(m_1+n\mathbf{Z}).
\end{aligned}$$

Hence $n\mathbf{Z}$ is identity element for addition.

(iii) **Existence of inverse:** If $m_1+n\mathbf{Z} \in \mathbf{Z}/n\mathbf{Z}$, then

$$(m_1+n\mathbf{Z})+(-m_1+n\mathbf{Z})=(m_1-m_1)+n\mathbf{Z}=n\mathbf{Z}$$

and

$$(-m_1+n\mathbf{Z})\ (m_1+n\mathbf{Z})=(-m_1+m_1)+n\mathbf{Z}=n\mathbf{Z}.$$

Hence every element of $\mathbf{Z}/n\mathbf{Z}$ is invertible. **Thus, $\mathbf{Z}/n\mathbf{Z}$ is a group.** We call it **group of integers modulo n.** Also, if $m_1+n\mathbf{Z}, m_2+n\mathbf{Z} \in \mathbf{Z}/n\mathbf{Z}$, then

$$\begin{aligned}(m_1+n\mathbf{Z})+(m_2+n\mathbf{Z})&=(m_1+m_2)+n\mathbf{Z}\\ &=(m_2+m_1)+n\mathbf{Z} \quad \text{(by commutativity in integers)}\\ &=(m_2+n\mathbf{Z})+(m_1+n\mathbf{Z}).\end{aligned}$$

Hence, $\mathbf{Z}/n\mathbf{Z}$ is an additive abelian group.

EXAMPLE 5.37

Let $B=\{0, 1\}$ be a set of bits and let the binary operation $*$ be defined by

$*$	0	1
0	0	1
1	1	0

Then $*$ is associative, 0 acts as identity and each element is its own inverse.

Definition 5.33

Let a and b be the elements of a group G. If there is an element $x \in G$ such that $a\,x=b$, then x is called **right quotient** of b by a. Similarly, if $x \in G$ is such that $x\,a=b$, then x is called **left quotient** of b by a.

Theorem 5.17

If a, b are elements of a group G, then the equations $a\,x=b$ and $y\,a=b$ have unique solutions in G, that is, the right and left quotients of b by a are unique.

Proof. Since

$$\begin{aligned}a\,(a^{-1}b)&=(aa^{-1})\,b && \text{by associativity}\\ &=e\,b && \text{since } aa^{-1}=e\\ &=b && \text{property of identity } e,\end{aligned}$$

it follows that $a^{-1}\,b$ is a solution of the equation $a\,x=b$.

To prove the uniqueness of the solution, let x_1 and x_2 be two solutions of $ax=b$. Then,

$$ax_1=b, \quad ax_2=b.$$

Thus,

$$ax_1=ax_2,$$

which implies

$$x_1=x_2 \qquad \text{by cancellation law in } G.$$

Hence the right quotient of b by a is unique. The second part of the theorem can be proved similarly.

Theorem 5.18

A non-empty set G with a binary operation $f: G\times G\to G$ defined by $f(a, b)=ab$ is a group if and only if

(i) $a(bc)=(ab)c$ for all $a, b, c\in G$ (associativity)
(ii) For every pair $a, b\in G$, there exist elements x and y in G such that $ax=b$ and $ya=b$.

Proof

The conditions are necessary: Let G be a group then (i) follows from the definition of G while (ii) follows from the theorem proved above.

The conditions are sufficient: Suppose that (i) and (ii) hold in G. Then by (ii), two elements e_1 and e_2 in G can be found such that for $a\in G$,

$$ae_1=a, \tag{1}$$
$$e_2a=a. \tag{2}$$

Let $b\in G$ be an arbitrary element of G. Then by (ii) there are x and y such that $b=ya$ and $b=ax$. Using (i), (1) and (2), we have

$$be_1=(ya)\,e_1=y\,(ae_1)=ya=b,$$
$$e_2b=e_2\,(ax)=(e_2\,a)\,x=ax=b.$$

If we take e_2 and e_1 in place of b, we get

$$e_2e_1=e_2, \qquad e_2e_1=e_1$$

and so $e_1=e_2$. If we put $e_1=e_2=$e, then

$$b\,e=b \text{ and } eb=b \text{ for all } b\in G.$$

Hence e is an identity element of G.

If $a, e\in G$, then condition (ii) implies that there are two elements x and y in G such that

$$a\,x=e, \qquad y\,a=e. \tag{3}$$

We have proved above that e is an identity of G.

Hence,

$$\begin{aligned} a\,x=e &\Rightarrow y\,(a\,x)=y\,e && \text{(pre-multiplication by } y)\\ &\Rightarrow (y\,a)\,x=y && \text{(by associativity and property of } e)\\ &\Rightarrow ex=y && \text{(by (3))}\\ &\Rightarrow x=y. \end{aligned}$$

Hence (3) becomes $ax=e$, $xa=e$, that is, $ax=xa=e$ showing that x is inverse of a.

Thus, all the postulates for a group are satisfied and hence G is a group.

Theorem 5.19

A **finite set** G with a binary composition $f: G\times G\to G$ defined by $f(a, b)=a\,b$ is a group if and only if

(i) $a(bc)=(ab)c$ for all $a, b, c\in G$ (associativity),
(ii) $a\,x=a\,y\Rightarrow x=y$ (left cancellation law)
$x\,a=y\,a\Rightarrow x=y$ (right cancellation law).

Proof.

The conditions are necessary: Let G be a group. The condition (i) follows from the definition of a group. Also cancellation laws hold good in a group and so (ii) is satisfied.

The conditions are sufficient: Let $G=\{x_1, x_2, \ldots, x_n\}$. If $x_i \in G$, then by (ii), all the products $x_i x_1$, $x_i x_2, \ldots, x_i x_n$ are different. Hence, every element x_m in group G can be expressed as

$$x_m = x_i x_l, \qquad l \leq m \leq n,$$

where x_l is some suitably chosen element. Thus, first part of the condition (ii) of Theorem 5.18 is satisfied. Similarly the second part also holds good. Hence G is a group.

5.9 SUBGROUPS

Definition 5.34

Let $(A, *)$ be a group and B a subset of A. Then $(B, *)$ is said to be a subgroup of $(A, *)$ if $(B, *)$ is also a group under the binary operation $*$.

For $(B, *)$ to be a group,

(i) $*$ should be a closed binary operation.
(ii) $*$ should be associative operation.
(iii) The identity element for $(A, *)$ should also be the identity element of $(B, *)$.
(iv) Every element in B should have its inverse in B.

Theorem 5.20

If H is a subgroup of a group G, then the identity element of G is also the identity element of H.

Proof. Let $h \in H$ and e_H be the identity element of H. Then,

$$e_H \cdot h = h$$
$$\Rightarrow (e_H h)\, h^{-1} = h\, h^{-1}$$
$$\Rightarrow e_H (h\, h^{-1}) = h\, h^{-1}$$
$$\Rightarrow e_H e = e \qquad (\because h \in H \Rightarrow h \in G \text{ and so } h\, h^{-1} = e)$$
$$\Rightarrow e_H = e \qquad (\because e_H \in H \Rightarrow e_H \in G \text{ and so } e_H e = e_H).$$

Hence identity of G is identity of subgroup H.

Theorem 5.21

A non-empty subset H of a group G is a subgroup of G if and only if

(i) $a, b \in H \Rightarrow a\, b \in$ H
(ii) $a \in H \Rightarrow a^{-1} \in H$

Proof. Let H be a subgroup of G, then

1. Since the binary operation is closed, it follows that a, $b \in H \Rightarrow a\, b \in H$.
2. Since H is a subgroup, each element $a \in H$ has inverse a^{-1} in H.

Conversely, suppose that H is a subset of a group G and let conditions (i) and (ii) be satisfied. Then

3. Condition (i) shows that binary operation in G is also a binary operation in H.
4. Since associative law holds good in G and H is a subset of G, therefore associative law also holds good for elements of H.
5. If $a \in H$, then (ii) implies that $a^{-1} \in H$. Thus, each element of H has a unique inverse.
6. Also, by (5), if $a \in$ H, then $a^{-1} \in H$ and then, by (i), $a\, a^{-1} \in H$. Therefore,

$$e = a\, a^{-1} \in H.$$

Thus, H has identity element also. Hence, H is a subgroup of G.

Theorem 5.22

A non-empty subset H of a group G is a subgroup of G if and only if

$$a, b \in H \Rightarrow a\, b^{-1} \in H.$$

Proof. Suppose first that H is a subgroup of G. Then, by Theorem 5.21, we must have

(i) $a, b \in H \Rightarrow a\, b \in H$,
(ii) $b \in H \Rightarrow b^{-1} \in H$.

Now by (ii) $b \in H \Rightarrow b^{-1} \in H$ and then by (i) $a \in H, b^{-1} \in H \Rightarrow a\, b^{-1} \in H$. Hence, $a, b \in H \Rightarrow a\, b^{-1} \in H$.

Conversely, let H be a subset of a group and let $a, b \in H \Rightarrow a\, b^{-1} \in H$ be given. Let $a \in H$. Then the given condition implies that $a\, a^{-1} \in H$, that is, $e \in H$. Thus, identity e is in H.

Let b be an arbitrary element of H. Then,

$$\begin{aligned} e \in H, b \in H &\Rightarrow e\, b^{-1} \in H \qquad \text{(using the given condition)} \\ &\Rightarrow b^{-1} \in H. \end{aligned}$$

Thus, each element of H is invertible.

Also, if $a \in H$, $b \in H$, then $b^{-1} \in H$. We have

$$\begin{aligned} a \in H, b^{-1} \in H &\Rightarrow a\, (b^{-1})^{-1} \in H \quad \text{(using given condition)} \\ &\Rightarrow a\, b \in H. \end{aligned}$$

Thus, we have shown that

(i) $a, b \in H \Rightarrow a\, b \in H$,
(ii) $b \in H \Rightarrow b^{-1} \in H$

and so conditions for a subset H to be a subgroup are satisfied. Hence, H is a subgroup of G.

Theorem 5.23

The intersection of any subgroups of a group G is again a subgroup of G.

Proof. Let H_1 and H_2 be two subgroups of a group G. They both have identity e in them. Hence, $H_1 \cap H_2$ is non-empty. Now, let $a \in H_1 \cap H_2$ and $b \in H_1 \cap H_2$. To prove that $H_1 \cap H_2$ is a subgroup of G it is sufficient to show that $a, b \in G$ imply that $a\, b^{-1} \in H_1 \cap H_2$.

We have

$$\begin{aligned} a \in H_1 \cap H_2 &\Rightarrow a \in H_1 \quad \text{and} \quad a \in H_2, \\ b \in H_1 \cap H_2 &\Rightarrow b \in H_1 \quad \text{and} \quad b \in H_2. \end{aligned}$$

Since H_1 is a subgroup,

$$a \in H_1, b \in H_1 \Rightarrow a\, b^{-1} \in H_1.$$

Since H_2 is a subgroup

$$a \in H_2, b \in H_2 \Rightarrow a\, b^{-1} \in H_2.$$

Hence $a\, b^{-1} \in H_1 \cap H_2$ and so $H_1 \cap H_2$ is a subgroup of G.

Note. Union of two subgroups of a group G need not be a subgroup of G.

For example,

$$\begin{aligned} H_1 &= \{0, \pm 2, \pm 4, \pm 6, + \ldots\}, \\ H_2 &= \{0, \pm 3, \pm 6, \pm 9, \pm \ldots\} \end{aligned}$$

are subgroups of the additive group $\mathbf{Z}$ of integers.

But,

$$H_1 \cup H_2 = 0, \pm 2, \pm 3, \pm 4, \pm 6, \ldots\}$$

is not a subgroup of **Z** because 2, 3, $\in H_1 \cup H_2$ and $2+3=5 \notin H_1 \cup H_2$.

Theorem 5.24

Let $(A, *)$ be a group and B be a subset of A. If B is a finite set, then $(B, *)$ is a subgroup of $(A, *)$ if $*$ is a closed operation on B.

Proof. Let a^m denote $\underbrace{a * a * a * \cdots * a}_{m \text{ times}}$ and let a be an element in B. If $*$ is a closed binary operation on B, the elements $a, a^2, a^3, \ldots$ are all in B. Since B is a finite set, according to pigeonhole principle, we must have $a^i = a^j$ for some i and j where $i<j$ which implies $a^i = a^i * a^{j-i}$. Thus, a^{j-i} acts as the identity of $(A, *)$ and it is included in the subset B.

On the other hand, if $j-i>1$, then

$$a^{j-i} = a * a^{j-i-1},$$

therefore a^{j-i-1} is the inverse of a ($\because a^{j-i}$ is identity element) and it is in the subset B. If $j-i=1$, that is, $j=i+1$, then $a^i = a^j$ gives

$$a^i = a^i * a. \tag{1}$$

Thus, in this case, a acts as identity element and is its own inverse. In fact if a is identity element of B, then it is identity element of $(A, *)$ also. Therefore, pre-multiplying (1) by $(a^i)^{-1}$ we get

$$(a^i)^{-1}(a^i) = (a^i)^{-1}(a^i) * a \quad \text{or} \quad a = a * a.$$

Hence, a is its own inverse. Thus, $(B, *)$ satisfies all the conditions to be a subgroup of $(A, *)$.

Definition 5.35

Let H_1 and H_2 be two subgroups of a group G. Then,

$$H_1 H_2 = \{x \in G: x = h_1 h_2, h_1 \in H_1, h_2 \in H_2\}$$

is called the **multiplication of H_1 and H_2**.

If H_1 is a singleton, say, $\{x\}$, then $H_1 H_2$ is denoted by $x H_2$. Similarly, if H_2 has only one element y, then $H_1 H_2$ is denoted by $H_1 y$.

Theorem 5.25

The product $H_1 H_2$ of two subgroups H_1 and H_2 of a group G is a subgroup of G is and only if $H_1 H_2 = H_2 H_1$.

Proof. Suppose first that $H_1 H_2$ is a subgroup of G. If $h_1 \in H_1$, $h_2 \in H_2$, then $h_1^{-1} \in H_1$, $h_2^{-1} \in H_2$. Therefore, by the definition of $H_1 H_2$, $h_1^{-1} h_2^{-1} \in H_1 H_2$. Then,

$$h_2 h_1 = (h_1^{-1} h_2^{-1})^{-1} \in H_1 H_2, \text{ since } H_1 H_2 \text{ is a subgroup.}$$

It follows that

$$H_2 H_1 \subseteq H_1 H_2. \tag{1}$$

Further, $H_1 H_2$ being a subgroup, $a \in H_1 H_2 \Rightarrow a^{-1} \in H_1 H_2$, and so

$$a^{-1} = h_1 h_2, \qquad h_1 \in H_1, h_2 \in H_2.$$

But then,

$$a=(a^{-1})^{-1}=(h_1 h_2)^{-1}=h_2^{-1} h_1^{-1} \in H_2 H_1.$$

Hence,

$$H_1 H_2 \subseteq H_2 H_1. \tag{2}$$

From (1) and (2), we have

$$H_1 H_2=H_2 H_1.$$

Conversely, suppose that $H_1 H_2=H_2 H_1$. We shall prove that $H_1 H_2$ is a subgroup of G. If $h_1 \in H_1$, $h_2 \in H_2$, then, by the definition of $H_1 H_2$, we can find two elements $h_1' \in H_1$ and $h_2' \in H_2$ such that $h_1 h_2=h_2' h_1'$. If $a=h_1 h_2 \in H_1 H_2$, $b=h_1' h_2' \in H_1 H_2$, then

$$a b=h_1 h_2 h_1' h_2'. \tag{3}$$

But $h_2 h_1' \in H_2 H_1 \Rightarrow h_2 h_1' \in H_1 H_2$ (since $H_1 H_2=H_2 H_1$)

$$\Rightarrow h_2 h_1'=h_1'' h_2'', \quad h_1'' \in H_1, h_2'' \in H_2.$$

Hence (3) reduces to

$$a b=h_1 h_1'' h_2'' h_2' \in H_1 H_2 \quad (\text{since } h_1 h_1'' \in H_1, h_2'' h_2' \in H_2).$$

Again, suppose that $a=h_1 h_2 \in H_1 H_2$. Then,

$$a^{-1}=(h_1 h_2)^{-1}=h_2^{-1} h_1^{-1} \in H_2 H_1=H_1 H_2.$$

Thus,

$$a \in H_1 H_2 \Rightarrow a^{-1} \in H_1 H_2.$$

We have proved therefore that

(i) $a, b \in H_1 H_2 \Rightarrow a b \in H_1 H_2$,
(ii) $a \in H_1 H_2 \Rightarrow a^{-1} \in H_1 H_2$.

Hence, $H_1 H_2$ is a subgroup of G.

Theorem 5.26

Let H be a subgroup of a group G. Then the relation

$\text{R}=\{(x, y): x, y \in G, \quad x^{-1} y \in H\}$

is an equivalence relation.

Proof.

(1) **Reflexivity:** Since H is a subgroup, identity $e \in H$. But,

$$e=x^{-1} x \quad \text{for all } x \in G.$$

Hence $(x, x) \in \text{R}$ for all $x \in G$.

(2) **Symmetry:** Suppose $(x, y) \in \text{R}$. Then, by the definition of R,

$$\begin{aligned} & x^{-1} y \in H \\ & \Rightarrow (x^{-1} y)^{-1} \in H, \text{ since } H \text{ is a subgroup} \\ & \Rightarrow y^{-1} x \in H \\ & \Rightarrow (y, x) \in R, \quad \text{by the definition of R.} \end{aligned}$$

Thus $(x, y) \in \text{R} \Rightarrow (y, x) \in \text{R}$.

(3) **Transitivity:** Let $(x, y), (y, z) \in R$. Therefore,

$$x^{-1} y \in H \quad \text{and} \quad y^{-1} z \in H.$$

Then,

$$x^{-1} z = (x^{-1} y)(y^{-1} z) \in H$$

and so $(x, z) \in R$.

Hence, the relation R is reflexive, symmetric and transitive and so is an equivalence relation. The relation R thus leads to a partition of G into equivalence classes. Let x be an element of G. Then,

$$\begin{aligned}[x] &= \{y: (x, y) \in R, \quad y \in G\} \\ &= \{y: x^{-1} y \in H\} \\ &= \{y: y \in x H\} \\ &= \{y: y = x h, h \in H\} = x H.\end{aligned}$$

These equivalence classes $x H, x \in G$ are called **left cosets of** H in G.

Similarly, we can show that the relation

$$R = \{(x, y): x y^{-1} \in H, x, y \in G\}$$

is an equivalence relation on G. The mutually disjoint equivalence classes of H in G are $H x, x \in G$. These equivalence classes are called **Right Cosets of H in** G.

Theorem 5.27

Let H be a subgroup of a group G. Then, H is cardinally equivalent to a left coset of H in G.

Proof. Let $x H, x \in G$ be an arbitrary left coset of H in G. Consider the mapping

$$f: H \to x H$$

defined by

$$f(h) = x h, h \in H.$$

If $h, k \in H$, then

$$\begin{aligned}f(h) = f(k) &\Rightarrow x h = x k \\ &\Rightarrow x^{-1}(x h) = x^{-1}(x k) \\ &\Rightarrow (x^{-1} x) h = (x^{-1} x) k \\ &\Rightarrow e h = e k \\ &\Rightarrow h = k.\end{aligned}$$

Hence, f is injective. Moreover, the range of f is

$$R(f) = \{x h: (h, x h) \in f\} = x H.$$

Thus, f is surjective. Hence, f is bijective and so H and $x H$ are cardinally equivalent. If H is finite, then H and $x H$ have the same number of elements.

Theorem 5.28

Let H be a subgroup of a finite group G. Then the number of left and right cosets of H in G is equal.

Proof. Let

$$\sum_l = \{xH : x \in G\} \quad \text{and} \quad \sum_r = \{Hx : x \in G\}$$

be the left and right cosets of H in G. Consider the mapping $f : \sum_l \to \sum_r$ defined by $f(x\,H) = H\,x^{-1}$, $x \in G$. We first prove that this mapping is well defined. To do so, we have to show that if $x\,H = y\,H$, then $H\,x^{-1} = H\,y^{-1}$. So, let $x\,H = y\,H$. Then $x = yh$, $h \in H$. Let $kx^{-1} \in Hx^{-1}$, $k \in H$. Then

$$\begin{aligned} k\,x^{-1} &= k\,(y\,h)^{-1} = k\,(h^{-1}\,y^{-1}) \\ &= (k\,h^{-1})\,y^{-1} \in H\,y^{-1} \end{aligned}$$

and so

$$H\,x^{-1} \subseteq H\,y^{-1}.$$

Similarly, we can show that

$$H\,y^{-1} \subseteq H\,x^{-1}.$$

Hence, $H\,x^{-1} = H\,y^{-1}$ and so f is well defined.

If $H\,y \in \sum_r$, then

$$f(y^{-1}H) = H\,(y^{-1})^{-1} = H\,y,$$

which implies that f is surjective. Further,

$$\begin{aligned} f(x\,H) = f(y\,H) &\Rightarrow H\,x^{-1} = H\,y^{-1} \\ &\Rightarrow x^{-1} = h\,y^{-1},\ h \in H \\ &\Rightarrow x = (h\,y^{-1})^{-1} = y\,h^{-1} \in y\,H \\ &\Rightarrow x\,H \subseteq y\,H. \end{aligned}$$

Similarly,

$$y\,H \subseteq x\,H.$$

Hence $x\,H = y\,H$ and f is injective and thus bijective. Since G is finite, it follows that the number of left cosets of H in G is equal to number of right cosets of H in G.

Definition 5.36

Let H be a subgroup of group G. Then the number of left cosets or right cosets of H in G is called the **index** of the subgroup H.

The index of the subgroup H of a group G is denoted by $[G: H]$.

Definition 5.37

The number of elements of a finite group G is called the **order of that group**. It is denoted by $O(G)$.

Definition 5.38

Let G be a group and a be an element of G. Then the least positive integer m such that $a^m = e$ (identity of G) is called the **order of** a. It is denoted by $O(a)$.

If no positive power of an element a equals to the identity, then the element a has order infinity.

Theorem 5.29

There exists a one-to-one correspondence between any two left cosets of a subgroup H in a group G.

Proof. Consider the mapping $f: xH \to yH$ defined by $f(xh_1)=yh_1$. The mapping f is clearly onto and

$$\begin{aligned} f(xh_1)=f(xh_2) &\Rightarrow yh_1=yh_2 \\ &\Rightarrow y^{-1}yh_1=y^{-1}yh_2 \\ &\Rightarrow eh_1=eh_2 \\ &\Rightarrow h_1=h_2 \\ &\Rightarrow xh_1=xh_2. \end{aligned}$$

Thus f is one-one. Hence, f is bijective.

Theorem 5.30 (Lagrange)

The order of a subgroup H of a finite group G is a divisor of the order of the group G.

Proof. Let G be a finite group of order n and let m be the order of its subgroup H. The group G can be decomposed into disjoint left cosets of H in G. Since G is finite, the number of left cosets is also finite. Let the number of left cosets of H in G be k. Since G is finite, the number of elements in H and a left coset of H in G are same. Hence each coset contains m elements. Therefore, the total number of elements in all the cosets of H in G is km which is equal to the total number of elements in G. Hence,

$$n=mk$$

and so $m \mid n$, that is, $O(H) \mid O(G)$.

Corollary 5.3

The index of every subgroup of a finite group is a divisor of the order of the group.

Proof. If n is the order of the finite group G and m is the order of the subgroup H of G, then, by the above theorem,

$$\begin{aligned} n&=mk \\ &= m\,[\text{number of left cosets of } H \text{ in } G] \\ &= m\,[G: H] \end{aligned}$$

and so

$$[G: H] \mid n.$$

Corollary 5.4

The order of each element of a finite group is a divisor of the order of the group.

Proof. Let G be a finite group and m the order of an element a in G. Therefore, $a^m=e$ and the elements

$$a, a^2, \ldots, a^{m-1}, a^m=e$$

are all different and form a subgroup of G whose order is m. Thus, the order of the element a is equal to the order of the subgroup generated by a. But, by Lagrange's theorem, order of a subgroup divides the order of the group. Hence

$$O(a) \mid O(G).$$

Corollary 5.5

If G is a finite group, then for any element a in G,

$$a^{O(G)} = e.$$

Proof. Let $O(a)=m$ and $O(G)=n$. By Corollary 5.4, $O(a) \mid O(G)$, that is, $m \mid n$ and therefore $n=m\,p$ for some positive integer p. Then,

$$a^{O(G)} = a^n = (a^m)^p = e^p = e,$$

which completes the proof.

Remark 5.5 It follows from the Lagrange's theorem that if G is a **group of prime order**, then **it cannot have any proper subgroup**.

For example, if G is a group of order 3 (prime), then a group having two elements cannot be a subgroup of G.

EXAMPLE 5.38

Let G be a group such that $(a\,b)^2 = a^2\,b^2$ for $a, b \in G$. Then, G is abelian.

Solution.

Since G is a group,

$$a \in G \Rightarrow a^{-1} \in G \quad \text{and} \quad a\,a^{-1} = e$$

and

$$b \in G \Rightarrow b^{-1} \in G \quad \text{and} \quad b\,b^{-1} = e.$$

Therefore,

$$\begin{aligned}(a\,b)^2 = a^2\,b^2 &\Rightarrow a\,b\,a\,b = a\,a\,b\,b\\ &\Rightarrow a^{-1}\,a\,b\,a\,b\,b^{-1} = a^{-1}\,a\,a\,b\,b\,b^{-1}\\ &\Rightarrow e\,b\,a\,e = e\,a\,b\,e\\ &\Rightarrow b\,a = a\,b.\end{aligned}$$

Hence, G is abelian.

EXAMPLE 5.39

If every elements of a group G is its own inverse, then G is abelian.

Solution.

Since G is a group, $a, b, \in G \Rightarrow a\,b \in G$.
Therefore,

$$\begin{aligned}&(a\,b)^{-1} = a\,b\\ &\Rightarrow b^{-1}\,a^{-1} = a\,b\\ &\Rightarrow b\,a = a\,b \quad (\text{since } b^{-1} = b,\ a^{-1} = a \text{ by hypothesis})\end{aligned}$$

Hence, G is abelian.

EXAMPLE 5.40

A group G each of whose elements other than identity is of order 2 is necessarily abelian.

Solution.

Let $a \neq e$, $b \neq e$ be two arbitrary elements of G. Then, according to the given hypothesis, we have

$$a^2 = e, b^2 = e \quad \text{and} \quad (a\,b)^2 = e.$$

But,

$$\begin{aligned}(a\,b)^2 = e \Rightarrow a\,b\,a\,b = e \Rightarrow a\,a\,b\,a\,b\,b &= a\,e\,b\\ \Rightarrow a^2\,b\,a\,b^2 &= a\,e\,b\\ \Rightarrow e\,b\,a\,e &= a\,b\\ \Rightarrow b\,a &= a\,b.\end{aligned}$$

Hence, G is abelian.

EXAMPLE 5.41

Let G be a group in which $(a\,b)^i = a^i\,b^i$ for three consecutive integers i for all $a, b \in G$. Then G is abelian.

Solution.

We are given that

$$(a\,b)^n = a^n\,b^n \qquad \text{(i)}$$
$$(a\,b)^{n+1} = a^{n+1}\,b^{n+1} \qquad \text{(ii)}$$
$$(a\,b)^{n+2} = a^{n+2}\,b^{n+2}. \qquad \text{(iii)}$$

From (ii), we have

$$\begin{aligned}&(a\,b)^{n+1} = a^{n+1}\,b^{n+1}\\ &\Rightarrow (a\,b)^n\,a\,b = a^{n+1}\,b^{n+1}\\ &\Rightarrow a^n\,b^n\,a\,b = a^n\,a\,b^n\,b && \text{(using (i))}\\ &\Rightarrow b^n\,a = a\,b^n && \text{(using cancellation law in } G\text{).} \qquad \text{(iv)}\end{aligned}$$

Now from (iii), we have

$$\begin{aligned}(a\,b)^{n+2} = (a\,b)^n\,a\,b\,a\,b = a^{n+2}\,b^{n+2} &\Rightarrow a^n\,b^n\,a\,b\,a\,b = a^n\,a^2\,b^n\,b^2 = a^n\,a^2\,b^n\,b\,b = a^n\,a^2\,b^{n+1}\,b\\ &\Rightarrow b^n\,a\,b\,a = a^2\,b^{n+1} && \text{(using cancellation law in } G\text{)}\\ &\Rightarrow a\,b^n\,b\,a = a^2\,b^{n+1} && \text{(using (iv))}\\ &\Rightarrow a\,b\,b^n\,a = a^2\,b\,b^n\\ &\Rightarrow b\,a\,b^n = a\,b\,b^n && \text{(using (iv) again)}\\ &\Rightarrow b\,a = a\,b && \text{(right cancellation law).}\end{aligned}$$

Hence, G is abelian.

EXAMPLE 5.42

If G is a group of even order, prove that it has an element $a \neq e$ satisfying $a^2 = e$.

Solution.

Since G is a group, it must contain the identity e. But we are given that the group is of even order. Therefore there exists $x \in G$ such that $x \neq e$. Now, $x \in G \Rightarrow x^{-1} \in G$. and also $x^{-1} \neq e$. Therefore, either $x^{-1} = x$ or $x^{-1} \neq x$. If $x^{-1} = x$, then

$$x\,x^{-1} = x^2$$
$$\Rightarrow e = x^2$$

If $x^{-1} \neq x$, then the group becomes of odd order which contradicts the hypothesis. Therefore, if $x^{-1} \neq x$, then there exists $y \in G$ such that $y \neq e, x, x^{-1}$. Now $y \in G \Rightarrow y^{-1} \in G$ and also $y^{-1} \neq e$. Therefore, either $y^{-1} = y$ or $y^{-1} \neq y$. If $y^{-1} = y$, then

$$y\,y^{-1} = y^2$$
$$\Rightarrow e = y^2.$$

If $y^{-1} \neq y$, then the group become of odd order. Thus we have proved that there exists an element $a \neq e$ satisfying $a^2 = e$ in a group of even order.

EXAMPLE 5.43

Prove that if G is an abelian group, then for all $a, b \in G$ and all integers n, $(a\,b)^n = a^n\,b^n$.

Solution.

We shall prove it by induction. It is obviously true for $n = 1$. For $n = 2$, we have

$$(a\,b)^2 = a\,b\,a\,b = a\,a\,b\,b = a^2\,b^2.$$

Let it be true for n. Then we have

$$(a\,b)^n = a^n\,b^n.$$

Now

$$(a\,b)^{n+1} = (a\,b)^n\,(a\,b) = a^n\,b^n\,(a\,b) = a^n\,b^n\,a\,b$$
$$= a^n\,a\,b^n\,b \quad (\text{since } b^n\,a = a\,b^n, \text{ see example 5.41})$$
$$= a^{n+1}\,b^{n+1}.$$

Hence the result is true by mathematical induction.

In case n is negative, we have

$$(a\,b)^{-n} = ((a\,b)^{-1})^n = ((b\,a)^{-1})^n = (a^{-1}\,b^{-1})^n = a^{-n}\,b^{-n}.$$

5.10 NORMAL SUBGROUP

Definition 5.39

A subgroup N of a group G is called a **normal subgroup** of G if $x\,N = N\,x$ for all $x \in G$.

It is denoted by $N \,\Delta\, G$.

It is clear from the definition of a normal subgroup N that every left coset of N in G is a right coset of N in G.

Further,

$$x\,N \subseteq N\,x \Rightarrow x^{-1}\,N \subseteq N\,x^{-1} \quad \text{for all } x \in G$$
$$\Rightarrow x\,x^{-1}\,N\,x \subseteq x\,N\,x^{-1}\,x$$
$$\Rightarrow N\,x \subseteq x\,N.$$

Hence, $x\,N = N\,x$. It follows, therefore, that if $x\,N \subseteq N\,x$, $x \in G$, then N is a normal subgroup of G.

Also, $x\,N \subseteq N\,x$ for all $x \in G$ can be written as $x\,N\,x^{-1} \subseteq N$ or $x\,a\,x^{-1} \in N$ for all $x \in G$, $a \in N$. Hence, we define normal subgroup as follows:

"A subgroup N of a group G is called a **normal subgroup** of G if $x\,a\,x^{-1} \in N$ for all $x \in G$, $a \in N$."

EXAMPLE 5.44

Let G be a group. Then, G itself and $\{e\}$ are normal subgroups of G.

EXAMPLE 5.45

Every subgroup N of an abelian group G is a normal subgroup of G because

$$x\,N = N\,x \quad \text{for all } x \in G.$$

EXAMPLE 5.46

Let G be a group and let

$$Z = \{a\colon a\,x = x\,a \quad \text{for all } x \in G\}.$$

Then, Z is a normal subgroup of G and is known as **Centre of the Group** G.

To prove that Z is a normal subgroup, let $a \in Z$. Then,

$$\begin{aligned} & a\,x = x\,a \quad \text{for all } x \in G \\ & \Rightarrow a^{-1}\,a\,x\,a^{-1} = a^{-1}\,x\,a\,a^{-1} \quad \forall\, x \in G \\ & \Rightarrow x\,a^{-1} = a^{-1}\,x \quad \forall\, x \in G \\ & \Rightarrow a^{-1} \in Z. \end{aligned} \tag{1}$$

Now, let $a, b \in Z$. Then

$$a\,x = x\,a \quad \text{and} \quad b\,x = x\,b \quad \text{for all } x \in G,$$

and so

$$\begin{aligned} (a\,b)\,x &= a\,b\,x = a\,(b\,x) \\ &= a\,(x\,b) = (a\,x)\,b \\ &= (x\,a)\,b = x\,(a\,b). \end{aligned}$$

Thus,

$$a\,b \in Z. \tag{2}$$

It follows from (1) and (2) that Z is a subgroup of G.
Further,

$$\begin{aligned} a \in Z &\Rightarrow x\,a = a\,x \quad \text{for all } x \in G \\ &\Rightarrow x\,a\,x^{-1} = a\,x\,x^{-1} \quad \forall\, x \in G \\ &\Rightarrow x\,a\,x^{-1} = a\,e = a \in Z. \end{aligned}$$

Hence, $x\,a\,x^{-1} \in Z$ for all $x \in G$, $a \in Z$. Hence, Z is a normal subgroup of G.

Definition 5.40

A group G that has no proper normal subgroup is called a **simple group**.

Thus, a simple group G has no normal subgroup except $\{e\}$ and G itself. For example, a group of prime order is simple.

Theorem 5.31

The intersection of any two normal subgroups of a group is a normal subgroup.

Proof. Let H_1 and H_2 be two normal subgroups of G. Let $a \in H_1 \cap H_2$. Then $a \in H_1$ and $a \in H_2$. Since H_1 and H_2 are normal subgroups, we have

$$x\,a\,x^{-1} \in H_1 \quad \text{and} \quad x\,a\,x^{-1} \in H_2.$$

Thus, $a \in H_1 \cap H_2$ implies $x\;a\;x^{-1} \in H_1 \cap H_2$ for all $x \in G$. Hence $H_1 \cap H_2$ is a normal subgroup.

5.11 QUOTIENT GROUP (FACTOR GROUP)

Let H be a normal subgroup of G. Then

$$x\,H = H\,x \qquad \text{for all } x \in G,$$
$$y\,H = H\,y \qquad \text{for all } y \in G.$$

We define the product of cosets of H in G by

$$(x\,H)\,(y\,H) = (x\,y)\,H.$$

Theorem 5.32

Let H be a normal subgroup of a group G. Then the set of cosets of H in G, denoted by,

$$G/H = \{x\,H\!: x \in G\} = \{H\,x\!: x \in G\}$$

is a multiplicative group under multiplication defined by

$$(x\,H)\,(y\,H) = (x\,y)\,H.$$

Proof. We observe that all the group postulates are satisfied.

(i) **Associativity:** If $x\,H, y\,H, z\,H \in G/H$, then

$$\begin{aligned}[(x\,H)\,(y\,H)]\,(z\,H) &= [(x\,y)\,H]\,(z\,H)\\ &= [(x\,y)\,z]\,H\\ &= [x\,(y\,z)]\,H \quad \text{by associativity in } G\\ &= (x\,H)\,(y\,z)\,H = (x\,H)\,[(y\,H)\,(z\,H)].\end{aligned}$$

(ii) **Existence of Identity:** If $x\,H \in G/H$, then

$$(e\,H)\,(x\,H) = (e\,x)\,H = x\,H,$$
$$(x\,H)\,(e\,H) = (x\,e)\,H = x\,H.$$

Hence, $e\,H = H$ is the identity in G/H.

(iii) **Existence of Inverse:** If $x \in G$, then

$$(x\,H)\,(x^{-1}\,H) = (x\,x^{-1})\,H = e\,H = H,$$
$$(x^{-1}\,H)\,(x\,H) = (x^{-1}\,x)\,H = e\,H = H.$$

Hence $x^{-1}\,H$ is the inverse of $x\,H$.

It follows therefore that G/H is a multiplicative group.

Definition 5.41

The group formed by the set of all cosets of a normal subgroup H of a group G under multiplication of complexes (cosets) is called a **Quotient Group** or a **factor group** of G by H. We denote this group by G/H.

Remarks 5.6

(i) If G an additive group, then the cosets of a subgroup H of G are of the form $x+H, x \in G$. Further, if H is a normal subgroup of G, then the cosets form a group under coset addition defined by

$$(x+H)+(y+H)=(x+y)+H.$$

(ii) Factor group of an abelian group G is abelian. In fact, if $x\,H$, $y\,H \in G/H$, then

$$\begin{aligned}(x\,y)\,(y\,H) &= (x\,y)\,H \\ &= (y\,x)\,H \quad \text{since } G \text{ is abelian} \\ &= (y\,H)\,(x\,H).\end{aligned}$$

5.12 HOMOMORPHISM OF GROUPS

Definition 5.42

Let $(G, *)$ and $(H, *')$ be two groups. Then a mapping $f: G \rightarrow H$ is called **group homomorphism** if

$$f(x * y)=f(x) *' f(y)$$

for all $x, y \in G$.

Definition 5.43

A homomorphism $f: G \rightarrow H$ is called **monomorphism** if f as a map is injecitve.

Definition 5.44

A homomorphism $f: G \rightarrow H$ is called **epimorphism** if f as a map is surjective.

Definition 5.45

A homomorphism which is monomorphism as well as epimorphism is called an **isomorphism**.

Further, if a group homomorphism $f: G \rightarrow H$ is onto, we say that H is a **homeomorphic image** of G.

Definition 5.46

A homomorphism of a group G into itself is called an **endomorphism**.

Definition 5.47

An isomorphism of a group G onto itself is called an **automorphism**.

EXAMPLE 5.47

Let **Z** be an additive group of integers and let $f: \mathbf{Z} \rightarrow \mathbf{Z}$ be defined by

$$f(x)=2x, \quad x \in \mathbf{Z}.$$

Then, we note that

$$\begin{aligned}f(x+y) &= 2(x+y), \quad x, y \in \mathbf{Z} \\ &= 2x+2y=f(x)+f(y)\end{aligned}$$

and

$$\begin{aligned}f(x)=f(y) &\Rightarrow 2x=2y \\ &\Rightarrow x=y.\end{aligned}$$

Thus, f is homomorphism and a one-one mapping. Also, f is clearly onto. Hence, f is an **isomorphism**.

EXAMPLE 5.48

Let **Z** be an additive group of integers and G be a multiplicative group. Suppose that G has an element a of order infinity. Consider the mapping

$$f: \mathbf{Z} \rightarrow G$$

defined by

$$f(n)=a^n, \quad n \in \mathbf{Z}.$$

Then,

$$f(m+n)=a^{m+n}, \quad m, n \in \mathbf{Z}$$
$$= a^m \cdot a^n = f(m) \cdot f(n)$$

and

$$f(m)=f(n) \Rightarrow a^m = a^n$$
$$\Rightarrow a^{m-n}=0$$
$$\Rightarrow m-n=0 \text{ since } O(a) \text{ is infinity}$$
$$\Rightarrow m=n.$$

Hence, f is an isomorphism.

EXAMPLE 5.49

Let H be a normal subgroup of a group G. Then the mapping $f: G \to G/H$ defined by

$$f(x)=x\,H, \quad x \in G$$

is a group homomorphism. In fact,

$$f(x\,y)=(x\,y)\,H, \quad x, y \in G$$
$$=(x\,H)\,(y\,H)=f(x)f(y).$$

EXAMPLE 5.50

Let $f: G \to H$ be a group homomorphism, then show that

(i) $f(e_G)=e_H$,

where e_G is the identity of G and e_H is the identity of H.

(ii) $(f(x))^{-1}=f(x^{-1})$, $x \in G$.

Solution.

(i) We note that

$$f(x)\,e_H=f(x)=f(x\,e_G)=f(x)\,f(e_G).$$

Therefore, by left cancellation law in the group H, we have $e_H=f(e_G)$.

(ii) Let $x \in G$. Since f is a homomorphism, we have

$$f(x\,x^{-1})=f(x)\,f(x^{-1})$$
$$\Rightarrow f(e_G)=f(x)\,f(x^{-1})$$
$$\Rightarrow e_H=f(x)\,f(x^{-1}).$$

Hence,

$$(f(x))^{-1}=f(x^{-1}).$$

Definition 5.48

Let $f: G \to H$ be a group homomorphism of G into H. Then the subset $\ker(f)$ defined by

$$\ker(f)=\{x: x \in G, \quad f(x)=e_H\}$$

is called the **Kernel** (**null space**) of f.

Theorem 5.33

Let $f: G \to H$ be a homomorphism of a group G onto a group H and let $\ker(f)$ be the kernel of f. Then $\ker(f)$ is a normal subgroup of G.

Proof. Let e_G and e_H be the identities of G and H, respectively. Then, as proved above, $e_H = f(e_G)$ and so $e_G \in \ker(f)$. Thus, $\ker(f) \neq \phi$.

Now let $x, y \in \ker(f)$. Then, by the definition of kernel,

$$f(x) = f(y) = e_H \tag{1}$$

and so

$$\begin{aligned} f(x\,y^{-1}) &= f(x)\,f(y^{-1}) \\ &= f(y)\,f(y^{-1}) \text{ using (1)} \\ &= f(y\,y^{-1}) \text{ since } f \text{ is homomorphism} \\ &= f(e_G) = e_H \text{ since } f(e_G) = e_H. \end{aligned}$$

Hence, $x\,y^{-1} \in \ker(f)$. Thus, $x, y \in \ker(f)$ imply $x\,y^{-1} \in \ker(f)$. Therefore, $\ker(f)$ is a subgroup of G. Further, if $x \in \ker(f)$, $g \in G$, then

$$\begin{aligned} f(g\,x\,g^{-1}) &= f(g\,x)\,f(g^{-1}) \\ &= [f(g)\,f(x)]\,f(g^{-1}) = [f(g)\,e_H]\,f(g^{-1}) \\ &= f(g)\,f(g^{-1}) = f(g\,g^{-1}) \\ &= f(e_G) = e_H. \end{aligned}$$

Hence, $g\,x\,g^{-1} \in \ker(f)$ and so $\ker(f)$ is a normal subgroup of G.

Theorem 5.34

Every normal subgroup H of a group G is kernel of some homomorphism.

Proof. Let $f: G \to G/H$ be the natural mapping defined by

$$f(x) = x\,H \quad \text{for all } x \in G.$$

If $x, y \in G$, then

$$f(x\,y) = (x\,y)\,H = (x\,H)\,(y\,H) = f(x)\,f(y).$$

Hence, f is a homomorphism from G onto G/H. Kernel of f is given by

$$\begin{aligned} \ker(f) &= \{x: x \in G, f(x) = \text{identity element of } G/H\} \\ &= \{x: x \in G, f(x) = H\} = \{x: x\,H = H\} \\ &= \{x: x \in H\} = H. \end{aligned}$$

Thus, the normal subgroup H is the kernel of homomorphism f defined above.

Theorem 5.35

Let $f: G \to H$ be a group homomorphism of G onto H. Then f is an isomorphism if and only if $\ker(f) = \{e_G\}$.

Proof. Suppose first that $\ker(f)=\{e_G\}$. We shall prove that f is one-to-one. So, let $x, y \in G$ and $f(x)=f(y)$. It suffices to show that $x=y$. To show this, we have

$$\begin{aligned} f(x)\,f(y)^{-1} &= f(y)\,(f(y))^{-1} = f(y)\,f(y^{-1}) \\ &= f(y\,y^{-1}), \quad \text{since } f \text{ is homomorphism} \\ &= f(e_G)=e_H, \quad \text{since } f(e_G)=e_H. \end{aligned}$$

Also,

$$f(x)\,(f(y))^{-1}=f(x)\,f(y^{-1})=f(x\,y^{-1}).$$

Hence,

$$f(x\,y^{-1})=e_H,$$

which implies that $x\,y^{-1} \in \ker(f)$ and so, by our assumption, $x\,y^{-1}=e_G$. Hence, $x=y$, proving that f is one-to-one. So f is bijective and homomorphism and is thus an isomorphism.

Conversely, suppose that f is an isomorphism and let $x \in \ker(f)$. To show that $\ker(f)$ consists of e_G only, we have to show that $x=e_G$. Towards this end, we observe that $x \in \ker(f)$ implies $f(x)=e_H$. Also, we know that if f is homomorphism, then $f(e_G)=e_H$. Hence,

$$f(x)=f(e_G). \tag{1}$$

Since f is one-one, (1) yields $x=e_G$ and so $\ker(f)=\{e_G\}$.

Theorem 5.36 (Fundamental Theorem of Isomorphism)

Let $f: G \to H$ be a group homomorphism of a group G onto another group H and let $\ker(f)$ be the kernel of f. Then,

$$G/\ker(f) \simeq H.$$

Proof. We know that $\ker(f)$ is a normal subgroup of G. So $G/\ker(f)$ is a factor group of G by $\ker(f)$. If $x \in G$, $y \in \ker(f)$, then

$$f(x\,y)=f(x)\,f(y)=f(x)\,e_H=f(x).$$

So, it is possible to define a mapping from $G/\ker(f)$ into H. Therefore, let $\overline{f}: G/\ker(f) \to H$ be a mapping defined by

$$\overline{f}\,(x\,K)=f(x),\ x \in G,$$

where K represents $\ker(f)$. Since

$$\begin{aligned} x\,K=y\,K &\Rightarrow x^{-1}\,y \in K \\ &\Rightarrow f(x^{-1}\,y)=e_H \\ &\Rightarrow f(x^{-1})\,f(y)=e_H \\ &\Rightarrow f(x)=f(y), \end{aligned}$$

it follows that $\overline{f}$ is well defined.

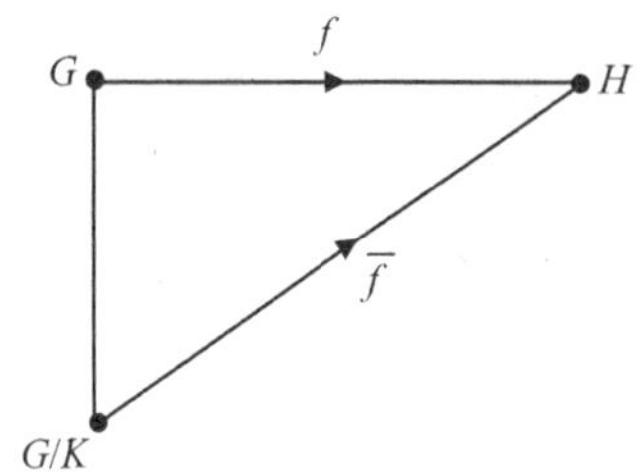

We note that

(i) If $xK, yK \in G/K$, then

$$\bar{f}\ (xK \cdot yK) = \bar{f}\ ((xy)K) = f(xy) = f(x)\, f(y) = \bar{f}\ (xK)\ \bar{f}\ (yK).$$

Thus, $\bar{f}$ is homomorphism.

(ii) If $h \in H$, then

$$h = f(x),\ x \in G, \text{ since } f \text{ is onto} = \bar{f}\ (xK).$$

Thus, for every element $h \in H$, there is an element $xK \in G/K$ such that $h = \bar{f}\ (xK)$. Hence, $\bar{f}$ is onto mapping. Moreover,

$$\begin{aligned}\bar{f}\ (xK) = \bar{f}\ (yK) &\Rightarrow f(x) = f(y)\\ &\Rightarrow f(x)\ (f(x))^{-1} = f(y)(f(x))^{-1}\\ &\Rightarrow f(xx^{-1}) = f(yx^{-1})\\ &\Rightarrow f(e_G) = f(yx^{-1})\\ &\Rightarrow e_H = f(yx^{-1})\\ &\Rightarrow yx^{-1} \in \ker(f) = K\\ &\Rightarrow xK = yK.\end{aligned}$$

Hence, $\bar{f}$ is one-to-one. It follows therefore, that $\bar{f}$ is an isomorphism and so

$$G/\ker(f) \simeq H.$$

5.13 CYCLIC GROUPS

Definition 5.49

A group capable of being generated by a single element is called a **cyclic group** or **monogenic**.

The generating element of a cyclic group is called its **generator**.

Let G be a cyclic group generated by a, then

$$G = \{a^n : n \in \mathbf{Z}\}.$$

We denote the cyclic group generated by a by <a>.

Since $a^n\ a^m = a^m a^n$ for all $a \in G$; $m, n \in \mathbf{Z}$, it follows that a **cyclic group is necessarily abelian**.

If all the elements of a cyclic group G are different, that is, if $a^m = a^n \Rightarrow m = n$, then G has infinite number of elements. Such a cyclic group is called **infinite cyclic group**.

EXAMPLE 5.51

The additive group $\mathbf{Z}$ of integers is an infinite cyclic group. The integers 1 and –1 are generators of this cyclic group. Thus

$$\mathbf{Z} = \{\ldots, -3, -2, -1, 0, 1, 2, 3, \ldots\}.$$

EXAMPLE 5.52

If ω is the cube root of unity, then the set $\{1, \omega, \omega^2\}$ is a finite multiplicative cyclic group. Since $\omega^3 = 1$, the order of this group is 3. The generator of this group is ω.

Theorem 5.37

The order of a cyclic group is equal to the order of its generator.

Proof. Let $<a>$ be cyclic group generated by a. If we can find a pair λ, μ of integers such that

$$a^\lambda = a^\mu, \quad \lambda \neq \mu,$$

then a positive integer h can be found such that $a^h = e$. If n is the smallest of these positive integers, then $O(a) = n$. In this case, the elements of the cyclic group shall be

$$e, a, a^2, \ldots, a^{n-1}$$

If $a^\lambda \in <a>$, then by Euclidean algorithm, we have

$$\lambda = q\,n + r, \quad 0 \leq r < n.$$

Therefore,

$$a^\lambda = a^{qn+r} = (a^n)^q \cdot a^r = e \cdot a^r = a^r.$$

Thus, n is also the order of cyclic group $<a>$.

Theorem 5.38

Let G be a finite cyclic group of order n. Then,

$$G \cong \mathbf{Z}/n\mathbf{Z}.$$

Proof. Let $G = <a>$ be a cyclic group generated by a. Then the mapping $f: \mathbf{Z} \rightarrow G$ defined by $f(n) = a^n$, $n \in \mathbf{Z}$ is a homomorphism. In fact,

$$f(m+n) = a^{m+n} = a^m \cdot a^n = f(m) \cdot f(n).$$

Further, if $x \in G = <a>$, then $x = a^m$ for some integer m, and so $f(m) = a^m = x$.
Thus to each $x \in G$, there is some $m \in \mathbf{Z}$ such that $x = f(m)$. Hence, f is onto.
Let K be the kernel of homomorphism f. Then, by the fundamental theorem of homomorphism,

$$\mathbf{Z}/K \cong G.$$

But K is a subgroup of $\mathbf{Z}$. Therefore

$$K = n\mathbf{Z}, \quad n \in \mathbf{Z}$$

and so

$$\mathbf{Z}/n\mathbf{Z} \cong G.$$

Theorem 5.39

A group of prime order is cyclic.

Proof. Let p be a prime and let G be a group of order p. If G consists of identity element only, then it is certainly cyclic. If order of G is greater than 1, then there is an element $a \in G, a \neq e$. Let,

$$H = \{a^n : n \in \mathbf{Z}\}$$

be a cyclic subgroup of G generated by a. Since $O(G)=p$, therefore by Lagrange's theorem, it cannot have any proper subgroup. Hence $H=\{e\}$ or $H=G$. But $H \neq \{e\}$ by supposition. Hence $H=G$ and so G is cyclic.

Theorem 5.40

Let H be a subgroup of a cyclic group $<a>$ and m be the least positive integer such that $a^m \in H$. If $a^n \in H$, then $m \mid n$.

Proof. By division algorithm, we have

$$n = qm + r,\ q, r \in \mathbf{Z},\ 0 \leq r < m.$$

Therefore,

$$a^r = a^{n-qm} = a^n(a^{-qm}) = a^n\ (a^{qm})^{-1} \in H.$$

Hence $r=0$, otherwise it will contradicts the fact that m is the least positive integer such that $a^m \in H$. Therefore,

$$n = qm$$

and so $m \mid n$. This completes the proof.

5.13.1 Generators of a Cyclic Group

Let $G=<a>$ be a cyclic group generated by a. Then a^{-1} will also be a generator of G. In fact, if $a^m \in G$, $m \in \mathbf{Z}$, then $a^m = (a^{-1})^{-m}$. The question arises which of the elements of G other than a and a^{-1} can be a generator of G. We consider the following two cases:

(i) G is an infinite cyclic group
(ii) G is a finite group

We discuss these cases in the form of the following theorems:

Theorem 5.41

An infinite cyclic group has exactly two generators.

Proof. Let a be a generator of an infinite cyclic group G. Then a is of infinite order and

$$G = \{\ldots, a^{-r}, \ldots, a^{-1}, e, a, a, \ldots, a^r, \ldots\}.$$

Let $a^t \in G$ be another generator of G, then

$$G = \{\ldots, a^{-2t}, a^{-t}, \mathrm{e}, a^t, a^{2t}, \ldots\}.$$

Since $a^{t+1} \in G$, therefore

$$a^{t+1} = a^{rt} \text{ for some integers } r.$$

Since G is infinite, we have

$$t + 1 = r\,t \quad \text{or} \quad (r-1)t = 1,$$

which holds only if $t = \pm 1$. Hence there exists only two generators a and a^{-1} of an infinite cyclic group $<a>$.

Theorem 5.42

Let $G=<a>$ be a cyclic group of order n. Then $a^m \in G$, $m \leq n$ is a generator of G if and only if gcd $(m, n)=1$

Proof. Let H be a subgroup of G generated by $a^m (m \leq n)$. If gcd $(m, n)=1$, then there exists two integers u, v such that

$$\begin{aligned} &um+vn=1 \\ &\Rightarrow a^{um+vn}=a \\ &\Rightarrow a^{um} \cdot a^{vn}=a \\ &\Rightarrow (a^m)^u \cdot (a^n)^v=a \\ &\Rightarrow (a^m)^u=a \quad (\therefore (a^n)^v=e \\ &\Rightarrow a \in H \quad (\therefore (a^m)^u \in H) \\ &\Rightarrow G \subseteq H. \end{aligned}$$

But, by supposition, $H \subseteq G$. Hence $G=H=<a^m>$, that is, a^m is a generator of G.

Conversely, let a^m $(m \leq n)$ be a generator of G. Then

$$G=\{a^{mn}: n \in \mathbf{Z}).$$

Therefore, we can find an integer u such that

$$\begin{aligned} &a^{mu}=a \\ &\Rightarrow a^{mu-1}=e \\ &\Rightarrow O(a) \mid (mu-1) \\ &\Rightarrow n \mid (mu-1). \end{aligned}$$

Hence, there exists an integer v such that

$$\begin{aligned} &nv=mu-1 \\ &\Rightarrow mu-nv=1 \\ &\Rightarrow \gcd(m, n)=1. \end{aligned}$$

This completes the proof of the theorem.

Theorem 5.43

Every subgroup H of a cyclic group G is cyclic.

Proof. If $H=\{e\}$, then H is obviously cyclic. So, let us suppose that $H \neq \{e\}$. If $a^\lambda \in H$, then $a^{-\lambda} \in H$. So, we can find a smallest positive integer m such that $a^m \in H$. Therefore,

$$<a^m> \subseteq H. \tag{i}$$

Moreover,

$$a^\lambda \in H \Rightarrow \lambda=q\,m, \quad q \in \mathbf{Z}.$$

Therefore,

$$\begin{aligned} &a^\lambda=a^{qm}=(q^m)^q \in <a^m> \\ &\Rightarrow <a^\lambda> \subseteq <a^m> \\ &\Rightarrow H \subseteq <a^m>. \end{aligned} \tag{ii}$$

It follows from (i) and (ii) that $H=<a^m>$, and hence H is cyclic.

Theorem 5.44

Let $G = <a>$ be a cyclic group of order n and H be a subgroup of G generated by a^m, $m \leq n$. Then,

$$O(H) = \frac{n}{\gcd(m,n)}.$$

Proof. We are given that $H = <a^m>$. Let gcd $(m, n) = d$, then we can find an integer q such that $m = q\,d$. Thus, $a^m = a^{qd}$. But $a^{qd} \in <a^d>$, where a^d is a subgroup generated by a^d. Therefore,

$$a^m \in <a^d>$$
$$\Rightarrow H = <a^m> \subseteq <a^d> \qquad \text{(i)}$$

Since gcd $(m, n) = d$, we can find $u, v \in \mathbf{Z}$ such that

$$d = un + vm$$
$$\Rightarrow a^d = a^{un+vm} = a^{un}\, a^{vm} = a^{vm} \; (\because a^{un} = e).$$

But $a^{vm} \in <a^m> = H$. Therefore

$$a^d \in H$$
$$\Rightarrow \quad <a^d> \subseteq H \qquad \text{(ii)}$$

From (i) and (ii), we have $H = <a^d>$ and so $O(H) = O(<a^d>)$. But

$$O(\langle a^d \rangle) = \frac{n}{d} \quad \text{since } (a^d)^{\frac{n}{d}} = e.$$

Hence,

$$O(H) = \frac{n}{\gcd(m,n)},$$

which completes the proof of the theorem.

Theorem 5.45

Any two cyclic groups of the same order are isomorphic. In particular, every infinite cyclic group is isomorphic to the additive group **Z** of integers.

Proof. Let G and H be two cyclic groups of same order. Consider the mapping $f: G \to H$ defined by $f(a^r) = b^r$. Then f is clearly a homomorphism. Also,

$$f(a^r) = f(a^s) \Rightarrow b^r = b^s.$$

If G and H are of infinite order, then $r = s$ and so $a^r = a^s$. If their order is finite, say n, then

$$b^r = b^s \Rightarrow b^{r-s} = e$$
$$\Rightarrow n \mid (r - s)$$
$$\Rightarrow n\,u = r - s, \; u \in \mathbf{Z}$$
$$\Rightarrow a^{r-s} = a^{nu} = (a^n)^u = e$$
$$\Rightarrow a^r = a^s.$$

Hence f is one-to-one mapping also. Therefore, $G \simeq H$.

Theorem 5.46

Every isomorphic image of a cyclic group is again cyclic.

Proof. Let $G=<a>$ be a cyclic group and let H be its image under isomorphism f. The elements of G are given by

$$G=\{\ldots, a^{-r}, \ldots, a^{-2}, a^{-1}, 1, a, a^2, a^3, \ldots, a^r, \ldots\}.$$

Let b be an arbitrary element of H. Since H is an isomorphic image of G, there exists $a^r \in G, r=0, 1, \ldots$ such that $b=f(a^r)$. Since f is homomorphism, we have

$$b=\underbrace{f(a)\cdot f(a)\ldots f(a)}_{r \text{ times}}=f(a)^r.$$

Thus, H is generated by $f(a)$ and hence is cyclic.

5.14 PERMUTATION GROUPS

The importance of the study of permutation groups lies in the fact that any finite group, whatsoever, is isomorphic to some permutation group. So by studying permutation groups, we are studying all possible types of finite group.

Definition 5.50

A one-to-one mapping of a finite set onto itself is called a **permutation**.

The set of all permutations of a set containing n elements will be denoted by S_n and is called the **symmetric set** of transformations of degree n. The number of elements in the set S_n is $n!$

If $S=\{x_1, x_2, \ldots, x_n\}$ is a finite set, then its permutation $f: S\to S$ is denoted by

$$\begin{pmatrix} x_1 & x_2 & \cdots & x_n \\ y_1 & y_2 & \cdots & y_n \end{pmatrix},$$

where $y_i=f(x_i)$, $i=1, 2, \ldots, n$.

EXAMPLE 5.53

Let $S=\{a, b, c\}$. Then S_3 consists of the following six elements:

$$\begin{pmatrix} a & b & c \\ a & b & c \end{pmatrix}, \begin{pmatrix} a & b & c \\ b & a & c \end{pmatrix}, \begin{pmatrix} a & b & c \\ a & c & b \end{pmatrix}$$
$$\begin{pmatrix} a & b & c \\ c & b & a \end{pmatrix}, \begin{pmatrix} a & b & c \\ b & c & a \end{pmatrix}, \begin{pmatrix} a & b & c \\ c & a & b \end{pmatrix}.$$

Definition 5.51

Let S be a finite set. A permutation which maps each element of S onto itself is called **identity permutation.**

Thus, the permutation $\begin{pmatrix} a & b & c \\ a & b & c \end{pmatrix}$ is the identity permutation of S_3 in Example 5.53.

Definition 5.52

Let

$$\alpha=\begin{pmatrix} x_1 & x_2 & \dots & x_n \\ y_1 & y_2 & \dots & y_n \end{pmatrix} \text{ and } \beta=\begin{pmatrix} y_1 & y_2 & \dots & y_n \\ z_1 & z_2 & \dots & z_n \end{pmatrix}$$

be two permutations of the set $\{x_1, x_2, \dots, x_n\}$ such that $y_i=\alpha(x_i)$, $z_i=\beta(y_i)$. Then the **composition** of α and β is defined by

$$\begin{aligned}\beta\circ\alpha &= \begin{pmatrix} y_1 & y_2 & \dots & y_n \\ z_1 & z_2 & \dots & z_n \end{pmatrix}\circ\begin{pmatrix} x_1 & x_2 & \dots & x_n \\ y_1 & y_2 & \dots & y_n \end{pmatrix} \\ &= \begin{pmatrix} x_1 & x_2 & \dots & x_n \\ z_1 & z_2 & \dots & z_n \end{pmatrix}.\end{aligned}$$

EXAMPLE 5.54

Let

$$\alpha=\begin{pmatrix} 1 & 2 & 3 \\ 2 & 3 & 1 \end{pmatrix} \text{ and } \beta=\begin{pmatrix} 1 & 2 & 3 \\ 3 & 1 & 2 \end{pmatrix}$$

be two permutations belonging to S_3. Then

$$\begin{aligned}\alpha\circ\beta &= \begin{pmatrix} 1 & 2 & 3 \\ 2 & 3 & 1 \end{pmatrix}\circ\begin{pmatrix} 1 & 2 & 3 \\ 3 & 1 & 2 \end{pmatrix} \\ &= \begin{pmatrix} 1 & 2 & 3 \\ 1 & 2 & 3 \end{pmatrix}\end{aligned}$$

and

$$\beta\circ\alpha=\begin{pmatrix} 1 & 2 & 3 \\ 1 & 2 & 3 \end{pmatrix}.$$

Thus $\alpha \circ \beta=\beta \circ \alpha$. Hence α and β commute with each other.

But the **composition of permutations is not always commutative**. For example, if we consider

$$\alpha=\begin{pmatrix} 1 & 2 & 3 \\ 3 & 2 & 1 \end{pmatrix},\quad \beta=\begin{pmatrix} 1 & 2 & 3 \\ 2 & 3 & 1 \end{pmatrix},$$

then

$$\beta\circ\alpha=\begin{pmatrix} 1 & 2 & 3 \\ 1 & 3 & 2 \end{pmatrix}$$

and

$$\alpha\circ\beta-\begin{pmatrix} 1 & 2 & 3 \\ 2 & 1 & 3 \end{pmatrix}.$$

Hence

$$\alpha \circ \beta\neq\beta \circ \alpha.$$

Definition 5.53

The set of permutations of a finite set having n elements form a group under function's composition. This group is known as **symmetric group** and is denoted by S_n. Its order is $n!$

The permutation $\begin{pmatrix} x_1 & \dots & x_n \\ x_1 & \dots & x_n \end{pmatrix}$ acts as **identity** of this group.

Definition 5.54

Let S be a finite set, $x \in S$ and $\alpha \in S_n$. Then α **fixes** x if $\alpha(x)=x$ otherwise α **moves** x.

Definition 5.55

Let $S=\{x_1, x_2, \dots, x_n\}$ be a finite set. If $\sigma \in S_n$ is such that

$$\sigma(x_i)=x_{i+1}, \quad i=1, 2, \dots, k-1$$
$$\sigma(x_k)=x_1$$

and

$$\sigma(x_j)=x_j, \quad j\neq 1, 2, \dots, k;$$

then σ is called a **cycle of length** k. We denote this cycle by

$$\sigma=(x_1, x_2, \dots, x_k).$$

Thus, the **length of a cycle is the number of objects permuted**.

For example, $\begin{pmatrix} a & b & c \\ b & c & a \end{pmatrix} \in S_3$ is a cyclic permutation because

$$f(a)=b, f(b)=c, f(c)=a.$$

In this case the length of the cycle is 3. We can denote this permutation by $(a\ b\ c)$.

Definition 5.56

A cyclic permutation of length 2 is called a **Transposition**.

For example, $\begin{pmatrix} 1 & 2 & 3 \\ 1 & 3 & 2 \end{pmatrix}$ is a transposition.

Definition 5.57

Two cycles are said to be **disjoint** if they have no object in common.

Definition 5.58

Two permutations $\alpha, \beta \in S_n$ are called **disjoint** if

$$\alpha(x)=x \Rightarrow \beta(x)\neq x,$$
$$\alpha(x)\neq x \Rightarrow \beta(x)=x,$$

for all $x \in S$.

In other words, α and β are disjoint if every $x \in S$ moved by one permutation is fixed by the other. Further, if α and β are disjoint permutations, then $\alpha\beta=\beta\alpha$. For example, if we consider

$$\alpha = \begin{pmatrix} 1 & 2 & 3 \\ 2 & 3 & 1 \end{pmatrix}, \quad \beta = \begin{pmatrix} 1 & 2 & 3 \\ 1 & 2 & 3 \end{pmatrix},$$

then $\alpha\beta = \beta\alpha$.

Definition 5.59

A permutation $\alpha \in S_n$ is said to be **regular** if either it is the identity permutation or it has no fixed point and is the product of disjoint cycles of the same length.

For example,

$$\begin{pmatrix} 1 & 2 & 3 & 4 & 5 & 6 \\ 2 & 3 & 1 & 5 & 6 & 4 \end{pmatrix} = (1 \ \ 2 \ \ 3)(4 \ \ 5 \ \ 6)$$

is a regular permutation.

Theorem 5.47

Every permutation can be expressed as a product of pairwise cycles.

Proof. Let $S = \{x_1, x_2, \ldots, x_n\}$ be a finite set having n elements and $f \in S_n$. If f is already a cycle, we are through. So, let us suppose that f is not a cycle. We shall prove this theorem by induction on n.

If $n=1$, the result is obvious. Let the theorem be true for a permutation of a set having less than n elements. Then there exists a positive integer $k<n$ and distinct elements $y_1, y_2, \ldots, y_k$ in $\{x_1, x_2, \ldots, x_n\}$ such that

$$\begin{aligned} &f(y_1)=y_2, \\ &f(y_2)=y_3, \\ &\_\_\_\_\_\_\_\_ \\ &\_\_\_\_\_\_\_\_ \\ &f(y_{k-1})=y_k, \\ &f(y_k)=y_1. \end{aligned}$$

Therefore $(y_1 \, y_2 \ldots y_k)$ is a cycle of length k. Next, let g be the restriction of f to

$$T = \{x_1, x_2, \ldots, x_n\} - \{y_1, y_2, \ldots, y_k\}.$$

Then, g is a permutation of the set T containing $n-k$ elements. Therefore, by induction hypothesis,

$$g = \alpha_1 \, \alpha_2 \ldots \alpha_m,$$

where $\alpha_1, \alpha_2, \ldots, \alpha_m$ are pairwise disjoint cycles. But

$$f = (y_1 \, y_2 \ldots y_k) \circ g = (y_1 \, y_2 \ldots y_k) \, \alpha_1 \, \alpha_2 \ldots \alpha_m.$$

Hence, every permutation can be expressed as a composition of disjoint cycles.

EXAMPLE 5.55

Let

$$f = \begin{pmatrix} 1 & 2 & 3 & 4 & 5 & 6 & 7 & 8 & 9 \\ 6 & 4 & 7 & 2 & 5 & 1 & 8 & 9 & 3 \end{pmatrix}$$

be a permutation. Here 5 is a fixed element. Therefore, (5) is a cycle of length 1. Cycles of length 2 are (1 6) and (2 4) whereas (3 7 8 9) is a cycle of length 4. Hence

$$f=(5)\,(1\ 6)\,(2\ 4)\,(3\ 7\ 8\ 9).$$

Theorem 5.48

Symmetric group S_n is generated by transpositions, that is, every permutation in S_n is a product of transpositions.

Proof. We have proved above that every permutation can be expressed as the composition of disjoint cycles. Consider the m-cycle $(x_1\ x_2\ \ldots\ x_m)$. A simple computation shows that

$$(x_1\ x_2\ \ldots\ x_m)=(x_1\ x_m)\,(x_1\ x_{m-1})\ \ldots\ (x_1\ x_3)\,(x_1\ x_2),$$

that is, every cycle can be expressed as a product of transposition. Hence every permutation $\alpha \in S_n$ can be expressed as a product of transpositions.

Remark 5.7 The above decomposition of a cycle as the product of transposition is not unique. For example,

$$(1\ 2\ 3)=(1\ 3)\,(1\ 2)=(3\ 2)\,(3\ 1).$$

However, it can be proved that the number of factors in the expression is **always even or always odd**.

Definition 5.60

A permutation is called **even** if it is product of an even number of transpositions.

Similarly, a permutation is called **odd** if it is a product of odd number of transpositions. Further,

(i) The product of two even permutations is even.
(ii) The product of two odd permutations is even.
(iii) The product of one odd and one even permutation is odd.
(iv) The inverse of an even permutation is an even permutation.

Theorem 5.49

If a permutation is expressed as a product of transpositions, then the number of transpositions is either even in both cases or odd in both cases.

Proof. Let a permutation σ be expressed as the product of transpositions as given below:

$$\sigma=\alpha_1\ \alpha_2\ \ldots\ \alpha_r=\beta_1\ \alpha_2\ \ldots\ \beta_s.$$

This yields

$$\begin{aligned} e&=\alpha_1\ \alpha_2\ \ldots\ \alpha_r\,\beta_s^{-1}\,\beta_{s-1}^{-1}\ \ldots\ \beta_2^{-1}\,\beta_1^{-1}\\ &=\alpha_1\ \alpha_2\ \ldots\ \alpha_r\,\beta_s\,\beta_{s-1}\ \ldots\ \beta_2\,\beta_1, \end{aligned}$$

since inverse of transposition is the transposition itself. The left-hand side, that is, identity permutation is even and therefore the right-hand side should also be an even permutation. Thus $r+s$ is even, which is possible if r and s are both even or both odd. This completes the proof of the theorem.

Theorem 5.50

The set A_n of all even permutations in S_n is a normal subgroup. Further $O(A_n)=\frac{n!}{2}$.

Proof. Let A_n be the subset of S_n consisting of all even permutations. Since

(i) The product of two even permutations is an even permutation,
(ii) The inverse of an even permutation is an even permutation,

it follows that A_n is a subgroup of S_n. To prove that A_n is a normal subgroup of S_n, we proceed as follows:

Let W be the group of real numbers 1 and -1 under multiplication. Define $f: S_n \to W$ by

$$f(a)=1 \text{ if } a \text{ is an even permutation,}$$
$$f(a)=-1 \text{ if } a \text{ is an odd permutation.}$$

Then it can be verified that f is a homomorphism of S_n onto W. The kernel (null space) of f is given by

$$\begin{aligned} K &= \{a \in S_n : f(a)=e_W=1\} \\ &= \{a \in S_n : f(a)=1\} \\ &= \{a \in S_n : a \text{ is even}\} = A_n. \end{aligned}$$

Now A_n, being the kernel of a homomorphism, is a normal subgroup of S_n and so Isomorphism theorem yields

$$\frac{S_n}{A_n} \cong W.$$

Therefore,

$$O(W) = O\left(\frac{S_n}{A_n}\right) = \frac{O(S_n)}{O(A_n)}.$$

But $O(W)=2$. Therefore,

$$2 = \frac{O(S_n)}{O(A_n)} \text{ which yields } O(A_n) = \frac{O(S_n)}{2} = \frac{n!}{2}.$$

This completes the proof of the theorem.

Definition 5.61

The normal subgroup A_n formed by all even permutation in S_n is called the **Alternating Group of degree** n.

We have shown above that order of A_n is $\frac{n!}{2}$.

Theorem 5.51 (Caley's Theorem)

Any finite group is isomorphic to a permutation group.

Proof. Let G be a finite group of order n and let $a \in G$. Consider the mapping $f_a : G \to G$ defined by $f_a(x)=a\,x$. We note that

(i) $f_a(x)=f_a(y) \Rightarrow a\,x=a\,y;\ x, y \in G \Rightarrow x=y$ by cancellation law in G.

and so f is one-one.

(ii) Since the range of f_a is whole of G, it follows that f is onto mapping.

Hence f_a is a permutation of the set G. Consider the set

$$G' = \{f_a : a \in G\}$$

of permutations of G. We note that

$$(f_a \circ f_b)(x) = f_a(f_b(x)) = f_a(b\,x)$$
$$= a\,(b\,x) = (a\,b)\,(x) = f_{ab}\,(x)$$

for all $x \in G$. Hence

$$f_a \circ f_b = f_{ab}.$$

We claim that G' is isomorphic to G. To show it, consider the mapping ϕ: $G \rightarrow G'$ defined by $\phi(a) = f_a$, $a \in G$. Then,

$$\phi(a\,b) = f_{ab} = f_a \circ f_b = \phi(a) \circ \phi(b)$$

and therefore ϕ is a homomorphism. Further,

$$\phi(a) = \phi(b) \Rightarrow f_a = f_b$$
$$\Rightarrow a\,x = b\,x \text{ for all } x \in G$$
$$\Rightarrow a = b,$$

and so ϕ is injective. Definition of ϕ shows that ϕ is onto also. Thus ϕ is an isomorphism and hence $G \cong G'$. This completes the proof.

EXAMPLE 5.56

Show that in S_3 there are four elements σ satisfying σ^2 = Identity and three elements satisfying σ^3 = Identity.

Solution.

Let $S = \{1, 2, 3\}$. Then permutations in S_3 are

$$\sigma_1 = \begin{pmatrix} 1 & 2 & 3 \\ 1 & 2 & 3 \end{pmatrix}, \quad \sigma_2 = \begin{pmatrix} 1 & 2 & 3 \\ 2 & 1 & 3 \end{pmatrix}$$

$$\sigma_3 = \begin{pmatrix} 1 & 2 & 3 \\ 3 & 2 & 1 \end{pmatrix}, \quad \sigma_4 = \begin{pmatrix} 1 & 2 & 3 \\ 1 & 3 & 2 \end{pmatrix}$$

$$\sigma_5 = \begin{pmatrix} 1 & 2 & 3 \\ 3 & 1 & 2 \end{pmatrix}, \quad \sigma_6 = \begin{pmatrix} 1 & 2 & 3 \\ 1 & 2 & 3 \end{pmatrix}.$$

The permutation σ_1 is the identity for S_3. The composition table for S_3 is given below:

	σ_1	σ_2	σ_3	σ_4	σ_5	σ_6
σ_1	σ_1	σ_2	σ_3	σ_4	σ_5	σ_6
σ_2	σ_2	σ_1	σ_5	σ_6	σ_3	σ_4
σ_3	σ_3	σ_6	σ_1	σ_5	σ_4	σ_2
σ_4	σ_4	σ_5	σ_6	σ_1	σ_2	σ_3
σ_5	σ_5	σ_4	σ_2	σ_3	σ_6	σ_1
σ_6	σ_6	σ_3	σ_4	σ_2	σ_1	σ_5.

From the composition table, we observe that there are four elements in S_3 satisfying σ^2=Identity. These elements are σ_1, σ_2, σ_3 and σ_4.

We further have

(i) $\sigma_1^3=\sigma_1$ since σ_1 is identity element in S_3.
(ii) $\sigma_5^2=\sigma_6$ and $\sigma_6 \cdot \sigma_5=\sigma_1$, therefore $\sigma_5^3=\sigma_6 \cdot \sigma_5=\sigma_1$.
(iii) $\sigma_6^2=\sigma_5$ and $\sigma_5 \cdot \sigma_6=\sigma_1$ and so $\sigma_6^3=\sigma_5 \cdot \sigma_6=\sigma_1$.

No other element satisfies this condition. Hence there are three elements in S_3 satisfying σ^3=Identity.

5.15 DIRECT PRODUCT AND DIRECT SUM OF GROUPS

Definition 5.62

Let $G_1, G_2, \ldots, G_k$ be a family of groups. Then the set $G_1 \times G_2 \times \ldots \times G_k=\{(g_1, g_2, \ldots, g_k): g_i \in G_i\}$ is called the **direct product** of the groups $G_1, G_2, \ldots, G_k$.

Theorem 5.52

Let G_1 and G_2 be two groups. Then

$$G_1 \times G_2=\{(g_1, g_2): g_1 \in G_1, g_2 \in G_2\}$$

is a group under the binary composition.

$$(g_1, g_2) \cdot (g_1', g_2')=(g_1 g_1', g_2 g_2').$$

Proof. Associativity of the binary composition is easy to establish. We further note that if $(g_1, g_2) \in G_1 \times G_2$, then

(i) $\left(e_{G_1}, e_{G_2}\right)(g_1, g_2)=\left(e_{G_1} g_1, e_{G_2} g_2\right)=(g_1, g_2)$,

and

$(g_1, g_2)\left(e_{G_1}, e_{G_2}\right)=\left(g_1 e_{G_1}, g_2 e_{G_2}\right)=(g_1, g_2)$.

Hence $\left(e_{G_1}, e_{G_2}\right)$ is identity element in $G_1 \times G_2$.

(ii) $(g_1, g_2)(g_1^{-1}, g_2^{-1})=(g_1 g_1^{-1}, g_2 g_2^{-1})=\left(e_{G_1}, e_{G_2}\right)$

and

$(g_1^{-1}, g_2^{-1})(g_1, g_2)=(g_1^{-1} g_1, g_2^{-1} g_2)=\left(e_{G_1}, e_{G_2}\right)$.

Hence (g_1^{-1}, g_2^{-1}) is the inverse of (g_1, g_2). It follows therefore that $G_1 \times G_2$ is a group under the binary composition defined in the statement of the theorem.

Remark 5.8 If the binary composition in the family $G_1, G_2, \ldots, G_k$ of groups is taken to be addition, then the direct product is called the **direct sum** and is denoted by $G_1 \oplus G_2 \oplus \ldots \oplus G_k$ Also, if each G_i is finite, then

$$O(G_1 \times G_2 \times \ldots \times G_k)=O(G_1) \cdot O(G_2) \ldots O(G_k).$$

5.16 GROUP AS DIRECT PRODUCT OF ITS SUBGROUPS

Definition 5.63

A group is said to be a **direct product of its subgroups H_1 and H_2** if

(i) Each element of H_1 commutes with each element of H_2.

(ii) Each element of G is uniquely expressible as product of an element of H_1 and an element of H_2.

If a group G is direct product of its subgroups H_1 and H_2, then we write $G=H_1 \times H_2$, whereas in case of direct sum, we write $G=H_1 \oplus H_2$.

Theorem 5.53

Necessary and sufficient conditions for a group G to be a direct product of its subgroups H_1 and H_2 are

(i) H_1 and H_2 are normal subgroups of G

(ii) $H_1 \cap H_2=\{e\}$

(iii) $G=H_1 H_2$.

Proof. **The conditions are necessary:** Suppose G is direct product of H_1 and H_2. Let a $\in$ G and $h \in H_1$. Since $G=H_1 H_2$, each element of G is expressible as product of an element of H_1 and an element of H_2. Therefore

$$a=a_1 a_2, \quad a_1 \in H_1, a_2 \in H_2.$$

Using condition (i) of Definition 5.63, we have

$$\begin{aligned} a h a^{-1} &= (a_1 a_2) h (a_1 a_2)^{-1}=(a_1 a_2) h (a_2^{-1} a_1^{-1}) \\ &= a_1(a_2 h a_2^{-1}) a_1^{-1} \\ &= a_1(h a_2 a_2^{-1}) a_1^{-1}, \text{ since } a_2h=ha_2 \\ &= a_1 h a_1^{-1} \in H_1. \end{aligned}$$

Hence, H_1 is a normal subgroup of G. Similarly, we can show that H_2 is a normal subgroup of G, that is, condition (i) is satisfied.

Suppose that $H_1 \cap H_2 \neq \{e\}$. Then $z \in H_1 \cap H_2$ will be an element such that $z=e\,z=z\,e$, which contradicts the fact that each element of G is uniquely expressible as a product of an element of H_1 and an element of H_2. Hence, $H_1 \cap H_2=\{e\}$ and so (ii) is satisfied. Finally, by definition of group as direct product of its subgroup, $G=H_1 H_2$ and so (iii) is satisfied.

The conditions are sufficient: Suppose that conditions (i) through (iii) are satisfied. Let $a_1 \in H_1$, $a_2 \in H_2$. Then

$$a_1 a_2 a_1^{-1} a_2^{-1}=a_1(a_2 a_1^{-1} a_2^{-1}) \in H_1, \text{ by normality of } H_1.$$

Similarly,

$$a_1 a_2 a_1^{-1} a_2^{-1}=(a_1 a_2 a_1^{-1}) a_2^{-1} \in H_2, \text{ by normality of } H_2.$$

But $H_1 \cap H_2=\{e\}$ and so

$$\begin{aligned} & a_1 a_2 a_1^{-1} a_2^{\ 1}=e \\ & \Rightarrow (a_1 a_2) (a_2 a_1)^{-1}=e \\ & \Rightarrow a_1 a_2=a_2 a_1. \end{aligned}$$

Thus each element of H_1 commutes with each element of H_2.

Since, by (iii) $G=H_1 H_2$, each element of G is expressible as a product of an element of H_1 and an element of H_2. We now establish uniqueness of the expression. If possible, let

$$a=a_1 a_2=b_1 b_2, a_1, b_1 \in H_1; a_2, b_2 \in H_2.$$

Then,

$$\begin{aligned} & b_1^{-1} a_1=b_2 a_2^{-1} \in H_1 \cap H_2 \\ & \Rightarrow b_1^{-1} a_1=e \text{ and } b_2 a_2^{-1}=e \\ & \Rightarrow a_1=b_1 \text{ and } a_2=b_2. \end{aligned}$$

Hence, the expression is unique.

Theorem 5.54

Let $G=H_1 \times H_2$, where H_1 and H_2 are subgroups of G. Then H_1 and H_2 are normal subgroups of G and

$$G/H_1 \cong H_2, \qquad G/H_2 \cong H_1.$$

Proof. Let $x \in G$. Since $G=H_1 H_2$, we have

$$x=x_1 x_2, \qquad x_1 \in H_1, \qquad x_2 \in H_2.$$

Consider the mapping $f: G \to H_1$ defined by $f(x)=x_1$ for $x \in G$. Let y be another element of G. Then,

$$y=y_1 y_2, \qquad y_1 \in H_1, \qquad y_2 \in H_2$$

and we have

$$\begin{aligned} x y &= (x_1 x_2)(y_1 y_2) \\ &= x_1(x_2 y_1) y_2 \\ &= x_1(y_1 x_2) y_2 && \text{since elements of } H_1 \text{ and } H_2 \text{ commute} \\ &= (x_1 y_1)(x_2 y_2), && x_1 y_1 \in H_1, x_2 y_2 \in H_2. \end{aligned}$$

Therefore,

$$f(x y)=x_1 y_1=f(x) \cdot f(y).$$

Thus, f is a homomorphism of G onto H_1. Kernel of f is given by

$$\begin{aligned} \ker(f) &= \{x \in G: f(x)=e_{H_1}=e_G\} \\ &= \{x \in G: f(x)=e_G\} \\ &= \{x_1 x_2: x_1=e_G\} \\ &= \{e_G x_2: x_2 \in H_2\}=H_2. \end{aligned}$$

Now, being the kernel of f, H_2 is a normal subgroup. Therefore, by the fundamental theorem of isomorphism,

$$G/H_2 \cong H_1.$$

Similarly it can be proved that $G/H_1 \cong H_2$.

5.17 RINGS

Definition 5.64

An algebraic system $(R, +, \cdot)$ consisting of a non-empty set R and two binary operations, denoted by $+$ and $\cdot$, is called a **Ring** if

(i) $(R, +)$ is an **abelian group** with respect to the binary operation $+$

(ii) $(R, \cdot)$ is a **semigroup** with respect to the binary operation $\cdot$

(iii) The binary operation $\cdot$ is distributive over the binary operation +, that is, for $x, y, z \in R$,

$$x \cdot (y+z)=x \cdot y+x \cdot z \quad \text{(right distributive law)}$$
$$(x+y) \cdot z=x \cdot z+y \cdot z \quad \text{(left distributive law)}.$$

Definition 5.65

Let $(R, +, \cdot)$ be a ring. If there exist an element 1 in R such that

$$x \cdot 1=1 \cdot x=x \quad \text{for all } x \in R,$$

then R is called a **ring with unit element**.

Definition 5.66

Let $(R, +, \cdot)$ be a ring. If

$$x \cdot y=y \cdot x \quad \text{for all } x, y \in R,$$

then the ring R is called a **commutative ring**.

Definition 5.67

Let $(R, +, \cdot)$ be a ring. If for $x, y \in R$,

$$x \cdot y=0 \Rightarrow x=0 \text{ or } y=0,$$

then R is called a **ring without zero divisor**.

EXAMPLE 5.57

Let **Z** be the set of all integers. Let + and $\cdot$ (addition and multiplication of integers) be the binary operations. Then,

(i) The operation + is closed binary operation because sum of two integers is again an integer.

(ii) For $x, y, z \in \mathbf{Z}$,

$$x+(y+z)=(x+y)+z \quad \text{(associative law for additon)}$$

(iii) For all $x \in \mathbf{Z}$

$$x+0=0+x=x \quad \text{(existence of identity)}.$$

Thus, 0 acts as additive identity.

(iv) For all $x \in \mathbf{Z}$

$$x+(-x)=(-x)+x)=0 \quad \text{(additive identity)}.$$

Therefore, $-x$ is inverse of x. Thus, all elements of **Z** have additive inverse.

(v) For all $x, y \in \mathbf{Z}$,

$$x+y=y+x \quad \text{(commutative law)}.$$

Thus (Z, +) becomes additive abelian group.

(vi) Product of two integers is an integer. Therefore $\cdot$ is a binary operation. Moreover, for $x, y, z \in \mathbf{Z}$,

$$x \cdot (y \cdot z)=(x \cdot y) \cdot z \quad \text{(associative law for multiplication of integer)}.$$

Thus (Z, $\cdot$) is a semigroup.

(vii) If $x, y, z \in \mathbf{Z}$, then

$$(x+y) \cdot z=x \cdot z+y \cdot z \quad \text{(left distribution law)},$$
$$x \cdot (y+z)=x \cdot y+x \cdot z \quad \text{(right distribution law)}.$$

Thus the binary operation · is distributive over the binary operation +. Hence (Z, +, ·) is a ring. Further, for $x, y \in \mathbf{Z}$

$$x \cdot y = y \cdot x.$$

Therefore (Z, +, ·) is a commutative ring.

Also, for $x \in \mathbf{Z}$, we have

$$x \cdot 1 = 1 \cdot x = x.$$

Therefore (Z, +, ·) is a commutative ring with unity.

EXAMPLE 5.58

Let R be the set of even integers under the usual operation of addition and multiplication. **Then R is a commutative ring.** But this ring has **no unit element** because

$$R = \{\ldots, -6, -4, -2, 0, 2, 4, 6, \ldots\}.$$

EXAMPLE 5.59

Let M be the set of $n \times n$ matrices. Then M becomes a ring with unity under the operations of addition of matrices and multiplication of matrices. The unit matrix

$$\begin{bmatrix} 1 & 0 & 0 & \ldots & 0 \\ 0 & 1 & 0 & \ldots & 0 \\ \ldots & \ldots & \ldots & \ldots & \ldots \\ \ldots & \ldots & \ldots & \ldots & \ldots \\ 0 & 0 & \ldots & \ldots & 1 \end{bmatrix}$$

acts as unit element for multiplication. The ring of matrices is **non-commutative because matrix multiplication is not commutative**.

EXAMPLE 5.60

Set of integer modulo n,

$$\mathbf{Z}/n\mathbf{Z} = \{m + n\mathbf{Z}: m \in \mathbf{Z}\}$$

is a **commutative ring**.

We know that $\mathbf{Z}/n\mathbf{Z}$ is abelian group under the operation of addition defined by

$$(m_1 + n\mathbf{Z}) + (m_2 + n\mathbf{Z}) = (m_1 + m_2) + n\mathbf{Z}.$$

We define multiplication in $\mathbf{Z}/n\mathbf{Z}$ by

$$(m_1 + n\mathbf{Z})\,(m_2 + n\mathbf{Z}) = m_1 m_2 + n\mathbf{Z}.$$

Then for **associativity of multiplication,** we have

$$\begin{aligned} &[(m_1 + n\mathbf{Z})\,(m_2 + n\mathbf{Z})](m_3 + n\mathbf{Z}) \\ &= (m_1\, m_2 + n\,\mathbf{Z})\,(m_3 + n\mathbf{Z}) \\ &= (m_1\, m_2)\, m_3 + n\mathbf{Z} \\ &= m_1(m_2\, m_3) + n\mathbf{Z} \qquad (\because \text{multiplication in } \mathbf{Z} \text{ is associative}) \\ &= (m_1 + n\mathbf{Z})\,(m_2\, m_3 + n\mathbf{Z}) \\ &= (m_1 + n\mathbf{Z})\,[(m_2 + n\mathbf{Z})\,(m_3 + n\mathbf{Z})]. \end{aligned}$$

Thus, $\mathbf{Z}/n\mathbf{Z}$ is a semigroup with respect to multiplication.

Further,

$$
\begin{aligned}
&(m_1+n\mathbf{Z})\,[(m_2+n\mathbf{Z})+(m_3+n\mathbf{Z})]\\
&\quad=(m_1+n\mathbf{Z})\,[(m_2+m_3)+n\mathbf{Z}]\\
&\quad=m_1(m_2+m_3)+n\mathbf{Z}\\
&\quad=(m_1\,m_2+m_1\,m_3)+n\mathbf{Z}\quad(\because \text{ multiplication is distributive over addition in } \mathbf{Z})\\
&\quad=(m_1\,m_2+n\mathbf{Z})+(m_1\,m_3+n\mathbf{Z})\\
&\quad=(m_1+n\mathbf{Z})\,(m_2+n\mathbf{Z})+(m_1+n\mathbf{Z})\,(m_3+n\mathbf{Z}).
\end{aligned}
$$

Thus multiplication in $\mathbf{Z}/n\mathbf{Z}$ is right distributive over addition in $\mathbf{Z}/n\mathbf{Z}$.

Similarly,

$$
\begin{aligned}
&[(m_1+n\mathbf{Z})+(m_2+n\mathbf{Z})]\,(m_3+n\mathbf{Z})\\
&\quad=(m_1+n\mathbf{Z})\,(m_3+n\mathbf{Z})+(m_2+n\mathbf{Z})\,(m_3+n\mathbf{Z}).
\end{aligned}
$$

Thus multiplication is left distributive also. Hence, multiplication operation in $\mathbf{Z}/n\mathbf{Z}$ is distributive over addition operation in $\mathbf{Z}/n\mathbf{Z}$. Thus, $\mathbf{Z}/n\mathbf{Z}$ is a ring. Also,

$$
\begin{aligned}
(m_1+n\mathbf{Z})\,(m_2+n\mathbf{Z})&=m_1m_2+n\mathbf{Z}\\
&=m_2m_1+n\mathbf{Z}\quad(\because \text{ multiplication in } \mathbf{Z} \text{ is commutative})\\
&=(m_2+n\mathbf{Z})\,(m_1+n\mathbf{Z}).
\end{aligned}
$$

Hence, $\mathbf{Z}/n\mathbf{Z}$ is a commutative ring.

Definition 5.68

An algebraic system $(R, +, \cdot)$ with two binary operations $+$ and $\cdot$ is called an **integral domain** if

(i) $(R, +)$ is an **abelian group**
(ii) $(R, \cdot)$ is a semigroup
(iii) The operation $\cdot$ is **commutative,** that is,

$x \cdot y = y \cdot x$ for all $x, y \in R$

(iv) The operation $\cdot$ is distributive over the operation $+$,
(v) For every $x \in R$, $1 \cdot x = x \cdot 1 = x$,
(vi) R is without zero divisor, that is $x \cdot y = 0 \Rightarrow x = 0$ or $y = 0$

Thus, a commutative ring with unit element is called an **integral domain** if it is without zero divisor.

EXAMPLE 5.61

Ring of integers is a commutative ring with unit element and also if $x, y \in \mathbf{Z}$, then

$$x\,y=0 \Rightarrow x=0 \text{ or } y=0.$$

Hence, ring of integers is an integral domain. Similarly, ring of rational numbers, ring of real numbers and ring of complex numbers are also integral domain.

Definition 5.69

A ring $(R, +, \cdot)$ with unity is called a **division ring (skew field)** if every non-zero element of R is invertible under multiplication $(\cdot)$.

It is clear from the definition that the set of non-zero elements of R form a multiplicative group, that is, $(R - \{0\}, \cdot)$ is a multiplicative group.

EXAMPLE 5.62

Let $D=\{a_0+a_1i+a_2j+a_3k;\ a_0,\ a_1,\ a_2,\ a_3 \in R\}$ and $i^2=j^2=k^2=-1$; $ij=-ji=k$, $jk=-kj=i$, $ki=-ik=j$. Then, $(D, +, \cdot)$ is a division ring under the operations of $+$ and $\cdot$ defined by

$$(a_0+a_1i+a_2j+a_3k)+(\beta_0+\beta_1i+\beta_2j+\beta_3k)=(a_0+\beta_0)+(a_1+\beta_1)i+(a_2+\beta_2)j+(a_3+\beta_3)k.$$

$$\begin{aligned}(a_0+a_1i+a_2j+a_3k)\,(\beta_0+\beta_1i+\beta_2j+\beta_3k)&=(a_0\beta_0-a_1\beta_1-a_2\beta_2-a_3\beta_3)\\ &+(a_0\beta_1-a_1\beta_0+a_2\beta_3-a_3\beta_2)i+(a_0\beta_2-a_2\beta_0-a_3\beta_1-a_1\beta_3)j\\ &+(a_0\beta_3+a_3\beta_0+a_1\beta_2-a_2\beta_1)k.\end{aligned}$$

If $x=a_0+a_1i+a_2j+a_3k\neq 0$ in D, then there exists an element $y=(a_0/\beta)-(a_1/\beta)i-(a_2/\beta)j-(a_3/\beta)k$ in D such that $xy=1$, where $\beta=a_0^2+a_1^2+a_2^2+a_3^2\neq 0$.

Definition 5.70

An algebraic system $(R, +, \cdot)$ with two binary operations $+$ and $\cdot$ is called a **Field** if

(i) $(R, +)$ is an abelian group
(ii) $(R - \{0\}, \cdot\}$ is an abelian group
(iii) The binary operation $\cdot$ is distributive over the binary operation $+$.

Thus, **a commutative division ring is called a field.**

Alternatively we may define a field as:

A commutative ring with unity in which every non-zero element has an inverse with respect to the binary operation $\cdot$ is called a **field**.

EXAMPLE 5.63

(i) Set of rational numbers is a field under the binary operation of usual addition and multiplication.
(ii) Set of real number is a field.
(iii) Set of complex number is a field.
(iv) Set of all integers do not form a field. For example, inverse of 2 is not in **Z**.

Theorem 5.55

Ring of integers mod p is an integral domain if and only if p is a prime.

Proof. Suppose first that $\mathbf{Z}/p\mathbf{Z}$ is an integral domain. We shall show that p is a prime. Suppose on the contrary that p is not a prime, that is,

$$p=m_1\, m_2,\quad m_1,\, m_2\neq p.$$

Then,

$$\begin{aligned}(m_1+p\mathbf{Z})\,(m_2+p\mathbf{Z})&=m_1\, m_2+p\mathbf{Z}\\ &=p+p\mathbf{Z}=p\mathbf{Z}.\end{aligned}$$

But neither $m_1+p\mathbf{Z}$ nor $m_2+p\mathbf{Z}$ is equal to $p\mathbf{Z}$ and so $\mathbf{Z}/p\mathbf{Z}$ is not without zero divisor. This contradicts the supposition that $\mathbf{Z}/p\mathbf{Z}$ is an integral domain. Hence, p is a prime.

Conversely, let p be a prime. Then for $m_1+p\mathbf{Z}$, $m_2+p\mathbf{Z} \in \mathbf{Z}/p\mathbf{Z}$, we have

$$\begin{aligned}(m_1+p\mathbf{Z})(m_2+p\mathbf{Z})=p\mathbf{Z} &\Rightarrow m_1m_2+p\mathbf{Z}=p\mathbf{Z}\\ &\Rightarrow m_1m_2 \in p\mathbf{Z}\\ &\Rightarrow p \mid m_1m_2\\ &\Rightarrow p \mid m_1 \text{ or } p \mid m_2 \text{ (since } p \text{ is a prime).}\end{aligned}$$

If $p \mid m_1$, then $m_1+p\mathbf{Z}=p\mathbf{Z}$ and if $p \mid m_2$, then $m_2+p\mathbf{Z}=p\mathbf{Z}$. Therefore

$$(m_1+p\mathbf{Z})(m_2+p\mathbf{Z})=p\mathbf{Z} \Rightarrow m_1+p\mathbf{Z}=p\mathbf{Z} \text{ or } m_2+p\mathbf{Z}=p\mathbf{Z}.$$

Hence, $\mathbf{Z}/p\mathbf{Z}$ is without zero divisor. Since $\mathbf{Z}/p\mathbf{Z}$ is a commutative ring with unity, it follows that $\mathbf{Z}/p\mathbf{Z}$ is an integral domain.

Theorem 5.56

Every field is without zero divisor.

Proof. Let F be a field and $x, y \in F$, $x \neq 0$. Then,

$$\begin{aligned}x\,y=0 &\Rightarrow x^{-1}\,x\,y=0\\ &\Rightarrow (x^{-1}\,x)\,y=0\\ &\Rightarrow e\,y=0\\ &\Rightarrow y=0.\end{aligned}$$

Similarly, if $y \neq 0$, then

$$\begin{aligned}x\,y=0 &\Rightarrow x\,y\,y^{-1}=0\\ &\Rightarrow x(y\,y^{-1})=0\\ &\Rightarrow x\,e=0\\ &\Rightarrow x=0.\end{aligned}$$

Hence $x\,y=0 \Rightarrow x=0$ or $y=0$. Hence, F is without zero divisor.

Corollary 5.6

Every field is an integral domain.

Proof. Since a field is a commutative ring with unity and is also without zero divisor, it follows that it is an integral domain.

The **converse of this corollary is not true**. For example, the ring of integers is an integral domain but it is not a field.

Theorem 5.57

A finite integral domain is a field.

Proof. Let D be an integral domain having n elements $a_1, a_2, \ldots, a_n$. To prove the result, it is sufficient to show that every non-zero element of D is invertible under multiplication. So, let x be a non-zero element of D. Then the elements $a_1x, a_2x, \ldots, a_nx$ are all in D. We claim that these elements are all distinct. Suppose on the contrary that

$$a_ix=a_jx, \qquad i\neq j, \tag{1}$$

Then

$$\begin{aligned}&(a_i-a_j)\,x=0\\ &\Rightarrow x=0 \text{ or } a_i-a_j=0,\end{aligned}$$

because D is without zero divisor. But, by supposition, $x \neq 0$ and so $a_i - a_j = 0$ implying $a_i = a_j$. This contradicts (1). Hence $a_1 x, \ldots, a_n x$ are n distinct elements lying in D, which has exactly n elements. Since $1 \in D$, there is an element $a \in D$ such that $ax = 1$, which implies a is inverse of x. Hence every non-zero element of D is invertible and so D is a field.

Corollary 5.7

$\mathbf{Z}/p\mathbf{Z}$, p is a prime, is a field.

Proof. We know that if p is a prime, then $\mathbf{Z}/p\mathbf{Z}$ is an integral domain. Moreover, this integral domain has elements

$$p\mathbf{Z},\ 1+p\mathbf{Z}, \ldots,\ 1+(p-1)\,\mathbf{Z}$$

and so it is a finite integral domain. Hence, by the above theorem, $\mathbf{Z}/p\mathbf{Z}$ is a field.

Definition 5.71

A subset S of a ring $(R, +, \cdot)$ is called a **subring** of R if S itself is a ring under binary operation $+$ and $\cdot$ in R.

Thus, a non-empty subset S of R is a subring if

(i) $a \in S, b \in S \Rightarrow a - b \in S$
(ii) $a, b \in S \Rightarrow a\,b \in S$.

For example, if R is a ring, then $\{0\}$ and R itself are subrings of R. Similarly, the set of even integers is a subring of the ring of integers, and the set of rational numbers is a subring of the ring of real numbers.

5.18 RING HOMOMORPHISM

Definition 5.72

Let $(R, +, \cdot)$ and $(S, \oplus, \cdot)$ be two rings. Then a mapping $f: R \to S$ is called a **ring homomorphism** if for $a, b \in R$,

(i) $f(a+b) = f(a) \oplus f(b)$,
(ii) $f(a \cdot b) = f(a) \cdot f(b)$.

If $f: R \to S$ is a ring homomorphism, then $(S, \oplus, \odot)$ is called a **homomorphic image** of $(R, +, \cdot)$.

If, in addition to (i) and (ii), we have

(iii) f is one-one and onto

then f is called a **ring isomorphism**.

EXAMPLE 5.64

Let $(R, +, \cdot)$ and $(S, +, \cdot)$ rings. Consider a mapping $f: R \to S$ defined by $f(a) = a$ for all $a \in R$. Then

$$f(a+b) = a+b = f(a) + f(b),$$
$$f(a \bullet b) = a\,b = f(a) \bullet f(b).$$

Hence, f is a ring homomorphism.

EXAMPLE 5.65

Let $(R, +, \cdot)$ be a ring. Then the mapping $f: R \to R$ defined by $f(a)=2a$ for all $a \in R$ is not a ring homomorphism. In fact,

$$f(a+b)=2(a+b)=2a+2b=f(a)+f(b),$$
$$f(a \cdot b)=2ab \neq f(a) \cdot f(b),$$

since for the given mapping $f(a) \cdot f(b)=(2a)(2b)=4ab$.

5.19 IDEALS AND QUOTIENT RINGS

Definition 5.73

Let S be a subring of ring R. If

$$x \in S, a \in R \Rightarrow ax \in S,$$

then S is called **left ideal** of R.

If $x \in S$, $a \in R \Rightarrow x\,a \in S$, then S is called the **right ideal** of R.

If $x \in S$, $a \in R \Rightarrow x\,a \in S$ and $a\,x \in S$, then S is called **two-sided ideal** or simply **ideal** of R.

We note that if R is a commutative ring, then all the three notions are same since in that case $ax=xa \in S$. Further, every ring has two trivial ideals:

(i) R itself and is called **unit ideal**

(ii) **Zero ideal** $\{0\}$, consisting of zero element only

Any other ideal except these two trivial ideals is called a **proper ideal**.

Theorem 5.58

The intersection of any two left ideals of a ring is again a left ideal of the ring.

Proof. Let S_1 and S_2 be two left ideals of R. Then S_1 and S_2 being subrings of R, $S_1 \cap S_2$ is also a subring of R. Again, let $x \in S_1 \cap S_2$ and let $a \in R$. Then, $x \in S_1$ and $x \in S_2$. Since S_1 and S_2 are left ideals,

$$a \in R, x \in S_1 \Rightarrow a\,x \in S_1,$$
$$a \in R, x \in S_2 \Rightarrow a\,x \in S_2,$$

which taken together imply $a\,x \in S_1 \cap S_2$. Hence, $S_1 \cap S_2$ is a left ideal.

Theorem 5.59

Let ker(T) be the kernel of a ring homomorphism $T: R \to S$. Then, ker (T) is a two-sided ideal of R.

Proof. Let $a, b \in$ ker (T). Then

$$T(a)=T(b)=0.$$

Therefore, by ring homomorphism,

$$T(a+b)=T(a)+T(b)=0+0=0,$$
$$T(a\,b)=T(a)\cdot T(b)=0\cdot 0=0,$$

which implies that $a+b$, $ab \in$ ker (T). Hence, ker (T) is a subring of R.

Now let $a \in$ ker (T) and $r \in R$. It suffices to prove that ar, $r\,a \in$ ker (T).

$$\begin{aligned} T(a\,r) &= T(a)\cdot T(r) \\ &= 0 \cdot T(r) \qquad (\because a \in \ker(T) \Rightarrow T(a)=0)=0 \end{aligned}$$

This implies that $a\,r \in \ker(T)$. Similarly,

$$T(r\,a)=T(r)\,T(a)=T(r)\cdot 0=0$$
$$\Rightarrow r\,a \in \ker(T).$$

Hence, ker (T) is an ideal of R.

Theorem 5.60

A field has no proper ideal.

Proof. Let us suppose that S is a proper ideal of a field F. Then $S \subseteq F$. If $x \in S$, then $x\,x^{-1} \in S$. But $x\,x^{-1}=1$. Therefore, $1 \in S$. As S is an ideal, $y \in F \Rightarrow y \cdot 1 \in S$. Thus, $y \in F \Rightarrow y \in S$. That is, $F \subseteq S$. Therefore, $F=S$. This contradicts our supposition. Hence F has no proper ideal.

Theorem 5.61

If a commutative ring R with unity has no proper ideal, then R is a field.

Proof. It suffices to prove that every non-zero element of R is invertible. Let a be a non-zero element of R. Consider the set

$$S=\{x\,a: x \in R\}.$$

We claim that S is an ideal of R. To show it, let $p, q \in S$. Then

$$p=x_1\,a,\ q=x_2\,a,\ x_1, x_2 \in R,$$
$$p+q=x_1\,a+x_2\,a=(x_1+x_2)\,a \in S, \quad \text{since } x_1+x_2 \in \text{R}.$$

Similarly

$$-p=-x_1\,a=(-x_1)\,a \in S.$$

Therefore, S is an additive subgroup of R.

Moreover, if $r \in R$, then

$$r\,p=r(x_1 a)=(r\,x_1)\,a \in S.$$

Since R is commutative, $r\,p \in S \Rightarrow p\,r \in S$. Hence S is an ideal of R. But by supposition, $S=\{0\}$ or $S=R$. Since $1 \in R$ implies $1 \cdot a \in S$, that is, $a \in S$, it follows that S is not equal to $\{0\}$. Hence $S=R$. By definition of S, $1=x\,a$, $x \in R$. Therefore, every non-zero element of R is invertible and hence R is a field.

5.19.1 Quotient Ring

Let A be an ideal of a ring R. Then R is an abelian group and A is an additive subgroup of R. But every subgroup of an abelian group is normal, therefore A is a normal subgroup of R. So we can define the set

$$R/A=\{r+A,\ r \in R\}.$$

We shall prove that R/A is a ring. This ring will be called a **quotient ring**.

Theorem 5.62

Let A be an ideal of R. Then the set

$$R/A=\{r+A,\ r \in R\}$$

is a ring.

Proof. We define addition and multiplication compositions as follows:

$$\left.\begin{aligned}(r+A)+(s+A)&=(r+s)+A\\(r+A)(s+A)&=rs+A\end{aligned}\right\}\quad \text{for } r,\ s\in R,.$$

We show first that the above-defined binary operations are well defined. Let

$$\left.\begin{aligned}r+A&=r_1+A\\s+A&=s_1+A\end{aligned}\right\}\quad \text{for } r_1,s_1\in R,$$

which implies $r-r_1\in A,\ s-s_1\in A$. Then

$$\begin{aligned}&(r+s)-(r_1+s_1)=(r-r_1)+(s-s_1)\in A\\&\Rightarrow (r+s)+A=(r_1+s_1)+A,\end{aligned}$$

which proves that addition is well defined.

Moreover,

$$\begin{aligned}r\,s-r_1\,s_1&=r\,s-r_1\,s+r_1\,s-r_1\,s_1\\&=(r-r_1)\,s+r_1(s-s_1)\in A.\end{aligned}$$

Therefore, $r\,s+A=r_1\,s_1+A$ and hence multiplication composition is also well defined. We now prove that these compositions satisfy all the properties of a ring.

(i) **Associativity of addition:** If $r+A,\ s+A,\ t+A\in R/A$, then

$$\begin{aligned}[(r+A)+(s+A)]+(t+A)&=[(r+s)+A]+(t+A)\\&=[(r+s)+t]+A\\&=(r+A)+[(s+t)+A]\\&=(r+A)+[(s+A)+(t+A)].\end{aligned}$$

(ii) **Existence of the identity of addition:** If $r+A\in R/A$, then

$$(0+A)+(r+A)=r+A$$

and

$$(r+A)+(0+A)=r+A.$$

Therefore $0+A=A$ is identity element of addition.

(iii) **Existence of additive inverse:** if $r+A\in R/A$, then

$$(r+A)+(-r+A)=[r+(-r)]+A=0+A=A$$

and

$$(-r+A)+(r+A)=[(-r)+r]+A=0+A=A,$$

which shows that $-r+A$ is the inverse of $r+A$.

(iv) **Commutativity of addition:** If $r+A,\ s+A\in R/A$, then

$$\begin{aligned}(r+A)+(s+A)&=(r+s)+A\\&=(s+r)+A,\ \text{by commutativity of}+\text{in } R.\\&=(s+A)+(r+A).\end{aligned}$$

(v) **Associativity of multiplication:** If $r+A$, $s+A$, $t+A \in R/A$, then

$$\begin{aligned}[(r+A)+(s+A)&=(r\,s+A)\,(t+A)\\ &=(r\,s)\,t+A\\ &=r\,(s\,t)+A \quad \text{by associativity of multiplication in } R.\\ &=(r+A)\,(s\,t+A)\\ &=(r+A)\,[(s+A)\,(t+A)].\end{aligned}$$

(vi) **Distributivity of multiplication over addition:** If $r+A$, $s+A$, $t+A \in R/A$, then

$$\begin{aligned}(r+A)\,[(s+A)+(t+A)]&=(r+a)\,[(s+t)+A]\\ &=r(s+t)+A=(r\,s+r\,t)+A.\\ &=(r\,s+A)+(r\,t+A)\\ &=(r+A)\,(s+A)+(r+A)\,(t+A).\end{aligned}$$

Similarly,

$$[(r+A)+(s+A)]\,(t+A)=(r+A)\,(t+A)+(s+A)\,(t+A).$$

Hence R/A is a ring.

Remark 5.9 If R is commutative, then R/A will be abelian since if $r+A$, $s+A \in R/A$, then by the commutativity of R, we have

$$\begin{aligned}(r+A)\,(s+A)&=r\,s+A=s\,r+A\\ &=(s+A)\,(r+A).\end{aligned}$$

In addition, if R has unit element then R/A has also identity $1+A$.

Theorem 5.63

Every ideal A of a ring R is a kernel of some ring homomorphism.

Proof. Let ϕ: $R \rightarrow R/A$ be a mapping defined by $\phi(r)=r+A$. This mapping is known as natural mapping. If r, $s \in R$, then

$$\begin{aligned}\phi(r+s)&=(r+s)+A\\ &=(r+A)+(S+A)=\phi(r)+\phi(s)\end{aligned}$$

and

$$\begin{aligned}\phi(r\,s)&=r\,s+A\\ &=(r+A)\,(s+A)=\phi(r)\,\phi(s).\end{aligned}$$

Therefore ϕ is a ring homomorphism. Kernel of this homomorphism, is given by

$$\begin{aligned}K(\phi)&=\{r\text{: } r \in R,\ \phi(r)=A\}\\ &=\{r\text{: } r \in R,\ r+A=A\}\\ &=\{r\text{: } r \in A\}=A,\end{aligned}$$

which proves the required result.

Theorem 5.64

Let ϕ: $R \rightarrow S$ be a ring homomorphism of R onto S. Then

$$R/K(\phi) \simeq S.$$

Proof. We know that $K(\phi)$ is an ideal of R. Therefore, $R/K(\phi)$ is defined. Elements of this set are cosets of $K(\phi)$ in R. Let $r+K \in R/K(\phi)$. Then

$$\begin{aligned}\phi(r+x) &= \phi(r)+\phi(x) \quad \text{for all } x \in K(\phi)\\ &= \phi(r)+0 \qquad (\because x \in K(\phi) \Rightarrow \phi(x)=0).\end{aligned}$$

Thus we can define a mapping $\psi(r+K)=\phi(r)$ for all $r \in R$. We shall prove ψ is an isomorphism. Let $r+K, s+K \in R/K(\phi)$. Then,

$$\begin{aligned}\psi[(r+K)+(s+K)] &= \psi[(r+s)+K]\\ &= \phi(r+s)=\phi(r)+\phi(s)\\ &= \psi(r+K)+\psi(s+K)\end{aligned}$$

and

$$\begin{aligned}\psi(r+K)\,(s+K)] &= \psi(r\,s+K)\\ &= \phi(r\,s)=\phi(r)\,\phi(s)\\ &= \psi(r+K)\,\psi(s+K).\end{aligned}$$

Therefore ψ is a ring homomorphism. If $x \in S$, then

$$\begin{aligned}x &= \phi(r),\ r \in R \quad (\text{ since } \phi \text{ is onto mapping})\\ &= \psi(r+K).\end{aligned}$$

Therefore to each element $x \in S$ there corresponds an element $r+K$ of $R/K(\phi)$ such that $\psi(r+K)=x$. Hence ψ is surjective.

Moreover,

$$\begin{aligned}\psi(r+K)=\psi(s+K) &\Rightarrow \phi(r)=\phi(s)\\ &\Rightarrow \phi(r-s)=0\\ &\Rightarrow r-s \in K\\ &\Rightarrow r+K=s+K.\end{aligned}$$

Therefore ψ is one-one mapping also. Hence ψ is an isomorphism, as a consequence of which

$$R/K(\phi) \simeq S.$$

Theorem 5.65

A homomorphic image of a ring R is also a ring.

Proof. Let $T: R \rightarrow S$ be a ring homomorphism. Then homomorphic image of R is

$$\text{Im}(T)=\{x: x \in S, \qquad x=T(r), r \in R\}.$$

We know that $T(0)=0$. Therefore, $\text{Im}(T)$ is non-empty. If $x, y \in \text{Im}(T)$, then there exists $r, s \in R$ such that

$$x=T(r), \qquad y=T(s).$$

Therefore,

$$x+y=T(r)+T(s),$$
$$= T(r+s) \in \text{Im}(T), \text{ since } T \text{ is a ring homomorphism}$$

and

$$x\,y=T(r)\,T(s)$$
$$= T(r\,s) \in \text{Im}(T), \text{ since } T \text{ is a ring homormorphism.}$$

Hence, Im(T) is a subring of S.

5.20 POLYNOMIAL RINGS

Definition 5.74

Let A be an arbitrary ring. By a polynomial over a ring A, is meant an ordered system $(a_0, a_1, a_2, \ldots, a_n, \ldots)$ of elements of A such that all except at the most a finite number of elements are zero.

Two polynomials $(a_0, a_1, a_2, \ldots, a_n, \ldots)$ and $(b_0, b_1, b_2, \ldots, b_n, \ldots)$ are said to be equal if and only if $a_n=b_n$, $n \in N$.

Let R be a ring and P be the set of all polynomials. Let $(a_0, a_1, \ldots, a_n, \ldots)$and $(b_0, b_1, b_2, \ldots, b_n, \ldots)$ be any two elements of P. If $a_n=0$ for all $n \geq j$ and $b_n=0$ for all $n \geq k$. Then,

$$a_n+b_n=0 \text{ for all } n \geq \max (j, k).$$

Thus all except at the most, a finite number of elements in the ordered system $(a_0+b_0, a_1+b_1, \ldots)$ are zero. Therefore, $(a_0+b_0, a_1+b_1, a_2+b_2, \ldots, a_n+b_n, \ldots) \in P$. Hence we can define addition composition in P by

$$(a_0, a_1, a_2, \ldots, a_n, \ldots)+(b_0, b_1, b_2, \ldots, b_n, \ldots)=(a_0+b_0, a_1+b_1, \ldots, a_n+b_n, \ldots).$$

Multiplication in P is defined by

$$(a_0, a_1, a_2, \ldots, a_n)\,(b_0, b_1, \ldots, b_n, \ldots)=(c_0, c_1, c_2, \ldots, c_n, \ldots),$$

where

$$c_0=a_0\,b_0,$$
$$c_1=a_0\,b_1+a_1\,b_0,$$
$$c_2=a_0\,b_2+a_1\,b_1+a_2\,b_0,$$
$$\cdots\cdots\cdots\cdots\cdots\cdots\cdots\cdots\cdots$$
$$c_n=a_0\,b_n+a_1\,b_{n-1}+\ldots+a_n \mathrm{b}_0=\sum_{m=0}^{n} a_m b_{n-m}\,.$$

If $a_n=0$ for all $n \geq j$ and $b_n=0$ for all $n \geq k$, then

$$c_n=0 \text{ for all } n \geq (j+k).$$

Thus, product of two polynomials is again a polynomial.

The set P of all polynomials over a ring R form a ring under these operations of addition and multiplication.

$$\text{Let } (a_0, a_1, a_2, \ldots, a_n, \ldots), (b_0, b_1, b_2, \ldots), (c_0, c_1, c_2, \ldots) \in P$$

Then

(i) $(a_0, a_1, a_2, \ldots)+[(b_0, b_1, b_2, \ldots)+(c_0, c_1, c_2, \ldots)]$
$= (a_0, a_1, a_2, \ldots)+[(b_0+c_0, b_1+c_1, b_2+c_2, \ldots)$
$= (a_0+b_0+c_0, a_1+b_1+c_1, a_2+b_2+c_2, \ldots)$
$= (a_0+b_0, a_1+b_1, a_2+b_2, \ldots)+(c_0, c_1, c_2, \ldots)$
$= [(a_0, a_1, a_2, \ldots)+(b_0, b_1, b_2, \ldots)]+(c_0, c_1, c_2, \ldots),$

(ii) $(a_0, a_1, a_2, \ldots)+(0, 0, 0, \ldots)=(a_0+0, a_1+0, a_2+0, \ldots)$
$= (a_0, a_1, a_2, \ldots)$

and

$$(0, 0, 0, \ldots)+(a_0, a_1, a_2, \ldots)=(0+a_0, 0+a_1, 0+a_2, \ldots) = (a_0, a_1, a_2, \ldots),$$

(iii) $(a_0, a_1, a_2, \ldots)+(-a_0, -a_1, -a_2, \ldots) = (a_0 - a_0, a_1 - a_1, a_2 - a_2, \ldots)$
$=(0, 0, 0, \ldots)$

and $(-a_0, -a_1, -a_2, \ldots)+(a_0, a_1, a_2, \ldots) = (0, 0, 0, \ldots)$

(iv) $(a_0, a_1, a_2, \ldots)+(b_0, b_1, b_2, \ldots)=(a_0+b_0, a_1+b_1, a_2+b_2, \ldots)$
$= (b_0+a_0, b_1+a_1, b_2+a_2, \ldots)$
$= (b_0, b_1, b_2, \ldots)+(a_0, a_1, a_2, \ldots),$

(v) $[(a_0, a_1, a_2, \ldots) (b_0, b_1, b_2, \ldots)] (c_0, c_1, c_2, \ldots)$
$= (d_0, d_1, d_2, \ldots) (c_0, c_1, c_2, \ldots)$ where $d_n = \sum_{j+k=n} a_j b_k$
$= (e_0, e_1, e_2, \ldots),$

where

$$e_m = \sum_{p+q=m} d_p c_q = \sum_{p+q=m} \left(\sum_{j+k=p} a_j b_k \right) c_q = \sum_{j+k+q=m} a_j b_k c_q.$$

Similarly, it can be shown that

$$(a_0, a_1, a_2, \ldots) [(b_0, b_1, b_2, \ldots) (c_0, c_1, c_2, \ldots)]=(f_0, f_1, f_2, \ldots),$$

where

$$f_m = \sum_{j+k+q=m} a_j b_k c_q.$$

Hence,

$(a_0, a_1, a_2, \ldots) [(b_0, b_1, \ldots) (c_0, c_1, c_2, \ldots)] = (a_0, a_1, a_2, \ldots) [(b_0, b_1, b_2, \ldots) (c_0, c_1, c_2, \ldots)]$

(vi) $(a_0, a_1, a_2, \ldots) [(b_0, b_1, b_2, \ldots)+(c_0, c_1, c_2, \ldots)] = (a_0, a_1, a_2, \ldots) (b_0+c_0, b_1+c_1, b_2+c_2, \ldots)$
$= (d_0, d_1, d_2, \ldots),$

where

$$d_m = \sum_{j+k=m} a_j(b_k c_k)$$
$$= \sum_{j+k=m} a_j b_k + \sum_{j+k=m} a_j c_k.$$

Also

$$(a_0, a_1, a_2, \ldots)(b_0, b_1, b_2, \ldots)=(f_0, f_1, f_2, \ldots),$$
$$(a_0, a_1, a_2, \ldots)(c_0, c_1, c_2, \ldots)=(g_0, g_1, g_2, \ldots).$$

Hence,

$$(a_0, a_1, a_2, \ldots)[(b_0, b_1, b_2, \ldots)+(c_0, c_1, c_2, \ldots)]$$
$$= (a_0, a_1, a_2, \ldots)(b_0, b_1, b_2, \ldots)+(a_0, a_1, a_2, \ldots)(c_0, c_1, \ldots).$$

Hence P is a ring. We call this ring of polynomials as **polynomial ring over *R*** and it is denoted by $R[x]$.

Let,

$$Q=\{(a, 0, 0, \ldots): a \in R\}.$$

Then a mapping $f: R \to Q$ defined by $f(a)=(a, 0, 0, \ldots)$ is an isomorphism. In fact,

$$f(a+b)=(a+b, 0, 0, \ldots)$$
$$= (a, 0, 0, \ldots)+(b, 0, 0, \ldots)$$
$$= f(a)+f(b),$$
$$f(a\,b)=(a\,b, 0, 0, \ldots)$$
$$= (a, 0, 0, \ldots)(b, 0, 0, \ldots)=f(a)f(b)$$

and

$$f(a)=f(b) \Rightarrow (a, 0, 0, \ldots)=(b, 0, 0, \ldots)$$
$$\Rightarrow a=b.$$

Hence

$$R \cong Q \qquad \text{(i)}$$

So we can identity the polynomial $(a, 0, 0, \ldots)$ with a.

If we represent $(0, 1, 0, \ldots)$ by x then we can see that

$$x^2=(0, 0, 1, 0, \ldots)$$
$$x^3=(0, 0, 0, 1, \ldots)$$

$$x^n = (\underbrace{0,0,0,\ldots 0}_{n\text{ terms}},1,0\ldots)$$

Therefore for $(a, 0, 0, \ldots) \in Q$ we have

$$\left.\begin{array}{l}(a,0,0,\ldots)x = (0,a,0,\ldots)\\(a,0,0,\ldots)x^2 = (0,0,a\ldots)\\\hline\\(a,0,0,\ldots)x^n = (\underbrace{0,0,0,\ldots0}_{n \text{ terms}},a,0\ldots)\end{array}\right\} \quad \text{(ii)}$$

If $(a_0, a_1, \ldots, a_n, 0, \ldots)$ be any arbitrary element of the polynomial ring P, then by (ii) we have

$$\begin{aligned}(a_0, a_1, \ldots, a_n, 0, \ldots) &= (a_0, 0, \ldots)+(0, a_1, \ldots)+\ldots +(\underbrace{0,0,0,\ldots,0}_{n\text{ terms}},a_n,0,\ldots)\\&= (a_0, 0, \ldots)+(a_1,0, \ldots)(0, 1, 0, \ldots)+\ldots\\&\quad + (a_n, 0, 0, \ldots)((\underbrace{0,0,0,\ldots,0,0}_{n\text{ terms}},1,0,\ldots)\\&= (a_0, 0, \ldots)+(a_1, 0, 0, \ldots)x+\ldots+(a_n, 0, 0, \ldots)x^n\\&= a_0+a_1 x+a_n x^n \qquad \text{(by (i))}.\end{aligned}$$

Hence every element $(a_0, a_1, a_2, \ldots)$ of P can be denoted by

$$a_0+a_1 x+a_2 x^2+\ldots+a_n x^n.$$

The number $a_0, a_1, \ldots, a_n$ are called **coefficients of the polynomial**. If the coefficient a_n of x^n is non-zero, then it is called **leading coefficient** of $a_0+a_1 x+\ldots+a_n x^n$.

A polynomial consisting of only one term a_0 is called **constant polynomial.**

5.20.1 Degree of Polynomial

Let $f(x)=a_0+a_1 x+\cdots+a_n x^n$ be a polynomial. If $a_n \neq 0$, then n is called the **degree** of $f(x)$. We denote it by $\deg f(x)=n$.

It is clear that degree of a constant polynomial is zero.

If

$$f(x)=a_0+a_1 x+a_2 x^2+\ldots+a_m x^m,\ a_m \neq 0$$

and

$$g(x)=b_0+b_1 x+\ldots+b_n x^n,\ b_n \neq 0$$

are two elements of $R[x]$, then $\deg f(x)=m$ and $\deg g(x)=n$ and

$$f(x)+g(x)=(a_0+a_1 x+a_2 x^2+\ldots+a_m x^m)+(b_0+b_1 x+b_2 x^2+\ldots+b_n x^n).$$

If $m=n$ and $a_m+b_n \neq 0$, then

$$f(x)+g(x)=(a_0+b_0)+(a_1+b_1)x+\ldots+(a_m+b_m)x^m.$$

Therefore in this case

$$\deg[f(x)+g(x)]=m.$$

It is also clear that if $m=n$ and $a_m+b_m=0$, then

$$\deg[f(x)+g(x)]<m.$$

If $m>n$, then

$$f(x)+g(x)=(a_0+b_0)+(a_1+b_1)x+\ldots+(a_n+b_n)x^n+a_{n+1}x^{n+1}+\ldots+a_m x^m.$$

Therefore in this situation

$$\deg[f(x)+g(x)]=m.$$

Similarly it can be seen that if $m<n$, then

$$\deg[f(x)+g(x)]=n.$$

It follows therefore that if $m\neq n$, then

$$\deg[f(x)+g(x)]=\max(m, n).$$

Also,

$$f(x)\,g(x)=a_0\,b_0+(a_0b_1+a_1b_0)x+\cdots+a_m b_n x^{m+n}.$$

Therefore,

$$\deg[f(x)g(x)=\begin{cases} m+n & \text{if } a_m b_n \neq 0 \\ < m+n & \text{where } a_m b_n = 0 \end{cases}$$

If R is without zero divisor, then

$$a_m\,b_n\neq 0 \qquad \text{since } a_m\neq 0,\ b_n\neq 0.$$

Hence for such a ring R we have

$$\deg[f(x)\,g(x)]=m+n=\deg f(x)+\deg g(x).$$

If R is without zero divisor and $f(x)$ and $g(x)$ are non-zero polynomial of $R[x]$, then

$$\deg f(x)\leq \deg\,[f(x)\,g(x)] \qquad (\because \deg g(x)\geq 0).$$

Theorem 5.66

If R is an integral domain, then so is also polynomial ring $R[x]$.

Proof. R is a commutative ring with unity. Therefore $R[x]$ is commutative with unit element. It suffices to prove that $R[x]$ is without zero divisor. Let

$$f(x)=\sum_{i=0}^{m} a_i x^i, \quad a_m\neq 0$$

and

$$g(x)=\sum_{i=0}^{n} b_j x^i, \quad b_n\neq 0,$$

be two non-zero polynomials of $R[x]$ and let m and n be their degrees, respectively.

Since R is an integral domain and $a_m\neq 0$, $b_n\neq 0$, therefore $a_m\,b_n\neq 0$. Hence $f(x)\,g(x)\neq 0$. Hence $R[x]$ is without zero divisor and therefore an integral domain.

5.21 DIVISION ALGORITHM FOR POLYNOMIALS OVER A FIELD

Theorem 5.67

Corresponding to any two polynomials $f(x)$ and $g(x)\neq 0$ belonging to $F[x]$, there exist uniquely two polynomials $q(x)$ and $r(x)$ also belonging to $F[x]$ such that

$$f(x)=g(x)\,q(x)+r(x),$$

where

$$r(x)=0 \quad \text{or} \quad \deg r(x)<\deg g(s).$$

Proof. Let

$$f(x)=\sum_{i=0}^{m} a_i x^i, \quad a_m\neq 0,$$

$$g(x)=\sum_{i=0}^{n} b_i x^i, \quad b_n\neq 0.$$

Then either

$$\deg f(x)<\deg g(x) \tag{i}$$

or

$$\deg f(x)\geq \deg g(x). \tag{ii}$$

In the first case we write

$$f(x)=g(x)\,0+f(x)$$

so that $q(x)=0$ and $r(x)=f(x)$.

In respect of the second case we shall prove the existence of $q(x)$ and $r(x)$ by mathematical induction on the degree of $f(x)$. If $\deg f(x)=1$, then the existence of $q(x)$ and $r(x)$ is obvious. Let us suppose that the result is true when $\deg f(x)\leq m-1$. If

$$h(x)=f(x)-\left(\frac{a_m}{b_n}\right)x^{m-n}g(x), \tag{iii}$$

then $\deg h(x)\leq m-1$.

Hence by supposition

$$h(x)=g(x)\,q_1(x)+r(x), \tag{iv}$$

where $r(x)=0$ or $\deg r(x)\leq \deg g(x)$.

From (iii) and (iv) we have

$$f(x)-\left(\frac{a_m}{b_n}\right)x^{m-n}g(x)=g(x)q_1(x)+r(x),$$

that is,

$$\begin{aligned} f(x)&=g(x)\left[q_1(x)\left(\frac{a_m}{b_n}\right)x^{m-n}\right]+r(x)\\ &=g(x)q(x)+r(x), \end{aligned}$$

where

$$q(x) = q_1(x) + \left(\frac{a_m}{b_n}\right) x^{m-n}.$$

Thus, existence of $q(x)$ and $r(x)$ is proved.

Now we shall prove the uniqueness of $q(x)$ and $r(x)$. Let us suppose that $q_1(x)$ and $r_1(x)$ are two polynomials belonging to $F(x)$ such that

$$f(x) = g(x)\, q_1(x) + r_1(x),$$

where $r_1(x) = 0$ or $\deg r_1(x) < \deg g(x)$.

But by the statement of the theorem, $q(x)$ and $r(x)$ are two elements of $F(x)$ such that

$$f(x) = g(x)\, q(x) + r(x),$$

where $$r(x) = 0 \text{ or } \deg r(x) < \deg g(x).$$

Hence

$$g(x)\, q(x) + r(x) = g(x)\, q_1(x) + r_1(x),$$

that is,

$$g(x)\,[q(x) - q_1(x)] = r_1(x) - r(x). \tag{v}$$

But,

$$\deg g(x)\,[q(x) - q_1(x)] \geq n$$

and

$$\deg[r_1(x) - r(x)] < n.$$

Hence (v) is possible only when

$$g(x)\,[q(x) - q_1(x)] = 0$$

and

$$r_1(x) - r(x) = 0,$$

that is, when

$$q(x) = q_1(x) \text{ and } r(x) = r_1(x).$$

Hence, $q(x)$ and $r(x)$ are unique.

5.22 ALGEBRAIC CODING THEORY

5.22.1 Hamming Distance and Encoding Function

In the section, our aim is to study the transmission of information represented by a string of the binary signals 0 and 1. A finite sequence of characters from a finite alphabet is called a *message*, which is the basic unit of information. In our study, we choose our finite alphabet as the set $B = \{0, 1\}$. A word is a sequence of 0's and 1's. We have shown in Example 5.37 that $B = \{0, 1\}$ is a group under the binary operation + (mod 2 addition) defined by the table.

+	0	1
0	0	1
1	1	0

In fact, + is associative, 0 acts as identity and *each element of B is its own inverse*. Now, let $B^n = B \times B \times \ldots \times B$ (*n* factors). Then $(B^n, \oplus)$ in a group under the operation $\oplus$ defined by

$$(x_1, x_2, \ldots, x_n) \oplus (y_1, y_2, \ldots, y_n) = (x_1 + y_1, x_2 + y_2, \ldots, x_n + y_n),$$

where $x_i + y_i$ is 1 if $x_i \neq y_i$ and 0 if $x_i = y_i$

For example, if $x = (0,0,1,0,1)$ and $y = (1,0,1,1,0)$, are two elements of B^5, then

$$x \oplus y = (1,0,0,1,1).$$

The identity of $(B^n, \oplus)$ is $\mathbf{0} = (0,0,\ldots,0)$ and every elements in $(B^n, \oplus)$ is its own inverse. Also $x \oplus y = y \oplus x$ for $x, y \in B^n$. Hence $(B^n, \oplus)$ is an abelian group. Further, B^n has 2^n elements and element $(x_1, x_2, \ldots, x_n) \in B^n$ can be written as $x_1 x_2 \ldots x_n$.

The *weight* of $x = (x_1, x_2, \ldots, x_n) \in B^n$ is defined as the number of 1's in *x*. For example, the weight of 10110001 in B^8 is 4, while the weight of 000 in B^3 is 0. The weight of $x \in B^n$ is denoted by $|x|$.

Let $x, y \in B^n$. Then the **Hamming distance** between *x* and *y*, denoted by $d(x,y)$, is defined to be the weight of $x \oplus y$. Thus,

$$d(x,y) = |x \oplus y|.$$

The definition of the operation $\oplus$ implies that the **distance *d*(*x*, *y*), between *x* and *y* is exactly the *number of positions at which they differ.***

For example, let $x = 0110010110$ and $y = 1001001001$ be in B^{10}. Then

$$d(x,y) = |x \oplus y| = 9.$$

The following theorem exhibits the properties of Hamming distance.

Theorem 5.68

Let $x, y, z \in B^n$. Then

(a) $d(x,y) \geq 0$

(b) $d(x,y) = 0$ if and only if $x = y$

(c) $d(x,y) = d(y,x)$

(d) $d(x,y) \leq d(x,z) + d(z,y)$ (triangle inequality)

Proof. Properties (a), (b) and (c) follow directly from the definition of (d). To prove (d), we note that

$$\begin{aligned} d(x,y) &= |x \oplus y| \\ &= |x \oplus \mathbf{0} \oplus y| \text{ since weight of } \mathbf{0} \text{ is zero} \\ &= |x \oplus z \oplus z \oplus y| \text{ since } z \oplus z = \mathbf{0} \\ &\leq |x \oplus z| + |z \oplus y| \\ &= d(x,z) + d(z,y). \end{aligned}$$

Hence (d) holds.

Let $n>m$ be an integer. Then a one-to-one function $e: B^m \to B^n$ is called an (m, n) ***encoding function.*** If $b \in B^n$, then $e(b)$ is called the ***code word*** representing the word b.

The ***minimum distance*** of (m, n) encoding function $e: B^m \to B^n$ is defined as

$$\min\{d(e(b),e(b')) : b,b' \in B^n, e(b) \neq e(b')\}.$$

EXAMPLE 5.66

Find the minimum distance of (3, 4) encoding function $e : B^3 \to B^4$ defined by

$$e(000)=1001,\quad e(001)=1000,\quad e(010)=0011,$$
$$e(011)=0110,\quad e(100)=0100,\quad e(101)=0010,$$
$$e(110)=0001,\quad e(111)=0000.$$

Solution.

The distances d between the code words are given in the distance table below:

d	1001	1000	0011	0110	0100	0010	0001	0000
1001	0	1	2	4	3	3	1	2
1000	1	0	3	3	2	2	2	1
0011	2	3	0	2	3	1	1	2
0110	4	3	2	0	1	1	3	2
0100	3	2	3	1	0	2	2	1
0010	3	2	1	1	2	0	2	1
0001	1	2	1	3	2	2	0	1
0000	2	1	2	2	1	1	1	0

Hence the minimum distance is

$$\min\{d(e(b),e(b')) : b,b' \in B^4, e(b) \neq e(b')\} = 1.$$

EXAMPLE 5.67

Find the minimum distance of the encoding function $e: B^2 \to B^3$ defined by $e(b_1,b_2)=(b_1,b_2,b_1+b_2)$.

Solution.

We observe that

$$e(0,0)=(0,0,0),\quad e(0,1)=(0,1,1),\quad e(1,0)=(1,0,1),\quad e(1,1)=(1,1,0).$$

Then the distance table is

d	000	011	101	110
000	0	2	2	2
011	2	0	2	2
101	2	2	0	2
110	2	2	2	0

Hence the minimum distance is 2.

5.22.2 Parity Check Code and Detection of Errors

Transmission of a code word by means of a transmission channel can now be illustrated as shown below:

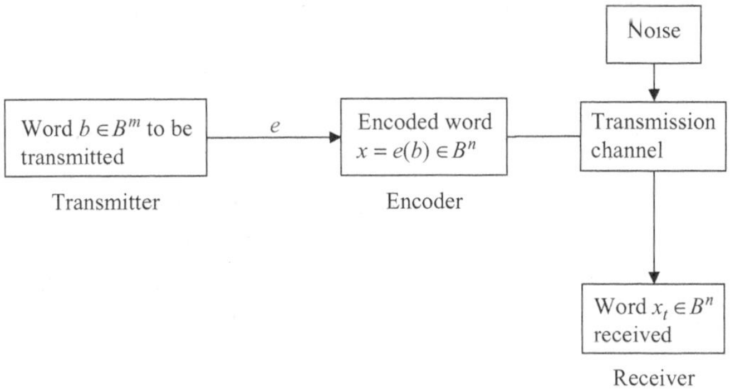

If the transmission channel is noiseless, then $x_t = x$ for all $x = e(b) \in B^n$. Hence x is received for each $b \in B^m$. Then the word b is identified using the definition of the encoding function e. But, in general, due to noise in transmission channel, errors do occur in transmission. If x and x_1 differ in at least 1 position but not more than in k position, then we say that the codeword x has been transmitted with k or fewer errors.

The encoding function $e: B^m \to B^{m+1}$ is called ***parity (m, m + 1) check code*** if for $b = b_1b_2 \ldots b_m$ in B^m,

$$e(b) = b_1b_2 \ldots b_m b_{m+1},$$

where

$$b_{m+1} = \begin{cases} 0 & \text{if } |b| \text{ is even} \\ 1 & \text{if } |b| \text{ is odd} \end{cases}$$

Clearly, for a parity check code, $e(b)$ *has even weight*. In case there is a single error in transmission of a codeword, the received word will be odd weight and therefore the error can be detected. Thus, any odd number of errors can be detected by party check code.

For example, consider the party check code $e: B^2 \to B^3$. Then

$$\begin{aligned} e(00) &= 000, \text{ since weight of } (00) \text{ is } 0 \text{ (even)}, \\ e(01) &= 011, \text{ since the weight of } (01) \text{ is } 1 \text{ (odd)}, \\ e(10) &= 101, \text{ since the weight of } (10) \text{ is } 1 \text{ (odd)}, \\ e(11) &= 110, \text{ since the weight of } (11) \text{ is } 2 \text{ (even)}. \end{aligned}$$

Suppose that $b=11$. Then $x=e(b)=110$. If the transmission channel transmit $x=e(b)$ as $x_t=111$, then $|x_t| = 3$ (odd). This implies that an odd number of errors (at least one) has occurred in transmission due to noise in the transmission channel.

It follows from the definition of the parity check code that it ***cannot detect even number of errors.*** However, the parity check code is widely used despite of the above said limitation.

An (m, n) encoding function $e: B^m \to B^n$ is said to **detect *k* or fewer errors** if whenever $x = e(b)$ is transmitted with k or fewer errors, then x_t is not a code word.

In other words, we can say that an encoding function $e: B^m \to B^n$ is said to be ***capable of correcting k or fewer errors*** if wherever the received word x_t is such that $d(x, x_t) \le k$, then x_t is not a code word.

Theorem 5.69

An (m, n) encoding function $e: B^m \to B^n$ can detect k or fewer code it and only if its minimum distance is at least $(k+1)$, that is, if and only if

$$\min\{d(e(b), e(b')) : b, b' \in B^m, e(b) \neq e(b')\} \geq k+1.$$

Proof. Suppose first that the minimum distance of (m, n) encoding function e is at least $(k+1)$. Let $b \in B^m$ and $x = e(b) \in B^n$. Then x is transmitted as x_t. If $x \neq x_t$, then $d(x, x_t) > k+1$. Therefore b is transmitted with $(k+1)$ or more errors. Hence if x is transmitted with k or fewer errors, then x_t cannot be a code word. Consequently e can detect k or fewer errors.

Conversely, suppose that minimum distance of e is less than $k+1$ that is, the minimum distance is $\leq k$. Let x and y be code words such that $d(x, y) \leq k$. If x has been transmitted and received as y, then errors transmitted are less than or equal to k. Since b is a code word, the error could not be detected. Hence it follows that if the minimum distance of the encoding function is not at least $k+1$, then it cannot detect k or fewer errors.

EXAMPLE 5.68

How many errors can be detected by the encoding function $e: B^2 \to B^4$ defined by

$$e(b_1 b_2) = (b_1, b_2, b_1 + b_2, b_1) \cdot$$

Solution.

We note that

$$e(00) = 0000$$
$$e(01) = 0110$$
$$e(10) = 1011$$
$$e(11) = 1101.$$

Therefore the distance table is

d	0000	0110	1011	1101
0000	0	2	3	3
0110	2	0	3	3
1011	3	3	0	2
1101	3	3	2	0

Hence the minimum weight of the encoding function is 2. Therefore this encoding function can detect at the most one error.

EXAMPLE 5.69

Consider the (2, 6) encoding function e defined by

$$e(00) = 000000$$
$$e(01) = 010101$$
$$e(10) = 101010$$
$$e(11) = 111111.$$

Find the minimum distance of e. How many errors will e detect?

Solution.

The distance table for the encoding function e is given below:

d	000000	010101	101010	111111
000000	0	3	3	6
010101	3	0	6	3
101010	3	6	0	3
111111	6	3	3	0

Therefore the minimum distance of e is 3 and this encoding function is capable of detecting 2 or fewer errors.

5.22.3 Group Codes

An (m, n) encoding function $e: B^m \to B^n$ is called a ***group code*** if $e(B^m) = \{e(b): b \in B^m\}$ is a subgroup of the group $(B^n, \oplus)$.

Since $(B^n, \oplus)$ is an *abelien group*, the subgroup $e(B^m)$ shall be a *normal subgroup* of $(B^n, \oplus)$, We therefore represent $e(B^m)$ by N.

EXAMPLE 5.70

Show that the encoding function $e: B^2 \to B^5$ defined by

$e(00) = 00000$, $e(01) = 01110$, $e(10) = 10101$, $e(11) = 11011$ is a group code.

Solution.

We have

$$e(B^2) = N = \{00000, 01110, 10101, 11011\}.$$

The addition table for N is

$\oplus$	00000	01110	10101	11011
00000	00000	01110	10101	11011
01110	01110	00000	11011	10101
10101	10101	11011	00000	01110
11011	11011	10101	01110	00000

We note that for every $x, y \in N, x \oplus y \in N$. Therefore $\oplus$ is binary operation. The element 00000 acts as identity for N. Also, $x \oplus x = 00000$ for every $x \in N$ showing that every element is inverse of itself. Hence N is a subgroup of B^5 and so the encoding function e is a group code.

EXAMPLE 5.71

Show that the encoding function $e: B^2 \to B^4$ defined by
$e(00) = 0000$, $e(01) = 0110$, $e(10) = 1011$, $e(11) = 1101$ is a group code.

Solution.

First method: The addition table for $e(B^2)$ is

$\oplus$	0000	0110	1011	1101
0000	0000	0110	1011	1101
0110	0110	0000	1101	1011
1011	1011	1101	0000	0110
1101	1101	1011	0110	0000

We observe that $x \oplus y \in e(B^2)$ for every $x, y \in e(B^2)$, the element 0000 acts as identity element for the binary operation $\oplus$ and each element of $e(B^2)$ is its own inverse. Hence the encoding function e is a group code.

Second method: The encoding function in the above example is nothing but the encoding function of Example 5.68 defined by

$$e(x_1x_2) = (x_1, x_2, x_1 + x_2, x_1).$$

We observe that if (x_1, x_2) and (y_1, y_2) are two elements of B^2, then

$$\begin{aligned} e((x_1, x_2) \oplus (y_1, y_2)) &= e(x_1 + y_1, x_2 + y_2) \\ &= (x_1 + y_1, x_2 + y_2, x_1 + y_1 + x_2 + y_2, x_1 + y_1) \\ &= (x_1, x_2, x_1 + x_2, x_1) + (y_1, y_2, y_1 + y_2, y_1) \\ &= e(x_1, x_2) + e(y_1, y_2). \end{aligned}$$

Hence e is a *group homomorphism*. Consequently the image $e(B^2)$ is a subgroup of B^4, proving that e is a group code.

Theorem 5.70

The minimum distance of a group code $e: B^m \to B^n$ is the minimum weight of a non-zero code word.

Proof. Let d be the minimum distance of the group code e. Then there exist *distinct* code words x and y such that $d = d(x, y)$. Let δ be the minimum weight of a non-zero code word, that is, $\delta = |z|$ for non-zero code word z. Since e is a group code, $x \oplus y$ is non-zero code word and therefore

$$d = d(x, y) = |x \oplus y| \geq \delta. \tag{1}$$

Also, the identity element $\mathbf{0}$ and z being district code words, we have

$$\delta = |z| = |z \oplus \mathbf{0}| = d(z, \mathbf{0}) \geq d. \tag{2}$$

Form (1) and (2), it follows that $d = \delta$. This proves the theorem.

Theorem 5.71

Let m and n be non-negative integers such that $m<n$ and let $r=n-m$. Suppose H is an $n\times r$ Boolean matrix. Then the function $f_H:B^n \to B^r$ defined by

$$f_H(x)=x*H, \quad x\in B^n$$

is a group homomorphism and $N=\{x\in B^n: x*H=\mathbf{0}\}$ is a normal subgroup of B^n.

(Here $*$ denotes mod – 2 Boolean product of the column matrix $[x_1x_2...x_n]^T$ and the matrix H).

Proof. Let $x, y\in B^n$. Then

$$\begin{aligned} f_H(x\oplus y) &= (x\oplus y)*H \text{ by definition } f_H \\ &= (x*H)\oplus(g*H) \text{ by } \textit{distributive property of} \oplus \text{ and } * \\ &= f_H(x)\oplus f_H(y). \end{aligned}$$

Thus f_H is a group homomorphism from the group B^n to the group B^r. Further, the *kernel (null space)* of f_H is

$$N=\{x\in B^n: x*H=\mathbf{0}\}.$$

Since null space of a group homomorphism is a normal group, it follows that N is a *normal subgroup* of B^n.

5.22.4 Parity Check Matrix

Let m and n be non-negative integers such that $m<n$ and $r=n-m$. Then an $n\times r$ Boolean matrix

$$H=\begin{bmatrix} h_{11} & h_{12} & \cdots & h_{1r} \\ h_{21} & h_{22} & \cdots & h_{2r} \\ \vdots & \vdots & & \vdots \\ h_{m1} & h_{m2} & \cdots & h_{mr} \\ 1 & 0 & \cdots & 0 \\ 0 & 1 & \cdots & 0 \\ \vdots & \vdots & & \vdots \\ 0 & 0 & \cdots & 1 \end{bmatrix},$$

whose last r ro ws constitute an $r\times r$ identity matrix, is called a ***parity check matrix.***

Let $e_H: B^m \to B^n$ be an encoding function defined by

$$e(b_1b_2\ldots b_m)=b_1b_2\ldots b_m x_1x_2\ldots x_n,$$

where

$$\begin{aligned} x_1 &= b_1h_{11}+b_2h_{21}+\cdots+b_mh_{m1} \\ x_2 &= b_1h_{12}+b_2h_{22}+\cdots+b_mh_{m2} \\ &\vdots \\ x_r &= b_1h_{1r}+b_2h_{2r}+\cdots+b_mh_{mr}. \end{aligned}$$

The following result holds for the function e_H.

Theorem 5.72

Let $r = n-m$ and $x = y_1 y_2 \cdots y_m x_1 x_2 \cdots x_n \in B^n$. Then $x * H = \mathbf{0}$ if and only if $x = e_H(b)$ for some $b \in B^m$.

Proof. Suppose first that $x * H = 0$. Then

$$[y_1 y_2 \ldots y_m \; x_1 x_2 \ldots x_r] * \begin{bmatrix} h_{11} & h_{12} & \cdots & h_{1r} \\ h_{21} & h_{22} & \cdots & h_{2r} \\ \vdots & & & \\ h_{m1} & h_{m2} & \cdots & h_{m2} \\ 1 & 0 & \cdots & 0 \\ 0 & 1 & \cdots & 0 \\ \vdots & & & \\ 0 & 0 & \cdots & 1 \end{bmatrix} = \begin{bmatrix} 0 \\ 0 \\ \vdots \\ 0 \\ 0 \\ 0 \\ \vdots \\ 0 \end{bmatrix}$$

or

$$\begin{aligned} & y_1 h_{11} + y_2 h_{21} + \cdots + y_m h_{m1} + x_1 = 0 \\ & y_1 h_{12} + y_2 h_{22} + \cdots + y_m h_{m2} + x_2 = 0 \\ & \vdots \\ & y_1 h_{1r} + y_2 h_{2r} + \cdots + y_m h_{mr} + x_r = 0. \end{aligned}$$

Putting $a_i = y_1 h_{1i} + y_2 h_{2i} + \cdots + y_m h_{mi}$, the above set of equation reduces to

$$\left.\begin{aligned} & a_1 + x_1 = 0 \\ & a_2 + x_2 = 0 \\ & \vdots \\ & a_r + x_r = 0 \end{aligned}\right\} \qquad (1)$$

But

$$a_i + (a_i + x_i) = a_i$$

or

$$(a_i + a_i) + x_i = a_i$$

or

$$0 + x_i = a_i \quad \text{since } a_i + a_i = 0.$$

or

$$x_i = a_i \text{ for } 1 \le i \le r.$$

or

$$x_i = y_1 h_{1i} + y_2 h_{2i} + \cdots + y_m h_{m3}.$$

Taking $y_1 = b_1, y_2 = b_2, \ldots, y_m = b_m$, we get

$$\begin{aligned} x_1 &= b_1h_{11} + b_2h_{21} + \cdots + b_mh_{m1} \\ x_2 &= b_1h_{12} + b_2h_{12} + \cdots + b_mh_{m2} \\ &\vdots \\ x_r &= b_1h_{1r} + b_2h_{2r} + \cdots + b_mh_{mr} \end{aligned}$$

or

$$x = e_H(b), \quad b \in B^m.$$

Conversely, let $x = e_H(b)$ for $b \in B^m$. Then, by definition of e_H, we have

$$\begin{aligned} x_1 &= b_1h_{11} + b_2h_{21} + \cdots + b_mh_{m1} \\ x_2 &= b_1h_{12} + b_2h_{12} + \cdots + b_mh_{m2} \\ &\vdots \\ x_r &= b_1h_{1r} + b_2h_{2r} + \cdots + b_mh_{mr} \end{aligned}$$

Adding $x_1, x_2, \ldots, x_n$ respectively to both sides of first, second, …, rth equation, and using the fact that $x_i + x_i = 0$, we get

$$\begin{aligned} & b_1h_{11} + b_2h_{21} + \cdots + b_mh_{m1} + x_1 = 0 \\ & b_1h_{12} + b_2h_{22} + \cdots + b_mh_{m2} + x_2 = 0 \\ & \vdots \\ & b_1h_{1r} + b_2h_{2r} + \cdots + b_mh_{mr} + x_r = 0 \end{aligned}$$

or

$$x * H = \mathbf{0}.$$

Theorem 5.73

The encoding function $e_H \colon B^m \to B^n$ defined by

$$x = e_H(b) = b_1b_2 \ldots b_m x_1 x_2 \ldots x_r,$$

where

$$x_i = b_1h_{1i} + b_2h_{2i} + \cdots + b_mh_{mi},$$

is a group code.

Proof. We have

$$\begin{aligned} e_H(B^m) &= \{e_H(b) : b \in B^m\} \\ &= \{x : x * H = \mathbf{0}\} \text{ (by Theorem 5.72)} \\ &= \ker(f_H) \text{ (see Theorem 5.71)}. \end{aligned}$$

Thus, being the kernel of homomorphism, $e_H(B^m)$ is a subgroup of B^n. Hence e_H is a group code.

Remark 5.10 Let N be the $m \times r$ sub-matrix of the parity check matrix M defined by

$$N = \begin{bmatrix} h_{11} & h_{12} & \cdots & h_{1r} \\ h_{21} & h_{22} & \cdots & h_{2r} \\ \vdots & & & \\ h_{m1} & h_{m2} & \cdots & h_{mr} \end{bmatrix}.$$

Then, using the Theorems 5.71 and 5.72, the ***group code*** e_H ***can be obtained by appending*** $b * N$ ***to*** b ***on the right for each*** $b \in B^m$.

EXAMPLE 5.72

Determine the encoding function $e_H: B^3 \to B^7$ corresponding to the parity check matrix

$$H = \begin{bmatrix} 1 & 0 & 0 \\ 0 & 1 & 1 \\ 1 & 1 & 1 \\ 1 & 0 & 0 \\ 0 & 1 & 0 \\ 0 & 0 & 1 \end{bmatrix}.$$

Solution.

We have

$$N = \begin{bmatrix} 1 & 0 & 0 \\ 0 & 1 & 1 \\ 1 & 1 & 1 \end{bmatrix}$$

and

$$B^3 = \{000, \quad 001, \quad 010, \quad 011, \quad 100, \quad 101, \quad 110, \quad 111\}.$$

Therefore

$$000 * \begin{bmatrix} 1 & 0 & 0 \\ 0 & 1 & 1 \\ 1 & 1 & 1 \end{bmatrix} = 000, \qquad 001 * \begin{bmatrix} 1 & 0 & 0 \\ 0 & 1 & 1 \\ 1 & 1 & 1 \end{bmatrix} = 111,$$

$$010 * \begin{bmatrix} 1 & 0 & 0 \\ 0 & 1 & 1 \\ 1 & 1 & 1 \end{bmatrix} = 011, \qquad 011 * \begin{bmatrix} 1 & 0 & 0 \\ 0 & 1 & 1 \\ 1 & 1 & 1 \end{bmatrix} = 100,$$

$$100 * \begin{bmatrix} 1 & 0 & 0 \\ 0 & 1 & 1 \\ 1 & 1 & 1 \end{bmatrix} = 100, \qquad 101 * \begin{bmatrix} 1 & 0 & 0 \\ 0 & 1 & 1 \\ 1 & 1 & 1 \end{bmatrix} = 011,$$

$$110 * \begin{bmatrix} 1 & 0 & 0 \\ 0 & 1 & 1 \\ 1 & 1 & 1 \end{bmatrix} = 111, \qquad 111 * \begin{bmatrix} 1 & 0 & 0 \\ 0 & 1 & 1 \\ 1 & 1 & 1 \end{bmatrix} = 000.$$

Hence the group code is

$$\begin{aligned} e_H(000) &= 000000, & e_H(001) &= 001111, \\ e_H(010) &= 010011, & e_H(011) &= 011100, \\ e_H(100) &= 100100, & e_H(101) &= 101011, \\ e_H(110) &= 110111, & e_H(111) &= 111000. \end{aligned}$$

EXAMPLE 5.73

Determine the group code $e_H: B^2 \to B^5$ corresponding to the parity check matrix

$$H = \begin{bmatrix} 1 & 1 & 1 \\ 0 & 1 & 1 \\ 1 & 0 & 0 \\ 0 & 1 & 0 \\ 0 & 0 & 1 \end{bmatrix}.$$

Also find the minimum distance of e_H. How many errors can e_H detect?

Solution.

We have

$$N = \begin{bmatrix} 1 & 1 & 1 \\ 0 & 1 & 1 \end{bmatrix}$$

and

$$B^2 = \{00, \quad 01, \quad 10, \quad 11\}.$$

Then,

$$00 * \begin{bmatrix} 1 & 1 & 1 \\ 0 & 1 & 1 \end{bmatrix} = 000, \qquad 01 * \begin{bmatrix} 1 & 1 & 1 \\ 0 & 1 & 1 \end{bmatrix} = 011,$$

$$10 * \begin{bmatrix} 1 & 1 & 1 \\ 0 & 1 & 1 \end{bmatrix} = 111, \qquad 11 * \begin{bmatrix} 1 & 1 & 1 \\ 0 & 1 & 1 \end{bmatrix} = 100.$$

Hence the required group code is

$$\begin{aligned} e_H(00) &= 00000, & e_H(01) &= 01011, \\ e_H(10) &= 10111, & e_H(11) &= 11100. \end{aligned}$$

The distance table for e_H is

d	00000	01011	10111	11100
00000	0	3	4	3
01011	3	0	3	4
10111	4	3	0	3
11100	3	4	3	0

Hence the minimum distance of the group code e_H is 3. This group code can detect 2 or fewer errors.

5.22.5 Decoding

Let $e : B^m \to B^n$ be an encoding function and let $x = e(b) \in B^n$ for $b \in b^m$ be received as the word x_t. Our aim now is to identify the original word b (the original message that was communicated). For this purpose, we define another function, called ***decoding function associated with the encoding function e.***

An *onto function* $d_e : B^n \to B^m$ is called an (n, m) decoding function associated with e if $d_e(x_t) = b' \in B^n$ is such that for noiseless transmission channel, $b' = b$, that is, $d_e \, o \, e = I$ (identity function in B^m).

Thus, the decoding function correctly decodes the errorless messages but messages received with errors (noise) may not be correctly decoded.

Taking decoding into consideration, the communication model can be represented as

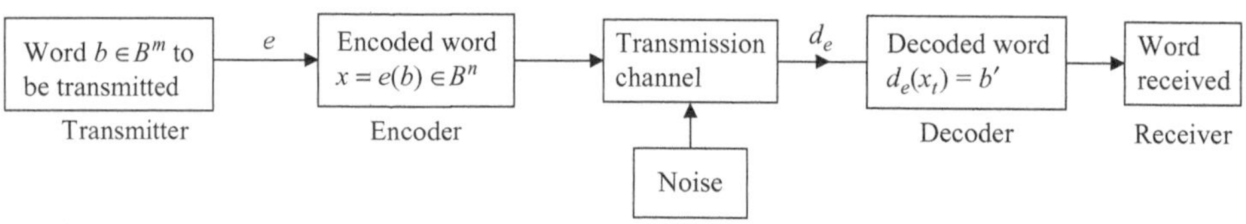

The pair (e, d_e) is said *correct k or fewer error* if whenever $x = e(b)$ is transmitted correctly or with k or fewer errors and x_t is the received word, then $d_e(x_t) = b$, that is, x_t is decoded as the correct message b.

Let $e : B^m \to B^n$ be an (m, n) encoding function and let the code words $e(B^m)$ be written in a fixed order

$$x^{(1)}, x^{(2)}, \ldots, x^{(2^m)} \qquad (1)$$

For the received word x_t, we find $d(x^{(i)}, x_t)$, $1 \le i \le 2^m$. Let $x^{(k)}$ be the first code from (1) such that

$$\min_{1 \le i \le 2^m} \{d(x^{(i)}, x_t\} = d(x^{(k)}, x_t).$$

This means $x^{(k)}$ is a code word closest to x_t and first in list (1).

If $x^{(s)} = e(b)$, $b \in B^m$, then the function d_e defined by $d_e(x_t) = b$ is called the ***maximum likelihood decoding function d_e*** *associated with e.*

Since d_e depends on the particular order of code words $e(B^m)$, it is not unique. For example, consider the encoding function $e : B^2 \to B^4$ defined by

$$e(b_1, b_2) = (b_1, b_2, b_1 + b_2, b_1).$$

Suppose that, we arrange the code words obtained as

0000, 0110, 1011, 1101.

Then the distance of these words from a received word $x_t = 0010$ are

$$d(0000, 0010) = 1$$
$$d(0110, 0010) = 1$$
$$d(1011, 0010) = 2$$
$$d(1101, 0010) = 4.$$

Here the first two distances are same. But 0000 being the first code in $e(B^2)$, we define the decoding function d_e as

$$d_e(x_t) = d_e(0010) = 0000$$

and not $d_e(0010) = 0110$.

The following theorem provides the condition for the pair (e, d_e) to correct k or fewer errors.

Theorem 5.74

Let $B^m \to B^n$ be an encoding function and $d_e: B^n \to B^m$ be a maximum likelihood decoding function associated with e. Then the pair (e, d_e) can correct k or fewer errors if and only if the minimum distance of the code e is at least $2k+1$.

Proof. Suppose first that the minimum distance of e is at least $2k+1$. Let $x = e(b) \in B^n$ for $b \in B^m$ and let x be transmitted as x_t with k or fewer errors. Therefore $d(x, x_t) \leq k$. Let z be another code word, then, using triangle inequality, we have

$$2k+1 \leq d(x,z) \leq d(x,x_t) + d(x_t,z) \leq k + d(x_t,z)$$

or

$$d(x_t,z) \geq 2k+1-k = k+1.$$

Since $z \neq x$ is arbitrary codeword, it follows that x is the unique code word that is closest to x_t and so $d_e(x_t) = b$. Hence (e, d_e) corrects k or fewer errors.

We prove the converse by contradiction. So assume that the minimum distance r of e is less than $2k+1$, that is, $r \leq 2k$. Let $x = e(b)$ and $x' = e(b')$ be two code words with $d(x, x') = r$. Suppose that x' precedes x in the order list of code words used in the definition of d_e. Let $x = x_1x_2 \ldots x_n$ and $x' = x_1'x_2' \ldots x_n'$. Since $d(x,x') = r$, x and x' differ at r positions. Assume that $x_i \neq x_i'$ for $1 \leq i \leq r$ but $x_i = x_i'$ for $i > r$.

Now consider the following two cases:

(i) Suppose that $r \leq k$. If x is transmitted as $x_t = x'$, then since x and x' differ at r positions, $r \leq k$ errors have been committed but as per our definition of maximum likelihood decoding function, $d_e(x_t) = b'$. Thus the pair (e,d_e) has not corrected the r errors

(ii) Let $(k+1) \le r \le 2k$ and let $y = x_1'x_2'\ldots x_k'x_{k+1}\ldots x_n$. If x is transmitted as $x_t = y$, then

$$d(x_t, x) \ge k \text{ and } d(x_t, x') \le r - k \le k .$$

This implies that x' is at least as close to x_t as x is. But x' precedes in the ordered list of codes. Therefore, $d_e(x_t) = b' \ne b$. Hence, transmission has committed k errors which the pair (e, d_e) has not corrected.

EXAMPLE 5.74

Let $e : B^3 \to B^5$ be the (3, 5) encoding function defined by

$$\begin{aligned} e(000) &= 00000, & e(001) &= 00110, \\ e(010) &= 01001, & e(011) &= 01111, \\ e(100) &= 10011, & e(101) &= 10101, \\ e(110) &= 11010, & e(111) &= 11100. \end{aligned}$$

and let d_e be an (5, 3) maximum likelihood decoding function associated with e. How many errors can the pair (e, d_e) correct?

Solution.

By Theorem 5.74, it is sufficient to find the minimum distance of e. The distance table for e is

d	00000	01001	10011	11010	00110	01111	10101	11100
00000	0	2	3	3	2	4	3	3
01001	2	0	3	3	3	3	3	3
10011	3	3	0	2	3	3	2	4
11010	3	3	2	0	3	3	4	2
00110	2	3	3	3	0	2	3	3
01111	4	3	3	3	2	0	3	3
10101	3	3	2	4	3	3	0	2
11100	3	3	4	2	3	3	2	0

Hence the minimum distance of e is 2. Therefore $2 \ge 2k+1$ yields $k \le \frac{1}{2}$ and so the pair (e, d_e) would not be capable of correcting any error.

EXAMPLE 5.75

Let e: $B^2 \to B^5$ be (2, 5) encoding function defined by

$$\begin{aligned} e(00) &= 00000, & e(01) &= 01110, \\ e(10) &= 10101, & e(11) &= 11011. \end{aligned}$$

and let (e, d_e) be maximum likelihood decoding function associated with e. Determine the number of errors that (e, d_e) will correct. Also decode the word 11001.

Solution.

The distance table for the encoding function e is

d	00000	01110	10101	11011
00000	0	3	3	4
01110	3	0	4	3
10101	3	4	0	3
11011	4	3	3	0

Hence the minimum distance of e is 3. Therefore $3 \geq 2k+1$ yields $k \leq 1$. Hence the pair (e, d_e) can correct one error.

To decode the word 11001, we first arrange the given code words as

00000 01110 10101 11011

Then

$$d(11001, 00000) = 3, \qquad d(11001, 01110) = 4$$
$$d(11001, 10101) = 2, \qquad d(11001, 11011) = 2.$$

We observe that 10101 and 11011 are equidistant from 11001. But 10101 comes first in our list of codes. Since $e(10) = 10101$, we decode 11001 as 10.

5.22.6 Determination of Maximum Likelihood Decoding Function Using Cosets

Let $e: B^m \to B^n$ be an encoding function and let $x = e(b)$ be received as x_t. Then the set N of code words in B^n is a normal subgroup of $(B^n, \oplus)$. The order of this subgroup is 2^m. So let

$$N = \{x^{(1)}, x^{(2)}, \ldots, x^{(2^m)}\}.$$

The left coset of x_t is

$$x_t \oplus N = \{\varepsilon_1, \varepsilon_2, \ldots, \varepsilon_{2^m}\},$$

where $\varepsilon_i = x_t \oplus x^{(i)}$, Further,

$$d(x_t, x^{(i)}) = |\varepsilon_i|.$$

Therefore if ε_j has smallest weight, then $x^{(j)}$ must be a code closest to x_t and

$$\begin{aligned} x^{(j)} &= \mathbf{0} \oplus x^{(j)} \\ &= x_t \oplus x_t \oplus x^{(j)} \quad \text{since } x_t \oplus x_t = \mathbf{0} \\ &= x_t \oplus \varepsilon^{(j)}, \end{aligned}$$

which will *provide* ***corrected*** **code** for x_t.

The coset member ε_j, having the smallest weight, is called a ***coset leader***. Since more than one coset members may have equal smallest weight, it follows that ***a coset leader need not be unique.***

To further simplify the procedure using parity check matrix, let H be the parity check matrix. Then it has already been established in Theorem 5.71 that the function $f_H : B^n \to B^r$ defined by $f_H(x) = x * H$ is a group homomorphism. Also f_H is onto mapping. Therefore, if N is kernel of f_H, then the mapping. $g: B^n/N \to B^r$ defined by

$$g(xN) = f_H(x) = x * H,$$

is an *isomorphism*.

The element $x * H$ is called the ***syndrome*** of x.

Theorem 5.75

Le $x, y \in B^n$. Then x and y lie in the same left coset of N in B^n if and only if

$$f_H(x) = f_H(y),$$

that is, if and only if $x * H = y * H$, that is, if and only if they have the same syndrome.

Proof. x and y lie in the same left coset of N if and only if $x \oplus y \in N$. Since N is kernel of f_H, $x \oplus y \in N$ if and only if

$$f_H(x \oplus y) = \mathbf{0} \quad \text{(identity elements of } B^r\text{)}$$

or

$$f_H(x) \oplus f_H(y) = \mathbf{0} \quad \text{since } f_H \text{ is homomorphism}$$

or

$$f_H(x) = f_H(y)$$

or

$$x * H = y * H$$

or if and only if x and y have the same syndrome.

In view of the above discussion, to decode a given word, we need only coset leaders and their syndromes. We shall follow the following procedure for decoding.

Step 1. To find the left cosets of N, we note that

(i) $\mathbf{0} \oplus N = N$ (the set of code words).
(ii) We take any element of B^n which is not in N and find the corresponding coset.
(iii) Choose an element of B^n which is not in any of the two former sets and find the corresponding coset.
(iv) Continue the procedure to get cosets of all elements of B^n.

Step 2. For each coset, find a coset leader and compute syndrome $x * H$ for each coset leader x and form the table of syndromes.

Step 3. Compute the syndrome $x_t * H$ for the received word x_t and find the coset leader ε having the same syndrome. Then $x_t \oplus \varepsilon = x$ (the code word $e_H(b)$ and $d_H(x_t) = b$.

EXAMPLE 5.76

Find the encoding function $e_H : B^2 \to B^5$ given by the party check matrix

$$H = \begin{bmatrix} 0 & 1 & 1 \\ 1 & 0 & 1 \\ 1 & 0 & 0 \\ 0 & 1 & 0 \\ 0 & 0 & 1 \end{bmatrix}$$

Decode 10100 and 11011.

Solution.

We have

$$N = \begin{bmatrix} 0 & 1 & 1 \\ 1 & 0 & 1 \end{bmatrix} \text{ and } B^2 = \{00, \quad 01, \quad 10, \quad 11\}.$$

Then

$$00 * \begin{bmatrix} 0 & 1 & 1 \\ 1 & 0 & 1 \end{bmatrix} = 000, \qquad 01 * \begin{bmatrix} 0 & 1 & 1 \\ 1 & 0 & 1 \end{bmatrix} = 101,$$

$$10 * \begin{bmatrix} 0 & 1 & 1 \\ 1 & 0 & 1 \end{bmatrix} = 011, \qquad 11 * \begin{bmatrix} 0 & 1 & 1 \\ 1 & 0 & 1 \end{bmatrix} = 110.$$

Hence the group code is

$$e_H(00) = 00000, \qquad e_H(01) = 01101,$$
$$e_H(10) = 10011, \qquad e_H(11) = 11110.$$

Therefore the set of code words is

$$N = \{00000, \quad 01101, \quad 10011, \quad 11110\}.$$

Now the cosets of the elements in B^5 which are not in N are given by

$$\begin{aligned}
00001 \oplus N &= \{00001, \quad 01100, \quad 10010, \quad 11111\} \\
00010 \oplus N &= \{00010, \quad 01111, \quad 10001, \quad 11100\} \\
00100 \oplus N &= \{00100, \quad 01001, \quad 10111, \quad 11010\} \\
01000 \oplus N &= \{01000, \quad 00101, \quad 11011, \quad 10110\} \\
10000 \oplus N &= \{10000, \quad 11101, \quad 00011, \quad 01110\} \\
00110 \oplus N &= \{00110, \quad 01011, \quad 10101, \quad 11000\} \\
01010 \oplus N &= \{01010, \quad 00111, \quad 11001, \quad 10100\}.
\end{aligned}$$

The coset leaders for these cosets are

$$00000, 00001, 00010, 00100, 01000, 10000, 00110, 01010.$$

Now to find syndromes for the coset leaders, we have

$$00000 * H = 000, \qquad 00001 * \begin{bmatrix} 0 & 1 & 1 \\ 1 & 0 & 1 \\ 1 & 0 & 0 \\ 0 & 1 & 0 \\ 0 & 0 & 1 \end{bmatrix} = 001,$$

$$00010 * \begin{bmatrix} 0 & 1 & 1 \\ 1 & 0 & 1 \\ 1 & 0 & 0 \\ 0 & 1 & 0 \\ 0 & 0 & 1 \end{bmatrix} = 010, \qquad 00100 * \begin{bmatrix} 0 & 1 & 1 \\ 1 & 0 & 1 \\ 1 & 0 & 0 \\ 0 & 1 & 0 \\ 0 & 0 & 1 \end{bmatrix} = 100,$$

$$01000 * \begin{bmatrix} 0 & 1 & 1 \\ 1 & 0 & 1 \\ 1 & 0 & 0 \\ 0 & 1 & 0 \\ 0 & 0 & 1 \end{bmatrix} = 101, \qquad 10000 * \begin{bmatrix} 0 & 1 & 1 \\ 1 & 0 & 1 \\ 1 & 0 & 0 \\ 0 & 1 & 0 \\ 0 & 0 & 1 \end{bmatrix} = 011$$

$$00110 * \begin{bmatrix} 0 & 1 & 1 \\ 1 & 0 & 1 \\ 1 & 0 & 0 \\ 0 & 1 & 0 \\ 0 & 0 & 1 \end{bmatrix} = 110, \qquad 01010 * \begin{bmatrix} 0 & 1 & 1 \\ 1 & 0 & 1 \\ 1 & 0 & 0 \\ 0 & 1 & 0 \\ 0 & 0 & 1 \end{bmatrix} = 111.$$

Hence the table for syndromes is

Coset Leader	*Syndrome*
00000	000
00001	001
00010	010
00100	100
01000	101
10000	011
00110	110
01010	111

(a) Further, the syndrome of the given code word 10100 is given by

$$10100 * \begin{bmatrix} 0 & 1 & 1 \\ 1 & 0 & 1 \\ 1 & 0 & 0 \\ 0 & 1 & 0 \\ 0 & 0 & 1 \end{bmatrix} = 111.$$

Since 111 is the syndrome of the code word 01010, therefore the given code 10100 is corrected as $01010 \oplus 10100 = 11110$ and is decoded as 11.

(b) The syndrome of the given code word 11011 is given by

$$11011 * \begin{bmatrix} 0 & 1 & 1 \\ 1 & 0 & 1 \\ 1 & 0 & 0 \\ 0 & 1 & 0 \\ 0 & 0 & 1 \end{bmatrix} = 101.$$

Since 101 is the syndrome of the code word 01000, therefore 11011 is corrected as $01000 \oplus 11011 = 10011$ and is decoded as10.

EXAMPLE 5.77

Let

$$H = \begin{bmatrix} 1 & 1 \\ 1 & 0 \\ 1 & 0 \\ 0 & 1 \end{bmatrix}$$

be a parity check matrix Determine the encoding function $e_H : B^2 \rightarrow B^4$ and decode 1110 and 1010.

Solution.

We have

$$N = \begin{bmatrix} 1 & 1 \\ 1 & 0 \end{bmatrix} \text{ and } B^2 = \{00, \ 01, \ 10, \ 11\}.$$

Then

$$00 * \begin{bmatrix} 1 & 1 \\ 1 & 0 \end{bmatrix} = 00, \qquad 01 * \begin{bmatrix} 1 & 1 \\ 1 & 0 \end{bmatrix} = 10$$
$$10 * \begin{bmatrix} 1 & 1 \\ 1 & 0 \end{bmatrix} = 11, \qquad 11 * \begin{bmatrix} 1 & 1 \\ 1 & 0 \end{bmatrix} = 01.$$

Therefore the encoding function e_H is defined as

$$e_H(00) = 0000, \qquad e_H(01) = 0110$$
$$e_H(10) = 1011, \qquad e_H(11) = 1101.$$

Therefore the set of code words is

$$N = \{0000, \ 0110, \ 1011, \ 1101\}.$$

Now the cosets of those elements of B^4 which are not in N are given by

$$0001 \oplus N = \{0001,\ 0111,\ 1010,\ 1100\}$$
$$0010 \oplus N = \{0010,\ 0100,\ 1001,\ 1111\}$$
$$1000 \oplus N = \{1000,\ 1110\ \ 0011,\ 0101\}.$$

The coset leaders of these cosets are

$$0000,\ 0001,\ 0010 \text{ or } 0100,\ 1000.$$

We now find syndromes for the coset leaders. We have

$$0000 * \begin{bmatrix} 1 & 1 \\ 1 & 0 \\ 1 & 0 \\ 0 & 1 \end{bmatrix} = 00, \quad 0001 * \begin{bmatrix} 1 & 1 \\ 1 & 0 \\ 1 & 0 \\ 0 & 1 \end{bmatrix} = 01,$$

$$0010 * \begin{bmatrix} 1 & 1 \\ 1 & 0 \\ 1 & 0 \\ 0 & 1 \end{bmatrix} = 10, \quad 1000 * \begin{bmatrix} 1 & 1 \\ 1 & 0 \\ 1 & 0 \\ 0 & 1 \end{bmatrix} = 11.$$

Hence the table for syndromes is

Coset Leader	*Syndrome*
0000	00
0001	01
0010 or 0100	10
1000	11

(a) The syndrome of the given code word 1110 is given by

$$1110 * \begin{bmatrix} 1 & 1 \\ 1 & 0 \\ 1 & 0 \\ 0 & 1 \end{bmatrix} = 11.$$

Since 11 is the syndrome of the code word 1000 (see table), the word 1110 is corrected as 1000+1110=0110 and is decoded as 01.

(b) The syndrome of the given code word 1010 is given by

$$1010 * \begin{bmatrix} 1 & 1 \\ 1 & 0 \\ 1 & 0 \\ 0 & 1 \end{bmatrix} = 01.$$

Since 01 is the syndrome of the code word 0001, the word is corrected as 0001 + 1010=1011 and is decoded as 10.

EXERCISES

1. Let $(M, *)$ be commutative monoid. Show that the set of idempotent elements of M form a submonoid.
2. Let $A=\{a, b\}$ and let multiplication table for a binary operation on A be

$*$	a	b
a	b	a
b	a	b

   Show that $(A, *)$ is a commutative monoid.
3. Let M be of the set of square matrices. Show that the determinant function defined on M is a semigroup homomorphism on $(M, \bullet)$ but it is not a semigroup homomorphism on $(M, +)$.
4. Show that the set $G=\{1, 5, 7, 11\}$ is a group under multiplication modulo 12.
5. Find the identity element in $(S, \odot_{14})$ if $S=\{2, 4, 8\}$ and $\odot_{14}$ denotes multiplication modulo 14.
6. Let G be a group with identity e. If $x^2=x$, $x \in G$, then show that $x=e$.
7. Let $\mathbf{Q}$ be the set of rational numbers. Show that $(\mathbf{Q}, +)$ is not a cyclic group.
8. Let U_m denote a reduced residue system modulo m which consists of those integers which are relatively prime to m. Then show that U_m is a group under multiplication (modulo m). Show that U_9 is a cyclic group. What are its generators?
9. Let G be a group and let $a, b \in G$. If $(ab)^n=a^n b^n$, $n>1$, show that

$$G^{(n)}=\{x^n: x \in G\}$$

   is a normal subgroup of G.
10. If $\alpha=\begin{pmatrix}1 & 2 & 3 & 4 & 5\\ 1 & 3 & 4 & 5 & 2\end{pmatrix}$,

$$\beta=\begin{pmatrix}1 & 2 & 3 & 4 & 5\\ 2 & 3 & 1 & 5 & 4\end{pmatrix}$$

   are elements of S_5, show that $\alpha \circ \beta \neq \beta \circ \alpha$.
11. Is A_n, $n=1, 2, 3$ simple? Give reasons.
12. Show that V_4 $\{(1), (1\ 2)(3\ 4), (1\ 3)(2\ 4), (1\ 4)(2\ 3)\}$ is a subgroup of alternating group A_4 (This subgroup is called Klien's Four Group).
13. Let G be a group and $a \in G$. Then the set

$$N(a)=\{x \in G: xa=ax\}$$

   is called **normalizer** of a in G. Show that $N(a)$ is a subgroup of G.
14. Let $\mathbf{Z}_6=(\bar{0}, \bar{1}, \bar{2}, \bar{3}, \bar{4}, \bar{5})$ be ring of integers modulo G. Find (i) the units of $\mathbf{Z}_6$ (ii) $-\bar{2}, -\bar{4}$ and $(\bar{3})^{-1}$.
15. Show that set of integer modulo 4 is a commutative ring with identity.
16. Let R be a ring with unit element 1 and let f be a ring homomorphism of R onto R'. Show that $f(1)$ is the unit element of R'.
17. Let R be a ring with unity. If $(xy)^2=x^2y^2$ for all $x, y \in R$, show that R is commutative.
18. Let R be a ring with unity e. If for some $x \in R$ there exists a unique element $y \in R$ such that $xy=e$, show that x is invertible.
19. Find the weight of the following words in B^5:
   (a) 10001 (b) 00110 (c) 11010
20. Find the Hamming distance between the words
   (a) 10110, 110100
   (b) 1010, 0111
   (c) 01111, 10101
21. Find the minimum distance of the (2, 5) encoding function $e: B^2 \to B^5$ defined by

$$e(00)=00000,\ e(01)=01110,$$
$$e(10)=00111,\ e(11)=11111$$

22. Show that the encoding function $e: B^3 \to B^6$ defined by

$$e(000)=000000, \quad e(001)=001100$$
$$e(010)=010011, \quad e(011)=011111$$
$$e(100)=100101, \quad e(101)=101001$$
$$e(110)=110110, \quad e(111)=111010$$

   is a group code.
23. Show that the function $e: B^3 \to B^5$ defined by $e(b_1, b_2, b_3)=b_1b_2b_1b_2b_1$ is not an encoding function.

Hint: we note that $e(001) = 00000$, $e(000) = 00000$.

Thus two elements 000 and 001 are mapping on to the same element. Hence e is not one-to-one function and so cannot be an encoding function.

24. Consider the (2, 5) encoding function $e: B^2 \to B^5$ defined by $e(00)=00000$, $e(01)=01110$, $e(10)=00111$, $e(11)=11111$. How many errors will e detect?
25. Consider the encoding function $e: B^2 \to B^3$ defined by $e(b_1, b_2) = (b_1, b_2, b_1 + b_2)$. How many errors will e detect?
26. Let

$$H = \begin{bmatrix} 0 & 1 & 1 \\ 0 & 1 & 1 \\ 1 & 0 & 0 \\ 0 & 1 & 0 \\ 0 & 0 & 1 \end{bmatrix}$$

be a parity check matrix. Determine the (2, 5) group code function $e_H: B^2 \to B^5$.

27. Let

$$H = \begin{bmatrix} 1 & 1 & 0 \\ 1 & 0 & 1 \\ 0 & 1 & 1 \\ 1 & 0 & 0 \\ 0 & 1 & 0 \\ 0 & 0 & 1 \end{bmatrix}$$

be a parity check matrix. Determine the encoding function $e_H: B^3 \to B^6$ and decode 001111.

28. Let

$$H = \begin{bmatrix} 1 & 0 & 0 \\ 1 & 1 & 0 \\ 0 & 1 & 1 \\ 1 & 0 & 0 \\ 0 & 1 & 0 \\ 0 & 0 & 1 \end{bmatrix}$$

be the party check matrix. Finding encoding function $e_H : B^2 \to B^6$, decode 101011.

6 Lattices

A lattice is a special kind of an ordered set with two binary operations. Lattice structures are used in computing and mathematical applications.

6.1 LATTICE

Definition 6.1

A **lattice** is a partially ordered set $(L, \leq)$ in which every subset $\{a, b\}$ consisting of two elements has a least upper bound and a greatest lower bound.

We denote LUB($\{a, b\}$) by $a \vee b$ and call it **join** or **sum of** a **and** b. Similarly, we denote GLB ($\{a, b\}$) by $a \wedge b$ and call it **meet** or **product of** a **and** b.

Other symbols used are

$$\text{LUB: } \oplus, +, \cup,$$
$$\text{GLB: } *, \cdot, \cap.$$

Thus **Lattice is** a mathematical structure with **two binary operations, join and meet**.

A totally ordered set is obviously a lattice but not all partially ordered sets are lattices.

EXAMPLE 6.1

Let A be any set and $P(A)$ be its power set. The partially ordered set $(P(A), \subseteq)$ is a lattice in which the meet and join are the same as the operations $\cap$ and $\cup$, respectively. If A has single element, say a, then $P(A)=\{\phi, \{a\}\}$ and

$$\text{LUB}(\{\phi, \{a\})=\{a\},$$
$$\text{GLB}(\{\phi, \{a\})=\phi.$$

The Hasse diagram of $(P(A), \subseteq)$ is a chain containing two elements ϕ and $\{a\}$ as shown in the Figure 6.1.

Figure 6.1

If A has two elements, say a and b. Then, $P(A)=\{\phi, \{a\}, \{b\}, \{a, b\}\}$. The Hasse diagram of $\{P(A), \subseteq)$ is then as shown in the Figure 6.2.

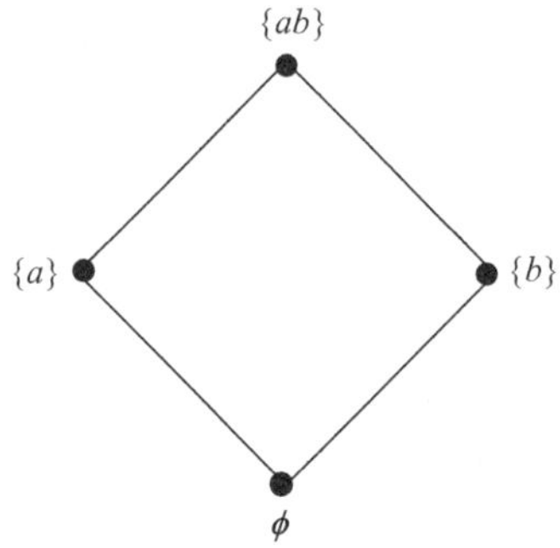

Figure 6.2

We note that

1. LUB exists for every two subsets and is $L \cup M$
2. GLB exists for every two subsets and is in $L \cap M$,

for $L, M \in P(A)$. Hence, $P(A)$ is a lattice.

EXAMPLE 6.2

Consider the poset ($\mathbf{N}$, $\leq$), where $\leq$ is relation of divisibility. Then $\mathbf{N}$ is a lattice in which
Join of a and $b=a \vee b=\text{lcm}(a, b)$,
Meet of a and $b=a \wedge b=\gcd(a, b)$ for $a, b \in \mathbf{N}$.

EXAMPLE 6.3

Let n be a positive integer and let D_n be the set of all positive divisors of n. Then D_n is a lattice under the relation of divisibility. The Hasse diagram of the lattices D_8, D_{20} and D_{30} are respectively shown in the Figures 6.3, 6.4 and 6.5.

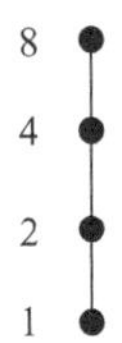

$D_8 = \{1, 2, 4, 8\}$

Figure 6.3

,

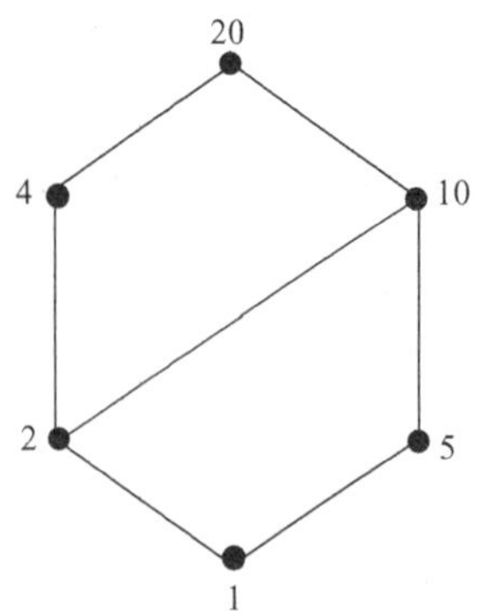

$D_{20} = \{1, 2, 4, 5, 10, 20\}$

Figure 6.4

and

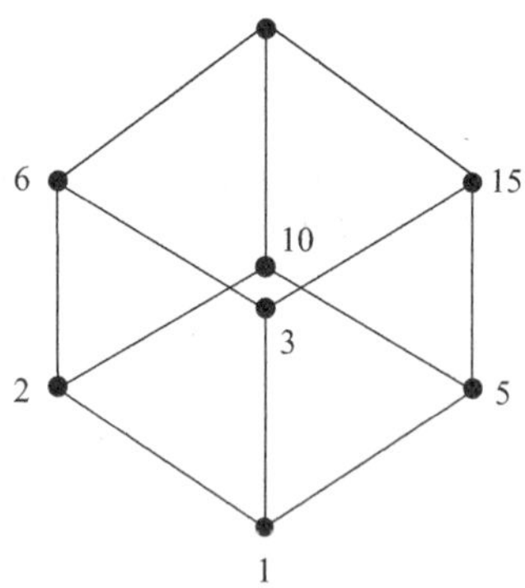

$D_{30} = \{1, 2, 3, 5, 6, 10, 15, 30\}$.

Figure 6.5

EXAMPLE 6.4

Which of the following Hasse diagrams represent lattices?

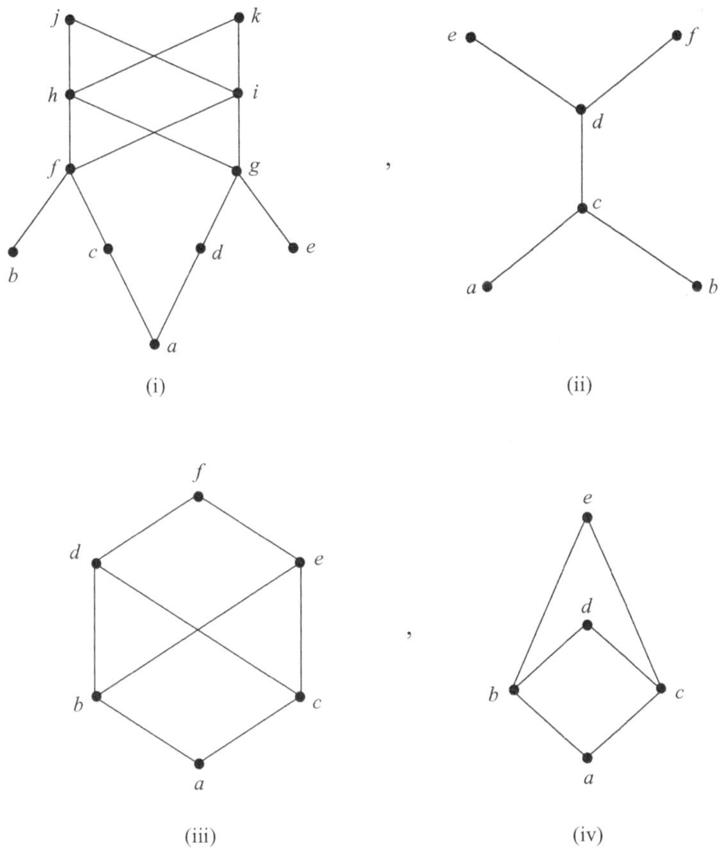

Figure 6.6

Solution.

The Hasse diagram (i) is **not a lattice** because $c \vee d$ and $b \wedge c$ do not exist.

The Hasse diagram (ii) is **not a lattice** because $a \wedge b$ does not exist, $e \vee f$ does not exist.

The Hasse diagram (iii) is **not a lattice** because neither $d \wedge e$ nor $b \vee c$ exists.

The Hasse diagram (iv) is **not a lattice** because $b \wedge c$ does not exist.

Definition 6.2

Let $(L, \leq)$ be a poset and let $(L, \geq)$ be the dual poset. If $(L, \leq)$ is a lattice, we can show that $(L, \geq)$ is also a lattice. In fact, for any a and b in L, the LUB of a and b in $(L, \leq)$ is equal to the GLB of a and b in $(L, \geq)$. Similarly, the GLB of a and b in $(L, \leq)$ is equal to LUB in $(L, \geq)$.

The operation $\vee$ and $\wedge$ are called **dual of each other**.

EXAMPLE 6.5

Let S be a set and $L=P(S)$. Then $(L, \subseteq)$ is a lattice and its **dual lattice** is $(L, \supseteq)$, where $\supseteq$ represents "contains". We note that in the poset $(L, \supseteq)$, the join $A \vee B$ is the set $A \cap B$ and the meet $A \wedge B$ is the set $A \cup B$.

Theorem 6.1

If $(L_1, \leq)$ and $(L_2, \leq)$ are lattices, then $(L, \leq)$ is a lattice, where $L = L_1 \times L_2$ and the partial order $\leq$ of L is the product partial order.

Proof. We denote the join and meet in L_1 by $\vee_1$, and $\wedge_1$ and the join and meet in L_2 by $\vee_2$ and $\wedge_2$, respectively. We know that Cartesian product of two posets is a poset. Therefore, $L = L_1 \times L_2$ is a poset. Thus, all we need to show is that if (a_1, b_1) and $(a_2, b_2) \in L$, then $(a_1, b_1) \vee (a_2, b_2)$ and $(a_1, b_1) \wedge (a_2, b_2)$ exist in L. Further, we know that

$$(a_1, b_1) \vee (a_2, b_2) = (a_1 \vee_1 a_2, b_1 \vee_2 b_2)$$

and

$$(a_1, b_1) \wedge (a_2, b_2) = (a_1 \wedge_1 a_2, b_1 \wedge_2 b_2).$$

Since L_1 is lattice, $a_1 \vee_1 a_2$ and $a_1 \wedge_1 a_2$ exist. Similarly, since L_2 is a lattice, $b_1 \vee_2 b_2$ and $b_1 \wedge_2 b_2$ exist. Hence $(a_1, b_1) \vee (a_2, b_2)$ and $(a_1, b_1) \wedge (a_2, b_2)$ both exist and therefore $(L, \leq)$ is a lattice, called **the direct product of $(L_1, \leq)$ and $(L_2, \leq)$.**

EXAMPLE 6.6

Let L_1 and L_2 be the lattices whose Hasse diagrams are given in the Figure 6.7.

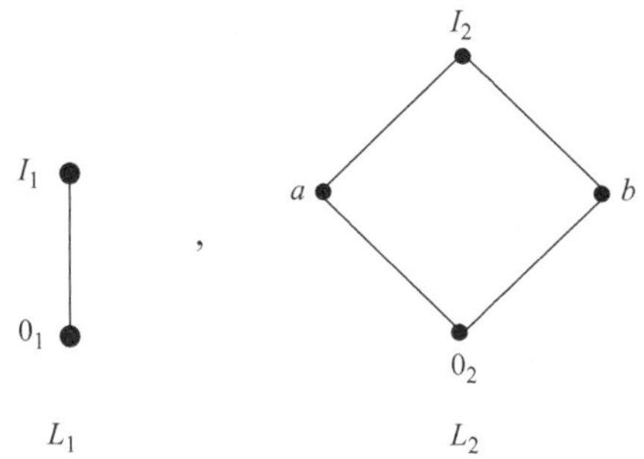

Figure 6.7

Then, $L = L_1 \times L_2$ is the lattice shown in the Figure 6.8.

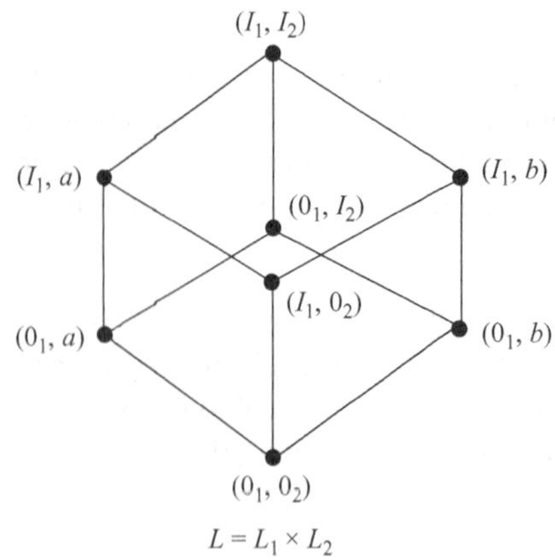

Figure 6.8

EXAMPLE 6.7

Let $L_1=\{1, 2, 4\}$ and $L_2=(1, 3, 9\}$ be the chains of divisors of 4 and 9 with partial order of divisibility (see Figure 6.9).

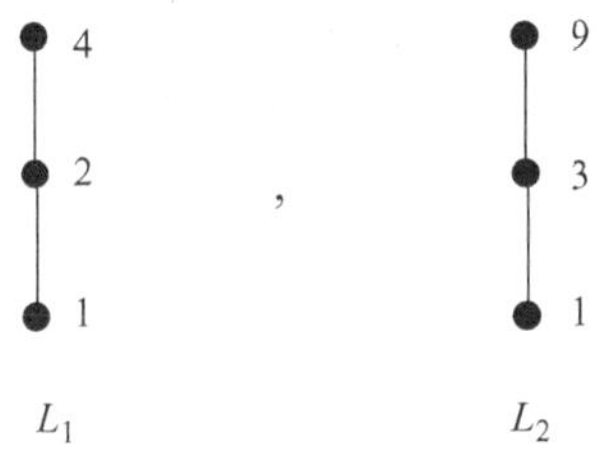

Figure 6.9

Then,

$$L_1 \times L_2=\{(1, 1), (1, 3), (1, 9), (2, 1), (2, 3), (2, 9), (4, 1), (4, 3), (4, 9)\}.$$

The lattice $L_1 \times L_2$ is shown in the Figure 6.10.

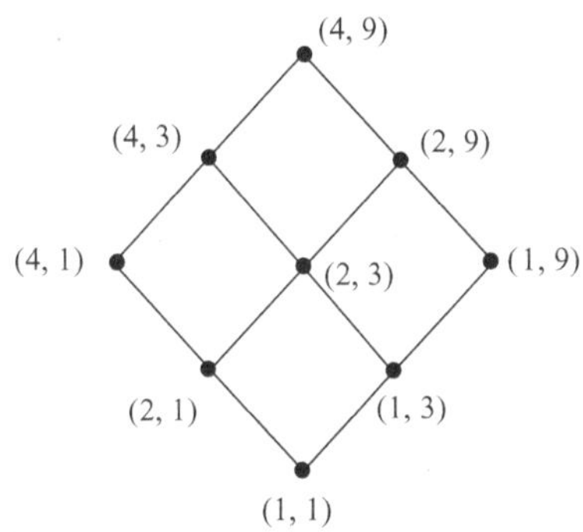

Figure 6.10

6.2 PROPERTIES OF LATTICES

Let $(L, \leq)$ be a lattice and let $a, b, c \in L$. Then, from the definition of $\vee$ (join) and $\wedge$ (meet) we have

(i) $a \leq a \vee b$ and $b \leq a \vee b$; $a \vee b$ is an upper bound of a and b.
(ii) If $a \leq c$ and $b \leq c$, then $a \vee b \leq c$; $a \vee b$ is the least upper bound of a and b.
(iii) $a \wedge b \leq a$ and $a \wedge b \leq b$; $a \wedge b$ is a lower bound of a and b.
(iv) If $c \leq a$ and $c \leq b$, then $c \leq a \wedge b$; $a \wedge b$ is the greatest lower bound of a and b.

Theorem 6.2

Let L be a lattice. Then for every a and b in L,

(i) $a \vee b=b$ if and only if $a \leq b$,
(ii) $a \wedge b=a$ if and only if $a \leq b$,
(iii) $a \wedge b=a$ if and only if $a \vee b=b$.

Proof.

(i) Let $a \vee b=b$. Since $a \leq a \vee b$, we have $a \leq b$.

Conversely, if $a \leq b$, then, since $b \leq b$, it follows that b is an upper bound of a and b. Therefore, by the definition of least upper bound, $a \vee b \leq b$. Also, $a \vee b$ being an upper bound, $b \leq a \vee b$. Hence, $a \vee b=b$.

(ii) Let $a \wedge b=$a. Since $a \wedge b \leq b$, we have $a \leq b$. Conversely, if $a \leq b$ and since $a \leq a$, a is a lower bound of a and b and so, by the definition of greatest lower bound, we have $a \leq a \wedge b$. Since $a \wedge b$ is a lower bound, $a \wedge b \leq a$. Hence $a \wedge b=a$.

(iii) From (ii), we have

$$a \wedge b=a \Leftrightarrow a \leq b. \tag{1}$$

From (i),

$$a \leq b \Leftrightarrow a \vee b=b. \tag{2}$$

Hence, combining (1) and (2), we have

$$a \wedge b=a \Leftrightarrow a \vee b=b.$$

EXAMPLE 6.8

Let L be a linearly (total) ordered set. Therefore, $a, b \in L$ implies either $a \leq b$ or $b \leq a$. Therefore, the above theorem implies that

$$a \vee b=b, \quad a \wedge b=a.$$

Thus for every pair of elements a, b in L, $a \vee b$ and $a \wedge b$ exist. Hence, **a linearly ordered set is a lattice**.

Theorem 6.3

Let $(L, \leq)$ be a lattice and let $a, b, c \in L$. Then we have

L_1: **Idempotent property**

(i) $a \vee a=a$,
(ii) $a \wedge a=a$.

L_2: **Commutative property**

(i) $a \vee b=b \vee a$,
(ii) $a \wedge b=b \wedge a$.

L_3: **Associative property**

(i) $a \vee (b \vee c)=(a \vee b) \vee c$,
(ii) $a \wedge (b \wedge c)=(a \wedge b) \wedge c$.

L_4: **Absorption property**

(i) $a \vee (a \wedge b)=a$,
(ii) $a \wedge (a \vee b)=a$.

Proof.

L_1: The idempotent property follows from the definition of LUB and GLB.
L_2: Commutativity follows from the symmetry of a and b in the definition of LUB and GLB.
L_3: (i) From the definition of LUB, we have

$$a \leq a \vee (b \vee c), \tag{1}$$

$$b \vee c \leq a \vee (b \vee c). \tag{2}$$

Also, $b \leq b \vee c$ and $c \leq b \vee c$ and so transitivity implies

$$b \leq a \vee (b \vee c) \tag{3}$$

and

$$c \leq a \vee (b \vee c) \tag{4}$$

Now, (1) and (3) imply that $a \vee (b \vee c)$ is an upper bound of a and b and hence by the definition of least upper bound, we have

$$a \vee b \leq a \vee (b \vee c). \tag{5}$$

Also by (4) and (5), $a \vee (b \vee c)$ is an upper bound of c and $a \vee b$. Therefore,

$$(a \vee b) \vee c \leq a \vee (b \vee c). \tag{6}$$

Similarly,

$$a \vee (b \vee c) \leq (a \vee b) \vee c. \tag{7}$$

Hence, by anti-symmetry of the relation $\leq$, (6) and (7) yield

$$a \vee (b \vee c) = (a \vee b) \vee c.$$

The proof of (ii) is analogous to the proof of part (i).
L_4: (i) Since $a \wedge b \leq a$ and $a \leq a$, it follows that a is an upper bound of $a \wedge b$ and a. Therefore, by the definition of least upper bound

$$a \vee (a \wedge b) \leq a. \tag{8}$$

On the other hand, by the definition of LUB, we have

$$a \leq a \vee (a \wedge b). \tag{9}$$

The expression (8) and (9) yields

$$a \vee (a \wedge b) = a.$$

(ii) Since $a \leq a \vee b$ and $a \leq a$, it follows that a is a lower bound of $a \vee b$ and a. Therefore, by the definition of GLB,

$$a \leq a \wedge (a \vee b). \tag{10}$$

Also, by the definition of GLB, we have

$$a \wedge (a \vee b) \leq a. \tag{11}$$

Then (10) and (11) imply

$$a \wedge (a \vee b) = a.$$

and the proof is completed.

In view of L_3, we can write $a \vee (b \vee c)$ and $(a \vee b) \vee c$ as $a \vee b \vee c$. Thus, we can express

$$\text{LUB}(\{a_1, a_2, \ldots, a_n\}) \quad \text{as} \quad a_1 \vee a_2 \vee \ldots \vee a_n,$$
$$\text{GLB}(\{a_1, a_2, \ldots, a_n\}) \quad \text{as} \quad a_1 \wedge a_2 \wedge \ldots \wedge a_n.$$

Remark 6.1 Using commutativity and absorption property, part (ii) of Theorem 6.2 can be proved as follows:
Let $a \wedge b=a$. We note that

$$b \vee (a \wedge b)=b \vee a=a \vee b \quad \text{(commutativity).}$$

But,

$$b \vee (a \wedge b)=b \quad \text{(absorption property).}$$

Hence $a \vee b=b$ and so by part (i) of Theorem 6.2, we have $a \leq b$. Hence $a \wedge b=a$ if and only if $a \leq b$.

Theorem 6.4

Let $(L, \leq)$ be a lattice. Then for any $a, b, c \in L$, the following properties hold:

1. **Isotonicity:** If $a \leq b$, then
   (i) $a \vee c \leq b \vee c$
   (ii) $a \wedge c \leq b \wedge c$
2. $a \leq c$ and $b \leq c$ if and only if $a \vee b \leq c$.
3. $c \leq a$ and $c \leq b$ if and only if $c \leq a \wedge b$.
4. If $a \leq b$ and $c \leq d$, then
   (i) $a \vee c \leq b \vee d$,
   (ii) $a \wedge c \leq b \wedge d$.

Proof.

1. (i) We know that

$$a \vee b=b \quad \text{if and only if} \quad a \leq b.$$

Therefore, to show that $a \vee c \leq b \vee c$, we shall show that

$$(a \vee c) \vee (b \vee c)=b \vee c.$$

We note that

$$\begin{aligned}(a \vee c) \vee (b \vee c) &= [(a \vee c) \vee b] \vee c \\ &= a \vee (c \vee b) \vee c \\ &= a \vee (b \vee c) \vee c \\ &= (a \vee b) \vee (c \vee c) \\ &= b \vee c \; (\because a \vee b=b \text{ and } c \vee c=c).\end{aligned}$$

(ii) This can be proved similarly.

2. If $a \leq c$, then 1(i) implies

$$a \vee b \leq c \vee b.$$

But,

$$\begin{aligned} b \leq c &\Leftrightarrow b \vee c=c \\ &\Leftrightarrow c \vee b=c \quad \text{(commutativity).}\end{aligned}$$

Hence, $a \leq c$ and $b \leq c$ if and only if $a \vee b \leq c$.

3. If $c \leq a$, then 1(ii) implies

$$c \wedge b \leq a \wedge b.$$

But,

$$c \leq b \Leftrightarrow c \wedge b = c.$$

Hence, $c \leq a$ and $c \leq b$ if and only if $c \leq a \wedge b$.

4. (i) We note that 1(i) implies that

$$\text{if } a \leq b, \quad \text{then} \quad a \vee c \leq b \vee c = c \vee b,$$
$$\text{if } c \leq d, \quad \text{then} \quad c \vee b \leq d \vee b = b \vee d.$$

Hence, by transitivity

$$a \vee c \leq b \vee d.$$

(ii) We note that 1(ii) implies that

$$\text{if } a \leq b, \quad \text{then} \quad a \wedge c \leq b \wedge c = c \wedge b,$$
$$\text{if } c \leq d, \quad \text{then} \quad c \wedge b \leq d \wedge b = b \wedge d.$$

Therefore, transitivity implies

$$a \wedge c \leq b \wedge d.$$

Theorem 6.5

Let $(L, \leq)$ be a lattice. If $a, b, c \in L$, then

(1) $a \vee (b \wedge c) \leq (a \vee b) \wedge (a \vee c)$,

(2) $a \wedge (b \vee c) \geq (a \wedge b) \vee (a \wedge c)$.

These inequalities are called **distributive inequalities**.

Proof. We have

$$a \leq a \vee b \quad \text{and} \quad a \leq a \vee c. \tag{i}$$

Also, by the above theorem, if $x \leq y$ and $x \leq z$ in a lattice, then $x \leq y \wedge z$. Therefore, (i) yields

$$a \leq (a \vee b) \wedge (a \vee c). \tag{ii}$$

Also,

$$b \wedge c \leq b \leq a \vee b$$

and

$$b \wedge c \leq c \leq a \vee c,$$

that is, $b \wedge c \leq a \vee b$ and $b \wedge c \leq a \vee c$ and so, by the above argument, we have

$$b \wedge c \leq (a \vee b) \wedge (a \vee c). \tag{iii}$$

Also, again by the above theorem if $x \leq z$ and $y \leq z$ in a lattice, then

$$x \vee y \leq z.$$

Hence, (ii) and (iii) yield

$$a \vee (b \wedge c) \leq (a \vee b) \wedge (a \vee c).$$

This proves (1).

The second distributive inequality follows by using the **principle of duality**.

Theorem 6.6 (Modular Inequality)

Let $(L, \leq)$ be a lattice. If $a, b, c \in L$, then

$$a \leq c \quad \text{if and only if} \quad a \vee (b \wedge c) \leq (a \vee b) \wedge c.$$

Proof. We know that

$$a \leq c \Leftrightarrow a \vee c = c. \tag{1}$$

Also, by distributive inequality,

$$a \vee (b \wedge c) \leq (a \vee b) \wedge (a \vee c).$$

Therefore, using (1), $a \leq c$ if and only if

$$a \vee (b \wedge c) \leq (a \vee b) \wedge c,$$

which proves the result.

The modular inequalities can be expressed also in the following way:

$$(a \wedge b) \vee (a \wedge c) \leq a \wedge [b \vee (a \wedge c)],$$
$$(a \vee b) \wedge (a \vee c) \geq a \vee [b \wedge (a \vee c)].$$

EXAMPLE 6.9

Let $(L, \leq)$ be a lattice and $a, b, c \in L$. If $a \leq b \leq c$, then (i) $a \vee b = b \wedge c$ and (ii) $(a \wedge b) \vee (b \wedge c) = (a \vee b) \wedge (a \vee c)$.

Solution.

(i) We know that

$$a \leq b \Leftrightarrow a \vee b = b$$

and

$$b \leq c \Leftrightarrow b \wedge c = b.$$

Hence, $a \leq b \leq c$ implies

$$a \vee b = b \wedge c.$$

(ii) Since $a \leq b$ and $b \leq c$, we have

$$a \wedge b = a \quad \text{and} \quad b \wedge c = b.$$

Thus,

$$\begin{aligned}(a \wedge b) \vee (b \wedge c) &= a \vee b \\ &= b, \quad \text{since } a \leq b \Leftrightarrow a \vee b = b.\end{aligned}$$

Also, $a \leq b \leq c \Rightarrow a \leq c$ by transitivity. Then,

$$a \leq b \quad \text{and} \quad a \leq c \Rightarrow a \vee b = b,\ a \vee c = c$$

and so

$$\begin{aligned}(a \vee b) \wedge (a \vee c) &= b \wedge c \\ &= b, \quad \text{since } b \leq c \Leftrightarrow b \wedge c = b.\end{aligned}$$

Hence,

$$(a \wedge b) \vee (b \wedge c)=b=(a \vee b) \wedge (a \vee c),$$

which proves (ii).

EXAMPLE 6.10

Show that a lattice with three or fewer elements is a chain.

Solution.

If a lattice consists of one element, then it is obviously a chain. If it consists of two elements, again it is a chain. Let the lattice consists of three elements. If it is not a chain, then its Hasse diagram should be as shown in the Figure 6.11.

Figure 6.11

In the first case, $b \vee c$ does not exist whereas in the second case $a \wedge b$ does not exist. Hence in that case, we cannot have a lattice. It follows, therefore, that a lattice with three or fewer elements is a chain.

6.3 LATTICES AS ALGEBRAIC SYSTEM

Definition 6.3

A **Lattice** is an algebraic system $(L, \vee, \wedge)$ with two binary operations $\vee$ and $\wedge$, called **join** and **meet**, respectively, on a non-empty set L which satisfies the following axioms for $a, b, c \in L$:

1. **Commutative Law:**

$$a \vee b=b \vee a \quad \text{and} \quad a \wedge b=b \wedge a.$$

2. **Associative Law:**

$$(a \vee b) \vee c=a \vee (b \vee c)$$

and

$$(a \wedge b) \wedge c=a \wedge (b \wedge c).$$

3. **Absorption Law:**

   (i) $a \vee (a \wedge b)=a$,
   (ii) $a \wedge (a \vee b)=a$.

We note that **Idempotent Law follows from axiom 3** above. In fact,

$$\begin{aligned} a \vee a &= a \vee [a \wedge (a \vee b)] \quad \text{using 3 (ii)} \\ &= a \quad \text{using 3 (i).} \end{aligned}$$

The proof of $a \wedge a=a$ follows by the principle of duality.

6.3.1 Partial Order Relations on a Lattice

A partial order relation on a lattice (L) follows as a consequence of the axioms for the binary operations $\vee$ and $\wedge$.

We define a relation $\leq$ on L such that for $a, b \in L$,

$$a \leq b \Leftrightarrow a \vee b = b$$

or analogously,

$$a \leq b \Leftrightarrow a \wedge b = a.$$

We note that

(i) For any $a \in L$,

$$a \vee a = a \quad \text{(idempotent law)},$$

therefore, $a \leq a$ showing that $\leq$ is **reflexive**.

(ii) Let $a \leq b$ and $b \leq a$. Therefore,

$$a \vee b = b \quad \text{and} \quad b \vee a = a.$$

But,

$$a \vee b = b \vee a \quad \text{(commutative law in lattice)}.$$

Hence $a = b$, showing that $\leq$ is **anti-symmetric**.

(iii) Suppose that $a \leq b$ and $b \leq c$. Therefore, $a \vee b = b$ and $b \vee c = c$. Then,

$$\begin{aligned} a \vee c &= a \vee (b \vee c) \\ &= (a \vee b) \vee c \quad \text{(associativity in lattice)} \\ &= b \vee c = c, \end{aligned}$$

showing that $a \leq c$ and hence $\leq$ is **transitive**. This shows that a **lattice is a partially ordered set.**

6.3.2 Least Upper Bounds and Greatest Lower Bounds in a Lattice

Let $(L, \vee, \wedge)$ be a lattice and let $a, b \in L$. We now show that LUB of $\{a, b\} \subseteq L$ with respect to the partial order introduced in Section 6.3.1 is $a \vee b$ and GLB of $\{a, b\}$ is $a \wedge b$.

From absorption law

$$a \wedge (a \vee b) = a, \qquad b \wedge (a \vee b) = b.$$

Therefore $a \leq a \vee b$ and $b \leq a \vee b$, showing that $a \vee b$ is upper bound for $\{a, b\}$. Suppose that there exists $c \in L$ such that $a \leq c$, $b \leq c$. Thus we have

$$a \vee c = c \quad \text{and} \quad b \vee c = c$$

and then

$$(a \vee b) \vee c = a \vee (b \vee c) = a \vee c = c,$$

implying that $a \vee b \leq c$. Hence, $a \vee b$ is the least upper bound of a and b.

Similarly, we can show that $a \wedge b$ is GLB of a and b.

The above discussion shows that ***Definitions 6.1 and 6.3 are equivalent.***

EXAMPLE 6.11

Let $\hat{C}$ be collection of sets with binary operations union and intersection of sets. Then, $(\hat{C}, \cup, \cap)$ is a lattice. In this lattice, the partial order relation is **set inclusion.** In fact, for $A, B \in \hat{C}$,

$$A \subseteq B \Leftrightarrow A \cup B = B$$

or

$$A \subseteq B \Leftrightarrow A \cap B = A.$$

For illustration, the diagram of lattice of subsets of $\{a, b\}$ is shown in the Figure 6.12.

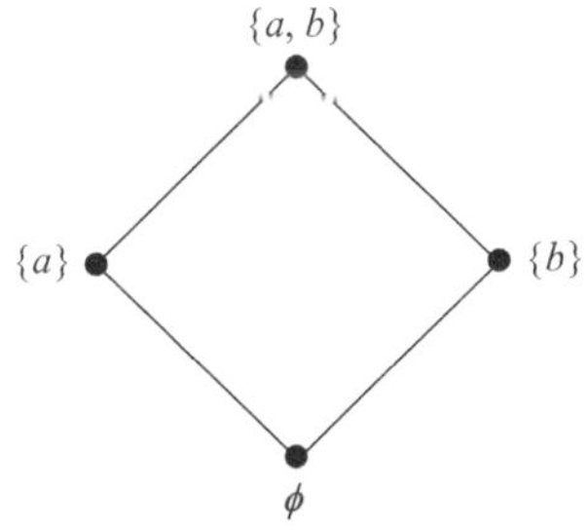

Figure 6.12

6.3.3 Sublattices

Definition 6.4

Let $(L, \leq)$ be a lattice. A non-empty subset S of L is called a **sublattice** of L if $a \vee b \in S$ and $a \wedge b \in S$ whenever $a \in S, b \in S$.

Equivalently, we may define sublattice as below:

Let $(L, \vee, \wedge)$ be a lattice and let $S \subseteq L$ be a subset of L. Then $(S, \vee, \wedge)$ is called a sublattice of $(L, \vee, \wedge)$ if and only if S is closed under both operations of join $(\vee)$ and meet $(\wedge)$.

It is clear from the definition that **sublattice itself is a lattice.**

However, **any subset of L which is a lattice need not be a sublattice.**

For example, consider the lattice shown in the Figure 6.13.

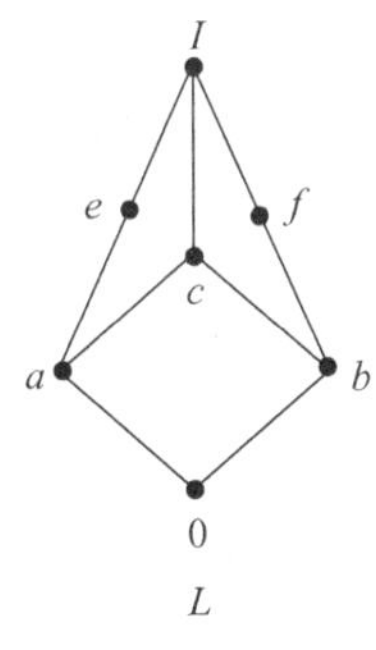

Figure 6.13

We note that

(i) The subset S shown in the Figure 6.14 is not a sublattice of L, since $a \wedge b \notin S$ and $a \vee b \notin S$.

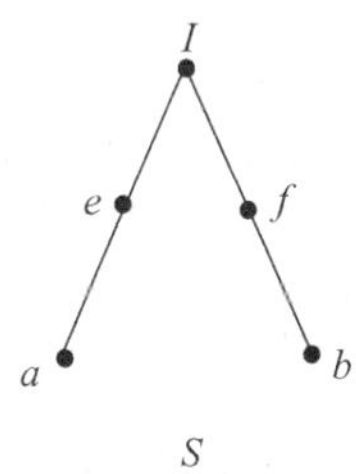

Figure 6.14

(ii) The set T shown in the Figure 6.15 is not a sublattice of L since $a \vee b \notin T$.

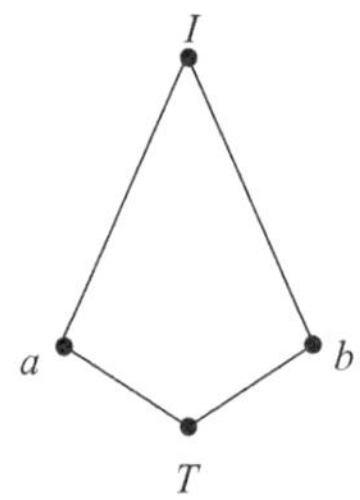

Figure 6.15

However, T is a lattice when considered as a poset by itself.

(iii) The subset U of L shown in the Figure 6.16 is a sublattice of L:

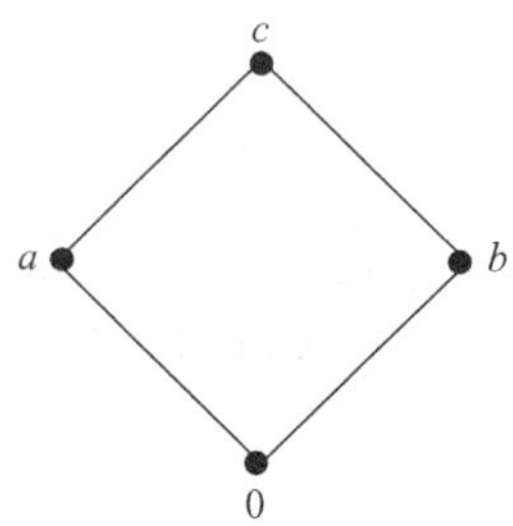

Figure 6.16

EXAMPLE 6.12

Let A be any set and $P(A)$ its power set. Then $(P(A), \vee, \wedge)$ is a lattice in which join and meet are union of sets and intersection of sets, respectively.

A family $\hat{C}$ of subsets of A such that $S \cup T$ and $S \cap T$ are in $\hat{C}$ for $S, T \in \hat{C}$ is a sublattice of $(P(A), \vee, \wedge)$. Such a family $\hat{C}$ is called a ring of subsets of A and is denoted by $(R(A), \vee, \wedge)$ (this is not a ring in the sense of algebra). Some authors call it lattice of subsets.

EXAMPLE 6.13

The lattice $(D_n, \leq)$ is a sublattice of $(\mathbf{N}, \leq)$, where $\leq$ is the relation of divisibility.

6.4 LATTICE ISOMORPHISM

Definition 6.5

Let $(L_1, \vee_1, \wedge_1)$ and $(L_2, \vee_2, \wedge_2)$ be two lattices. A mapping $f: L_1 \to L_2$ is called a **lattice homomorphism** from the lattice $(L_1, \vee_1, \wedge_1)$ to $(L_2, \vee_2, \wedge_2)$ if for any $a, b \in L_1$,

$$f(a \vee_1 b) = f(a) \vee_2 f(b) \quad \text{and} \quad f(a \wedge_1 b) = f(a) \wedge_2 f(b).$$

Thus, here both the binary operations of join and meet are preserved. **There may be mappings which preserve only one of the two operations. Such mappings are not lattice homomorphism.**

Let $\leq_1$ and $\leq_2$ be partial order relations on $(L_1, \vee_1, \wedge_1)$ and $(L_2, \vee_2, \wedge_2)$, respectively. Let $f: L_1 \to L_2$ be lattice homomorphism. If $a, b \in L_1$, then

$$a \leq_1 b \Leftrightarrow a \vee_1 b = b$$

and so

$$\begin{aligned} f(b) &= f(a \vee_1 b) \\ &= f(a) \vee_2 f(b) \\ &\Leftrightarrow f(a) \leq_2 f(b). \end{aligned}$$

Thus,

$$a \leq_1 b \Leftrightarrow f(a) \leq_2 f(b).$$

Thus, **order relations are also preserved** under lattice homomorphism.

If a lattice homomorphism $f: L_1 \to L_2$ is one-one and onto, then it is called **lattice isomorphism.**

If there exists an isomorphism between two lattices, then the lattices are called **isomorphic**.

Since lattice isomorphism preserves order relation, therefore **isomorphic lattices can be represented by the same diagram in which nodes are replaced by images.**

Theorem 6.7

Let $A = \{a_1, a_2, \ldots, a_n\}$ and $B = \{b_1, b_2, \ldots, b_n\}$ be any two finite sets with n elements. Then the lattices $(P(A), \subseteq)$ and $(P(B), \subseteq)$ are isomorphic and so have identical Hasse diagram.

Proof. Consider the mapping $f: P(A) \to P(B)$ defined by

$$f(\{a_n\} = \{b_n\}, f(\{a_1, a_2, \ldots, a_m\}) = \{b_1, b_2, \ldots, b_m\} \text{ for } m \leq n.$$

Then f is bijective mapping and $L \subseteq M \Leftrightarrow f(L) \subseteq f(M)$ for subsets L and M of $P(A)$. Hence $P(A)$ and $P(B)$ are isomorphic.

For example, let $A = \{a, b, c\}$, $B = \{2, 3, 5\}$. The Hasse diagrams of $P(A)$ and $P(B)$ are then given in the Figure 6.17.

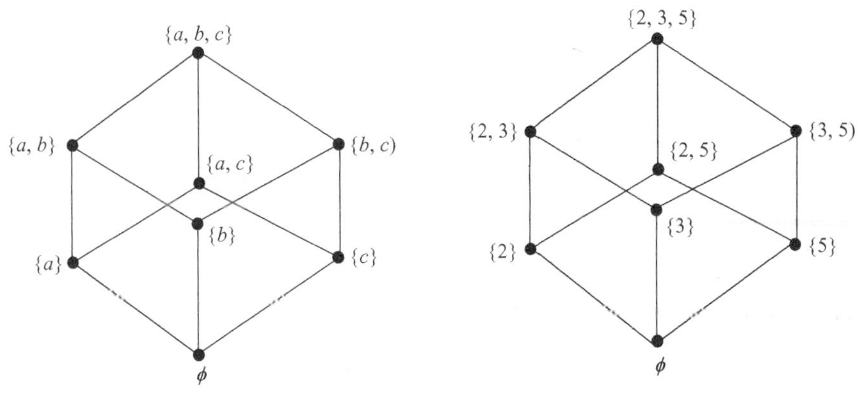

Figure 6.17

Define a mapping $f: P(A) \rightarrow P(B)$ by

$$f(\phi)=\phi, f(\{a\})=\{2\}, f(\{b\})=\{3\}, f(\{c\})=\{5\},$$
$$f(\{a, b\})=\{2, 3\}, f(\{b, c\})=\{3, 5\}, f(\{a, c\})=\{2, 5\},$$

and

$$f(\{a, b, c\})=\{2, 3, 5\}.$$

This is a bijective mapping satisfying the condition that if S and T are subsets of A, then $S \subseteq T$ if and only if $f(S) \subseteq f(T)$. Hence f is isomorphism and $(P(A), \subseteq)$ and $(P(B), \subseteq)$ are isomorphic.

Thus, for each $n=0, 1, 2, \ldots$, there is only one type of lattice and this lattice depends only on n, the number of elements in the set A, and not on A. It has 2^n elements. Also, we know that if A has n elements, then all subsets of A can be represented by sequences of 0's and 1's of length n. We can therefore label the Hasse diagram of a lattice $(P(A), \subseteq)$ by such sequence of 0's and 1's.

For example, lattices of $P(A)$ and $P(B)$ of the last example can be labelled as in the Figure 6.18.

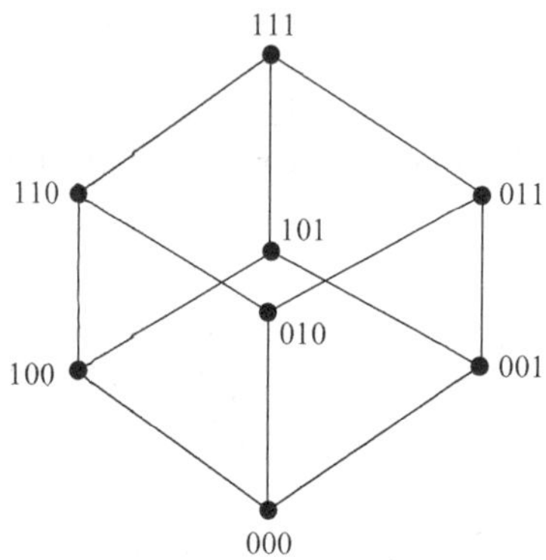

Figure 6.18

The lattice so obtained is named B_n. The properties of the partial order in B_n can be described directly as follows:

Let $x=a_1\, a_2 \ldots a_n$ and $y=b_1\, b_2 \ldots b_n$ be any two elements of B_n. Then

1. $x \leq y$ if and only if $a_k < b_k$, $k = 1, 2, \ldots, n$, where a_k and b_k are 0 or 1.
2. $x \wedge y=c_1\, c_2 \ldots c_n$, where $c_k=\min(a_k, b_k)$.
3. $x \vee y=d_1\, d_2 \ldots d_n$, where $d_k=\max(a_k, b_k)$.
4. x has a complement $x'=z_1\, z_2 \ldots z_n$ where $z_k=1$ if $x_k=0$ and $z_k=0$ if $x_k=1$.

Remark 6.2 $(B_n, \leq)$ under the partial order $\leq$ defined above is isomorphic to $(P(A), \subseteq)$, when A has n elements. In such a case $x \leq y$ corresponds to $S \subseteq T$, $x \vee y$ corresponds to $S \cup T$ and x' corresponds to A^c.

EXAMPLE 6.14

Let $D_6=\{1, 2, 3, 6\}$, set of divisors of 6. Then D_6 is isomorphic to B_2. In fact, $f: D_6 \rightarrow B_2$ defined by

$$f(1)=00, \quad f(2)=10, \quad f(3)=01, \quad f(6)=11$$

is an **isomorphism** (see Figure 6.19).

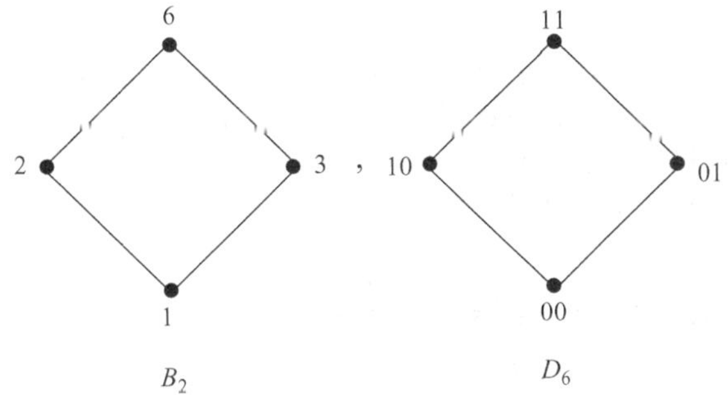

Figure 6.19

EXAMPLE 6.15

Let $A=\{a, b\}$ and $P(A)=\{\phi, \{a\}, \{b\}, \{a, b\}\}$. Then the lattice $(P(A), \subseteq)$ is isomorphic to lattice $(D_6, 1)$ with divisibility as the partial order relation. In fact, we define a mapping $f: D_6 \to P(A)$ by

$$f(1)=\phi, \quad f(2)=\{a\}, \quad f(3)=\{b\}, \quad f(6)=\{a, b\} \text{ (see Figure 6.20).}$$

Then, f is bijective and we note that

$$1 \mid 2 \Leftrightarrow \{\phi\} \subseteq \{a\} \Leftrightarrow f(1) \subseteq f(2),$$
$$2 \mid 6 \Leftrightarrow \{a\} \subseteq \{a, b\} \Leftrightarrow f(2) \subseteq f(6)$$

and so on. Hence f is isomorphism.

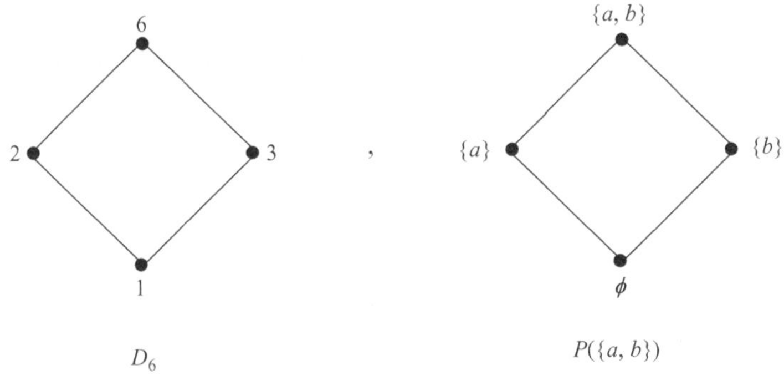

Figure 6.20

Definition 6.6

Let $(L, \vee, \wedge)$ be a lattice. Then lattice homomorphism $f: L \to L$ is called an **endomorphism**.

Definition 6.7

Let $(L, \vee, \wedge)$ be a lattice. Then the lattice isomorphism $f: L \to L$ is called an **automorphism**.

If $f: L \to L$ is an endomorphism, then the image set of f is sublattice of L.

Definition 6.8

Let $(A, \leq)$ and $(B, \leq')$ be two partially ordered sets. A mapping $f: A \to B$ is called order preserving relative to the ordering $\leq$ in A and $\leq'$ in B if and only if for $a, b \in A$,

$$a \leq b \Rightarrow f(a) \leq' f(b).$$

If A and B are lattices and $f: A \to B$ is a lattice homomorphism, then f is order preserving.

Definition 6.9

Two partially ordered sets $(A, \leq)$ and $(B, \leq')$ are said to be **order isomorphic** if there exists a mapping $f: A \to B$ which is bijective and both f and f^{-1} are order preserving.

For lattices $(A, \leq)$ and $(B, \leq')$, an order isomorphism is equivalent to lattice isomorphism. Hence lattices which are order-isomorphic as partially ordered sets are isomorphic.

6.5 BOUNDED, COMPLEMENTED AND DISTRIBUTIVE LATTICES

Let $(L, \vee, \wedge)$ be a lattice and let $S = \{a_1, a_2, \ldots, a_n\}$ be a finite subset of L. Then,

$$\text{LUB of } S \text{ is represented by } a_1 \vee a_2 \vee \ldots \vee a_n,$$
$$\text{GLB of } S \text{ is represented by } a_1 \wedge a_2 \wedge \ldots \wedge a_n.$$

Definition 6.10

A lattice is called **complete** if each of its non-empty subsets has a least upper bound and a greatest lower bound.

Obviously, every finite lattice is complete.

Also, every complete lattice must have a least element, denoted by 0 and a greatest element, denoted by I.

The least and greatest elements if exist are called **bound (units, universal bounds)** of the lattice.

Definition 6.11

A lattice L is said to be **bounded** if it has a greatest element I and a least element 0.

For the lattice $(L, \vee, \wedge)$ with $L = \{a_1, a_2, \ldots, a_n\}$,

$$a_1 \vee a_2 \vee \ldots \vee a_n = I \quad \text{and} \quad a_1 \wedge a_2 \wedge \ldots \wedge a_n = 0.$$

EXAMPLE 6.16

The lattice $\mathbf{Z}^+$ of all positive integers under partial order of divisibility is not a bounded lattice since it has a least element (the integer 1) but no greatest element.

EXAMPLE 6.17

The lattice $\mathbf{Z}$ of integers under partial order $\leq$ (less than or equal to) is not bounded since it has neither a greatest element nor a least element.

EXAMPLE 6.18

Let A be a non-empty set. Then the lattice $(P(A), \subseteq)$ is bounded. Its greatest element is A and the least element is empty set ϕ.

If $(L, \leq)$ is a bounded lattice, then for all $a \in L$,

$$0 \leq a \leq I,$$
$$a \vee 0 = a, \quad a \wedge 0 = 0,$$
$$a \vee I = I, \quad a \wedge I = a.$$

Thus, 0 acts as identity of the operation $\vee$ and I acts as identity of the operation $\wedge$.

Definition 6.12

Let $(L \vee, \wedge, 0, I)$ be a bounded lattice with greatest element I and the least element 0. Let $a \in L$. Then an element $b \in L$ is called a **complement** of a if

$$a \vee b=I \quad \text{and} \quad a \wedge b=0.$$

It follows from this definition that

0 and I are complement of each other.

Further, I is the only complement of 0. Suppose that $c \neq I$ is a complement of 0 and $c \in L$, then

$$0 \vee c=I \quad \text{and} \quad 0 \wedge c=0.$$

But $0 \vee c=c$. Therefore $c=I$ which contradicts $c \neq I$. Similarly, 0 is the only complement of I.

Definition 6.13

A lattice $(L, \vee, \wedge, 1, 0)$ is called **complemented** if it is bounded and if every element of L has at least one complement.

EXAMPLE 6.19

The lattice $(P(A), \subseteq)$ of the power set of any set A is a bounded lattice, where meet and join operations on $P(A)$ are $\cap$ and $\cup$, respectively. Its bounds are ϕ and A. The lattice $(P(A), \subseteq)$ is complemented in which the complement of any subset B of A is $A-B$.

EXAMPLE 6.20

Let L^n be the lattice of n tuples of 0 and 1, where partial ordering is defined for $a=(a_1, a_2, \ldots, a_n)$, $b=(b_1, b_2, \ldots, b_n) \in L^n$ by

$$a \leq_n b \Leftrightarrow a_i \leq b_i \quad \text{for all } i=1, 2, \ldots, n,$$

where $\leq$ means less than or equal to. Then $(L^n, \leq_n)$ is lattice which is bounded.

For example, the bounds are (0, 0, 0) and (1, 1, 1) for L^3.

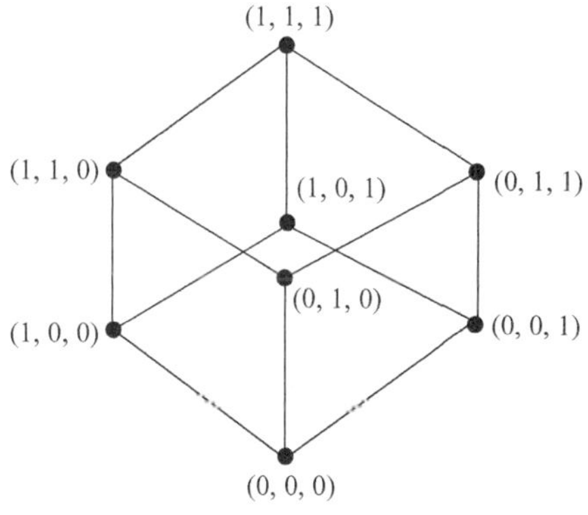

Figure 6.21

The complement of an element of L^n can be obtained by interchanging 1 by 0 and 0 by 1 in the n-tuple representing the element. For example, complement of (1, 0, 1) in L^3 is (0, 1, 0) (see Figure 6.21).

Definition 6.14

A lattice $(L, \vee, \wedge)$ is called a **distributive lattice** if for any elements a, b and c in L,

(1) $a \wedge (b \vee c)=(a \wedge b) \vee (a \wedge c)$,

(2) $a \vee (b \wedge c)=(a \vee b) \wedge (a \vee c)$.

Properties (1) and (2) are called **distributive properties**.

Thus, in a distributive lattice, the operations $\wedge$ and $\vee$ are distributive over each other.

We further note that, by the principle of duality, the condition (1) holds if and only if (2) holds. Therefore it is sufficient to verify any one of these two equalities for all possible combinations of the elements of a lattice.

If a lattice L is not distributive, we say that L is **non-distributive**.

EXAMPLE 6.21

For a set S, the lattice $(P(S), \subseteq)$ is distributive. The meet and join operation in $P(S)$ are $\cap$ and $\cup$, respectively. Also we know, by set theory, that for $A, B, C \in P(S)$,

$$A \cap (B \cup C)=(A \cap B) \cup (A \cap C),$$
$$A \cup (B \cap C)=(A \cup B) \cap (A \cup C).$$

EXAMPLE 6.22

The **five-element** lattices given in the Figure 6.22 are **non-distributive**.

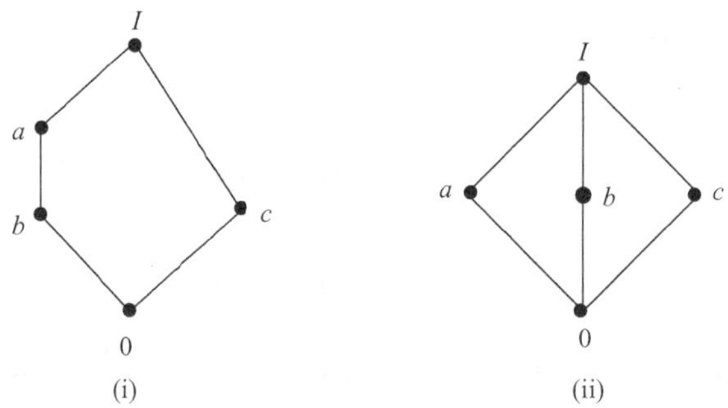

Figure 6.22

In fact, for lattice (i), we note that

$$a \wedge (b \vee c)=a \wedge I=a,$$

while

$$(a \wedge b) \vee (a \wedge c)=b \vee 0=b.$$

Hence,

$$a \wedge (b \vee c)\neq(a \wedge b) \vee (a \wedge c),$$

showing that (i) is non-distributive.

For the lattice (ii), we have

$$a \wedge (b \vee c) = a \wedge I = a,$$

while

$$(a \wedge b) \vee (a \wedge c) = 0 \vee 0 = 0.$$

Hence,

$$a \wedge (b \vee c) \neq (a \wedge b) \vee (a \wedge c),$$

showing that (ii) is also non-distributive.

EXAMPLE 6.23

The lattice shown in the Figure 6.23 is **distributive**.

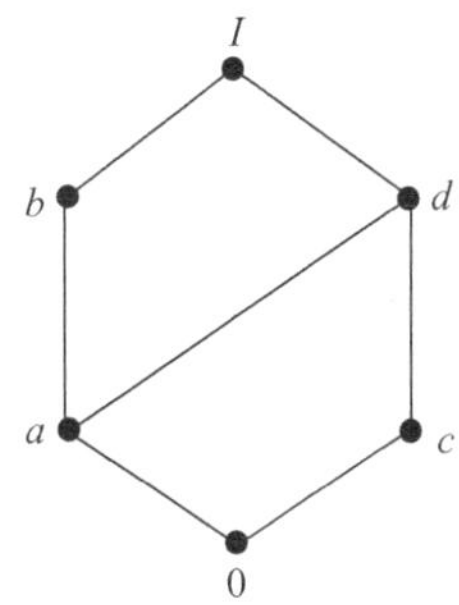

Figure 6.23

The distributive properties are satisfied for any ordered triplet chosen from the given elements.

Theorem 6.8

A lattice L is **non-distributive** if and only if it contains a sublattice isomorphic to any one of the following two five-element lattices of Figure 6.24.

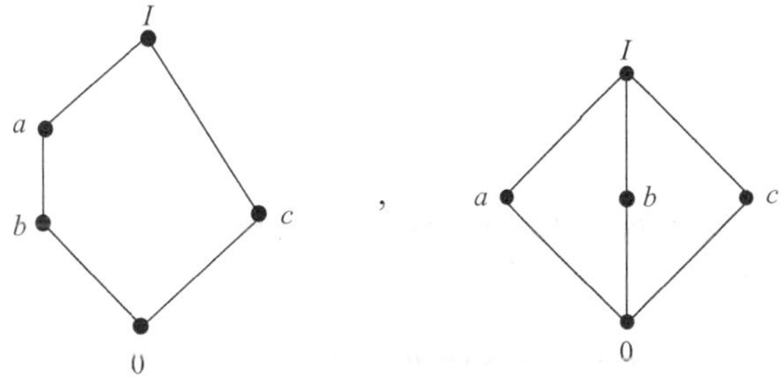

Figure 6.24

(The proof of this theorem is out of the scope of this book.)

EXAMPLE 6.24

Is the lattice in the Figure 6.25 a distributive lattice?

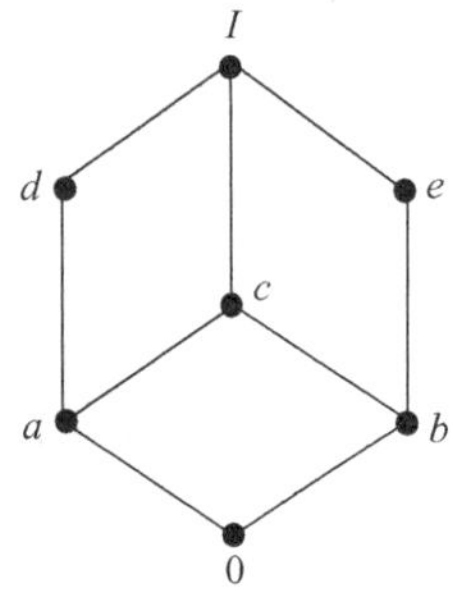

Figure 6.25

Solution.

The given lattice is **not a distributive lattice** since $\{0, a, d, e, I\}$ is a sublattice which is isomorphic to the five-element lattice shown in the Figure 6.26.

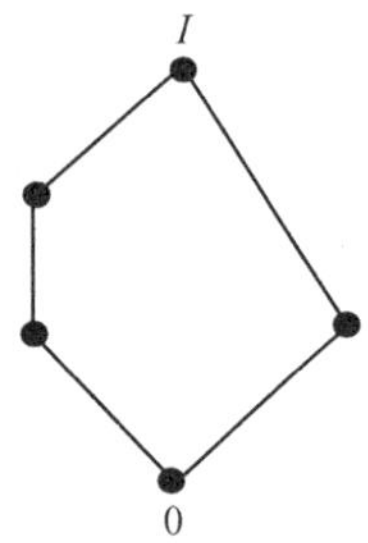

Figure 6.26

Theorem 6.9

Every chain is a distributive lattice.

Proof. Let $(L, \leq)$ be a chain and $a, b, c \in L$. We shall show that distributive law holds for any $a, b, c \in L$. Two cases arise:

Case I. Let $a \leq b$ or $a \leq c$. In this case

$$a \wedge (b \vee c) = a \quad \text{and} \quad (a \wedge b) \vee (a \wedge c) = a.$$

Hence,

$$a \wedge (b \vee c) = (a \wedge b) \vee (a \wedge c).$$

Also, by Principle of Duality,

$$a \vee (b \wedge c) = (a \vee b) \wedge (a \vee c).$$

Case II. Let $b \leq a$ or $c \leq a$. Then we have

$$a \wedge (b \vee c) = (b \vee c) \quad \text{and} \quad (a \wedge b) \vee (a \wedge c) = (b \vee c).$$

Hence,

$$a \wedge (b \vee c) = (b \vee c) = (a \wedge b) \vee (a \wedge c).$$

Hence distributive law holds for any $a, b, c \in L$.

Theorem 6.10

The direct product of any two distributive lattices is a distributive lattice.

Proof. Let $(L_1, \leq_1)$ and $(L_2, \leq_2)$ be two lattices in which meet and join are $\wedge_1, \vee_1$ and $\wedge_2, \vee_2$, respectively. Then meet and join in $L_1 \times L_2$ are defined by

$$(a_1, b_1) \wedge (a_2, b_2) = (a_1 \wedge_1 a_2, b_1 \wedge_2 b_2) \tag{1}$$

and

$$(a_1, b_1) \vee (a_2, b_2) = (a_1 \vee_1 a_2, b_1 \vee_2 b_2). \tag{2}$$

Since L_1 is distributive,

$$a_1 \wedge_1 (a_2 \vee_1 a_3) = (a_1 \wedge_1 a_2) \vee_1 (a_1 \wedge_1 a_3). \tag{3}$$

Since L_2 is distributive,

$$b_1 \wedge_2 (b_2 \vee_2 b_3) = (b_1 \wedge_2 b_2) \vee_2 (b_1 \wedge_2 b_3). \tag{4}$$

Therefore

$$\begin{aligned}(a_1, b_1) \wedge [(a_2, b_2) \vee (a_3, b_3)] &= (a_1, b_1) \wedge [(a_2 \vee_1 a_3, b_2 \vee_2 b_3)] \\ &= [a_1 \wedge_1 (a_2 \vee_1 a_3), b_1 \wedge_2 (b_2 \vee_2 b_3)] \\ &= [(a_1 \wedge_1 a_2) \vee_1 (a_1 \wedge_1 a_3), (b_1 \wedge_2 b_2) \vee_2 (b_1 \wedge_2 b_3)] \\ &\qquad \text{(using (3) and (4))}\end{aligned}$$

and using (1) and (2), we have

$$\begin{aligned}[(a_1, b_1) \wedge (a_2, b_2)] \vee [(a_1, b_1) \wedge (a_3, b_3)] &= (a_1 \wedge_1 a_2, b_1 \wedge_2 b_2) \vee (a_1 \wedge_1 a_3, b_1 \wedge_2 b_3) \\ &= [(a_1 \wedge_1 a_2) \vee_1 (a_1 \wedge_1 a_3), (b_1 \wedge_2 b_2) \vee_2 (b_1 \wedge_2 b_3)].\end{aligned}$$

Hence,

$$(a_1, b_1) \wedge [(a_2, b_2) \vee (a_3, b_3)] = [(a_1, b_1) \wedge (a_2, b_2)] \vee [(a_1, b_1) \wedge (a_3, b_3)],$$

proving that $L_1 \times L_2$ is distributive.

Theorem 6.11

Let L be a bounded distributive lattice. If a complement of any element exists, it is unique.

Proof. Suppose on the contrary that b and c are complements of the element $a \in L$. Then

$$\begin{aligned} a \vee b &= I, \qquad & a \vee c &= I, \\ a \wedge b &= 0, \qquad & a \wedge c &= 0. \end{aligned}$$

Using distributive law, we have

$$\begin{aligned} b &= b \vee 0 = b \vee (a \wedge c) \\ &= (b \vee a) \wedge (b \vee c) \\ &= (a \vee b) \wedge (b \vee c) \\ &= I \wedge (b \vee c) = b \vee c. \end{aligned}$$

Similarly,

$$\begin{aligned} c &= c \vee 0 = c \vee (a \wedge b) \\ &= (c \vee a) \wedge (c \vee b) \\ &= (a \vee c) \wedge (c \vee b) \\ &= I \wedge (c \vee b) \\ &= I \wedge (b \vee c) = b \vee c. \end{aligned}$$

Hence $b=c$.

Definition 6.15

Let $(L, \wedge, \vee)$ be a lattice. An element $a \in L$ is said to be **join-irreducible** if it cannot be expressed as the join of two distinct elements of L.

In other words, $a \in L$ is join-irreducible if for any $b, c \in L$

$$a = b \vee c \Rightarrow a = b \quad \text{or } a = c.$$

For example, prime numbers under multiplication have this property. In fact, if p is a prime number, then $p = a\,b \Rightarrow p = a$ or $p = b$.

Clearly 0 is join-irreducible.

Further, suppose a has at least two immediate predecessors, say b and c as in the Figure 6.27.

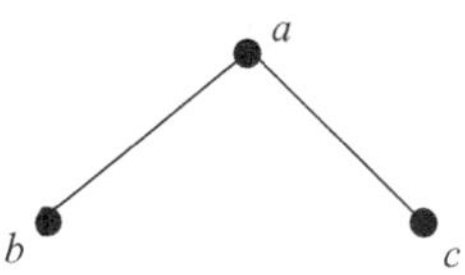

Figure 6.27

Then, $a = b \vee c$ and so a **is not join-irreducible.**

On the other hand, if a has a unique immediate predecessor c, then $a \neq \sup(b_1, b_2) = b_1 \vee b_2$ for any other elements b_1 and b_2 because c would lie between b_1, b_2 and a (see Figure 6.28).

Figure 6.28

In other words, $a \neq 0$ **is join-irreducible if and only if** a **has a unique predecessor**.

Definition 6.16

Those elements which immediately succeed 0 are called **atoms**.

From the above discussion, it follows that the **atoms are join-irreducible**. For illustration, in the Figure 6.29, the atoms are a and b.

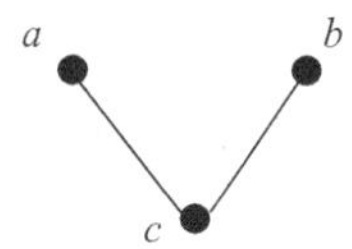

Figure 6.29

However, lattices can have other join-irreducible elements. For example, the element c in five-element lattice shown in the Figure 6.30 is not an atom, even then it is join-irreducible because it has only **one immediate predecessor, namely** a.

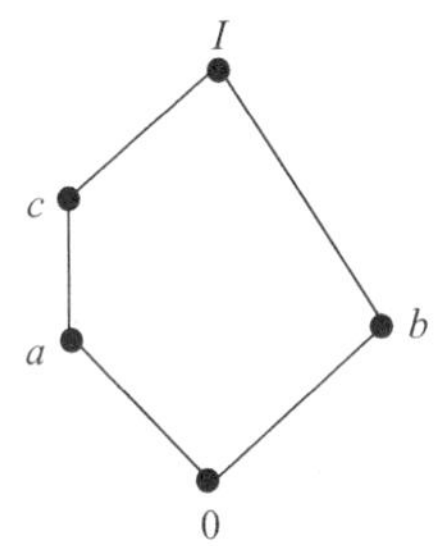

Figure 6.30

Let a be an element of a finite lattice which is not join-irreducible, then we can write $a=b \vee c$. If b and c are not join-irreducible, then we can write them as the join of other elements. Since L is finite we shall finally have

$$a=d_1 \vee d_2 \vee d_3 \vee \ldots \vee d_n, \tag{1}$$

where d_i, $i=1, 2, \ldots, n$ are join-irreducible. If d_i precedes d_j, then $d_i \vee d_j=d_j$, so we delete d_i from the expression. Thus d's are irredundant, that is, no d precedes any other d.

The expression (1) need not be unique. For example, in five-element lattice, shown above

$$I=a \vee b \quad \text{and} \quad I=b \vee c.$$

Theorem 6.12

Let $(L, \wedge, \vee)$ be a finite distributive lattice. Then every a in L can be written uniquely (except for order) as the join of irredundant join-irreducible elements.

Proof. Let $a \in L$. Since L is finite, we can express a as the join of irredundant join-irreducible elements (as discussed above). To prove uniqueness let

$$a=b_1 \vee b_2 \vee \ldots \vee b_n=c_1 \vee c_2 \vee \ldots \vee c_m,$$

where b_i are irredundant join-irreducible and c_i are irredundant and join-irreducible. For any given i, we have

$$b_i \leq (b_1 \vee b_2 \vee \ldots \vee b_n)=c_1 \vee c_2 \vee \ldots \vee c_m.$$

Hence,

$$b_i = b_i \wedge (c_1 \vee c_2 \vee \ldots \vee c_m)$$
$$= (b_i \wedge c_1) \vee (b_i \wedge c_2) \vee \ldots \vee (b_i \wedge c_m).$$

Since b_i is irredundant join-irreducible, there exists j such that $b_i = b_i \wedge c_j$ and so $b_i \leq c_j$.

Similarly, for c_i there exists a b_k such that $c_j \leq b_k$. Hence,

$$b_i \leq c_j \leq b_k,$$

which gives $b_i = c_j = b_k$ since b_i are irredundant. Hence b_i and c_i may be paired off. Hence the representation for a is unique except for order.

Theorem 6.13

Let L be a complemented lattice with unique complements. Then the join-irreducible elements of L, other than 0, are its atoms.

Proof. Suppose a is join-irreducible and is not an atom. Then a has a unique immediate predecessor $b \neq 0$. Let b' be the complement of b (complement exists since L is complemented). Since $b \neq 0$, $b' \neq I$. If a precedes b', then $b \leq a \leq b'$, and so $b \vee b' = b'$ which is impossible since $b \vee b' = I$. Thus, a does not precede b' and so $a \wedge b'$ must strictly precede a. Since b is the unique immediate predecessor of a, we also have that $a \wedge b'$ precedes b. But $a \wedge b'$ precedes b' (see Figure 6.31).

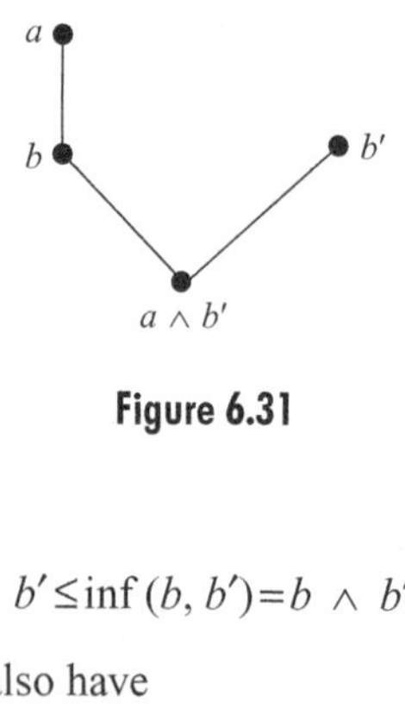

Figure 6.31

Hence

$$a \wedge b' \leq \inf(b, b') = b \wedge b' = 0.$$

Thus, $a \wedge b' = 0$. Since $a \vee b = a$, we also have

$$a \vee b' = (a \vee b) \vee b' = a \vee (b \vee b') = a \vee I = I.$$

Therefore b' is a complement of a. Since complements are unique, $a = b$. This contradicts the assumption that b is an immediate predecessor of a. Thus the only join-irreducible elements of L are its atoms.

Combining this result with the above-proved theorems, we have the following theorem.

Theorem 6.14

Let L be a finite complemented distributive lattice. Then every element a in L is the join of a unique set of atoms.

EXERCISES

1. Let D_4 and D_6 be two lattices. Draw the Hasse diagram of $D_4 \times D_6$. Is it a lattice?
2. Does the Hasse diagram 6.32 represent lattice?

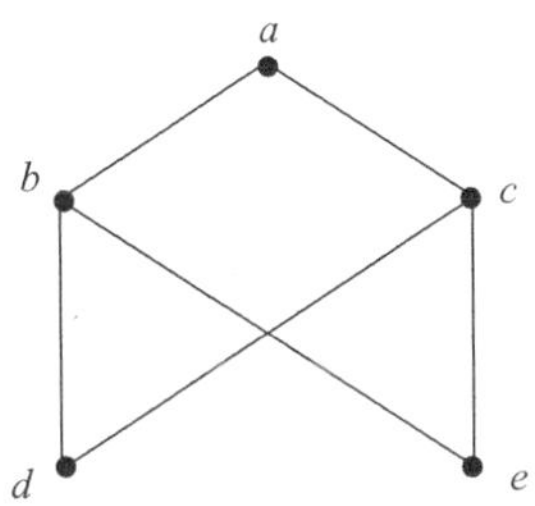

Figure 6.32

3. Show that the lattice D_{30} is complemented.
4. Let B_3 be the lattice with Hasse diagram 6.33. Find (i) complements of (1, 0, 1), (0, 0, 1), (0, 0, 0); (ii) join-irreducible elements and (iii) atoms.

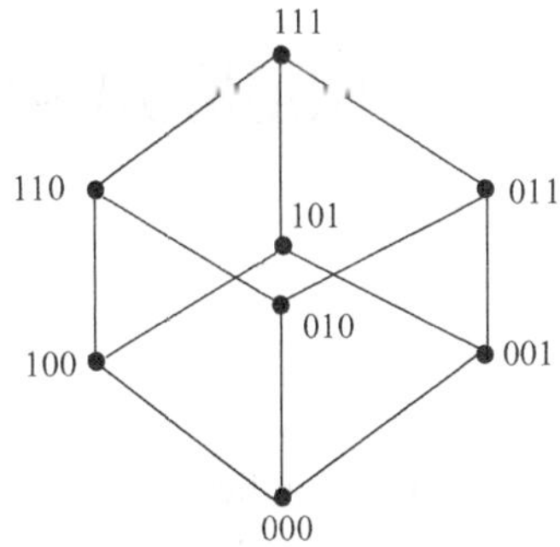

Figure 6.33

5. Find the atoms and join-irreducible elements in D_{24}.
6. Show that D_{12} is distributive whereas $(A, \leq)$, where $A = \{1, 2, 3, 4, 12\}$ and $\leq$ is partial order of divisibility, is not distributive.
7. Show that a subset of a linearly ordered poset is a sublattice.
8. Find the sublattices of D_{12}.

7 Boolean Algebra

Boolean algebra is a significant tool for the analysis and design of electronic computers. It has wide applications to switching theory and logical design of electronic circuits.

7.1 DEFINITIONS AND BASIC PROPERTIES

Definition 7.1

A non-empty set B with two binary operations $\vee$ and $\wedge$, a unary operation $'$, and two distinct elements 0 and I is called a **Boolean Algebra** if the following axioms holds for any elements $a, b, c \in B$:

$[B_1]$: **Commutative Laws:**

$$a \vee b = b \vee a \quad \text{and} \quad a \wedge b = b \wedge a.$$

$[B_2]$: **Distributive Laws:**

$$a \wedge (b \vee c) = (a \wedge b) \vee (a \wedge c) \quad \text{and} \quad a \vee (b \wedge c) = (a \vee b) \wedge (a \vee c).$$

$[B_3]$: **Identity Laws:**

$$a \vee 0 = a \quad \text{and} \quad a \wedge I = a.$$

$[B_4]$: **Complement Laws:**

$$a \vee a' = I \quad \text{and} \quad a \wedge a' = 0.$$

We shall call 0 as zero element, 1 as unit element and a' the complement of a.

We denote a Boolean algebra by $(B, \vee, \wedge, ', 0, I)$.

EXAMPLE 7.1

Let A be a non-empty set and $P(A)$ be its power set. Then the set algebra $(P(A), \cup, \cap, -, \phi, A)$ is a Boolean algebra. In fact, we have for $L, M, N \in P(A)$,

1. $L \cup M = M \cup L$ and $L \cap M = M \cap L$
2. $L \cap (M \cup N) = (L \cap M) \cup (L \cap N)$
   $L \cup (M \cap N) = (L \cup M) \cap (L \cup N)$
3. $L \cup \phi = L$ and $L \cap A = L$
4. $L \cup [A - L] = A$ and $L \cap [A - L] = \phi$.

If A has n elements, then $P(A)$ has 2^n elements and the diagram of the Boolean algebra is an n cube. The partial order relation on $P(A)$ corresponding to the operations $\cap$ and $\cup$ is set inclusion $\subseteq$.

If A has one element, two elements, three elements, then the corresponding Boolean algebras are shown by the following diagrams (Figure 7.1):

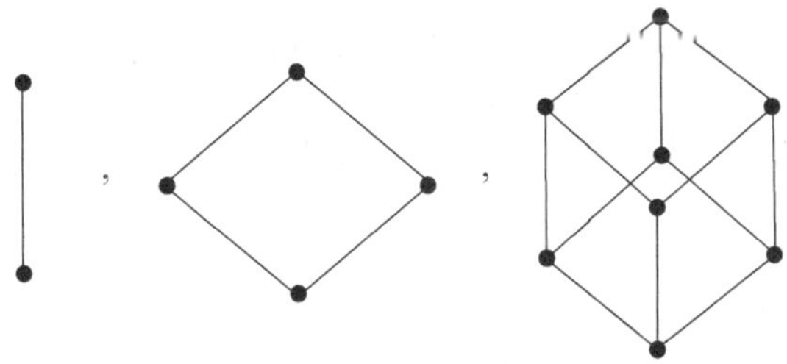

Boolean algebra for singleton set   Boolean algebra for set having two elements   Boolean algebra for set having three elements

Figure 7.1

EXAMPLE 7.2

Let $B=\{0, 1\}$ be the set of bits (binary digits) with the binary operations $\vee$ and $\wedge$ and the unary operation $'$ defined by the following tables:

$$\begin{array}{c|cc} \vee & 1 & 0 \\ \hline 1 & 1 & 1 \\ 0 & 1 & 0 \end{array}, \qquad \begin{array}{c|cc} \wedge & 1 & 0 \\ \hline 1 & 1 & 0 \\ 0 & 0 & 0 \end{array}, \qquad \begin{array}{c|cc} ' & 1 & 0 \\ \hline & 0 & 1 \end{array}$$

Here the operations $\vee$ and $\wedge$ are logical operations and complement of 1 is 0 whereas complement of 0 is 1. Then $(B, \vee, \wedge, ', 0, 1)$ is a Boolean algebra. It is the simplest example of a two-element algebra.

Further, a two-element Boolean algebra is the only Boolean algebra whose diagram is a chain.

EXAMPLE 7.3

Let B_n be the set of n tuples whose members are either 0 or 1. Let $a=(a_1, a_2, \ldots, a_n)$ and $b=(b_1, b_2, \ldots, b_n)$ be any two members of B_n. Then we define

$$a \vee_1 b=(a_1 \vee b_1, a_2 \vee b_2, \ldots, a_n \vee b_n),$$
$$a \wedge_1 b=(a_1 \wedge b_1, a_2 \wedge b_2, \ldots, a_n \wedge b_n),$$

where $\vee$ and $\wedge$ are logical operations on $\{0, 1\}$, and

$$a'=(\sim a_1, \sim a_2, \ldots, \sim a_n),$$

where $\sim 0=1$ and $\sim 1=0$.

If 0_n represents $(0, 0, \ldots, 0)$ and $1_n=(1, 1, \ldots, 1)$, then $(B_n, \vee_1, \wedge_1, ', 0_n, 1_n)$ is a Boolean algebra.

This algebra is known as **switching algebra** and represents a switching network with n inputs and one output.

EXAMPLE 7.4

The poset $D_{30}=\{1, 2, 3, 5, 6, 10, 15, 30\}$ has eight elements. Define $\vee$, $\wedge$ and $'$ on D_{30} by

$$a \vee b=\text{lcm}(a, b), \quad a \wedge b=\gcd(a, b) \quad \text{and} \quad a'=\frac{30}{a}.$$

Then D_{30} is a Boolean algebra with 1 as the zero element and 30 as the unit element.

EXAMPLE 7.5

Let S be the set of statement formulas involving n statement variables. The algebraic system $(S, \wedge, \vee, \sim, F, T)$ is a Boolean algebra in which $\wedge, \vee, \sim$ denote the operations of conjunction, disjunction and negation, respectively. The elements F and T denote the formulas which are contradictions and tautologies, respectively. The partial ordering corresponding to $\wedge$, $\vee$ is implication $\Rightarrow$.

We have seen that B_n is a Boolean algebra. Using this fact, we can also define Boolean algebra as follows:

Definition 7.2

A finite lattice is called a **Boolean algebra** if it is isomorphic with B_n for some non-negative integer n.

For example, D_{30} is isomorphic to B_3. In fact, the mapping $f: D_{30} \to B_3$ defined by

$$f(1)=000,\ f(2)=100,\ f(3)=010,\ f(5)=001,$$
$$f(6)=110,\ f(10)=101,\ f(15)=011,\ f(30)=111$$

is an isomorphism. Hence D_{30} is a Boolean algebra.

If a finite lattice L does not contain 2^n elements for some non-negative integer n, then L cannot be a Boolean algebra (this is a consequence of Corollary 7.1).

For example, consider $D_{20}=\{1, 2, 4, 5, 10, 20\}$ that has 6 elements and $6 \neq 2^n$ for any integer $n \geq 0$. Therefore, D_{20} is not a Boolean algebra.

If $|L| = 2^n$, then L may or may not be a Boolean algebra. If L is isomorphic to B_n, then it is Boolean algebra, otherwise it is not.

For large value of n, we use the following theorem for determining whether D_n is a Boolean algebra or not.

Theorem 7.1

Let $n=p_1 p_2 \ldots p_k$, where p_i are distinct primes, known as set of atoms. Then D_n is a Boolean algebra.

Proof. Let $A=\{p_1, p_2, \ldots, p_k\}$. If $B \subseteq A$ and a_B is the product of primes in B, then $a_B \mid n$. Also, any divisor of n must be of the form a_B for some subset B of A, where we assume that $a_\phi=1$. Further, if C and B are subsets of A, then $C \subseteq B$ if and only if $a_C \mid a_B$. Also,

$$a_{C \cap B}=a_C \wedge a_B=\gcd(a_C, a_B)$$

and

$$a_{C \cup B}=a_C \vee a_B=\text{lcm}(a_C, a_B).$$

Thus the function $f: P(A) \to D_n$ defined by $f(B)=a_B$ is an isomorphism. Since $P(A)$ is a Boolean algebra, it follows that D_n is also a Boolean algebra.

For example, consider $D_{20}, D_{30}, D_{210}, D_{66}, D_{646}$. We notice that

(i) 20 cannot be represented as product of distinct primes and so D_{20} is not a Boolean algebra.
(ii) $30=2 \cdot 3 \cdot 5$, where 2, 3, 5 are distinct primes. Hence, D_{30} is a Boolean algebra.
(iii) $210=2 \cdot 3 \cdot 5 \cdot 7$ (all distinct primes) and so D_{210} is a Boolean algebra.
(iv) $66=2 \cdot 3 \cdot 11$ (product of distinct primes) and so D_{66} is a Boolean algebra.
(v) $646=2 \cdot 17 \cdot 19$ (product of distinct primes) and so D_{646} is a Boolean algebra.

7.1.1 Duality

The **dual of any statement** in a Boolean algebra B is obtained **by interchanging** $\vee$ and $\wedge$ and interchanging the zero element and unit element in the original statement.

For example, the dual of $a \wedge 0=0$ is $a \vee I=I$.

Principle of duality. The dual of any theorem in a Boolean algebra is also a theorem.

(Thus, dual theorem is proved by using the dual of each step of the proof of the original statement).

Theorem 7.2

Let a, b and c be any elements in a Boolean algebra $(B, \vee, \wedge, ', 0, I)$. Then

1. **Idempotent Laws.**

   (i) $a \vee a = a$ (ii) $a \wedge a = a$.

2. **Boundedness Laws:**

   (i) $a \vee I = I$ (ii) $a \wedge 0 = 0$.

3. **Absorption Laws:**

   (i) $a \vee (a \wedge b) = a$ (ii) $a \wedge (a \vee b) = a$.

4. **Associative Laws:**

   (i) $(a \vee b) \vee c = a \vee (b \vee c)$ (ii) $(a \wedge b) \wedge c = a \wedge (b \wedge c)$.

Proof. It is sufficient to prove first part of each law since second part follows from the first by principle of duality.

1. (i) We have

$$\begin{aligned} a &= a \vee 0 && \text{(by identity law in a Boolean algebra)} \\ &= a \vee (a \wedge a') && \text{(by complement law)} \\ &= (a \vee a) \wedge (a \vee a') && \text{(by distributive law)} \\ &= (a \vee a) \wedge I && \text{(complement law)} \\ &= a \vee a && \text{(identity law),} \end{aligned}$$

which proves 1(i).

2. (i) We have

$$\begin{aligned} a \vee I &= (a \vee I) \wedge I && \text{(identity law)} \\ &= (a \vee I) \wedge (a \vee a') && \text{(complement law)} \\ &= a \vee (I \wedge a') && \text{(distributive law)} \\ &= a \vee a' && \text{(identity law)} \\ &= I && \text{(complement law).} \end{aligned}$$

3. (i) we note that

$$\begin{aligned} a \vee (a \wedge b) &= (a \wedge I) \vee (a \wedge b) && \text{(identity law)} \\ &= a \wedge (I \vee b) && \text{(distributive law)} \\ &= a \wedge (b \vee I) && \text{(commutativity)} \\ &= a \wedge I && \text{(identity law)} \\ &= a && \text{(identity law).} \end{aligned}$$

4. (i) Let

$$L = (a \vee b) \vee c, \quad R = a \vee (b \vee c).$$

Then,

$$\begin{aligned} a \wedge L &= a \wedge [(a \vee b) \vee c] \\ &= [a \wedge (a \vee b)] \vee (a \wedge c) && \text{(distributive law)} \\ &= a \vee (a \wedge c) && \text{(absorption law)} \\ &= a && \text{(absorption law)} \end{aligned}$$

and

$$\begin{aligned} a \wedge R &= a \wedge [a \vee (b \vee c)] \\ &= (a \wedge a) \vee (a \wedge (b \vee c)] && \text{(distributive law)} \\ &= a \vee (a \wedge (b \vee c)] && \text{(idempotent law)} \\ &= a && \text{(absorption law).} \end{aligned}$$

Thus, $a \wedge L = a \wedge R$ **and so, by duality,** $a \vee L = a \vee R$.
Further,

$$\begin{aligned} a' \wedge L &= a' \wedge [(a \vee b) \vee c] \\ &= [a' \wedge (a \vee b)] \vee (a' \wedge c) && \text{(distributive law)} \\ &= [(a' \wedge a) \vee (a' \wedge b)] \vee (a' \wedge c) && \text{(distributive law)} \\ &= [0 \vee (a' \wedge b)] \vee (a' \wedge c) && \text{(complement law)} \\ &= (a' \wedge b)] \vee (a' \wedge c) && \text{(identity law)} \\ &= a' \wedge (b \vee c) && \text{(distributive law).} \end{aligned}$$

On the other hand,

$$\begin{aligned} a' \wedge R &= a' \wedge [a \vee (b \vee c)] \\ &= (a' \wedge a) \vee [a' \wedge (b \vee c)] && \text{(distributive law)} \\ &= 0 \vee [a' \wedge (b \vee c)] && \text{(complement law)} \\ &= a' \wedge (b \vee c)] && \text{(identity law).} \end{aligned}$$

Hence,

$$a' \wedge L = a' \wedge R \text{ and so by duality } a' \vee L = a' \vee R.$$

Therefore,

$$\begin{aligned} L &= (a \vee b) \vee c \\ &= 0 \vee [(a \vee b) \vee c] = 0 \vee L && \text{(identity law)} \\ &= (a \wedge a') \vee [(a \vee b) \vee c] = (a \wedge a') \vee L && \text{(complement law)} \\ &= (a \vee L) \wedge (a' \vee L) && \text{(distributive law)} \\ &= (a \vee R) \wedge (a' \vee R) \text{ (using } A \vee L = a \vee R \text{ and } a' \vee L = a' \vee R] \\ &= (a \wedge a') \vee R && \text{(distributive law)} \\ &= 0 \vee R && \text{(complement law)} \\ &= R && \text{(identity law).} \end{aligned}$$

Hence

$$(a \vee b) \vee c = a \vee (b \vee c),$$

which completes the proof of the theorem.

Theorem 7.3

Let a be any element of a Boolean algebra B. Then

(i) Complement of a is unique **(uniqueness of complement)**,
(ii) $(a')' = a$ **(involution law)**,
(iii) $0' = 1$ and $1' = 0$.

Proof

(i) Let a' and x be two complements of $a \in B$. Then

$$a \vee a' = I \quad \text{and} \quad a \wedge a' = 0 \qquad \text{(i)}$$

$$a \vee x = I \quad \text{and} \quad a \wedge x = 0 \qquad \text{(ii)}$$

and we have

$$\begin{aligned} a' &= a' \vee 0 && \text{(identity law)} \\ &= a' \vee (a \wedge x) && \text{by (ii)} \\ &= (a' \vee a) \wedge (a' \vee x) && \text{(distributive law)} \\ &= I \wedge (a' \vee x) && \text{by (i)} \\ &= a' \vee x && \text{(identity law).} \end{aligned}$$

Also,

$$\begin{aligned} x &= x \vee 0 && \text{(identity law)} \\ &= x \vee (a \wedge a'), && \text{by (i)} \\ &= (x \vee a) \wedge (x \vee a') && \text{(distributive law)} \\ &= I \wedge (x \vee a'), && \text{(by (ii))} \\ &= x \vee a' = a' \vee x && \text{(identity and commutative law).} \end{aligned}$$

Hence $a' = x$ and so complement of any element in B is unique.

(ii) Let a' be a complement of a. Then,

$$a \vee a' = I, \quad \text{and} \quad a \wedge a' = 0$$

or, by commutativity,

$$a' \vee a = I \quad \text{and} \quad a' \wedge a = 0.$$

This implies that a is complement of a', that is, $a = (a')'$.

(iii) By boundedness law,

$$0 \vee I = I$$

and by identity law

$$0 \wedge I = 0.$$

These two relations imply that I is the complement of 0, that is, $I = 0'$.
By principle of duality, we have then $0 = I'$.

Theorem 7.4

Let a, b be elements of a Boolean algebra. Then, $(a \vee b)' = a' \wedge b'$ and $(a \wedge b)' = a' \vee b'$.

Proof. We have

$$\begin{aligned} (a \vee b) \vee (a' \wedge b') &= (b \vee a) \vee (a' \wedge b') && \text{(commutativity)} \\ &= b \vee (a \vee (a' \wedge b')) && \text{(associativity)} \\ &= b \vee [(a \vee a' \wedge (a \vee b')] && \text{(distributivity)} \\ &= b \vee [I \wedge (a \vee b') && \text{(complement law)} \\ &= b \vee (a \vee b') && \text{(identity law)} \\ &= b \vee (b' \vee a) && \text{(commutativity)} \\ &= (b \vee b') \vee a && \text{(associative law)} \\ &= I \vee a && \text{(complement law)} \\ &= I && \text{(identity law).} \end{aligned}$$

Also,

$$\begin{aligned}(a \vee b) \wedge (a' \wedge b') &= [(a \vee b) \wedge a'] \wedge b' && \text{(associative law)}\\ &= [a \wedge a') \vee (b \wedge a')] \wedge b' && \text{(distributive law)}\\ &= [0 \vee (b \wedge a')] \wedge b' && \text{(complement law)}\\ &= (b \wedge a') \wedge b' && \text{(identity law)}\\ &= b \wedge b' \wedge a' = 0 \wedge a' = 0.\end{aligned}$$

Hence $a' \wedge b'$ is complement of $a \vee b$, i.e. $(a \vee b)' = a' \wedge b'$.

The second part follows by principle of duality.

We have proved already that Boolean algebra $(B, \vee, \wedge, ', 0, I)$ satisfies associative laws, commutative law and absorption law. Hence every Boolean algebra is a lattice with join as $\vee$ and meet as $\wedge$. Also boundedness law holds in Boolean algebra. Thus, Boolean algebra becomes a bounded lattice. Also, Boolean algebra obeys distributive law and is complemented. Conversely, every bounded, distributive and complemented lattice satisfies all the axioms of a Boolean algebra. Hence we can define Boolean algebra as.

Definition 7.3

A Boolean algebra is a **bounded, distributive and complemented lattice**.

Now, being a lattice, a Boolean algebra must have a partial ordering. Recall that in case of lattice, we had defined partial ordering $\le$ by $a \le b$ if $a \vee b = b$ or $a \wedge b = a$.
The following result yields much more than these required conditions:

Theorem 7.5

If a, b are in a Boolean algebra, then the following are equivalent:

1. $a \vee b = b$,
2. $a \wedge b = a$,
3. $a' \vee b = I$,
4. $a \wedge b' = 0$.

Proof. (1)$\Leftrightarrow$(2), proved already in Theorem 6.2.

(1)$\Rightarrow$(3): Suppose $a \vee b = b$, then

$$\begin{aligned}a' \vee b &= a' \vee (a \vee b)\\ &= (a' \vee a) \vee b && \text{(associativity)}\\ &= I \vee b = I && \text{(complement and boundedness).}\end{aligned}$$

Conversely, suppose $a' \vee b = I$, then

$$\begin{aligned}a \vee b = 1 \wedge (a \vee b) &= (a' \vee b) \wedge (a \vee b) && \text{(by assumption of (3))}\\ &= (a' \wedge a) \vee b && \text{(distributivity)}\\ &= 0 \vee b = b && \text{(complement and identity).}\end{aligned}$$

Thus, (1)$\Leftrightarrow$(3).

Now we show that (3)$\Leftrightarrow$(4). Suppose first that (3) holds. Then, using De Morgan law and involution, we have

$$\begin{aligned}0 = I' = (a' \vee b)' &= a'' \wedge b'\\ &= a \wedge b' && \text{(involution).}\end{aligned}$$

Conversely, if (4) holds, then

$$I=0'=(a \wedge b')'=a' \vee b''=a' \vee b.$$

Thus, (3)$\Leftrightarrow$(4). Hence all the four condition are equivalent.

EXAMPLE 7.6

Show that the lattice whose diagram is shown in the Figure 7.2 is not a Boolean algebra.

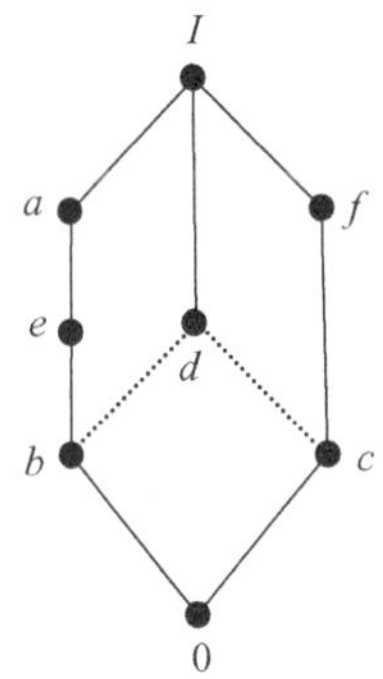

Figure 7.2

Solution.

Elements a and e are both complements of c since $c \vee a=I$, $c \wedge a=0$ and $c \vee e=I$, $c \wedge e=0$. But in a Boolean algebra, complement of an element is unique. Hence the given lattice is not a Boolean algebra.

Definition 7.4

Let $(B, \vee, \wedge, ', 0, I)$ be a Boolean algebra and $S \subseteq B$. If S contains the elements 0 and I and is closed under the operations $\vee$, $\wedge$ and $'$, then $(S, \wedge, \vee, ', 0, I)$ is called **sub-Boolean algebra**.

In practice, it is sufficient to check closure with respect to the set of operations $(\wedge, ')$ or $(\vee, ')$ for proving a subset S of B as the sub-Boolean algebra.

The definition of sub-Boolean algebra implies that it is a Boolean algebra.

A subset of Boolean algebra can be a Boolean algebra, but not necessarily a Boolean sub-algebra because it is not closed with respect to the operations in B.

For any Boolean algebra $(B, \wedge, \vee, ', 0, I)$, the subsets $\{0, I\}$ and the set B are both sub-Boolean algebras.

In addition to these sub-Boolean algebras, consider now any element $a \in B$ such that $a \neq 0$ and $a \neq 1$ and consider the set $\{a, a', 0, I\}$. Obviously this set is a sub-Boolean algebra of the given Boolean algebra.

For example $D_{70}=\{1, 2, 5, 7, 10, 14, 35, 70\}$ is a Boolean algebra and $\{1, 2, 35, 70\}$ is a sub-algebra of D_{70}.

Every element of a Boolean algebra generates a sub-Boolean algebra. More generally, any subset of B generates a sub-Boolean algebra.

EXAMPLE 7.7

Consider the Boolean algebra given in the Figure 7.3.

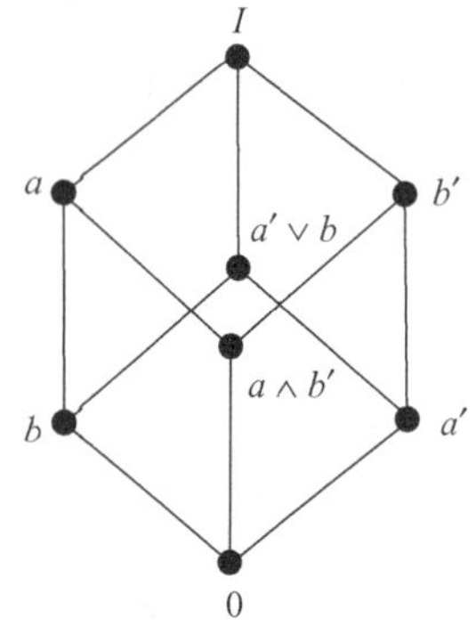

Figure 7.3

Verify whether the following subsets are Boolean algebras or not:

$$S_1=\{a, a', 0, I\},$$
$$S_2=\{a' \vee b, a \wedge b', 0, I\},$$
$$S_3=\{a \wedge b', b', a, I\},$$
$$S_4=\{b', a \wedge b', a', 0\},$$
$$S_5=\{a, b', 0, I\}.$$

Solution.

The subset S_1 and S_2 are sub-Boolean algebras. The subsets S_3 and S_4 are Boolean algebras but not sub-Boolean algebras of the given Boolean algebra. The subset S_5 is not even a Boolean algebra.

Definition 7.5

Let $(B_1, \wedge_1, \vee_1, ', 0_1, I_1)$ and $(B_2, \wedge_2, \vee_2, '', 0_2, I_2)$ be two Boolean algebras. The **direct product** of the two Boolean algebras is defined to be a Boolean algebra, denoted by $(B_1 \times B_2, \wedge_3, \vee_3, ''', 0_3, I_3)$, in which the operations are defined for any (a_1, b_1) and $(a_2, b_2) \in B_1 \times B_2$ as

$$(a_1, b_1) \wedge_3 (a_2, b_2) = (a_1 \wedge_1 a_2, b_1 \wedge_2 b_2),$$
$$(a_1, b_1) \vee_3 (a_2, b_2) = (a_1 \vee_1 a_2, b_1 \vee_2 b_2),$$
$$(a_1, b_1)''' = (a_1', b_1''),$$
$$0_3 = (0_1, 0_2) \quad \text{and} \quad I_3 = (I_1, I_2).$$

Thus, from a Boolean algebra B, we can generate $B^2 = B \times B$, $B^3 = B \times B \times B$, etc.

Definition 7.6

Let $(B, \wedge, \vee, ', 0, I)$ and $(P, \cap, \cup, —, \alpha, \beta)$ be two Boolean algebras. A mapping $f: B \to P$ is called a **Boolean homomorphism** if all the operations of the Boolean algebra are preserved, that is, for any $a, b \in B$

$$f(a \wedge b) = f(a) \cap f(b),$$
$$f(a \vee b) = f(a) \cup f(b),$$
$$f(a') = \overline{f(a)},$$
$$f(0) = \alpha,$$
$$f(1) = \beta.$$

The above definition of homomorphism can be simplified by asserting that $f: B \to P$ preserves either the operations $\wedge$ and $'$ or the operations $\vee$ and $'$.

We now consider a mapping $g: B \to P$ in which the operations $\wedge$ and $\vee$ are preserved. Thus, g is a lattice homomorphism. Naturally g preserves the order and hence it maps the bounds 0 and I into the least and the greatest element, respectively, of the image set $g(B) \subseteq P$. It is, however, not necessary that $g(0)=\alpha$ and $g(1)=\beta$. The complements, if defined in terms of $g(0)$ and $g(1)$ in $g(B)$, are preserved, and $(g(B), \cap, \cup, —, g(0), g(1))$ is a Boolean algebra. Note that $g: B \to P$ is not a Boolean homomorphism, although $g: B \to g(B)$ is a Boolean homomorphism.

In any case, **for any mapping from a Boolean algebra which preserves the operations $\wedge$ and $\vee$, the image set is a Boolean algebra**.

A Boolean homomorphism is called **Boolean isomorphism** if it is bijective.

7.2 REPRESENTATION THEOREM

Let B be a **finite** Boolean algebra. We know that an element a in B is called an **atom (minterm)** if a immediately succeeds the **least element** 0. Let A be the set of atoms of B and let $P(A)$ be the Boolean algebra of all subsets of the set A of atoms. Then, by Theorem 6.14, each $x \neq 0$ in B can be expressed uniquely (except for order) as the join of atoms (i.e., elements of A). So, let

$$x=a_1 \vee a_2 \vee \ldots \vee a_n.$$

Consider the function $f: B \to P(A)$ defined by

$$f(x)=\{a_1, a_2, \ldots, a_n\} \quad \text{for each} \quad x=a_1 \vee a_2 \vee \ldots \vee a_n.$$

Theorem 7.6 (Stone's Representation)

Any Boolean algebra is isomorphic to a power set algebra $(P(S), \cap, \cup, \sim, \phi, S)$ for some set S.

Restricting our discussion to finite Boolean algebra B, the presentation theorem can be stated as follows:

Theorem 7.7

Let B be a finite Boolean algebra and let A be the set of atoms of B. If $P(A)$ is the Boolean algebra of all subsets of the set A of atoms, then the mapping $f: B \to P(A)$ is an isomorphism.

Proof. Suppose B is finite Boolean algebra and $P(A)$ is the Boolean algebra of all subsets of the set A of atoms of B. Consider the mapping $f: B \to P(A)$ defined by

$$f(x)=\{a_1, a_2, \ldots, a_n\},$$

where $x=a_1 \vee a_2 \vee \ldots \vee a_n$ is the unique representation of $x \in B$ as the join of atoms $a_1, a_2, \ldots, a_n \in A$. If a_i are atoms, then we know that $a_i \wedge a_i=a_i$ but $a_i \wedge a_j=0$ for $a_i \neq a_j$.

Let x and y be in the Boolean algebra B and suppose

$$x=a_1 \vee \ldots \vee a_r \vee b_1 \vee \ldots \vee b_s,$$
$$y=b_1 \vee \ldots \vee b_s \vee c_1 \vee \ldots \vee c_t,$$

where $A=\{a_1, a_2, \ldots, a_r, b_1, b_2, \ldots, b_s, c_1, \ldots, c_t, d_1 \ldots, d_k\}$ is the set of atoms of B. Then

$$x \vee y=a_1 \vee \ldots \vee a_r \vee b_1 \vee \ldots \vee b_s \vee c_1 \ldots \vee c_t,$$
$$x \wedge y=b_1 \vee \ldots \vee b_s.$$

Hence,

$$\begin{aligned} f(x \vee y) &= \{a_1, a_2, \ldots, a_r, b_1, b_2, \ldots, b_s, c_1, c_2, \ldots, c_t\} \\ &= \{a_1, \ldots, a_r, b_1, \ldots, b_s\} \cup \{b_1, b_2, \ldots, b_s, c_1, c_2, \ldots, c_t\} \\ &= f(x) \cup f(y) \end{aligned}$$

and

$$f(x \wedge y) = \{b_1, \ldots, b_s\}$$
$$= \{a_1, a_2, \ldots, a_r, b_1, \ldots, b_s\} \cap \{b_1, \ldots, b_s, c_1, \ldots, c_t\}$$
$$= f(x) \cap f(y).$$

Let

$$y = c_1 \vee \ldots \vee c_t \vee d_1 \vee \ldots \vee d_k.$$

Then,

$$x \vee y = I \quad \text{and} \quad x \wedge y = 0$$

and so $y = x'$. Thus,

$$f(x') = f(y) = \{c_1 \ldots c_t, d_1, \ldots, d_k = \{a_1, a_2, \ldots, a_r, b_1, b_2, \ldots, b_s\}^c = (f(x))^c.$$

Since the representation is unique, f is one-one and onto. Hence f is a Boolean algebra isomorphism. **Thus, every finite Boolean algebra is structurally the same as a Boolean algebra of sets.**

If a set A has n elements, then its power set $P(A)$ has 2^n elements. Thus, we have the following:

Corollary 7.1

A finite Boolean algebra has 2^n elements for some positive integer n.

EXAMPLE 7.8

Consider the Boolean algebra

$$D_{70} = \{1, 2, 5, 7, 10, 14, 35, 70\}.$$

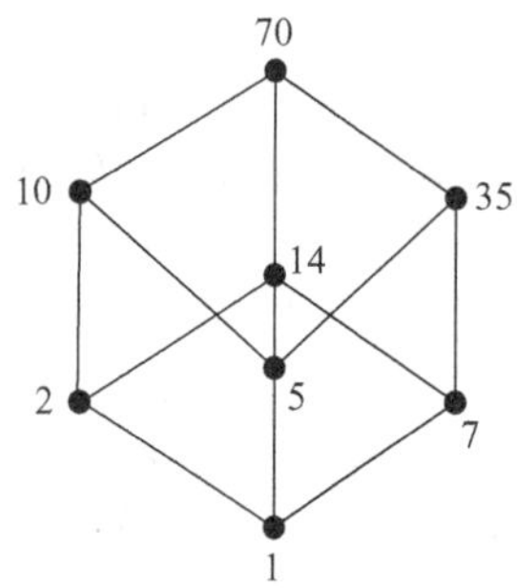

Figure 7.4 (D_{70})

Then the set of atoms of D_{70} is

$$A = \{2, 5, 7\}.$$

The unique representation of each non-atom by atoms is

$$10 = 2 \vee 5,$$
$$14 = 2 \vee 7,$$
$$35 = 5 \vee 7,$$
$$70 = 2 \vee 5 \vee 7.$$

The diagram of the Boolean algebra of the power set $P(A)$ of the set A of atoms is given below (Figure 7.5):

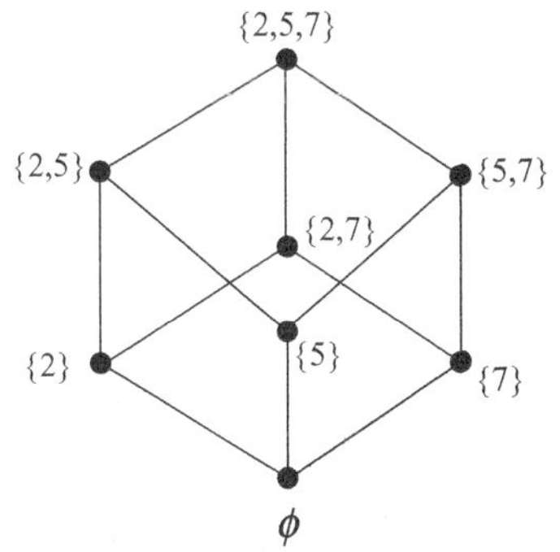

Figure 7.5

We note that the diagram for D_{70} and $P(A)$ are structurally the same.

7.3 BOOLEAN EXPRESSIONS

Definition 7.7

Let $x_1, x_2, \ldots, x_n$ be a set of n variables (or letters or symbols). **A Boolean polynomial (Boolean expression, Boolean form or Boolean formula)** $p(x_1, x_2, \ldots, x_n)$ in the variables $x_1, x_2, \ldots, x_n$ is defined recursively as follows:

1. The symbols 0 and I are Boolean polynomials
2. $x_1, x_2, \ldots, x_n$ are all Boolean polynomials
3. If $p(x_1, x_2, \ldots, x_n)$ and $q(x_1, x_2, \ldots, x_n)$ are two Boolean polynomials, then so are
$$p(x_1, x_2, \ldots, x_n) \vee q(x_1, x_2, \ldots, x_n)$$
and
$$p(x_1, x_2, \ldots, x_n) \wedge q(x_1, x_2, \ldots, x_n)$$
4. If $p(x_1, x_2, \ldots, x_n)$ is a Boolean polynomial, then so is
$$(p(x_1, x_2, \ldots, x_n))'$$
5. There are no Boolean polynomials in the variables $x_1, x_2, \ldots, x_n$ other than those obtained in accordance with rules 1 through 4.

Thus, Boolean expression is an expression, built from the variables given using Boolean operations $\vee$, $\wedge$ **and** $'$.

For example, for variables x, y, z, the expressions

$$p_1(x, y, z) = (x \vee y) \wedge z,$$
$$p_2(x, y, z) = (x \vee y') \vee (y \wedge 1),$$
$$p_3(x, y, z) = (x \vee (y' \wedge z)) \vee (x \wedge (y \wedge 1))$$

are Boolean expressions.

Notice that a Boolean expression in n variables may or may not contain all the n variables. Obviously, an infinite number of Boolean expressions may be constructed in n variables.

Definition 7.8

A **literal** is a variable or complemented variable such as x, x', y, y', and so on.

Definition 7.9

A **fundamental product** is a literal or a product of two or more literals in which no two literals involve the same variable.

For example,

$$x \wedge z', x \wedge y' \wedge z, x, y', x' \wedge y \wedge z$$

are fundamental products whereas

$$x \wedge y \wedge x' \wedge z \quad \text{and} \quad x \wedge y \wedge z \wedge y$$

are not fundamental products.

Remark 7.1 Fundamental product is also called a **minterm** or **complete product.** In what follows we shall denote $x \wedge y$ by xy.

Any product of literals can be reduced to either 0 or a fundamental product.

For example, consider $x\, y\, x'\, z$. Since $x \wedge x' = 0$ by complement law, we have $x\, y\, x'\, z = 0$.

Similarly, if we consider $x\, y\, z\, y$, then since $y \wedge y = y$ (idempotent law), we have $x\, y\, z\, y = x\, y\, z$, which is a fundamental product.

Definition 7.10

A fundamental product P_1 is said to be **contained in** (or **included in**) another fundamental Product P_2 if the literals of P_1 are also literals of P_2.

For example, $x'\, z$ is contained in $x'\, y\, z$ but $x'\, z$ is not contained in $x\, y'\, z$ since x' is not a literal of $x\, y'\, z$.

Observe that, if P_1 is contained in P_2, say $P_2 = P_1 \wedge Q$, then, by the absorption law,

$$P_1 \vee P_2 = P_1 \vee (P_1 \wedge Q) = P_1.$$

For example,

$$x'\, z \vee x'\, y\, z = x'\, z.$$

Definition 7.11

A Boolean expression E is called **a sum-of-products expression (disjunctive normal form or DNF)** if E is a fundamental product or the sum (join) of two or more fundamental products none of which is contained in another.

Definition 7.12

Two Boolean expression $P(x_1, x_2, \ldots, x_n)$ and $Q(x_1, x_2, \ldots, x_n)$ are called **equivalent (or equal)** if one can be obtained from the other by a finite number of applications of the identities of a Boolean algebra.

Definition 7.13

Let E be any Boolean expression. A **sum of product form** of E is an equivalent Boolean sum-of-products expression.

EXAMPLE 7.9

Consider the expression

$$E_1(x, y, z) = x\, z' + y'\, z + x\, y\, z'.$$

Although the expression E_1 is a sum of products, it is not a sum-of-products expression because, the product $x\, z'$ is contained in the product $x\, y\, z'$. But, by absorption law, E_1 can be expressed as

$$E_1(x, y, z) = x\, z' + y'\, z + x\, y\, z' = x\, z' + x\, y\, z' + y'\, z = x\, z' + y'\, z,$$

which is a sum-of-product form for E_1.

7.3.1 Algorithm for Finding Sum-of-Products Forms

The input is a Boolean expression E. The output is a sum-of-products expression equivalent to E.

Step 1. Use De Morgan's law and involution to move the complement operation into any parenthesis until finally the complement operation only applies to variables. Then E will consist only sums and products of literals.

Step 2. Use the distributive operation to next transform E into a sum of products.

Step 3. Use the commutative, idempotent, and complement laws to transform each product in E into 0 or a fundamental product.

Step 4. Use the absorption law and identity law to finally transform E into a sum-of-products expression.

For example, we apply the above algorithm to the Boolean expression.

$$E=((x\,y)'\,z)'\,((x'+z)\,(y'+z')')$$

Step 1. Using De Morgan's laws and involution, we obtain

$$\begin{aligned} E&=((x\,y)'' \vee z')\,((x' \vee z)' \vee (y' \vee z'))' \\ &= (x\,y \vee z') \wedge [(x \wedge z') \vee (y \wedge z)]. \end{aligned}$$

Thus, E consists only of sum and products of literals.

Step 2. Using the distributive laws, we obtain

$$\begin{aligned} E&=(x\,y+z')\,x\,z'+(x\,y+z')\,yz \\ &=x\,y\,x\,z'+x\,z'\,z'+x\,y\,y\,z+y\,z\,z'. \end{aligned}$$

Thus, E is now a sum of products.

Step 3. Using commutative, idempotent and complement laws, we obtain

$$E=x\,y\,z'+x\,z'+x\,y\,z+0.$$

Thus each term in E is a fundamental product or 0.

Step 4. Using absorption law

$$x\,z'+x\,y\,z'=x\,z'+(x\,z' \wedge y)=x\,z'.$$

Hence,

$$E=x\,z'+x\,y\,z+0.$$

Step 5. Now using identity law

$$E=x\,z'+x\,y\,z,$$

which is the required sum-of-products expression.

EXAMPLE 7.10

Show that

$$(x_1'\,x_2'\,x_3'\,x_4')+(x_1'\,x_2'\,x_3'\,x_4)+(x_1'\,x_2'\,x_3\,x_4)+(x_1'\,x_2'\,x_3\,x_4')=x_1'\,x_2'.$$

Solution.

We have

$$\begin{aligned} (x_1'\,x_2'\,x_3'\,x_4')+(x_1'\,x_2'\,x_3'\,x_4)&=(x_1'\,x_2'\,x_3'), \\ (x_1'\,x_2'\,x_3\,x_4')+(x_1'\,x_2'\,x_3\,x_4')&=(x_1'\,x_2'\,x_3). \end{aligned}$$

Hence the given formula is equal to

$$(x_1'\,x_2'\,x_3')+(x_1'\,x_2'\,x_3)=x_1'\,x_2'.$$

Definition 7.14

A Boolean expression $E(x_1, x_2, \ldots, x_n)$ is said to be a **complete sum-of-product expression (or full disjunctive normal form or disjunctive canonical form, or the minterm canonical form)** if E is a sum-of-products expression where each product involves all the n variables.

A fundamental product which involves all the variables is called a **minterm** and there is a maximum of 2^n such products for n variables.

It can be seen that "**every non-zero Boolean expression $E(x_1, x_2, \ldots, x_n)$ is equivalent to a complete sum-of-product expression and such a representation is unique.**"

7.3.2 Algorithm for Obtaining Complete Sum-of-Product Expression

The input is a Boolean sum-of-products expression $E(x_1, x_2, \ldots, x_n)$. The output is a complete sum-of-products expression equivalent to E.

Step 1. Find a product P in E which does not involve the variable x_i and then multiply P by (x_i+x_i') deleting any repeated products (This is possible since $x+x'=I$ and $P+P=P$).

Step 2. Repeat step 1 until every product in E is a minterm, i.e. every product P involves all the variables.

EXAMPLE 7.11

Express $x_1 \wedge x_2$ in its complete sum-of-products form in three variables x_1, x_2, x_3.

Solution.

Here the variable x_3 does not involve in the given sum-of-product. Multiplying $x_1\,x_2$ by (x_3+x_3'), we get

$$x_1\,x_2 = x_1\,x_2\,(x_3+x_3') = (x_1\,x_2\,x_3) + (x_1\,x_2\,x_3').$$

EXAMPLE 7.12

Express $x_1 \vee x_2$ in its complete sum-of-products form in three variables x_1, x_2, x_3.

Solution.

We have, using the above stated algorithm,

$$\begin{aligned} x_1+x_2 &= [x_1(x_2+x_2')]+[x_2\,(x_1+x_1')] \\ &= x_1\,x_2+x_1\,x_2'+x_2\,x_1+x_1'\,x_2 \\ &= x_1\,x_2+x_1\,x_2'+x_1'\,x_2 \\ &= x_1\,x_2(x_3+x_3')+x_1\,x_2'(x_3+x_3')+x_1'\,x_2(x_3+x_3') \\ &= x_1\,x_2\,x_3+x_1\,x_2\,x_3'+x_1\,x_2'\,x_3+x_1\,x_2'\,x_3'+x_1'\,x_2\,x_3+x_1'\,x_2\,x_3', \end{aligned}$$

which is the complete sum-of-products form in x_1, x_2, x_3.

EXAMPLE 7.13

Express $E(x, y, z) = x(y'\,z)'$ in its complete sum-of-products form.

Solution.

We have

$$\begin{aligned} E &= x(y''+z') && \text{using De Morgan law} \\ &= x\,(y+z') && \text{(involution law)} \\ &= x\,y+x\,z' && \text{(distribution law),} \end{aligned}$$

which is sum-of-products form.

Now we convert this sum-of-product form into complete sum-of-product form. We have

$$\begin{aligned} E &= x\,y+x\,z' \\ &= x\,y\,(z+z')+x\,z'\,(y+y') \\ &= x\,y\,z+x\,y\,z'+x\,z'\,y+x\,z'\,y' \\ &= x\,y\,z+x\,y\,z'+x\,z'\,y', \end{aligned}$$

which is the required complete sum-of-product form.

EXAMPLE 7.14

Express $(x_1+x_2)'\,x_3$ in complete sum-of-product form.

Solution.

We have

$$(x_1+x_2)'\,x_3 = x_1'\,x_2'\,x_3,$$

which is the required complete sum-of-product form.

In a similar fashion, we can convert a given Boolean expression into product-of-sums canonical forms:

EXAMPLE 7.15

Obtain product of sums canonical form of Boolean expression $x_1\,x_2$.

Solution.

$$\begin{aligned} x_1\,x_2 &= [x_1+(x_2\,x_2')]\,[x_2+(x_1\,x_1')] \\ &= (x_1+x_2)\,(x_1+x_2')\,(x_2+x_1)\,(x_2+x_1') \\ &= (x_1+x_2)\,(x_1+x_2')(x_2+x_1') \\ &= [(x_1+x_2)+(x_3+x_3')]\,[(x_1+x_2')+(x_3+x_3')]\,[(x_2+x_1')+(x_3+x_3')] \\ &= [(x_1+x_2+x_3)\,(x_1+x_2+x_3')]\,[(x_1+x_2'+x_3)\,(x_1+x_2'+x_3')][(x_2+x_1'+x_3)\,(x_1+x_2'+x_3'], \end{aligned}$$

which is the required form.

Obviously, it is easier to obtain sum-of-products form than the product-of-sums canonical form in some cases. Any how, if one of the canonical forms is known, then the other canonical form can be obtained directly.

7.3.3 Minimal Sum-of-Products

Consider a Boolean sum-of-products expression E. Let E_L denote the number of literals in E (counted according to multiplicity) and let E_S denote the number of summands in E. For example, let

$$E = x\,y\,z'+x'\,y'\,z+x\,y'\,z'\,t+x'\,y\,z\,t.$$

Then

$$E_L = 3+3+4+4 = 14 \quad \text{and} \quad E_S = 4.$$

Let E and F be equivalent Boolean sum-of-products expressions. Then E is called **simpler** than F if

$$\text{(i) } E_L < F_L \text{ and } E_S \le F_L \quad \text{or} \quad \text{(ii) } E_L \le F_L \text{ and } E_S < F_L.$$

Definition 7.15

A Boolean sum-of-product expression is called **minimal** if there is no equivalent sum-of-product expression which is simpler than E.

There can be more than one equivalent minimal sum-of-products expressions.

Definition 7.16

A fundamental product P is called **prime implicants** of a Boolean expression E if $P+E=E$ but no other fundamental product contained in P has this property.

For example, suppose

$$E=x\,y'+x\,y\,z'+x'\,y\,z'.$$

Then, we find first the complete-sum-of-products form of $x\,z'$. Towards this end, we have

$$x\,z'=x\,z'(y+y')=x\,z'\,y+x\,z'\,y'. \tag{1}$$

Also, we know that the complete sum-of-products form is unique, $A+E=E$, where $A\neq 0$ if and only if the summands in the complete sum-of-products form for A are among the summands in the complete sum-of-products form for E. We observe that summands $x\,y\,z'$ and $x\,y'\,z'$ in (1) are in the complete form of E given below:

$$\begin{aligned}E&=x\,y'(z+z')+x\,y\,z'+x'\,y\,z'\\&=x\,y'\,z+x\,y'\,z'+x\,y\,z'+x'\,y\,z'.\end{aligned}$$

Therefore, by the above argument,

$$x\,z'+E=E.$$

Also, the complete sum-of-products form of x is

$$\begin{aligned}x&=x(y+y')\,(z+z')=(x\,y+x\,y')\,(z+z')\\&=x\,y\,z+x\,y\,z'+x\,y'\,z+x\,y'\,z'.\end{aligned}$$

The summand $x\,y\,z$ of x is not a summand of E. Hence,

$$x+E\neq E.$$

Similarly, the complete sum-of-product form of z' is

$$\begin{aligned}z'&=z'(x+x')\,(y+y')=(z'\,x+z'\,x')\,(y+y')\\&=z'\,x\,y+z'\,x\,y'+z'\,x'\,y+z'\,x'\,y'.\end{aligned}$$

The summand $x'\,y'\,z'$ of z' is not a summand of E. Hence,

$$z'+E\neq E.$$

Thus the fundamental products x and z' contained in $x\,z'$ do not have the property $P+E=E$ whereas $x\,z'$ has this property. Hence, $x\,z'$ is a prime implicant of E.

It can be seen that "**a minimal sum-of-products form for a Boolean expression E is a sum of prime implicants of E**".

7.3.4 Consensus of Fundamental Products

Let P_1 and P_2 be fundamental products such that exactly one variable, say x_k, appears uncomplemented in **one of** P_1 and P_2 and complemented in the other. Then the **consensus of P_1 and P_2** is the product (without repetitions) of the literals of P_1 and P_2 after x_k and x_k' are deleted. (**we do not define the consensus of $P_1=x$ and $P_2=x'$**).

Lemma 7.1

Suppose Q is the consensus of P_1 and P_2. Then, $P_1+P_2+Q=P_1+P_2$.

Proof. Since the literals commute, we can assume without loss of generality that

$$P_1=a_1\,a_2\ldots a_r\,t,\quad P_2=b_1\,b_2\ldots b_s\,t'$$
$$Q=a_1\,a_2\ldots a_r\,b_1\,b_2\ldots b_s.$$

Now $Q=Q(t+t')=Q\,t+Q\,t'$. Because $Q\,t$ contains P_1, $P_1+Q\,t=P_1$; and because $Q\,t'$ contains P_2,

$$P_2+Qt'=P_2.$$

Hence,

$$P_1+P_2+Q=P_1+P_2+Q\,t+Q\,t'$$
$$=(P_1+Q\,t)+(P_2+Q\,t')=P_1+P_2.$$

EXAMPLE 7.16

Find the consensus Q of P_1 and P_2, where

(i) $P_1=x\,y\,z'\,s,\ P_2=x\,y'\,t,$
(ii) $P_1=x\,y',\ P_2=y,$
(iii) $P_1=x'\,y\,z,\ P_2=x'\,y\,t,$
(iv) $P_1=x'\,y\,z,\ P_2=x\,y\,z'.$

Solution.

(i) $P_1=x\,y\,z'\,s,\ P_2=x\,y'\,t$. Delete y and y' and then multiply the literals of P_1 and P_2 (without repetition) to obtain $Q=x\,z'\,s\,t$.

(ii) $P_1=x\,y',\ P_2=y$. Delete y and y' then multiply the literals of P_1 and P_2 (without repetition) to obtain $Q=x$.

(iii) $P_1=x'\,y\,z,\ P_2=x'\,y\,t$. In this case, no variable appears uncomplemented in one of the products and complemented in the other. Hence P_1 and P_2 have no consensus.

(iv) $P_1=x'\,y\,z,\ P_2=x\,y\,z'$. In this case, each x and z appear complemented in one of the products and uncomplemented in the other. Hence deleting x and z and multiplying the literals of P_1 and P_2 without repetition, we get $Q=y$.

7.3.5 Consensus Method for Finding Prime Implicants

The following algorithm, known as **consensus method** is used to find the prime implicants of a Boolean expression.

Algorithm (Consensus Method)

The input is a Boolean expression $E=P_1+P_2+\ldots+P_m$, where P_m are fundamental products. The output expresses E as a sum of its prime implicants.

Step 1. Delete any fundamental product P_i which includes any other fundamental product P_j (this is permissible by the absorption law).

Step 2. Add the consensus of any P_i and P_j providing Q does not include any of the P_i (this is permissible by the lemma $P_1+P_2+\ldots+P_n+Q=P_1+\ldots+P_n$).

Step 3. Repeat Step 1 or Step 2 until neither can be applied.

EXAMPLE 7.17

Let

$$E(x,y,z)=x\,y\,z+x'\,z'+x\,y\,z'+x'\,y'\,z+x'\,y\,z'.$$

Then,

$$E=x\,y\,z+x'\,z'+x\,y\,z'+x'\,y'\,z \qquad (\because x'yz' \text{ include } x'z')$$
$$=x\,y\,z+x'\,z'+x\,y\,z'+x'\,y'\,z+x\,y \qquad (\text{consensus } xy \text{ of } xyz,\ xyz' \text{ added})$$
$$=x'\,z'+x'\,y'\,z+x\,y \qquad (\because xyz \text{ and } xyz' \text{ include } xy)$$
$$=x'\,z'+x'\,y'\,z+x\,y+x'\,y' \qquad (\text{consensus } x'y' \text{ of } x'z' \text{ and } x'y'z \text{ added})$$
$$=x'\,z'+x\,y+x'\,y' \qquad (\because x'y'z \text{ include } x'y')$$
$$=x'\,z'+x\,y+x'\,y'+y\,z' \qquad (\text{consensus of } x'z' \text{ and } xy, \text{ which is } yz', \text{ added}).$$

After this, none of the step in the consensus method will change E. Thus, E is the sum of its prime implicants $x'z'$, xy, $x'y'$ and yz'.

7.3.6 Use of Consensus Method for Finding Minimal Sum-of-Products Form

We have seen that consensus method can be used to express a Boolean expression E as a sum of all its prime implicants. Using such a sum, we can find a minimal sum-of-products form for E as follows:

Algorithm

The input is a Boolean expression $E=P_1+P_2+\ldots+P_m$, where P_i are all prime implicants of E. The output expresses E as a minimal sum-of-products.

Step 1. Express each prime implicant P as a complete sum-of-products.
Step 2. Delete one by one those prime implicants whose summands appear among the summands of the remaining prime implicants.

EXAMPLE 7.18

Consider Boolean expression E expressed as the sum of prime implicants in the above example. We have

$$E=x'\,z'+x\,y+x'\,y'+y\,z'.$$

We first convert each prime implicant into complete sum-of-products form. We have

$$x'\,z'=x'z'(y+y')=x'\,z'\,y+x'\,z'\,y',$$
$$x\,y=x\,y(z+z')=x\,y\,z+x\,y\,z',$$
$$x'\,y'=x'\,y'(z+z')=x'\,y'\,z+x'\,y'\,z',$$
$$y\,z'=y\,z'(x+x')=y\,z'\,x+y\,z'\,x'.$$

The summands of $x'\,z'$ appear in the summands of $x'\,y'$ and $y\,z'$. So we delete $x'\,z'$ and get

$$E=x\,y+x'\,y'+y\,z'. \tag{1}$$

The summands of no other prime implicant appear among the summands of the remaining prime implicants. Hence expression (1) is a minimal sum-of-product form for E. In other words, none of the remaining prime implicants is superfluous, that is, none can be deleted without changing E.

7.4 LOGIC GATES AND CIRCUITS

Definition 7.17

Logic circuit (or **logic networks**) are structures which are built up from certain elementary circuits called **logical gates.**

Each logic circuit may be treated as **a machine** L **which contains one or more input devices and exactly one output device.** Each input device in L sends a signal, specifically, a bit (binary digit), 0 or 1 to the circuit L, and L processes the set of bits to yield an output bit. Accordingly, an n bit sequence may be assigned to each input device, and L process the input sequences, one bit at a time, to produce an n-bit output sequence.

7.4.1 Logic Gates

There are three basic logic gates. The lines (wires) entering the gate symbol from the left are input lines and the single line on the right is the output line.

1. **OR Gate:** An OR gate has input x and y and output $z=x \vee y$ or $z=x+y$, where addition (or join) is defined by the truth table. In this case, the output z equals zero only when both inputs x and y are zero.

The symbol and the truth table for OR gate are shown in the diagram below:

x	y	$x+y$
1	1	1
1	0	1
0	1	1
0	0	0

(Truth table for OR gate)

Thus, the OR gate only yields 0 when both input bits are 0. An OR gate may have more than two inputs. For example, if it has three inputs x, y, z, then the symbol and truth table for OR gates are

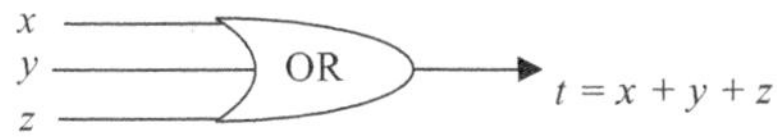

(OR gate with three inputs)

x	y	z	$t=x+y+z$
1	1	1	1
1	1	0	1
1	0	0	1
1	0	1	1
0	1	0	1
0	1	1	1
0	0	1	1
0	0	0	0

(Truth table for OR gate with three inputs)

If the input data for the OR gate are the following 6-bit sequences

$$x=100010, \quad y=100101, \quad z=110001,$$

then the output t will yield 0 only when all input bits are 0. This occurs only in third position (from the left). Thus, $t=110111$.

2. **AND Gate:** In this gate the inputs are x and y and output is $x \wedge y$ or x·y or xy, where multiplication is defined by the truth table.

x	y	$z=x \wedge y$
1	1	1
1	0	0
0	1	0
0	0	0

(Truth table for AND gate)

Thus, output is 1 only when $x=1$, $y=1$, otherwise it is zero.

The AND gate may have more than two inputs. The output in such a case will be 1 if all the inputs are 1.

3. **NOT Gate (Inverter):** The diagram below shows NOT gate with input x and output $y=x'$, where inversion, denoted by the prime, is defined by the truth table:

(NOT gate)

x	$y=x'$
1	0
0	1

(Truth table for NOT gate)

For example, if $x=10101$, then output x' in NOT gate shall be $x'=01010$.

EXAMPLE 7.19

Draw logic circuit for $a\,b'+a'\,b$.

Solution.

The logic circuit for the given expression is shown in the Figure 7.6.

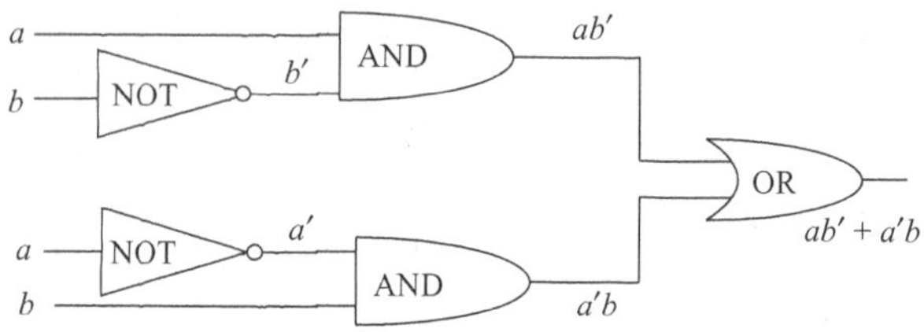

Figure 7.6

7.4.2 Logic Circuits as a Boolean Algebra

The truth tables for OR, AND and NOT gates are respectively identical to the truth tables for the propositions $p \vee q$ (disjunction, "p or q"), $p \wedge q$ (conjunction, "p and q") and $\sim p$ (negation, "not p"). The only difference is that 0 and 1 are used instead of F (contradiction) and T (tautology). Thus the logic circuits satisfy the same laws as do propositions and hence they form a Boolean algebra. Hence, we have established the following:

Theorem 7.8

Logic circuits form a Boolean algebra.

In view of this theorem, all terms used with Boolean algebras such as, complements, literals, fundamental products, minterm, sum-of-products and complete sum-of-products may be used with logic circuits.

7.4.3 AND-OR Circuits

The logic circuit L which corresponds to Boolean sum-of-products expression is called an **AND-OR circuit.** Such a circuit L has several inputs, where

(i) Some of inputs or their complements are fed into each AND gate.
(ii) The outputs of all the AND gates are fed into a single OR gate.
(iii) The output of the OR gate is the output for the circuit L.

EXAMPLE 7.20

Express the output t as a Boolean expression in the inputs x, y, z in the AND-OR circuit shown in the Figure 7.7.

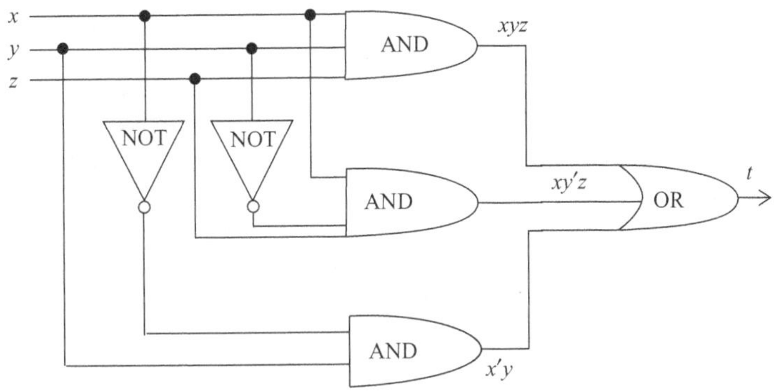

Figure 7.7

We notice that

(i) The inputs of first AND gate are x, y, z and so $x\,y\,z$ is the output.
(ii) The inputs of the second AND gate are x, y', z and hence $x\,y'\,z$ is the output.
(iii) The inputs of the third AND gate are x' and y so $x'\,y$ is the output.
(iv) These outputs of AND gates are inputs for the OR gate and so the output is $t = x\,y\,z + x\,y'\,z + x'\,y$.

Hence, the Boolean expression for the output t is

$$t = x\,y\,z + x\,y'\,z + x'\,y.$$

EXAMPLE 7.21

Express the output z as a Boolean expression of the logic circuit, shown in the Figure 7.8, for which a, b, c are inputs:

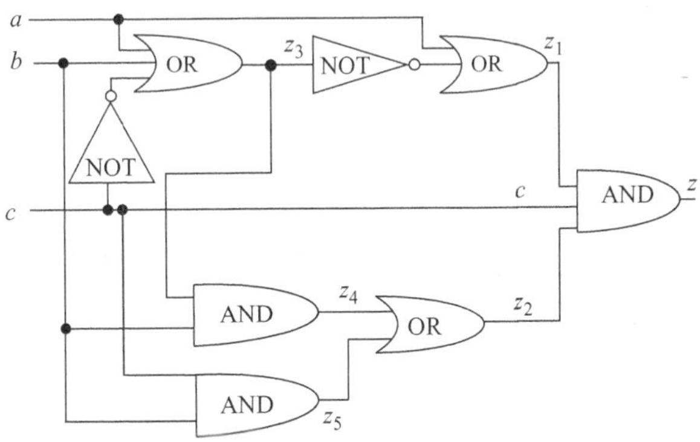

Figure 7.8

We note that

$$\begin{aligned} z &= z_1\,c\,z_2 \\ &= (a+z_3')\,c\,(z_4+z_5) \\ &= (a+(a+b+c')')\,c\,(b\,z_3+b\,c) \\ &= (a+(a+b+c')')\,c\,[b\,(a+b+c')+b\,c]. \end{aligned}$$

We simplify this expression further by using Boolean identity and have

$$\begin{aligned} z &= (a+a'\,b'\,c\,)\,c\,b\,[a+b+c'+c] && \text{(using De Morgan's law)} \\ &= (a\,c\,b+a'\,b'\,b\,c)\,(a+b+I) && \text{(using complement law)} \\ &= (a\,c\,b)\,I && \text{(using absorption law)} \\ &= a\,c\,b = a\,b\,c. \end{aligned}$$

EXAMPLE 7.22

Express the output of the logic circuit in the Figure 7.9 as a Boolean expression. (Here small circle represents complement (NOT))

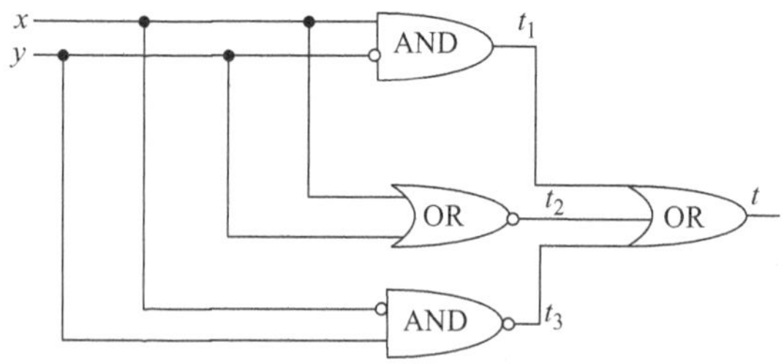

Figure 7.9

Solution.

We note that

$$t_1 = xy', \quad t_2 = (x+y)', \quad t_3 = (x'y)'$$

and so we have

$$\begin{aligned} t = t_1 + t_2 + t_3 &= x\,y' + (x+y)' + (x'\,y)' \\ &= xy' + x'y' + x + y'. \end{aligned}$$

7.4.4 NAND and NOR Gates

NAND and NOR gates are frequently used in computers.

1. NAND gate: It is equivalent to AND gate followed by a NOT gate. Its symbol is

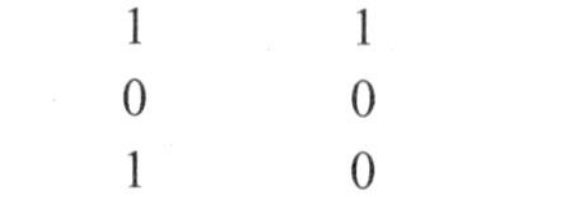

NAND gate

Its truth table is

x	y	$x\,y$	$z=(x\,y)'$
1	1	1	0
1	0	0	1
0	1	0	1
0	0	0	1

Thus, the **output of a NAND gate is 0 if and only if all the inputs are 1.**

2. NOR gate: This gate is equivalent to OR gate followed by a NOT gate. Its symbol is

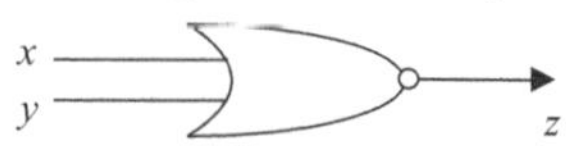

NOR gate

Its truth table is as shown below:

x	y	$x+y$	$(x+y)'$
1	1	1	0
1	0	1	0
0	1	1	0
0	0	0	1

Thus, **the output of NOR gate is 1 if and only if all inputs are 0.**

EXAMPLE 7.23

Find the value of

$$x_1 x_2 [(x_1 x_4)+x_2'+(x_3 x_1')]$$

for $x_1=a$, $x_2=1$, $x_3=b$ and $x_4=I$, where $a, b, I \in B$ and the Boolean algebra is

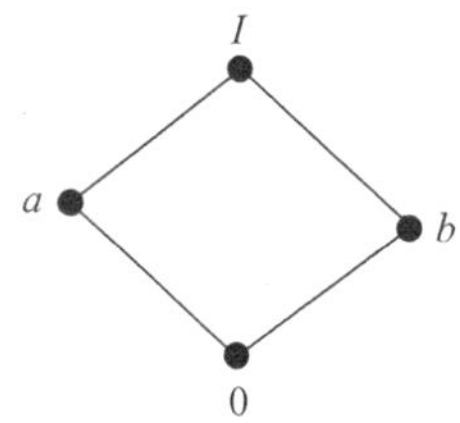

Figure 7.10

Solution.

The given expression is equal to

$$x_1 x_2 x_1 x_4+x_1 x_2 (x_2'+(x_3 x_1') = x_1 x_2 x_4+x_1 x_2 x_2'+x_1 x_2 x_3 x_1'$$
$$= x_1 x_2 x_4+0+0=x_1 x_2 x_4=a \cdot I \cdot I=a.$$

7.5 BOOLEAN FUNCTION

We know that ordinary polynomials could produce functions by substitution. For example, the polynomial $x\,y+y\,z^3$ produces a function f: $\mathbf{R}^3 \to \mathbf{R}$ by letting $f(x, y, z)=xy+yz^3$. Thus, $f(3, 4, 2)=3 \cdot 4+4 \cdot 2^3=44$. In a similar way, Boolean polynomials involving n variables produce functions from B_n to B.

Definition 7.18

Let $(B, \cdot, +, ', 0, I)$ be a Boolean algebra. A function $f: B_n \to B$ which is associated with a Boolean expression (polynomial) in n variables is called a **Boolean function**.

Thus a Boolean function is completely determined by the Boolean expression $\alpha(x_1, x_2, \ldots, x_n)$ because it is nothing but the evaluation function of the expression. It may be mentioned here that every function g: $B_n \to B$ need not be a Boolean function.

If we assume that the Boolean algebra B is of order 2^m for $m \geq 1$, then the number of function from B_n to B is greater than 2^{2n} showing that there are functions from B_n to B which are not Boolean functions. On the other hand, for $m=1$, that is, for a two element Boolean algebra, the number of functions from B_n to B is 2^{2n} which is same as the number of distinct Boolean expressions in n variables. Hence, in this case, every function from B_n to B is a Boolean function.

EXAMPLE 7.24

Consider the Boolean polynomial

$$p(x_1, x_2, x_3) = (x_1 \wedge x_2) \vee (x_1 \vee (x_2' \wedge x_3)).$$

Construct the truth table for the Boolean function $f: B_3 \to B$ determined by this Boolean polynomial.

Solution.

The Boolean function is described by substituting all the 2^3 ordered triple of values from B for x_1, x_2 and x_3. The truth table is as shown below:

x_1	x_2	x_3	x_1x_2	x_2'	$x_2'x_3$	$x_1+x_2'x_3$	$p(x_1, x_2, x_3)$
1	1	1	1	0	0	1	1
1	1	0	1	0	0	1	1
1	0	1	0	1	1	1	1
1	0	0	0	1	0	1	1
0	1	1	0	0	0	0	0
0	1	0	0	0	0	0	0
0	0	1	0	1	1	1	1
0	0	0	0	1	0	0	0

EXAMPLE 7.25

Show that the following Boolean expressions are equivalent to one another. Obtain their sum-of-product canonical form.

(a) $(x+y)\,(x'+z)\,(y+z)$
(b) $(xz)+(x'y)+(yz)$
(c) $(x+y)\,(x'+z)$
(d) $xz+x'y$.

Solution.

The binary valuation of the expression are

x	y	z	$x+y$	$x'+z$	$y+z$	(a)	(c)	xz	$x'y$	yz	(b)	(d)
0	0	0	0	1	0	0	0	0	0	0	0	0
0	0	1	0	1	1	0	0	0	0	0	0	0
0	1	0	1	1	1	1	1	0	1	0	1	1
0	1	1	1	1	1	1	1	0	1	1	1	1
1	0	0	1	0	0	0	0	0	0	0	0	0
1	0	1	1	1	1	1	1	1	0	0	1	1
1	1	0	1	0	1	0	0	0	0	0	0	0
1	1	1	1	1	1	1	1	1	0	1	1	1

Since the values of the given Boolean expression are equal over every triple of the two-element Boolean algebra, they are equal.

To find the sum-of-product **canonical (complete)** form, we note that (d) is in sum-of-product form. Therefore, to find complete sum-of-product form, we have

$$\begin{aligned}(d) &= (x\,z)+(x'\,y)\\ &= x\,z(y+y')+(x'\,y)\,(z+z')\\ &= x\,z\,y+x\,z\,y'+x'\,y\,z+x'\,y\,z'.\end{aligned}$$

7.6 METHOD TO FIND TRUTH TABLE OF A BOOLEAN FUNCTION

Consider a logic circuit consisting of three input devices x, y, z. Each assignment of a set of three bits to the input x, y, z yields an output bit for t. There are $2^n=2^3=8$ possible ways to assign bits to the input as follows:

000, 001, 010, 011, 100, 101, 110, 111.

The assumption is that the sequence of first bits is assigned to x, the sequence of second bits to y, and the sequence of third bits to z. Thus, the above set of inputs may be rewritten in the form

$$x=00001111, \quad y=00110011, \quad z=01010101.$$

These three sequences (of 8 bits) contain the eight possible combinations of the input bits.

The truth table $T=T(L)$ of the circuit L consists of the output t that corresponds to the input sequences x, y, z.

The truth table is same as we generally have written in vertical columns. The difference is that here we write x, y, z and t horizontally.

Consider a logic circuit L with n input devices. There are many ways to form n input sequences $x_1, x_2, \ldots, x_n$ so that they contain 2^n different possible combinations of the input bits (Each sequence must contain 2^n bits).

The assignment scheme is as follows:

x_1: Assign 2^{n-1} bits which are 0 followed by 2^{n-1} bits which are 1.
x_2: Assign 2^{n-2} bits which are 0 followed by 2^{n-2} bits which are 1.
x_3: Assign 2^{n-3} bits which are 0 followed by 2^{n-3} bits which are 1.

and so on.

The sequence obtained in this way is called "**special sequence**". Replacing 0 with 1 and 1 with 0 in the special sequences yield the **complements** of the special sequences.

EXAMPLE 7.26

Suppose a logic circuit L has $n=4$ input devices—x, y, z and t. Then $2^n=2^4=16$ bit special sequences for x, y, z, t are

$x=0000000011111111$ ($2^{n-1}=2^3=8$ zeros followed by 8 ones)
$y=0000111100001111$ ($2^{n-2}=2^{4-2}=4$ zeros followed by 4 ones)
$z=0011001100110011$ ($2^{n-3}=2^{4-3}=2$ zeros followed by 2 ones)
$t=0101010101010101$ ($2^{n-4}=2^{4-4}=2^0=1$ zeros followed by 1 one)

7.6.1 Algorithm for Finding Truth Table for a Logic Circuit L where Output T is Given by a Boolean Sum-of-Product Expression in the Inputs

The input is a Boolean sum-of-products expression $t(x_1, x_2, \ldots)$.

Step 1. Write down the special sequences for the inputs x_1, x_2, ... and their complements.
Step 2. Find each product appearing in $t(x_1, x_2, \ldots)$ keeping in mind that $x_1\,x_2 \ldots=1$ is a position if and only if all $x_1\,x_2 \ldots$ have 1 in the position.

Step 3. Find the sum t of the products keeping in mind that $x_1+x_2+\ldots=0$ in a position if and only if all $x_1\, x_2\, \ldots$ have 0 in the position.

EXAMPLE 7.27

Find the truth table for a circuit whose Boolean sum-of-product expression is $t=x\,y\,z+x\,y'\,z+x'\,y$.

Solution.

Here $n=3$. So the special sequences for x, y, z are

$$x=00001111, \quad y=00110011, \quad z=01010101.$$

So,

$$x'=11110000, \quad y'=11001100.$$

Hence,

$$x\,y\,z=00000001,$$
$$x\,y'\,z=00000100,$$
$$x'\,y=00110000.$$

The sum is $t=00110101$. Hence $T(L)$ is given by

$$T(L)=T(00001111, 00110011, 01010101)=00110101.$$

7.6.2 Representation of Boolean Functions

A Boolean function may be expressed in four ways:

1. Boolean function may be represented by one of the Boolean expressions to which the function corresponds. For example,

$$f(x, y, z)=x\,y\,z+x\,y'\,z+x'\,y.$$

2. Boolean function may be represented by its truth table. But this representation becomes very cumbersome for expressions of five or more variables.
3. Boolean functions may be represented by n-space representation. For example, consider the Boolean function

$$f(x, y, z)=x(y+z').$$

There are three input variables, therefore there are $2^3=8$ input combinations. **These correspond to the vertices of a three-dimensional cube (Figure 7.11).**

Each vertex is labelled to indicate to which input triple it corresponds to and each vertex that corresponds to an input triple which has a functional value 1 is denoted by a **bold dot**. For example, for the triple 100, we have

$$x=1, y=0, z=0 \quad \text{and} \quad z'=1 \text{ and } y+z'=1.$$

Hence, $f(x, y, z)=x\,(y+z')=1\cdot 1=1$.

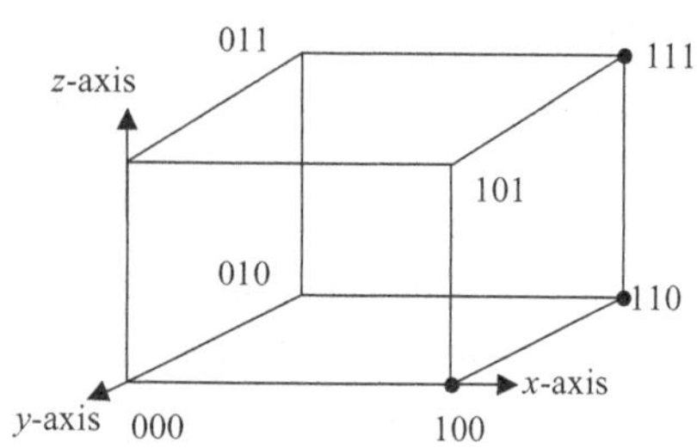

Figure 7.11

4. **Karnaugh Map**

A Karnaugh map is a graphical procedure to represent Boolean function as an "or" combination of minterms where minterms are represented by squares. This procedure is easy to use with functions $f: B_n \rightarrow B$, if n is not greater than 6. We shall discuss this procedure for $n=2$, 3 and 4.

A Karnaugh map structure is an area which is subdivided into 2^n cells, one for each possible input combination for a Boolean function of n variables. Half of the cells are associated with an input value of 1 for one of the variables and the other half are associated with an input value of 0 for the same variable. This association of cell is done for each variable, with the splitting of the 2^n cells yielding a different pair of halves for each distinct variable.

Case of 1 variable: In this case, the Karnaugh map consists of $2^1=2$ squares. The variable x is represented by the right square and its complement x' by the left square (see Figure 7.12).

0	1
x'	x

Figure 7.12

Case of 2 variables: For $n=2$, the Boolean function is of two variable, say x and y. We have $2^2=4$ squares, that is, a 2×2 matrix of squares. Each square contains one possible input from B_2.

The variable x appears in the first row of the matrix as x' whereas x appears in the second row as x. Similarly, y appears in the first column as y' and as y in the second column.
In this case, x is represented by the points in lower half of the map and y is represented by the points in the right half of the map (see Figure 7.13).

	0	1
0	00	01
1	10	11

	y'	y
x'	$x'y'$	$x'y$
x	xy'	xy

(Two variables Karnaugh map)

Figure 7.13

Definition 7.19

Two fundamental products are said to be **adjacent** if they have the same variables and if they differ in exactly one literal. Thus, there must be an uncomplemented variable in one product which is complemented in the other.

For example, if $P_1=x\,y\,z'$ and $P_2=x\,y'\,z'$, then they are adjacent.

The sum of two such adjacent products will be a fundamental product with one less literal.

For example, in case of the above-mentioned adjacent products,

$$P_1+P_2=x\,y\,z'+x\,y'\,z'=x\,z'(y+y')=x\,z'(I)=x\,z'.$$

We note that two squares in Karnaugh map above are adjacent if and only if squares are geometrically adjacent, that is, have a side in common.

We know that a complete sum-of-products Boolean expression $E(x, y)$ is a sum of minterms and hence can be represented in the Karnaugh map by placing checks in the appropriate square. A prime implicant of $E(x, y)$ will be either a **pair of adjacent squares** in E or an isolated square (a square which is not adjacent to other square of $E(x, y)$). A minimal sum of products for $E(x, y)$ will consist of a minimal number of prime implicants which cover all the squares of $E(x, y)$.

EXAMPLE 7.28

Find the prime implicants and a minimal sum-of-products form from each of the following complete sum-of-products Boolean expression:

(a) $E_1=x\,y+x\,y'$
(b) $E_2=x\,y+x'\,y+x'\,y'$
(c) $E_3=x\,y+x'\,y'$.

Solution.

(a) The Karnaugh map for E_1 is shown in the Figure 7.14.

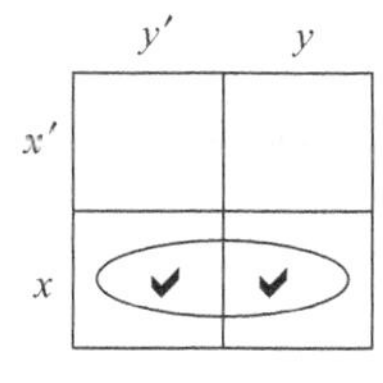

Figure 7.14

Check the squares corresponding to $x\,y$ and $x\,y'$. We note that E_1 consists of one prime implicant, the two adjacent square designated by the loop. The pair of adjacent square represents the variable x. So x is the only prime implicant of E_1. Consequently $E_1 = x$ is its minimal sum.

(b) The Karnaugh map for E_2 is shown in the Figure 7.15.

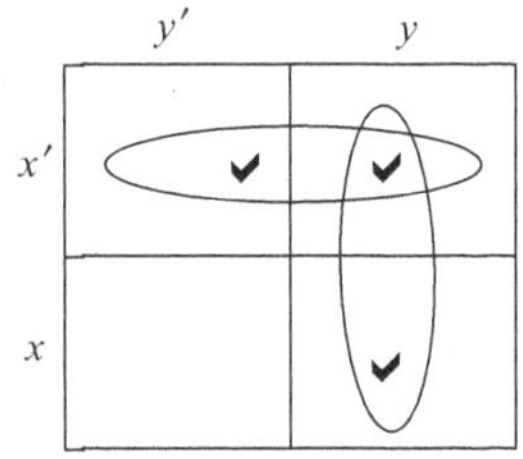

Figure 7.15

Check the squares corresponding to $x\,y$, $x'\,y$, $x'\,y'$. The expression E_2 contains two pairs of adjacent squares (designated by two loops) which include all the squares of E_2. The vertical pair represents y and the horizontal pair x'. Hence, y and x' are the prime implicants of E_2. Thus,

$$E_2(x, y) = x' + y$$

is a minimal sum.

(c) The Karnaugh map for E_3 is shown in the Figure 7.16.

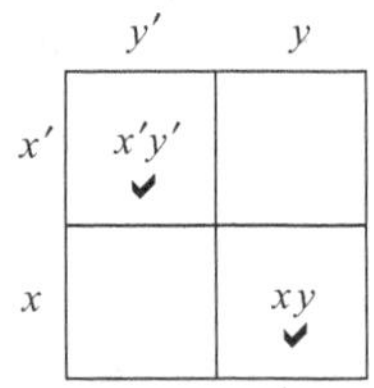

Figure 7.16

Check (tick) the squares corresponding to xy and $x'y'$. The expression E_3 consists of two isolated squares which represent xy and $x'y'$. Hence $x\,y$ and $x'\,y'$ are the prime implicants of E_3 and so $E_3 = x\,y + x'\,y'$ is its minimal sum.

Case of 3 variables: We now turn to the case of a function $f: B_3 \to B$ which is function of x, y and z. The Karnaugh map corresponding to Boolean expression $E(x, y, z)$ is shown in the Figure 7.17.

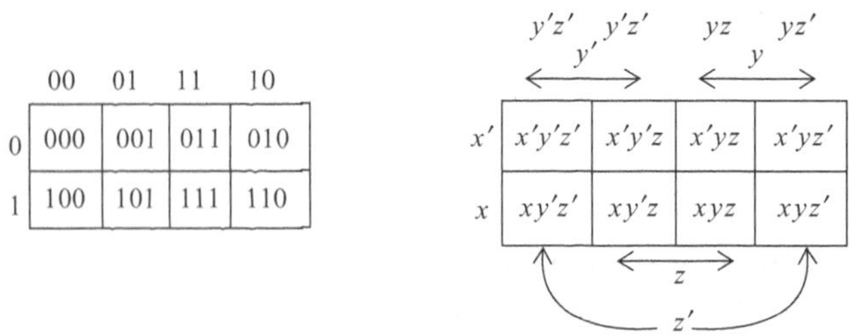

Figure 7.17

Here x, y, z are, respectively, represented by lower half, right half and middle two quarters of the map.

Similarly, x', y', z' are, respectively, represented by upper half, left half, and left and right quarters of the map.

Definition 7.20

By a **basic rectangle** in the Karnaugh map with three variables, we mean a square, two adjacent squares or four squares which form a one by four, or a two by two rectangle. These basic rectangles correspond to fundamental products of three, two and one literal respectively.

Further, the fundamental product represented by a basic rectangle is the product of just those literals that appear in every square of the rectangle.

Let a complete sum of products Boolean expression $E(x, y, z)$ is represented in the Karnaugh map by placing checks in the appropriate squares. A prime implicant of E will be a maximal basic rectangle of E, that is, a basic rectangle contained in E which is not contained in any larger basic rectangle in E.

A minimal sum-of-products form for E will consist of a minimal cover of E, that is, a minimal number of maximal basic rectangles of E which together include all the squares of E.

EXAMPLE 7.29

Find the prime implicants and a minimal sum-of-products form for each of the following complete sum of products Boolean expressions:

(a) $E_1 = x\,y\,z + x\,y\,z' + x'\,y\,z' + x'\,y'\,z$,

(b) $E_2 = x\,y\,z + x\,y\,z' + x\,y'\,z + x'\,y\,z + x'\,y'\,z$,

(c) $E_3 = x\,y\,z + x\,y\,z' + x'\,y\,z' + x'\,y'\,z + x'y'z'$.

Solution.

(a) The Karnaugh map for E_1 is shown in the Figure 7.18.

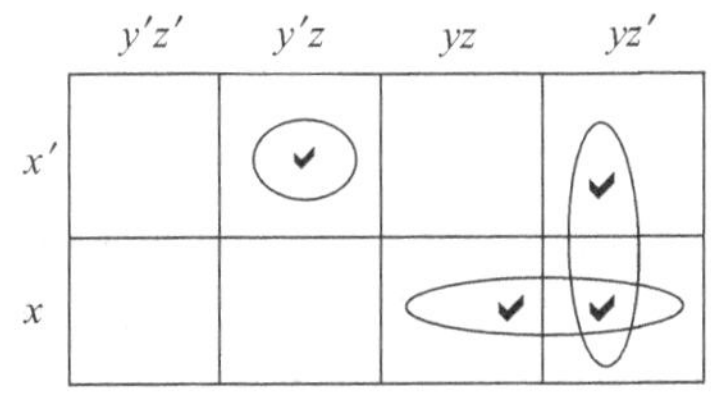

Figure 7.18

We check the four squares corresponding to four summands in E_1. Here E_1 has three prime implicants (maximal basic rectangles) which are encircled. These are $x\,y$, $y\,z'$ and $x'\,y'\,z$. All three are needed to cover E_1. Hence minimal sum for E_1 is

$$E_1 = x\,y + y\,z' + x'\,y'\,z.$$

(b) The Karnaugh map for E_2 is shown in the Figure 7.19.

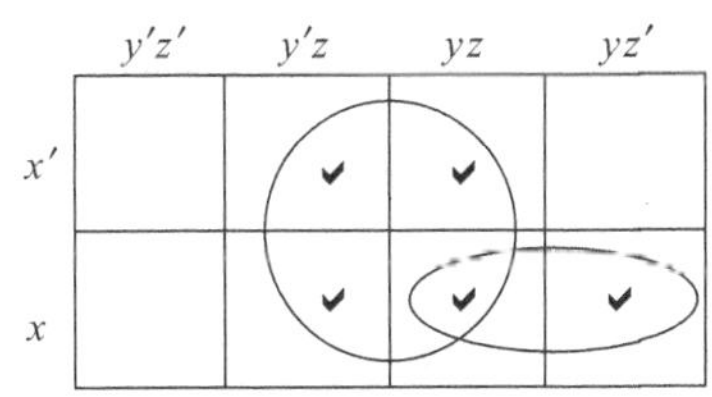

Figure 7.19

Check the squares corresponding to the five summands. E_2 has two prime implicants which are circled. One is the two adjacent squares which represent $x\,y$, and the other is the two-by-two square which represents z. Both are needed to cover E_2 so the minimal sum for E_2 is

$$E_2 = x\,y + z.$$

(c) The Karnaugh map for E_3 is shown in the Figure 7.20.

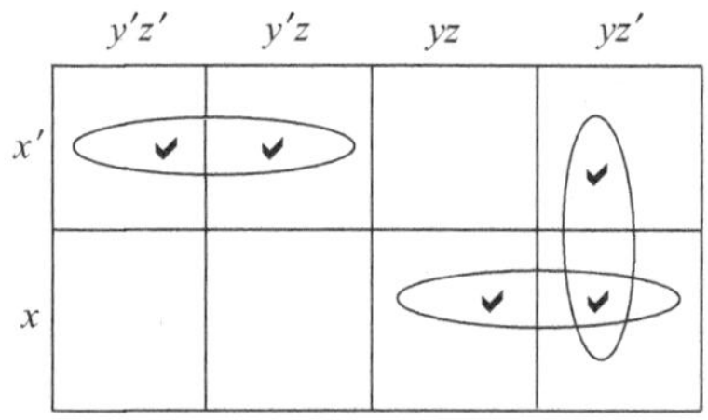

Figure 7.20

Check the squares corresponding to the five summands. Here E_3 has three prime implicants $x\,y$, yz', $x'\,y'$. All these are needed in a minimal cover of E_3. Hence E_3 has minimal sum as

$$E_3 = x\,y + y\,z' + x'\,y'.$$

Remark 7.2 **To find the fundamental product represented by a basic rectangle, find literals which appear in all the squares of the rectangle.**

Case of 4 Variables: We consider a Boolean function $f: B_4 \to B$, considered as a function of x, y, z and t. Each of the 16 squares (2^4) corresponds to one of the minterms with four variables.

$$x\,y\,z\,t,\; x\,y\,z\,t',\; \ldots,\; x'\,y\,z'\,t.$$

We consider first and last columns to be adjacent, and first and last rows to be adjacent, both by Wrap around, and we look for rectangles with sides of length some power of 2, so the length is 1, 2 or 4. The expression for such rectangles is given by intersecting the large labelled rectangles.

	00	01	11	10
00	0000	0001	0011	0010
01	0100	0101	0111	0110
11	1100	1101	1111	1110
10	1000	1001	1011	1010

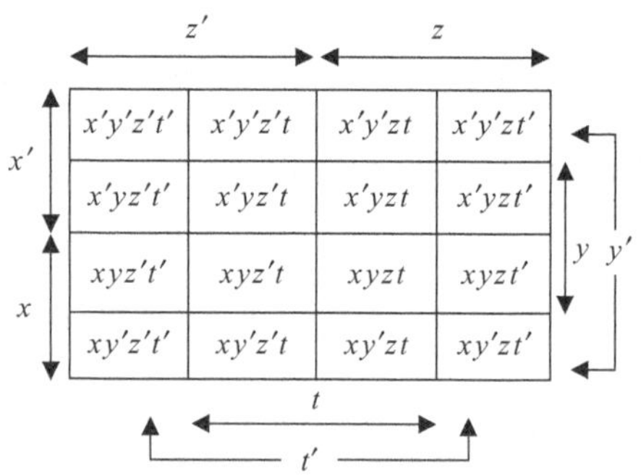

Figure 7.21

A basic rectangle in a four-variable Karnaugh map is a square, two adjacent squares, four squares which form a one-by-four or two-by-two rectangle or eight square squares which form a two-by-four rectangle. These rectangles correspond to fundamental product with four, three, two and one literal respectively. Maximal basic rectangles are prime implicants.

EXAMPLE 7.30

Find the fundamental product P represented by the basic rectangle in the Karnaugh map given in the Figure 7.22.

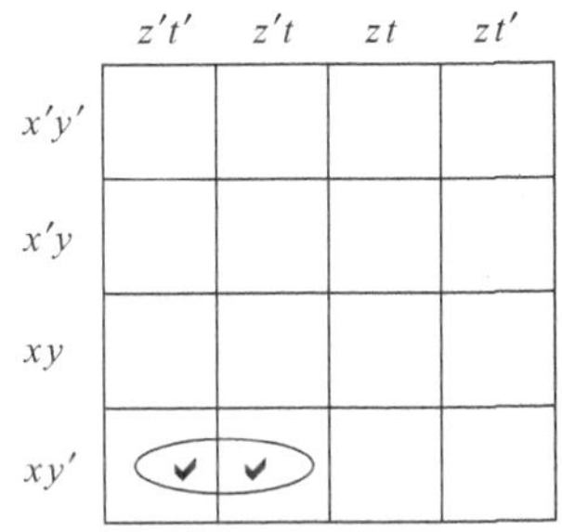

Figure 7.22

Solution.

We find the literals which appear in all the squares of the basic rectangle. Then P will be the product of such literals.

Here x, y', z' appear in both squares. Hence, $P=x\,y'\,z'$ is the fundamental product represented by the basic rectangle in question.

EXAMPLE 7.31

Use a Karnaugh map to find a minimal sum-of-products form for

$$E=x\,y'+x\,y\,z+x'\,y'\,z'+x'\,y\,z\,t'.$$

Solution.

The Karnaugh map for E is shown in the Figure 7.23.

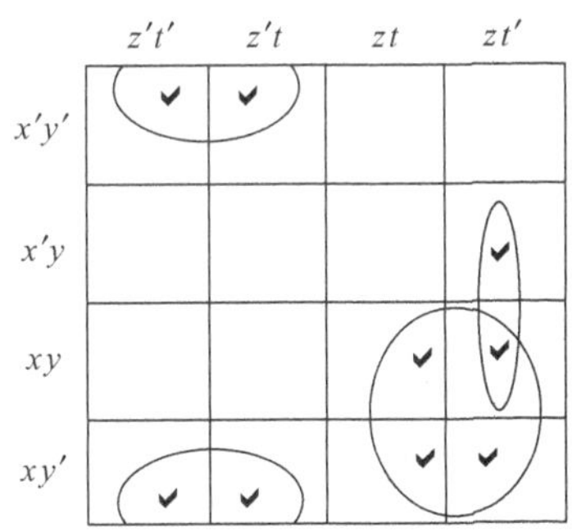

Figure 7.23

Check all the (four) squares representing xy', check all the (two) squares representing xyz, check all the (two) squares representing $x'y'z'$ and check the square representing $x'yzt'$. Two-by-two square represents xz (shown on the right side of the figure). Further, considering the first row and the last row as adjacent, the four square represent $y'z'$ Similarly, two adjacent squares on the right side of the figure represent yzt'. Hence,

$$E=x\,z+y'\,z'+y\,z\,t'.$$

7.7 EXPRESSING BOOLEAN FUNCTIONS AS BOOLEAN POLYNOMIALS

Let $f: B_n \to B$ be a Boolean function and let

$$S(f)=\{b \in B_n: f(b)=1\}.$$

Then we have the following result:

Theorem 7.9

Let f, f_1 and f_2 be three functions from B_n to B. Then

(i) If $S(f)=S(f_1) \cup S(f_2)$, then

$$f(b)=f_1(b) \vee f_2(b) \quad \text{for all } b \in B_n.$$

(ii) If $S(f)=S(f_1) \cap S(f_2)$, then

$$f(b)=f_1(b) \wedge f_2(b) \quad \text{for all } b \in B_n,$$

where $\vee$ denotes lub and $\wedge$ denotes glb in B.

Proof.

(i) Let $b \in B_n$. If $b \in S(f)$, then by definition $f(b)=1$. Since $S(f)=S(f_1) \cup S(f_2)$, therefore either $b \in S(f_1)$ or $b \in S(f_2)$ or b belongs to both. But in every case, $f_1(b) \vee f_2(b)=1$. If $b \notin S(f_1)$ then by definition, $f(b)=0$. Since $S(f)=S(f_1) \cup S(f_2)$, therefore, $b \notin S(f_1)$ and $b \notin S(f_2)$ and so $f_1(b)=f_2(b)=0$. Hence, $f_1(b) \vee f_2(b)=0$. Thus, for all $b \in B_n$, we have

$$f(b)=f_1(b) \vee f_2(b).$$

(ii) The proof is completely analogous to part (i).

EXAMPLE 7.32

Let $f_1: B_2 \to B$ be produced by the expression $E(x, y)=x'$ and let $f_2: B_2 \to B$ be produced by the expression $E(x, y)=y'$. Let $f: B_2 \to B$ be the function whose truth table is

x	y	$f(x, y)$
0	0	1
0	1	1
1	0	1
1	1	0

Show that f is produced by the Boolean expression $x' \vee y'$.

Solution.

The truth tables for f_1 and f_2 are respectively

x	y	$f_1(x, y)$
0	0	1
0	1	1
1	0	0
1	1	0

, and

x	y	$f_2(x, y)$
0	0	1
0	1	0
1	0	1
1	1	0

.

We note that

$$S(f_1)=\{(0,0), (0,1)\}, S(f_2)=\{(0,0), (1, 0)\}$$
$$S(f)=\{(0, 0), (0, 1), (1, 0)\},$$

Thus,

$$S(f)=S(f_1) \cup S(f_2).$$

Hence, by the above theorem, $f=f_1 \vee f_2$ and so f is produced by $x' \vee y'$.

EXAMPLE 7.33

Let $f: B_2 \to B$ be the function with truth table given below:

x	y	$f(x, y)$
0	0	0
0	1	0
1	0	1
1	1	0

Find the Boolean expression which produces f.

Solution.

The function is equal to 1 only when $x=1, y=0$. Thus, $S(f)=\{(0, 1)\}$. This is also true for $E(x, y)=x \wedge y'$. Hence, f is produced by

$$E(x, y)=x \wedge y'.$$

EXAMPLE 7.34

Let $f: B_3 \to B$ be the function with the truth table

x	y	z	$f(x, y, z)$
0	0	0	0
0	0	1	0
0	1	0	0
0	1	1	1
1	0	0	0
1	0	1	0
1	1	0	0
1	1	1	0

Find the Boolean expression which produces f.

Solution.

Here $S(f)=\{(0, 1, 1)\}$. Thus, $f(x, y, z)=1$ only for $x=0$, $b=1$, $z=1$. This is also true for $E(x, y, z)= x' \wedge y \wedge z$. Hence, f is produced by

$$E(x, y, z)=x' \wedge y \wedge z.$$

Theorem 7.10

Any function $f: B_n \to B$ is produced by a Boolean expression.

Proof. Let $S(f)=\{b_1, b_2, \ldots, b_k\}$ and let for each $i, f_i: B_n \to B$ be the function defined by

$$f_i(b_i)=1,$$
$$f_i(b)=0 \quad \text{if } b \neq b_i.$$

Then, $S(f_i)=\{b_i\}$. Hence,

$$S(f)=S(f_1) \cup S(f_2) \cup \ldots \cup S(f_k)$$

and so,

$$f=f_1 \vee f_2 \vee \ldots \vee f_n.$$

Remark 7.3 As we have seen in examples given above, each f_i is produced by the minterm E_{bi}. Thus f is produced by the Boolean expression

$$E_{b_1} \vee E_{b_2} \vee \ldots \vee E_{b_n}.$$

EXAMPLE 7.35

Find Boolean expression for the function $f: B_3 \to B$ whose truth table is given below:

x	y	z	$f(x, y, z)$
0	0	0	0
0	0	1	0
0	1	0	0
0	1	1	1
1	0	0	0
1	0	1	0
1	1	0	0
1	1	1	1

Solution.

From the truth table, we note that

$$S(f)=\{(0, 1, 1), (1, 1, 1)\}.$$

Hence Boolean expression for the function f is

$$\begin{aligned} E(x, y, z) &= E_{b_1} \vee E_{b_2} \\ &= E_{(0,1,1)} \vee E_{(1,1,1)} \\ &= (x' \wedge y \wedge z) \vee (x \wedge y \wedge z) \\ &= (x' \vee x) \wedge (y \wedge z) \quad \text{(distributive law)} \\ &= I \wedge (y \wedge z) = y \wedge z. \end{aligned}$$

EXAMPLE 7.36

Find the Boolean expression $E(x, y, z)$ corresponding to the truth table

$$T(E)=01001001.$$

Solution.
The truth table is

x	y	z	$E(x, y, z)$
0	0	0	0
0	0	1	1
0	1	0	0
0	1	1	0
1	0	0	1
1	0	1	0
1	1	0	0
1	1	1	1

From the truth table, we note that

$$S(f)=\{(0, 0, 1), (1, 0, 0), (1, 1, 1)\}.$$

Hence, the Boolean expression for the function f is

$$\begin{aligned} E(x, y, z) &= E_{(0,0,1)} \vee E_{(1,0,0)} \vee E_{(1,1,1)} \\ &= (x' \wedge y' \wedge z) \vee (x \wedge y' \wedge z') \vee (x \wedge y \wedge z) \\ &= x' y' z + x y' z' + x y z. \end{aligned}$$

EXAMPLE 7.37

Find the Boolean expression $E(x, y, z)$ corresponding to the truth table

$$T(E)=00010001.$$

Solution.
The truth table is

x	y	z	$E(x, y, z)$
0	0	0	0
0	0	1	0
0	1	0	0
0	1	1	1
1	0	0	0
1	0	1	0
1	1	0	0
1	1	1	1

From the truth table we note that

$$S(f)=\{(0, 1, 1), (1, 1, 1)\}.$$

Hence, the Boolean expression is

$$E(x, y, z)=E_{(0,1,1)} \vee E_{(1,1,1)}=x' y z+x y z.$$

EXAMPLE 7.38

Design a three-input-minimal AND-OR circuit with the following truth table:

$$T=[A, B, C; L]=[00001111, 00110011, 01010101, 11001101].$$

Solution.

Special sequences for this circuit are

$$A=00001111, \quad B=00110011, \quad C=01010101, \quad L=11001101.$$

The complete sum-of-products form for L is

$$L=E_{(0,0,0)} \vee E_{(0,0,1)} \vee E_{(1,0,0)} \vee E_{(1,0,1)} \vee E_{(1,1,1)}$$
$$= A'B'C'+A'B'C+AB'C'+AB'C+ABC.$$

The Karnaugh map for L is shown in the Figure 7.24.

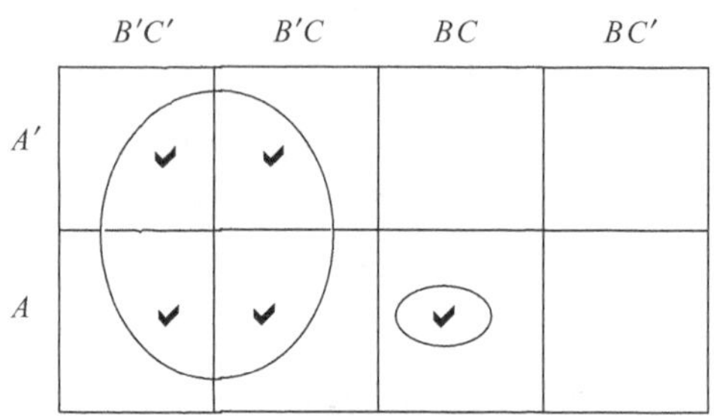

Figure 7.24

Check the squares corresponding to the five summands. Here L has two prime implicant B' and ABC.
Hence the sum is

$$L=B'+ABC=B'+(ACB)=(B'+AC)\,(B'+B)=B'+AC.$$

Thus, the minimal cover is $L=B'+AC$.
The minimal AND-OR circuit for L is therefore

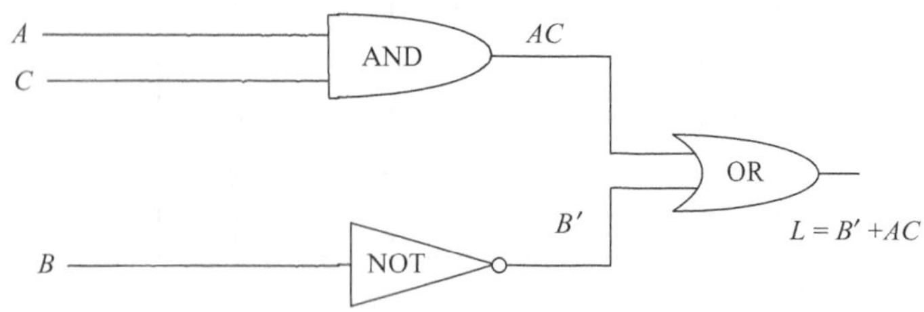

Figure 7.25

EXAMPLE 7.39

Consider the function f whose truth table is

x	y	z	$f(x, y, z)$
0	0	0	1
0	0	1	1
0	1	0	0
0	1	1	1
1	0	0	1
1	0	1	0
1	1	0	0
1	1	1	0

Find Boolean expression using Karnaugh map.

Solution.

Karnaugh map corresponding to f is given in the Figure 7.26.

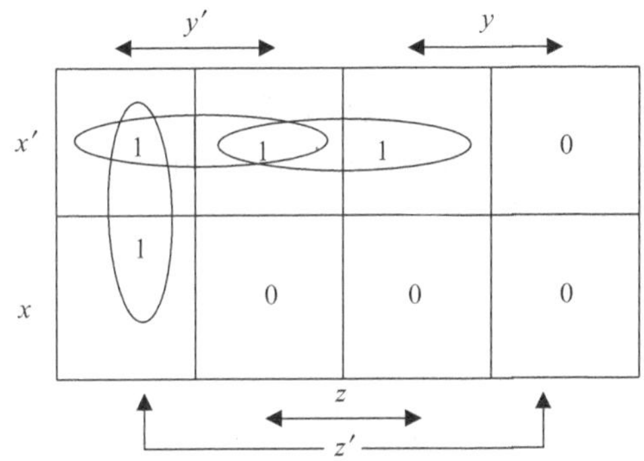

Figure 7.26

The Boolean expression has three prime implicants (three maximal basic rectangles). These are $y'z'$, $x'y'$ and yz.

Hence the Boolean expression for f is

$$E(x, y, z) = (y' \wedge z') \vee (x' \wedge y') \vee (x' \wedge z)$$
$$= y'z' + x'y' + y'z,$$

EXAMPLE 7.40

Consider the function whose truth table is

$$T(E) = 10101011.$$

Find Boolean expression using Karnaugh map.

Solution.

The truth table of the given function is given below:

x	y	z	$f(x, y, z)$
0	0	0	1
0	0	1	0
0	1	0	1
0	1	1	0
1	0	0	1
1	0	1	0
1	1	0	1
1	1	1	1

The corresponding Karnaugh map is

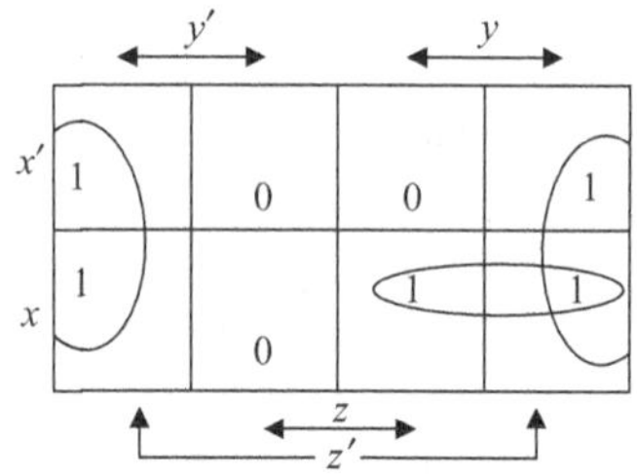

Figure 7.27

Taking into account that first and last columns are adjacent representing z', the Boolean expression has two prime implicants z' and xy. Hence,

$$E = z' + xy.$$

EXAMPLE 7.41

Consider the function $f: B_4 \rightarrow B$ where truth table is

x	y	z	t	$f(x, y, z, t)$
0	0	0	0	1
0	0	0	1	0
0	0	1	0	1
0	0	1	1	0
0	1	0	0	0
0	1	0	1	1
0	1	1	0	0
0	1	1	1	1
1	0	0	0	1
1	0	0	1	0
1	0	1	0	1
1	0	1	1	0
1	1	0	0	0
1	1	0	1	1
1	1	1	0	0
1	1	1	1	1

Find the Boolean expression using Karnaugh map.

Solution.

The corresponding Karnaugh map is shown in the Figure 7.28.

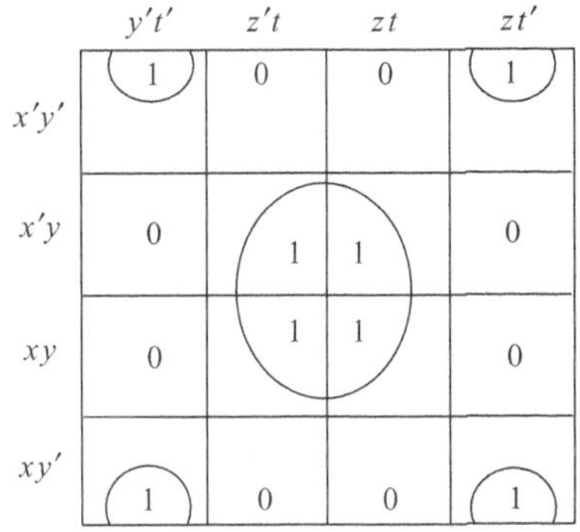

Figure 7.28

The central 2×2 square represents the Boolean expression $y \wedge t$. The four corners also form a square of side 2 since the right and left edges and the top and bottom edges are considered adjacent which represents $y' \wedge t'$.
Hence the Boolean expression is

$$E(x, y, z, t) = (y \wedge t) \vee (y' \wedge t')$$
$$= yt + y't' \quad \text{Ans.}$$

7.8 ADDITION OF BINARY DIGITS

We know that computers generally work on binary numbers. Thus the digits used are 0 and 1 which are called **bits.** We use **binary operation of addition modulo 2** in case of binary system. We denote this binary operation by ⊕. Thus, we have the truth table for addition in binary system as

a_i	b_i	c_i	$S_i(a_i \oplus b_i)$
0	0	0	0
0	1	0	1
1	0	0	1
1	1	1	0

,

where a_i, b_i are bits 0 or 1, c_i is carry bit and S_i is sum bit of bits a_i and b_i. Thus, in case of two 1's, a two digit sequence is required to express their sum, i.e.

$$\begin{array}{r} 1 \\ 1 \\ \hline 10 \end{array}$$

↗ Carry

The left digit in this sum is called **carry**. In hand calculation, the carry bit creates no problem, but in fixed length computer words, a carry may require an extra bit position which does not exist. Such a condition is called an **overflow**.

From the truth table the Boolean functions for the two outputs can be obtained as follows:

(i) We have

$$S(S_i)=\{(0, 1), (1, 0)\}.$$

Hence,

$$E(a_i, b_i)=E_{(0, 1)} \vee E_{(1, 0)}$$
$$= a_i'b_i+a_ib_i'.$$

We can also find Boolean expression for the sum bit using Karnaugh map. Here we have $2^2=4$ squares (see Figure 7.29).

Thus,

$$a_i \oplus b_i=a_i'b_i+a_ib_i'$$

	b_i'	b_i
a_i'	0	1
a_i	1	0

Figure 7.29

(ii) For the carry bit

$$S(c_i)=\{(1, 1)\}.$$

Hence Boolean expression for c_i is

$$c_i=a_ib_i.$$

7.8.1 Half-Adder

The most basic digital arithmetic function is the addition of two binary digits. A combinational circuit that performs the addition of two bits is called a **half-adder**.

The input variables of a half-adder are called the augend and addend bits. The output variables are called the **sum** and **carry**. It is necessary to specify two output variables because the sum of 1+1 is binary 10, which has two digits. We assign symbols a_i and b_i to the two input variables and S_i (for the sum function) and C_i (for carry) to the two output variables. The truth table for the **half-adder** is thus

Inputs		Outputs	
a_i	b_i	C_i	S_i
0	0	0	0
0	1	0	1
1	0	0	1
1	1	1	0

As derived earlier, the Boolean expressions for S_i and C_i are

$$S_i= \bar{a}_i b_i+a_i \bar{b}_i =a_i \oplus b_i,$$
$$C_i=a_i b_i,$$

where $\oplus$ denotes binary operation of addition modulo 2.

The logic diagram of half-adder consists of an exclusive OR gate and AND gate as shown below:

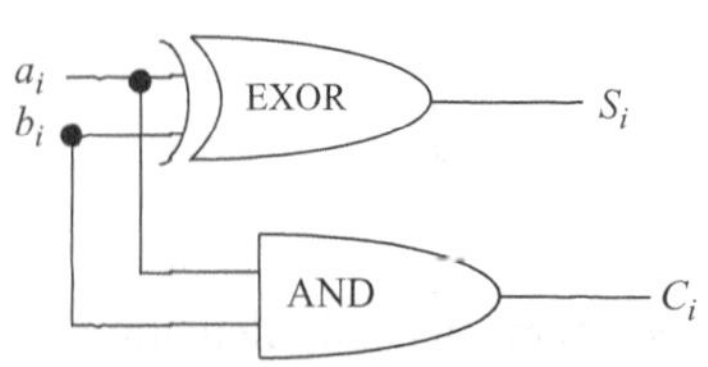

Figure 7.30

7.8.2 Full-Adder

A full-adder is a combination circuit that forms the **arithmetic sum of three input bits.** It consists of **three inputs and two outputs.** Two of the input variables, denoted by a_i and b_i, represent the two significant bits to be added. The third input c_{i-1} represents the carry from the previous lower significant position. Two outputs are necessary because the arithmetic sum of three binary digit ranges in value from 0 to 3, and the decimal numbers 2 and 3 need two binary digits. The two outputs are designated by S_i (For sum) and C_i (for carry). The binary variable C_i gives the output carry. The truth table of the full-adder consist of the eight rows under the input variable designate all possible combinations of 1's and 0's that these variables may have. The 1's and 0's for the output variables are determined from the arithmetic sum of the input bits. When all input bits are 0's the output is 0. The S_i output is equal to 1 when only one input is equal to 1. The C_i output has carry of 1 if two or more inputs are equal to 1.

Inputs			Outputs	
a_i	b_i	c_{i-1}	C_i	S_i
0	0	0	0	0
0	0	1	0	1
0	1	0	0	1
0	1	1	1	0
1	0	0	0	1
1	0	1	1	0
1	1	0	1	0
1	1	1	1	1

The Karnaugh map of C_i and S_i are

	$b_i'c_{i-1}'$	$b_i'c_{i-1}$	b_ic_{i-1}	b_ic_{i-1}'
a_i'	0	0	1	0
a_i	0	1	1	1

(Karnaugh map for C_i)

Figure 7.31

	$b_i'c_{i-1}'$	$b_i'c_{i-1}$	b_ic_{i-1}	b_ic_{i-1}'
a_i'	0	1	0	1
a_i	1	0	1	0

(Karnaugh map for S_i)

Figure 7.32

Hence the Boolean expression for S_i and C_i are

$$S_i = a_i'b_i'c_{i-1} + a_i'b_ic_{i-1}' + a_ib_i'c_{i-1}' + a_ib_ic_{i-1}'$$
$$= a_i \oplus b_i \oplus c_{i-1}$$
$$C_i = a_ib_i + a_ic_{i-1} + b_ic_{i-1}$$
$$= a_i\,b_i + a_i\,c_{i-1}\,(b_i + b_i') + b_i\,c_{i-1}(a_i + a_i')$$
$$= a_i\,b_i + a_i\,c_{i-1}\,b_i + a_i\,c_{i-1}\,b_i' + b_i\,c_{i-1}a_i + b_i\,c_{i-1}a_i'$$
$$= a_i\,b_i + a_i\,b_i\,c_{i-1} + a_i\,b_i'\,c_{i-1} + a_i'\,b_i\,c_{i-1} \quad \text{(idempotent law)}$$
$$= a_i\,b_i + (a_i\,b_i' + a_i'\,b_i)\,c_{i-1} \quad \text{(absorption law).}$$

EXERCISES

1. Is D_{60} a Boolean algebra? Give reasons.
2. Is the lattice D_{72} a Boolean algebra? Give reasons.
3. Let $A=\{2, 3, 5\}$. Show that D_{30} and $P(A)$ are structurally the same Boolean algebra.
4. Find the disjunctive normal form for the function f whose truth table is

x	y	z	$f(x,y,z)$
1	1	1	1
1	1	0	1
1	0	1	0
1	0	0	1
0	1	1	1
0	1	0	1
0	0	1	0
0	0	0	0

5. Find the truth table for the Boolean function $f: B_3 \to B$ defined by

$$f(x_1, x_2, x_3)=x_1 \wedge (x_2' \vee x_3).$$

6. Find the Boolean expression for the following logic diagram:

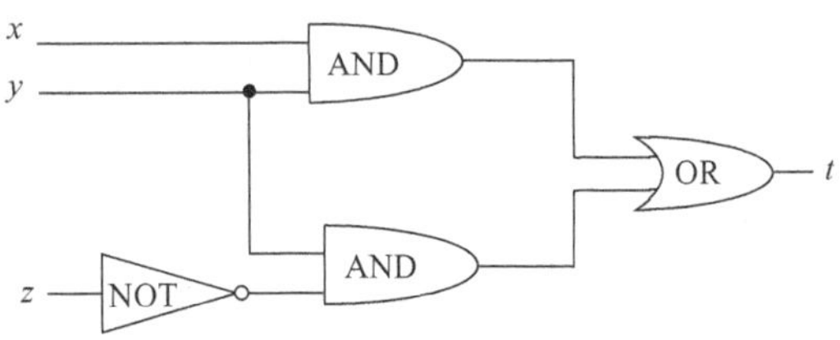

Figure 7.33

7. Using Karnaugh map redesign the logic circuit given below to get a minimal circuit.

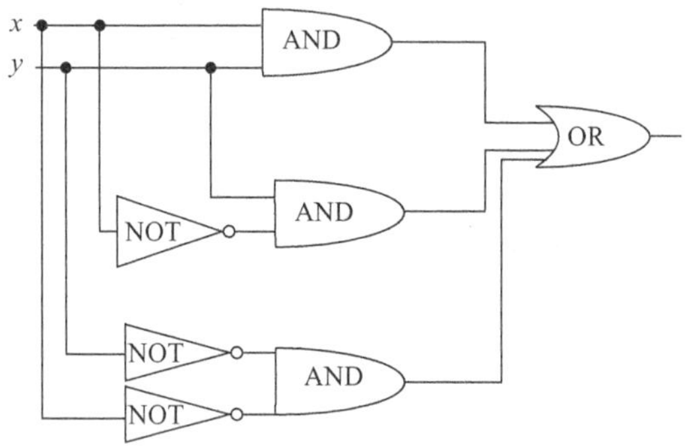

Figure 7.34

8. Find minimal sum of products form for the Boolean expression:

$$E=xy'+x'z't+xyzt'+x'y'zt'.$$

9. Using Karnaugh map, find the Boolean expression for the function $f: B_3 \to B$ whose truth table is

$$T(00001111, 00110011, 01010101) = 10001001.$$

10. Design logic circuit using NAND gates to compute the function $f(x,y)=x \vee y$.
11. Represent the expression $x'y'z+x'yz'+xyz'$ by Karnaugh map. Also find minimal expression.

8 Graphs

The interest in the study of graph theory has increased due to its applicability in so many fields like artificial intelligence, electrical engineering, transportation system, scheduling problems, economics, chemistry and operations research.

8.1 DEFINITIONS AND BASIC CONCEPTS

Definition 8.1

A **graph** $G=(V, E)$ is a mathematical structure consisting of two finite sets V and E. The elements of V are called **vertices (or nodes)** and the elements of E are called **edges**. Each edge is associated with a set consisting of **either one** or **two vertices** called its **endpoints**.

The correspondence from edges to endpoints is called **edge-endpoint function**. This function is generally denoted by γ. Due to this function, some authors denote graph by $G=(V, E, \gamma)$.

Definition 8.2

A graph consisting of one vertex and no edges is called a **trivial graph**.

Definition 8.3

A graph whose vertex and edge sets are empty is called a **null graph**.

Definition 8.4

An edge with just one endpoint is called a **loop** or a **self-loop**.

Thus, a loop is an edge that joins a single endpoint to itself.

Definition 8.5

An edge that is not a self-loop is called a **proper edge**.

Definition 8.6

If two or more edges of a graph G have the same vertices, then these edges are said to be **parallel** or **multi-edges**.

Definition 8.7

Two vertices that are connected by an edge are called **adjacent**.

Definition 8.8

An endpoint of a loop is said to be **adjacent to itself**.

Definition 8.9

An edge is said to be **incident** on each of its endpoints.

Definition 8.10

Two edges incident on the same endpoint are called **adjacent edges**.

Definition 8.11

The number of edges in a graph G which are incident on a vertex is called the **degree of that vertex**.

Definition 8.12

A vertex of degree zero is called an **isolated vertex**.

Thus, a vertex on which no edges are incident is called **isolated**.

Definition 8.13

A graph without multiple edges **(parallel edges)** and loops is called **simple graph**.

Notations: In pictorial representations of a graph, the vertices will be denoted by dots and edges by line segments.

EXAMPLE 8.1

(a) Let

$$V=\{1, 2, 3, 4\} \quad \text{and} \quad E=\{e_1, e_2, e_3, e_4, e_5\}.$$

Let γ be defined by

$$\gamma(e_1)=\gamma(e_5)=\{1, 2\}, \quad \gamma(e_2)=\{4, 3\}, \gamma(e_3)=\{1, 3\}, \quad \gamma(e_4)=\{2, 4\}.$$

We note that both edges e_1 and e_5 have same endpoints $\{1, 2\}$. The endpoints of e_2 are $\{4, 3\}$, the endpoints of e_3 are $\{1, 3\}$ and endpoints of e_4 are $\{2, 4\}$. Thus the graph is as shown in the Figure 8.1.

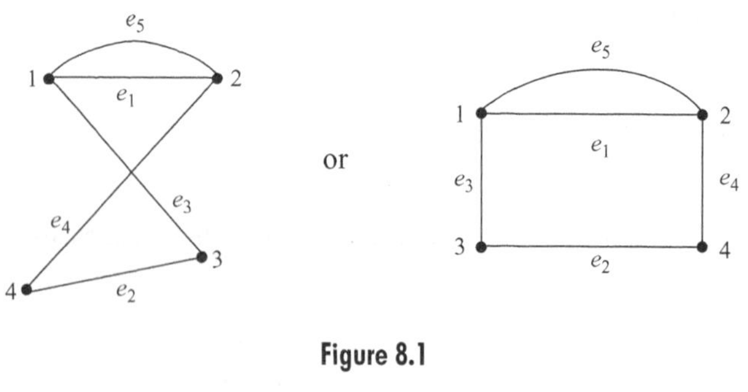

Figure 8.1

(b) For the graph pictured in the Figure 8.2,
  (i) Write down the degree of the vertices A, B, D.
  (ii) Which of the edges are parallel?
  (iii) Which vertex is isolated?
  (iv) Point out whether it is a simple graph or a multi-graph.
  (v) Point out one pair of adjacent vertices and one pair of adjacent edges.

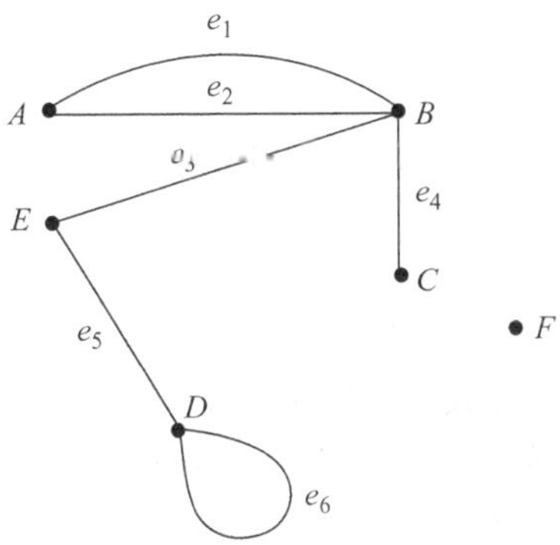

Figure 8.2

In this graph,

(i) We notice that

Degree of the vertex $A=2$

Degree of the vertex $B=4$

Degree of the vertex $D=3$ (because loop e_6 has degree 2).

The vertex D has degree 3 (because e_6 is a loop and so has degree 2)

(ii) The edges e_1 and e_2 are parallel

(iii) The vertex F is isolated because no edge is incident on F

(iv) It is a multi-graph because it has multi-edges and a loop.

(v) A and B are adjacent vertices because they are connected by the edge e_2 or e_1 whereas A and E are not adjacent.

The edges e_2 and e_3 are adjacent edges because they are incident on the same vertex B.

(c) Consider the graph with the vertices A, B, C, D and E pictured in Figure 8.3.

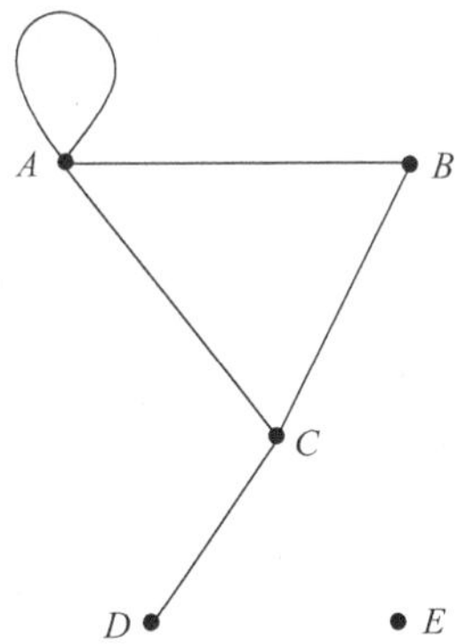

Figure 8.3

In this graph, we note that

Number of edges $=5$

Degree of vertex $A=4$

Degree of vertex $B=2$

Degree of vertex $C=3$

Degree of vertex $D=1$

Degree of vertex $E=0$

Sum of the degree of vertices $=4+2+3+1+0=10$.

Thus, we observe that

$$\sum_{i=1}^{5} \deg(v_i) = 2e,$$

where $\deg(v_i)$ denotes the degree of vertex v_i and e denotes the number of edges.

Euler's Theorem 8.1 (The First Theorem of Graph Theory)

The sum of the degrees of the vertices of a graph G is equal to twice the number of edges in G. **Thus, total degree of a graph is even.**

Proof. Each edge in a graph contributes a count of 1 to the degree of two vertices (endpoints of the edge), that is, each edge contributes 2 to the degree sum. Therefore, the sum of degrees of the vertices is equal to twice the number of edges.

Corollary 8.1

There can be only an even number of vertices of odd degree in a given graph G.

Proof. We know, by the fundamental theorem, that

$$\sum_{i=1}^{n} \deg(v_i) = 2 \times \text{number of edges}.$$

Thus the right-hand side is an even number. Hence to make the left-hand side an even number there can be only even number of vertices of odd degree.

Remarks 8.1

(i) A vertex of degree d is also called a ***d*-valent vertex**.

(ii) The degree (or valence) of a vertex v in a graph G is the number of proper edges incident on v plus twice the number of self-loops.

Theorem 8.2

A nontrivial simple graph G must have at least one pair of vertices whose degrees are equal.

Proof. Let the graph G has n vertices. Then there appear to be n possible degree values, namely $0, 1, \ldots, n-1$. But there cannot be both a vertex of degree 0 and a vertex of degree $n-1$ because if there is a vertex of degree 0 then each of the remaining $n-1$ vertices is adjacent to at most $n-2$ other vertices. Hence the n vertices of G can realize at most $n-1$ possible values for their degrees. Hence, the pigeonhole principle implies that at least two of the vertices have equal degree.

EXAMPLE 8.2

Is there a graph with eight vertices of degree 2, 2, 3, 6, 5, 7, 8, 4?

Solution.

The answer is No. In such a graph, there will be three vertices of odd degree which is impossible.

We can also argue as follows: Total degree of the graph (if possible) is equal to $2+2+3+6+5+7+8+4=37$, which is odd. This contradicts the first theorem of graph theory, according to which, total degree of a graph is always even.

8.2 SPECIAL GRAPHS

Definition 8.14

A graph G is said to **simple** if it has no parallel edges or loops. In a simple graph, an edge with end-points v and w is denoted by $\{v, w\}$.

Definition 8.15

For each integer $n \geq 1$, let D_n denote the graph with n **vertices** and **no edges**. Then D_n is called the **discrete graph on n vertices**.

For example, we have

D_3

and

D_5

Definition 8.16

Let $n \geq 1$ be an integer. Then a simple graph with n vertices in which there is an edge between each pair of distinct vertices is called the **complete graph** on n vertices. It is denoted by K_n.

For example, the complete graphs K_2, K_3 and K_4 are shown in Figure 8.4.

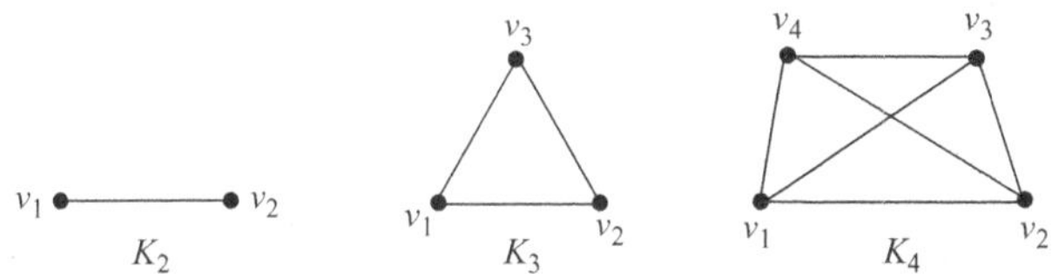

Figure 8.4

Definition 8.17

If each vertex of a graph G has the same degree as every other vertex, then G is called a **regular graph**.

A **k-regular graph** is a regular graph whose common degree is k.

For example, consider K_3. The degree of each vertex in K_3 is 2. Hence K_3 is regular. Similarly, K_4 is regular. Also the graph shown in Figure 8.5 is regular because degree of each vertex here is 2.

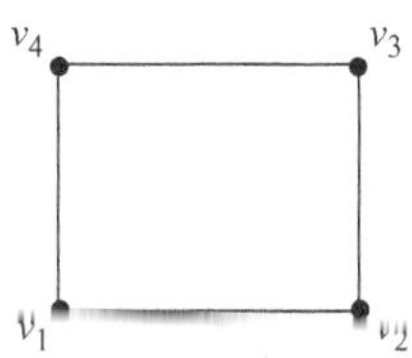

2-regular graph

Figure 8.5

But this graph is not complete because v_2 and v_4 have not been connected through an edge. Similarly, v_1 and v_3 are not connected by any edge.

Thus, **a complete graph is always regular** but **a regular graph need not be complete**.

EXAMPLE 8.3

The oxygen molecule O_2, made up of two oxygen atoms linked by a double bond can be represented by the regular graph shown in Figure 8.6.

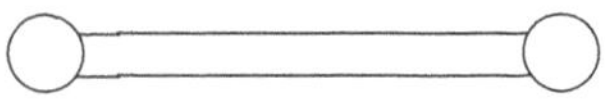

Figure 8.6

Definition 8.18

Let $n \geq 1$ be an integer. Then a graph L_n with n vertices $\{v_1, v_2, \ldots, v_n\}$ and with edges $\{v_i, v_{i+1}\}$ for $1 \leq i < n$ is called a **linear graph on** n **vertices**.

For example, the linear graphs L_2 and L_4 are shown in Figure 8.7.

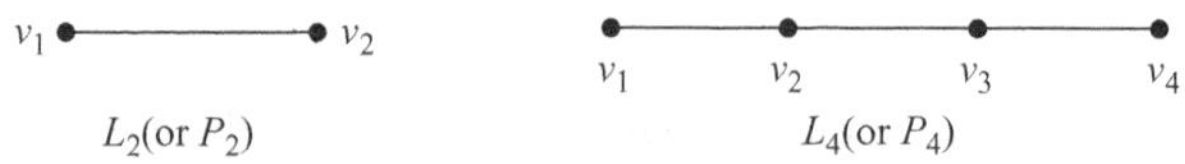

Figure 8.7

It is also called **path graph** denoted by P_n.

Definition 8.19

A **bipartite graph** G is a graph whose vertex set V can be partitioned into two subsets U and W, such that each edge of G has one endpoint in U and one endpoint in W.

The pair (U, W) is called a **vertex bipartition of** G and U and W are called the bipartition subsets. Obviously, a bipartite graph cannot have any self-loop.

EXAMPLE 8.4

If vertices in U are solid vertices and vertices in W are hollow vertices, then the graphs shown in Figure 8.8 are bipartite graphs.

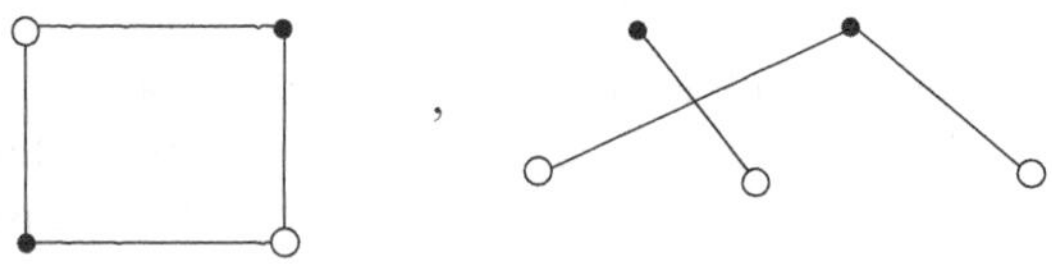

Figure 8.8

EXAMPLE 8.5

The smallest possible simple graph that is not bipartite is the complete graph K_3 shown in Figure 8.9.

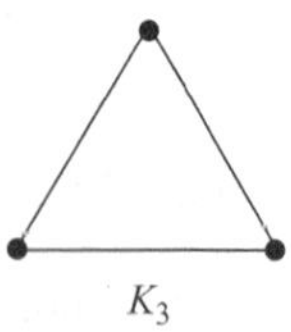

Figure 8.9

Definition 8.20

A complete bipartite graph G is a simple graph whose vertex set V can be partitioned into two subsets $U=\{v_1, v_2, \ldots, v_m\}$ and $W=\{w_1, w_2, \ldots, w_n\}$ such that for all i, k in $\{1, 2, \ldots, m\}$ and j, l in $\{1, 2, \ldots, n\}$

(i) There is an edge from each vertex v_i to each vertex w_j.
(ii) There is not an edge from any vertex v_i to any other vertex v_k.
(iii) There is not an edge from any vertex w_j to any other vertex w_l.

A complete bipartite graph on (m, n) vertices is denoted by $K_{m,n}$.

EXAMPLE 8.6

The complete bipartite graphs $K_{3,2}$ and $K_{3,4}$ are shown in Figure 8.10.

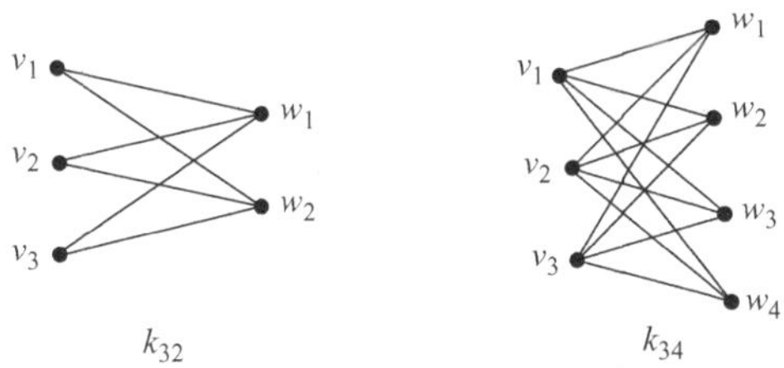

Figure 8.10

Definition 8.21

A **cycle graph** is a single vertex with a self-loop or a simple connected graph C with $|V_c|=|E_c|$, that can be drawn so that all of its vertices and edges lie on a circle.

An n-vertex cycle graph is denoted by C_n. The cycle graphs C_1, C_2, and C_4 are shown in Figure 8.11.

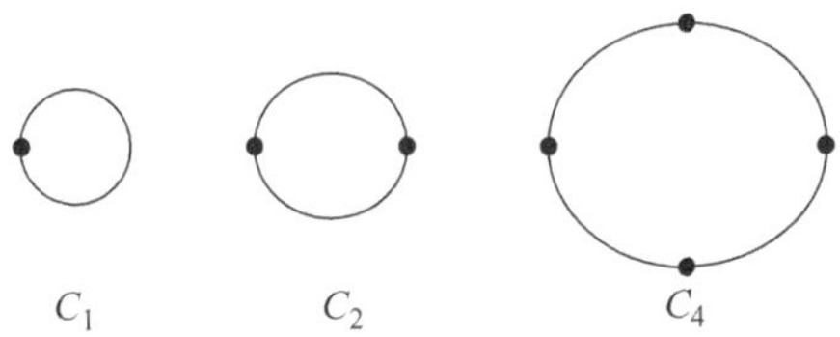

Figure 8.11

Definition 8.22

The **Petersen graph** is the 3-regular graph shown in Figure 8.12.

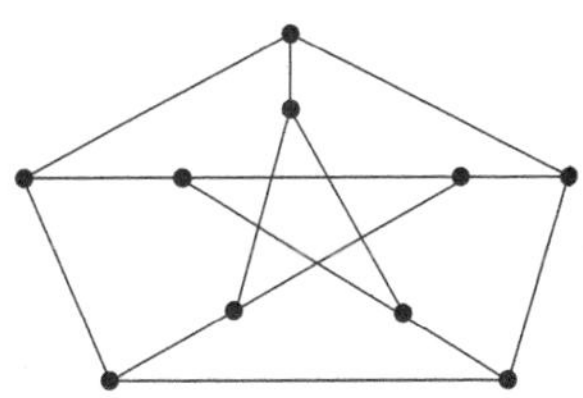

(Petersen graph)

Figure 8.12

This graph is frequently used to establish theorems and to test conjectures.

Definition 8.23

The **hypercube graph** Q_n is the n-regular graph whose vertex set is the set of bit string of length n and such that there is an edge between two vertices if and only if they differ in exactly one bit.

For example, 8-vertex cube graph is a hypercube graph Q_3 shown in Figure 8.13

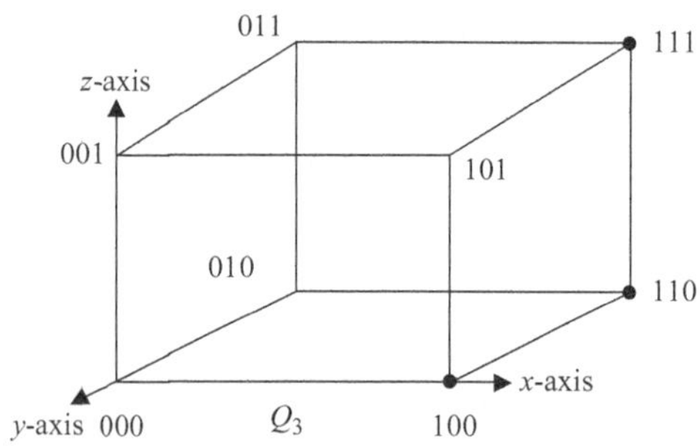

Figure 8.13

Definition 8.24

A graph consisting of two concentric n-cycles in which each of the n pairs of the corresponding vertices is joined by an edge is called the **circular ladder graph** CL_n.

For example, the graphs CL_2 and CL_4 are shown in the Figure 8.14

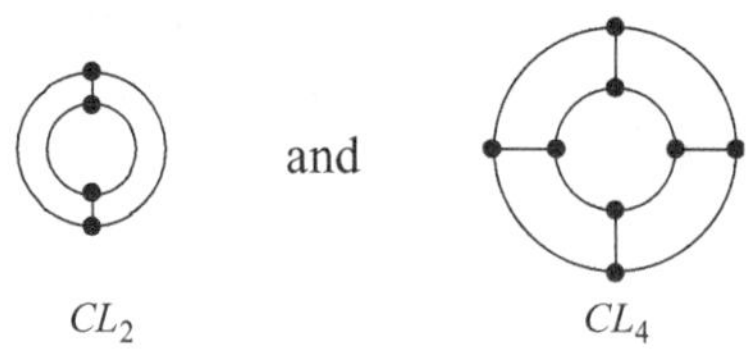

Figure 8.14

Definition 8.25

A graph consisting of a single vertex having n loops is called a **bouquet** and is denoted by B_n.

For example, the graphs shown in Figure 8.15 represent bouquet B_2 and bouquet B_4.

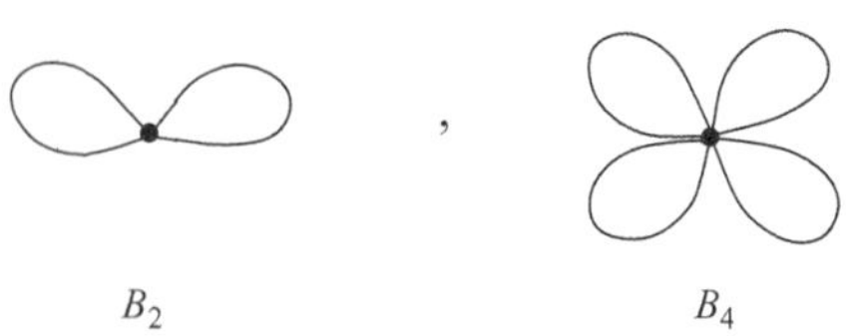

Figure 8.15

Definition 8.26

A graph consisting of two vertices and n edges joining them is called a **dipole**. It is denoted by D_n.

For example, Figure 8.16 represents D_2, D_3 and D_4.

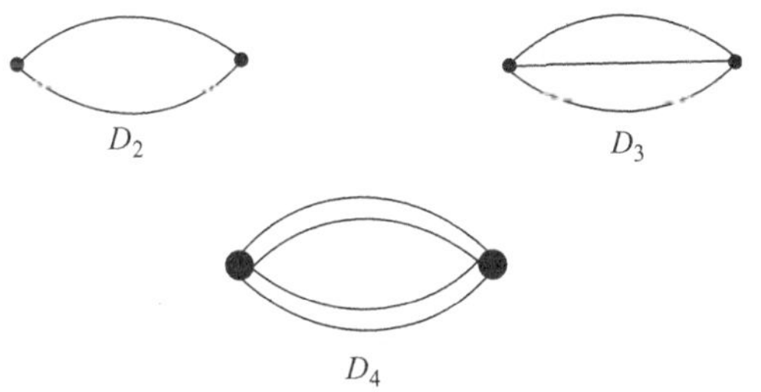

Figure 8.16

EXAMPLE 8.7

Draw all simple graphs with vertices $\{u, v, x, y\}$ and two edges, one of which is $\{u, v\}$.

Solution.

In a simple graph, an edge corresponds to a subset of two vertices. Since there are four vertices, we can have $4c_2 = 6$ such subsets in all and they are $\{u, v\}$, $\{u, x\}$, $\{u, y\}$, $\{v, x\}$, $\{v, y\}$ and $\{x, y\}$. But one edge of the graph is specified to be $\{u, v\}$. Hence the possible simple graphs are as shown in Figure 8.17.

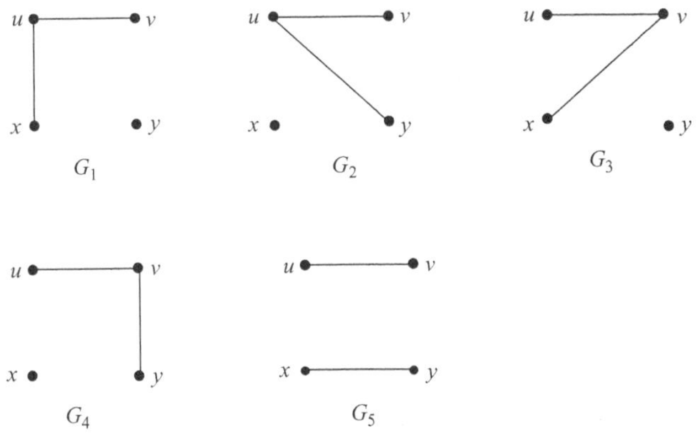

Figure 8.17

8.3 SUBGRAPHS

Definition 8.27

A graph H is said to be a subgraph of a graph G if and only if every vertex in H is also a vertex in G, every edge in H is also an edge in G and every edge in H has the same endpoints as in G.
We may also say that G is a supergraph of H. For example, the graphs (Figure 8.18)

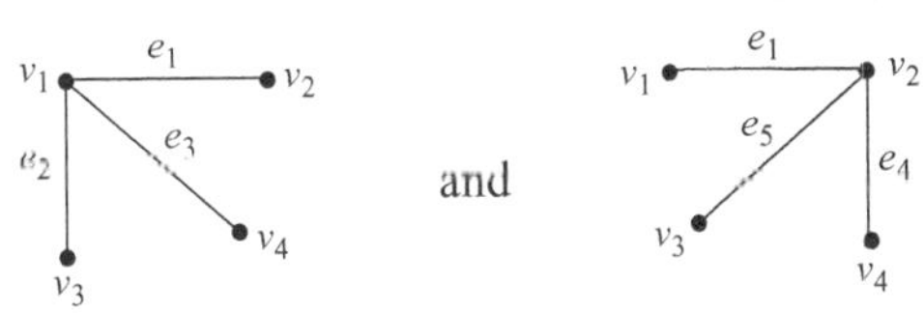

Figure 8.18

are subgraphs of the graph given in Figure 8.19.

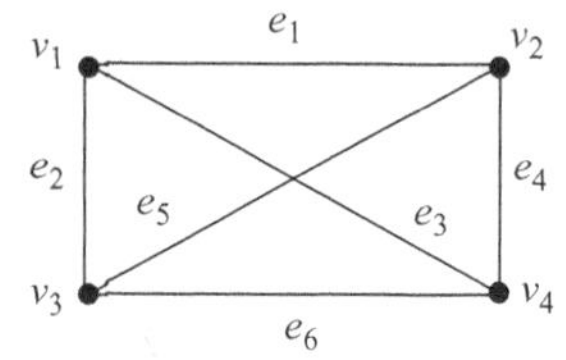

Figure 8.19

Similarly, the graph (Figure 8.20)

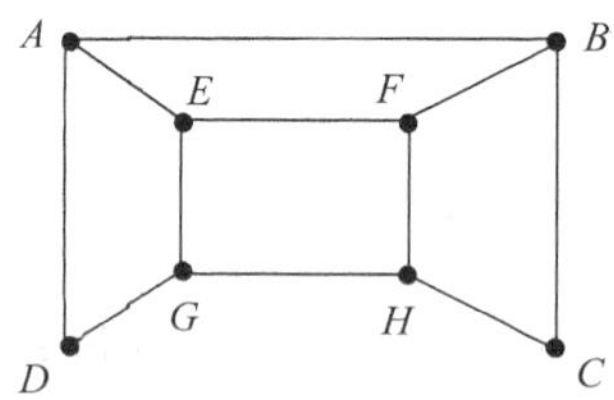

Figure 8.20

is a subgraph of the graph given in the Figure 8.21.

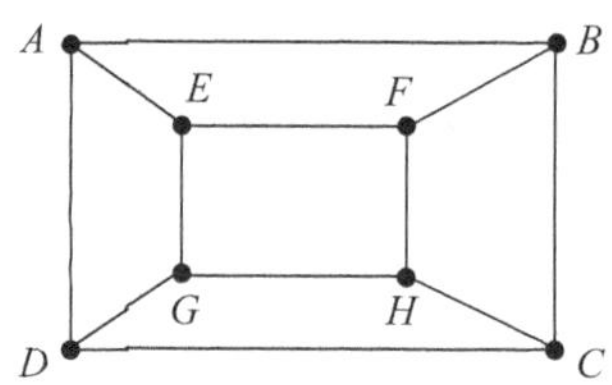

Figure 8.21

Definition 8.28

A subgraph H is said to be a **proper subgraph** of a graph G if vertex set V_H of H is a proper subset of the vertex set V_G of G or edge set E_H is a proper subset of the edge set E_G.

For example, the subgraphs in the above examples are proper subgraphs of the given graphs.

Definition 8.29

A subgraph H is said to **span** a graph G if $V_H = V_G$.

Thus H is a spanning subgraph of graph G if it contains all the vertices of G.

For example, the subgraph (Figure 8.22)

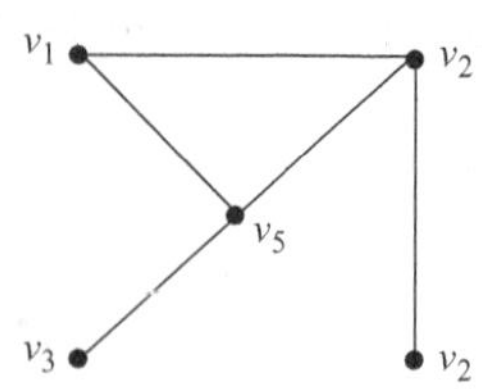

Figure 8.22

spans the graph give in Figure 8.23.

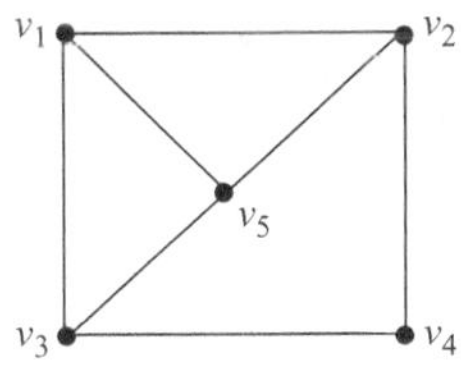

Figure 8.23

Definition 8.30

Let $G=(V, E)$ be a graph. Then the **complement of a subgraph** $G'=(V', E')$ with respect to the graph G is another subgraph $G''=(V'', E'')$ such that $E''=E-E'$ and V'' contains only the vertices with which the edges in E'' are incident.

For example, the subgraph (Figure 8.24)

Figure 8.24

is the complement of the subgraph (Figure 8.25)

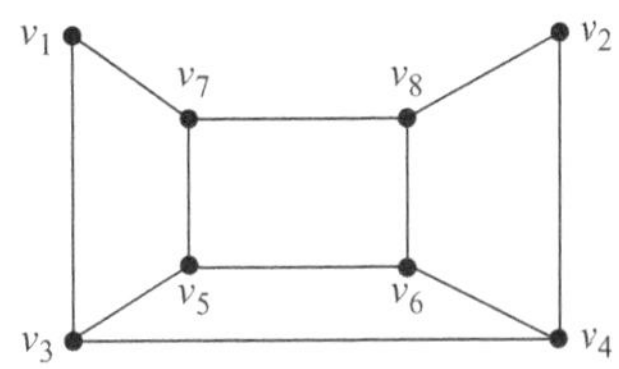

Figure 8.25

with respect to the graph G shown in Figure 8.26.

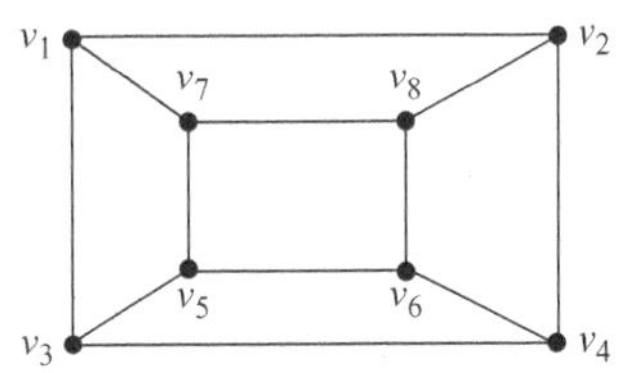

Figure 8.26

Definition 8.31

If G is a simple graph, the **complement of** G **(edge complement)**, denoted by G' or G^c is a graph such that

(i) The vertex set of G' is identical to the vertex set of G, that is, $V_{G'}=V_{G}$.

(ii) Two distinct vertices v and w of G' **are connected by** an edge if and only if v and w **are not connected** by an edge in G.

For example, consider the graph G shown in Figure 8.27.

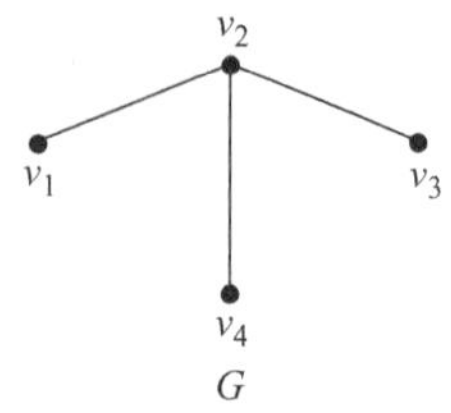

Figure 8.27

Then complement G' of G is the graph shown in Figure 8.28.

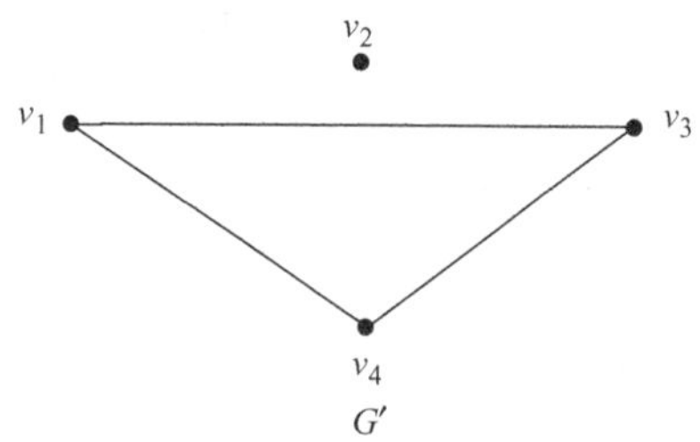

Figure 8.28

EXAMPLE 8.8

Find the complement of the graphs:

(a)

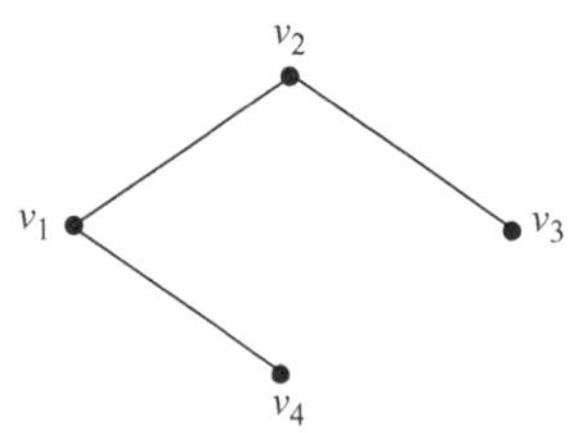

Figure 8.29

(b)

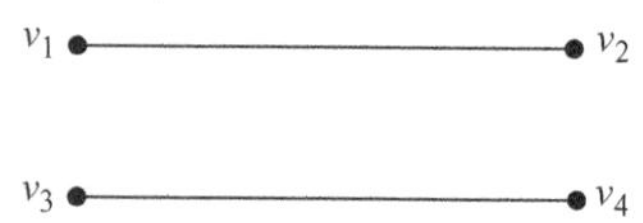

Figure 8.30

(c) Complete graph K_4:

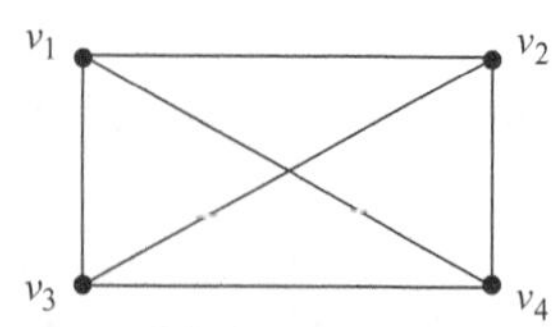

Figure 8.31

Solution.

(a) 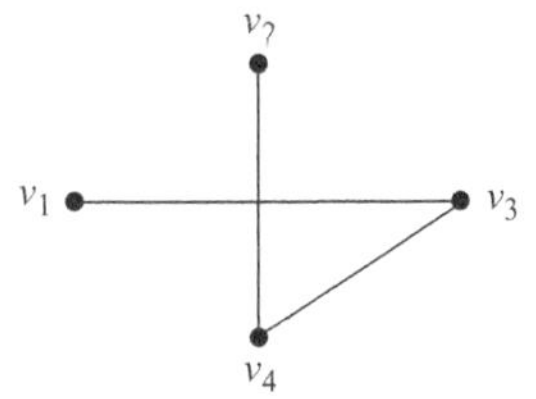

Figure 8.32

(b) 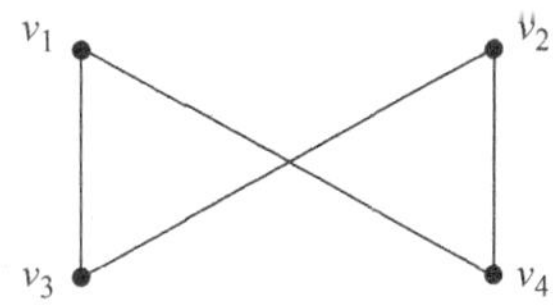

Figure 8.33

(c) Null graph.

EXAMPLE 8.9

Find the edge complement of the graph G shown in Figure 8.34.

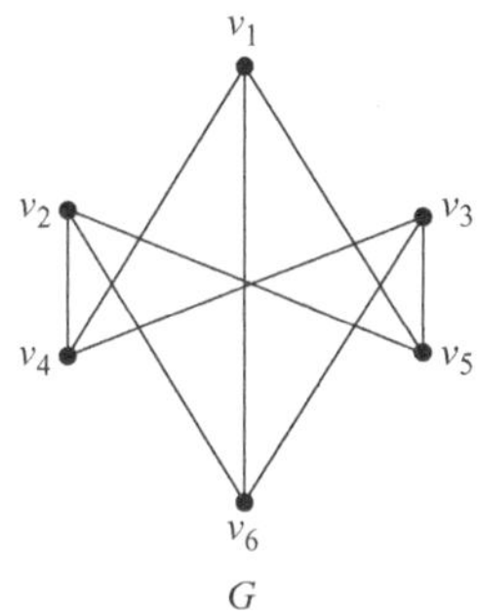

Figure 8.34

Solution.

The edge complement of G is the following graph G^c (Figure 8.35)

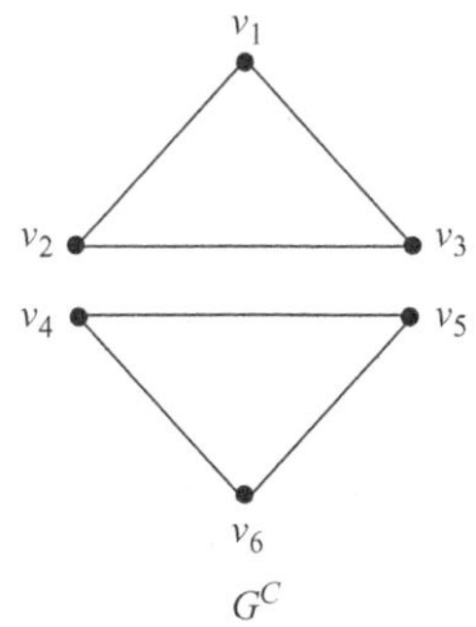

Figure 8.35

Definition 8.32

If a new vertex v is joined to each of the pre-existing vertices of a graph G, then the resulting graph is called the **join of G and v** or the **suspension of G from v**. It is denoted by $G+v$.

Thus, a graph obtained by joining a new vertex v to each of the vertices of a given graph G is called the **join of G and v** or the **suspension of G from v**.

For example, consider the graph given below (Figure 8.36)

Figure 8.36

Let v be a vertex. Then the graph shown in the Figure 8.37 is the join of G to v.

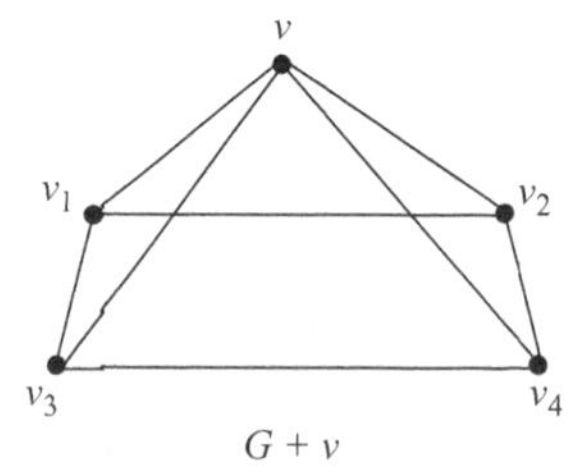

Figure 8.37

8.4 ISOMORPHISMS OF GRAPHS

We know that shape or length of an edge and its position in space are not part of specification of a graph. For example, the Figures 8.38 represent the same graph.

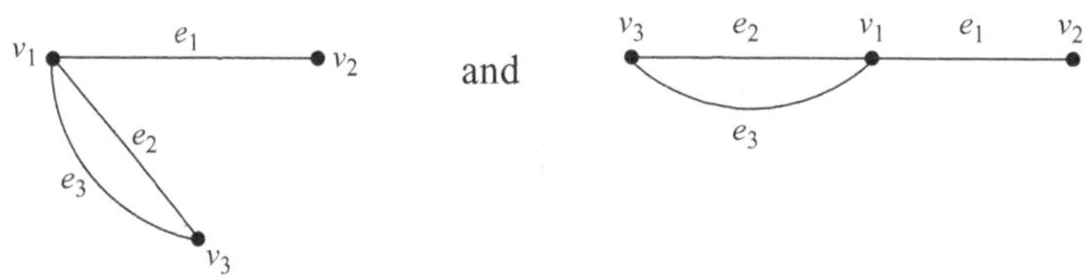

Figure 8.38

Definition 8.33

Let G and H be graphs with vertex sets $V(G)$ and $V(H)$ and edge sets $E(G)$ and $E(H)$, respectively. Then G **is said to isomorphic to** H if there exist one-to-one correspondences g: $V(G) \to V(H)$ and h: $E(G) \to E(H)$ such that for all $v \in V(G)$ and $e \in E(G)$,

$$v \text{ is an endpoint of } e \Leftrightarrow g(v) \text{ is an endpoint of } h(e).$$

Definition 8.34

The property of mapping endpoints to endpoints is called **preserving incidence** or **the continuity rule** for graph mappings.

As a consequence of this property, a self-loop must map to a self-loop.

Thus, two isomorphic graphs are same except for the labelling of their vertices and edges.

For example, the graphs shown in the Figure 8.39 are isomorphic.

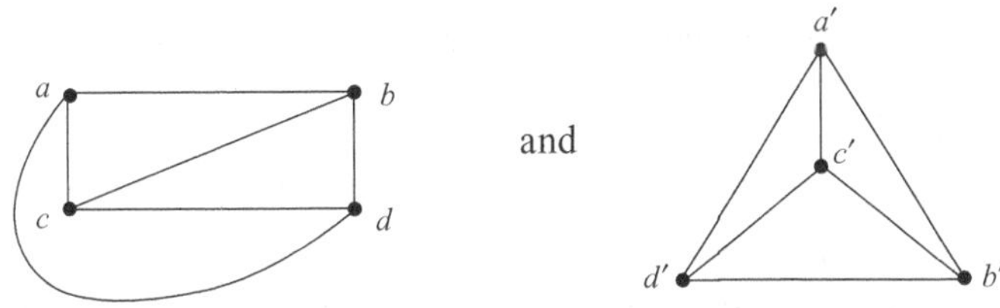

Figure 8.39

Similarly, the hypercube graph Q_3 and circular ladder CL_4 shown in Figure 8.40 are isomorphic.

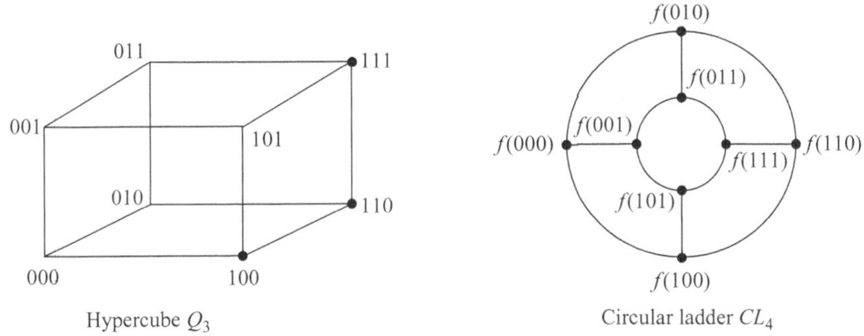

Figure 8.40

EXAMPLE 8.10

Show that the graphs shown in Figure 8.41 are isomorphic.

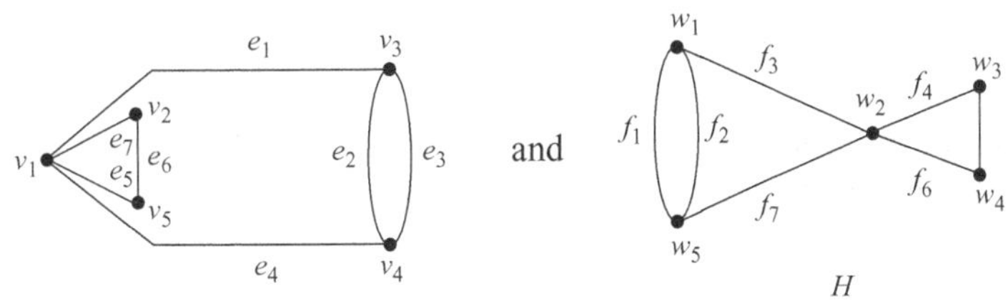

Figure 8.41

Solution.

To solve this problem, we have to find g: $V(G) \to V(H)$ and h: $E(G) \to E(H)$ such that for all $v \in V(G)$ and $e \in E(G)$,

$$v \text{ is an endpoint of } e \Leftrightarrow g(v) \text{ is an endpoint of } h(e).$$

Since e_2 and e_3 are parallel (have the same endpoints), $h(e_2)$ and $h(e_3)$ must also be parallel. Thus we have

$$h(e_2)=f_1 \text{ and } h(e_3)=f_2 \quad \text{or} \quad h(e_2)=f_2 \text{ and } h(e_3)=f_1.$$

Also the endpoints of e_2 and e_3 must correspond to the endpoints of f_1 and f_2 and so

$$g(v_3)=w_1 \text{ and } g(v_4)=w_5 \quad \text{or} \quad g(v_3)=w_5 \text{ and } g(v_4)=w_1.$$

Further, we note that v_1 is the endpoint of four distinct edges e_1, e_7, e_5 and e_4 and so $g(v_1)$ should be the endpoint of four distinct edges. We observe that w_2 is the vertex having four edges and so $g(v_1)=w_2$. If $g(v_3)=w_1$, then since v_1 and v_3 are endpoints of e_1 in G, $g(v_1)=w_2$ and $g(v_3)=w_1$ must be endpoints of $h(e_1)$ in H. This implies that $h(e_1)=f_3$.

Continuing this way, we can find g and h to define the isomorphism between G and H.

One such pair of functions (of course there exist several) is shown in Figure 8.42.

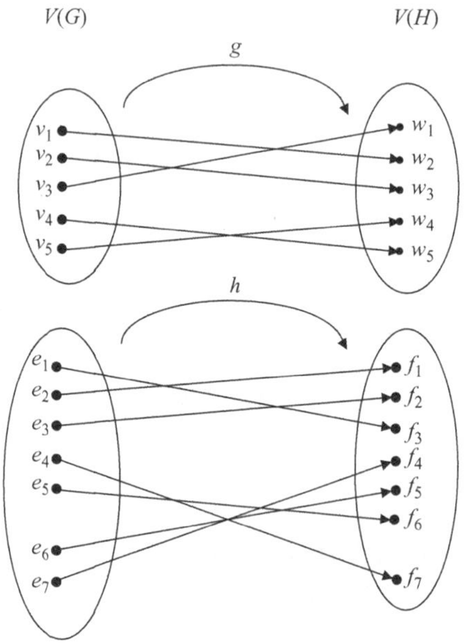

Figure 8.42

Remark 8.2 Each of the following properties of a graph is invariant under graph isomorphism, where n, m and k are all non-negative integers:

1. Has n vertices
2. Has m edges
3. Has a vertex of degree k
4. Has m vertices of degree k
5. Has a circuit of length k
6. Has a simple circuit of length k
7. Has m simple circuits of length k,
8. Is connected
9. Has an Euler circuit
10. Has a Hamiltonian circuit

} to be studied later on in this chapter (items 5–10)

EXAMPLE 8.11

Examine for isomorphism

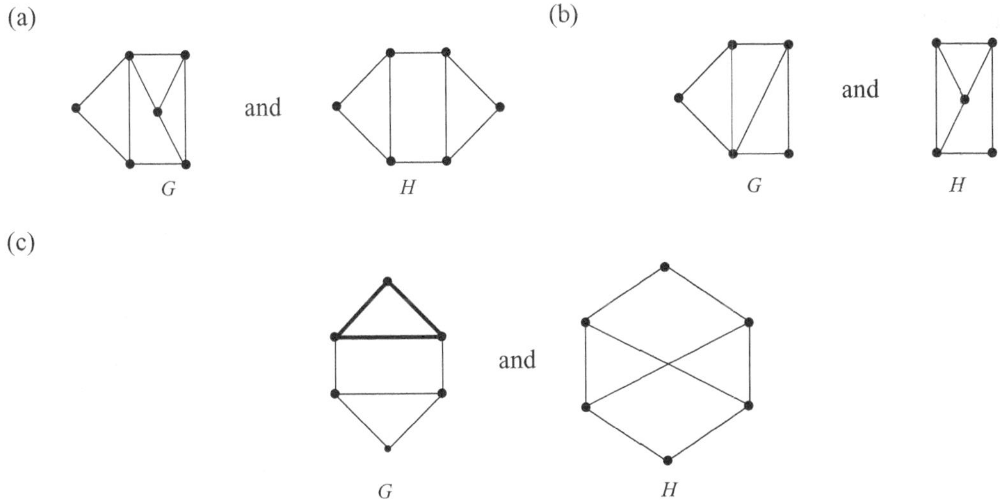

Figure 8.43

Solution.

We note in Figure 8.43 that

(a) G has nine edges whereas H has only eight. Hence, G is not isomorphic to H.
(b) G has a vertex v of degree 4, whereas H has no vertex of degree 4. Hence, G is not isomorphic to H
(c) G has a simple circuit of length 3 (**marked dark**) whereas H does not have. Hence, G is not isomorphic to H.

8.5 WALKS, PATHS AND CIRCUITS

Definition 8.35

In a graph G, **a walk from vertex** v_0 **to vertex** v_n is a finite alternating sequence $\{v_0, e_1, v_1, e_2, \ldots, v_{n-1}, e_n, v_n\}$ of vertices and edges such that v_{i-1} and v_i are the endpoints of e_i.

The **trivial walk** from a vertex v to v consists of the single vertex v.

Definition 8.36

In a graph G, a **path** from the vertex v_0 to the vertex v_n is a walk from v_0 to v_n that does not contain a repeated edge.

Thus a **path** from v_0 to v_n is a walk of the form

$$\{v_0, e_1, v_1, e_2, v_2, \ldots, v_{n-1}, e_n, v_n\},$$

where all the edges e_i are distinct.

Definition 8.37

In a graph, a **simple path** from v_0 to v_n is a path that does not contain a repeated vertex.

Thus a simple path is a walk of the form

$$\{v_0, e_1, v_1, e_2, v_2, \ldots, v_{i-1}, e_n, v_n\},$$

where all the e_i and all the v_i are distinct.

Definition 8.38

A walk in a graph G that starts and ends at the same vertex is called a **closed walk**.

Definition 8.39

A closed walk that does not contain a repeated edge is called a **circuit**.

Thus, a closed path is called a circuit (or a **cycle**) if it is a walk of the form

$$\{v_0, e_1, v_1, e_2, v_2, \ldots, v_{n-1}, e_n, v_n\},$$

where $v_0 = v_n$ and all the e_i are distinct.

Definition 8.40

A **simple circuit** is a circuit that does not have any other repeated vertex except the first and the last.

Thus, a simple circuit is a walk of the form

$$\{v_0, e_1, v_1, e_2, \ldots, v_{n-1}, e_n, v_n\},$$

where all the e_i are distinct and all the v_j are distinct except that $v_0 = v_n$.

EXAMPLE 8.12

In the graph given in the Figure 8.44, determine whether the following walks are paths, simple paths, closed walks, circuits, simple circuits or are just walks.

1. $v_1\, e_2\, v_2\, e_3\, v_3\, e_4\, v_4\, e_5\, v_2\, e_2\, v_1\, e_1\, v_0$,
2. $v_5\, v_4\, v_2\, v_1$,
3. $v_2\, v_3\, v_4\, v_5\, v_2$,
4. $v_4\, v_2\, v_3\, v_4\, v_5\, v_2\, v_4$,
5. $v_2\, v_1\, v_5\, v_2\, v_3\, v_4\, v_2$,
6. $v_0\, v_5\, v_2\, v_3\, v_4\, v_2\, v_1$.

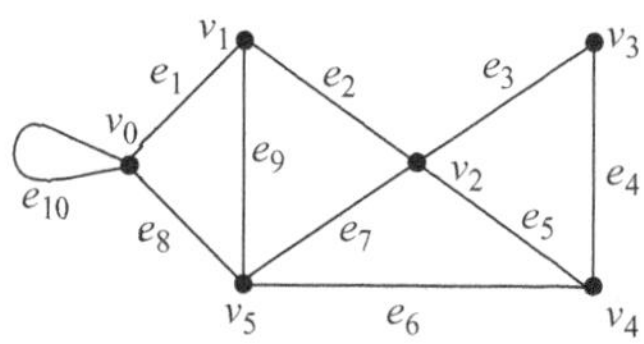

Figure 8.44

Solution.

1. It is **just a walk** because it has repeated edges and repeated vertices
2. It is a **simple path**, because here no vertex is repeated and no edge is repeated.
3. It is **closed walk** in which no edge is repeated and hence it is a **circuit**. It starts and ends at the same vertex, and therefore is a **simple circuit**.
4. It is a closed walk in which no edge is repeated but vertices v_2 and v_4 are repeated. Hence it is a **circuit**.
5. It is a closed walk in which no edge is repeated. Hence it is a circuit. Only one vertex is repeated twice, hence it is **not a simple circuit**.
6. It is a walk in which no edge is repeated but the vertex v_2 is repeated. Hence it is a **path**.

EXAMPLE 8.13

Consider the graph shown in Figure 8.45.

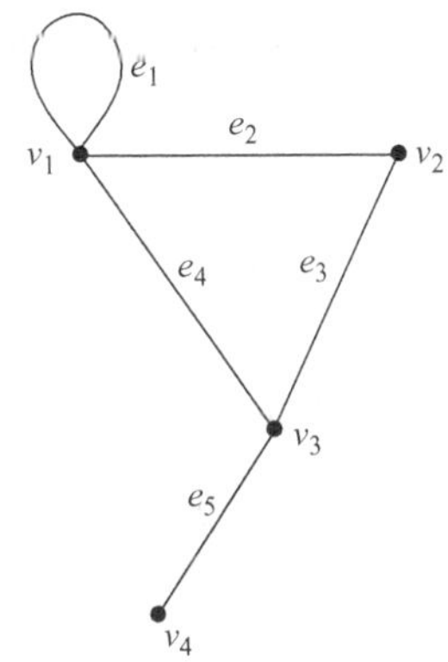

Figure 8.45

We note that e_3, e_5 is a path. The walk e_1, e_2, e_3, e_5 is a path but it is not a simple path because the vertex v_1 is repeated (e_1 being a self-loop). The walk e_2, e_3, e_4 is a circuit. The walk e_2, e_3, e_4, e_1 is a circuit but it is not simple circuit because vertex v_1 repeats twice (or we may write that v_1 is met twice).

Definition 8.41

In a graph the number of edges in the path $\{v_0, e_1, v_1, e_2, \ldots, e_n, v_n\}$ from v_0 to v_n is called the **length of the path**.

Definition 8.42

A cycle with k-edges is called a **k-cycle** or **cycle of length** k.

For example, loop is a cycle of length 1. On the other hand, a pair of parallel edges e_1 and e_2, shown in the Figure 8.46, is a cycle of length 2

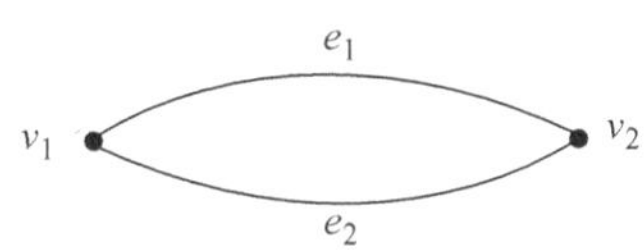

Figure 8.46

EXAMPLE 8.14

Consider the graph shown in Figure 8.47.

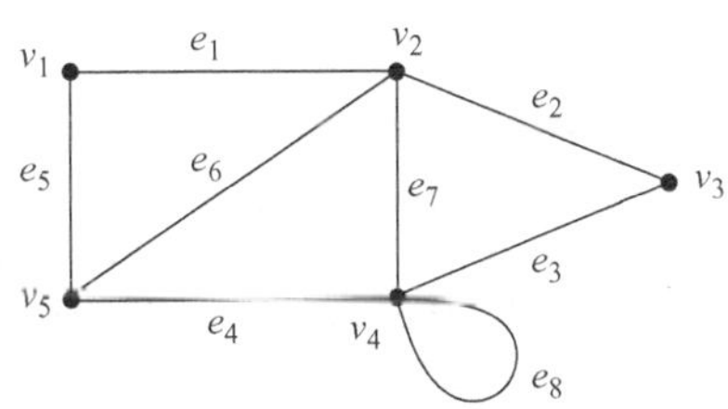

Figure 8.47

In this graph, we observe that

(i) The path $v_4\, e_8\, v_4$ is a cycle of length 1.
(ii) The path $v_5\, e_4\, v_4\, e_8\, v_4\, e_3\, v_3$ from v_5 to v_3 (Figure 8.48) is a path of length 3.

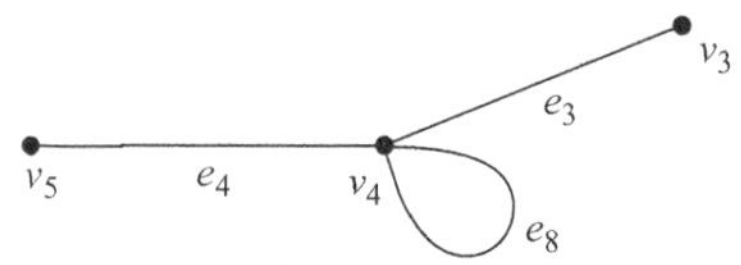

Figure 8.48

(iii) The path $v_1\, e_1\, v_2\, e_6\, v_5\, e_5\, v_1$ (Figure 8.49) is a cycle of length 3, since it consists of three edges.

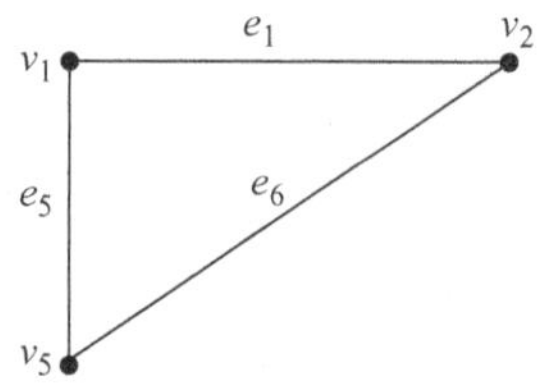

Figure 8.49

Definition 8.43

A graph is said to be **acyclic** if it contains no cycle.

For example, the graphs shown in Figure 8.50 are acyclic.

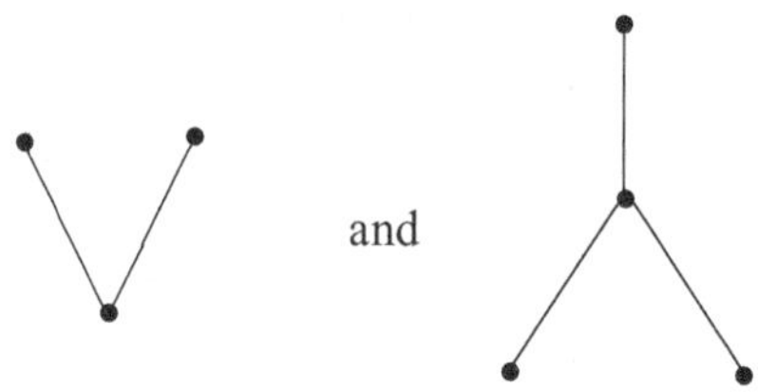

Figure 8.50

Theorem 8.3

If there is a path from vertex v_1 to v_2 in a graph with n vertices, then there does not exist a simple path of more than $n-1$ edges from vertex v_1 to v_2.

Proof. Suppose there is a path from v_1 to v_2. Let $v_1, \ldots, v_i, \ldots, v_2$ be the sequence of vertices which the path meets between the vertices v_1 and v_2. Let there be m edges in the simple path. Then there will be $m+1$ vertices in the sequence. Therefore if $m>n-1$, then there will be more than n vertices in the sequence. But the graph is with n vertices. Therefore some vertex, say v_k, appears more than once in the sequence. So the sequence of vertices shall be $v_1, \ldots, v_i, \ldots, v_k, \ldots, v_k, \ldots, v_2$. Deleting the edges in the path that lead v_k back to v_k we have a path from v_1 to v_2 that has fewer edges than the original one. This argument is repeated until we get a path that has $n-1$ or fewer edges.

Definition 8.44

Two vertices v_1 and v_2 of a graph G are said to be **connected** if and only if there is a walk from v_1 to v_2.

Definition 8.45

A graph G is said to be **connected** if and only if given any two vertices v_1 and v_2 in G, there is a walk from v_1 to v_2.

Thus, a graph G is connected if there exists a walk between every two vertices in the graph.

Definition 8.46

A graph which is not connected is called **disconnected graph**.

EXAMPLE 8.15

Which of the graphs shown below are connected?

(a)

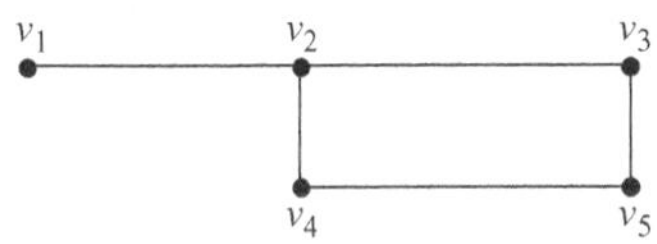

(b)

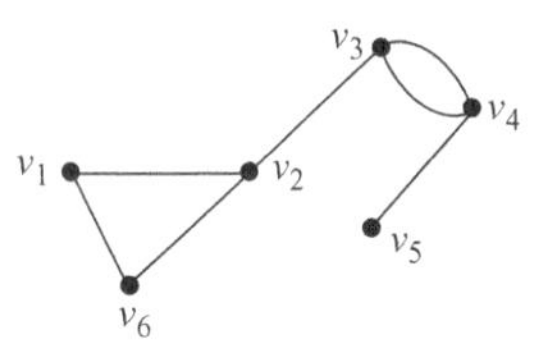

(c)

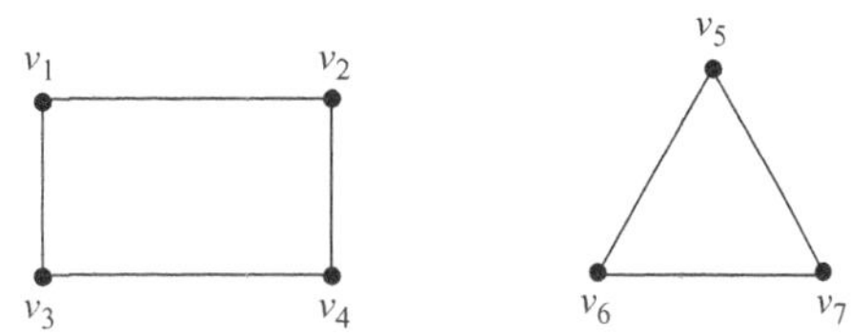

(d)

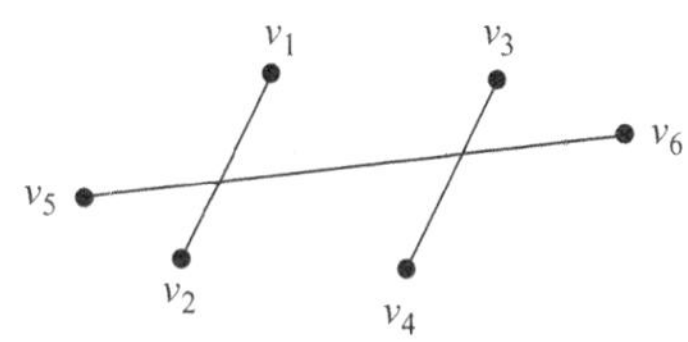

Figure 8.51

Solution.

In Figure 8.51, we observe that

(a) Since every two vertices in the given graph are connected by a path, therefore the given graph is connected.
(b) This graph is also connected.
(c) The graph in (c) is disconnected. It has two connected components.
(d) In this graph, the edge (v_5, v_6) cross the edges (v_1, v_2) and (v_3, v_4) at points which are not vertices. Therefore the graph can be redrawn as shown in Figure 8.52.

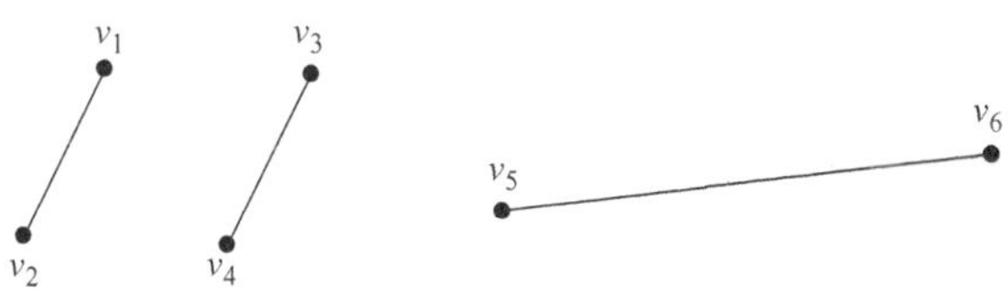

Figure 8.52

There is no path from v_2 to v_4, etc. Hence the given graph is disconnected and has three connected components.

EXAMPLE 8.16

Which of the graphs shown in Figure 8.53 are connected?

(a) (b)

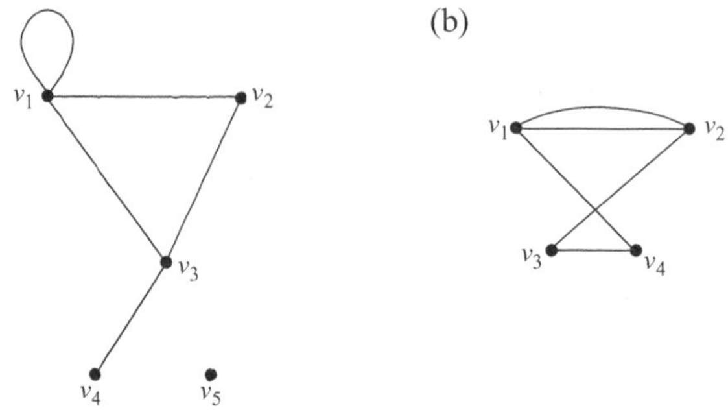

Figure 8.53

Solution.

Graph (a) is not connected as there is no walk from any of v_1, v_2, v_3, v_4 to the vertex v_5.

The graph (b) is clearly connected.

Definition 8.47

If a graph G is disconnected, then the various connected pieces of G are called the **connected components of the graph**.

EXAMPLE 8.17

Find the connected components of the graphs given below (Figure 8.54)

(a)

(b)

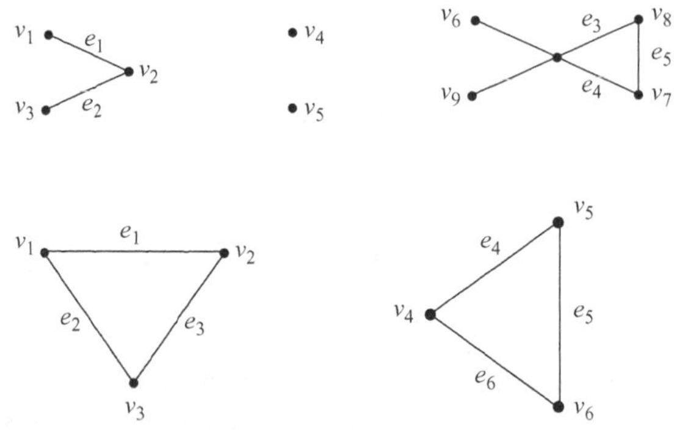

Figure 8.54

Solution.

(a) The graph in (a) has four connected components (Figures 8.55 and 8.56)

II_1: 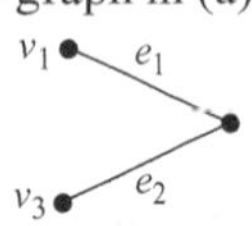

with vertex set $\{v_1, v_2, v_3)$ and edge set $\{e_1, e_2)$

Figure 8.55

H_2: with vertex set $\{v_4\}$ and edge set ϕ
H_3: with vertex set $\{v_5\}$ and edge set ϕ

H_4: 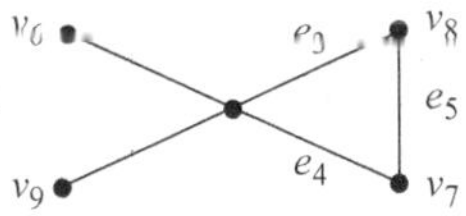 with vertex set $\{v_6, v_7, v_8, v_9\}$ and edge set $\{e_3, e_4, e_5\}$

Figure 8.56

(b) The graph in (b) is disconnected and have two connected components (Figures 8.57(a), 8.57(b))

H_1:

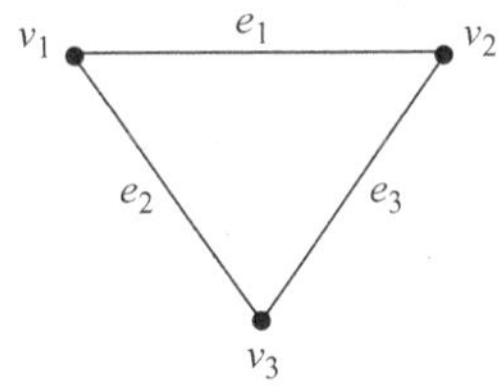

Figure 8.57(a)

with vertex set $\{v_1, v_2, v_3\}$ and edge set $\{e_1, e_2, e_3\}$

H_2:

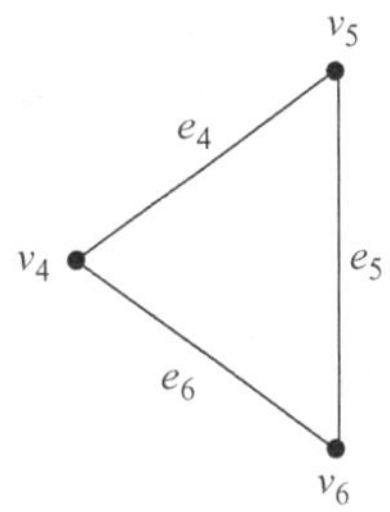

Figure 8.57(b)

with vertex set $\{v_4, v_5, v_6\}$ and edge set $\{e_4, e_5, e_6\}$.

EXAMPLE 8.18

Find the number of connected components in the graph shown below (Figure 8.58)

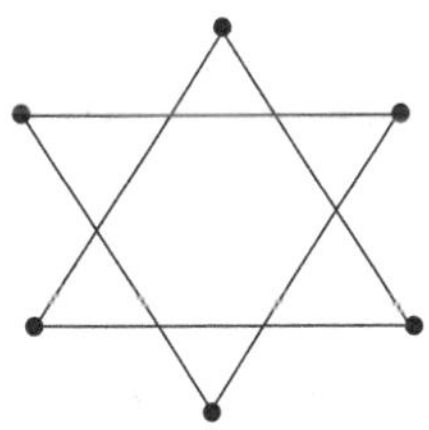

Figure 8.58

Solution.

The connected components are shown in Figure 8.59.

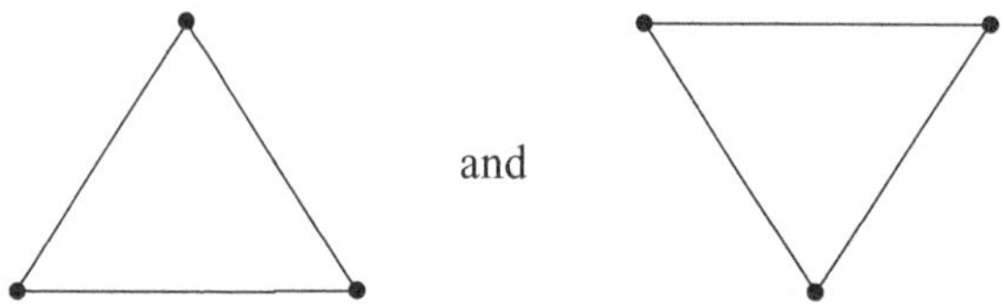

Figure 8.59

Remark 8.3 If a connected component has n vertices, then degree of any vertex cannot exceed $n-1$.

8.6 EULERIAN PATHS AND CIRCUITS

Definition 8.48

A path in a graph G is called an **Euler path** if it includes **every edge exactly once**.

Definition 8.49

A circuit in a graph G is called an **Euler circuit** if it includes every edge exactly once. Thus, an Euler circuit (Eulerian trail) for a graph G is a sequence of adjacent vertices and edges in G that starts and ends at the same vertex, uses **every vertex of G at least once**, and uses **every edge of G exactly once**.

Definition 8.50

A graph is called **Eulerian graph** if there exists an Euler circuit for that graph.

EXAMPLE 8.19

Find which of the following are Euler paths and Euler circuits in the graph given below (Figure 8.60)

(a) $v_1\, e_1\, v_2\, e_2\, v_3\, e_3\, v_4\, e_4\, v_5\, e_5\, v_2$

(b) $v_1\, e_1\, v_2\, e_2\, v_3\, e_3\, v_4\, e_4\, v_5\, e_5\, v_2\, e_1\, v_1$

(c) $v_2\, e_2\, v_3\, e_3\, v_4\, e_4\, v_5\, e_5\, v_2$

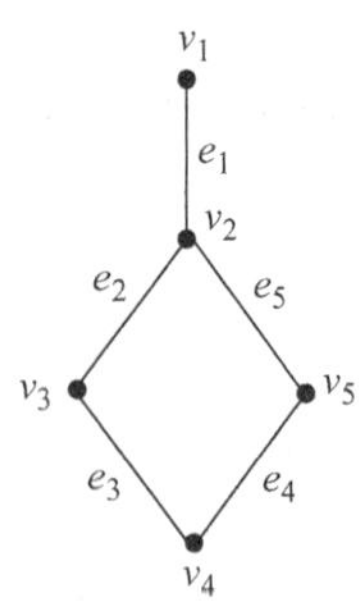

Figure 8.60

Solution.

(a) The walk in (a) is an Euler path but it is not Euler circuit because it is not closed.
(b) It is not an Euler circuit because the edge e_1 is covered twice.
(c) It is neither an Euler path nor an Euler circuit because the vertex v_1 of G has not been used.

Theorem 8.4

If a graph has an Euler circuit, then every vertex of the graph has even degree.

Proof. Let G be a graph which has an Euler circuit. Let v be a vertex of G. We shall show that degree of v is even. By definition, Euler circuit contains every edge of graph G. Therefore the Euler circuit contains all edges incident on v. We start a journey beginning in the middle of one of the edges adjacent to the start of Euler circuit and continue around the Euler circuit to end in the middle of the starting edge. Since Euler circuit uses every edge exactly once, the edges incident on v occur in entry/exist pair and hence the degree of v is a multiple of 2. Therefore, the degree of v is even. This completes the proof of the theorem.

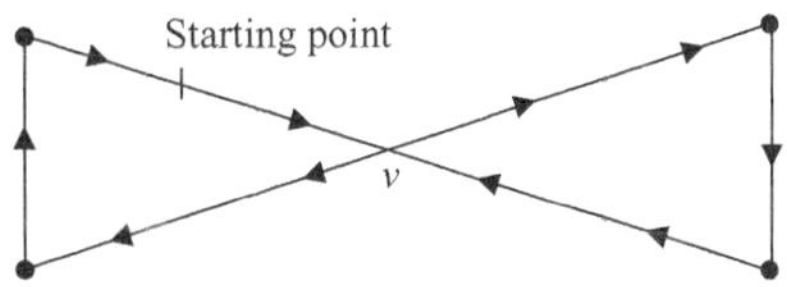

Figure 8.61

We know that ***contrapositive of a conditional statement is logically equivalent to the statement***. Thus, Theorem 8.4 is equivalent to the following:

Theorem 8.5

If a vertex of a graph is not of even degree, then it does not have an Euler circuit.
Thus,
If some vertex of a graph has odd degree, then that graph does not have an Euler circuit.

EXAMPLE 8.20

Show that the graphs shown in Figure 8.62 do not have Euler circuits.

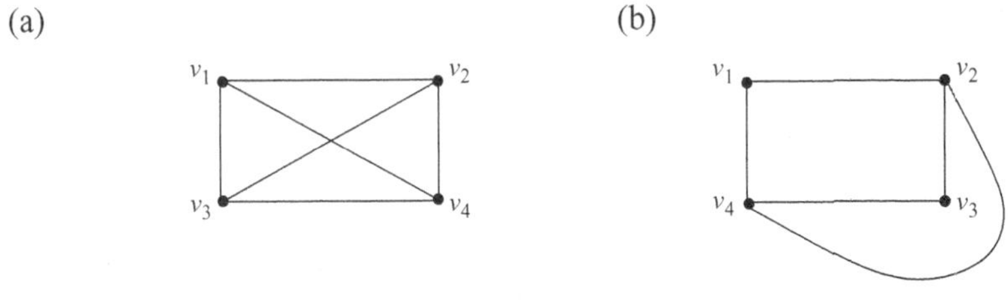

Figure 8.62

Solution.

In graph (a), degree of each vertex is 3. Hence this **does not** have an Euler circuit.

In graph (b), we have

$$\deg(v_2)=3 \quad \text{and} \quad \deg(v_4)=3.$$

Since there are vertices of odd degree in the given graph, therefore it **does not** have an Euler circuit.

Remark 8.4 The converse of Theorem 8.4 is not true. There exist graphs in which every vertex has even degree but the Euler circuits do not exist.

For example,

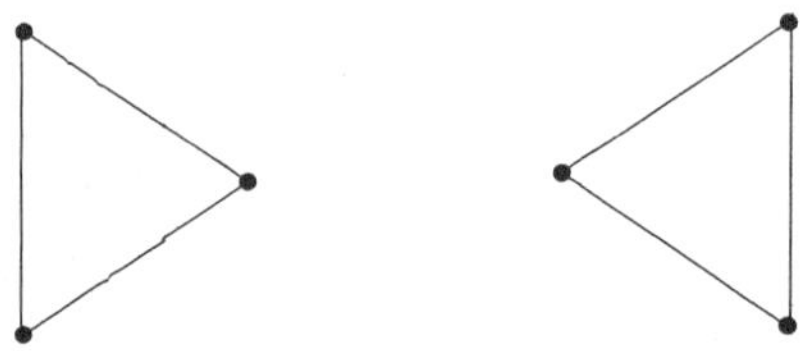

and

Figure 8.63

are graphs (Figure 8.63) in which each vertex has degree 2 but these graphs do not have Euler circuits since there is no path which uses each vertex at least once.

Theorem 8.6

If G is a connected graph and every vertex of G has even degree, then G has an Euler circuit.

Proof. Let every vertex of a connected graph G has even degree. If G consists of a single vertex v, the trivial walk from v to v is an Euler circuit. So suppose that G consists of more than one vertices. We start from any vertex v of G. Since the degree of each vertex of G is even, if we reach each vertex other than v by travelling on one edge, the same vertex can be reached by travelling on another previously unused edge. Thus a sequence of distinct adjacent edges can be produced indefinitely as long as v is not reached. Since the number of edges of the graph is finite (by definition of graph), the sequence of distinct edges will terminate. Thus the sequence must return to the starting vertex. We thus obtain a sequence of adjacent vertices and edges starting and ending at v without repeating any edge. Thus we get a circuit C.

If C contains every edge and vertex of G, then C is an Euler circuit.

If C does not contain every edge and vertex of G, remove all edges of C from G and also any vertices that become isolated when the edges of C are removed. Let the resulting subgraph be G'. We note that when we removed edges of C, an even number of edges from each vertex have been removed. Thus degree of each remaining vertex remains even.

Further, since G is connected, there must be at least one vertex common to both C and G'. Let it be w (in fact there are two such vertices). Pick any sequence of adjacent vertices and edges of G' starting and ending at w without repeating an edge. Let the resulting circuit be C'.

Join C and C' together to create a new circuit C''. Now, we observe that if we start from v and follow C all the way to reach w and then follow C' all the way to reach back to w. Then continuing travelling along the untravelled edges of C, we reach v.

If C'' contains every edge and vertex of C, then C'' is an Euler circuit. If not, then we again repeat our process. Since the graph is finite, the process must terminate.

The process followed has been described in the graph G shown below (Figure 8.64).

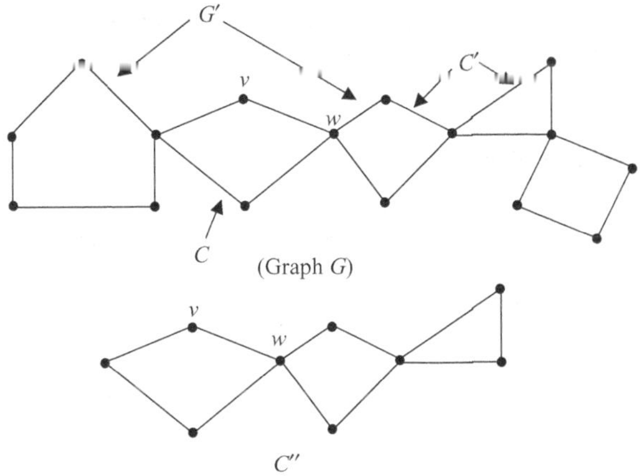

Figure 8.64

Theorems 8.4 and 8.6 taken together imply:

Theorem 8.7 (Euler's Theorem)

A finite connected graph G has an Euler circuit if and only if every vertex of of G has even degree. Thus, finite connected graph is Eulerian if and only if each vertex has even degree.

Theorem 8.8

If a graph G has more than two vertices of odd degree, then there can be no Euler path in G.

Proof. Let v_1, v_2 and v_3 be vertices of odd degree. Since each of these vertices had odd degree, any possible Euler path must leave (arrive at) each of v_1, v_2, v_3 with no way to return (or leave). One vertex of these three vertices may be the beginning of Euler path and another the end, but this leaves the third vertex at one end of an untravelled edge. Thus, there is no Euler path.

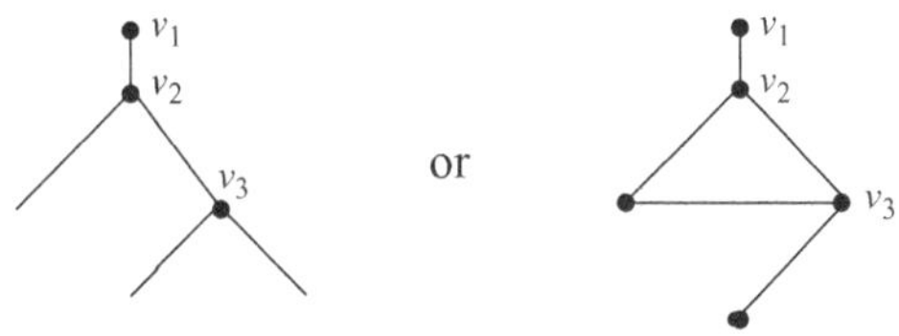

(Graphs having more than two vertices of odd degree).

Figure 8.65

Theorem 8.9

If G is a connected graph and has exactly two vertices of odd degree, then there is an Euler path in G. Further, any Euler path in G must begin at one vertex of odd degree and end at the other.

Proof. Let u and v be two vertices of odd degree in the given connected graph G (see Figure 8.66)

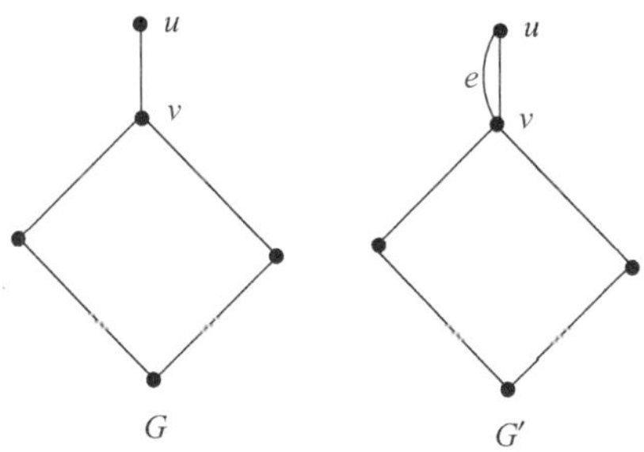

Figure 8.66

If we add the edge e to G, we get a connected graph G' all of whose vertices have even degree. Hence there will be an Euler circuit in G'. If we omit e from Euler circuit, we get an Euler path beginning at u (or v) and ending at v (or u).

EXAMPLE 8.21

Has the graph given in Figure 8.67 an Eulerian path?

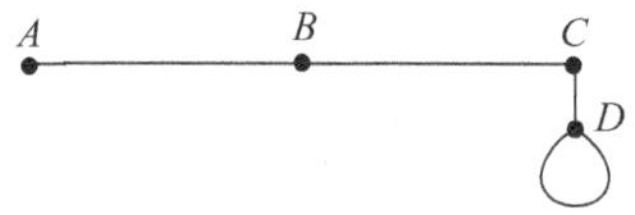

Figure 8.67

Solution.

In the given graph,

$$\deg(A)=1, \quad \deg(B)=2,$$
$$\deg(C)=2, \quad \deg(D)=3.$$

Thus the given connected graph has exactly two vertices of odd degree. Hence, it has an Eulerian path.

If it starts from A (vertex of odd degree), then it ends at D (vertex of odd degree). If it starts from D (vertex of odd degree), then it ends at A (vertex of odd degree).

But, on the other hand, let the graph be as shown below (Figure 8.68)

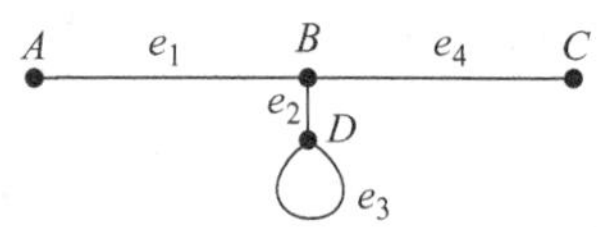

Figure 8.68

Then $\deg(A)=1$, $\deg(B)=3$, $\deg(C)=1$, degree of $D=3$ and so we have four vertices of odd degree. Hence it does not have an Euler path.

EXAMPLE 8.22

Does the graph given in Figure 8.69 possess an Euler circuit?

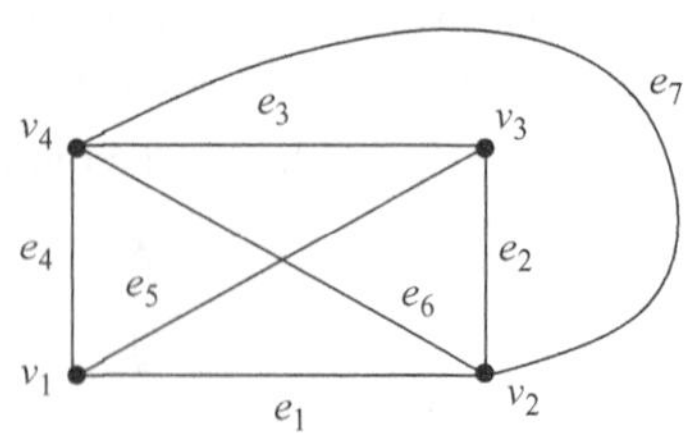

Figure 8.69

Solution.

The given graph is connected. Further

$$\deg(v_1)=3, \quad \deg(v_2)=4,$$
$$\deg(v_3)=3, \quad \deg(v_4)=4.$$

Since this connected graph has vertices with odd degree, it cannot have Euler circuit. But this graph has Euler path, since it has exactly two vertices of odd degree. For example, $v_3\, e_2\, v_2\, e_7\, v_4\, e_6\, v_2\, e_1\, v_1\, e_4\, v_4\, e_3\, v_3\, e_5\, v_1$ is an Euler path.

EXAMPLE 8.23

Consider the graph given below (Figure 8.70):

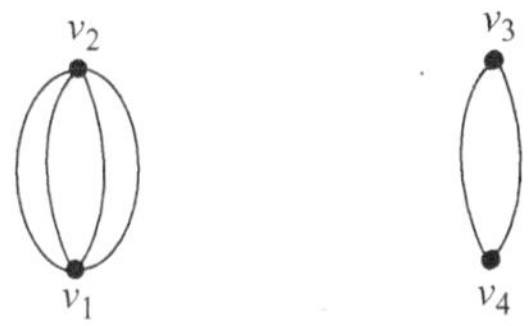

Figure 8.70

Here, $\deg(v_1)=4$, $\deg(v_2)=4$, $\deg(v_3)=2$, $\deg(v_4)=2$. Thus degree of each vertex is even. But the graph is not Eulerian since it is **not connected**.

EXAMPLE 8.24

Is it possible to trace the graph in Figure 8.71 without lifting the pencil?

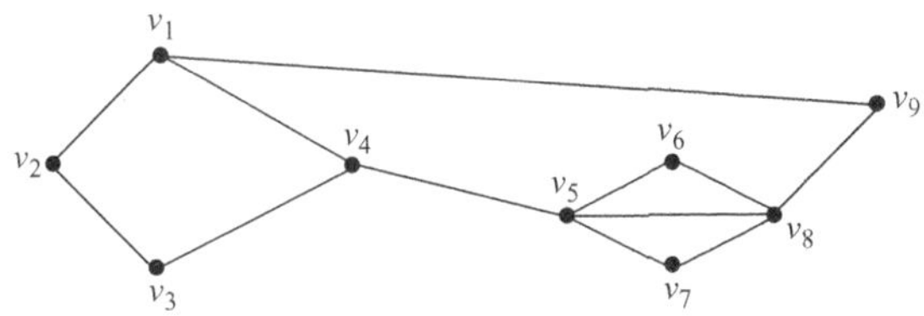

Figure 8.71

Solution.

The problem is equivalent to say that "Is it possible for this graph to have an Eulerian circuit"? We observe that

$$\deg(v_1)=3,\ \deg(v_2)=2,\ \deg(v_3)=2,\ \deg(v_4)=3,\ \deg(v_5)=4,$$
$$\deg(v_6)=2,\ \deg(v_7)=2,\ \deg(v_8)=4,\ \deg(v_9)=2.$$

Since the graph contains vertex of odd degree, therefore it cannot have Euler's circuit and therefore can not be traced without lifting the pencil.

EXAMPLE 8.25

Find the Euler's circuit for the graph given below (Figure 8.72):

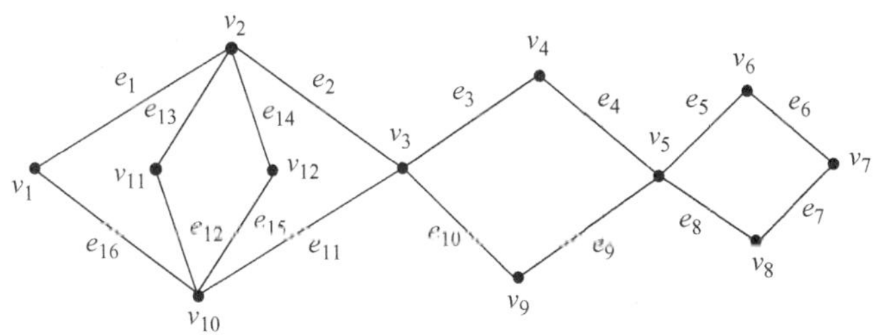

Figure 8.72

Solution.

The given connected graph has 12 vertices and degree of each vertex is even. Hence, by Euler's theorem, this has an Euler's circuit. For example, we observe that $v_1\ e_1\ v_2\ e_{13}\ v_{11}\ e_{12}\ v_{10}\ e_{15}\ v_{12}\ e_{14}\ v_2\ e_2\ v_3\ e_3\ v_4\ e_4\ v_5\ e_5\ v_6\ e_6\ v_7\ e_7\ v_8\ e_8\ e_9\ v_9\ e_{10}\ v_3\ e_{11}\ v_{10}\ e_{16}\ v_1$ is an Euler's circuit. In short we can represent it by $e_1, e_{13}, e_{12}, e_{15}, e_{14}, e_2, e_3, e_4, e_5, e_6, e_7, e_8, e_9, e_{10}, e_{11}, e_{16}$.

EXAMPLE 8.26 (The bridges of Konigsberg)

The graph theory began in 1736 when Leonhard Euler solved the problem of seven bridges on Pregel River in the town of Konigsberg in Prussia (now Kaliningrad in Russia). The two islands and seven bridges are shown in Figure 8.73.

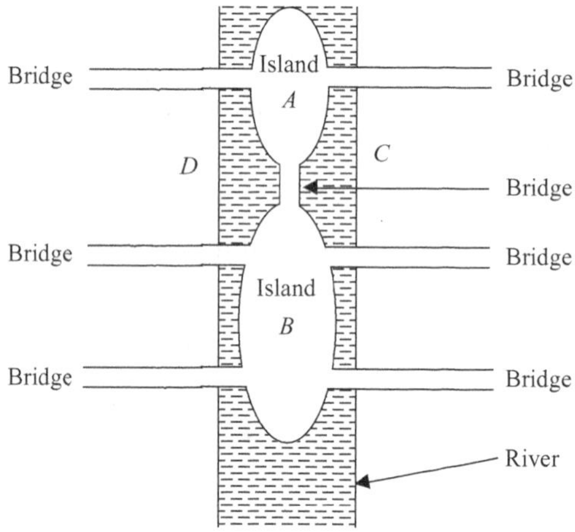

Figure 8.73

The people of Konigsberg posed the following question to famous Swiss Mathematician Leonhard Euler:

"Beginning anywhere and ending any where, can a person walk through the town of Konigsberg crossing all the seven bridges exactly once?"

Euler showed that such a walk is impossible. He replaced the islands A, B and the two sides (banks) C and D of the river by vertices and the bridges as edges of a graph. We note then that

$$\deg(A)=3, \quad \deg(B)=5,$$
$$\deg(C)=3, \quad \deg(D)=3.$$

Thus the graph of the problem is as shown in Figure 8.74.

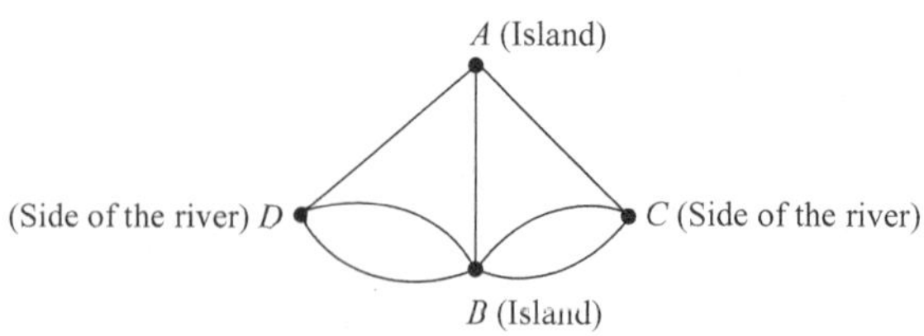

(Euler's graphical representation of seven bridges problem)

Figure 8.74

The problem then reduces to

"Is there any Euler's path in the above diagram?".

To find the answer, we note that there are more than two vertices having odd degree. Hence there exists no Euler path for this graph.

EXAMPLE 8.27

The floor plan shown in Figure 8.75 is for a house that is open for public viewing. Each room is connected to every room with which it has a common wall and to the outside along each wall. Is it possible to begin in a room or outside and take a walk which goes through each door exactly once?

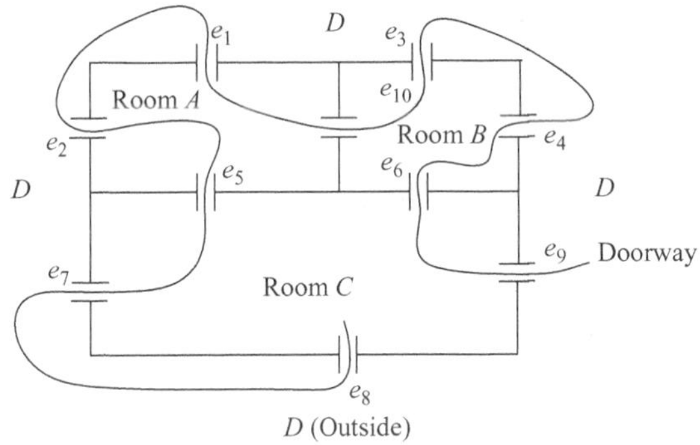

Figure 8.75

If each room and out side constitute a vertex and each door corresponds to an edge, then the floor plan converts into the graph. There are two edges e_1 and e_2 from A to D, two edges e_3 and e_4 from B to D, three edges e_7, e_8, e_9 from C to D, one edge e_{10} from A to B, one edge e_5 from A to C, one edge e_6 from B to C. Hence the graph is as shown below (Figure 8.76):

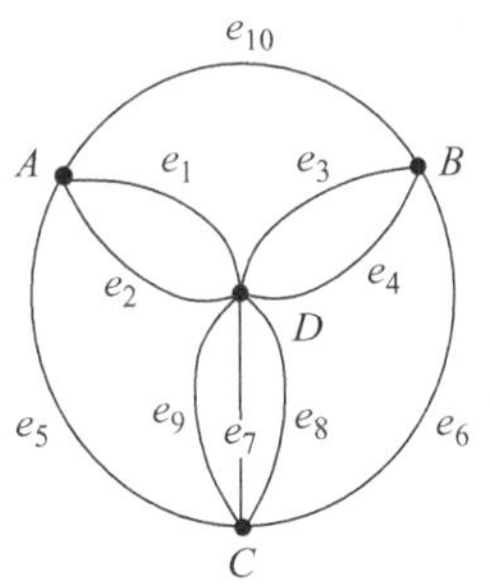

Figure 8.76

We note that

$$\deg(A)=\deg(B)=4, \quad \deg(C)=5, \quad \deg(D)=7.$$

Since the graph is connected and degree of exactly two vertices C and D is odd, there exists an Euler's path and that path should start from one vertex of odd degree and end to the other vertex of odd degree. Therefore, either the path will begin from C and end at D or it will begin from D and end at C. An Euler's path is shown (which begins from C).

Definition 8.51

An edge in a connected graph is called a **bridge** or a **cut edge** if deleting that edge creates a disconnected graph.

For example, consider the graph shown below (Figure 8.77):

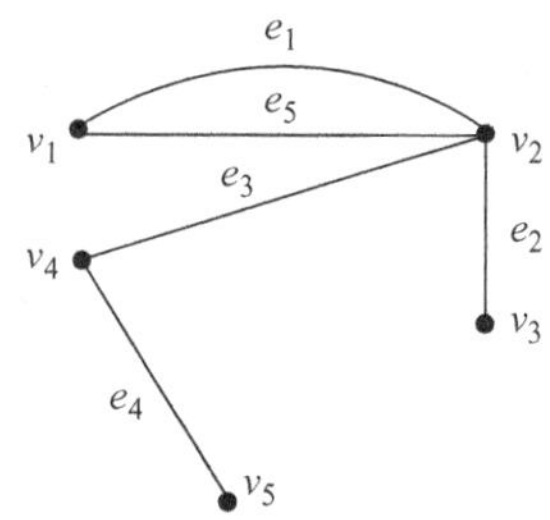

Figure 8.77

In this graph, if we remove the edge e_3, then the graph breaks into two connected components given below (Figure 8.78):

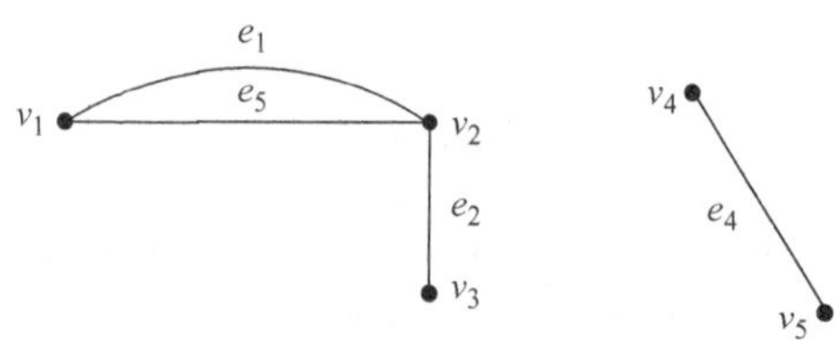

Figure 8.78

Hence the edge e_3 is a bridge in the given graph.

8.6.1 Methods for Finding Euler Circuit

Method 1. We know that if every vertex of a non-empty connected graph has even degree, then the graph has an Euler circuit. We shall make use of this result to find an Euler path in a given graph.

Consider the graph shown in Figure 8.79

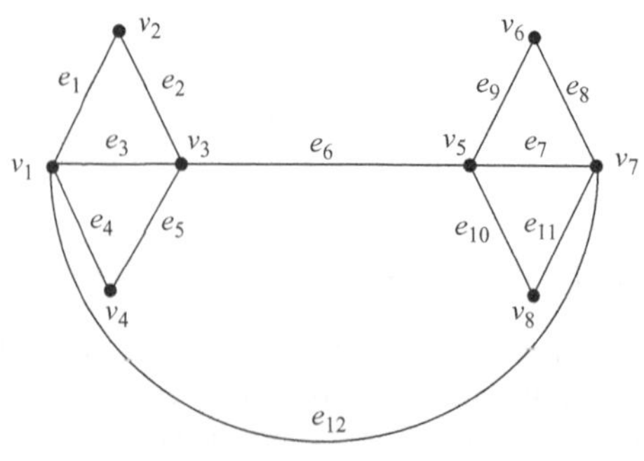

Figure 8.79

We note that

$$\deg(v_2)=\deg(v_4)=\deg(v_6)=\deg(v_8)=2,$$
$$\deg(v_1)=\deg(v_3)=\deg(v_5)=\deg(v_7)=4.$$

Hence all vertices have even degree. Also the given graph is connected. Hence the given graph has an Euler circuit. We start from the vertex v_1 and let C be

$$C: v_1\ v_2\ v_3\ v_1.$$

Then C is not an Euler circuit for the given graph but C intersects the rest of the graph at v_1 and v_3.

Let C' be

$$C': v_1 v_4\ v_3\ v_5\ v_7\ v_6\ v_5\ v_8\ v_7\ v_1.$$

(In case we start from v_3, then C' will be $v_3\ v_4\ v_1\ v_7\ v_6\ v_5\ v_7\ v_8\ v_5$).

Path C' into C and obtain

$$C'': v_1 v_2\ v_3\ v_1\ v_4\ v_3\ v_5\ v_7\ v_6\ v_5\ v_8\ v_7\ v_1,$$

that is,

$$C'': e_1 e_2\ e_3\ e_4\ e_5\ e_6\ e_7\ e_8\ e_9\ e_{10}\ e_{11}\ e_{12}.$$

(If we had started from v_2, then C'': $v_1 v_2\ v_3\ v_4\ v_1\ v_7\ v_6\ v_5\ v_7\ v_8\ v_5\ v_3\ v_1$ **or** $e_1 e_2\ e_5\ e_4\ e_{12}\ e_8\ e_9\ e_7\ e_{11}\ e_{10}\ e_6\ e_3$). In C'' all edges are covered exactly once. Also every vertex has been covered at least once. Hence C'' is an Euler circuit.

Method 2: Fleury's Algorithm: This algorithm is used to find Euler's circuit for a connected graph with no vertices of odd degree.

We shall illustrate the method with the help of the following example:

Let G be the connected graph as shown in the Figure 8.80

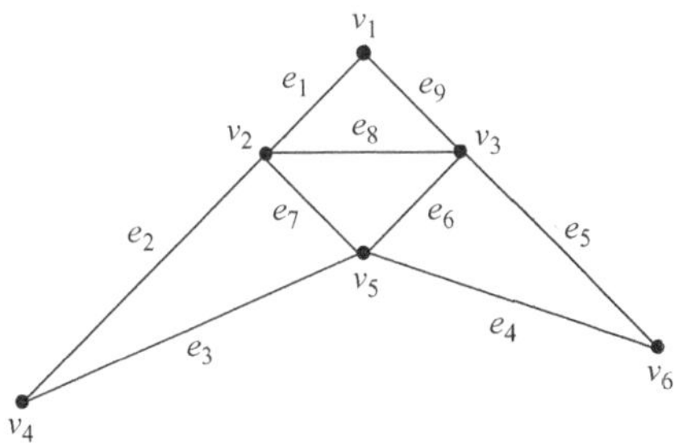

Figure 8.80

Step 1. We begin from any vertex, say v_1

Step 2. We then construct the following table:

Current Path	*Next Edge*	*Reason*
v_1	$\{v_1, v_2\}$	No edge from v_1 is a bridge, so choose any edge, say $\{v_1, v_2\}$
$v_1\ v_2$	$\{v_2, v_4\}$	(v_2, v_5) a bridge, so choose $\{v_2, v_4\}$
$v_1\ v_2\ v_4$	$\{v_4, v_5\}$	only one edge $\{v_4, v_5\}$from v_4 remains

(Continued)

Current Path	*Next Edge*	*Reason*
$v_1\ v_2\ v_4\ v_5$	$\{v_5, v_6\}$	Since $\{v_2, v_5\}$ and $\{v_5, v_3\}$ are bridges, choose $\{v_5, v_6\}$
$v_1\ v_2\ v_4\ v_5\ v_6$	$\{v_6, v_3\}$	only one edge $\{v_6, v_3\}$ remains
$v_1\ v_2\ v_4\ v_5\ v_6\ v_3$	$\{v_3, v_5\}$	Since $\{v_1, v_3\}$ is a bridge, choose either $\{v_3, v_2\}$or $\{v_3, v_5\}$
$v_1\ v_2\ v_4\ v_5\ v_6\ v_3\ v_5$	$\{v_5, v_2\}$	only one edge $\{v_5, v_2\}$ remains
$v_1\ v_2\ v_4\ v_5\ v_6\ v_3\ v_5 v_2$	$\{v_2, v_3\}$	only one edge $\{v_2, v_3\}$ remains
$v_1\ v_2\ v_4\ v_5\ v_6\ v_3\ v_5\ v_2\ v_3$	$\{v_3, v_1\}$	only one edge $\{v_3, v_1\}$ remains

Hence one possible Euler circuit is

$$C: v_1\ v_2\ v_4\ v_5\ v_6\ v_3\ v_5\ v_2\ v_3\ v_1.$$

In term of edges, this Euler circuit can be expressed as

$$e_1\ e_2\ e_3\ e_4\ e_5\ e_6\ e_7\ e_8\ e_9.$$

8.7 HAMILTONIAN CIRCUITS

Definition 8.52

A **Hamiltonian path** for a graph G is a sequence of adjacent vertices and distinct edges in which every vertex of G appears exactly once.

Definition 8.53

A **Hamiltonian circuit** for a graph G is a sequence of adjacent vertices and distinct edges in which every vertex of G appears exactly once, except for the first and the last which are the same.

Definition 8.54

A graph is called **Hamiltonian** if it admits a Hamiltonian circuit.

EXAMPLE 8.28

The wooden graph shown in Figure 8.81 and constructed by William Hamilton in the shape of a regular dodecahedron is a Hamiltonian circuit.

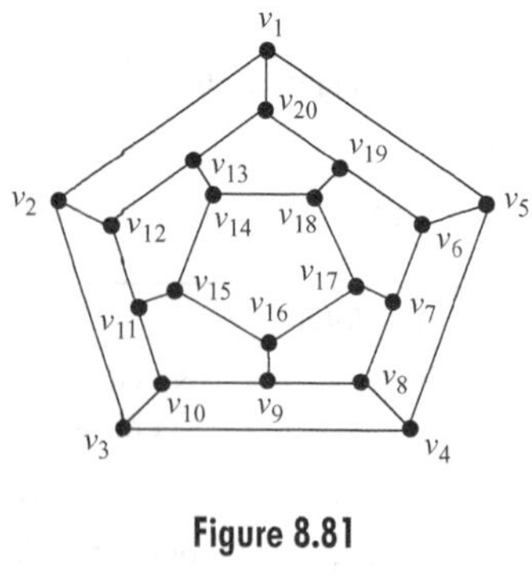

Figure 8.81

The Hamilton circuit is $v_1\ v_2\ v_3 \ldots v_{18}\ v_{19}\ v_{20}\ v_1$.

EXAMPLE 8.29

A complete graph K_n has a Hamiltonian circuit. In particular, the graphs shown in Figure 8.82 are Hamiltonian.

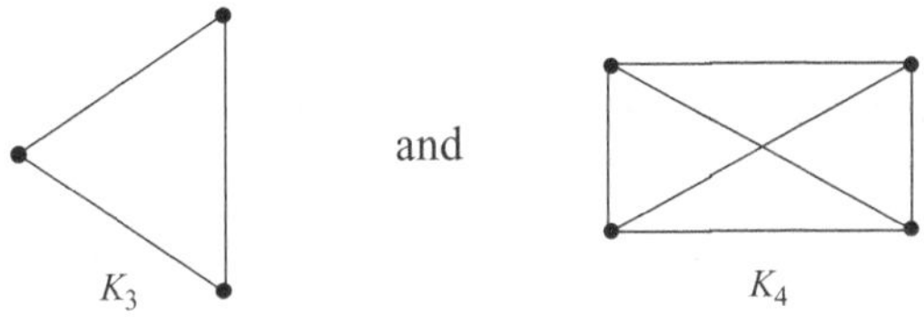

Figure 8.82

EXAMPLE 8.30

The graph shown below does not have a Hamiltonian circuit.

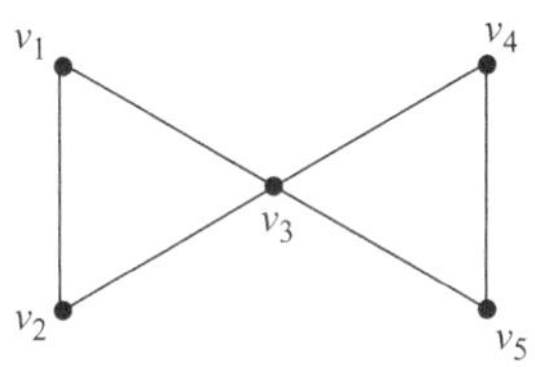

Figure 8.83

EXAMPLE 8.31

The graph shown in Figure 8.84 does not have a Hamiltonian circuit

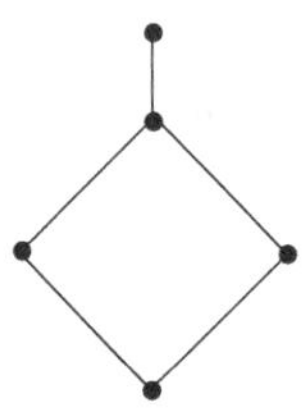

Figure 8.84

Remark 8.5 It is clear that **only connected graphs can have Hamiltonian circuit**. However, there is no simple criterion to tell us whether or not a given graph has Hamiltonian circuit. The following results give us some sufficient conditions for the existence of Hamiltonian circuit/path.

Theorem 8.10

Let G be a linear graph of n vertices. If the sum of the degrees for each pair of vertices in G is greater than or equal to $n-1$, then there exists a Hamiltonian path in G.

Theorem 8.11

Let G be a connected graph with n vertices. If $n \geq 3$ and $\deg(v) \geq n$ for each vertex v in G, then G has a Hamiltonian circuit.

Theorem 8.12

Let G be a connected graph with n vertices and let u and v be two vertices of G that are not adjacent. If $\deg(u)+\deg(v) \geq n$, then G has a Hamiltonian circuit.

Corollary 8.2

Let G be a connected graph with n vertices. If each vertex has degree greater than or equal to $n/2$, then G has a Hamiltonian circuit.

Proof. It is given that degree of each vertex is greater than or equal to $n/2$. Hence the sum of the degree of any two vertices is greater than or equal to $n/2+n/2=n$. So, by the above theorem, the graph G has a Hamiltonian circuit.

Theorem 8.13

Let n be the number of vertices and m be the number of edges in a connected graph G. If $m \geq \frac{1}{2}(n^2-3n+6)$, then G has a Hamiltonian circuit.

The following example shows that the above conditions are not necessary for the existence of Hamiltonian path.

EXAMPLE 8.32

Let G be the connected graph shown in the Figure 8.85.

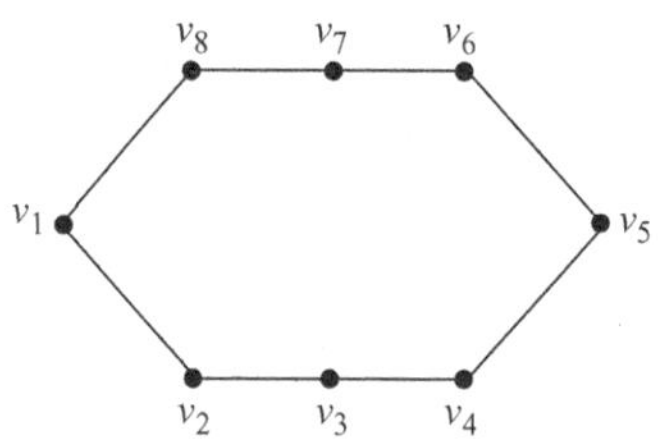

Figure 8.85

We note that

$$n = \text{Number of vertices in } G = 8,$$
$$m = \text{Number of edges in } G = 8,$$
$$\text{Degree of each vertex} = 2.$$

Thus, if u and v are non-adjacent vertices, then

$$\deg(u)+\deg(v)=2+2=4 \ngeq 8.$$

Also,

$$\frac{1}{2}(n^2-3n+6)=\frac{1}{2}(64-24+6)=23.$$

Clearly $m \ngeq \frac{1}{2}(n^2-3n+6)$. Therefore the above two theorems fail. But the given graph has Hamiltonian circuit. For example, $v_1, v_2, v_3, v_4, v_5, v_6, v_7, v_8, v_1$ is an Hamiltonian circuit for the graph.

Proposition 8.1

Let G be a graph with at least two vertices. If G has a Hamiltonian circuit, then G has a subgraph H with the following properties:

1. H contains every vertex of G
2. H is connected
3. H has the same number of edges as vertices
4. Every vertex of H has degree 2

The contrapositive of this proposition is

"If a graph G with at least two vertices does not have a subgraph H satisfying (1) – (4), then G does not have a Hamiltonian circuit".

Also we know that contrapositive of a statement is logically equivalent to the statement. Therefore the above result can be used to show non-existence of a Hamiltonian circuit.

EXAMPLE 8.33

Consider the graph G shown below (Figure 8.86):

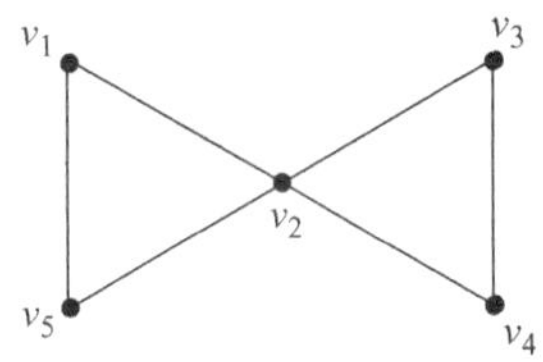

Figure 8.86

Let us verify whether this graph has a Hamiltonian circuit.

Suppose that G has a Hamiltonian circuit. Then it has a subgraph H such that

1. H contains every vertex of G, i.e., H has 5 vertices v_1, v_2, v_3, v_4, v_5
2. H is connected
3. H has 5 edges
4. Every vertex of H has degree 2

Since the degree of v_2 in G is 4 and every vertex of H has degree 2, two edges incident on v_2 must be removed from G to create H. But we note that the edge $\{v_1, v_2\}$ cannot be removed, because the removal of $\{v_1, v_2\}$ will decrease the degree of v_1 to 1, which is less than 2. Similarly the edge $\{v_2, v_3\}$, $\{v_2, v_4\}$ and $\{v_2, v_5\}$ cannot be removed. Hence the degree of v_2 in H must stand at 4 which contradicts the condition that every vertex in H has degree 2 in H. Therefore, no such subgraph H exists. Hence G does not have a Hamiltonian circuit.

EXAMPLE 8.34

Does the graph G given below (Figure 8.87) have Hamiltonian circuit?

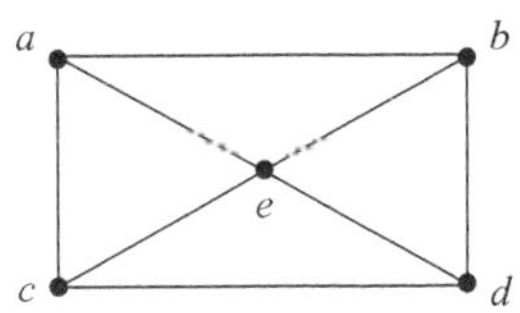

Figure 8.87

Solution.

The given graph has

n = Number of vertices = 5

m = Number of edges = 8

$\deg(a) = \deg(b) = \deg(c) = \deg(d) = 3$

$\deg(e) = 4$

We observe that

(i) Degree of each vertex is greater than $n/2$

(ii) The sum of degrees of any non-adjacent pair of vertices is greater than n

(iii) $\frac{1}{2}(n^2 - 3n + 6) = \frac{1}{2}(25 - 15 + 6) = 8.$

Thus the condition $m \geq \frac{1}{2}(n^2 - 3n + 6)$ is satisfied.

(iv) The sum of degrees of each pair of vertices in the given graph is greater than $n - 1 = 5 - 1 = 4$.

Thus four sufficiency conditions are satisfied (whereas one condition out of these four conditions is sufficient for the existence of Hamiltonian path/circuit). Hence the graph has a Hamiltonian circuit.

For example, the following circuits in G are Hamiltonian (Figure 8.88).

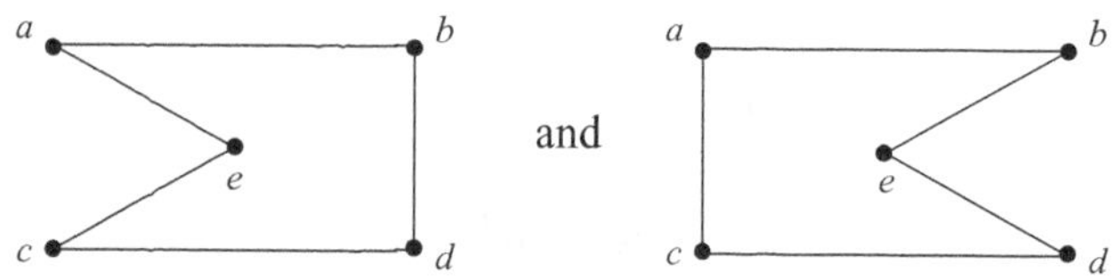

Figure 8.88

EXAMPLE 8.35

Does the graph shown below (Figure 8.89) has Hamiltonian circuit?

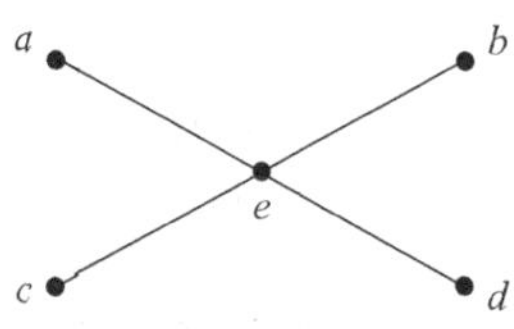

Figure 8.89

Solution.

Here

Number of vertices $(n) = 5$

Number of edges $(m) = 4$

$\deg(a) = \deg(b) = \deg(c) = \deg(d) = 1$

$\deg(e) = 4$

We note that

(i) $\deg(a) = \deg(b) = \deg(c) = \deg(d) \ngeq \frac{5}{2}$

(ii) $\deg(a) + \deg(b) = 2 \ngeq 5$, that is sum of any non-adjacent pair of vertices is not greater than 5

(iii) $\frac{1}{2}(n^2-3n+6)=\frac{1}{2}(25-15+6)=8$

Therefore the condition $m \geq \frac{1}{2}(n^2-3n+6)$ is **not satisfied**,

(iv) $\deg(a)+\deg(b)=2 \ngeq 4$, i.e., the condition that sum of degrees of each pair of vertices in the graph is not greater than or equal to $n-1$.

Hence no sufficiency condition is satisfied. So we try the Proposition 8.1. Suppose that G has a Hamiltonian circuit. Then G should have a subgraph which contains every vertex of G, and number of vertices and number of edges in H should be same. Thus, H should have five vertices a, b, c, d, e and five edges. Since G has only four edges, H cannot have more than four edges. Hence no such subgraph is possible. Hence, the given graph does not have Hamiltonian circuit.

EXAMPLE 8.36

Does the graph shown below (Figure 8.90) possess a Hamiltonian circuit?

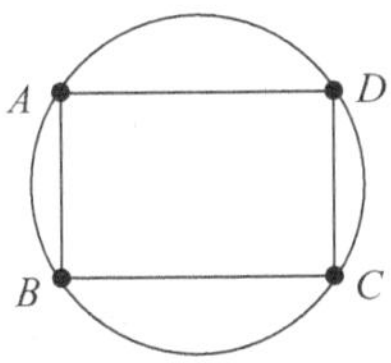

Figure 8.90

Solution.

In the given graph

Number of vertices $(n)=4$

Number of edges $(m)=8$

Degree of each vertex $=4$

Thus we see that

(i) Degree of each vertex is greater than $n/2$

(ii) Sum of degree of each pair of vertices is greater than $n-1$,

(iii) $\frac{1}{2}(n^2-3n+6)=\frac{1}{2}(16-12+6)=5$ and so $m \geq \frac{1}{2}(n^2-3n+6)$.

Hence the given graph has a Hamiltonian path. For example, $ABCDA$ is a Hamiltonian path in the given graph (see Figure 8.91).

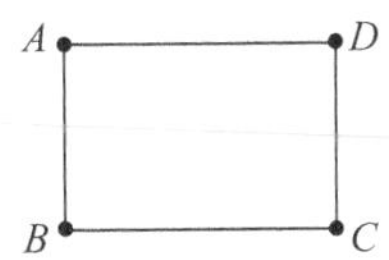

Figure 8.91

EXAMPLE 8.37

Is the graph shown in Figure 8.92 a Hamiltonian?

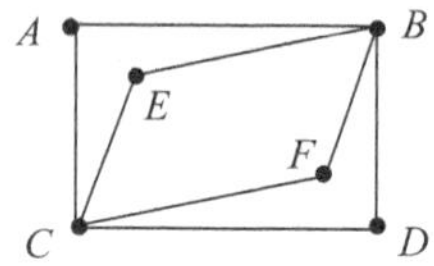

Figure 8.92

Solution.

We note that

$$\deg(A)=\deg(D)=\deg(E)=\deg(F)=2,$$
$$\deg(B)=\deg(C)=4.$$

Further,

Number of vertices $(n)=6$

Number of edges $(m)=8$.

So,

$$\frac{1}{2}(n^2-3n+6)=\frac{1}{2}(36-18+6)=12.$$

Thus the conditions

(i) Degree of each vertex is greater than or equal to $n/2$ is not satisfied

(ii) $m \geq \frac{1}{2}(n^2-3n+6)$ is not satisfied

(iii) $\deg(A)+\deg(E)=2+2=4$ and so the condition that "the sum of degrees of non-adjacent vertices is greater than or equal to n" is not satisfied

(iv) "The sum of degrees of any pair of vertices is greater than or equal to $n-1$" is not satisfied.

So suppose that G has a Hamiltonian circuit. Then it should have a connected subgraph H containing six vertices, six edges and degree of each vertex should be 2. To have degree of each vertex equal to 2, we should remove two edges from C and two edges from B.

For example, now, if we remove CE from C, then $\deg(E)\neq 2$, If we remove CF, then $\deg(F)\neq 2$. So, we cannot remove CE and CF. If we remove CD, then $\deg(D)\neq 2$. If we remove CA, then $\deg(A)\neq 2$. Hence no such subgraph H exists. So, G cannot have a Hamiltonian circuit.

Remark 8.6 Since the degree of each vertex in the above graph is even, it has Eulerian circuit. Thus the graph in Example 8.37 is Eulerian but not Hamiltonian.

EXAMPLE 8.38

Is the graph given in Figure 8.93 Hamiltonian?

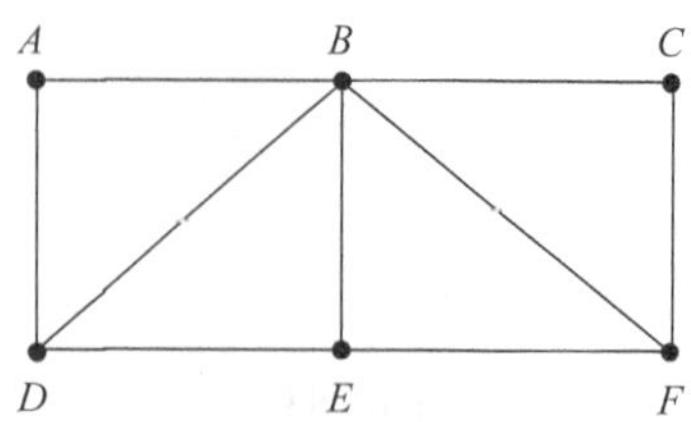

Figure 8.93

Solution.

In this graph,

Number of vertices $(n)=6$

Number of edges $(m)=9$.

So,

$$\frac{1}{2}(n^2-3n+6)=\frac{1}{2}(36-18+6)=12,$$
$$\deg(A)=\deg(C)=2,$$
$$\deg(D)=\deg(F)=3=\deg(E),$$
$$\deg(B)=5.$$

Thus, no sufficient condition is satisfied in this case. **But, the graph has Hamiltonian circuit** *ADEFCBA* shown in the Figure 8.94.

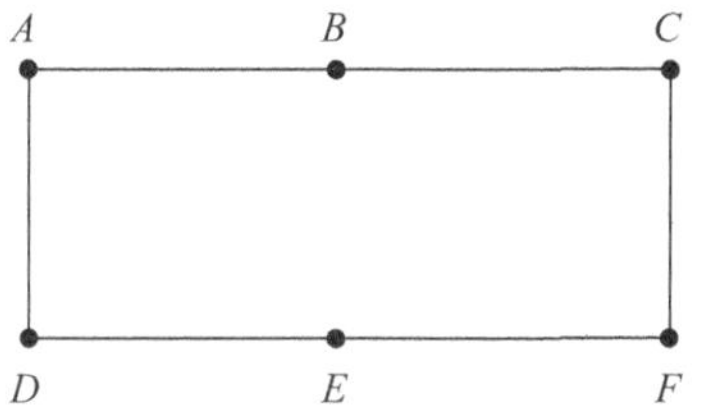

Figure 8.94

EXAMPLE 8.39

Show that the graph G shown in Figure 8.95 is not Hamiltonian.

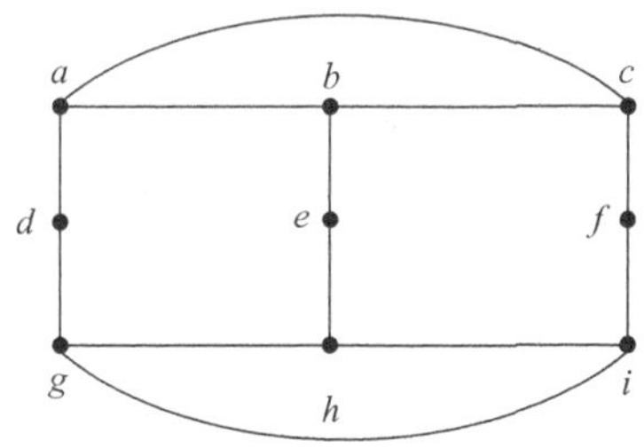

Figure 8.95

Solution.

If G is Hamiltonian, then it has a subgraph H that

1. Contains every vertex of G
2. Is connected
3. Has the same number of edges as vertices
4. Is such that every vertex has degree 2

Thus, if such a graph exists, then it will have vertices $(a, b, c, d, e, f, g, h, i)$, will be connected, will have six edges and the degree of each vertex shall be 2. We note that $\deg(d)=\deg(e)=\deg(f)=2$. The degree of the vertex h in G is 3, one edge incident on h must be deleted from G to create H. The edge $\{h, e\}$ is required since otherwise $\deg(e)\neq 2$. So we have to delete either $\{g, h\}$ or $\{h, i\}$. If we delete $\{h, i\}$ and retain $\{g, h\}$, then $\{g, i\}$ has to be deleted to keep g of degree 2. In such a case, there is only one edge $\{i, f\}$ on i and so there is a contradiction to (4). So we cannot remove $\{h, i\}$. Similarly, we see that $\{g, h\}$ cannot be deleted.

Hence no such subgraph H exists and so G does not have a Hamiltonian circuit.

Definition 8.55

A **weighted graph** is a graph for which each edge or each vertex or both is (are) labelled with a numerical value, called its **weight**.

For example, if vertices in a graph denote recreational sites of a town and weights of edges denote the distances in kilometers between the sites, then the graph shown in Figure 8.96 is a weighted graph.

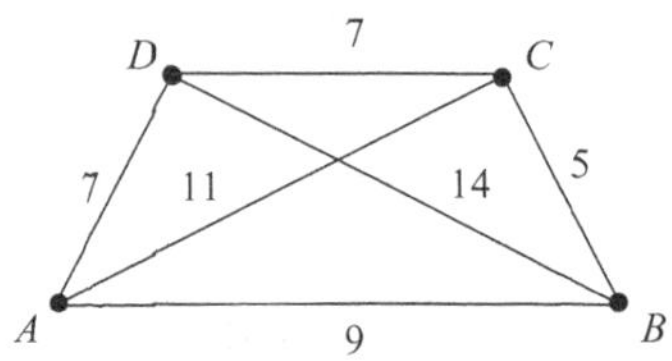

Figure 8.96

Definition 8.56

The **weight of an edge** (v_i, v_j) is called **distance between the vertices** v_i **and** v_j.

Definition 8.57

A vertex u is a **nearest neighbour** of vertex v in a graph if u and v are adjacent and no other vertex is joined to v by an edge of lesser weight than (u, v).

For example, in the above example, B is the nearest neighbour of C, whereas A and C are both nearest neighbours of the vertex D. **Thus, nearest neighbour of a set of vertices is not unique.**

Definition 8.58

A vertex u is a nearest neighbour of a set of vertices $\{v_1, v_2, \ldots, v_n\}$ in a graph if u is adjacent to some member v_i of the set and no other vertex adjacent to member of the set is joined by an edge of lesser weight than (u, v_i).

In the above example, if we have set of vertices as $\{B, D\}$, C is the neatest neighbour of $\{B, D\}$ because the edge (C, B) has weight 5 and no other vertex adjacent to $\{B, D\}$ is linked by an edge of lesser weight than (C, B).

Definition 8.59

The **length of a path** in a graph is the sum of lengths of edges in the path.

Definition 8.60

Let $G=(V, E)$ be a graph and let l_{ij} denote the length of edge (v_i, v_j) in G. Then a **shortest path** from v_i to v_k is a path such that the sum $l_{12}+l_{23}+\cdots+l_{k-1,k}$ of lengths of its edges is **minimum**, that is, total edge weight is minimum.

8.7.1 Travelling Salesperson Problem

This problem requires the determination of a **shortest Hamiltonian circuit** in a given graph of cities and lines of transportation to minimize the total fare for a travelling person who wants to make a tour of n cities visiting each city exactly once before returning home.

The weighted graph model for this problem consists of vertices representing cities and edges with weight as distances (fares) between the cities. The salesman starts and ends his journey at the same city and visits each of $n-1$ cities once and only once. We want to find minimum total distance.

We discuss the case of five cities and so consider the weighted graph shown in the Figure 8.97.

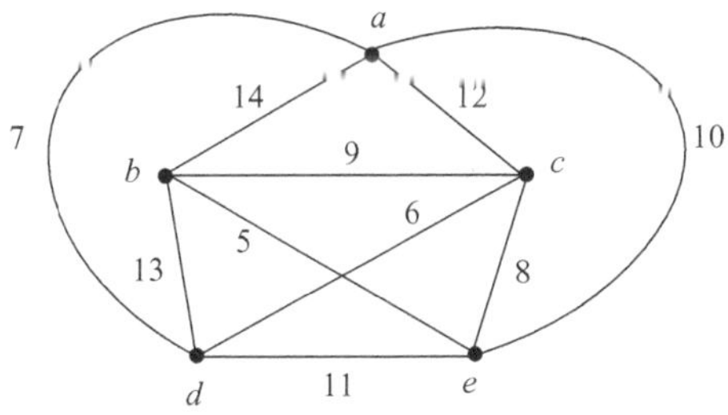

Figure 8.97

We shall use **nearest neighbour** algorithm to solve the problem:
Algorithm: Nearest neighbour (closest insertion)
Input: A weighted complete graph G.
Output: A sequence of labelled vertices that forms a Hamiltonian cycle.

Start at any vertex v.
Initialize $l(v)=0$.
Initialize $i=0$.

While there are unlabelled vertices

$i:=i+1.$

Traverse the cheapest edge that join v to an unlabelled vertex, say w

Set $l(w)=i$.
$v:=w$.

For the present example,

(i) Let us choose a as the starting vertex. Then d is the nearest vertex and then (a, d) is the corresponding edge. Thus we have the figure

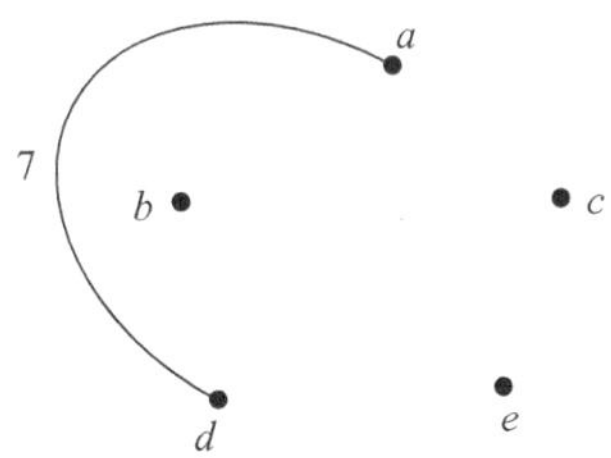

(ii) From d, the nearest vertex is c, so we have a path shown below:

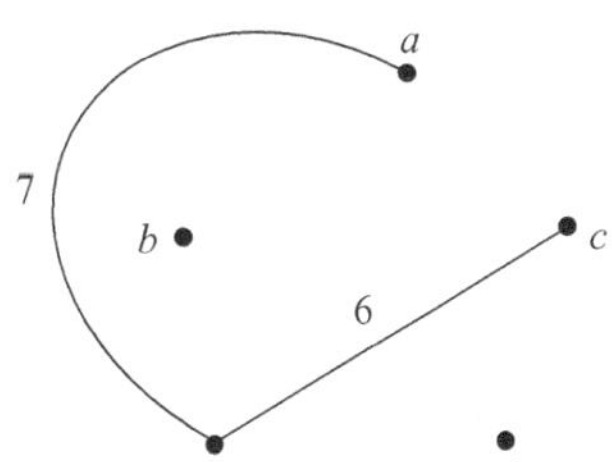

(iii) From c, the nearest vertex is e. So we have the path as shown below:

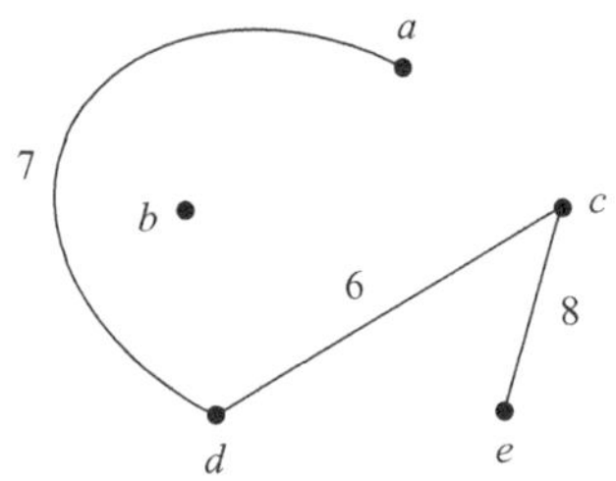

(iv) From e, the nearest vertex is b and so we have the path

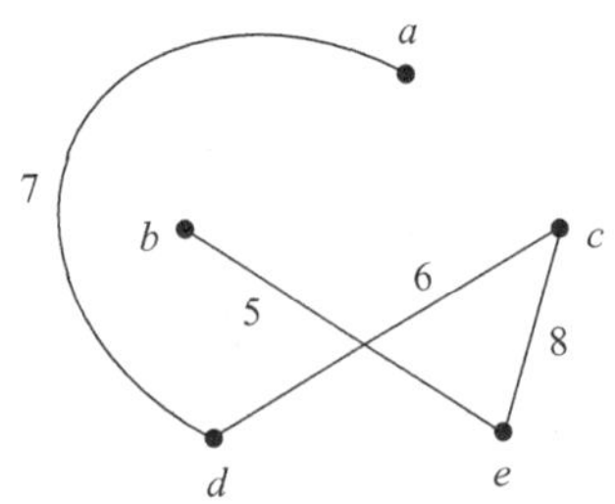

(v) Now, from b, the only vertex to be covered is a to **form Hamiltonian circuit.** Thus we have a Hamiltonian circuit as given below. The length of this Hamiltonian circuit is

$$7+6+8+5+14=40.$$

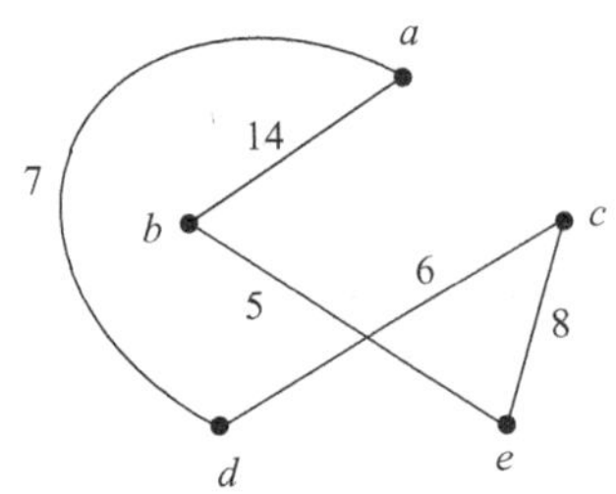

However, **this is not Hamiltonian circuit of minimal length.**
The **total distance** of a **minimum Hamiltonian** circuit (Figure 8.98) **is 37**.

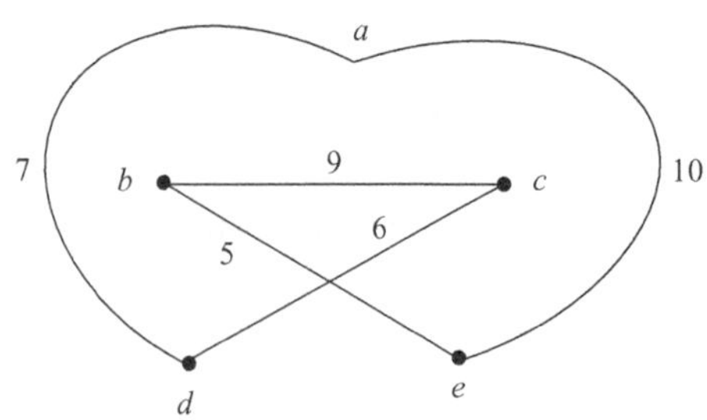

Figure 8.98

Total length $=7+6+9+5+10=37$.

Remark 8.7 Unless otherwise stated, try to start from a vertex of largest weight.

EXAMPLE 8.40

Find a Hamiltonian circuit of minimal weight for the graph given below (Figure 8.99)

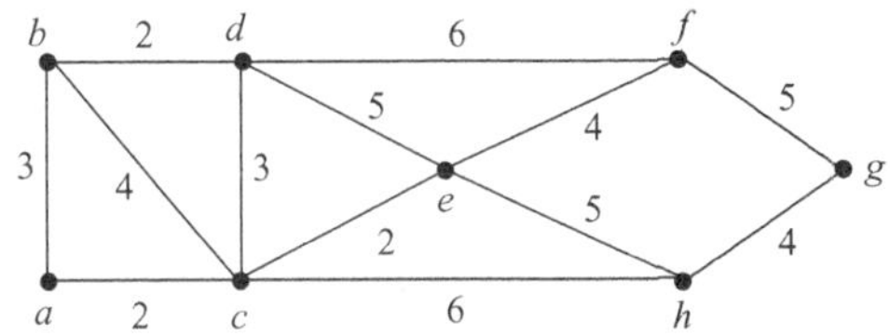

Figure 8.99

Solution.

Starting from c and applying nearest neighbour method, we have the required Hamiltonian circuit as $c\ a\ b\ d\ e\ f\ g\ h\ c$ with total length as

$$2+3+2+5+4+5+4+6=31.$$

EXAMPLE 8.41

Find a Hamiltonian circuit of minimal weight for the graph shown below (Figure 8.100)

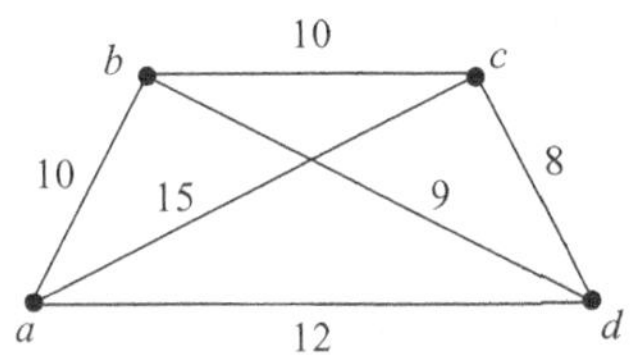

Figure 8.100

Solution.

Starting from the point a and using nearest neighbour method, we have the required Hamiltonian circuit as $a\ b\ c\ d\ a$ with total length as

$$10+10+8+12=40.$$

Definition 8.61

A k-**factor of a graph** is a spanning subgraph of the graph **with the degree of its vertices being** k.

Consider the graph shown in Figure 8.101.

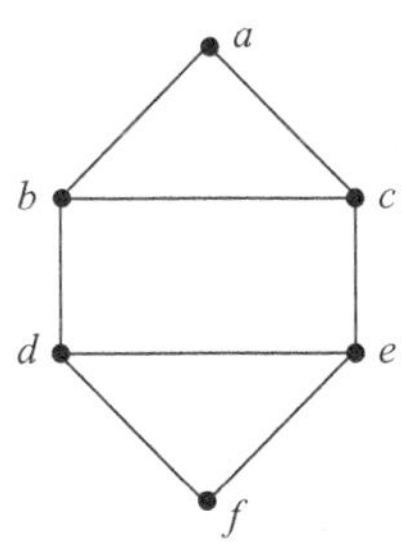

Figure 8.101

Then the graph shown in Figure 8.102 shows a 1-factor of the given graph.

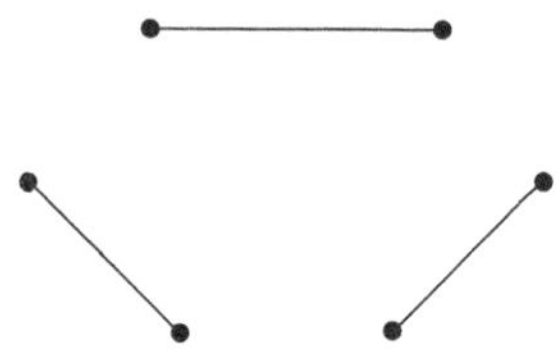

Figure 8.102

Also then, (see Figure 8.103)

Figure 8.103

is a 2-factor of the given graph.

8.8 MATRIX REPRESENTATION OF GRAPHS

A graph can be represented inside a computer by using the adjacency matrix or the incidence matrix of the graph.

Definition 8.62

Let G be a graph with n ordered vertices $v_1, v_2, \ldots, v_n$. Then the **adjacency matrix of** G is the $n \times n$ matrix $A(G) = (a_{ij})$ over the set of non-negative integers such that

$$a_{ij} = \text{the number of edges connecting } v_i \text{ and } v_j \text{ for all } i, j = 1, 2, \ldots, n.$$

We note that if G has no loop, then there is no edge joining v_i to v_i, $i = 1, 2, \ldots, n$. Therefore, in this case, all the entries on the main diagonal will be 0.

Further, if G has no parallel edge, then the entries of $A(G)$ are either 0 or 1.

It may be noted that adjacent matrix of a graph is symmetric.

Conversely, given a $n \times n$ symmetric matrix $A(G) = (a_{ij})$ over the set of non-negative integers, we can associate with it a graph G, whose adjacency matrix is $A(G)$, by letting G have n vertices and joining v_i to vertex v_j by a_{ij} edges.

EXAMPLE 8.42

Find the adjacency matrix of the graph shown below (Figure 8.104):

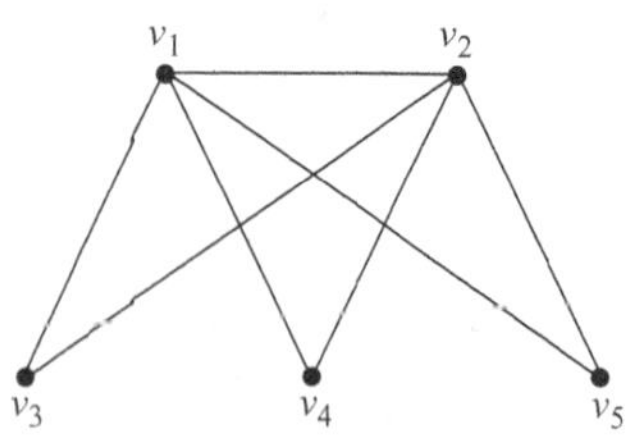

Figure 8.104

Solution.

The adjacency matrix $A(G)=(a_{ij})$ is the matrix such that

$$a_{ij}=\text{Number of edges connecting } v_i \text{ and } v_j.$$

So, for the given graph, the adjacency matrix is

$$A(G)=\begin{bmatrix} 0 & 1 & 1 & 1 & 1 \\ 1 & 0 & 1 & 1 & 1 \\ 1 & 1 & 0 & 0 & 0 \\ 1 & 1 & 0 & 0 & 0 \\ 1 & 1 & 0 & 0 & 0 \end{bmatrix}.$$

EXAMPLE 8.43

Find the adjacency matrix of the graph given below:

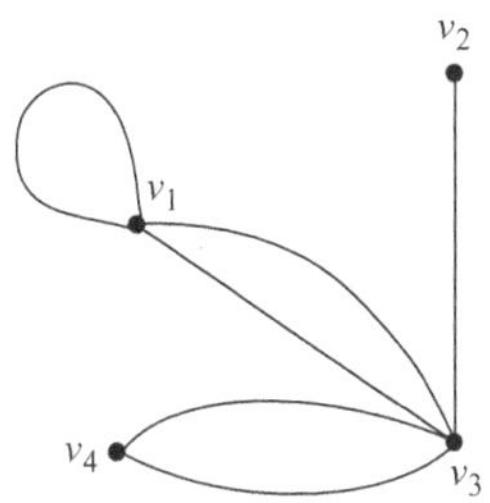

Figure 8.105

Solution.

We note that there is a loop at v_1 and parallel edges between v_1, v_3, and v_3, v_4. So the adjacency matrix $A(G)$ is given by the following 4×4 matrix:

$$A(G)=\begin{bmatrix} 1 & 0 & 2 & 0 \\ 0 & 0 & 1 & 0 \\ 2 & 1 & 0 & 2 \\ 0 & 0 & 2 & 0 \end{bmatrix}.$$

EXAMPLE 8.44

Find the graph that has the following adjacency matrix.

$$\begin{bmatrix} 1 & 2 & 1 & 2 \\ 2 & 0 & 2 & 1 \\ 1 & 2 & 1 & 0 \\ 2 & 1 & 0 & 0 \end{bmatrix}.$$

Solution.

We note that there is a loop at v_1 and a loop at v_3. There are parallel edges between v_1, v_2; v_1, v_4; v_2, v_1; v_2, v_3; v_3, v_2; v_4, v_1. Thus the graph is as shown in Figure 8.106.

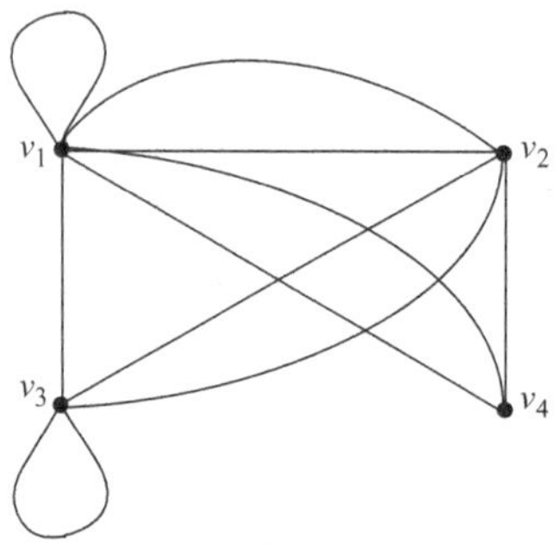

Figure 8.106

The following theorem is stated without proof.

Theorem 8.14

Let G be a graph with n vertices $v_1, v_2, \ldots, v_n$ and let $A(G)$ denote the matrix of G. If $B=(b_{ij})$ is the matrix

$$B=A+A^2+\cdots+A^{n-1},$$

then the graph G is a connected graph if and only if $b_{ij}\neq 0$ for $i\neq j$, that is, B has no zero entry off the main diagonal.

EXAMPLE 8.45

A graph G has the following adjacency matrix. Verify whether it is connected.

$$A=\begin{bmatrix} 0 & 0 & 1 & 0 & 0 \\ 0 & 0 & 0 & 1 & 0 \\ 1 & 0 & 0 & 0 & 1 \\ 0 & 1 & 0 & 0 & 1 \\ 0 & 0 & 1 & 1 & 0 \end{bmatrix}.$$

Solution.

We have

$$A^2=\begin{bmatrix} 1 & 0 & 0 & 0 & 1 \\ 0 & 1 & 0 & 0 & 1 \\ 0 & 0 & 2 & 1 & 0 \\ 0 & 0 & 1 & 2 & 0 \\ 1 & 1 & 0 & 0 & 2 \end{bmatrix}, \quad A^3=\begin{bmatrix} 0 & 0 & 2 & 1 & 0 \\ 0 & 0 & 1 & 2 & 0 \\ 2 & 1 & 0 & 0 & 3 \\ 1 & 2 & 0 & 0 & 3 \\ 0 & 0 & 3 & 3 & 0 \end{bmatrix} \quad A^4=\begin{bmatrix} 2 & 1 & 0 & 0 & 3 \\ 1 & 2 & 0 & 0 & 3 \\ 0 & 0 & 5 & 4 & 0 \\ 0 & 0 & 4 & 5 & 0 \\ 3 & 3 & 0 & 0 & 6 \end{bmatrix}.$$

Therefore,

$$B = A + A^2 + A^3 + A^4 = \begin{bmatrix} 3 & 1 & 3 & 1 & 4 \\ 1 & 3 & 1 & 3 & 4 \\ 3 & 1 & 7 & 5 & 4 \\ 1 & 3 & 5 & 7 & 4 \\ 4 & 4 & 4 & 4 & 8 \end{bmatrix}.$$

Since B has no zero entry off the main diagonal, the graph is connected.

Definition 8.63

Suppose a graph G has n vertices $v_1, v_2, \ldots, v_n$ and t edges $e_1, e_2, \ldots, e_t$. The **incidence matrix** $B(G)$ of G is the $n \times t$ matrix $B(G)=(b_{ij})$, where

b_{ij}=the number of times that the vertex v_i is incident with the edge e_j,

that is,

$$b_{ij} = \begin{cases} 0 & \text{if } v_i \text{ is not end of } e_j \\ 1 & \text{if } v_i \text{ is an end of the non-loop } e_j \\ 2 & \text{if } v_i \text{ is an end of the loop } e_j \end{cases}$$

8.9 PLANAR GRAPHS

Definition 8.64

A graph which can be drawn in the plane so that its edges do not cross is said to be **planar.** For example, the graph shown in Figure 8.107 is planar:

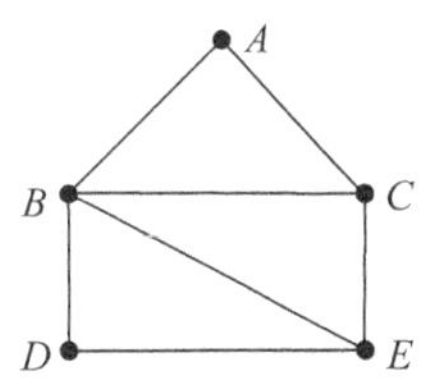

Figure 8.107

Also the complete graph K_4 shown below (Figure 8.108) is planar.

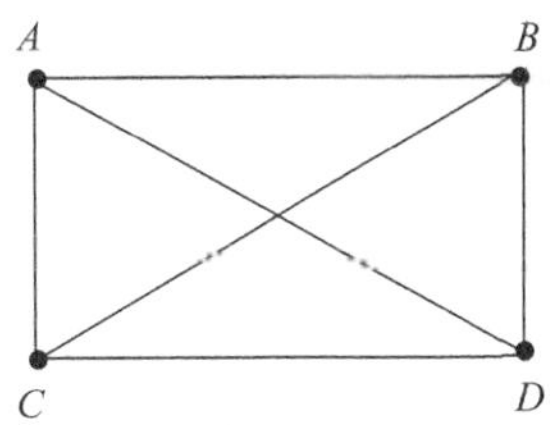

Figure 8.108

In fact, it can be redrawn as shown below in Figure 8.109 so that no edges cross.

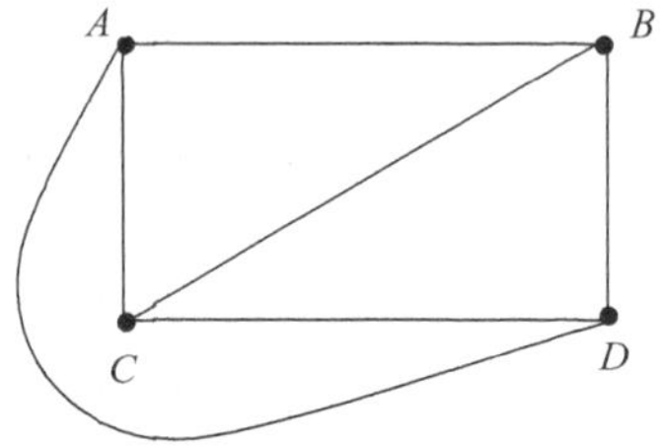

Figure 8.109

But the map K_5 is not planar because in this case, the edges cross each others (see Figure 8.110).

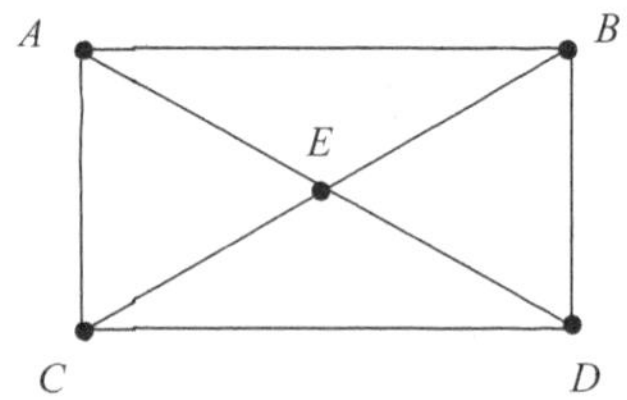

Figure 8.110

Definition 8.65

An area of the plane that is bounded by edges of the planar graph and is not further subdivided into subareas is called a **region** or **face** of the planar graph.

A face is characterized by the cycle that forms its boundary.

Definition 8.66

A region is said to be **finite** if its area is finite and **infinite** if its area is infinite. Clearly a planar graph has exactly one infinite region.

For example, consider the graphs shown in Figure 8.111.

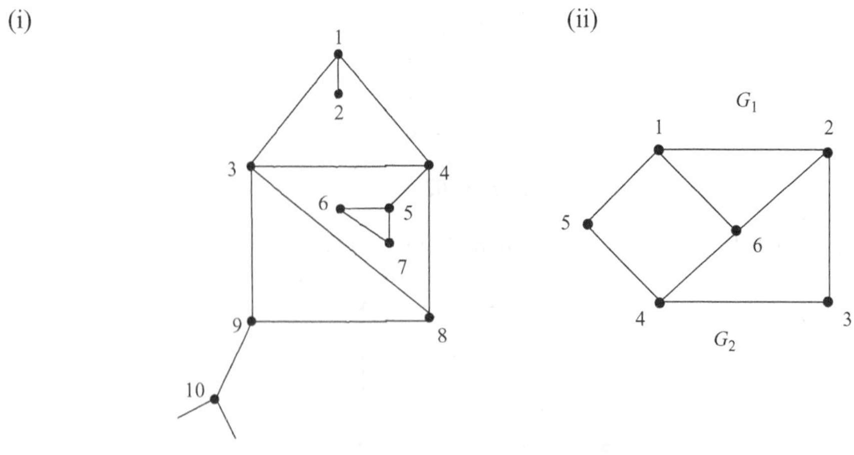

Figure 8.111

We note that in the graph G_1, there are five regions **A, B, C, D, E** as shown below:

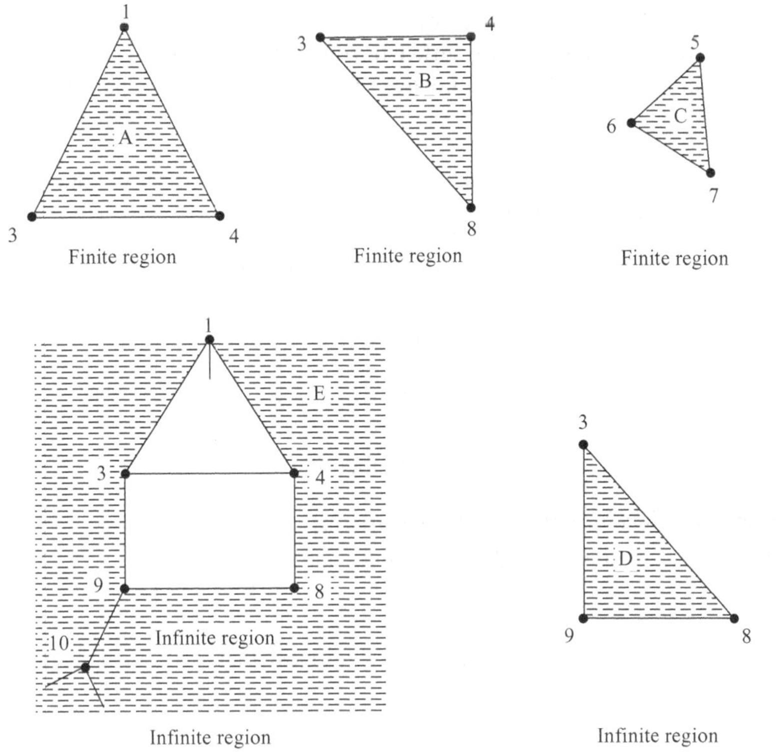

In graph G_2, there are four region **A, B, C, D** (Figure 8.112)

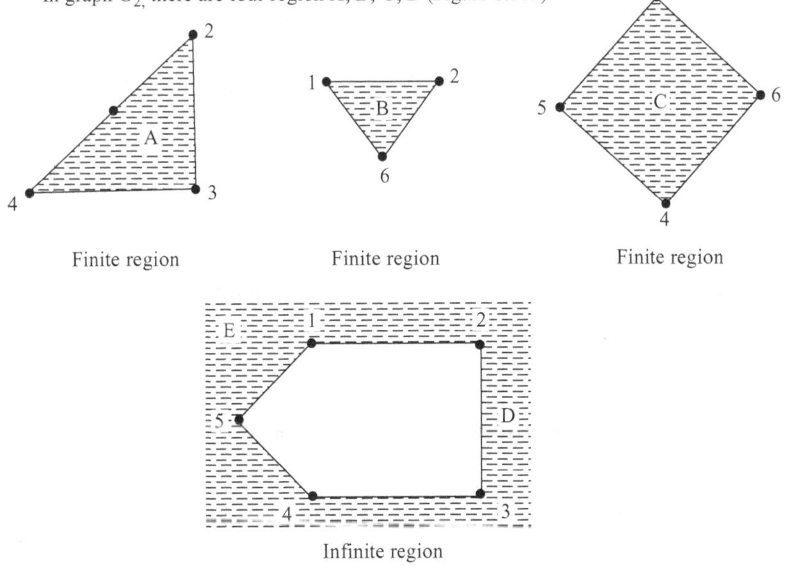

Figure 8.112

Definition 8.67

Let f be a face (region) in a planar graph. The length of the cycle (or closed walk) which borders f is called the **degree of the region** f. It is denoted by deg(f).

In a planar graph we note that **each edge either borders two regions or is contained in a region and will occur twice in any walk along the border of the region.** Thus we have the following:

Theorem 8.15

The sum of the degrees of the regions of a map is equal to twice the number of edges.

For example, in the graph G_2, discussed above, we have

$$\deg(A)=4, \quad \deg(B)=3, \quad \deg(C)=4, \quad \deg(d)=5.$$

The sum of degrees of all regions $=4+3+4+5=16$.

$$\text{Number of edges in } G_2=8.$$

Hence the sum of degrees of regions is twice the number of edges.

Theorem 8.16 (Euler's Formula for Connected Planar Graphs)

If G is a connected planar graph with e edges, v vertices and r regions, then

$$v-e+r=2.$$

Proof. We shall use induction on the number of edges. Suppose that $e=0$. Then the graph G consists of a single vertex, say P. Thus, G is as shown below:

$$\bullet\, P$$

and we have $e=0$, $v=1$, $r=1$. Thus $1-0+1=2$ and the formula holds in this case.

Suppose that $e=1$. Then the graph G is one of the two graphs shown below:

$e=1,\ v=2,\ r=1$

$e=1,\ v=1,\ r=2$

We see that, in either case, the formula holds.

Suppose that the formula holds for connected planar graph with n edges. We shall prove that this holds for graph with $n+1$ edges. So, let G be the graph with $n+1$ edges. Suppose first that G contains no cycles. Choose a vertex v_1 and trace a path starting at v_1. Ultimately, we will reach a vertex a with degree 1, that we cannot leave (see Figure 8.113).

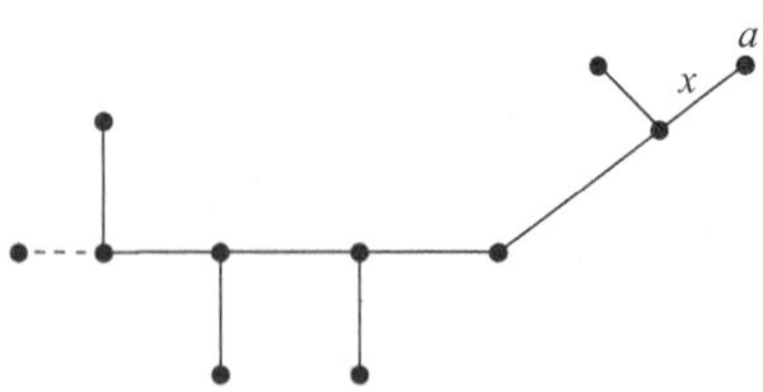

Figure 8.113

We delete "a" and the edge x incident on "a" from the graph G. The resulting graph G' (Figure 8.114) has n edges and so by induction hypothesis, the formula holds for G'. Since G has one more edge than G', one more vertex than G' and the same number of faces as G', it follows that the formula $v-e+r=2$ holds also for G.

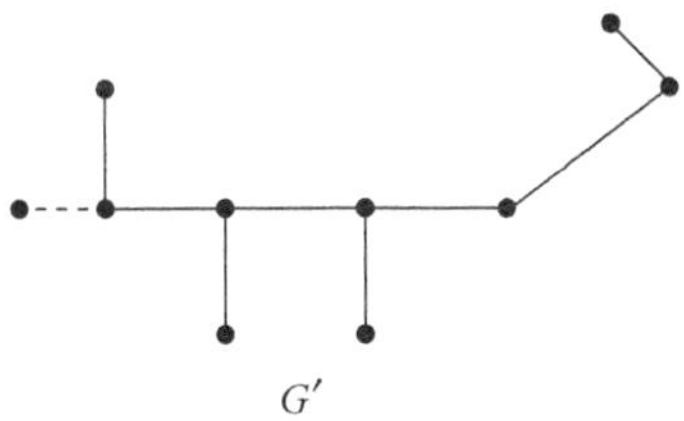

Figure 8.114

Now suppose that G contains a cycle. Let x be an edge in a cycle as shown in Figure 8.115.

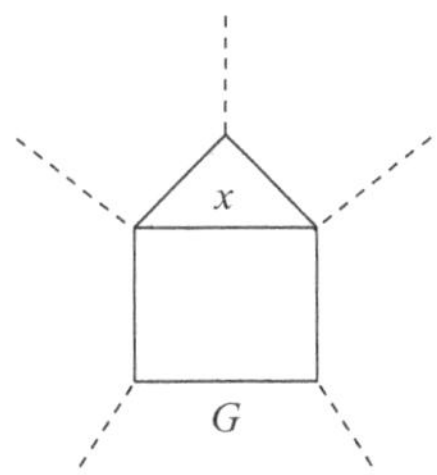

Figure 8.115

Now the edge x is part of a boundary for two faces. We delete the edge x but no vertices to obtain the graph G' as shown in Figure 8.116.

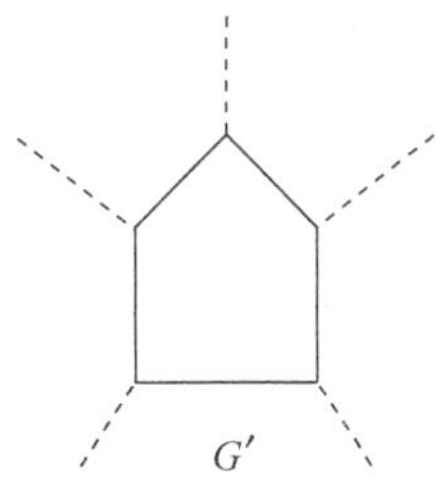

Figure 8.116

Thus G' has n edges and so by induction hypothesis the formula holds. Since G has one more face (region) than G', one more edge than G' and the same number of vertices as G', it follows that the formula $v-e+r=2$ also holds for G. Hence, by mathematical induction, the theorem is true.

Remark 8.8 Planarity of a graph is not affected if

(i) An edge is divided into two edges by the insertion of new vertex of degree 2 (Figure 8.117).

Figure 8.117

(ii) Two edges that are incident with a vertex of degree 2 are combined as a single edge by the removal of that vertex (Figure 8.118).

Figure 8.118

Definition 8.68

Two graphs G_1 and G_2 are said to be **isomorphic to within vertices of degree 2 (or homeomorphic)** if they are isomorphic or if they can be transformed into isomorphic graphs by repeated insertion and/or removal of vertices of degree 2.

Definition 8.69

The repeated insertion/removal of vertices of degree 2 is called **sequence of series reduction**.

For example, the graphs shown in Figure 8.119 are isomorphic to within vertices of degree 2.

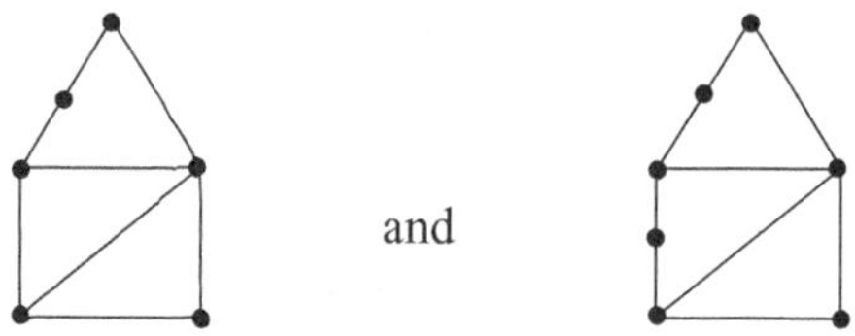

Figure 8.119

If we define a relation R on the set of graphs by $G_1 \, R \, G_2$ if G_1 and G_2 are homeomorphic, then R is an equivalence relation. Each equivalence class consists of a set of mutually homeomorphic graphs.

EXAMPLE 8.46

Show that the graph $K_{3,3}$, given in Figure 8.120 below, is not planar.

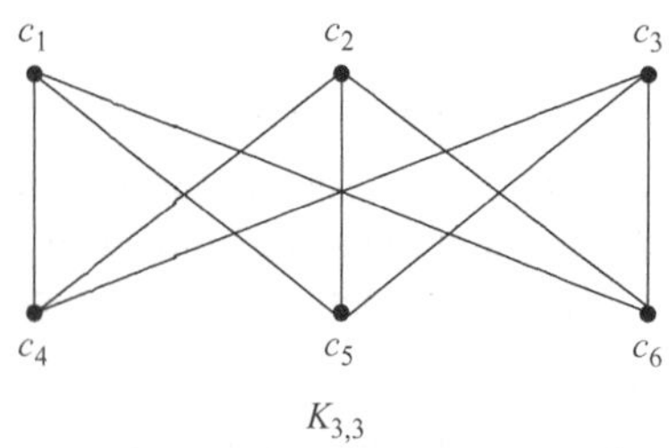

Figure 8.120

A problem based on this example can be stated as "Three cities c_1, c_2 and c_3 are to be directly connected by express ways to each of three cities c_4, c_5 and c_6. Can this road system be designed so that the express ways do not cross?" This example shows that it cannot be done.

Solution.

Suppose that $K_{3,3}$ is planar. Since every cycle in $K_{3,3}$ has at least four edges, each face (region) is bounded by at least four edges. Thus the number of edges that bound regions is at least $4r$. Also, in a planar graph each edge belongs to at most two bounding cycles. Therefore,

$2e \geq 4r$ (sums of degrees of region is equal to twice the number of edges)

But, by Euler's formula for planar graph,

$$r = e - v + 2.$$

Hence,

$$2e \geq 4(e - v + 2). \tag{1}$$

In case of $K_{3,3}$ we have $e=9$, $v=6$ and so (1) yields

$$18 \geq 4\ (9-6+2) = 20,$$

which is a contradiction. Therefore $K_{3,3}$ is not planar.

Remark 8.9 By a argument similar to the above example, we can show that the graph K_5 (Figure 8.121) is not planar.

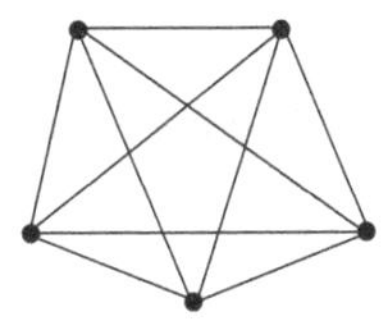

(Non-planar graph K_5)

Figure 8.121

We observe that if a graph contains $K_{3,3}$ or K_5 as a subgraph, then it cannot be planar.

The following theorem, which we state without proof, gives necessary and sufficient condition for a graph to be planar.

Kuratowski's Theorem 8.17

A graph G is planar if and only if G does not contain a **subgraph** homeomorphic to $K_{3,3}$ or K_5.

The complete graph K_5 and the complete bipartite graph $K_{3,3}$ are called the **Kuratowski graphs.**

EXAMPLE 8.47

Using Kuratowski's theorem show that the graph G, shown in Figure 8.122, is not planar.

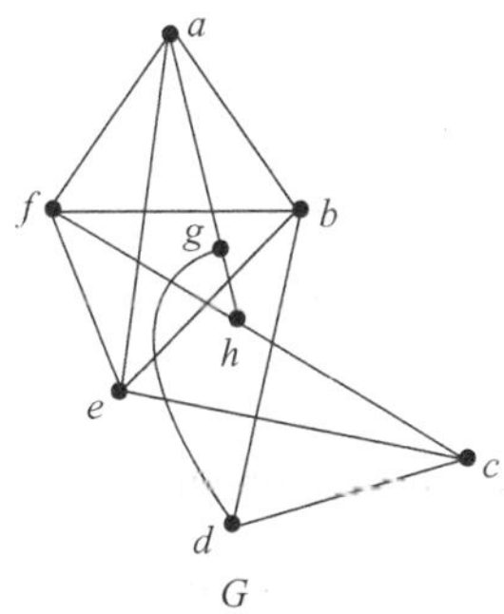

Figure 8.122

Solution.

Let us try to find $K_{3,3}$ in the graph G. We know that in $K_{3,3}$, each vertex has degree 3. But we note that in G, the degree of a, b, f and e each is 4. So we eliminate the edges (a, b) and (f, e) so that all vertices have degree 3. If we eliminate one more edge, we will obtain two vertices of degree 2 and we can then carry out series reduction. The resulting graph will have nine edges

Also we know that $K_{3,3}$ **has nine edges**. So this approach seems promising. Using trial and error, we find that the edge (g, h) should be removed. Then g and h have degree 2.

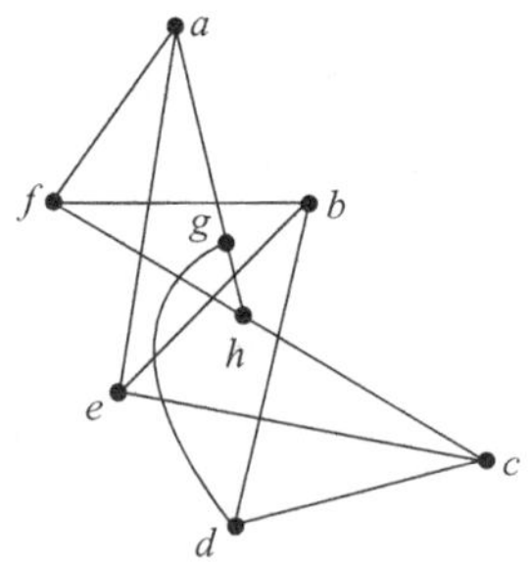

(Graph obtained by deleting edges (a, b) and (f, e)).

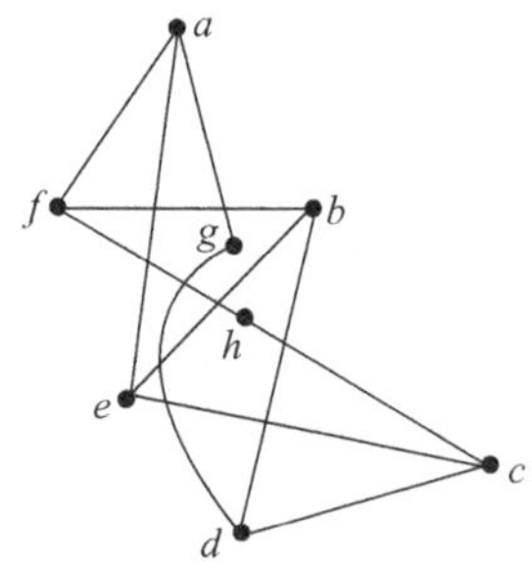

(Graph obtained by eliminating the edge (g, h).

Performing series reduction now, we obtain an isomorphic copy of $K_{3,3}$ (Figure 8.123).

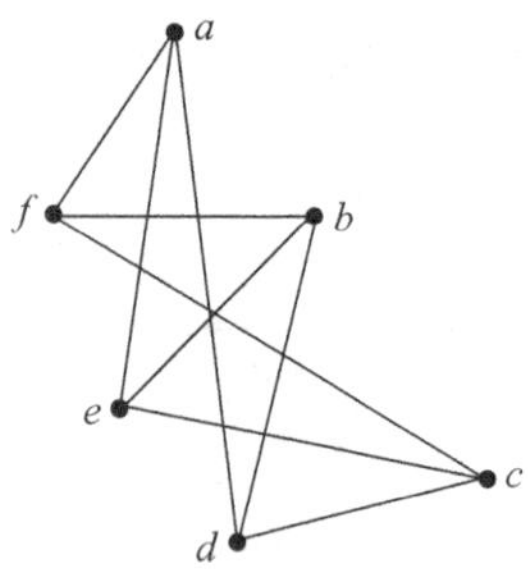

(Isomorphic copy of $K_{3,3}$, obtained by series reduction)

Figure 8.123

Hence, by Kurtowski's theorem, the given graph G is not planar.

EXAMPLE 8.48

If the graph given in Figure 8.124 is planar, redraw it so that no edges cross, otherwise find a subgraph homeomorphic to either K_5 or $K_{3,3}$.

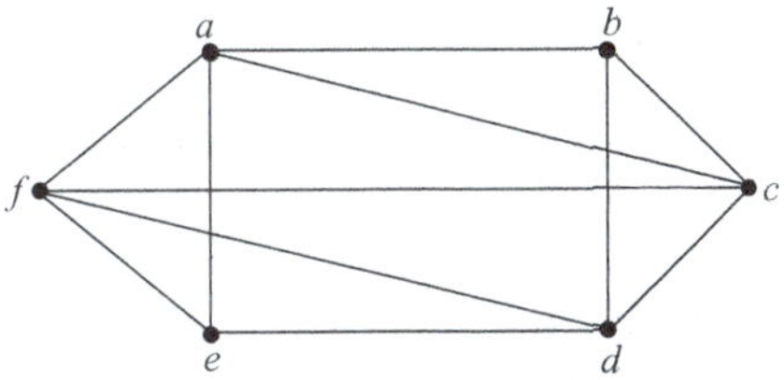

Figure 8.124

Solution.

The given graph can be redrawn as shown in the Figure 8.125.

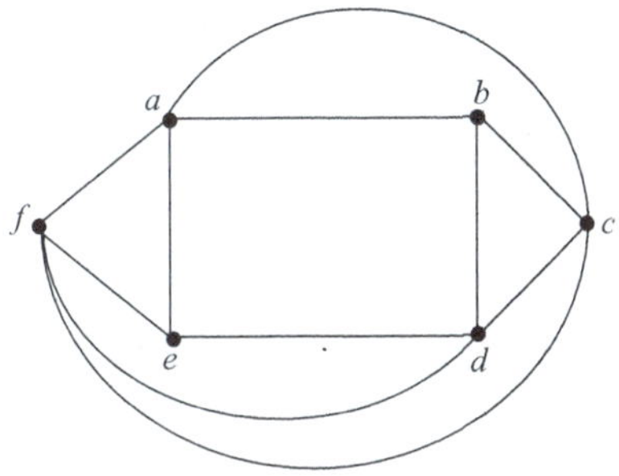

Figure 8.125

We then observe that no edges cross. Hence the given graph is planar.

Theorem 8.18

Let G be a connected planar graph with v vertices, e edges, where $v \geq 3$. Then $e \leq 3v-6$ or $6 \leq 3v - e$.

(Note that the theorem is not true for K_1, where $v=1$ and $e=0$ and is not true for K_2, where $v=2$ and $e=1$).

Proof. Let r be the number of regions in a planar representation of the graph G. The sum of the degrees of the regions is equal to $2e$. But each region has degree greater than or equal to 3. Hence $2e \geq 3r$, that is, $r \leq \frac{2e}{3}$. But, by Euler's formula,

$$v-e+r=2.$$

So,

$$2=v-e+r \leq v-e+\frac{2e}{3}=v-\frac{e}{3}.$$

Hence,

$$6 \leq 3v-e \quad \text{or} \quad e \leq 3v-6$$

Remark 8.10 Using above theorem, let us consider planarity of K_5. Here $v=5$, $e=10$. Suppose K_5 is planar, then by the above theorem,

$$6 \leq 3(5) - 10 = 5,$$

which is impossible. Hence K_5 is non-planar.

Definition 8.70

Any graph homeomorphic either to K_5 or to $K_{3,3}$ is called **Kuratowski subgraph**.

EXAMPLE 8.49

Find Kuratowski subgraph of the following graph (Figure 8.126):

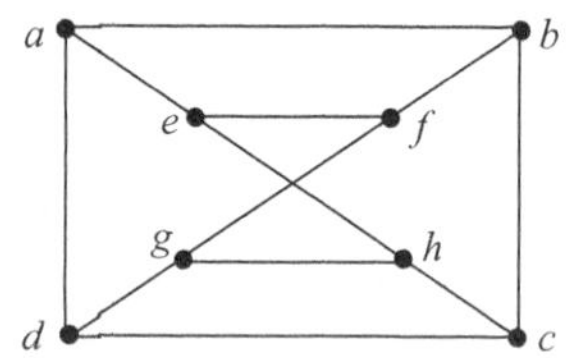

Figure 8.126

Solution.

In the given graph,

Number of vertices $=8$

Number of edges $=12$.

Firstly, we remove the edge (g, h). Then $\deg(g)=2$, $\deg(h)=2$ and so the vertices g and h can be removed. Then, we have

Number of vertices $=6$

Number of edges $=9$

and the graph reduces to

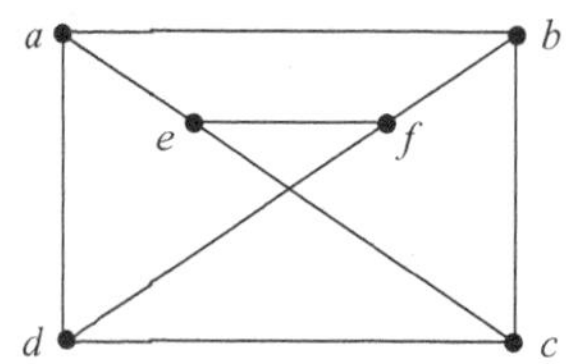

,

which is homeomorphic to $K_{3,3}$ (Kuratowski graph) shown in Figure 8.127.

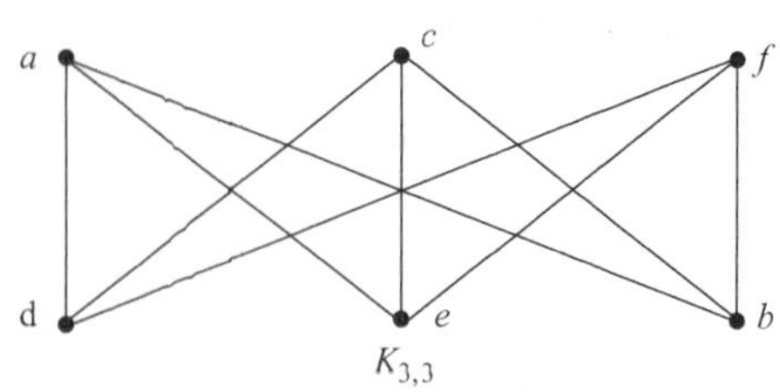

Figure 8.127

Thus the given graph is non-planar by Kuratowski theorem.

EXAMPLE 8.50

Find a subgraph homeomorphic to either K_5 or $K_{3,3}$ in the graph given below (Figure 8.128):

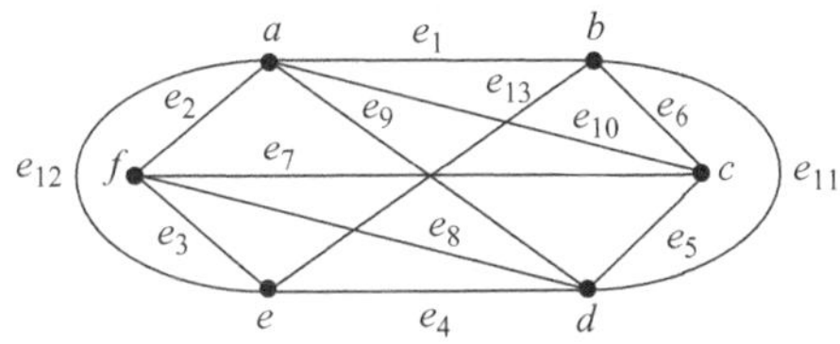

Figure 8.128

Solution.

Here,

Number of vertices = 6

Number of edges = 13

$$\deg(a)=5, \quad \deg(f)=4, \quad \deg(c)=4, \quad \deg(d)=5$$

But, in $K_{3,3}$, the degree of each vertex is 3. We delete the edges (a, c), (f, d), (a, e), (b, d), then degree of each vertex a, f, c and d becomes 3. As such, the subgraph becomes as given below (Figure 8.129), which has six vertices each with degree 3 and has 9 edges.

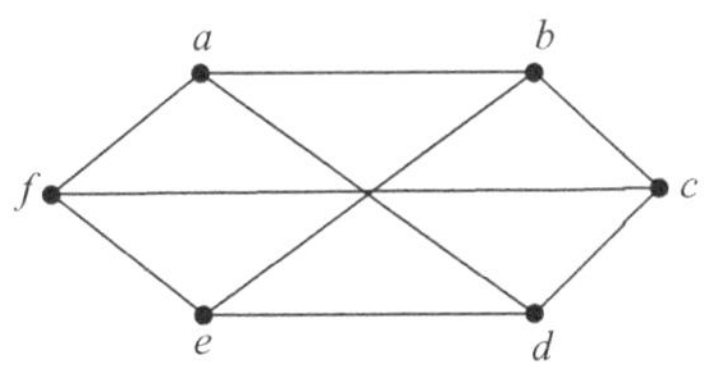

Figure 8.129

This subgraph is homeomorphic to $K_{3,3}$ as shown in Figure 8.130.

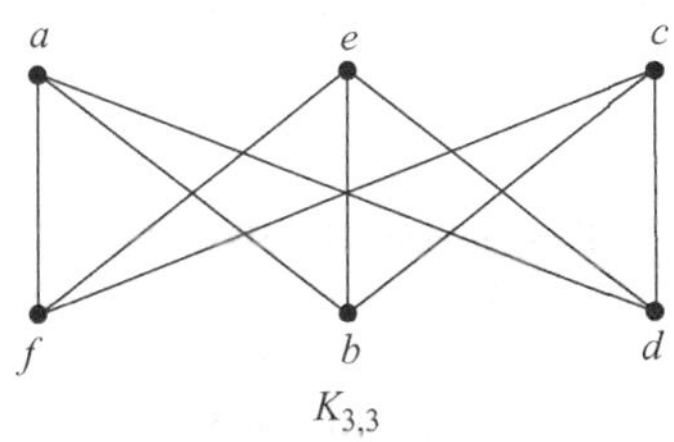

Figure 8.130

This also shows, by Kuratowski theorem, that the given graph is not planar.

8.10 COLOURING OF GRAPH

Definition 8.71

Let G be a graph. The assignment of colours to the vertices of G, one colour to each vertex, so that the adjacent vertices are assigned different colours is called **vertex colouring** or **colouring of the graph** G.

Definition 8.72

A graph G is **n-colourable** if there exists a colouring of G which uses n colours.

Definition 8.73

The minimum number of colours required to paint (colour) a graph G is called the **chromatic number of** G and is denoted by $\chi(G)$.

EXAMPLE 8.51

Find the chromatic number for the graph shown in the Figure 8.131.

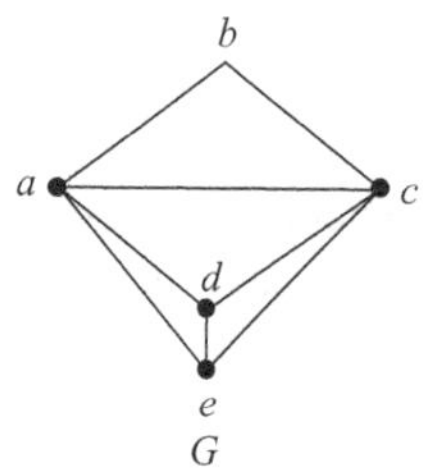

Figure 8.131

Solution.

The triangle $a\ b\ c$ needs three colours. Suppose that we assign colours c_1, c_2, c_3 to a, b and c respectively. Since d is adjacent to a and c, d will have different colour than c_1 and c_3. So we paint d by c_2. Then e must be painted with a colour different from those of a, d and c, that is, we cannot colour e with c_1, c_2 or c_3. Hence, we have to give e a fourth colour c_4. Hence

$$\chi(G)=4.$$

Welsh–Powell algorithm to determine upper bound to the chromatic number of a given graph.

The input is a given graph G.

1. Order the vertices of G according to decreasing degree.
2. Assign the first colour, say c_1, to the first vertex and then, in sequential order, assign c_1 to each vertex, which is not adjacent to a previous vertex assigned c_1.
3. Repeat Step 2 with a second colour c_2 and the subsequence of the remaining non-painted vertices.
4. Repeat Step 3 with a third colour c_3, then a fourth colour c_4 and so on until all vertices are coloured.
5. Exit.

EXAMPLE 8.52

Use Welsh-Powell algorithm to determine an upper bound to the chromatic number of the "Wheel" graph shown in Figure 8.132.

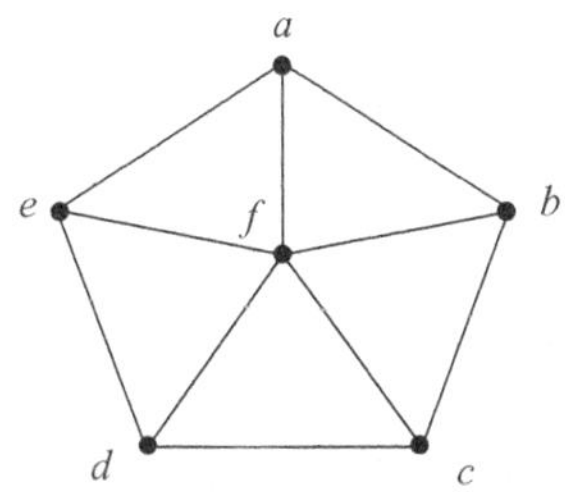

Figure 8.132

Solution.

We note that

$$\deg(f)=5,$$
$$\deg(a)=\deg(b)=\deg(c)=\deg(d)=\deg(e)=3.$$

Step 1. Ordering the vertices according to decreasing degree yields

$$f, a, b, c, d, e$$

Step 2. Paint f with colour c_1.
Step 3. Paint a, d with colour c_2.
Step 4. Paint b, e with colour c_3.
Step 5. Paint c with colour c_4.

Hence $\chi(G) \le 4$. Also since there is a triangle in the given graph so we have $\chi(G) \ge 3$. Thus $3 \le \chi(G) \le 4$. But we do not yet know exactly what the chromatic number is. We try to build a 3-colouring of G.

Let us start colouring the triangle abf with the colours c_1, c_2, c_3, respectively. Since c is adjacent to the vertices b and f of colour c_2 and c_3, respectively, c is forced to be coloured c_1 and then d is forced to be c_2. However, now the adjacent vertices a and e cannot both have colour c_1. Thus the graph cannot be 3-coloured. But using a fourth colour c_4 for e gives us 4-colouring of G. Hence, $\chi(G)=4$.

EXAMPLE 8.53

Use Welsh-Powell algorithm to colour the graph shown in Figure 8.133.

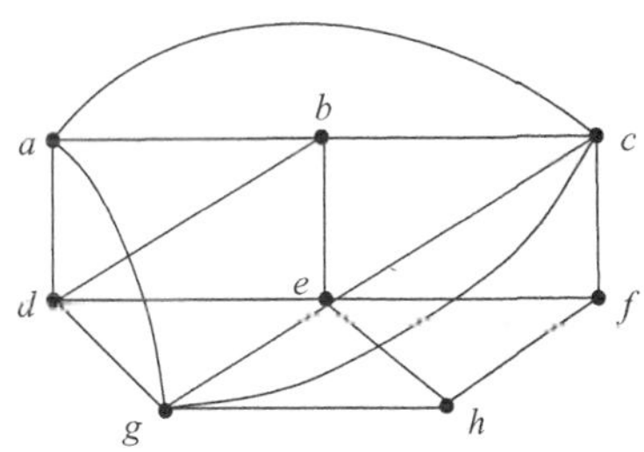

Figure 8.133

Solution.

Ordering the vertices according to decreasing degrees, we get the sequence e, c, g, a, b, d, f, h.

Use the colour c_1 to colour (paint) e and a.
Use the colour c_2 to paint c, d and h.
Use third colour c_3 to paint vertices g, b and f.

Thus all the vertices are painted such that no adjacent vertices get the same colour. Hence $\chi(G)=3$.

Some rules for colouring a graph

1. $\chi(G) \leq |V|$, where $|V|$ is the number of vertices of G.
2. A triangle always requires three colours, that is, $\chi(K_3)=3$. Similarly $\chi(K_n)=n$, where K_n is the complete graph of n vertices.
3. If some subgraph of G requires k-colours, then $\chi(G) \geq k$.
4. If $\deg(v)=n$, then at most n colours are required to colour the vertices adjacent to v.
5. $\chi(G)=\max\{\chi(C): C \text{ is a connected component of } G\}$.
6. Every n-colourable graph has at least n vertices v such that $\deg(v) \geq n-1$.
7. $\chi(G) \leq 1+\Delta(G)$, where $\Delta(G)$ is the largest degree of any vertex of G.
8. The following statements are equivalent:
   (i) A graph G is 2-colourable
   (ii) G is bipartite
   (iii) Every cycle of G has even length
9. If $\delta(G)$ is the smallest degree of any vertex of G, then

$$\chi(G) \geq \frac{|v|}{|v|-\delta(G)}.$$

EXAMPLE 8.54

Find chromatic number of the graphs shown in Figure 8.134.

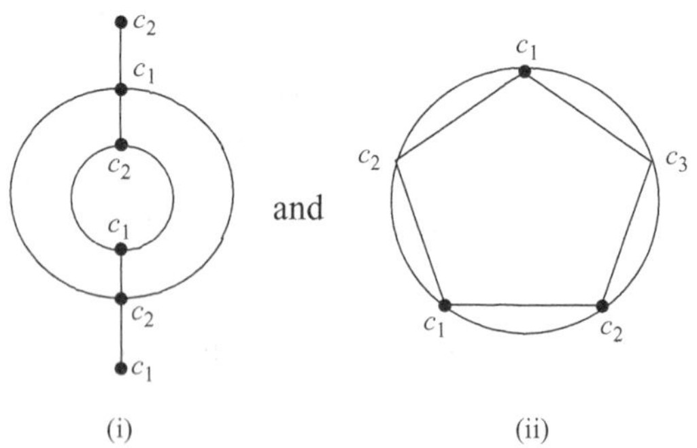

Figure 8.134

Solution.

The graph (i) is 2-colourable whereas the graph (ii) is 3-colourable.

Theorem 8.19

Bipartite graph is 2-colourable unless G is edgeless.

Proof. A two colouring is obtained by assigning one colour to every vertex in one of the bipartition parts and another colour to every vertex in the other partition part as shown in Figure 8.135.

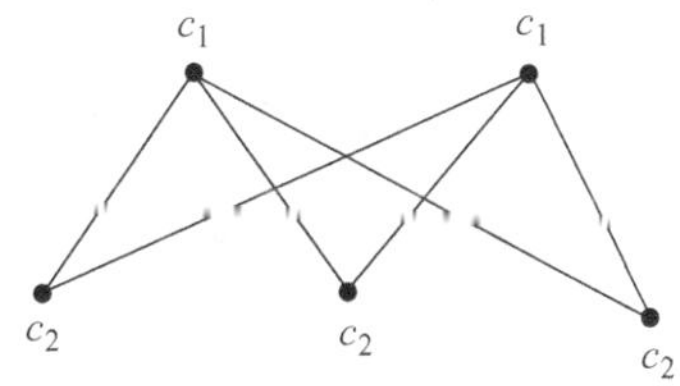

Figure 8.135

Corollary 8.3

Even cycle graphs C_{2n} have $\chi(C_{2n})=2$.

Proof. An even cycle graph is bipartite and therefore, by Theorem 8.19, its chromatic number is 2.

For example consider the graph shown in Figure 8.136.

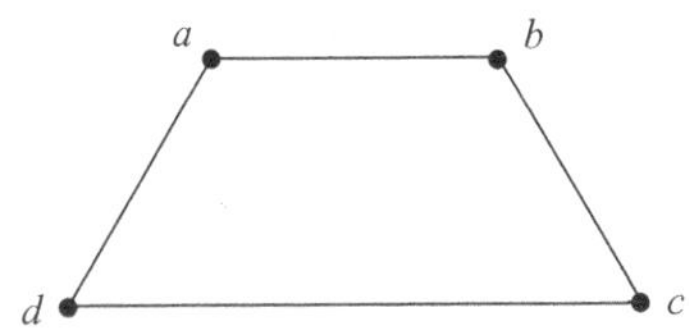

Figure 8.136

It is equivalent to the following bipartite graph shown in Figure 8.137.

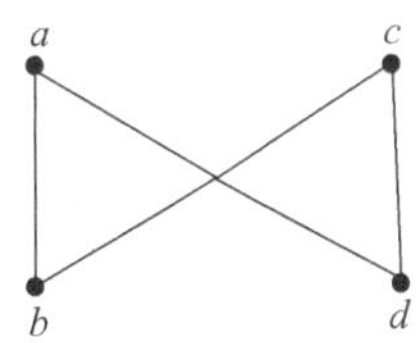

Figure 8.137

Therefore, it is 2-colourable.

Proposition 8.2

Odd cycle graph C_{2n+1} has $\chi(C_{2n+1})=3$.

Proof. Let $v_1, v_2, \ldots, v_{2n}, v_{2n+1}$ be vertices of a cycle graph c_{2n+1}. If two colours were sufficient, then they would have to alternate around the cycle. Thus, the odd subscripted would have to be one colour and the even subscripted vertices have to be second colour but vertex v_{2n+1} is adjacent to v_1, and so according to this scheme two adjacent vertices v_1 and v_{2n+1} have the same colour. This is a contradiction and so C_{2n+1} is not 2-colourable but 3-colourable.

It follows, therefore, that "The chromatic number of a cycle is either two or three, depending on whether its length is even or odd."

Definition 8.74

The **join** $G+H$ of the graphs G and H is obtained from the graph $G \cup H$ by adding an edge between each vertex of H.

For example, if we have a graph as shown below (Figure 8.138):

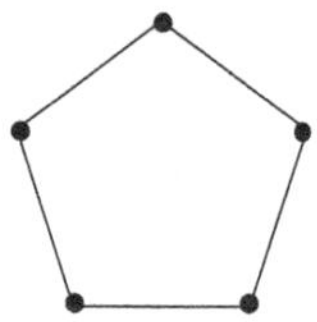

(G)

Figure 8.138

and a graph

$$\bullet K_1,$$

then the join $G+K_1$ of the graphs G and K_1 is the graph as shown below (Figure 8.139):

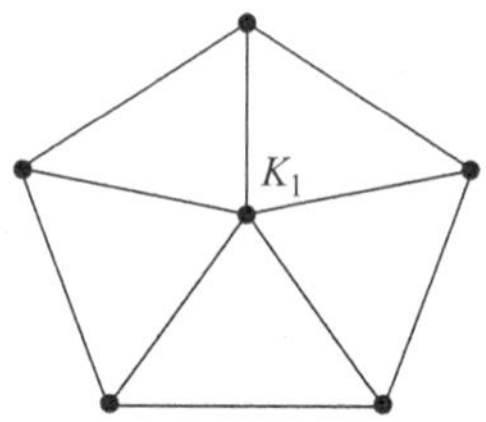

Figure 8.139

Regarding the chromatic number of the join $G+H$ of the graphs G and H we have the following:

Proposition 8.3

The join of a graph G and H has chromatic number

$$\chi(G+H)=\chi(G)+\chi(H).$$

Definition 8.75

The n vertex wheel graph W_n is called an **odd-order wheel** if n is odd and an **even-order wheel** if n is even.

Proposition 8.4

Odd-order wheel graph has

$$\chi(W_{2m+1})=3, \quad \text{for all } m \geq 1.$$

Proof. Using the fact that the wheel graph W_{2m+1} is the join $C_{2m}+K_1$, it follows that

$$\chi(W_{2m+1})=\chi(C_{2m})+\chi(K_1)=2+1=3.$$

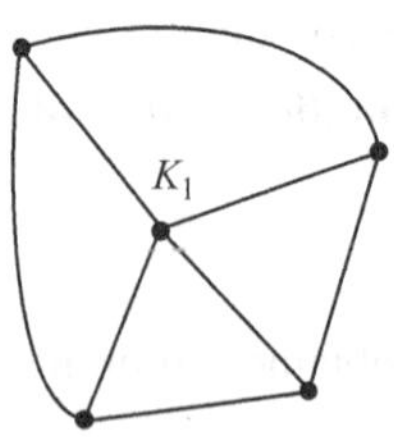

Figure 8.140 (W_5)

Proposition 8.5

Even-order wheel graph has $\chi(W_{2m})=4$.

Proof. Using the fact that $W_{2m}=c_{2m-1}+K_1$, it follows that

$$\chi(W_{2m})=\chi(c_{2m-1})+\chi(K_1)=3+1=4.$$

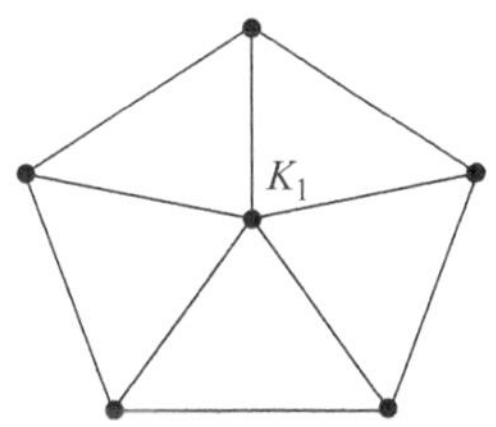

Figure 8.141 (W_6)

EXAMPLE 8.55

Find chromatic number of the graph shown in Figure 8.142:

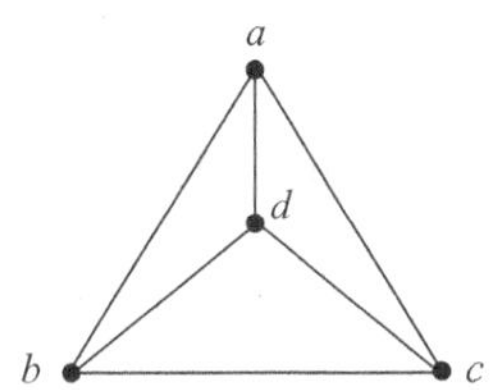

Figure 8.142

Solution.

Here degree of each vertex is 3. Therefore, $\chi(G)\leq 1+3=4$. Since it has a triangle, $\chi(G)\geq 3$. Thus, $3\leq \chi(G)\leq 4$.

We first colour the triangle *abd* with colours c_1, c_2, c_3, respectively. Since *c* is adjacent to each of *a*, *b* and *d*, therefore a different colour c_4 have to be given to it. Hence the graph is 4-colourable and so $\chi(G)=4$.

Note: If we apply rule 9, then $\chi \geq \frac{4}{(4-3)} = 4$. Hence $\chi(G) = 4$.

Theorem 8.20 (Four-Colour Theorem of Appel and Haken)

Any planar graph is 4-colourable.

(Proof of the theorem is out of the scope of this book).

8.11 DIRECTED GRAPHS

Definition 8.76

A **directed graph** or **digraph** consists of two finite sets:

(i) A set *V* of vertices (or nodes or points).

(ii) A set *E* of directed edges (or arcs), where each edge is associated with an ordered pair (*v*, *w*) of vertices called its endpoints. If edge *e* is associated with the ordered pair (*v*, *w*), then *e* is said to be **directed edge from** *v* **to** *w*.

The directed edges are indicated by arrows.

We say that edge *e*=(*v*, *w*) is incident from *v* and is incident into *w*.

The vertex *v* is called **initial vertex** and the vertex *w* is called the **terminal vertex** of the directed edge (*v*, *w*).

Definition 8.77

Let G be a directed graph. The **outdegree of a vertex** v **of** G is the number of edges beginning at v. It is denoted by outdeg(v).

Definition 8.78

Let G be a directed graph. The **indegree of a vertex** v **of** G is the number of edges ending at v. It is denoted by indeg(v).

EXAMPLE 8.56

Consider the directed graph shown below (Figure 8.143):

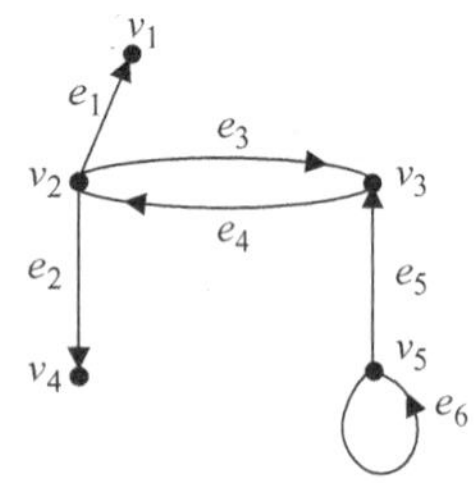

Figure 8.143

Here edge e_1 is (v_2, v_1) whereas e_6 is denoted by (v_5, v_5) and is called a loop. The indegree of v_2 is 1, outdegree of v_2 is 3.

Definition 8.79

A vertex with 0 indegree is called a **source**, whereas a vertex with 0 outdegree is called a **sink**.
For instance, in the above example, v_1 is a sink.

Definition 8.80

If the edges and/or vertices of a directed graph G are labelled with some type of data, then G is called a **labelled directed graph.**

Definition 8.81

Let G be a **directed graph** with ordered vertices $v_1, v_2, \ldots, v_n$. **The adjacency matrix of** G **is the matrix** $A=(a_{ij})$ over the set of non-negative integers such that

$$a_{ij}=\text{the number of arrows from } v_i \text{ to } v_j,\ i, j=1, 2, \ldots, n.$$

EXAMPLE 8.57

Find the adjacency matrices for the graphs given below (Figure 8.144):

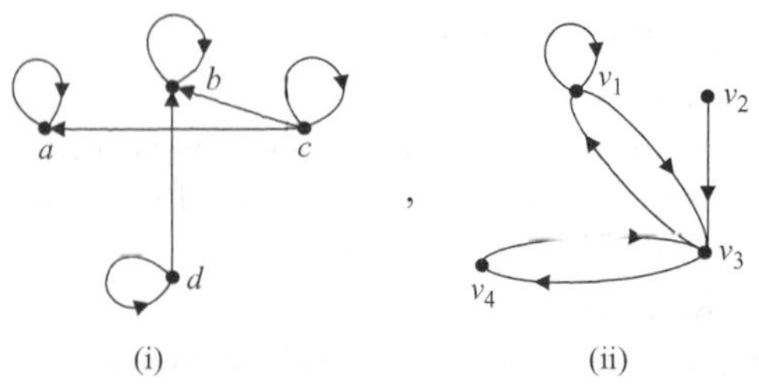

Figure 8.144

Solution.

(i) The edges in the directed graph are (a, a), (b, b), (c, c), (d, d), (c, a), (c, b) and (d, b). Therefore the adjacency matrix $A=(a_{ij})$ is

$$\begin{bmatrix} 1 & 0 & 0 & 0 \\ 0 & 1 & 0 & 0 \\ 1 & 1 & 1 & 0 \\ 0 & 1 & 0 & 1 \end{bmatrix}.$$

(ii) The edges in the graph in (ii) are (v_2, v_3), (v_1, v_1), (v_1, v_3), (v_3, v_1), (v_3, v_4), (v_4, v_3). Hence the adjacency matrix is

$$\begin{bmatrix} 1 & 0 & 1 & 0 \\ 0 & 0 & 1 & 0 \\ 1 & 0 & 0 & 1 \\ 0 & 0 & 1 & 0 \end{bmatrix}.$$

EXAMPLE 8.58

Find the directed graph represented by the adjacency matrix:

$$\begin{bmatrix} 0 & 1 & 0 & 0 & 0 \\ 0 & 0 & 1 & 0 & 0 \\ 0 & 0 & 0 & 1 & 1 \\ 1 & 1 & 0 & 0 & 0 \\ 0 & 0 & 0 & 0 & 0 \end{bmatrix}.$$

Solution.

We observe that $a_{12}=1$, $a_{23}=1$, $a_{34}=1$, $a_{35}=1$, $a_{41}=1$, $a_{42}=1$. Hence the digraph is as shown in Figure 8.145.

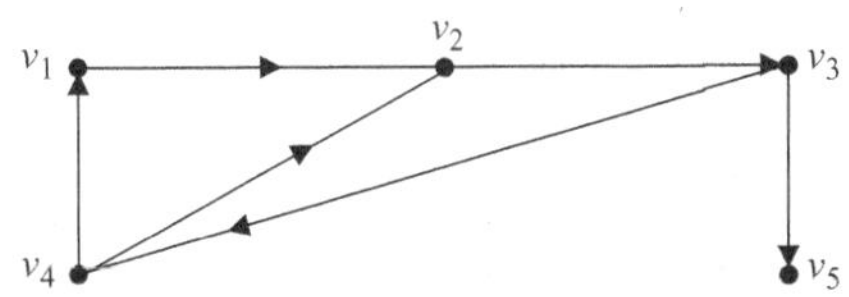

Figure 8.145

Definition 8.82

In a directed graph, if there is no more than one directed edge in a particular direction between a pair of vertices, then it is called **simple directed graph**.

For example, the graph shown in Figure 8.146 is a simple directed graph.

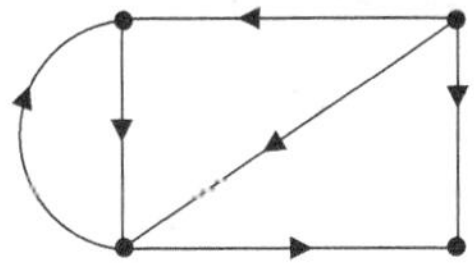

Figure 8.146

A directed graph which is not simple is called **directed multi-graph**.

Definition 8.83

A simple digraph is said to be **strongly connected** if for each pair of vertices v, w, there are path, from v to w and w to v.

For example, the graphs show in Figure 8.147 are strongly connected.

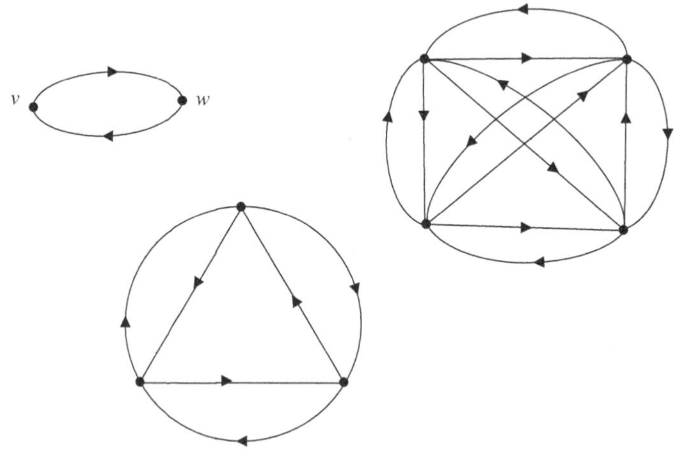

Figure 8.147

Definition 8.84

Let G be a simple digraph with n vertices. Then a $n \times n$ matrix $P=(p_{ij})$ such that

$$p_{ij} = \begin{cases} 1 & \text{if there is a path from } v_i \text{ to } v_j \\ 0 & \text{otherwise} \end{cases}$$

is called **path matrix** or **reachability matrix** of G.

The path matrix can be given in term of adjacency matrix:

$$P = A \vee A^{(2)} \vee A^{(3)} \vee \cdots \vee A^{(n)}$$

and

$$A^{(2)} = A \odot A, \quad A^{(r)} = A^{(r-1)} \odot A,$$

where $\vee$ represents Boolean addition and $\odot$ denotes the Boolean matrix multiplication.

Warshall's algorithm for finding path matrix from adjacency matrix

We discuss this algorithm with the help of the following example.

EXAMPLE 8.59

Find the adjacency matrix and the path matrix for the digraph shown in Figure 8.148.

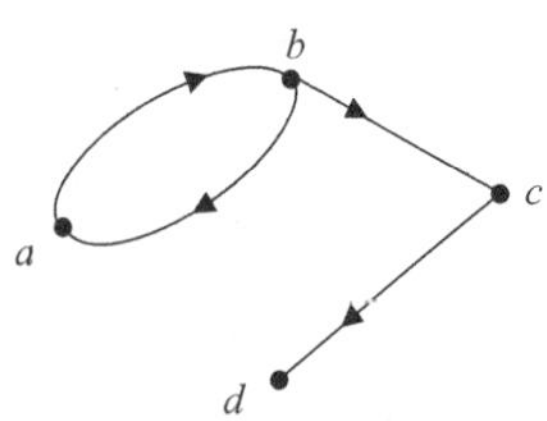

Figure 8.148

Solution.

The adjacency matrix of this digraph is:

$$A=\begin{bmatrix}0&1&0&0\\1&0&1&0\\0&0&0&1\\0&0&0&0\end{bmatrix}.$$

Then the path matrix $P=(p_{ij})$ is found as follows: We take

$$P_0=\begin{bmatrix}0&1&0&0\\1&0&1&0\\0&0&0&1\\0&0&0&0\end{bmatrix}=A \text{ itself.}$$

Now we find P_1 by consulting column 1 and row 1. We note that P_0 has 1 in location **2** of column 1 and location **2** of row 1. Thus the element p_{22} is replaced by 1 in P_0. Thus we have

$$P_1=\begin{bmatrix}0&1&0&0\\1&1&1&0\\0&0&0&1\\0&0&0&0\end{bmatrix}.$$

To find P_2 we consult column 2 and row 2 of P_1. We note that P_1 has 1 in locations 1 and 2 of column 2 and locations 1, 2 and 3 of row 2. Thus, to obtain P_2 we should put 1 in positions $p_{11}, p_{12}, p_{13}, p_{21}, p_{22}$ and p_{23} of matrix P_1. We thus have

$$P_2=\begin{bmatrix}1&1&1&0\\1&1&1&0\\0&0&0&1\\0&0&0&0\end{bmatrix}.$$

We now consult column 3 and row 3 of P_2. The matrix P_2 has 1 in locations 1 and 2 of column 3 and 1 in location 4 of row 3. Thus we shall obtain P_3 by putting 1 in positions p_{14}, p_{24}. Thus

$$P_3=\begin{bmatrix}1&1&1&1\\1&1&1&1\\0&0&0&1\\0&0&0&0\end{bmatrix}.$$

Finally, we consult column 4 and row 4 of P_3. We have 1 in location 1, 2, 3 of column 4 and there is no 1 in row 4. Hence $P_4=P_3$. Thus the path matrix is

$$P=(p_{ij})=\begin{bmatrix}1&1&1&1\\1&1&1&1\\0&0&0&1\\0&0&0&0\end{bmatrix}.$$

Remark 8.11 Given an adjacency matrix of a graph G, we first draw digraph of G and then apply definition of path matrix. We will obtain path matrix.

Now, we state a result without proof.

Theorem 8.21

Let A be the adjacency matrix of a graph G with n vertices and let $B_n = A + A^2 + A^3 + \cdots + A^n$. Then the path matrix P and B_n have the same non-zero entries.

Let G be a strongly connected directed graph. Then for any pair of vertices v and w in G, there is a path from v to w and from w to v. Accordingly, G is **strongly connected if and only if the path matrix P of G has no zero entries**.

Thus, in view of the above theorem, we have:

Theorem 8.22

Let A be adjacency matrix of a graph G with n vertices and let $B_n = A + A^2 + \ldots + A^n$. Then G is strongly connected if and only if B_n has no zero entries.

EXAMPLE 8.60

A directed graph has the following adjacency matrix. Check whether it is strongly connected.

$$A(G) = \begin{bmatrix} 0 & 1 & 1 \\ 1 & 0 & 1 \\ 1 & 1 & 0 \end{bmatrix}.$$

Solution.

To find the path matrix, we first consider column 1 and row 1 of

$$P_0 = \begin{bmatrix} 0 & 1 & 1 \\ 1 & 0 & 1 \\ 1 & 1 & 0 \end{bmatrix} = A(G).$$

In column 1, the entry 1 is located at 2 and 3 positions and 1 is at locations 2 and 3 in row 1. Thus put 1 at $p_{22}, p_{23}, p_{32}, p_{33}$ and get

$$P_1 = \begin{bmatrix} 0 & 1 & 1 \\ 1 & 1 & 1 \\ 1 & 1 & 1 \end{bmatrix}.$$

We note that 1 is located in column 2 of P_1 at position 1, 2, 3, and 1 is located at position 1, 2, 3 of row 2. Thus put 1 at $p_{11}, p_{12}, p_{13}, p_{21}, p_{22}, p_{23}, p_{31}, p_{32}, p_{33}$, and get

$$P_2 = \begin{bmatrix} 1 & 1 & 1 \\ 1 & 1 & 1 \\ 1 & 1 & 1 \end{bmatrix}.$$

The next step will not change the position. Hence,

$$P = \begin{bmatrix} 1 & 1 & 1 \\ 1 & 1 & 1 \\ 1 & 1 & 1 \end{bmatrix} \quad \text{(all non-zero entries).}$$

Therefore, G is strongly connected.

The graph of G is as shown below (Figure 8.149):

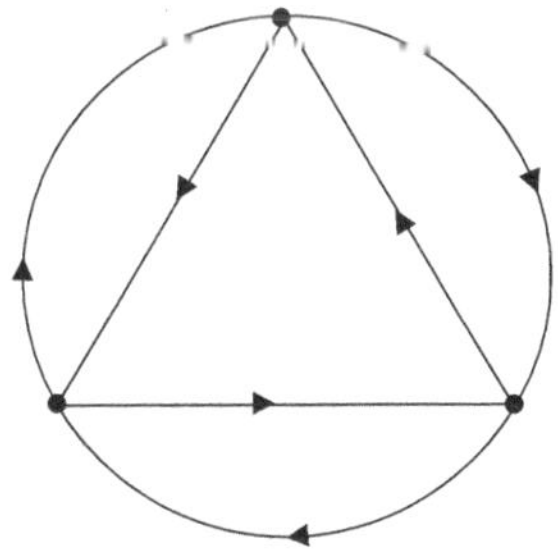

Figure 8.149

Remark 8.12 In the above example, we also note that

$$B_3 = A + A^2 + A^3$$

$$= \begin{bmatrix} 0 & 1 & 1 \\ 1 & 0 & 1 \\ 1 & 1 & 0 \end{bmatrix} + \begin{bmatrix} 2 & 1 & 1 \\ 1 & 2 & 1 \\ 1 & 1 & 2 \end{bmatrix} + \begin{bmatrix} 2 & 3 & 3 \\ 3 & 2 & 3 \\ 3 & 3 & 2 \end{bmatrix} = \begin{bmatrix} 4 & 5 & 5 \\ 5 & 4 & 5 \\ 5 & 5 & 4 \end{bmatrix} \quad \text{(all non-zero entries)}$$

showing that G is strongly connected.

8.12 TREES

Definition 8.85

A graph is said to be a **tree** if it is a connected acyclic graph.

A **trivial tree** is a graph that consists of a single vertex. An **empty tree** is a tree that does not have any vertices or edges.

For example, the graphs shown in Figure 8.150 are all trees.

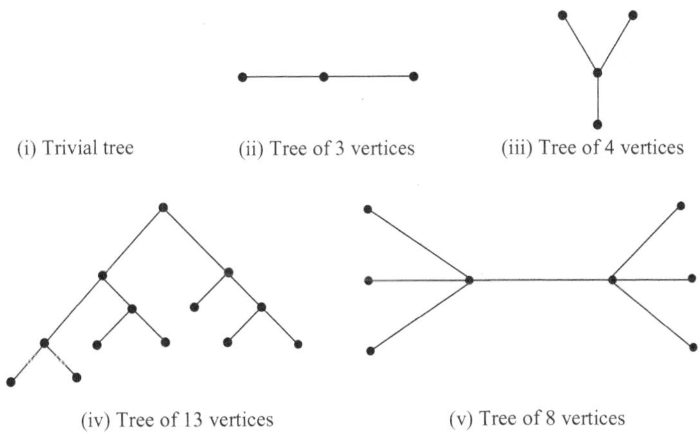

(i) Trivial tree (ii) Tree of 3 vertices (iii) Tree of 4 vertices

(iv) Tree of 13 vertices (v) Tree of 8 vertices

Figure 8.150

But the graphs shown in Figure 8.151 are not trees:

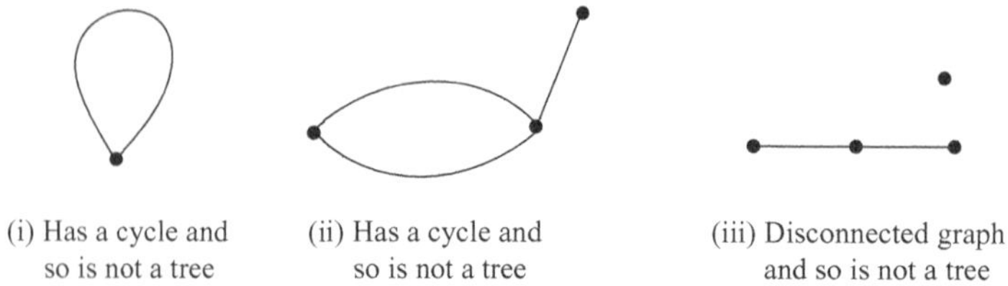

Figure 8.151

Definition 8.86

A collection of disjoint trees is called a **forest**.

Thus a graph is a forest if and only if it is circuit free.

Definition 8.87

A vertex of degree 1 in a tree is called a **leaf** or a **terminal node** or a **terminal vertex**.

Definition 8.88

A vertex of degree greater than 1 in a tree is called a **branch node** or **internal node** or **internal vertex**.

Consider the tree shown in Figure 8.152.

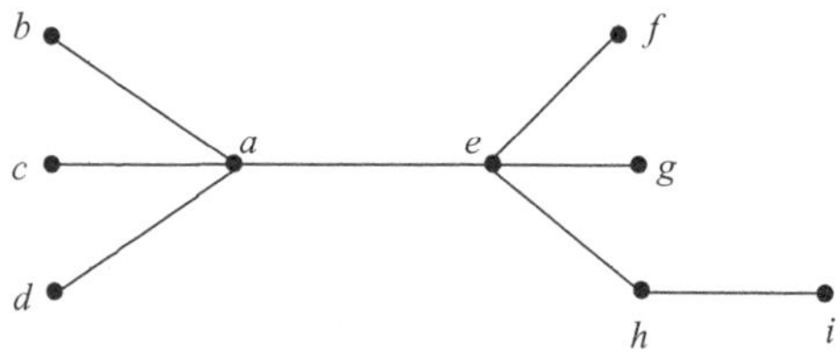

Figure 8.152

In this tree, the vertices b, c, d, f, g, i are leaves whereas the vertices a, e, h are branch nodes.

8.12.1 Characterization of Trees

We have the following interesting characterization of trees:

Lemma 8.1

A tree that has more than one vertex has at least one vertex of degree 1.

Proof. Let T be a particular but arbitrary chosen tree having more than one vertex (Figure 8.153).

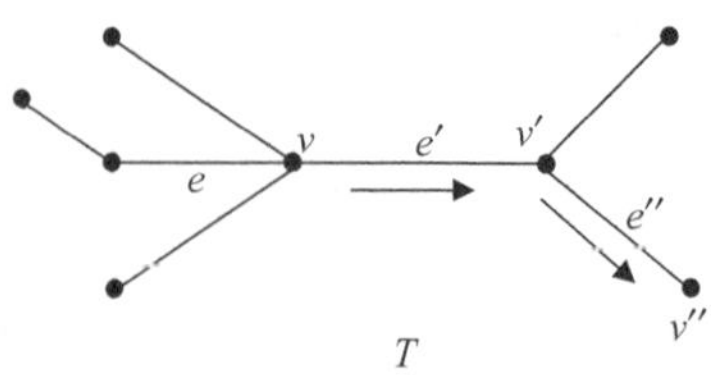

Figure 8.153

1. Choose a vertex v of T. Since T is connected and has at least two vertices, v is not isolated and there is an edge e incident on v.
2. If $\deg(v) > 1$, there is an edge $e' \neq e$ because there are, in such a case, at least two edges incident on v. Let v' be the vertex at the other end of e'. This is possible because e' is not a loop by the definition of a tree.
3. If $\deg(v') > 1$, then there are at least two edges incident on v'. Let e'' be the other edge different from e' and v'' be the vertex at other end of e''. This is again possible because T is acyclic.
4. If $\deg(v'') > 1$, repeat the above process. Since the number of vertices of a tree is finite and T is circuit free, the process must terminate and we shall arrive at a vertex of degree 1.

Remark 8.13 In the proof of the Lemma 8.1, after finding a vertex of degree 1, if we return to v and move along a path outward from v starting with e, we shall reach to a vertex of degree 1 again. Thus it follows that **"Any tree that has more than one vertex has at least two vertices of degree 1".**

Lemma 8.2

There is a unique path between every two vertices in a tree.

Proof. Suppose on the contrary that there are more than one path between any two vertices in a given tree T. Then T has a cycle which contradicts the definition of a tree because T is acyclic. Hence the lemma is proved.

Lemma 8.3

The number of vertices is one more than the number of edges in a tree.
Equivalently, we may state this lemma as

"For any positive integer n, a tree with n vertices has $n-1$ edges".

Proof. We shall prove the lemma by mathematical induction.

Let T be a tree with **one** vertex. Then T has no edges, that is, T has 0 edge. But $0=1-1$. Hence the lemma is true for $n=1$.

Suppose that the lemma is true for $k > 1$. We shall show that it is then true also for $k+1$. Since the lemma is true for k, the tree has k vertices and $k-1$ edges. Let T be a tree with $k+1$ vertices. Since k is +ve, $k+1 \geq 2$ and so T has more than one vertex. Hence, by Lemma 8.1, T has a vertex v of degree 1. Also there is another vertex w and so there is an edge e connecting v and w. Define a subgraph T' of T so that

$$V(T') = V(T) - \{v\},$$
$$E(T') = E(T) - \{e\}.$$

Then number of vertices in $T' = (k+1) - 1 = k$ and since T is circuit free and T' has been obtained on removing one edge and one vertex, it follows that T' is acyclic. Also T' is connected. Hence T' is a tree having k vertices and therefore by induction hypothesis, the number of edges in T' is $k-1$.

But then

$$\text{Number of edges in } T = \text{number of edges in } T' + 1$$
$$= k - 1 + 1 = k.$$

Thus the Lemma is true for tree having $k+1$ vertices. Hence the lemma is true by mathematical induction.

Corollary 8.4

Let $C(G)$ denote the number of components of a graph. Then a forest G on n vertices has $n-C(G)$ edges.

Proof. Apply Lemma 8.3 to each component of the forest G.

Corollary 8.5

Any graph G on n vertices has at least $n-C(G)$ edges.

Proof. If G has cycle-edges, remove them one at a time until the resulting graph G^* is acyclic. Then G^* has $n-C(G^*)$ edges by Corollary 8.4. Since we have removed only circuit, $C(G^*)=C(G)$. Thus G^* has $n-C(G)$ edges. Hence G has at least $n-C(G)$ edges.

Lemma 8.4

A graph in which there is a unique path between every pair of vertices is a tree

(This lemma is converse of Lemma 8.2).

Proof. Since there is a path between every pair of points, therefore the graph is connected. Since a path between every pair of points is unique, there does not exist any circuit because existence of circuit implies existence of distinct paths between pair of vertices. Thus the graph is connected and acyclic and so is a tree.

Lemma 8.5 (Converse of Lemma 8.3)

A connected graph G with $e=v-1$ is a tree

First Proof. The given graph is connected and $e=v-1$. To prove that G is a tree, it is sufficient to show that G is acyclic. Suppose on the contrary that G has a cycle. Let m be the number of vertices in this cycle. Also, we know that **number of edges in a cycle is equal to number of vertices in that cycle**. Therefore number of edges in the present case is m. Since the graph is connected, every vertex of the graph which is not in cycle must be connected to the vertices in the cycle (see Figure 8.154).

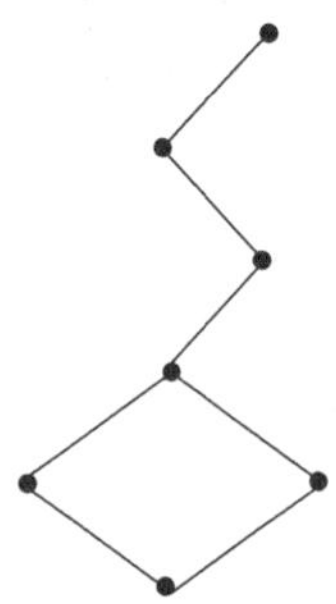

Figure 8.154

Now each edge of the graph that is not in the cycle can connect only one vertex to the vertices in the cycle. There are $v-m$ vertices that are not in the cycle. So the graph must contain at least $v-m$ edges that are not in the cycle. Thus we have $e\geq v-m+m=v$, which is a contradiction to our hypothesis. Hence there is no cycle and so the graph is a tree.

Second Proof. We shall show that a connected graph with v vertices and $v-1$ edges is a tree. It is sufficient to show that G is acyclic. Suppose on the contrary that G is not circuit free and has a nontrivial circuit C. If we remove one edge of C from the graph G, we obtain a graph G' which is connected (see Figure 8.155).

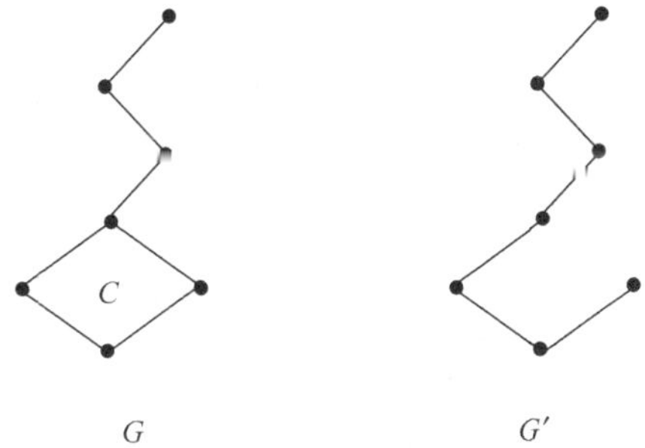

Figure 8.155

If G' still has a nontrivial circuit, we repeat the above process and remove one edge of that circuit obtaining a new connected graph. Continuing this process, we obtain a connected graph G^* which is circuit free. Hence G^* is a tree. Since no vertex has been removed, the tree G^* has v vertices. Therefore, by Lemma 8.3, G^* has $v-1$ edges. But at least one edge of G has been removed to form G^*. This means that G^* has not more than $v-1-1=v-2$ edges. Thus, we arrive at a contradiction. Hence our supposition is wrong and G has no cycle. Therefore G is connected and cycle free and so is a tree.

Lemma 8.6

A graph G with $e=v-1$, that has no circuit is a tree.

Proof. It is sufficient to show that G is connected. Suppose G is not connected and let G', G'', … be the connected components of G. Since each of G', G'', … is connected and has no cycle, they all are tree. Therefore, by Lemma 8.3,

$$e'=v'-1,$$
$$e''=v''-1,$$
$$\overline{\qquad\qquad}$$
$$\overline{\qquad\qquad},$$

where e', e'', … are the number of edges and v', v'', … are the number of vertices in G', G'', …, respectively. We have, on adding

$$e'+e''+\cdots=(v'-1)+(v''-1)+\cdots$$

Since

$$e=e'+e''+\cdots,$$
$$v=v'+v''+\cdots,$$

we have

$$e<v-1,$$

which contradicts our hypotheses. Hence G is connected. So G is connected and acyclic and is therefore a tree.

EXAMPLE 8.61

Construct a graph that has six vertices and five edges but is not a tree.

Solution.

We have

$$\text{Number of vertices}=6$$
$$\text{Number of edges}=5$$

So the condition $e=v-1$ is satisfied. Therefore, to construct graph with six vertices and five edges that is not a tree, we should keep in mind that the graph should not be connected. The graph shown in Figure 8.156 has six vertices and five edges but is not connected. Hence it is not a tree.

Figure 8.156

Definition 8.89

A directed graph is said to be a **directed tree** if it becomes a tree when the direction of edges are ignored.
For example, the graph shown in Figure 8.157 is a directed tree.

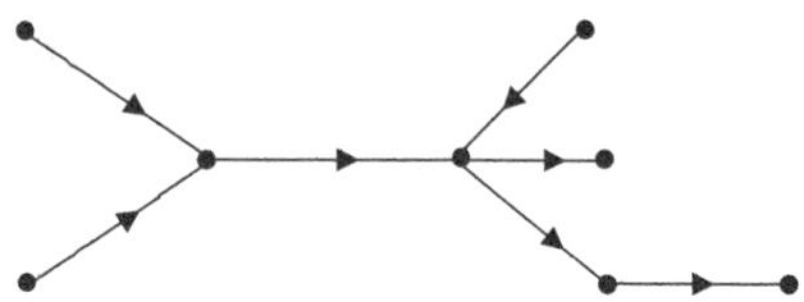

Figure 8.157

Definition 8.90

A directed tree is called a **rooted tree** if there is exactly one vertex whose incoming degree is 0 and the incoming degrees of all other vertices are 1.
The vertex with incoming degree 0 is called the **root** of the rooted tree.
A tree T with root v_0 will be denoted by (T, v_0).

Definition 8.91

In a rooted tree, a vertex, whose outgoing degree is 0 is called a **leaf** or **terminal node**, whereas a vertex whose outgoing degree is non-zero is called a **branch node** or an **internal node**.

Definition 8.92

Let u be a branch node in a rooted tree. Then a vertex v is said to be **child** (**son** or **offspring**) of u if there is an edge from u to v. In this case, u is called **parent** (**father**) of v.

Definition 8.93

Two vertices in a rooted tree are said to be **siblings** (**brothers**) if they are both children of same parent.

Definition 8.94

A vertex v is said to be a **descendent** of a vertex u if there is a unique directed path from u to v.
In this case, u is called the **ancestor** of v.

Definition 8.95

The **level** (or **path length**) of a vertex u in a rooted tree is the number of edges along the unique path between u and the root.

Definition 8.96

The **height** of a rooted tree is the maximum level to any vertex of the tree.

As an example of these terms, consider the rooted tree shown in Figure 8.158:

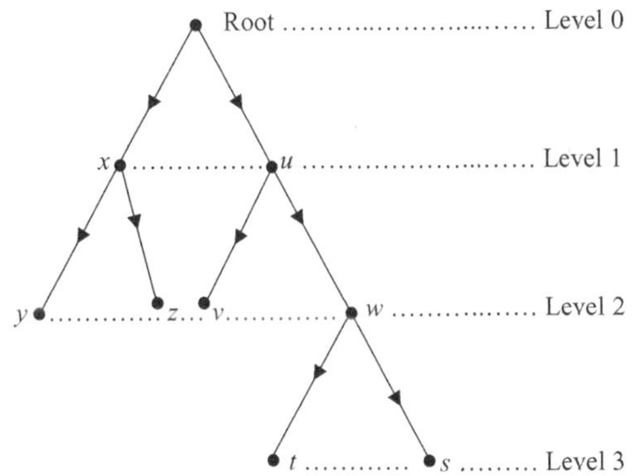

Figure 8.158

Here y is a child of x; x is the parent of y and z. Thus y and z are siblings. The descendents of u are v, w, t and s. Levels of vertices are shown in the figure. The height of this rooted tree is 3.

Definition 8.97

Let u be a branch node in the tree $T=(V, E)$. Then the subgraph $T'=(V', E')$ of T such that the vertices set V' contains u and all of its descendents and E' contains all the edges in all directed paths emerging from u is called a **subtree** with u as the root.

Definition 8.98

Let u be a branch node. By a subtree of u, we mean a subtree that has child of u as root.

In the above example, we note that the Figure 8.159 is a subtree of T,

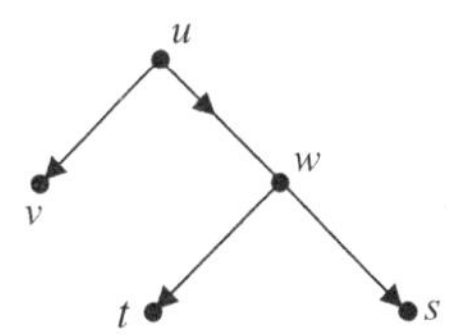

Figure 8.159

whereas the Figure 8.160 is a subtree of the branch node u.

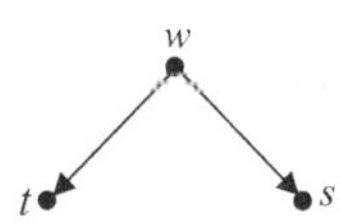

Figure 8.160

EXAMPLE 8.62

Let

$$V = \{v_1, v_2, v_3, v_4, v_5, v_6, v_7, v_8\}$$

and let

$$E = \{(v_2, v_1), (v_2, v_3), (v_4, v_2), (v_4, v_5), (v_4, v_6), (v_6, v_7), (v_5, v_8)\}.$$

Show that (V, E) is rooted tree. Identify the root of this tree.

Solution.

We note that

Incoming degree of $v_1 = 1$,
Incoming degree of $v_2 = 1$,
Incoming degree of $v_3 = 1$,
Incoming degree of $v_4 = 0$,
Incoming degree of $v_5 = 1$,
Incoming degree of $v_6 = 1$,
Incoming degree of $v_7 = 1$,
Incoming degree of $v_8 = 1$.

Since incoming degree of the vertex v_4 is 0, it follows that v_4 is root.

Further,

Outgoing degree of $v_1 = 0$,
Outgoing degree of $v_3 = 0$,
Outgoing degree of $v_7 = 0$,
Outgoing degree of $v_8 = 0$.

Therefore, v_1, v_3, v_7, v_8 are leaves. Also,

Outgoing degree of $v_2 = 2$,
Outgoing degree of $v_4 = 3$,
Outgoing degree of $v_5 = 1$,
Outgoing degree of $v_6 = 1$.

Now the root v_4 is connected to v_2, v_5 and v_6. So, we have the following diagram:

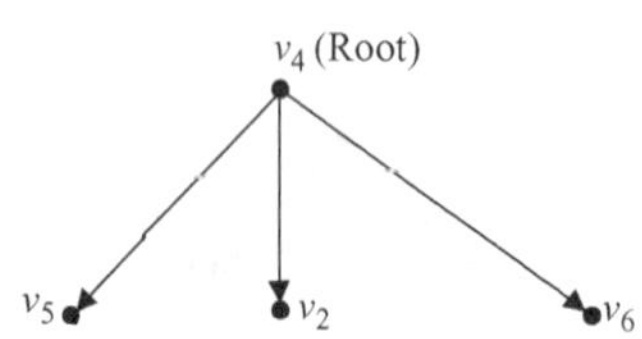

Now v_2 is connected to v_1 and v_3, v_5 is connected to v_8, v_6 is connected to v_7. Thus, we have the graph shown in Figure 8.161.

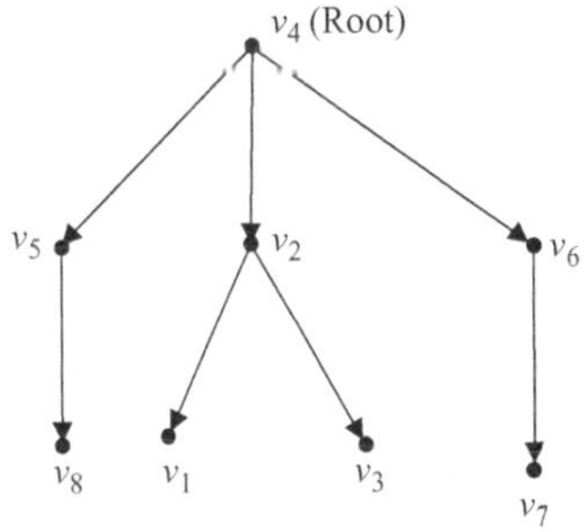

Figure 8.161

We thus have a connected acyclic graph and so (V, E) is a rooted tree with root v_4.

EXAMPLE 8.63

Let F be the set of all female descendent of a lady v_0. We define a relation T on F as follows: if $v_1, v_2, \in F$, then $v_1 \; T \; v_2$ if v_1 is mother of v_2. Show that the relation T on F is a rooted tree with root v_0. Is this relation an equivalence relation?

Solution.

The graph of T is cycle free, because descendents cannot be parents of their parents. The graph is also connected. Hence T is rooted tree. We further note that

(i) Since v_1 is not mother of itself, $(v_1, v_1) \notin T$.

(ii) If $(v_1, v_2) \in T$, then v_1 is mother of v_2. But then v_2 cannot be mother of v_1. Hence $(u_1, v_2) \in T$ does not imply $(v_2, v_1) \in T$.

(iii) If $(v_1, v_2) \in T$, $(v_2, v_3) \in T$, then

$$v_1 \text{ is mother of } v_2 \quad \text{and} \quad v_2 \text{ is mother of } v_3.$$

This does not imply that v_1 is mother of v_3, i.e., $(v_1, v_3) \notin T$. Thus T is irreflexive, asymmetric and is not transitive. Hence T is not an equivalence relation.

Consider the rooted trees T and T' shown in the Figure 8.162, where T is family tree of a man who has two sons, with the elder son having no child and younger having three children.

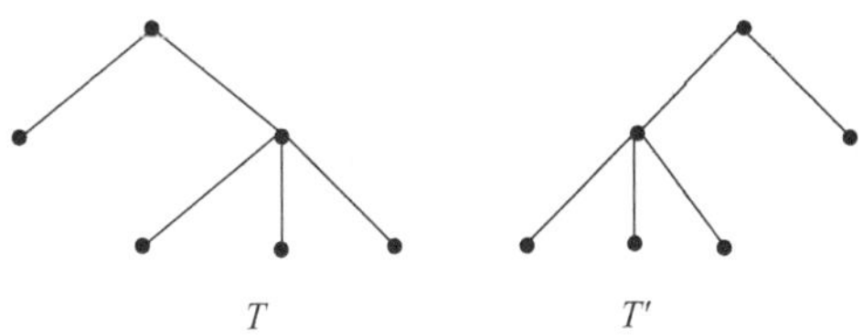

Figure 8.162

Although T'(as a graph) is isomorphic to T, it could be the family tree of another man whose elder son has three children and whose younger son has no child.

The above discussion motivates the following definitions:

Definition 8.99

A rooted tree in which the edges incident from each branch node are labelled with integers 1, 2, 3, … is called an **ordered tree**.

8.13 ISOMORPHISM OF TREES

Definition 8.100

Two ordered trees are said to be **isomorphic** if (i) there exists a one-to-one correspondence between their vertices and edges and that preserves the incident relation (ii) labels of the corresponding edges match.

In view of this definition, the ordered trees shown in Figure 8.163 are not isomorphic.

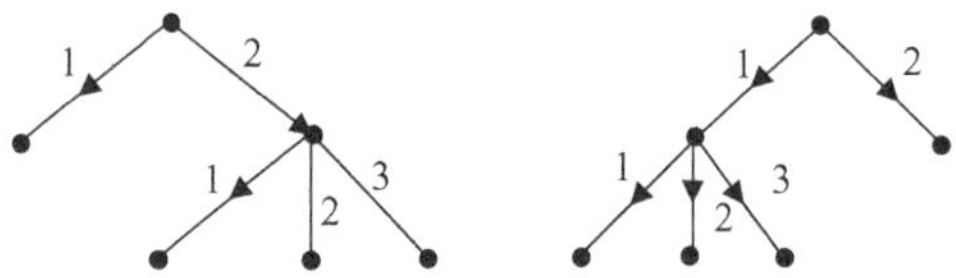

Figure 8.163

EXAMPLE 8.64

Show that the tree T_1 and T_2 shown in the diagram below are isomorphic.

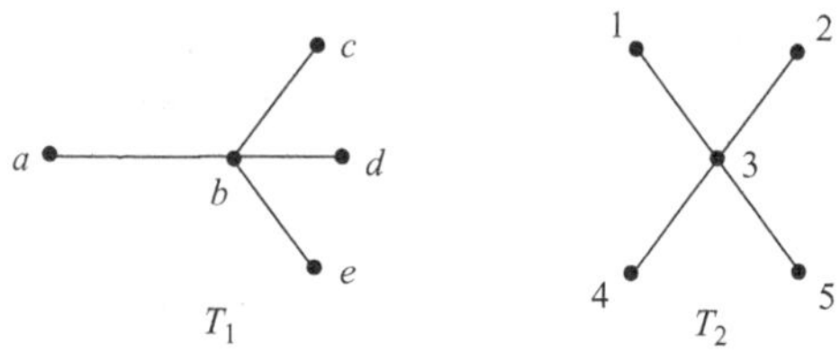

Figure 8.164

Solution.

We observe that in the tree T_1,

$$\deg(b)=4.$$

In the tree T_2,

$$\deg(3)=4.$$

Further, $\deg(a)=\deg(1)=1, \deg(c)=\deg(2), \deg(d)=\deg(4)=\deg(e)=1=\deg(5)$ and $\deg(\text{b})=\deg(3)=4$. Thus we may define a function f from the vertices of T_1 to the vertices of T_2 by

$$f(a)=1, \quad f(b)=3, \quad f(c)=2, \quad f(d)=4, \quad f(e)=5.$$

This is a one-to-one and onto function. Also, adjacency relation is preserved because if v_i and v_j are adjacent vertices in T_1, then $f(v_i)$ and $f(v_j)$ are adjacent vertices in T_2. Hence T_1 is isomorphic to T_2.

EXAMPLE 8.65

Show that the tree T_1 and T_2, shown in the Figure 8.165 are isomorphic

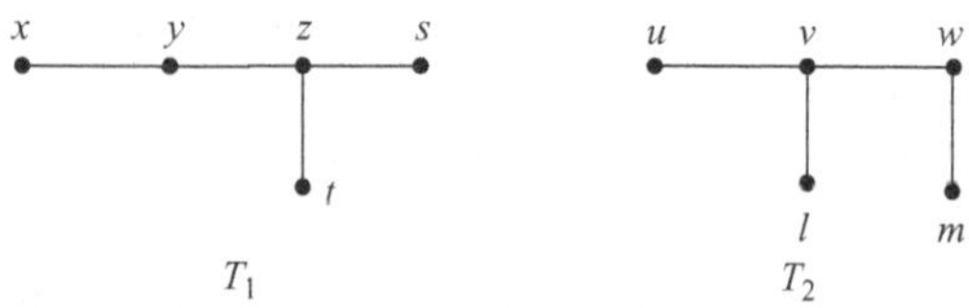

Figure 8.165

Solution.

Let f be a function defined by

$$f(z)=v,\quad f(y)=w,\quad f(x)=m,$$
$$f(s)=u,\quad f(t)=l.$$

Then f is an one-one, onto mapping which preserves adjacency. Hence, T_1 and T_2 are isomorphic.

Remark 8.14 There are three non-isomorphic trees (Figure 8.166) with five vertices:

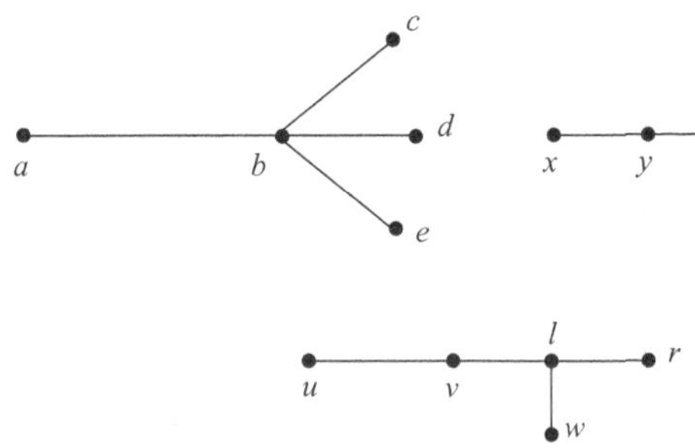

Figure 8.166

Definition 8.101

Let T_1 and T_2 be rooted trees with roots r_1 and r_2, respectively. Then T_1 and T_2 are **isomorphic** if there exists a one-one, onto function f from the vertex set of T_1 to the vertex set of T_2 such that

(i) Vertices v_i and v_j are adjacent in T_1 if and only if the vertices $f(v_i)$ and $f(v_j)$ are adjacent in T_2,

(ii) $f(r_1)=r_2$.

The function f is then called an **isomorphism**.

EXAMPLE 8.66

Show that the trees T_1 and T_2 shown in Figure 8.167 are isomorphic.

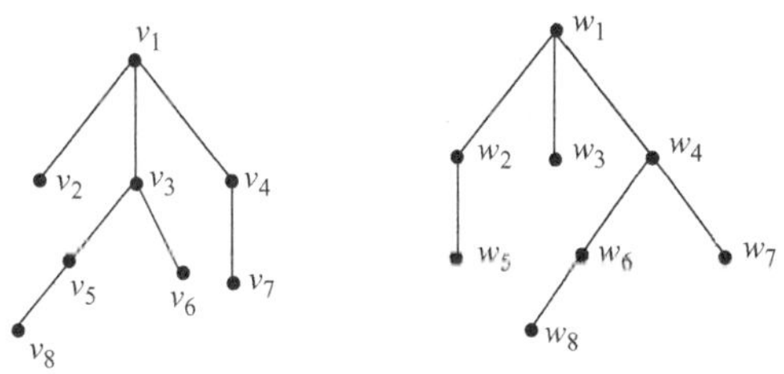

Figure 8.167

Solution.

We observe that T_1 and T_2 are rooted trees.
Define f: (vertex set of T_1)→(vertex set of T_2) by

$$f(v_1)=w_1,\quad f(v_2)=w_3,\quad f(v_3)=w_4$$
$$f(v_4)=w_2,\quad f(v_5)=w_6,\quad f(v_6)=w_7$$
$$f(v_7)=w_5,\quad f(v_8)=w_8.$$

Then f is one-to-one and adjacency relation is preserved. Hence f is an isomorphism and so the rooted trees T_1 and T_2 are isomorphic

EXAMPLE 8.67

Show that the rooted trees shown in Figure 8.168 are not isomorphic:

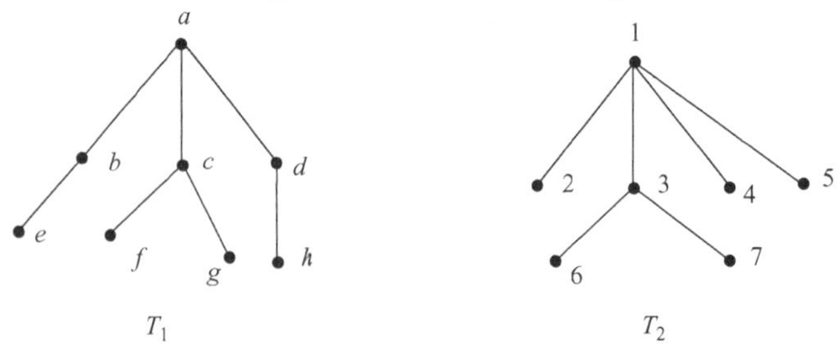

Figure 8.168

Solution.

We observe that the degree of root in T_1 is 3, whereas the degree of root in T_2 is 4. Hence T_1 is not isomorphic to T_2.

Definition 8.102

An ordered tree in which every branch node has at most n offspring is called a **n-ary tree** (or **n-tree**).

Definition 8.103

An n-ary tree is said to be **fully n-ary tree** (**complete n-ary tree** or **regular n-ary tree**) if every branch node has exactly n offspring.

Definition 8.104

An ordered tree in which every branch node has at most two offspring is called a **binary tree** (or **2-tree**).

Definition 8.105

A binary tree in which every branch node (internal vertex) has exactly two offspring is called a **fully binary tree**.

For example, the tree given in Figure 8.169 is a binary tree,

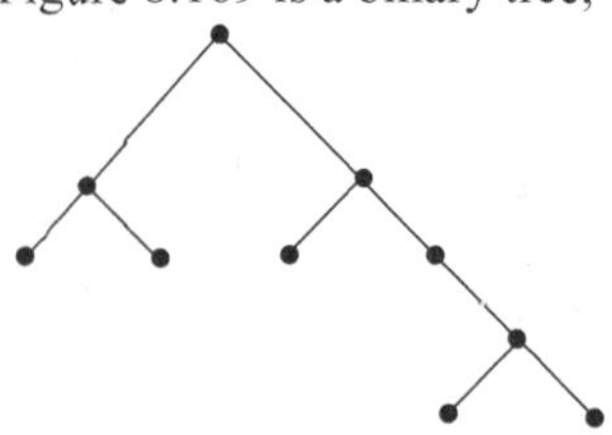

Figure 8.169

whereas the tree shown in Figure 8.170 is a fully binary tree.

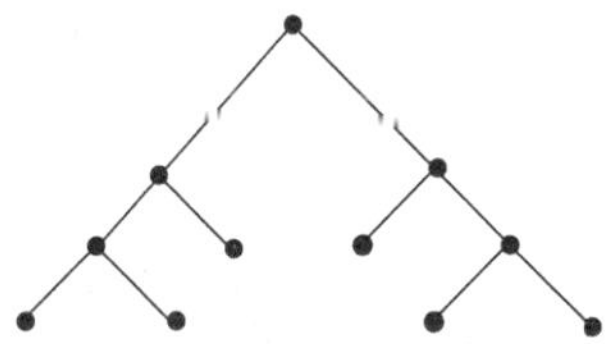

Figure 8.170

Definition 8.106

Let T_1 and T_2 be binary trees with roots r_1 and r_2, respectively. Then T_1 and T_2 are **isomorphic** if there is a one-one, onto function f from the vertex set of T_1 to the vertex set of T_2 satisfying

(i) Vertices v_i and v_j are adjacent in T_1 if and only if the vertices $f(v_i)$ and $f(v_j)$ are adjacent in T_2
(ii) $f(r_1)=r_2$
(iii) v is a left child of w in T_1 if and only if $f(v)$ is a left child of $f(w)$ in T_2
(iv) v is a right child of w in T_1 if and only if $f(v)$ is a right child of $f(w)$ in T_2

The function f is then called an **isomorphism** between binary trees T_1 and T_2.

EXAMPLE 8.68

Show that the trees given in Figure 8.171 are isomorphic.

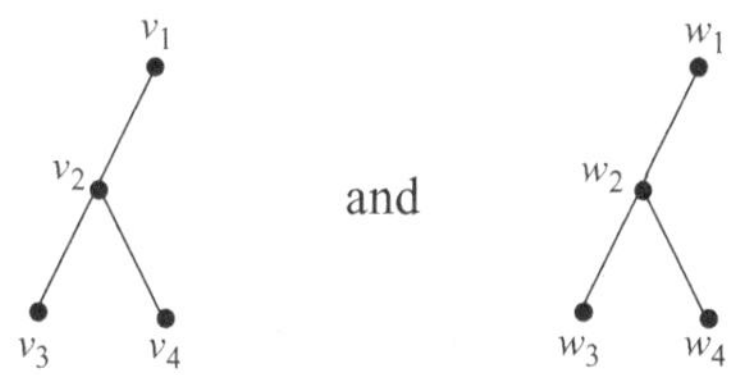

Figure 8.171

Solution.

Define f by $f(v_i)=w_i$, $i=1, 2, 3, 4$. Then f satisfies all the properties for isomorphism. Hence T_1 and T_2 are isomorphic.

EXAMPLE 8.69

Show that the trees given in Figure 8.172 are not isomorphic.

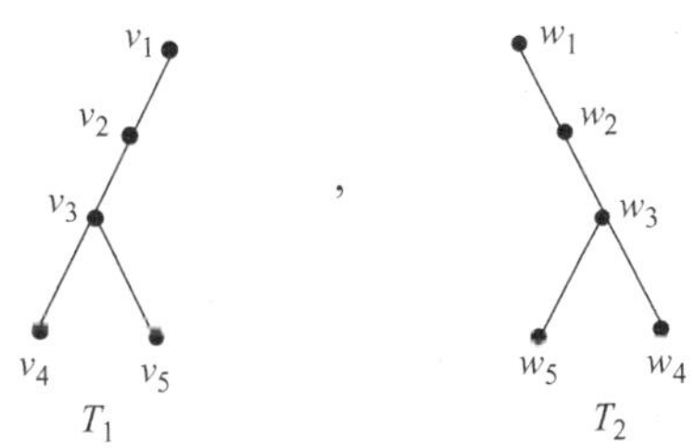

Figure 8.172

Solution.

Since the root v_1 in T_1 has a left child but the root w_1 in T_2 has no left child, the binary trees are not isomorphic.

Definition 8.107

Let v be a branch node of a binary tree T. The **left subtree** of v is the binary tree whose root is the left child of v, whose vertices consists of the left child of v and all its descendents and whose edges consists of all those edges of T that connects the vertices of the left subtree together.

The **right subtree** can be defined analogously.

For example, the left subtree and the right subtree of v in the tree (Figure 8.173)

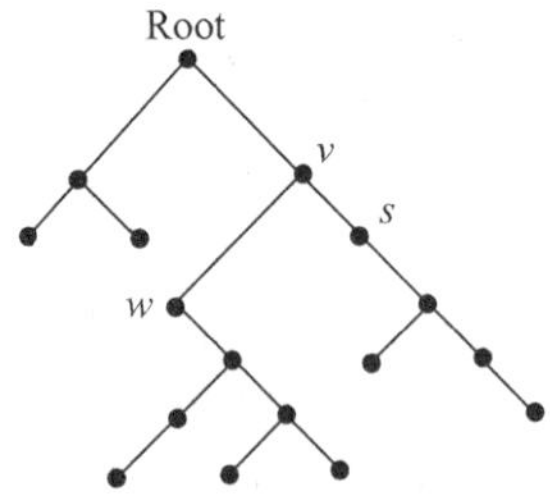

Figure 8.173

are, respectively the thees shown in Figure 8.174.

(Left subtree of v) (Right subtree of v)

Figure 8.174

8.14 REPRESENTATION OF ALGEBRAIC EXPRESSIONS BY BINARY TREES

Binary trees are used in computer science to represent algebraic expressions involving parentheses. For example, the binary trees (Figure 8.175).

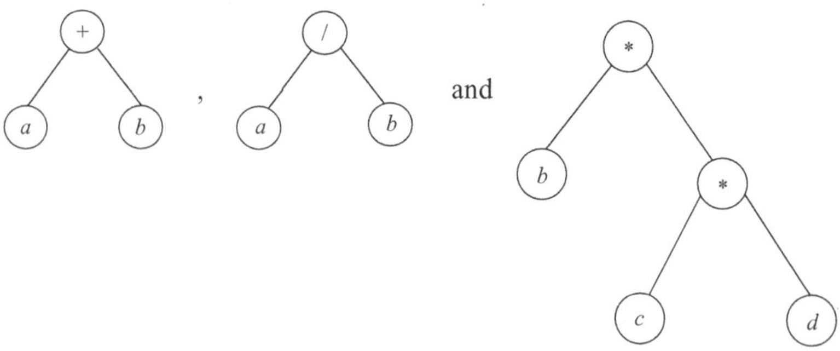

Figure 8.175

represent the expressions, $a+b$, a/b and $b * (c * d)$, respectively.

Thus, the **central operator acts as root of the tree.**

EXAMPLE 8.70

Draw a binary tree to represent

(i) $(2-(3\times x))+((x-3)-(2+x))$
(ii) $ab-(c/(d+e))$.

Solution.

(i) In this expression + is the central operator. Therefore the root of tree is + . The binary tree is shown in Figure 8.176.

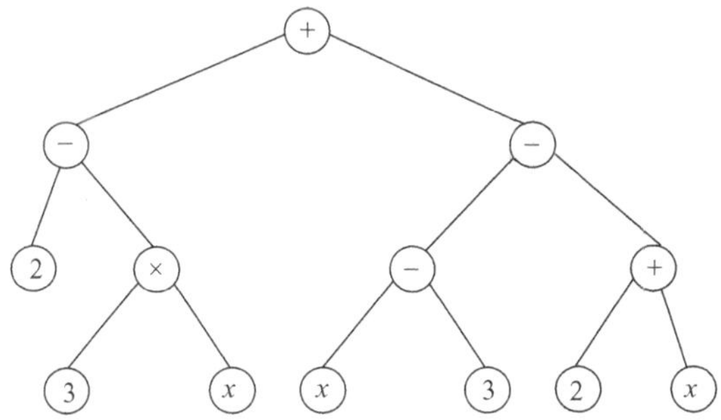

Figure 8.176

(ii) Here the central operator is –. Therefore it is the root of the tree. We have the following binary tree (Figure 8.177) to represent this expression:

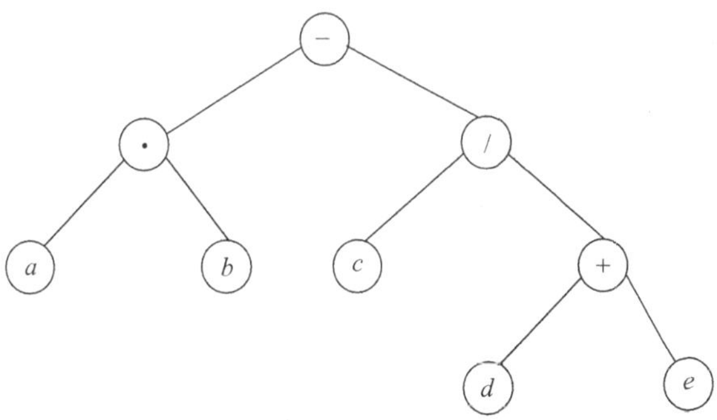

Figure 8.177

EXAMPLE 8.71

Consider a single elimination tournament with eight players in a wrestling game. How many rounds will be there to declare the champion and how many games are to be played?

Solution.

In this example, the leaves of the tree shown in Figure 8.178 will represent player, the branch nodes will represent winners in the first round and the second round. Then the tree showing the tournament shall be as given below: The graph of single elimination tournament is a full binary tree. The champion shall be at the root.

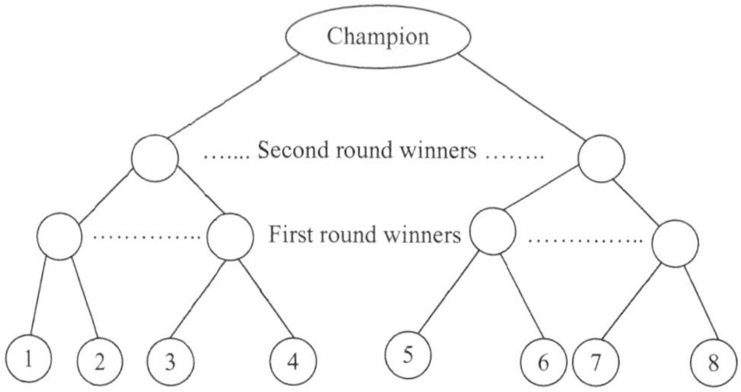

Figure 8.178

We note that there shall be three rounds to declare the champion. Let

i=Number of games played=number of internal vertices
p=Number of leaves=number of players.

Then, we see that

$$i=p-1=7.$$

In general, for complete n-ary tree, we will have

$$(n-1)\,i=p-1.$$

To derive this formula we first prove the following result:

Theorem 8.23

If T is a full binary tree with i internal vertices, then T has $i+1$ terminal vertices (leaves) and $2i+1$ total vertices.

Proof. The vertices of T consist of the vertices that are children (of some parent) and the vertices that are not children (of any parent). There is a non-child (the root). Since there are i internal vertices, each having two children, there are $2i$ children. Thus the total number of vertices of T is $2i+1$ and the number of terminal vertices is

$$(2i+1)-i=i+1.$$

This completes the proof.

In the context of above example, we have

$$\text{Number of leaves}=p=i+1$$
$$\Rightarrow i=p-1.$$

Remark 8.15 In case of full n-ary tree, if i denotes the number of branch nodes, then total number of vertices of T is $ni+1$ and the number of terminal vertices is

$$ni+1-i=i(n-1)+1.$$

If p is the number of terminal vertices, then

$$p=i(n-1)+1,$$

which yields

$$(n-1)\,i=p-1.$$

EXAMPLE 8.72

Find the minimum number of extension cords, each having four outlets, required to connect 22 bulbs to a single electric outlet.

Solution.

Clearly, the graph of the problem is a regular quaternary tree with 22 leaves.
Let i denote the internal vertices and p denote the number of leaves, then using the expression $(n-1)i = p-1$, we have

$$(4-1)i = 22-1 \quad \text{or} \quad i = \frac{21}{3} = 7.$$

Thus seven extension cords, as shown in Figure 8.179, are required.

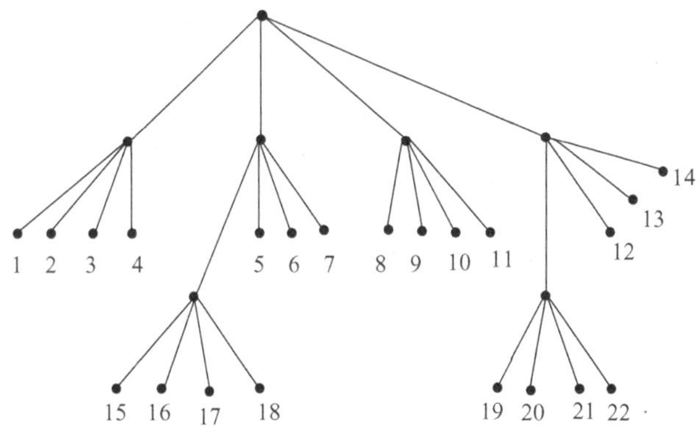

Figure 8.179

EXAMPLE 8.73

A computing machine has been given instruction which computes the sum of three numbers. How many times the addition instruction will be executed to perform the sum of 11 numbers?

Solution.

In this problem, we will have a regular ternary tree with 11 leaves. Thus $n=3, p=11$ and so the relation $(n-1)i = p-1$ yields

$$i = \frac{p-1}{n-1} = \frac{10}{2} = 5.$$

Theorem 8.24

If a binary tree of height h has p leaves, then

$$\log_2 p \le h \quad \text{or} \quad p \le 2^h.$$

Proof. It is sufficient to prove that

$$p \le 2^h, \tag{1}$$

because then taking logarithm to the base 2 of both sides, we will get

$$\log_2 p \le h.$$

We shall prove the inequality (1) by induction on h.

If $h=0$, then the binary tree consists of a single vertex and $p=1$ and so (1) is satisfied.

Now suppose that (1) holds for a binary tree whose height is less than h. We shall prove that it holds for a tree of height h. So let T be a tree with height $h>0$ with p leaves. We consider the following two cases:

(a) Suppose that the root of T has only one child (Figure 8.180).

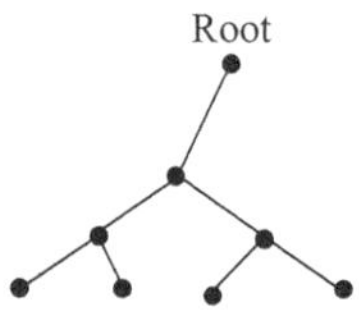

Figure 8.180

Eliminating the root and the edge incident on the root, we get a tree T' of height $h-1$ in which number of leaves is same as in T. Using induction hypothesis on T', we get $p\leq 2^{h-1}$. Since $2^{h-1}<2^h$, we have $p\leq 2^h$ and the inequality holds in this case.

(b) Suppose the root of T has offspring's v and w (Figure 8.181).

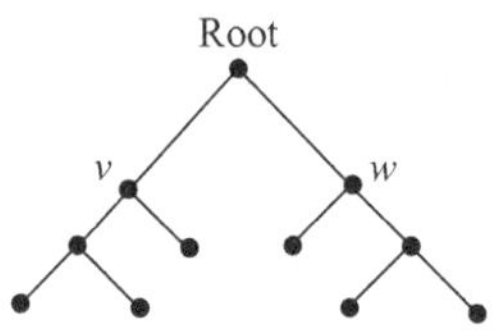

Figure 8.181

Let T_1 be the subtree rooted at v whose height is h_1 and has p_1 leaves. Similarly, let T_2 be the subtree rooted at w whose height is h_2 and has p_2 leaves. Using mathematical induction, to these trees, we have $p_1 \leq 2^{h_1}$ and $p_2 \leq 2^{h_2}$. But the leaves of T consist of the leaves of T_1 and T_2. Hence,

$$p = p_1 + p_2 \leq 2^{h_1} + 2^{h_2}$$

$$\leq 2^{h-1}+2^{h-1}=2^{h-1}(1+1)=2^h.$$

Hence, by mathematical induction, the result holds.

It can be proved on the same lines that an **n-ary tree of height h has atmost n^h leaves.**

EXAMPLE 8.74

Find the number of terminal vertices in a regular binary tree of height 4.

Solution.

Let p be the number of terminal vertices. We are given that height (h) of the tree is 4. So, we must have $p\leq 2^4=16$. Since the tree is regular, we have $p=16$.

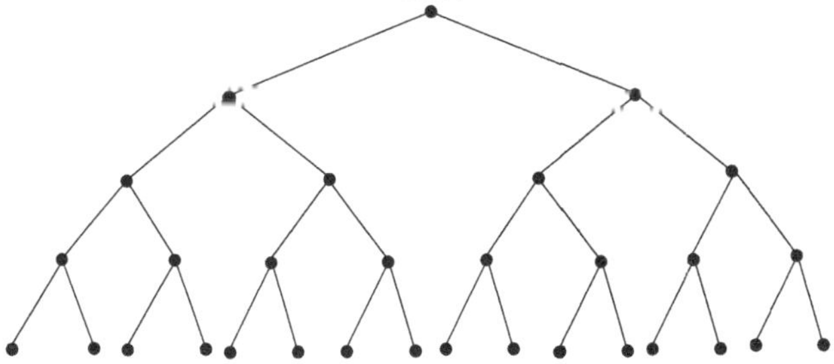

(A regular binary tree of height 4 with 16 leaves)

Figure 8.182

EXAMPLE 8.75

Does there exist a full binary tree with 12 internal vertices and 15 leaves?

Solution.

We know that if i is the number of branch nodes in a full binary tree, then the number of leaves is $i+1$. Therefore for a tree with 12 branch nodes, the number of leaves should be 13 and not 15. Hence such tree does not exist.

EXAMPLE 8.76

Is there a binary tree with height 6 and 65 leaves?

Solution.

No, since

$$p \leq 2^h = 2^6 = 64.$$

We now state some results without proof on n-ary tree.

Theorem 8.25

The number b_n of different binary trees on n vertices is given by

$$b_n = \frac{1}{n+1}\binom{2n}{n}.$$

Theorem 8.26

A regular n-ary tree of height h has **at least** $n+(n-1)(h-1)$ leaves.

Theorem 8.27

If in a rooted regular n-ary tree,

I = Sum of the path length of all branch nodes

E = Sum of the path length of all leaves

i = Number of branch nodes

then

$$E=(n-1)I+n\,i$$

Thus, in particular, for a full binary tree,

$$E=I+2i.$$

EXAMPLE 8.77

How many binary trees are possible on three vertices?

Solution.

We know that the number of different trees on n vertices is given by

$$b_n = \frac{1}{n+1}\binom{2n}{n}.$$

So we have

$$b_n = \frac{1}{3+1}\binom{6}{3} = \frac{1}{4}\left(\frac{6\times5\times4\times3\times2\times1}{3\times2\times1\times3\times2\times1}\right) = 5 \text{ trees}$$

and those are shown in Figure 8.183.

Figure 8.183

8.15 SPANNING TREE OF A GRAPH

Definition 8.108

Let G be a graph, then a subgraph of G which is a tree is called **tree of the graph**.

Definition 8.109

A **spanning tree** for a graph G is a subgraph of G that contains every vertex of G and is a tree.

Thus,

"A **spanning tree** for a graph G is a spanning subgraph of G which is a tree".

EXAMPLE 8.78

Determine a tree and a spanning tree for the connected graph given below (Figure 8.184):

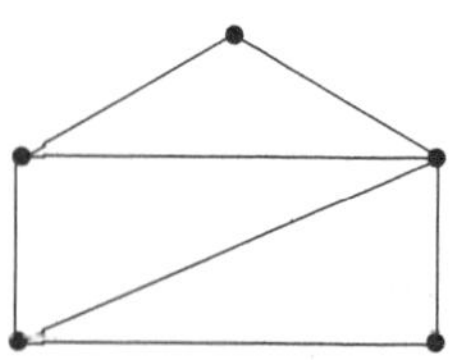

G

Figure 8.184

Solution.

The given graph G contains circuits and we know that removal of the circuits gives a tree. So, we note that the Figure 8.185 is a tree.

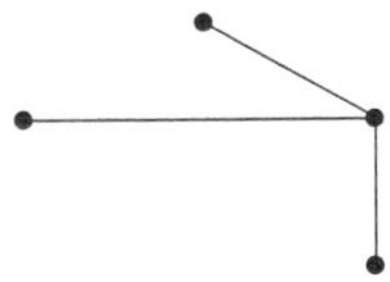

Figure 8.185

And the Figure 8.186 is a spanning tree of the graph G.

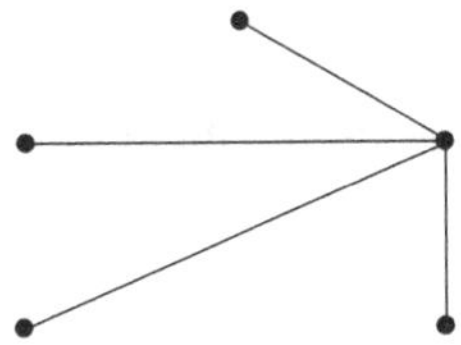

Figure 8.186

EXAMPLE 8.79

Find all spanning trees for the graph G shown below (Figure 8.187).

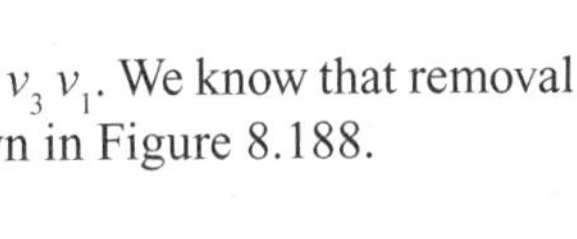

Figure 8.187

Solution.

The given graph G has a circuit $v_1 v_2 v_3 v_1$. We know that removal of any edge of the circuit gives a tree. So the spanning trees of G are shown in Figure 8.188.

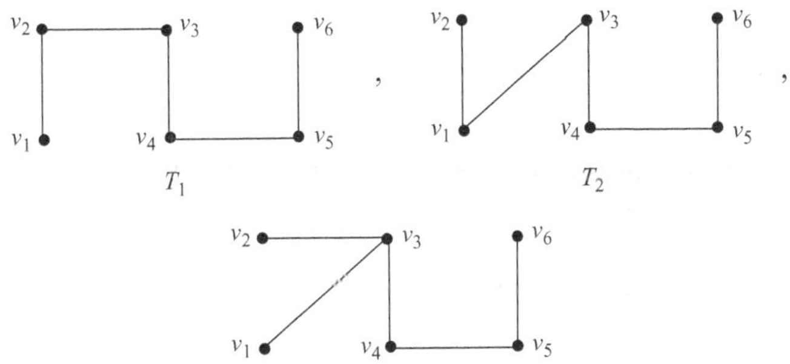

Figure 8.188

Remark 8.16 We know that a tree with n vertices has exactly $n-1$ edges. Therefore if G is a connected graph with n vertices and m edges, a spanning tree of G must have $n-1$ edges. Hence the number of edges that must be removed before a spanning tree is obtained must be

$$m-(n-1)=m-n+1.$$

For illustration, in Example 8.79, $n=6$, $m=6$, so, we had to remove one edge to obtain a spanning tree.

Definition 8.110

A **branch** of a tree is an edge of the graph that is in the tree.

Definition 8.111

A **chord** (or a **link**) of a tree is an edge of the graph that is not in the tree.

It follows from the above remark that the number of chords in a tree is equal to $m-n+1$, where n is the number of vertices and m is the number of edges in the graph related to the tree.

Definition 8.112

The set of the chords of a tree is called the **complement of the tree**.

EXAMPLE 8.80

Consider the graph discussed in the above example. We note that the edge (v_2, v_3) is a branch of the tree T_1, whereas (v_1, v_3) is a chord of the tree T_1.

Theorem 8.28

A graph G has a spanning tree if and only if G is connected.

Proof. Suppose first that a graph G has a spanning tree T. If v and w are vertices of G, then they are also vertices in T and since T is a tree there is a path from v to w in T. This path is also a path in G. Thus every two vertices are connected in G. Hence G is connected.

Conversely, suppose that G is connected. If G is acyclic, then G is its own spanning tree and we are done. So suppose that G contains a cycle C_1. If we remove an edge from the cycle, the subgraph of G so obtained is also connected. If it is acyclic, then it is a spanning tree and we are done. If not, it has at least one circuit, say C_2. Removing one edge from C_2, we get a subgraph of G which is connected. Continuing in this way, we obtain a connected circuit free subgraph T of G. Since T contains all vertices of G, it is a spanning tree of G.

Theorem 8.29 (Cayley's Formula)

The number of spanning trees of the complete graph K_n, $n\geq 2$ is n^{n-2}.

(Proof of this formula is out of scope of this book)

EXAMPLE 8.81

Find all the spanning trees of K_4 (Figure 8.189).

Solution.

According to Cayley's formula, K_4 has $4^{4-2}=4^2=16$ different spanning trees.

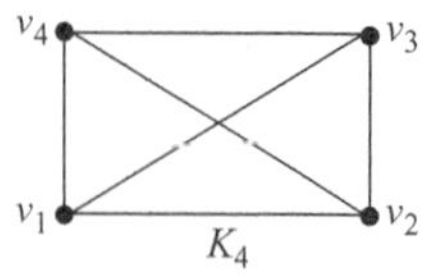

Figure 8.189

Here $n=4$, so the number of edges in any tree should be $n-1=4-1=3$. But here number of edges is equal to 6. So to get a tree, we have to remove three edges of K_4. The 16 spanning trees so obtained are shown in Figure 8.190.

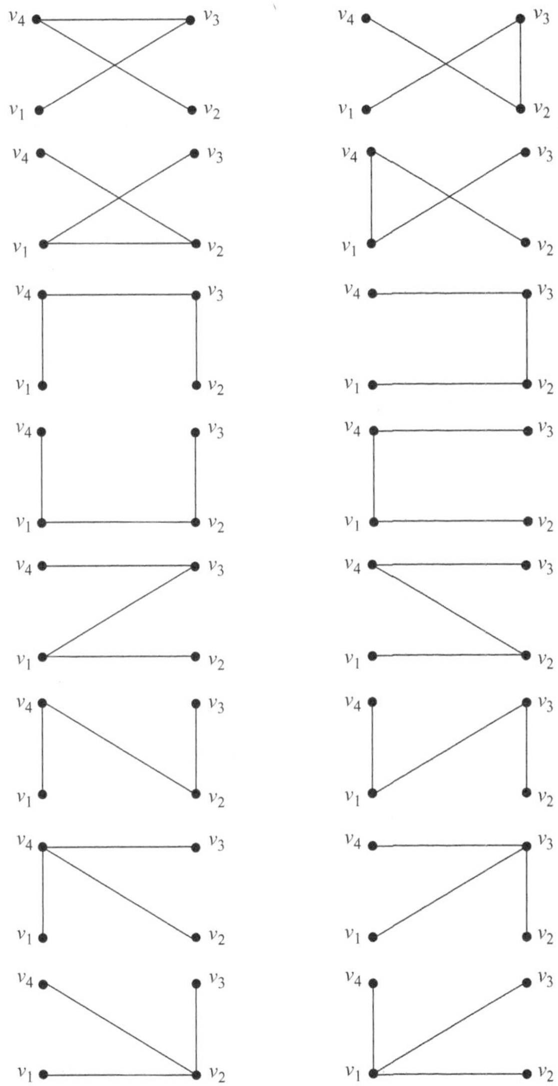

Figure 8.190

8.16 SHORTEST PATH PROBLEM

Let s and t be two vertices of a connected weighted graph G. Shortest path problem is to find a **path from s to t whose total edge weight is minimum.** We now discuss algorithm due to $E \cdot W$. Dijkstra which efficiently solves the shortest path problem. The idea is to grow a Disjkstra tree, starting at the vertex s, by adding, at each iteration, a frontier edge, whose non-tree endpoint is as close as possible to s. The algorithm involves assigning labels to vertices.

For each tree vertex x, let dist $[x]$ denote the distance from vertex s to x and for each edge e in the given weighted graph G, let $w(e)$ be its edge-weight.

After each iteration, the vertices in the Dijkstra tree (the labelled vertices) are those to which the shortest paths from s have been found.

Priority of the Frontier Edges: Let e be a frontier edge and let its P value be given by

$$P(e) = \text{dist }[x] + w(e),$$

where x is the labelled endpoint of e and $w(e)$ is the edge-weight of e. Then,

(i) The edge with the smallest P value is given the **highest priority**.

(ii) The P value of this highest priority edge e gives the distant from the vertex s to the unlabelled endpoint of e.

We are now in a position to describe Dijkstra's shortest path algorithm.

8.16.1 Dijkstra's Shortest Path Algorithm

Input: A connected weighted graph G with non-negative edge-weights and a vertex s of G.

Output: A spanning tree T of G, rooted at the vertex s, whose path from s to each vertex v is a shortest path from s to v in G and a vertex labelling giving the distance from s to each vertex.

Initialize the Dijkstra tree T as vertex s.
Initialize the set of frontier edges for the tree T as empty.

$$\text{dist: }[s] = 0.$$

Write label 0 on vertex s.
While Dijkstra tree T does not yet span G.
For each frontier edge e for T,
Let x be the labelled endpoint of edge e.
Let y be the unlabelled endpoint of edge e.
Set $P(e) = \text{dist }[x] + w(e)$
Let e be a frontier edge for T that has smallest P value
Let x be the labelled endpoint of edge e
Let y be the unlabelled endpoint of edge e
Add edge e (and vertex y) to tree T

$$\text{dist }[y]\text{: } P(e)$$

Write label dist $[y]$ on vertex y.

Return Dijkstra tree T and its vertex labels.

EXAMPLE 8.82

Apply Dijkstra algorithm to find shortest path from s to each other vertex in the graph given in Figure 8.191.

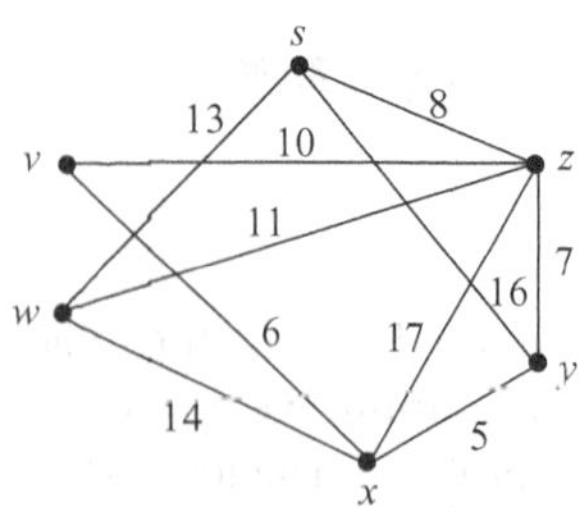

Figure 8.191

If t is the labelled endpoint of edge e, then P values are given by

$$P(e)=\text{dist}\ [t]+w(e),$$

where dist $[t]$=distance from s to t and $w(e)$ is the edge weight of edge e. For each vertex v, dist $[v]$ appears in the parenthesis. Iteration tree at the end of each iteration is **drawn in dark line**

Iteration 1

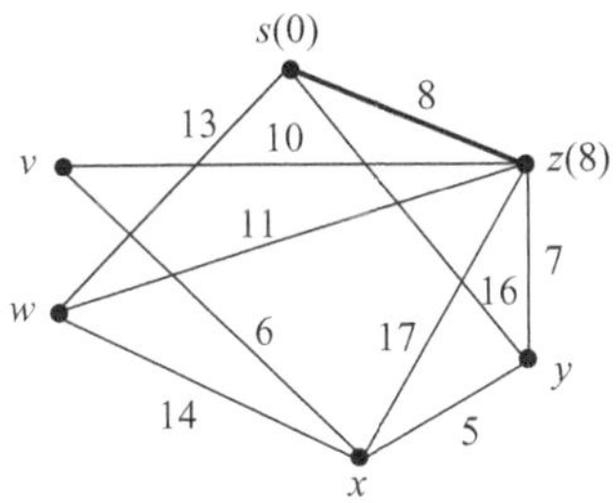

dist $[s]=0$ $\quad P(sw)=13$ (minimum)

dist $[z]=8$ $\quad P(zy)=8+7=15$

$P(sy)=16$

$P(zv)=8+10=18$

$P(zw)=8+11=19$

$P(zx)=8+17=25$

Iteration 2

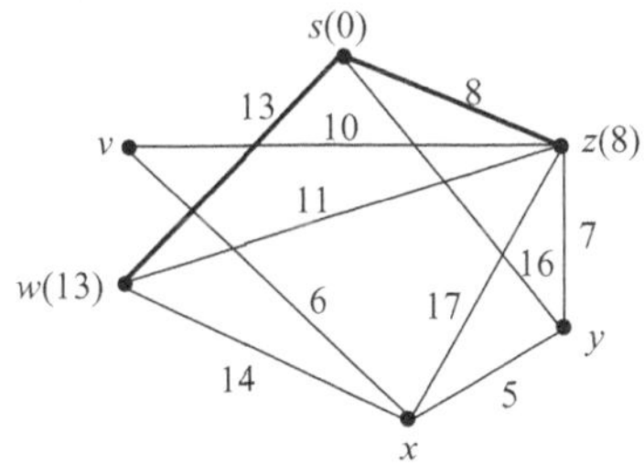

dist $[s]=0$ $\quad P(zy)=8+7=15$ (minimum)

dist $[z]=8$ $\quad P(zx)=8+17=25$

dist $[w]=13$ $\quad P(zv)=8+10=18$

$P(sy)=16$

$P(wx)=13+14=27$

Iteration 3

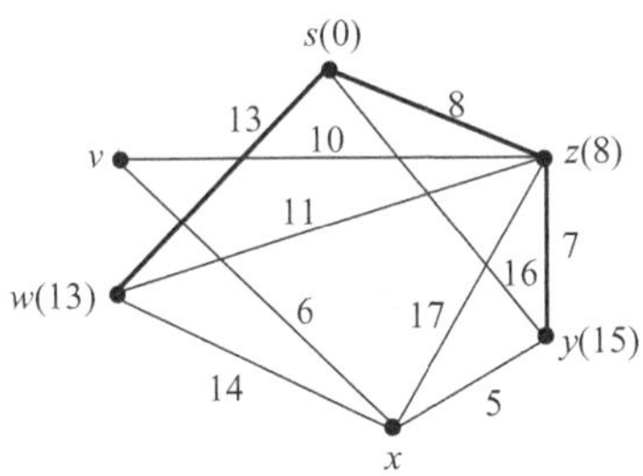

dist $[s]=0$ $\quad P(zv)=18$ (minimum)

dist $[z]=8$ $\quad P(zx)=8+17=25$

dist $[w]=13$ $\quad P(wx)=13+14=27$

dist $[y]=15$ $\quad P(yx)=15+5=20$

Iteration 4

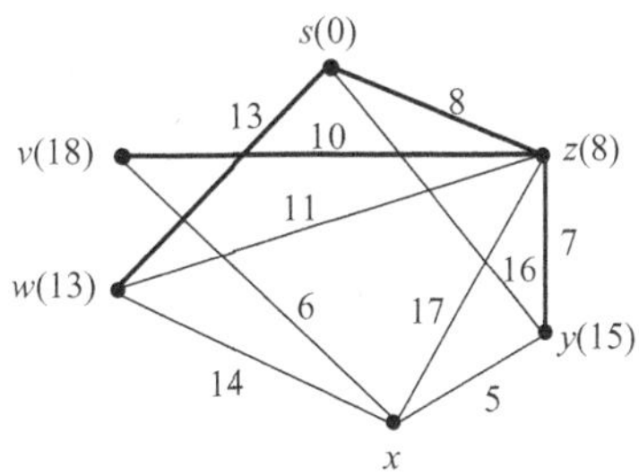

Dist $[s]=0$ $\quad P(yx)=20$ (minimum)

Dist $[z]=8$ $\quad P(zx)=8+17=25$

Dist $[w]=13$ $\quad P(vx)=18+6=24$

Dist $[y]=15$ $\quad P(wx)=13+14=27$

Dist $[v]=18$

Iteration 5

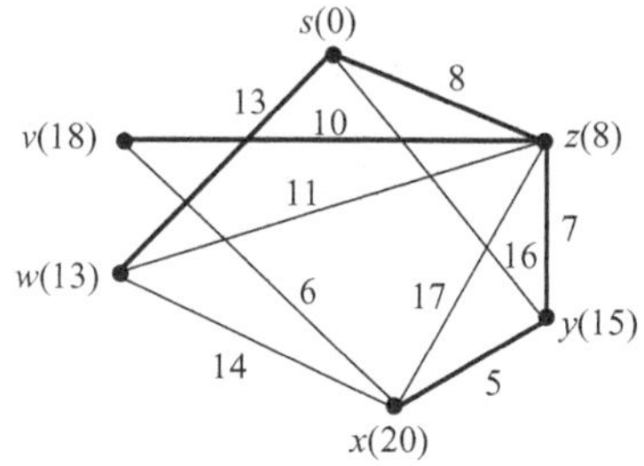

$$\text{dist }[s]=0$$
$$\text{dist }[z]=8$$
$$\text{dist }[w]=13$$
$$\text{dist }[y]=15$$
$$\text{dist }[v]=18$$
$$\text{dist }[x]=20,$$

which are the required shortest paths from s to any other point. The Dijkstra tree is shown in dark lines in Figure 8.192.

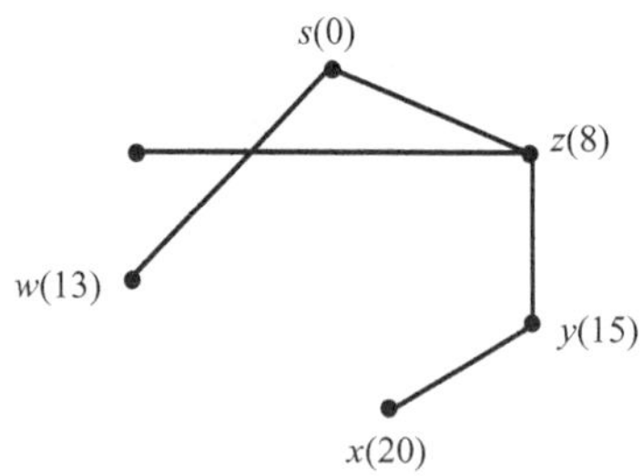

Figure 8.192

EXAMPLE 8.83

Apply Dijkstra algorithm to find the shortest path from s to t in the graph given below (Figure 8.193).

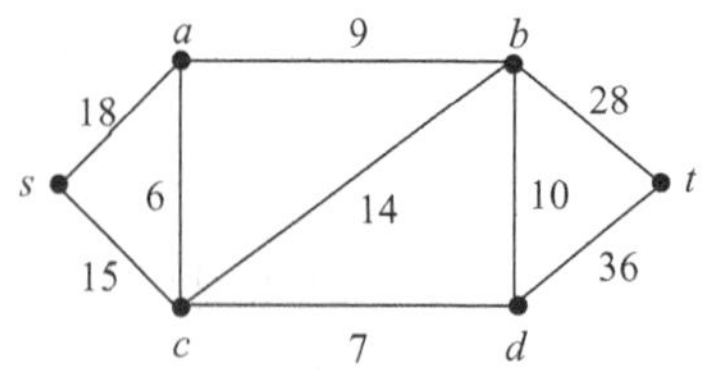

Figure 8.193

Solution.

Let x be the labelled endpoint of the edge e, then P values are given by $P(e)=\text{dist }[x]+w(e)$, where dist $[x]$ denotes the distance of x from s and $w(e)$ is the edge weight of e.

For each vertex v, dist $[v]$ appears in the bracket. Iteration tree at the end of each iteration is shown by dark lines:

Iteration 1

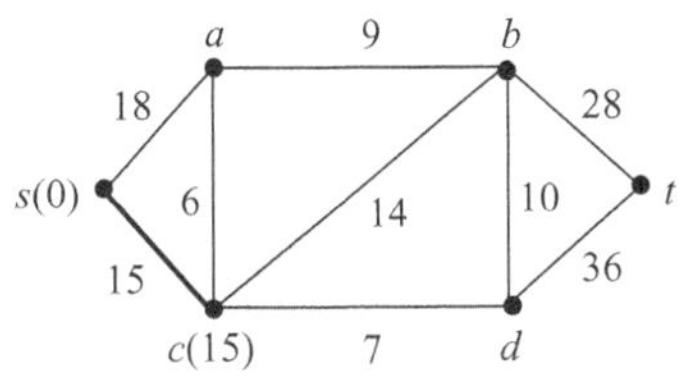

dist $[s]=0$ $\quad P(s\ a)=18$ (minimum)

dist $[c]=15$ $\quad P(c\ a)=15+6=21$

$P(c\ b)=15+14=29$

$P(c\ d)=15+7=22$

$P(c\ t)=\infty$

Iteration 2

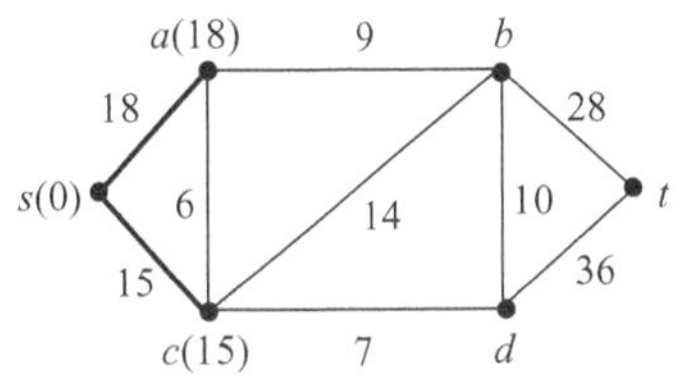

dist $[s]=0$ $\quad P(a\ b)=18+9=27$

dist $[c]=15$ $\quad P(c\ b)=15+14=29$

dist $[a]=18$ $\quad P(c\ d)=15+7=22$ (minimum)

$P(c\ t)=\infty$

$P(a\ t)=\infty$

Iteration 3

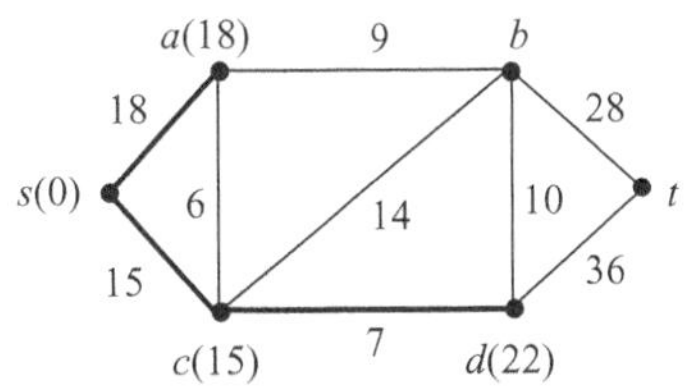

dist $[s]=0$ $\quad P(c\ b)=15+14=29$

dist $[c]=15$ $\quad P(a\ b)=18+9=27$ (minimum)

dist $[a]=18$ $\quad P(d\ b)=22+10=32$

dist $[d]=22$ $\quad P(d\ t)=22+36=58$

Iteration 4

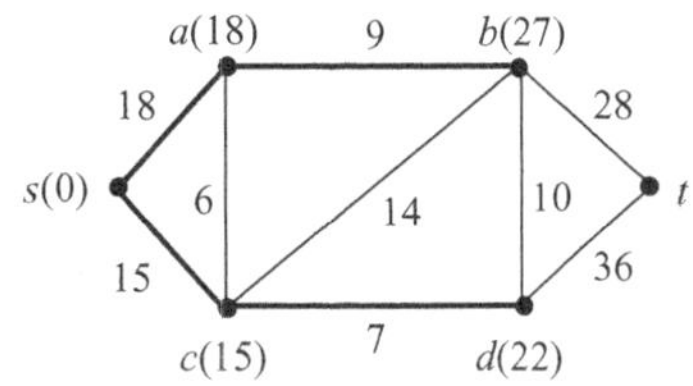

dist $[s]=0$ $\quad P(d\ t)=22+36=58$

dist $[c]=15$ $\quad P(b\ t)=27+28=55$ (minimum)

dist $[a]=18$

dist $[d]=22$

dist $[b]=27$

Iteration 5

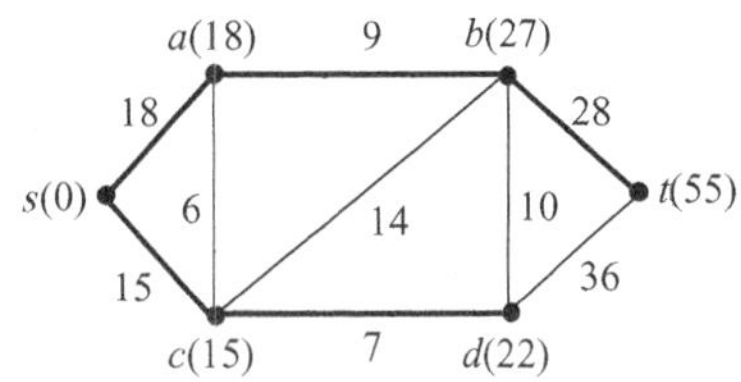

dist $[s]=0$

dist $[c]=15$

dist $[a]=18$

dist $[s]=0$

dist $[c]=15$

dist $[a]=18$

dist $[d]=22$

dist $[b]=27$

dist $[t]=55$

Hence, **the length of the shortest path** (s, a, b, t) **from** s **to** t **is 55** and the Dijkstra's tree is shown in Figure 8.194.

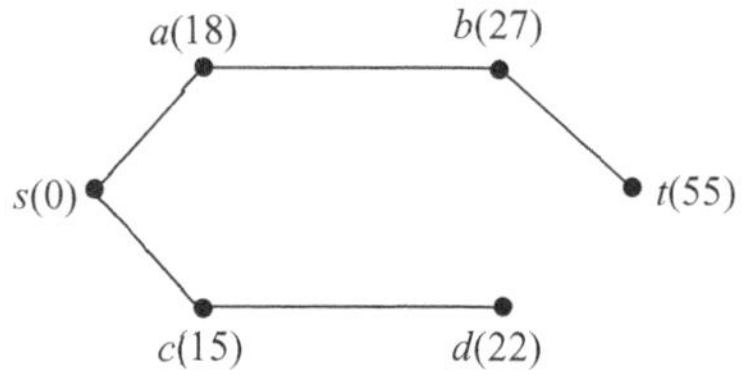

Figure 8.194

EXAMPLE 8.84

Find a shortest path from s to t and its length for the graph given below:

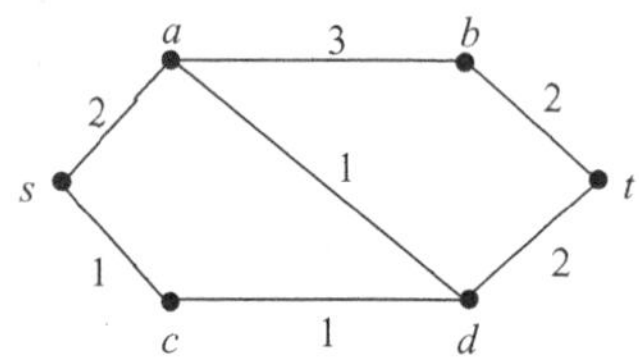

Figure 8.195

Solution.

Let x be the labelled endpoint of edge e, then P values are given by

$$P(c) = \text{dist }[x] + w(e),$$

where dist $[x]$ denotes the distance from s to x and $w(e)$ is the weight of the edge e.

For each vertex v, dist $[v]$ appears in the bracket. Iteration tree at the end of each iteration is shown in dark lines.

Iteration 1

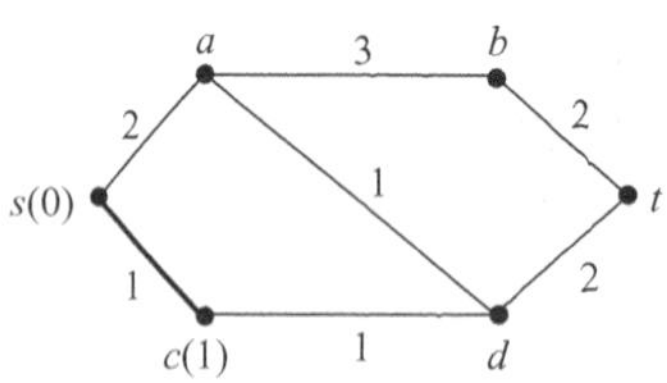

dist $[s]=0$ $\quad$ $P(c\ d)=2$

dist $[c]=1$ $\quad$ $P(s\ a)=2$ (minimum)

$P(s\ b)=\infty$

$P(s\ t)=\infty$

$P(d\ t)=\infty$

$P(b\ t)=\infty$

Iteration 2

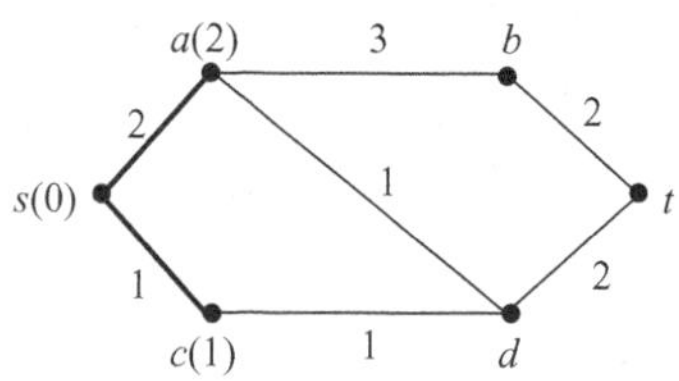

dist $[s]=0$ $\quad$ $P(a\ b)=2+3=5$

dist $[c]=1$ $\quad$ $P(c\ d)=1+1=2$ (minimum)

dist $[a]=2$ $\quad$ $P(d\ t)=\infty$

$P(b\ t)=\infty$

$P(a\ d)=3$

Iteration 3

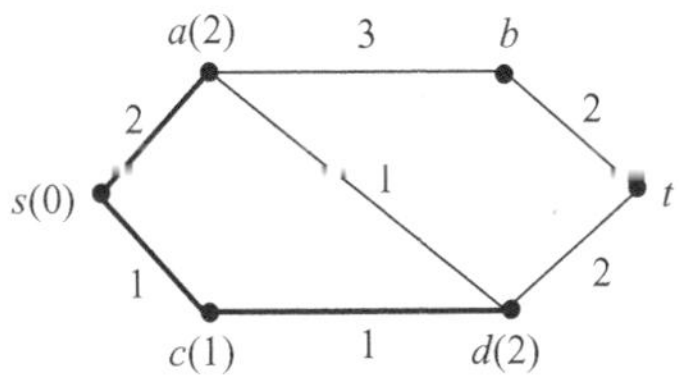

dist $[s]=0$ $P(d\ t)=2$ (minimum)

dist $[c]=1$ $P(a\ b)=5$

dist $[a]=2$ $P(b\ t)=\infty$

dist $[d]=2$

dist $[s]=0$

dist $[c]=1$

dist $[a]=2$

dist $[d]=2$

dist $[t]=4$

dist $[b]=5$

Iteration 4

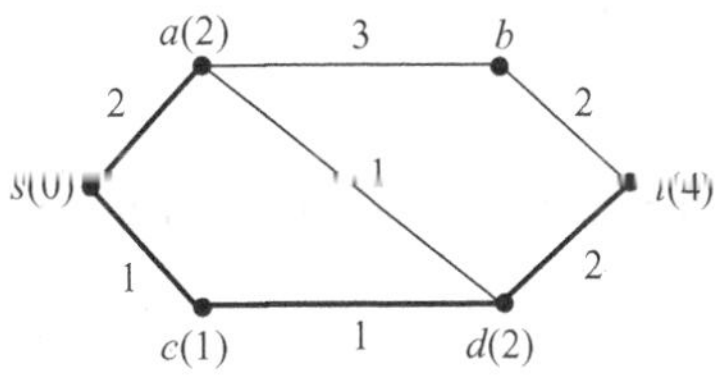

dist $[s]=0$ $P(a\ b)=5$ (minimum)

dist $[c]=1$ $P(b\ t)=\infty$

dist $[a]=2$

dist $[d]=2$

dist $[t]=4$

Iteration 5

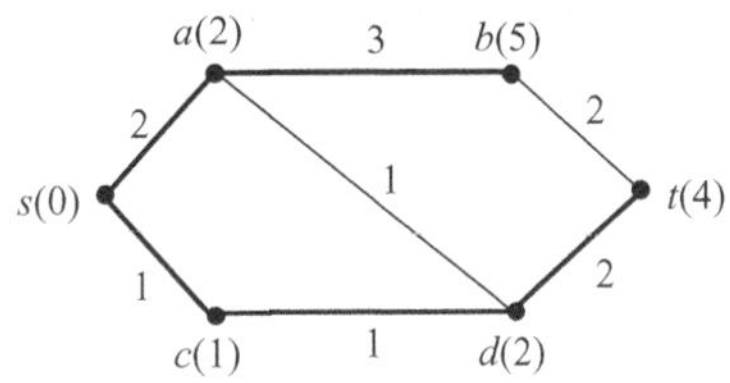

Thus, the Dijkstra tree is

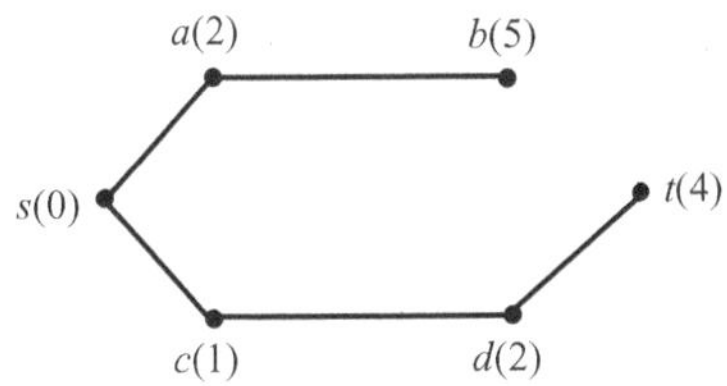

Figure 8.196

Thus the shortest path is *scdt* and its length is 4.

8.16.2 Shortest Path if All Edges Have Length 1

If all edges in a connected graph G have length 1, then a **shortest path** $v_1 \to v_k$ is the path that has the smallest number of edges among all paths $v_1 \to v_k$ in the given graph G.

Moore's Breadth First Search Algorithm

This method of finding shortest path in a connected graph G from a vertex s to a vertex t is used when all edges have length 1.

Input: Connected graph $G=(V, E)$, in which one vertex is denoted by s and one by t and each edge (v_i, v_j) has length 1.

Initially all vertices are unlabelled.

Output: A shortest path $s \to t$ in $G=(V, E)$.

1. Label s with 0.
2. Set $v_i = 0$.
3. Find all unlabelled vertices **adjacent to a vertex labelled** v_i.
4. Label the vertices just found with v_{i+1}.
5. If vertex t is labelled, then "**back tracking**" gives the shortest path. If k is label of t(i.e., $t=v_k$), then

Output: $v_k, v_{k-1}, \ldots, v_1, 0$.

Else increase i by 1. Go to Step 3.
End Moore.

Remark 8.17 **There could be several shortest paths from s to t.**

EXAMPLE 8.85

Use BFS algorithm to find shortest path from s to t in the connected graph G given in Figure 8.197.

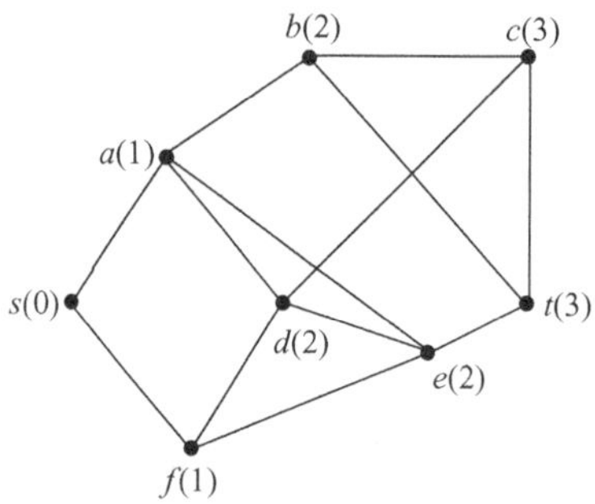

Figure 8.197

Solution.

Label s with 0 and then label the adjacent vertices with 1. Thus two vertices have been labelled by 1. Now label the adjacent vertices of all vertices labelled by 1 with label 2. Thus three vertices have been labelled with 2. Label the vertices adjacent to these vertices (labelled by 2) with 3. Thus two vertices have been labelled with 3. We have reached t. Now backtracking yields the following three shortest paths:

(i) $t(3), e(2), f(1), s(0)$, that is, $s\,f\,e\,t$,
(ii) $t(3), b(2), a(1), s(0)$, that is $s\,a\,b\,t$,
(iii) $t(3), e(2), a(1), s(0)$, that is, $s\,a\,e\,t$.

Thus there are three possible shortest paths of length 3.

EXAMPLE 8.86

Find a shortest path $s \to t$ in the graph given in Figure 8.198 (using Moore's BFS algorithm).

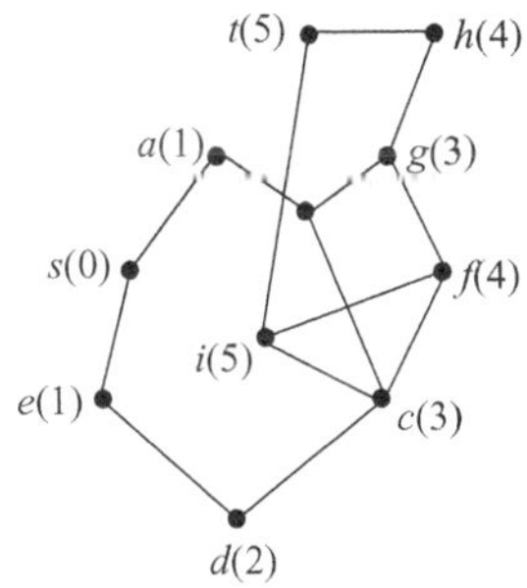

Figure 8.198

Solution.

Label *s* with 0 and then label the adjacent vertices with 1. Thus two vertices have been marked as 1. Label the adjacent vertices of al vertices labelled as 1 with 2. Thus two vertices have been marked as 2. Now label the adjacent vertices of all vertices labelled as 2 with 3. Thus two vertices have been marked as 3. Label the adjacent vertices of all vertices labelled as 3 with 4. Thus three vertices have been marked as 4. Now there is only one vertex *t* to be labelled. Label it with 5. We have reached at *t*.

Now backtracking gives the following shortest paths

t h g b a s	or	*s a b g h t*	with length 5,
t i c d e s	or	*s e b c i t*	of length 5,
t i c b a s	or	*s a b c i t*	of length 5.

Thus, there are three possible shortest paths of length 5.

8.17 MINIMAL SPANNING TREE

Definition 8.113

Let *G* be a weighted graph. A spanning tree of *G* with minimum weight is called **minimal spanning tree of** *G*.

We discuss two algorithms to find a minimal spanning tree for a weighted graph *G*.

8.17.1 Prim Algorithm

Prim algorithm builds a minimal spanning tree *T* by expanding outward in connected links from some vertex. In this algorithm, one edge and one vertex are added at each step. The edge added is the one of least weight that connects the vertices already in *T* with those not in *T*.

Input: A connected weighted graph *G* with *n* vertices.

Output: The set of edges *E* in a minimal spanning tree.

1. Choose a vertex v_1 of G. Let $V=\{v_1\}$ and $E=\{\ \}$.
2. Choose a nearest neighbour v_i of V that is adjacent to v_j, $v_j \in V$ and for which the edge (v_i, v_j) does not form a cycle with members of E. Add v_i to V and add (v_i, v_j) to E.
3. Repeat Step 2 till number of edges in T is $n-1$. Then V contains all n vertices of G and E contains the edges of a minimal spanning tree for G.

Definition 8.114

A **greedy algorithm** is an algorithm that optimizes the choice at each iteration without regard to previous choices.

For example, **Prim algorithm is a greedy algorithm.**

EXAMPLE 8.87

Find a minimal spanning tree for the graph shown in Figure 8.199.

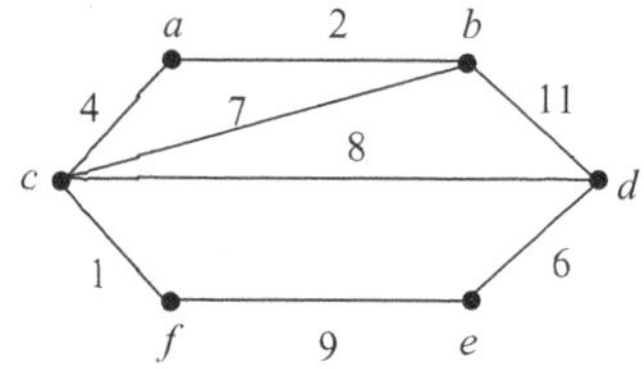

Figure 8.199

Solution.

We shall use **Prim algorithm to find the required minimal spanning tree.** We note that number of vertices in this connected weighted graph is 6. Therefore the tree will have five edges.

We start with any vertex, say c. The nearest neighbour of c is f and (c, f) does not form a cycle. Therefore (c, f) is the first edge selected.

Now we consider the set of vertices $V=\{c, f\}$. The vertex a is nearest neighbour to $V=\{c, f\}$ and the edge (c, a) does not form a cycle with the member of set of edges selected so far. Thus,

$$E=\{(c, f), (c, a)\} \quad \text{and} \quad V=\{c, f, a\}.$$

The vertex b is now nearest neighbour to $V=\{c, f, a\}$ and the edge (a, b) do not form a cycle with the member of $E=\{(c, f), (c, a)\}$. Thus,

$$E=\{(c, f), (c, a), (a, b)\} \quad \text{and} \quad V=\{c, f, a, b\}.$$

Now the edge (b, c) cannot be selected because it forms a cycle with the members of E. We note that d is the nearest point to $V=\{c, f, a, b\}$ and (c, d) is the edge which does not form a cycle with members of $E=\{(c, f), (c, a), (a, b)\}$. Thus we get

$$E=\{(c, f), (c, a), (a, b)\}, (c, d)\}, \quad V=\{c, f, a, b, d\}.$$

The nearest vertex to V is now e and (d, e) is the corresponding edge. Thus,

$$E=\{(c, f), (c, a), (a, b), (c, d), (d, e)\}, \quad V=\{c, f, a, b, d, e\}$$

Since number of edges in the Prim tree is 5, the process is complete. The minimal spanning tree is shown in Figure 8.200:

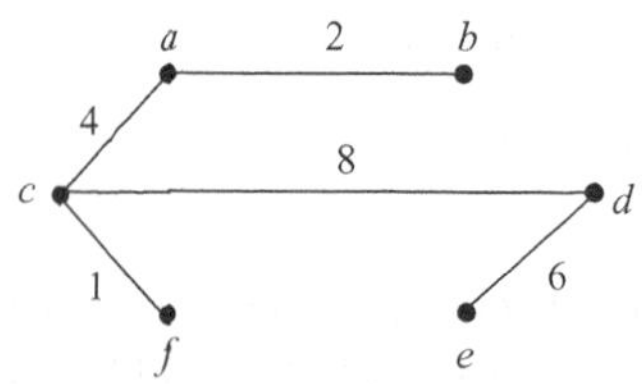

Figure 8.200

The length of the tree is $1+4+2+8+6=21$

EXAMPLE 8.88

Build the lowest-cost road system that will connect the cities in the graph given in Figure 8.201.

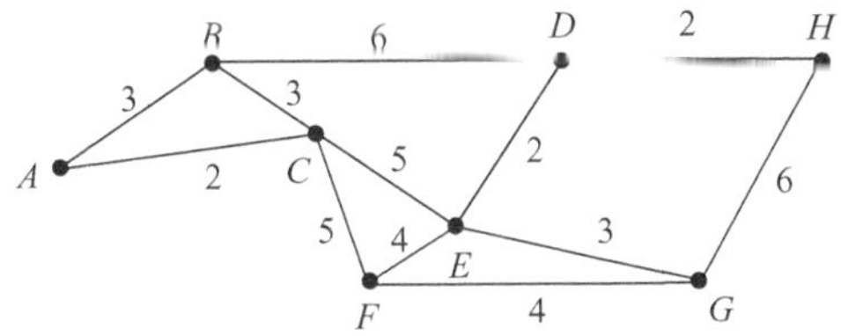

Figure 8.201

Solution.

We shall use Prim algorithm to build this system. Here the number of vertices (Cities) is 8. So, the minimal spanning tree would have seven edges.

We start from any vertex, say A and have at the initial stage:

$$V=\{A\},\quad E=\{\ \}.$$

The nearest neighbour of $V=\{A\}$ is C and the corresponding edge is (A, C). So we have

$$E=\{(A, C)\},\quad V=\{A, C\}.$$

Now B is the nearest neighbour of $V=\{A, C\}$ and edge (C, B) does not form cycle with the edges in $E=\{(A, C)\}$. Here the edge (A, B) can also be chosen. Thus, choosing arbitrarily one out of (A, B) and (C, B), we have

$$E=\{(A, C), (C, B)\},\quad V=\{A, C, B\}.$$

Now, E and F are both nearest neighbour to V and (C, E) or (C, F) do not form cycle with the members of $E=\{(A, C), (C, B)\}$. So we may choose any of E or F. Let us choose F and the corresponding edge (C, F). Then,

$$E=\{(A, C), (C, B), (C, F)\},\quad V=\{A, C, B, F\}.$$

Now E and G are nearest neighbours of $V=\{A, C, B, F\}$ and both (F, E) and (F, G) do not form cycle with the members of $E=\{(A, C), (C, B), (C, F)\}$. Therefore we may pick up any of these vertices. We pick arbitrarily E.

Then,

$$E=\{(A, C), (C, B), (C, F), (F, E)\},\quad V=\{A, C, B, F, E\}.$$

Now D is the nearest neighbour of $V=\{A, C, B, F, E\}$ and (E, D) is the corresponding edge to be added to the tree. So, we have

$$E=\{(A, C), (C, B), (C, F), (F, E), (E, D)\},\quad V=\{A, C, B, F, E, D\}.$$

Now H is the nearest neighbour and (D, H) is the corresponding edge and we have

$$E=\{(A, C), (C, B), (C, F), (F, E), (E, D), (D, H)\},$$
$$V=\{A, C, B, F, E, D, H\}.$$

Now G is the nearest neighbour and (E, G) is the corresponding edge. So, we have

$$E=\{(A, C), (C, B), (C, F), (F, E), (E, D), (D, H), (E, G)\},$$
$$V=\{A, C, B, F, E, D, H, G\}$$

Since the number of edges in the Prim tree is 7, the process is complete. The minimal spanning tree is shown in Figure 8.202:

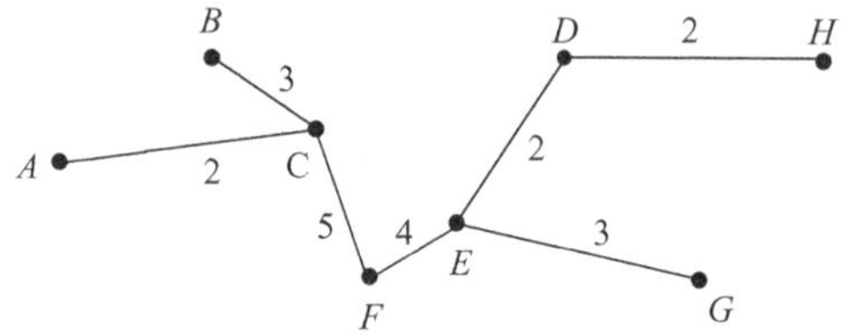

Figure 8.202

The total length of the roads is

$$2+3+5+4+2+2+3=21.$$

EXAMPLE 8.89

Using Prim algorithm, find the minimal spanning tree of the following graph (Figure 8.203):

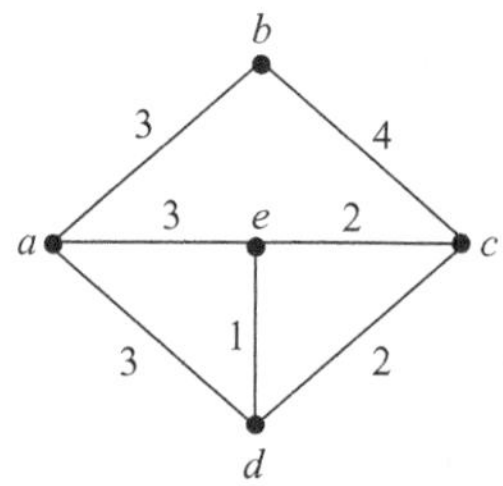

Figure 8.203

Solution.

Pick up the vertex a. Then

$$E=\{\ \} \quad \text{and} \quad V=\{a\}.$$

The nearest neighbour of V is b, d or e and the corresponding edges are (a, b) or (a, d) or (a, e). We choose arbitrarily (a, b) and have

$$E=\{(a, b)\}, \quad V=\{a, b\}.$$

Now d is the nearest neighbour of $V=\{a, b\}$ and the corresponding edge (a, d) does not form cycle with (a, b). Thus we get

$$E=\{(a, b), (a, d)\}, \quad V=\{a, b, d\}.$$

Now e is the nearest neighbour of $\{a, b, d\}$ and (d, e) does not form cycle with $\{(a, b), (a, d)\}$.

Hence,

$$E=\{(a, b), (a, d), (d, e)\}, \quad V=\{a, b, d, e\}.$$

Now c is the nearest neighbour of $V=\{a, b, d, e\}$ and the corresponding edges are (e, c), (d, c). Thus we have, choosing (e, c),

$$E=\{(a, b), (a, d), (d, e), (e, c)\}, \quad V=\{a, b, d, e, c\}$$

Total weight$=3+3+1+2=9$.

(If we choose (d, c), then total weight is $3+3+1+2=9$).

The minimal tree is shown in Figure 8.204.

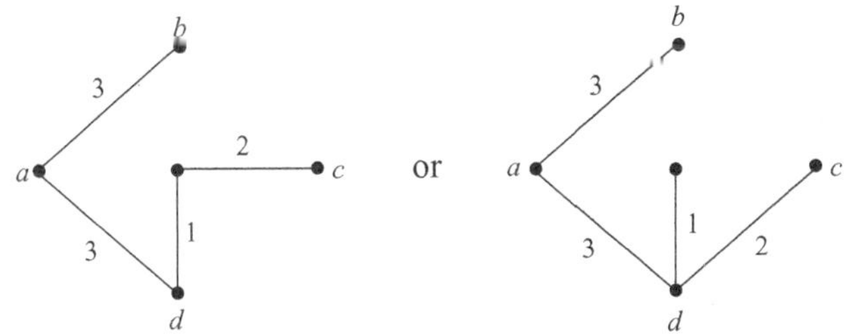

Figure 8.204

8.17.2 Kruskal's Algorithm

In Kruskal's algorithm, the edges of a weighted graph are examined one by one in order of increasing weight. At each stage, an edge with least weight out of edge-set remaining at that stage is added provided this additional edge does not create a circuit with the members of existing edge set at that stage. After $n-1$ edges have been added, these edges together with the n vertices of the connected weighted graph form a minimal tree.

Algorithm

Input: A connected weighted graph G with n vertices and the set $E=\{e_1, e_2, \ldots, e_k\}$ of weighted edges of G.

Output: The set of edges in a minimal spanning tree T for G.

Step 1. Initialize T to have all vertices of G and no edges.

Step 2. Choose an edge e_1 in E of least weight. Let

$$E^*=\{e_1\}, \quad E=E-\{e_1\}.$$

Step 3. Select an edge e_i in E of least weight that does not form circuit with members of E^*. Replace

$$E^* \text{ by } E^* \cup \{e_i\} \quad \text{and} \quad E \text{ with } E-\{e_i\}.$$

Step 4. Repeat Step 3 until number of edges in E^* is equal to $n-1$.

EXAMPLE 8.90

Use Kruskal's algorithm to determine a minimal spanning tree for the connected weighted graph G shown in Figure 8.205.

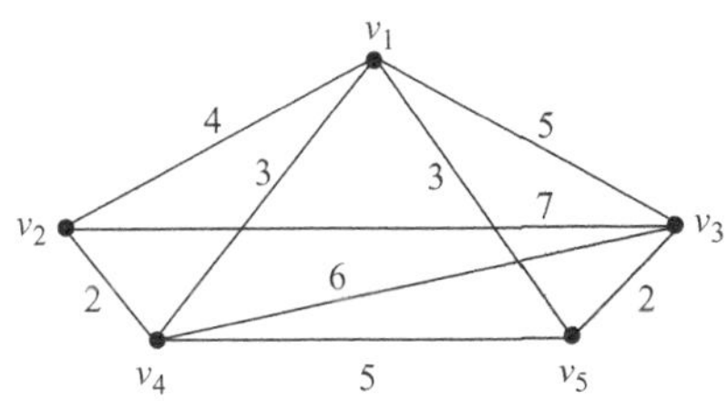

Figure 8.205

Solution.

The given weighted graph has five vertices. The minimal spanning tree would, therefore, have four edges.

Let

$$E=\{(v_1, v_2), (v_1, v_4), (v_1, v_5), (v_2, v_3), (v_1, v_3), (v_2, v_4), (v_4, v_5), (v_5, v_3), (v_3, v_4)\}.$$

The edges (v_2, v_4) and (v_3, v_5) have minimum weight. We choose arbitrarily one of these, say (v_2, v_4). Thus

$$E^*=\{(v_2, v_4)\},$$
$$E=E-\{(v_2, v_4)\}.$$

The edge (v_3, v_5) has minimum weight, so we pick it up. We have thus

$$E^*=\{(v_2, v_4), (v_3, v_5)\},$$
$$E=E-\{(v_2, v_4), (v_3, v_5)\}.$$

The edges (v_1, v_4) and (v_1, v_5) have minimum weight in the remaining edge set. We pick (v_1, v_4) say, as it does not form a cycle with E^*. Thus

$$E^*=\{(v_2, v_4), (v_3, v_5), (v_1, v_4),$$
$$E=E-\{(v_2, v_4), (v_3, v_5), (v_1, v_4)\}.$$

Now the edge (v_1, v_5) has minimum weight in $E-\{(v_1, v_4), (v_3, v_5), (v_1, v_4)\}$ and it does not form a cycle with E^*. So, we have

$$E^*=\{(v_2, v_4), (v_3, v_5), (v_1, v_4), (v_1, v_5)\}$$

and

$$E=E-\{(v_2, v_4), (v_3, v_5), (v_1, v_4), (v_1, v_5)\}.$$

Thus all the four edges have been selected. The minimal tree has the edges.

$$(v_2, v_4), (v_3, v_5), (v_1, v_4), (v_1, v_5)$$

and is shown in the Figure 8.206.

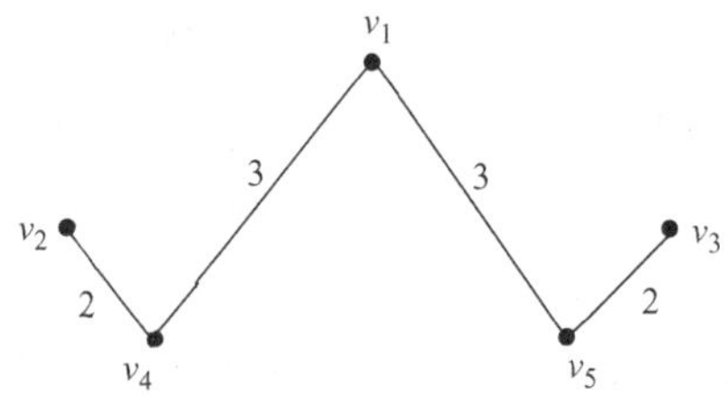

(Minimal spanning tree)

Figure 8.206

EXAMPLE 8.91

Using Kruskal's algorithm, find the minimal spanning tree for the graph given in the Figure 8.207.

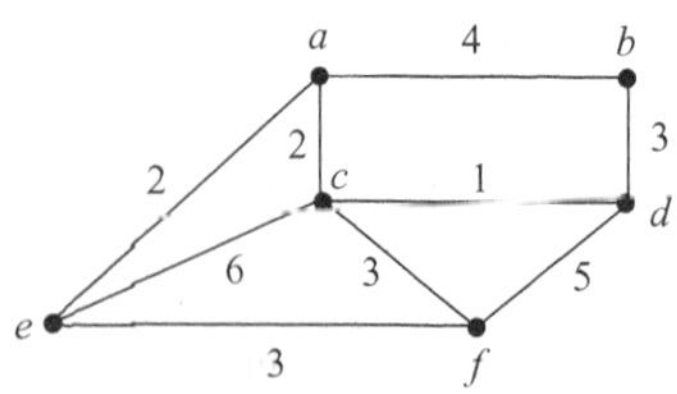

Figure 8.207

Solution.

The given graph has six vertices, therefore the minimal spanning tree would have 6 – 1 = 5 edges

Let

$$E = \{(a, b), (a, c), (b, d), (c, f), (d, f), (c, d), (a, e), (e, f), (e, c)\}.$$

Choose the edge (c, d) first because it has minimum weight. Therefore,

$$E^* = \{(c, d)\}, \quad E = E - \{(c, d)\}.$$

Now choose the edge (a, c) because it has minimum weight. So,

$$E^* = \{(c, d), (a, c), \quad E = E - \{(c, d), (a, c)\}.$$

Now choose the edge (a, e), since it has also minimum weight in $E - \{(c, d), (a, c)\}$ and it does not form cycle with E^*. So

$$E^* = \{(c, d), (a, c), (a, e)\}, \quad E = E - \{(c, d), (a, c), (a, e)\}.$$

Now choose (b, d) since it has minimum weight and it does not form circuit with E^*. Thus

$$E^* = \{(c, d), (a, c), (a, e), (b, d)\},$$
$$E = E - \{(c, d), (a, c), (a, e), (b, d)\}.$$

Now choose (c, f) or (e, f) because they have equal and minimal weight. Let us choose (c, f). Then

$$E^* = \{(c, d), (a, c), (a, e), (b, d), (c, f)\},$$
$$E = E - \{(c, d), (a, c), (a, e), (b, d), (c, f)\}.$$

We have thus obtained all the five required edges. The minimal spanning tree is shown in the Figure 8.208.

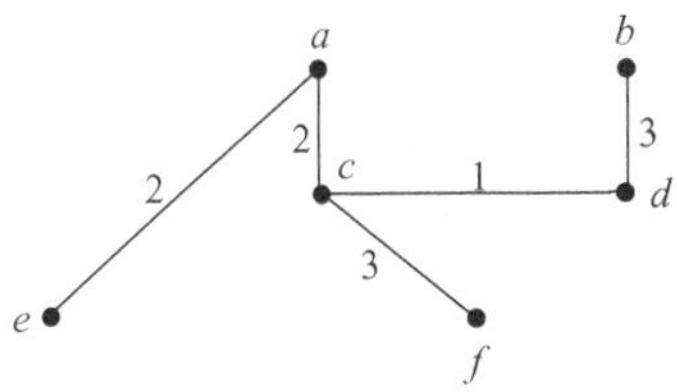

(Minimal spanning tree)

Figure 8.208

Remark 8.18 In the above example, if we had chosen (e, f) in place of (c, f) in the last step, then the minimal spanning tree would have been as shown in the Figure 8.209.

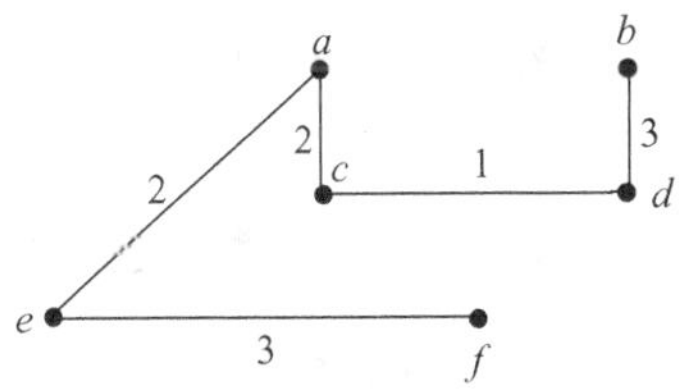

(Minimal spanning tree)

Figure 8.209

EXAMPLE 8.92

Use Kruskal's algorithm to find a shortest spanning tree for the graph G given in the Figure 8.210.

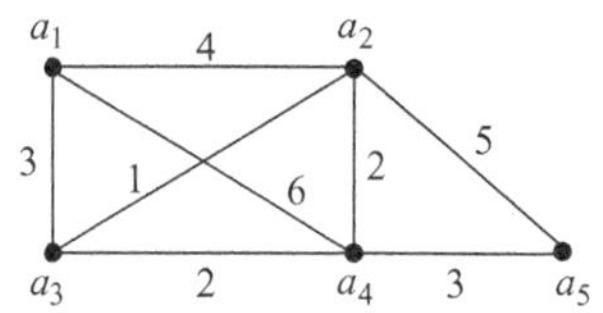

Figure 8.210

Solution.

The given graph has five vertices. So the required spanning tree would have $5-1=4$ edges. Let

$$E=\{(a_1, a_2), (a_1, a_3), (a_3, a_4), (a_2, a_4), (a_2, a_5), (a_4, a_5), (a_2, a_3), (a_1, a_4)\}.$$

Choose the edge (a_2, a_3) since it has minimum weight. Then

$$\begin{aligned} E^* &= \{(a_2, a_3)\}, \\ E &= E-\{a_2, a_3\} \\ &= \{(a_1, a_2), (a_1, a_3), (a_3, a_4), (a_2, a_4), (a_2, a_5), (a_4, a_5), (a_1, a_4)\}. \end{aligned}$$

Now choose the edge (a_2, a_4) or (a_3, a_4) having minimum weight in E. Let us take (a_2, a_4), say. Then

$$E^* = \{(a_2, a_3), (a_2, a_4)\}$$

and

$$\begin{aligned} E &= E-\{(a_2, a_3), (a_2, a_4)\} \\ &= \{(a_1, a_2), (a_1, a_3), (a_3, a_4), (a_2, a_5), (a_4, a_5), (a_1, a_4)\}. \end{aligned}$$

We cannot choose (a_3, a_4) now because it will form circuit with already chosen edges (a_2, a_3), (a_2, a_4). Therefore we can choose (a_1, a_3) or (a_4, a_5). Suppose we choose (a_1, a_3). Then

$$\begin{aligned} E^* &= \{(a_2, a_3), (a_2, a_4), (a_1, a_3)\}, \\ E &= E-\{(a_2, a_3), (a_2, a_4), (a_1, a_3)\} \\ &= \{(a_1, a_2), (a_3, a_4), (a_2, a_5), (a_4, a_5), (a_1, a_4)\}. \end{aligned}$$

Now choose (a_4, a_5) as it has minimal weight in E. Then

$$E^* = \{(a_2, a_3), (a_2, a_4), (a_1, a_3), (a_4, a_5)\}$$

and

$$\begin{aligned} E &= E-\{(a_2, a_3), (a_2, a_4), (a_1, a_3), (a_4, a_5)\} \\ &= \{(a_1, a_2), (a_3, a_4), (a_2, a_5), (a_1, a_4)\}. \end{aligned}$$

Thus we have obtained four edges and therefore the process is complete.

The Kruskal minimal spanning tree is shown in the Figure 8.211.

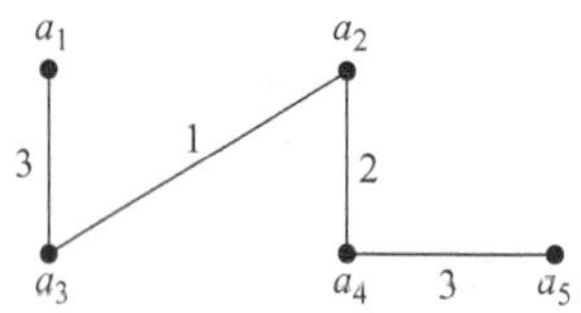

Figure 8.211

Its length is $3+1+2+3=9$.

8.18 CUT SETS

Let G be a connected graph. We know that the distance between two vertices v_1 and v_2, denoted by $d(v_1, v_2)$, is the length of the shortest path

Definition 8.115

The **diameter** of a connected graph G, denoted by diam (G), is the maximum distance between any two vertices in G.

For example, in graph G shown in the Figure 8.212, we have

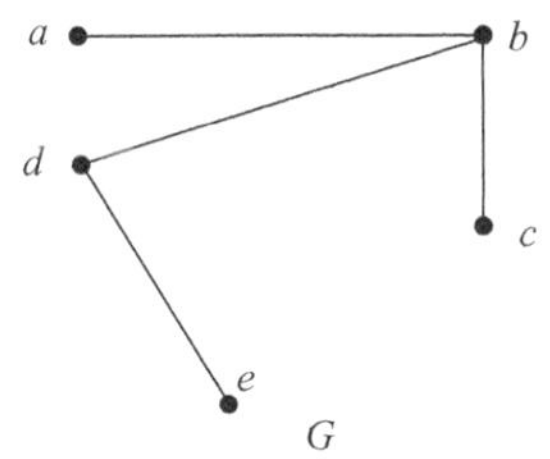

Figure 8.212

$d(a, e)=3$, $d(a, c)=2$, $d(b, e)=2$ and diam $(G)=3$.

Definition 8.116

A vertex v in a connected graph G is called a **cut point** if $G-v$ is disconnected, where $G-v$ is the graph obtained from G by deleting v and all edges containing v.

For example, in the above graph, d is a cut point.

Definition 8.117

An edge e of a connected graph G is called a **bridge** (or **cut edge**) if $G-e$ is disconnected, where $G-e$ is the graph obtained by deleting the edge e.

For example, consider the graph G shown in the Figure 8.213.

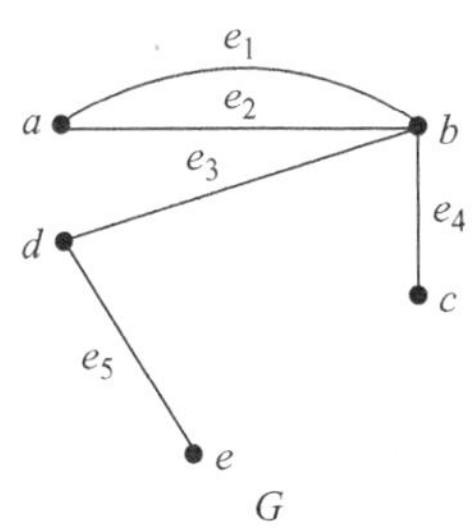

Figure 8.213

We observe that $G-e_3$ is disconnected. Hence the edge e_3 is a bridge.

Definition 8.118

A minimal set C of edges in a connected graph G is said to be a **cut set (or minimal edge-cut)** if the subgraph $G-C$ has more connected components than G has.

For example, in the above graph, if we delete the edge $(b, d)=e_3$, the resulting subgraph $G-e_3$ has two connected components shown in the Figure 8.214.

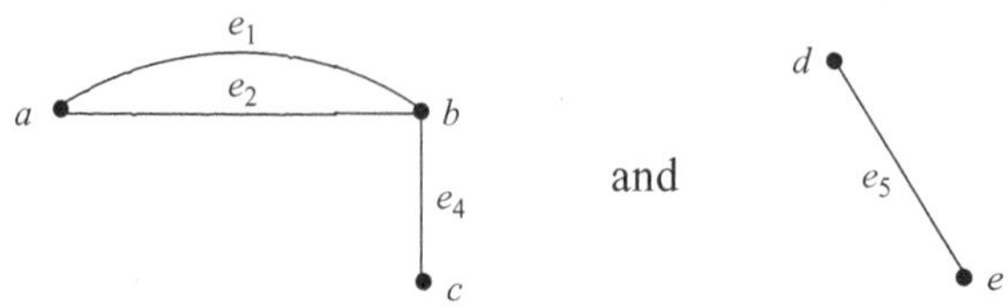

Figure 8.214

So, in this example, the cut set consists of single edge $(b, d)=e_3$, which is called cut edge or bridge.

EXAMPLE 8.93

Find a cut set for the graph given below in the Figure 8.215.

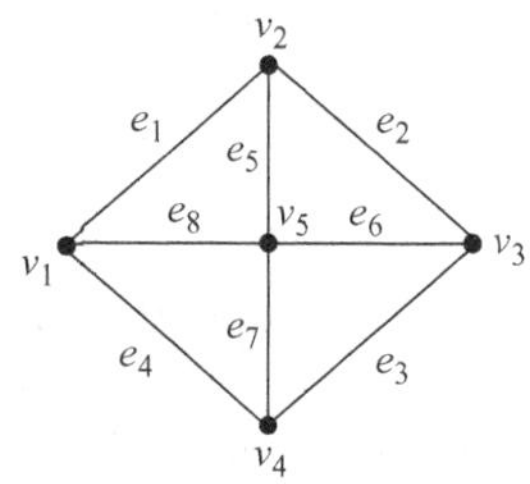

Figure 8.215

Solution.

The given graph is connected. It is sufficient to reduce the graph into two connected components. To do so we have to remove the edges e_1, e_4, e_5, e_6, e_7. The two connected components are shown in the Figure 8.216.

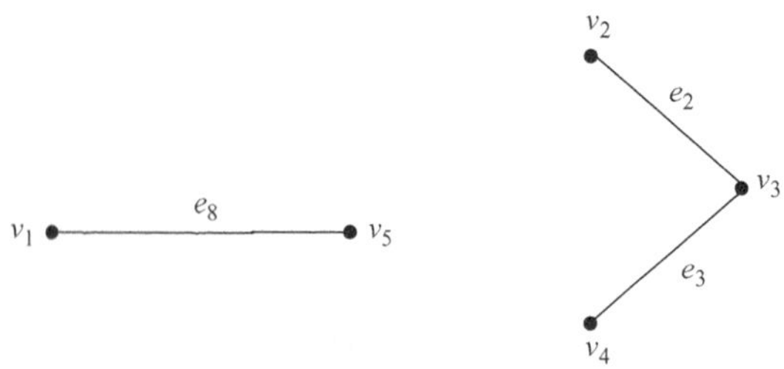

Figure 8.216

But, if we remove any proper subset of $\{e_1, e_4, e_5, e_6, e_7\}$, then there is no increase in connected components of G. Hence $\{e_1, e_4, e_5, e_6, e_7\}$ is a cut set.

EXAMPLE 8.94

Find a cut set for the graph given in the Figure 8.217.

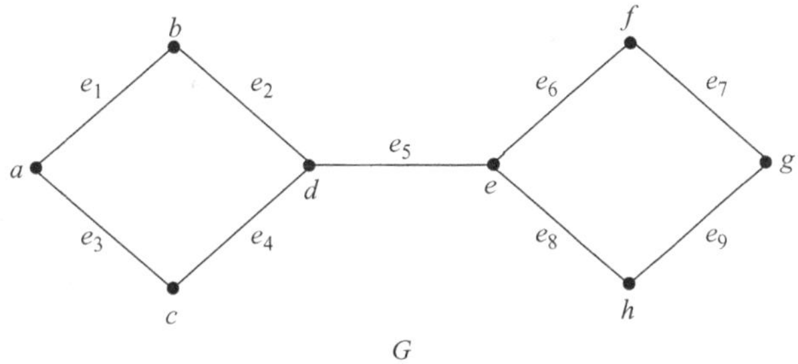

Figure 8.217

The given graph G is connected. Clearly $G-e_5$ has two connected components give in the Figure 8.218.

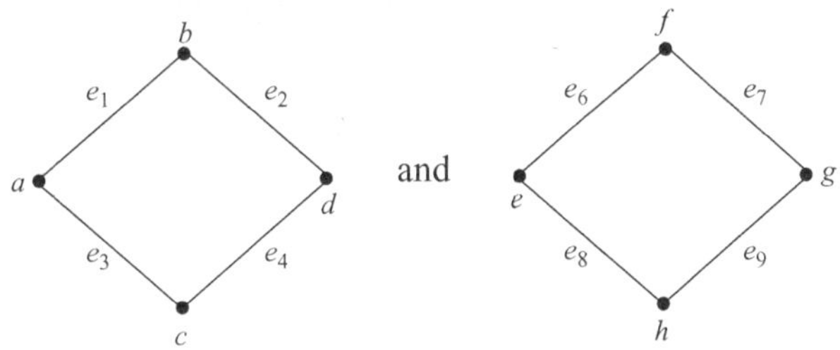

Figure 8.218

Hence a cut set for this graph is $\{e_5\}$. Some other cut sets for G are $\{e_1, e_3\}$, $\{e_3, e_2\}$, $\{e_6, e_9\}$, $\{e_6, e_8\}$

EXAMPLE 8.95

Find a cut set for the graph represented by the Figure 8.219.

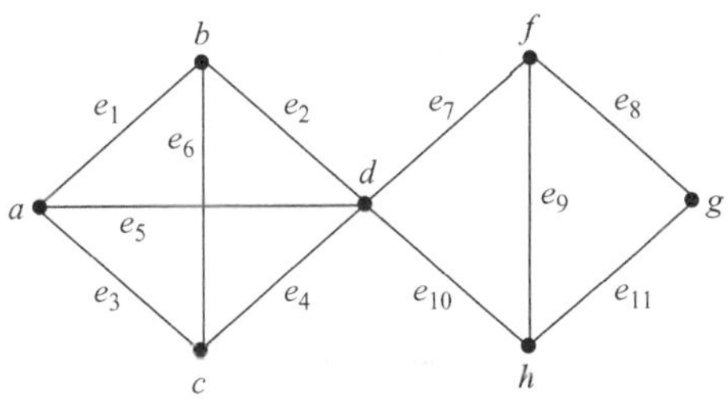

Figure 8.219

Solution.

The given graph is a connected graph. We note that removal of the edges e_7 and e_{10} creates two connected components of G shown in Figure 8.220.

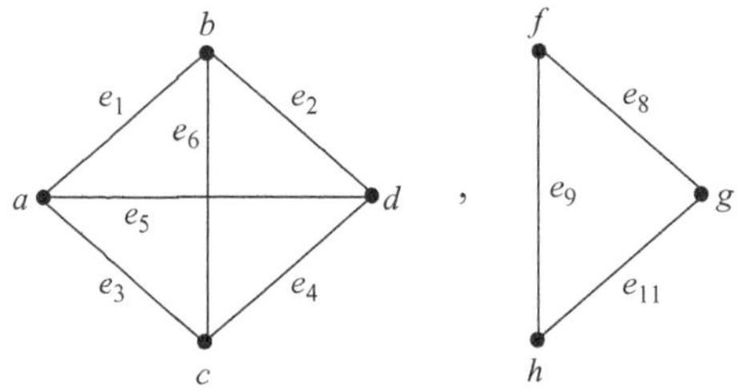

Figure 8.220

Hence the set $\{e_7, e_{10}\}$ is a cut set for the given graph G.

EXAMPLE 8.96

Find a cut set (minimal edge cut) for the following graph (Figure 8.221).

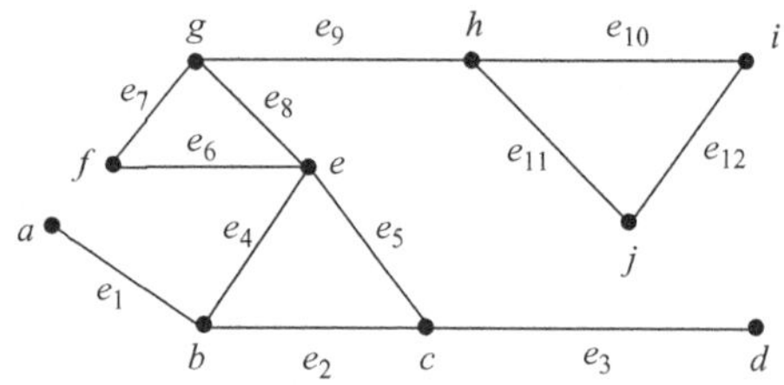

Figure 8.221

Solution.

We note that $\{e_4, e_5\}$ is a cut set (minimal edge cut) for the given graph because removal of e_4 and e_5 gives rise to two components shown in the Figure 8.222.

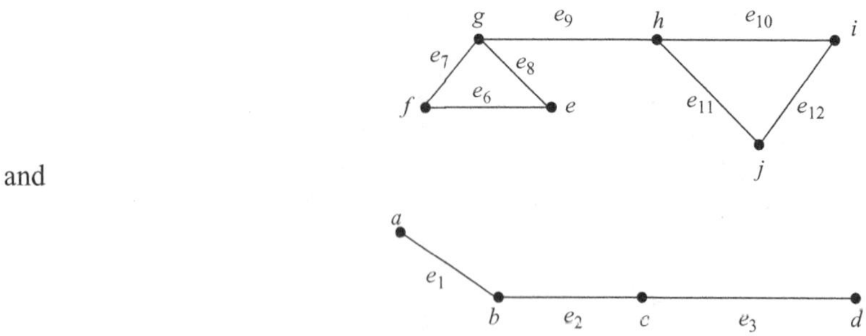

Figure 8.222

Similarly, $\{e_9\}$ forms a cut set (bridge) for this graph.

Theorem 8.30

Let G be a connected graph with n vertices. Then G is a tree if and only if every edge of G is a bridge (cut edge).

(This theorem asserts that every edge in a tree is a bridge).

Proof. Let G be a tree. Then it is connected and has $n - 1$ edges (proved already). Let e be an arbitrary edge of G. Since $G-e$ has $n-2$ edges, and also we know that a graph G with n vertices has at least $n-C(G)$ edges, it follows that $n-2 \geq n-C(G-e)$. Thus, $G-e$ has at least two components. Thus removal of the edge e created more components than in the graph G. Hence e is a cut edge. This proves that every edge in a tree is a bridge.

Conversely, suppose that G is connected and every edge of G is a bridge. We have to show that G is a tree. To prove it, we have only to show that G is circuit-free. Suppose on the contrary that there exists a cycle between two points x and y in G (Figure 8.223). Then any edge on this cycle is not a cut edge which contradicts the fact that every edge of G is a cut edge. Hence G has no cycle. Thus G is connected and acyclic and so is a tree.

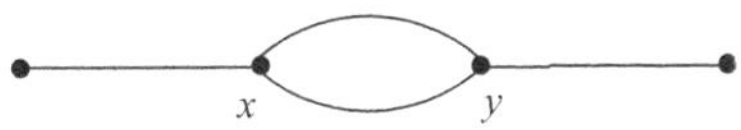

Figure 8.223

8.18.1 Relation Between Spanning Trees, Circuits and Cut Sets

A spanning tree contains a unique path between any two vertices in the graph. Therefore, addition of a chord to the spanning tree yields a subgraph that contains exactly one circuit. For example, consider the graph G shown below in the Figure 8.224.

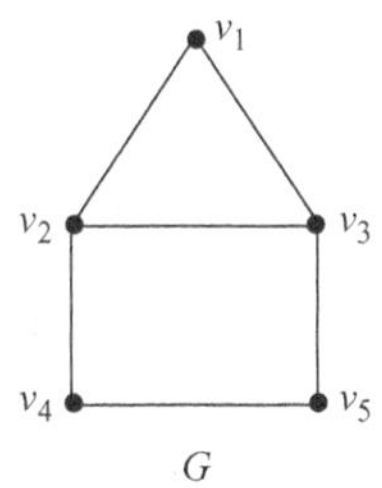

Figure 8.224

For this graph, the Figure 8.225 is a spanning tree.

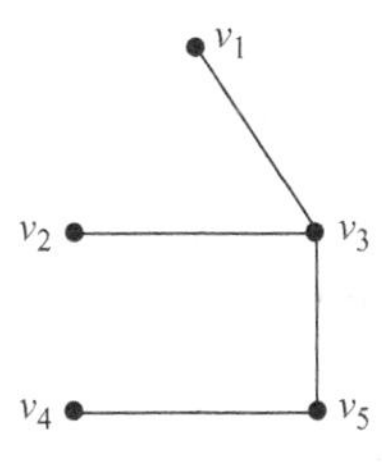

Figure 8.225 (Spanning tree)

The chords of this tree are (v_1, v_2) and (v_2, v_4). If we add (v_1, v_2) to this spanning tree, we get a circuit $v_1 v_2 v_3 v_1$. Similarly addition of (v_2, v_4) gives one more circuit $v_2 v_3 v_5 v_4 v_2$. If there are v vertices and e edges in a graph, then there are $e-v+1$ chords in a spanning tree. Therefore, if we add all the chords to the spanning tree, there will be $e-v+1$ circuits in the graph.

Definition 8.119

Let v be the number of vertices and e be the number of edges in a graph G. Then the set of $e-v+1$ circuits obtained by adding $e-v+1$ chords to a spanning tree of G is called the **fundamental system of circuits relative to the spanning tree**.

A circuit in the fundamental system is called a **fundamental circuit**.

For example, $\{v_1, v_2, v_3, v_1\}$ is the fundamental circuit corresponding to the chord (v_1, v_2).

On the other hand, since each branch of a tree is cut edge, removal of any branch from a spanning tree breaks the spanning tree into two trees. For example, if we remove (v_1, v_3) from the above figured spanning tree, the resulting components are shown in the Figure 8.226.

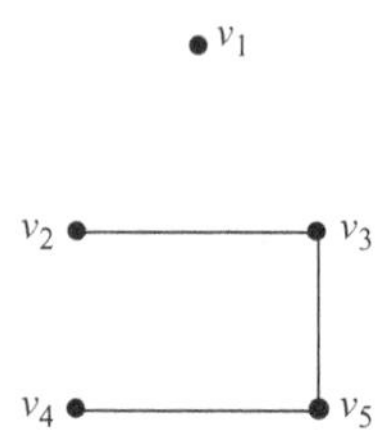

Figure 8.226

Thus, **to every branch in a spanning tree, there is a corresponding cut set.** But, in a spanning tree, there are $v-1$ branches. Therefore, **there are $v-1$ cut sets corresponding to $v-1$ branches.**

Definition 8.120

The set of $v-1$ cut sets corresponding to $v-1$ branches in a spanning tree of a graph with v vertices is called the **fundamental system of cut sets relative to the spanning tree.**

A cut set in the fundamental system of cut sets is called a **fundamental cut set.**

For example, the fundamental cut sets in the spanning tree (figured above) are

$$\{(v_1, v_2), (v_1, v_3)\}, \{(v_1, v_3), (v_2, v_3), (v_3, v_4)\},$$
$$\{(v_3, v_5), (v_4, v_5)\}, \{v_2, v_4), (v_4, v_5)\}.$$

Theorem 8.31

A circuit and the complement of any spanning tree must have at least one edge in common.

Proof. We recall that the set of all chords of a tree is called the complement of the tree. Suppose on the contrary that a circuit has no common edge with the complement of a spanning tree. This means the circuit is wholly contained in the spanning tree. This contradicts the fact that a tree is acyclic (circuit- free). Hence a circuit has at least one edge in common with complement of a spanning tree.

Theorem 8.32

A cut set and any spanning tree must have at least one edge in common.

Proof. Suppose on the contrary that there is a cut set which does not have a common edge with a spanning tree. Then removal of cut set has no effect on the tree, that is, the cut set will not separate the graph into two components. But this contradicts the definition of a cut set. Hence the result.

Theorem 8.33

Every circuit has an even number of edges in common with every cut set.

Proof. We know that a cut set divides the vertices of the graph into two subsets each being set of vertices in one of the two components. Therefore a path connecting two vertices in one subset must traverse the edges in the cut set an even number of times. Since a circuit is a path from some vertex to itself, it has an even number of edges in common with every cut set.

8.19 TREE SEARCHING

Let T be a binary tree of height $h \geq 1$ and root v. Since $h \geq 1$, v has at least one child : v_L and/or v_R. Now v_L and v_R are the roots of the left and right subtrees of v called T_L and T_R, respectively.

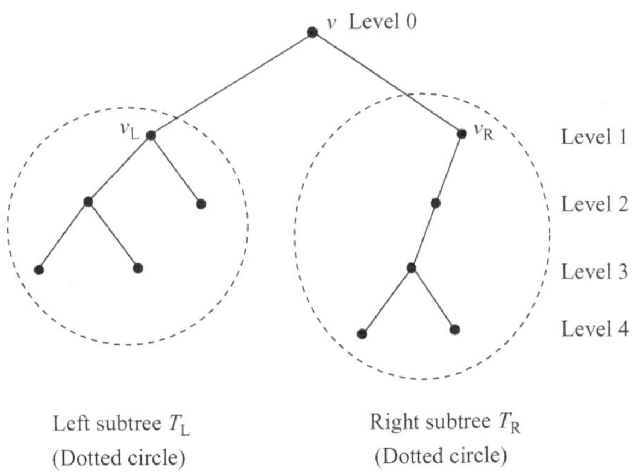

Figure 8.227

Definition 8.121

Performing appropriate tasks at a vertex is called **visiting the vertex**.

Definition 8.122

The process of visiting each vertex of a tree in some specified order is called **searching the tree** or **walking** or **traversing the tree**.

We now discuss methods of searching a tree.

1. Pre-order Search Method

Input: The root v of a binary tree.

Output: Vertices of a binary tree using preorder traversal

1. **Visit** v
2. If v_L (left child of v) exists, then apply the algorithm to $(T(v_L), v_L)$
3. If v_R (right child of v) exists, then apply this algorithm to $(T(v_R), v_R)$

End of algorithm pre-order.

In other words, pre-order search of a tree consists of the following steps:

Step 1. Visit the root.

Step 2. Search the left subtree if it exists.

Step 3. Search the right subtree if it exists.

EXAMPLE 8.97

Find the order of the vertices of the following tree T processed using pre-order traversal.

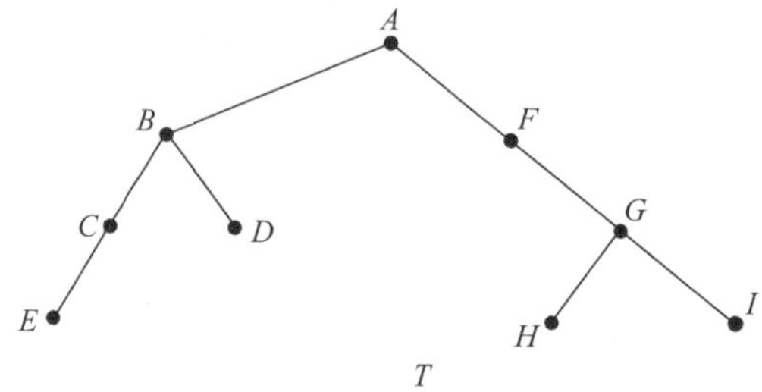

Figure 8.228

Solution.

The root of the given tree is A. It has two children B and F. The left subtree and right subtree are shown in Figure 8.229.

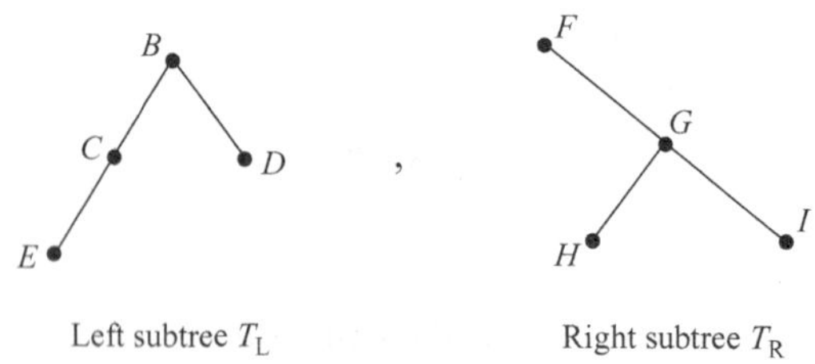

Figure 8.229

If we apply pre-order algorithm to the given tree T, we visit the root and print A. Using pre-order to left subtree, we visit the root B and print B and then search T_L printing C and E and then D. Up to this point, the search has yielded the string $A\ B\ C\ E\ D$. Thus the search of T_L is complete. We now go to subtree T_R. Using the same procedure, we get the string $F\ G\ H\ I$. Thus the complete search of the tree T is

A B C E D F G H I.

EXAMPLE 8.98

Find binary tree representation of the expression

$$(a-b)\times(c+(d\div e))$$

and represent the expression in string form using pre-order traversal.

Solution.

In the given expression, $\times$ is the central operator and therefore shall be the root of the binary tree. Then the operator $-$ acts as v_L and the operator $+$ acts as v_R. Thus the tree representation of the given expression is as shown in the Figure 8.230.

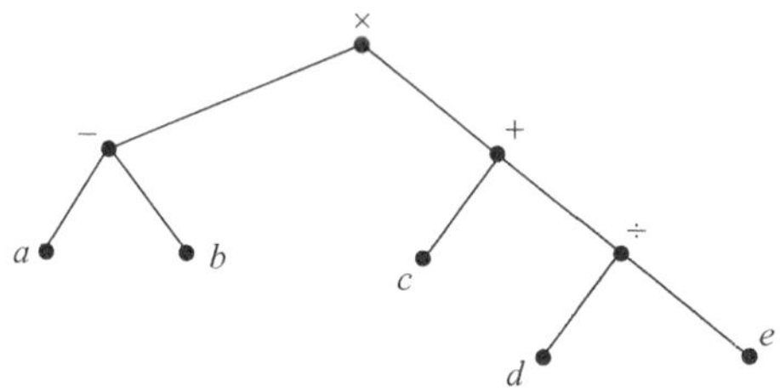

Figure 8.230

The result of the pre-order traversal to this binary tree is the string

$$\times - a\ b + c \div d\ e.$$

This form of the expression is called **prefix form** or **polish form** of the expression

$$(a-b)\times(c+(d\div e)).$$

In a polish form, the variables a, b, c,...are called **operands** and $-,+,\times,\div$ are called **operators**. We observe that, **in polish form, the operands follow the operator**.

EXAMPLE 8.99

Represent the expression $(A+B) * (C-D)$ as a binary tree and write the prefix form of the expression.

Solution.

Here $*$ is the central operator. Further $+$ and $-$ operators are v_L and v_R. Hence the binary tree for the given experssion is as shown in the Figure 8.231.

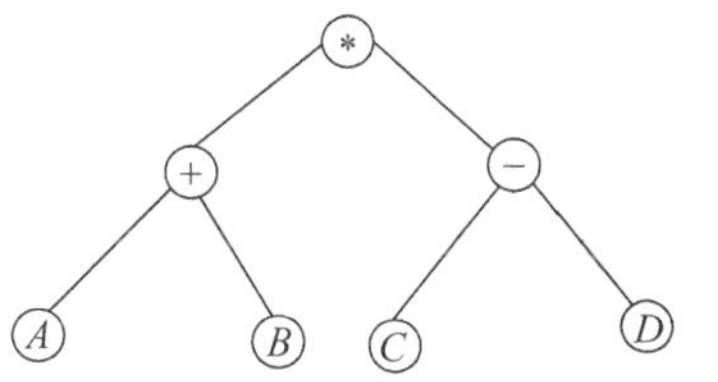

Figure 8.231

Using pre-order traversal, the prefix expression for it is

$$* + A\ B - C\ D.$$

8.19.1 Procedure to Evaluate an Expression Given in Polish Form

To find the value of a polish form, we proceed as follows:

Move from left to right until we find a string of the form K x y, where K is operator and x, y are operands.

Evaluate x K y and substitute the answer for the string K x y. Continue this procedure until only one number remains.

EXAMPLE 8.100

Find parenthesized form of the polish expression

$$-+A\ B\ C.$$

Solution.

The parenthesized form of the given polish expression is derived as follows:

$$-(A+B)\ C,$$
$$(A+B)-C.$$

The corresponding binary tree is shown in the Figure 8.232.

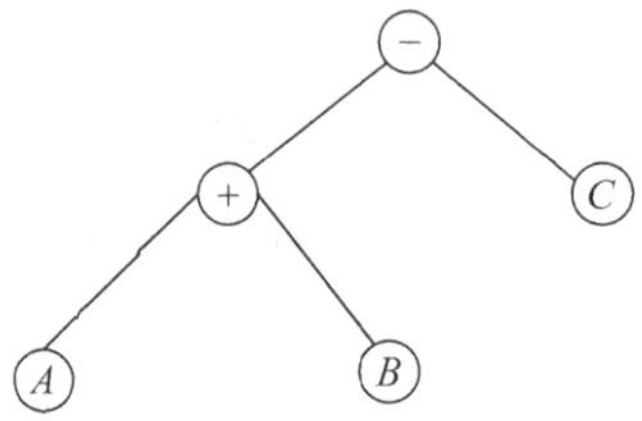

Figure 8.232

EXAMPLE 8.101

Evaluate the polish form

$$\times-64+5\div22$$

Solution.

We have the following steps in this regard:

1. $\times(6-4)+5\div22$
2. $\times2+5\div22$
3. $\times2+5\ (2\div2)$
4. $\times2+51$
5. $\times2\ (5+1)$
6. $\times26$
7. 2×6
8. 12, which is the required value of the expression.

2. Post-order Search Method

Algorithm

Step 1. Search the left subtree if it exists
Step 2. Search the right subtree if it exists
Step 3. Visit the root

End of algorithm

EXAMPLE 8.102

Represent the expression

$$(A+B) * (C-D)$$

as a binary tree and write the result of post-order search for that tree.

Solution.

The binary tree expression (as shown earlier) of the given algebraic expression is given in Figure 8.233.

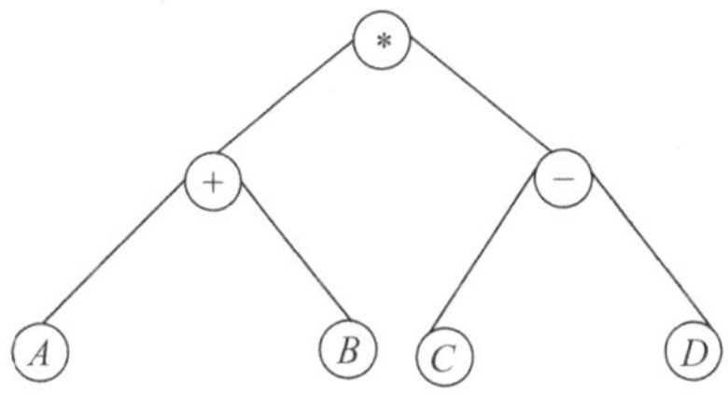

Figure 8.233

The result of post-order search of this tree is

$$A\ B + C\ D - *$$

This form of the expression is called **postfix form** of the expression or **reverse polish form** of the expression.

In postfix form, the operator follows its operands.

EXAMPLE 8.103

Find the parenthesized form of the postfix form

$$A\ B\ C * * C\ D\ E + / -.$$

Solution.

We have

1. $A\ B\ C * * C\ D\ E+/-$
2. $A\ (B * C) * C\ (D+E)\ /-$
3. $(A * (B * C))\ (C\ /\ (D+E))-$
4. $(A * (B * C))-(C\ /\ (D+E))$.

The corresponding binary tree is shown in the Figure 8.234.

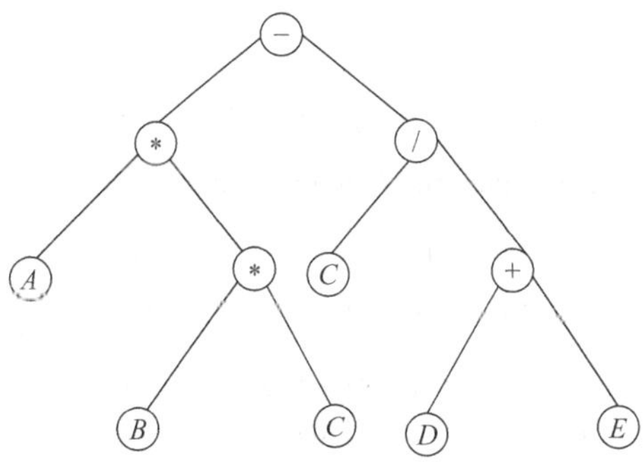

Figure 8.234

EXAMPLE 8.104

Evaluate the postfix form

$$21-342\div+\times.$$

Solution.

We have

$$\begin{aligned}21-342\div+\times &= (2-1)\ 342\div+\times\\ &= 13\ (4\div2)+\times\\ &= 132+\times\\ &= 1\ (3+2)\times\\ &= 15\times\\ &= 1\times5=5.\end{aligned}$$

8.20 TRANSPORT NETWORKS

An important application of weighted directed graph is to model transport networks like oil pipeline, water supply pipeline, communication network, electric power distribution system and highway system. The purpose of a network is to implement the flow of oil, water, electricity, messages, traffic, etc and we desire that the flow through the network should be largest possible value. Such a flow will be called maximum flow.

Definition 8.123

A ***transport network*** or simply a ***network*** is a simple weighted directed graph with the following properties:

(i) There is a designated vertex (node), called the ***source,*** that has no incoming edge.
(ii) There is a designated vertex, called the ***sink,*** that has no outgoing edge.
(iii) There is a non-negative weight c_{ij} on the directed edge (i, j), called the ***capacity*** of the edge (i, j).

Since the graph is simple, it follows that if the edge (i, j) is in the network, then (j, i) is not there.

Definition 8.124

A ***flow*** in a network is a function that assigns to each (i, j) of the network a non-negative number f_{ij} such that

(i) $0 \le f_{ij} \le c_{ij}$, where c_{ij} is the capacity of the edge (i, j)
(ii) For each vertex j, which is neither the source nor the sink,

$$\sum_i f_{ij} = \sum_k f_{jk}.$$

The non-negative number f_{ij} is called the ***flow*** in the edge (i, j). For any vertex (distinct from source and sink) j, the sum $\sum_i f_{ij}$ is called the ***flow into*** j and the sum $\sum_k f_{jk}$ is called the ***flow out of*** j. Thus the condition (ii) implies that material cannot accumulate, be created, dissipate or lost at any vertex other than the source or the sink. Hence the equation $\sum_i f_{ij} = \sum_k f_{jk}$ is called ***conservation of flow***. As a consequence of conservation, it follows that the ***sum of flows leaving the source must be equal to the sum of the flows entering the sink.*** This sum is called the ***value of the flow*** and is denoted by value (F). A flow F in a network is represented by labelling each edge (i, j) with the pair (c_{ij}, f_{ij}).

For example consider the oil pipeline network shown in Figure 8.235

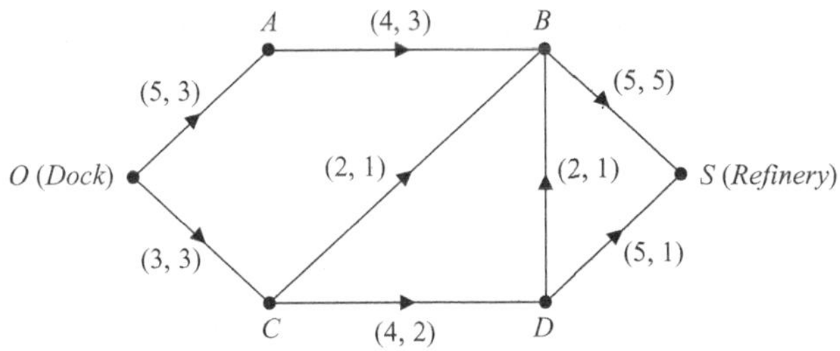

Figure 8.235

In this network, the crude oil unloaded at the dock O (source) is being pumped to the refinery S (sink). The capacities of the flow are

$$c_{OA} = 5, \quad c_{AB} = 4, \quad c_{BS} = 5, \quad c_{OC} = 3, \quad c_{CB} = 2, \; c_{CD} = 4, \; c_{DB} = 2, \quad c_{DS} = 5.$$

The flows in the edges are

$$f_{OA} = 3, \quad f_{AB} = 3, \quad f_{BS} = 5, \quad f_{OC} = 3, \quad f_{CB} = 1, \quad f_{CD} = 2, \quad f_{DB} = 1, \quad C_{DS} = 1.$$

We note that

Flow into the vertex B=3+1+1=5

Flow out of the vertex B=5.

Therefore, ***conservation of flow*** in satisfied. Further, value (F)=3+3=5+1=6.

EXAMPLE 8.105

Consider the water pipeline network shown in Figure 8.236 for two towns A and B in which water is supplied from four borewells w_1, w_2, w_3, w_4 and a, b, c, d represent intermediate pumping stations. Form an equivalent network with unique source and unique sink and determine the value of the flow.

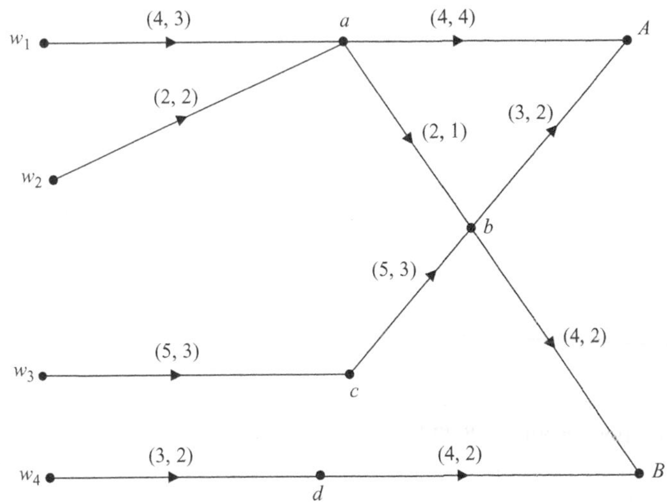

Figure 8.236

Solution.

To obtain the required equivalent network with designated source and designated sink, we tie together the given sources w_1, w_2, w_3 and w_4 into a ***super source*** O and the towns A and B into a ***super sink*** S as shown in Figure 8.237.

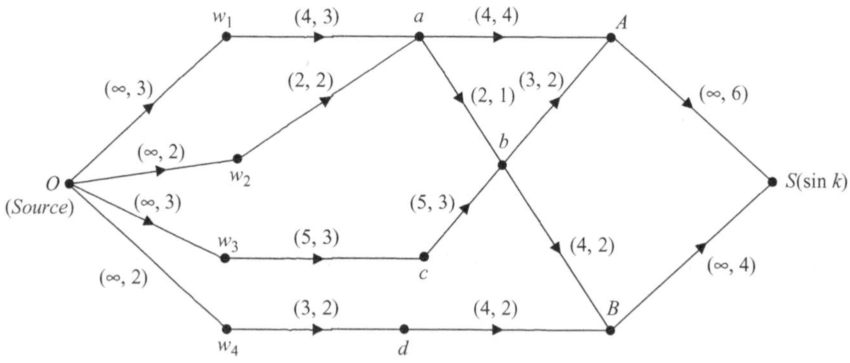

Figure 8.237

The value of the flow is

$$\text{Value }(F) = 3+2+3+2 = 6+4 = 10.$$

Definition 8.125

A ***maximal flow*** in a network N is a flow with maximum value.

In general, there will be several flows having the same maximum value. For example, consider the network shown in Figure 8.238.

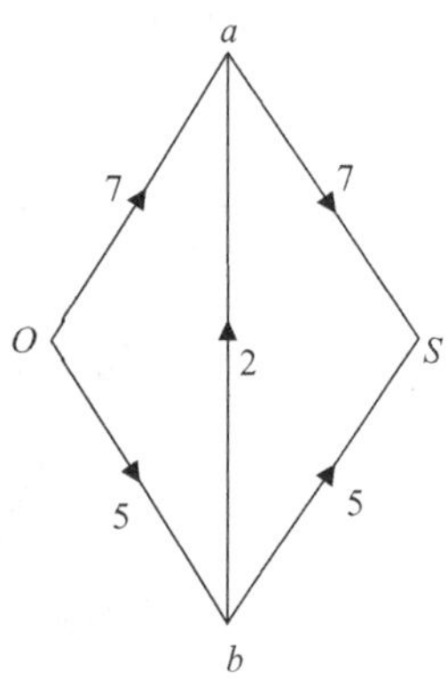

Figure 8.238

Consider the following two flows for this network:

The value of the flow in Figure 8.239(a) is 10 and three of the five edges in this network attain their maximum capacity. But for the same network, the value of the flow in Figure 8.239(b) is 12 and four out of five edges are carrying maximum capacity. Therefore, the flow function in the second case is better than the first one.

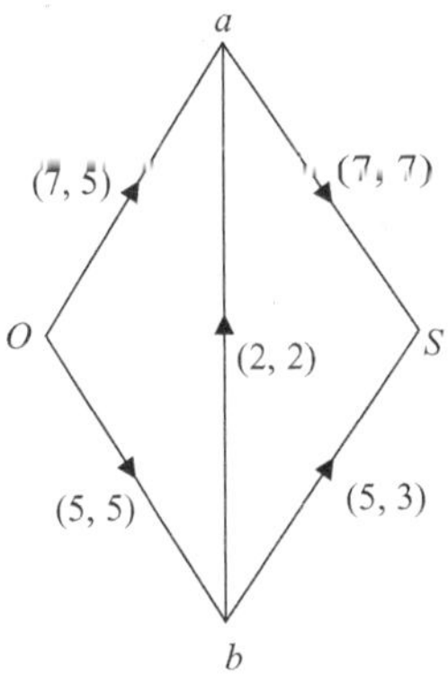

Figure 8.239(a)

and

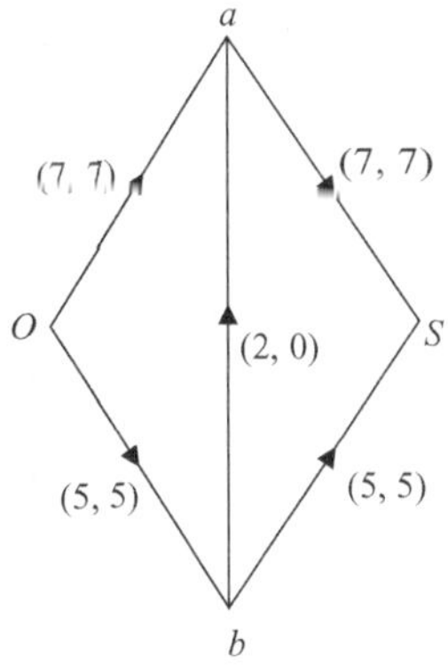

Figure 8.239(b)

EXAMPLE 8.106

Find the maximum possible increase in the following network (Figure 8.240).

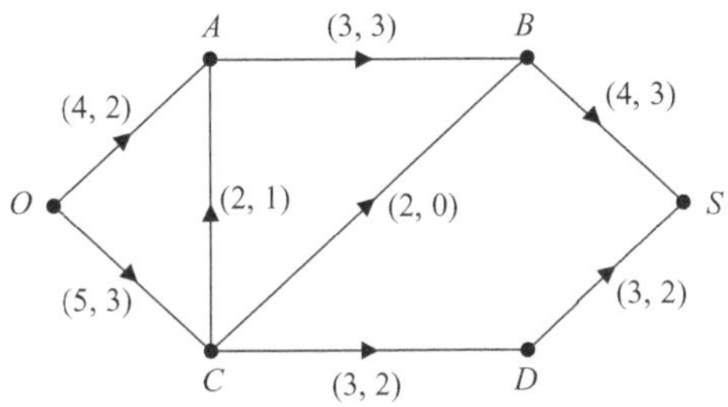

Figure 8.240

Solution.

We cannot increase the flow in the path *OABS* because the capacity and the flow of the edge *AB* are equal. But if increase flow in the path *OCDS* by 1, we get the network shown in Figure 8.241.

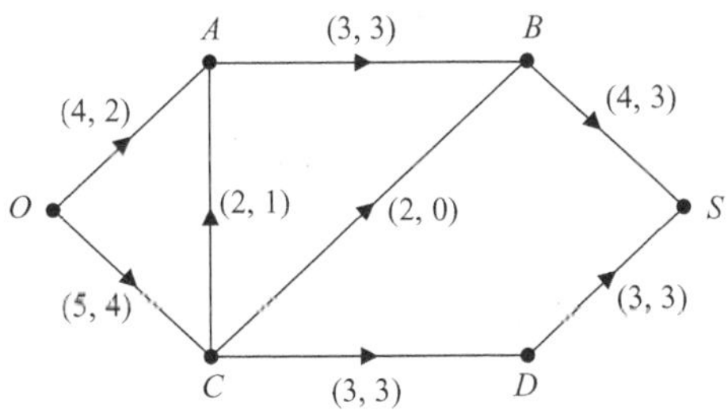

Figure 8.241

Further, if we increase the flow in the path *OCBS* by 1, we get the network shown in Figure 8.242.

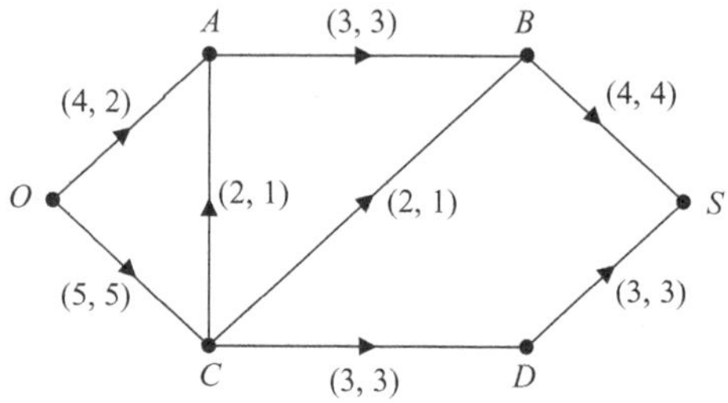

Figure 8.242

No further increase in flow is possible since the capacities of all the edges leading to the sink have been exhausted. The value of the flow is 7.

EXAMPLE 8.107

Find a maximal flow in the network shown in Figure 8.243.

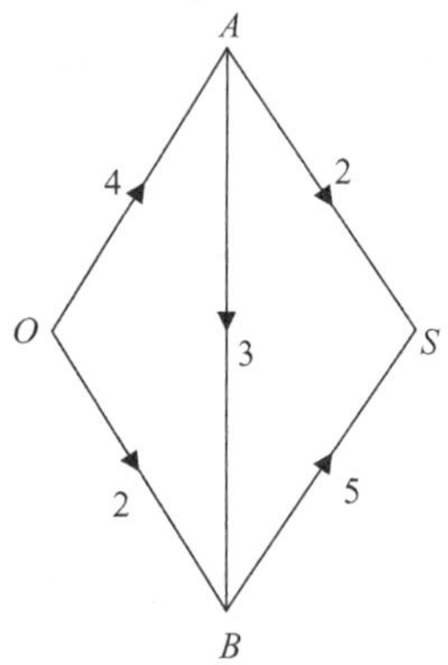

Figure 8.243

Solution.

The capacity of each edge is shown in Figure 8.243. Let the initial flow be 0 in each edge. Thus we have the network as shown in Figure 8.244(a).

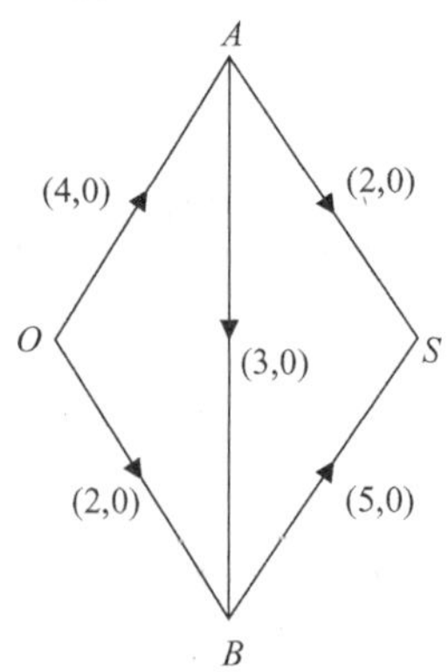

Figure 8.244(a)

Increase the flow by 2 in the path *OAS* to get the network shown in the Figure 8.244(b).

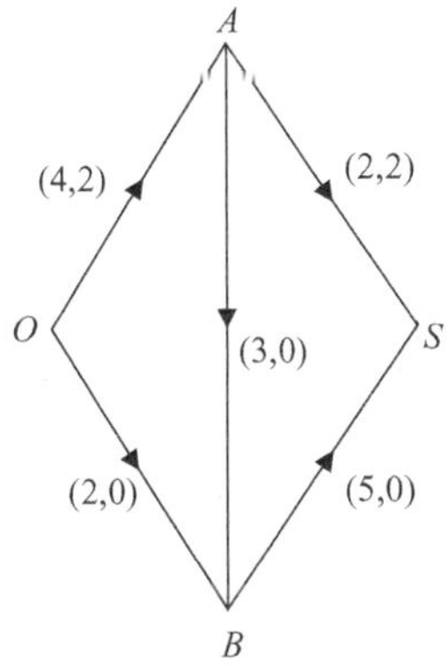

Figure 8.244(b)

Now increase the flow by 2 in the path *OBS* to get the network shown in Figure 8.244(c).

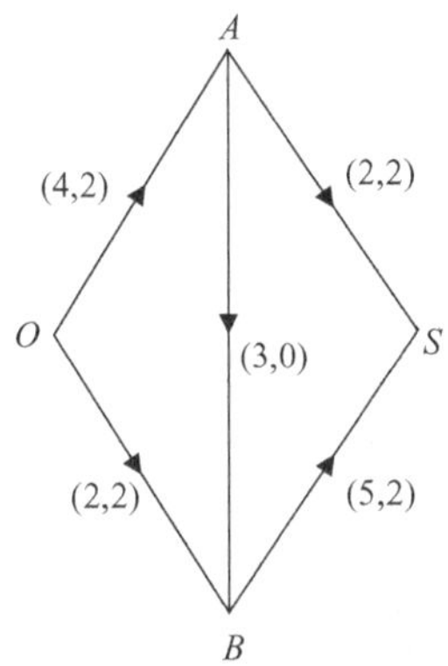

Figure 8.244(c)

Now increase the flow by 2 in the path *OABS* to get the network shown in Figure 8.244(d).

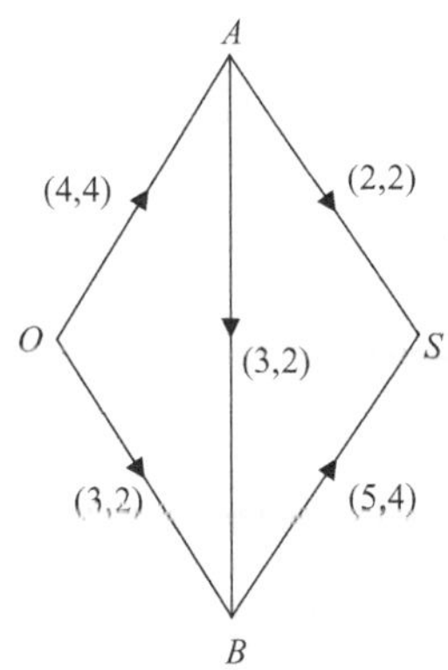

Figure 8.244(d)

Any further increase in the flow is not possible. Thus maximal flow in the network has been reached. Thus, value $(F)=2+4=6$.

EXAMPLE 8.108

Find a maximal flow in the network shown in Figure 8.245.

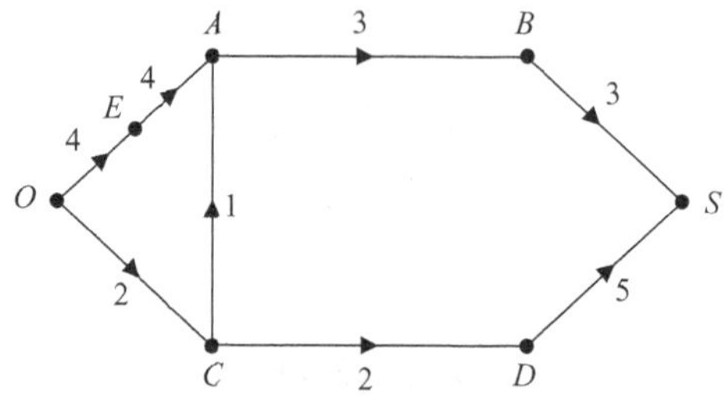

Figure 8.245

Solution.

The initial labelling yields the network shown in Figure 8.246(a).

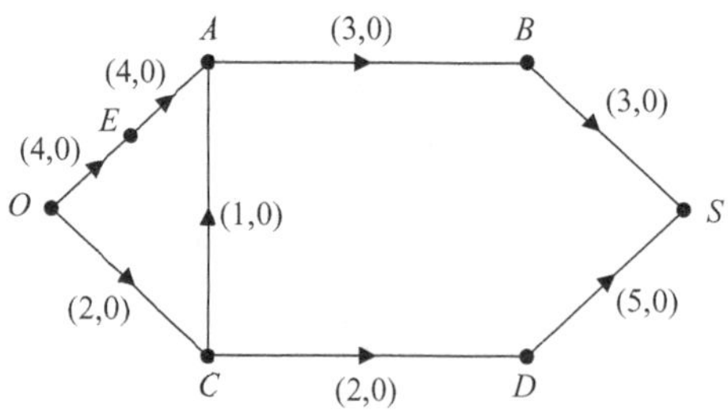

Figure 8.246(a)

Increase the flow by 2 in the path *OCDS* by to get the network shown in Figure 8.246(b).

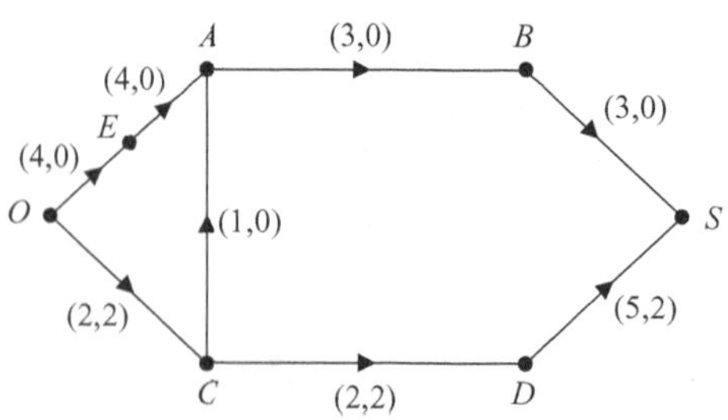

Figure 8.246(b)

Now increase the flow by 3 in the path *OEABS* to get the network shown in Figure 8.246(c).

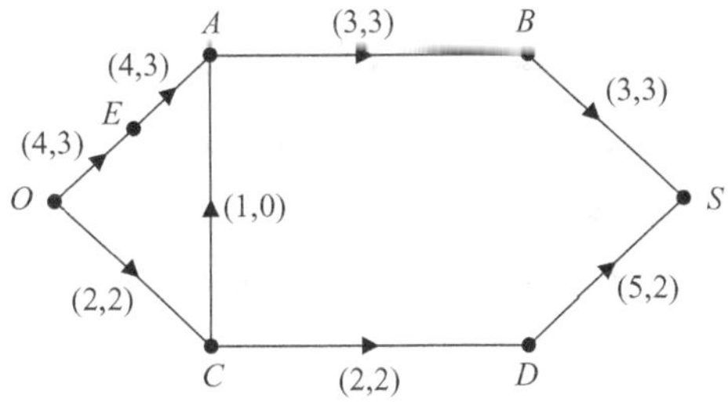

Figure 8.246(c)

No further increase in flow in possible since the capacities in the edges (*A*, *B*) and (*C*, *D*) has been exhausted. Hence value (*F*)=5.

Definition 8.126

Let *N* be a network. A ***cut*** in *N* is a partition $\left(P, \overline{P}\right)$ of the set of vertices in *N* such that the source $O \in P$ and the sink $S \in \overline{P}$.

Thus, a cut $\left(P, \overline{P}\right)$ in a network *N* is a set *K* of edges $(v, w), v \in P, w \in \overline{P}$ such that every path from source to sink contains at least one edge from *K*. In fact, a cut does cut a directed graph into two pieces, one containing the source and the other containing the sink. ***Nothing can flow from source to sink if edges of a cut are removed.*** We indicate a cut by drawing a dashed line to partition the vertex set. As an illustration, consider the network of Example 8.106. The dashed line divides the vertex set into the sets $P = \{O, A\}$ and $\overline{P} = \{C, B, D, S\}$.

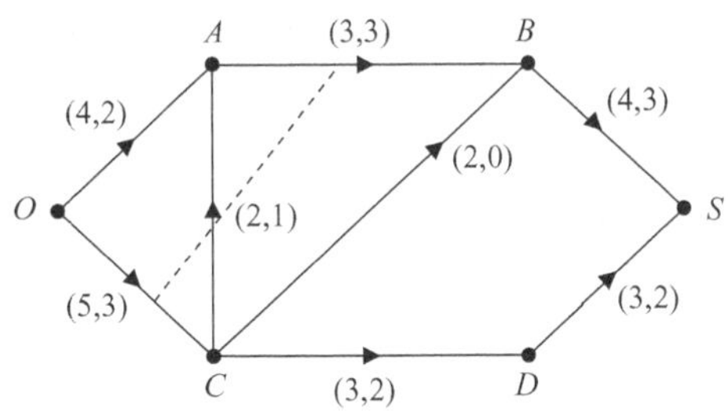

Figure 8.247

Definition 8.127

The ***capacity of a cut*** $K = (P, \overline{P})$ is the sum of the capacities of all edges in *K* and is denoted by *C(K)* or $C(P, \overline{P})$.

For example, the capacity of the cut in Figure 8.247 is

$$C(P, \overline{P}) = C_{OC} + C_{AC} + C_{AB} = 5 + 2 + 3 = 10.$$

EXAMPLE 8.109

Find the capacity of the cut, shown by dashed line, in the network shown in Figure 8.248.

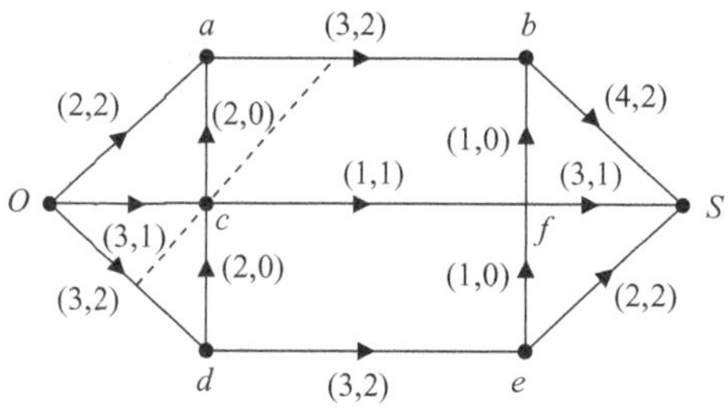

Figure 8.248

Solution.

The cut $(P, \overline{P})$ is given by

$$P = \{o, a, c\}, \quad \overline{P} = \{d, e, f, b, S\}.$$

Therefore the capacity of the given cut is

$$C(P, \overline{P}) = c_{Od} + c_{cf} + C_{ab} = 3 + 1 + 3 = 7.$$

Definition 8.128

A ***minimal cut*** is a cut having minimum capacity.

Theorem 8.34 (The Max Flow Min Cut Theorem)

A maximum flow F in a network has value equal to the capacity of a minimum cut of the network.

Proof. Let F be any flow and K be any cut in a network. Since all parts of F must pass through the edges of the cut K and since $C(K)$ is the maximum amount that can pass through the edges of K, it follows that $\textit{\textbf{value}}(F) \le C(K)$. The equality will hold if the flow F uses the full capacity of all edges in the cut K. Thus equality holds if the flow is maximum. Further, K must be a minimum capacity cut since every cut must have capacity at least equal to ***value*** (F). This completes the proof of the theorem.

EXAMPLE 8.110

In the network shown in Figure 8.249, find a maximum flow, give its value and prove that it is maximum by appealing to the max flow min cut theorem.

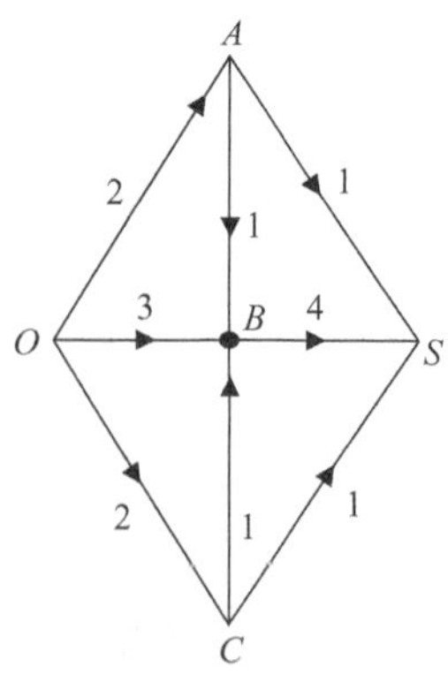

Figure 8.249

Solution.

The initial labelling yields the following network shown in Figure 8.250(a):

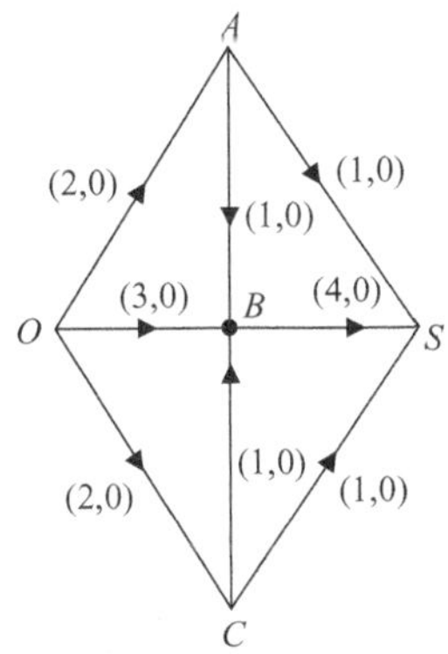

Figure 8.250(a)

Increase the flow by 1 in the path *OAS* to get the network shown in Figure 8.250(b).

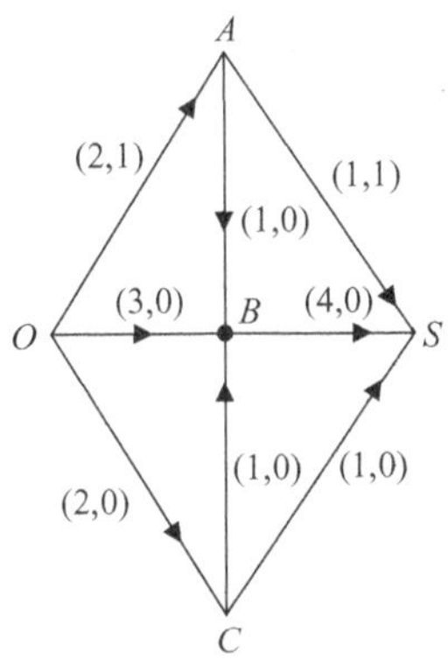

Figure 8.250(b)

Now increase the flow by 1 in the path *OCS* to get the network shown in Figure 8.250(c).

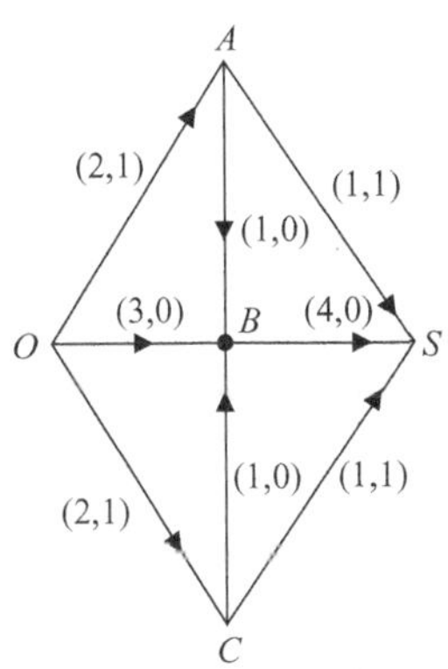

Figure 8.250(c)

Further, increase in flow by 1 in the path *OABS* to get the network of Figure 8.250(d).

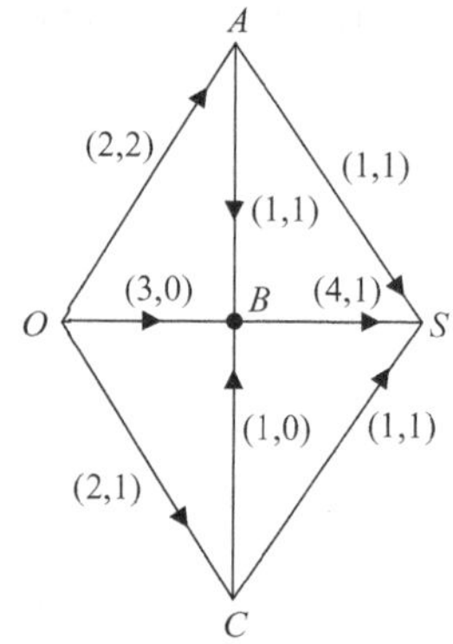

Figure 8.250(d)

Now increase the flow by 1 in the path *OCBS* and get the network shown in Figure 8.250(e).

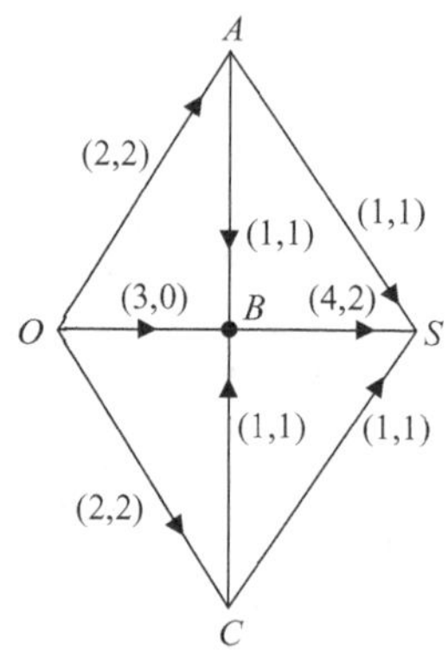

Figure 8.250(e)

Lastly, we increase the flow by 2 in the path *OBS* and get the network with maximum flow as shown in Figure 8.250(f). In fact the capacities in the edges leading to the sink have been completely exhausted.

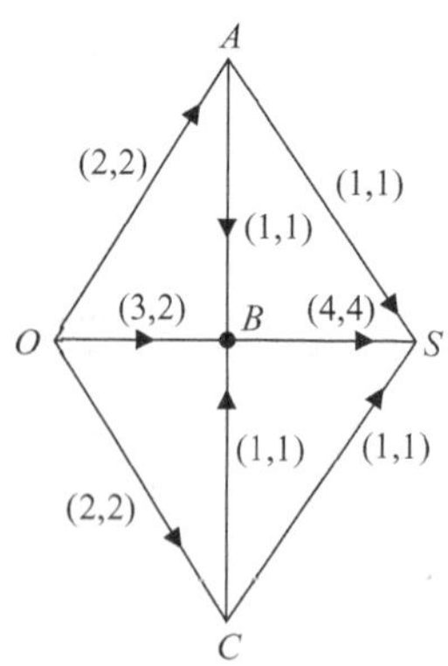

Figure 8.250(f)

We note that value ($F=2+2+2=1+4+1=6$). To check whether it is maximum flow, we appeal to max flow min cut theorem. We consider the following cuts:

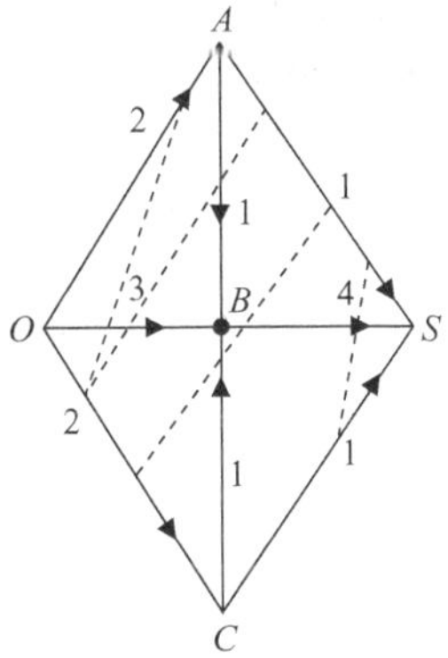

Figure 8.250(g)

(i) $K_1 = \{OC, OB, OA\}$
with $\text{cap}(K_1) = C_{OC} + C_{OB} + C_{OA} = 2+3+2 = 7$,

(ii) $K_2 = \{OC, BC, BS, AS\}$
with $\text{cap}(K_2) = C_{OC} + C_{BC} + C_{BS} + C_{AS} = 2+1+4+1 = 8$,

(iii) $K_3 = \{CS, BS, AS\}$
with $\text{cap}(K_3) = C_{CS} + C_{BS} + C_{AS} = 1+4+1 = 6$,

(iv) $K_4 = \{OC, OB, AS, AB\}$
with cap $(K_4) = C_{OC} + C_{OB} + C_{AS} + C_{AB} = 2+3+1+1 = 7$.

We note that capacity of the minimum cut in 6. Hence, the maximum flow is 6.

EXERCISES

1. Is there a non-empty simple graph with twice as many edges as vertices?
2. Is the graph given below a bipartite graph?

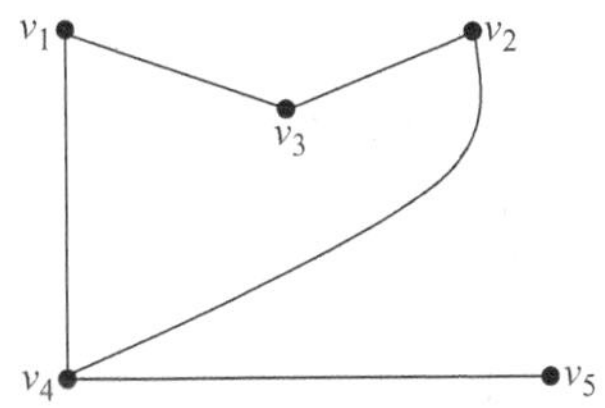

Figure 8.251

3. Find a formula for the number of edges in complete bipartite graph K_{mn}.
4. A graph G has vertices of degrees 1, 4, 3, 7, 3 and 2. Find the number of edges in the graph.
5. How many edges are there in an n-cube?
6. When does the complete graph K_n possess an Euler circuit?
7. Does the graph given below have an Euler circuit?

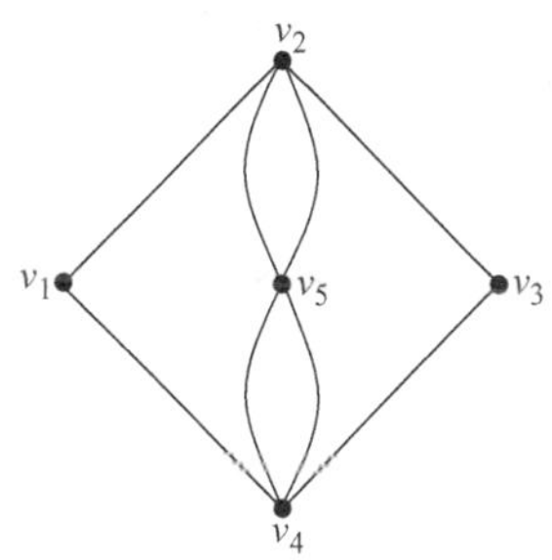

Figure 8.252

8. When does the complete bipartite graph K_{mn} contain Euler cycle?
9. In seven bridges problem, was it possible for a citizen of Konigsberg to make a tour of the city and cross each bridge exactly twice? Give reasons.
10. Show that the graph given below does not contain a Hamiltonian cycle.

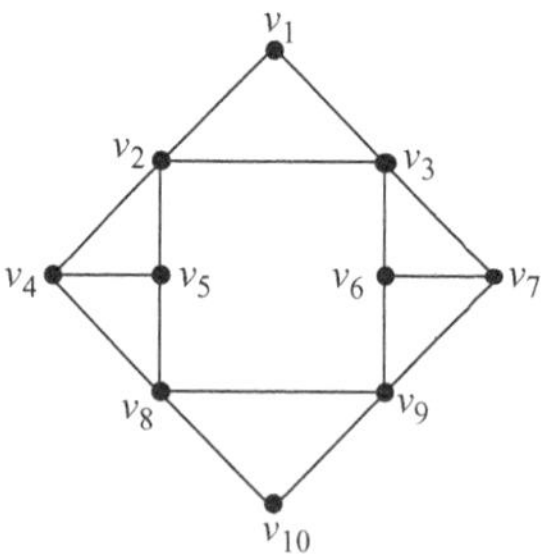

Figure 8.253

11. Give example of a graph that has Euler circuit but not Hamiltonian circuit.
12. Give example of a graph that has an Hamiltonian circuit but not an Euler circuit.
13. Give example of a graph that has both an Euler circuit and an Hamiltonian circuit
14. Use Dijkstra's shortest path algorithm to find the shortest path from a to f in the graph given below:

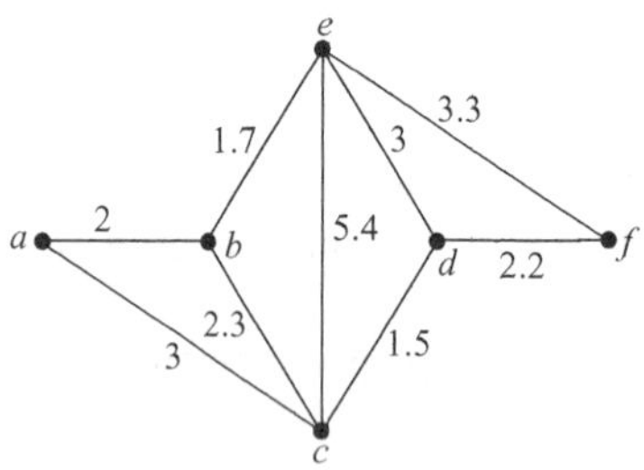

Figure 8.254

15. Find the adjacency matrix of the graph

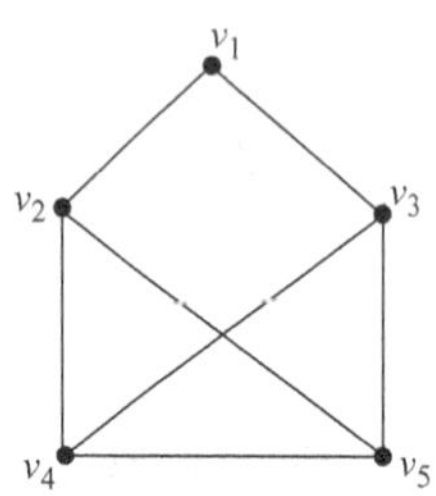

Figure 8.255

16. Find directed graph that have the following adjacency matrix.

$$\begin{bmatrix} 0 & 0 & 1 & 1 \\ 1 & 0 & 2 & 0 \\ 2 & 1 & 0 & 0 \\ 1 & 1 & 0 & 0 \end{bmatrix}.$$

17. A graph G has the adjacency matrix given below. Verify whether it is connected.

$$A = \begin{bmatrix} 0 & 0 & 1 & 0 & 0 \\ 0 & 0 & 0 & 1 & 0 \\ 0 & 0 & 0 & 0 & 1 \\ 0 & 1 & 0 & 0 & 1 \\ 0 & 0 & 1 & 1 & 0 \end{bmatrix}.$$

18. Defining an isomorphism. Show that the graphs shown below are isomorphic.

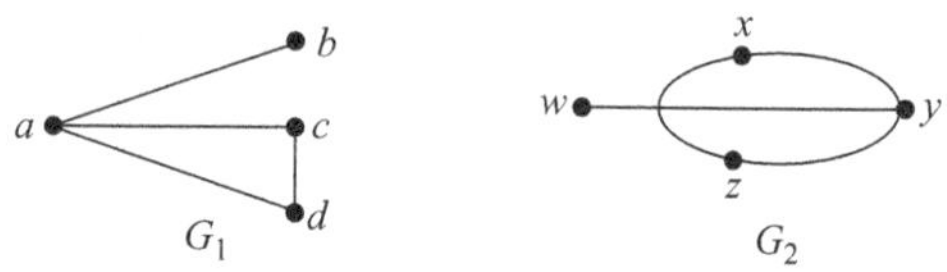

(Clearly, the graph G_2 can be drawn as)

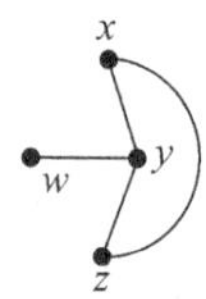

Figure 8.256

19. Show that the graphs shown below are not isomorphic. Give reasons.

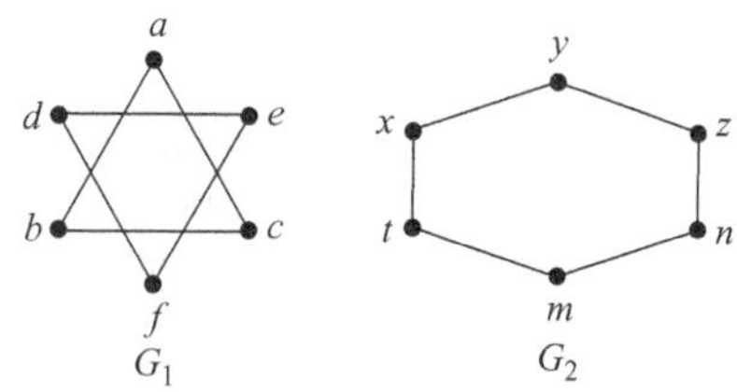

Figure 8.257

20. Show that the graphs shown below are not isomorphic.

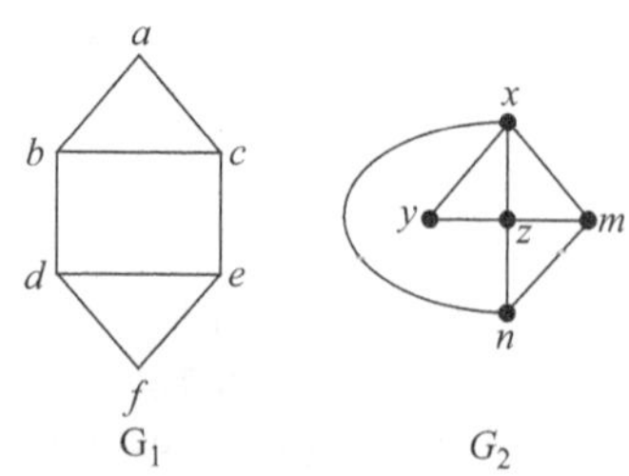

Figure 8.258

21. Find the complement of the graph shown below.

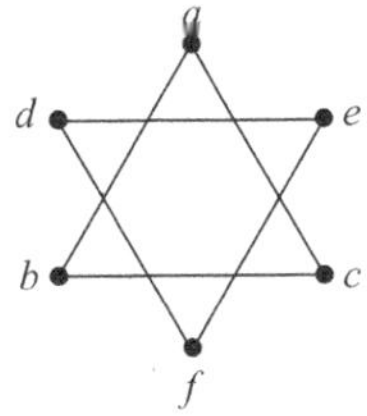

Figure 8.259

22. Let G be a graph and V be its vertex set. Then

(i) Eccentricity of $v \in V$ is defined as

$$e(v) = \max\{d(u, v): u \in V, u \neq v\}.$$

(ii) The radius of the graph is defined by

$$\text{rad}(G) = \min\{e(v): v \in V\}.$$

(iii) The diameter of G is defined by

$$\text{diam}(G) = \max\{e(v): v \in V\}.$$

$$= \max[d(u, v): u \in V, v \in V, u \neq v\}.$$

(iv) v is a central point if

$$e(v) = \text{rad}(G).$$

(v) The centre of G is the set of all central points.

Find the centre of the graph given below:

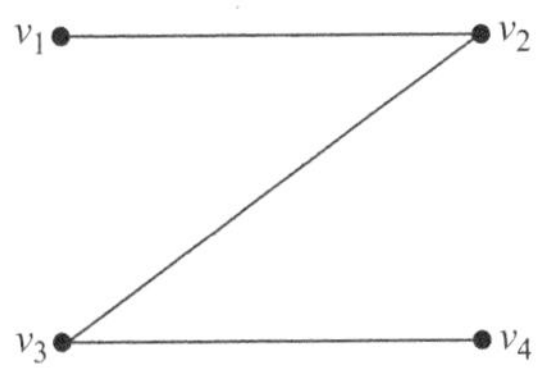

Figure 8.260

23. Show that any graph having n vertices, $n = 1, 2, 3, 4$ is planar.

24. Show that the colouring of map given below requires at least three colours

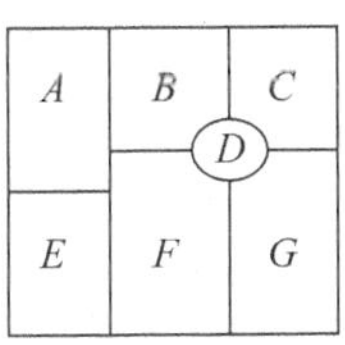

Figure 8.261

25. How many colours are required to paint the graph given below?

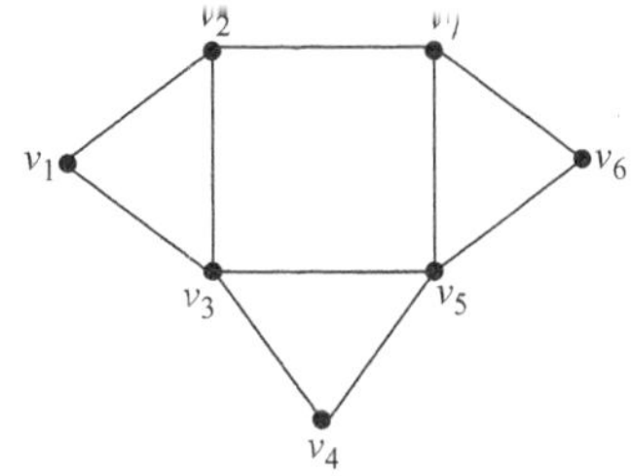

Figure 8.262

26. For which values of m and n is the complete bipartite graph K_{mn} a tree?

27. Let G be a graph with n vertices and $n-2$ or fewer edges. Show that G is not connected.

28. Represent the weighted graph given below in the matrix form.

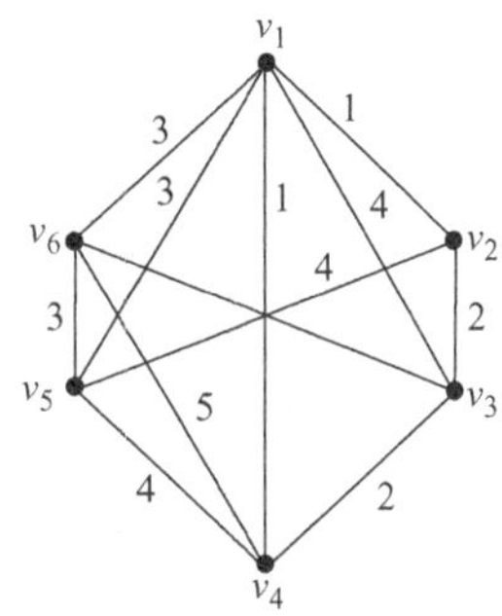

Figure 8.263

29. Represent the expression given below by a binary tree and obtain the prefix form and postfix form of the expression.

$$a \cdot b - (c/(d+e)).$$

30. Convert the following postfix form of a binary tree into prefix form and parenthesized infix form and usual infix form

$$ab + c -.$$

31. Draw a tree for the algebraic expression

$$(3x + y)(4m + 5n)^2$$

and find its prefix polish form.

32. If $\uparrow$ denotes exponentiation, evaluate the polish form

$$* \uparrow + 233 - 24.$$

33. Find all cut sets for the graph G given below.

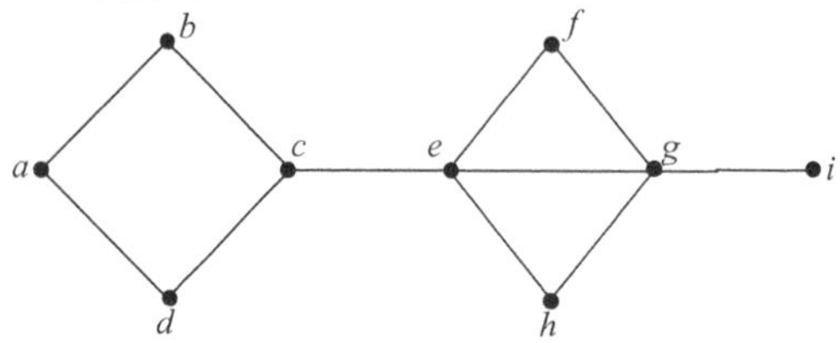

Figure 8.264

34. Find a minimal spanning tree for the graph shown below:

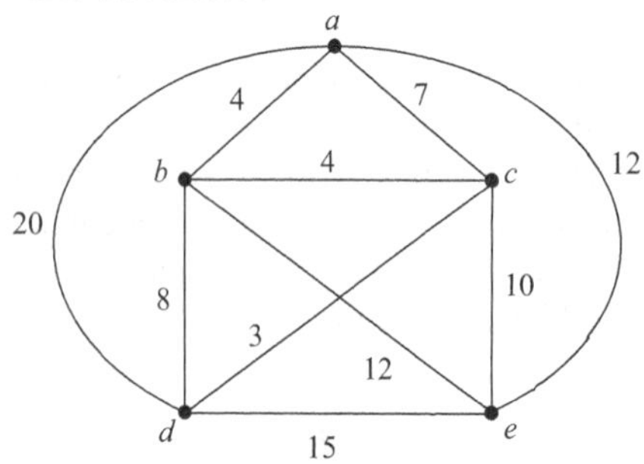

Figure 8.265

35. Find a minimal tree for the graph of Exercise 34 by deleting one by one those costliest edges whose deletion does not disconnect the graph.

36. Find all spanning trees of the graph shown below:

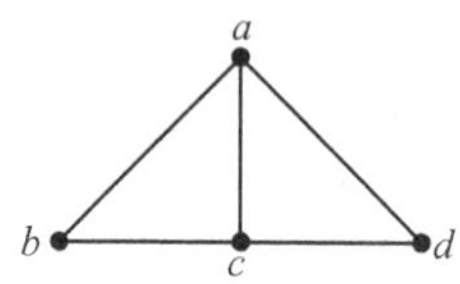

Figure 8.266

37. Find the value of the maximum flow in the network of Example 8.109 and verify the value obtained by max flow min cut theorem.

38. Find the value of the maximum flow in the network shown in Figure 8.267 and verify your answer using max flow min cut theorem.

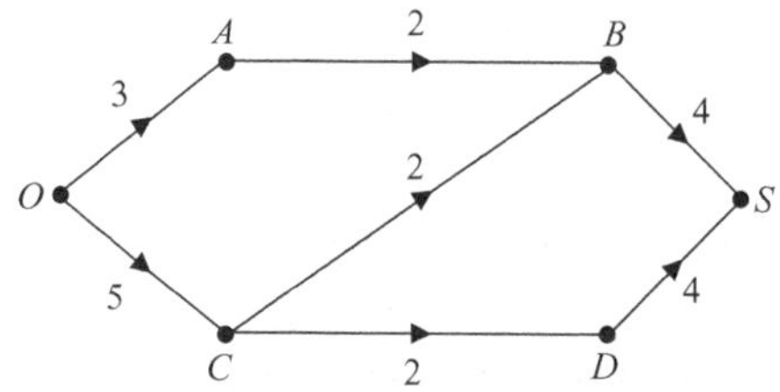

Figure 8.267

9 Finite State Automata

In this chapter, we study abstract models of machines that can accept input, produce output and have primitive internal memory to keep information about previous inputs. In these machines, the output depends not only on the input but also on the state of the system at the time the input is introduced.

9.1 FINITE STATE MACHINES

Definition 9.1

A **finite state machine** (**complete sequential machine**) is an abstract model of a machine with a primitive internal memory. A finite state machine M consists of

1. A finite set I of **input symbols**
2. A finite set S of "**internal**" states
3. A finite set O of **output** symbols
4. An initial state s_0 in S
5. A next-state function $f: S\times I\to S$
6. An output function $g: S\times I\to O$

A finite state machine M is denoted by $M=M(I, S, O, s_0, f, g)$.

EXAMPLE 9.1

Let us take

$$I=\{a, b\},$$
$$S=\{s_0, s_1, s_2),$$
$$O=\{x, y, z\}.$$
$$\text{Initial state: } s_0.$$

Next state function $f: S\times I\to S$ defined by

$$f(s_0, a)=s_1, f(s_1, a)=s_2, f(s_2, a)=s_0,$$
$$f(s_0, b)=s_2, f(s_1, b)=s_1, f(s_2, b)=s_1.$$

Output function $g: S\times I\to O$ defined by

$$g(s_0, a)=x, g(s_1, a)=x, g(s_2, a)=z,$$
$$g(s_0, b)=y, g(s_1, b)=z, g(s_2, b)=y.$$

Then $M=M(I, S, O, s_0, f, g)$ is a finite state machine.

9.1.1 Transition (State) Table and Transition (State) Diagram

There are two ways of representing a finite state machine M in a compact form:

(a) **Transition (state) table:** In this table, the functions f and g are represented by a table. Thus, in case of Example 9.1, the transition table is

	f		g	
$S \backslash I$	a	b	a	b
s_0	s_1	s_2	x	y
s_1	s_2	s_1	x	z
s_2	s_0	s_1	z	y

(b) **Transition (state) diagram:** A transition diagram of a finite state machine M is a labelled directed graph in which there is a node for each state symbol in S and each node is labelled by a state symbol with which it is associated. The initial state is indicated by an arrow. Moreover, if $f(s_i, a_j)=s_k$ and $g(s_i, a_j)=O_r$, then there is an arrow (arc) from s_i to s_k which is labelled with the pair a_j, O_r. We usually put the input symbol a_j near the base of the arrow (near s_i) and the output symbol O_r near the centre of the arrow **(Also, we can represent it by a/O_i near the centre of the arrow).** Thus, the transition diagram of the finite state machine in Example 9.1 is as given in the Figure 9.1.

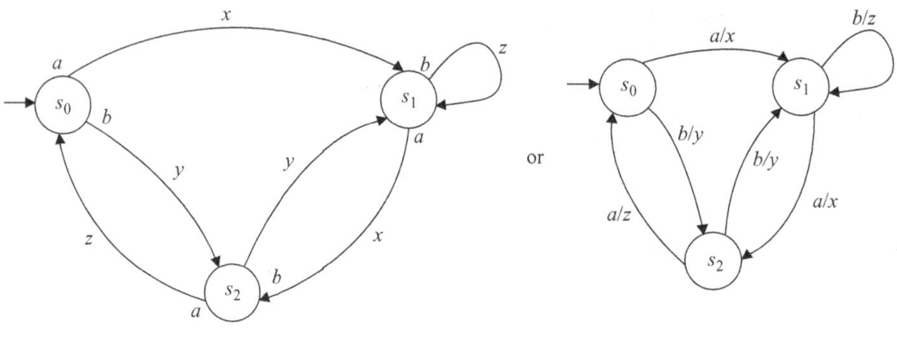

Figure 9.1

EXAMPLE 9.2

Let $I=\{a, b\}$, $O=\{0, 1\}$ and $S=\{s_0, s_1\}$. Let s_0 be the initial state. Define $f: S\times I\rightarrow S$ by

$$f(s_0, a)=s_0, f(s_0, b)=s_1, f(s_1, a)=s_1, f(s_1, b)=s_1$$

and define $g: S\times I\rightarrow O$ by

$$g(s_0, a)=0, g(s_0, b)=1, g(s_1, a)=1, g(s_1, b)=0.$$

Then $M=M(I, S, O, s_0, f, g)$ is a finite state machine. Its transition table representation is given below:

	f		g	
$S \backslash I$	a	b	a	b
s_0	s_0	s_1	0	1
s_1	s_1	s_1	1	0

The transition diagram for this finite state machine is drawn in the Figure 9.2.

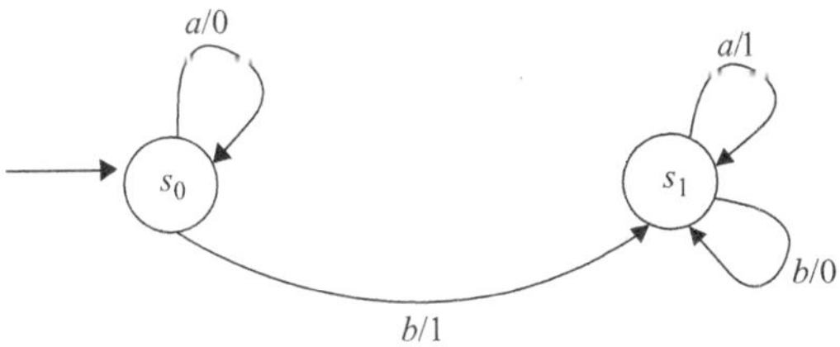

Figure 9.2

Remark 9.1 We can regard the finite state machine $M=M(I, S, O, s_0, f, g)$ as a simple computer. We begin in state S, input a string over I, and produce a string of output.

9.1.2 Input and Output Strings

Let $M=M(I, S, O, s_0, f, g)$ be a finite state machine. An **input string** for M is a string over I.

The string $y_1\, y_2 \ldots y_n$ is the **output string** for M corresponding to the input string $x_1\, x_2 \ldots x_n$ if there exist states $s_0, s_1, \ldots, s_n \in S$ such that

$$s_i=f(s_{i-1}, x_i) \quad \text{for } i=1, 2, \ldots, n,$$
$$y_i=g(s_{i-1}, x_i) \quad \text{for } i=1, 2, \ldots, n.$$

EXAMPLE 9.3

In Example 9.2, we had taken

$$I=\{a, b\}, O=\{0, 1\} \quad \text{and} \quad S\ \{s_0, s_1\}$$

with

$$f(s_0, a)=s_0, f(s_0, b)=s_1, f(s_1, a)=s_1, f(s_1, b)=s_1$$

and

$$g(s_0, a)=0, g(s_0, b)=1, g(s_1, a)=1, g(s_1, b)=0.$$

We had shown that $M=M(I, S, O, s_0, f, g)$ is a FSM. We want to find the output string to the input string $a\,a\,b\,a\,b\,b\,a$ for this machine. Initially, we are in a state s_0. The first symbol input is a. Therefore, the output is $g(s_0, a)=0$. The edge points out to s_0. Next symbol input is again a. So we again have $g(s_0, a)=0$ as the output and the edge points out to s_0. Next, b is the input symbol and so $g(s_0, b)=1$ as the output and there is a state of change s_1. Next symbol is a, so $g(s_1, a)=1$ as the output and the state is s_1. Now b is input and so $g(s_1, b)=0$ as the output. Again b is input and s_1 is the state, so $g(s_1, b)=0$. The last input symbol is a and the state is s_1. Therefore $g(s_1, a)=1$ as the output symbol. Thus the output string is 0 0 1 1 0 0 1.

EXAMPLE 9.4

Consider the FSM of Example 9.1. Let the input string be *abaab*. We begin by taking s_0 as the initial state. Using state diagram, we have

$$S_0 \xrightarrow{a/x} S_1 \xrightarrow{b/z} S_1 \xrightarrow{a/x} S_2 \xrightarrow{a/z} S_0 \xrightarrow{b/y} S_2.$$

Hence, the output string is *xzxzy*.

EXAMPLE 9.5

Draw the transition diagram of the finite state machine (I, S, O, s_0, f, g), where $I=\{a, b\}$, $S=\{s_0, s_1\}$, $O=\{0, 1\}$ and the transitional table is

	f		g	
$S \backslash I$	a	b	a	b
s_0	s_1	s_0	0	1
s_1	s_0	s_1	1	0

Also, find the output string for the input string $b\ b\ a\ a$.

Solution.

From the transition table, we observe that

$$\begin{aligned} &f(s_0, a)=s_1, \quad f(s_0, b)=s_0,\\ &f(s_1, a)=s_0, \quad f(s_1, b)=s_1,\\ &g(s_0, a)=0, \quad g(s_0, b)=1,\\ &g(s_1, a)=1, \quad g(s_1, b)=0. \end{aligned}$$

Therefore, the transition diagram for the given finite state machine is as shown in the Figure 9.3.

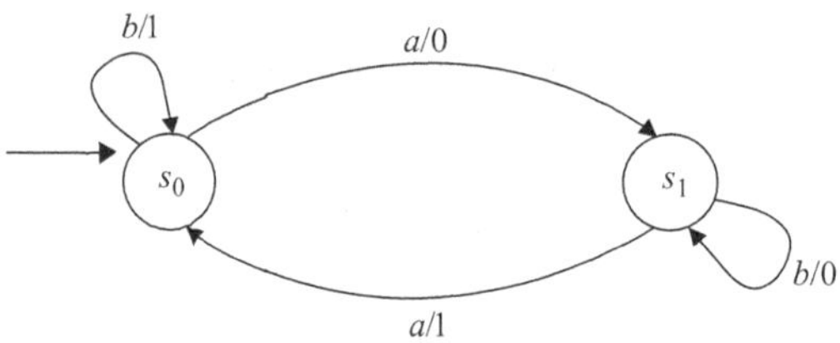

Figure 9.3

The input string is $b\ b\ a\ a$. Since the initial state is s_0, the use of transition diagram yields

$$s_0 \xrightarrow{b/1} s_0 \xrightarrow{b/1} s_0 \xrightarrow{a/0} s_1 \xrightarrow{a/1} s_0.$$

Hence, the output string is 1101.

9.1.3 Binary Addition

We want to describe a finite state machine M which can perform binary addition. Suppose that the machine is given the input

$$\begin{array}{r} 1\,1\,0\,1\,0\,1\,1 \\ +\,0\,1\,1\,1\,0\,1\,1, \\ \hline \end{array}$$

then we want to have the output to be the binary sum

$$1\,0\,1\,0\,0\,1\,1\,0.$$

Thus, the input is the string of pairs of digits to be added:

$$11, 11, 00, 11, 01, 11, 10, b,$$

where b denotes blank spaces and the output should be the string

$$0\,1\,1\,0\,0\,1\,0\,1.$$

We also want the machine to enter a state called "stop" when the machine finishes the addition. The input symbols are

$$I = \{00, 01, 10, 11, b\}$$

and the output symbols are

$$O = \{0, 1, b\}.$$

The machine that we construct will have three states:

$$S = \{\text{carry}(c), \text{ no carry } (nc), \text{ stop}(s)\}.$$

In this case, nc is the initial state.
In fact, given an input xy, we take one of three actions:

(a) We add x and y if carry bit is 0
(b) We add x, y and 1 if carry bit is 1
(c) We stop.

Next, we consider the possible inputs at each vertex. For examples if 00 is input to nc, we should output 0 and remain in the state nc. Thus, nc has a loop labelled 00/0. Further, if 11 is input to c, we compute $1+1+1=11$. In this case we output 1 and remain in the state c. Thus c has a loop labelled 11/1. As a final example, if we are in state nc and 11 is input, we should output 0 and move to the state c. By considering all possibilities, we arrive at the transition diagram given in the Figure 9.4.

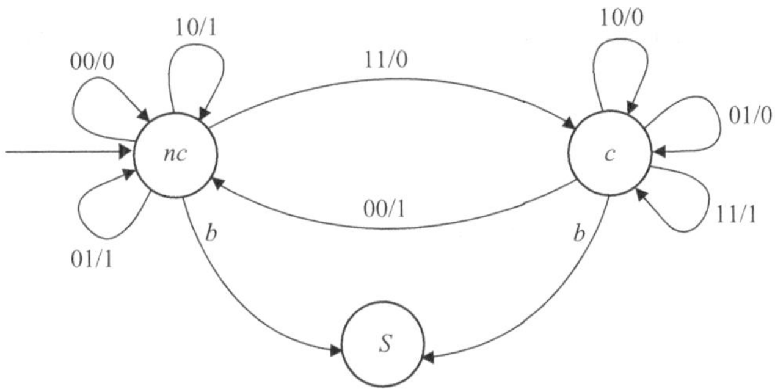

Figure 9.4

Limitation of Machines: There is no finite state machine which can perform binary multiplication.

9.1.4 Generalization of f and g in the Definition of FSM

Consider a sequence $x_0 x_1 \ldots$ of input symbols. Let s_0 be the initial state. Then the next state s_1 of the machine for the input x_0 is given by $s_1 = f(s_0, x_0) = f_1(s_0, x_0)$ say, where $f = f_1 : S \times I \to S$. Next consider the change in state due to second input symbol x_1 and the next state is $s_2 = f(s_1, x_1) = f(f_1(s_0, x_0), x_1) = f_2(s_0, x_0 x_1)$, where $f_2 : S \times I^2 \to S$. The next state due to third input symbol x_2 is $s_3 = f(s_2, x_2) = f(f_2(s_0, x_0 x_1), x_2) = f_3(s_0, x_0 x_1 x_2)$, where $f_3 : S \times I^3 \to S$. Continuing in this fashion, we can define a function $f_n : S \times I^n \to S$ such that

$$s_n = f(f_{n-1}(s_0, x_0 x_1 \ldots x_{n-2}), x_{n-1}) = f_n(s_0, x_0 x_1 \ldots x_{n-1}).$$

Similarly, the output symbol $O_0, O_1, \ldots$ can be described with the help of g as shown below:

$$O_0 = g(s_0, x_0) = g_1(s_0, x_0),$$
$$O_1 = g(s_1, x_1) = g(f_1(s_0, x_0), x_1) = g_2(s_0, x_0 x_1),$$

$$O_{n-1} = g(s_{n-1}, x_{n-1}) = g_n(s_0, x_0 x_1 \ldots x_{n-1}).$$

9.1.5 Equivalence of Finite State Machines

The aim of this section is to obtain an equivalent minimal machine for some given machine. First, we treat equivalent states. Intuitively, two states are equivalent if and only if they produce the same output for any input sequence. Thus, we can make the following definition:

Definition 9.2

Let $M=M\{I, S, O, s_0, f, g)$ be a finite state machine. Two states $s_i, s_j \in S$ are said to be **equivalent**, written as $s_i \equiv s_j$, if and only if

$$g(s_i, x)=g(s_j, x) \text{ for every word } x \in I^*,$$

where I^* denotes the set of words on the input alphabets.

It can be seen that the relation $\equiv$ is an **equivalence relation.**

Theorem 9.1

Let s be any state in a finite state machine and let x and y be any words. Then,

$$f(s, x\,y)=f(f(s, x), y)$$

and

$$g(s, x\,y)=g(f(s, x), y).$$

Proof. We shall prove the theorem by induction on length of y. Let $y=a$. Then,

$$f(s, x\,a)=f(f(s, x), a).$$

Assume that the equation is true for any y of length n, that is,

$$f(s, x\,y)=f(f(s, x), y).$$

We want to show that it is true for y having $n+1$ symbols. From the generalized definition, we can write

$$f(s, x\,y\,a)=f(f(s, x\,y), a)=f(f(f(s, x), y), a),$$

by the induction hypothesis. Taking $s'=f(s, x)$, we have

$$f(f(f(s, x), y), a)=f(f(s', y), a)=f(s', y\,a)$$
$$=f(f(s, x), y\,a).$$

The result regarding g may be established similarly.

Theorem 9.2

Let $M=M(I, S, O, s_0, f, g)$ be a finite state machine. If the states s_i and s_j are equivalent, then for any input sequence x,

$$f(s_i, x)\equiv f(s_j, x),$$

that is, if two states are equivalent, then their next states are also equivalent.

Proof. Since $s_i \equiv s_j$, it follows by definition that

$$g(s_i, x\,y)=g(s_j, x\,y) \qquad (1)$$

for any input word $x\,y$. Then, by Theorem 9.1, (1) reduces to

$$g(f(s_i, x)\,y)=g\,(f(s_j, x), y)$$

for any y belonging to the set of words I^*, which in term of definition of equivalence of states implies

$$f(s_i, x)\equiv f(s_j, x),$$

that is, the next states are equivalent.

Definition 9.3

Let $M=(I, S, O, s_0, f, g)$ be a finite state machine. Then for some positive integer k, s_i is said to be **k-equivalent to s_j** if and only if

$$g(s_i, x)=g(s_j, x) \text{ for all } x \text{ such that } |x|\leq k.$$

Obviously, equivalence of states is *a* generalization of k-equivalence of states for all k, that is,

$$s_i \equiv s_j \Rightarrow s_i \equiv s_j$$

but not conversely.

Definition 9.4

Let $M=(I,S,O, s_i, f, g)$ and $M'=(I,S',O,s_i', f', g')$ be finite state machines. Then M is said to be **equivalent to M'**, written as $M\equiv M'$ if and only if for all $s_i \in S$, there exists an $s_j' \in S'$ such that

$$s_i \equiv s_j'$$

and for all $s_j' \in S'$, there exists an $s_i \in S$ such that

$$s_i \equiv s_j'.$$

The relation $\equiv$ is an equivalence relation.
For example, consider two finite state machines whose transition tables are

	f		g	
S \ I	0	1	0	1
s_0	s_5	s_3	0	1
s_1	s_1	s_4	0	0
s_2	s_1	s_3	0	0
s_3	s_1	s_2	0	0
s_4	s_5	s_2	0	1
s_5	s_4	s_1	0	1

$M(I, S, O, s_i, f, g)$

and

	f'		g'	
S' \ I	0	1	0	1
s_0'	s_3'	s_2'	0	1
s_1'	s_1'	s_0'	0	0
s_2'	s_1'	s_2'	0	0
s_3'	s_0'	s_1'	0	1

$M'(I, S', O, s_i', f', g')$

Observe that s_0' in M' is equivalent to s_0 and s_4 in M; s_1' in M' is equivalent to s_1 in M; s_2' in M' is equivalent to s_2 and s_3 in M, and s_3' in M' is equivalent to s_5 in M. Also note that the functions g and g' are same for the indicated correspondence, but this is only a necessary condition for equivalence, not a sufficient one.

Definition 9.5

A finite state machine $M=(I, S, O, s_i, f, g)$ is said to be **reduced** if and only if $s_i \equiv s_j$ implies that $s_i = s_j$ for all states $s_i, s_j \in S$.

Thus, a reduced finite state machine is one in which each state is equivalent to itself and to no other. The partition of S in such a machine has all its equivalence classes consisting of a single element.

9.1.6 Construction of a Reduced Finite State Machine Which is Equivalent to Some Given Machine

Let M be a given machine. Let the set of states S be partitioned in a set of equivalence classes $[s]$ such that partition $P = U[s]$. Let ϕ be the function defined on the partition P such that $\phi([s]) = s'$, where s' is an arbitrary fixed element of $[s]$, called a representative. Clearly $s' \equiv s$ in M. Let S' in M' be defined as

$$S' = \{s': \text{there exists } s \in S \text{ such that } \phi([s]) = s'\}$$

and let $I' = I$ and $O' = O$, that is, both machines will have the same input and output alphabets. The functions f' and g' are defined as follows:

$$f'(s', a) = \phi([f(s', a)])$$

and

$$g'(s', a) = g(s', a),$$

where s' is both in S and S'. Therefore, the reduced machine is $M' = (I, S', O, s_i', f', g')$.

Remark 9.2 Applying this procedure to the machines in the above example, we see that M' is equivalent reduced machine of the machine M.

Theorem 9.3

Let $M = M(I, S, O, s_i, f, g)$ be a finite state machine. Then there exists an equivalent machine M' with a set of states S' such that $S' \subseteq S$ and M' is reduced.

(Proof of this theorem is out of the scope of this book).

Definition 9.6

Let $M=(I, S, O, s_i, f, g)$ and $M'=(I, S', O, s_i', f', g')$ be two finite state machines. Let ϕ be a mapping from S into S'. Then ϕ is called a finite state **homomorphism** if

$$\left.\begin{aligned} \phi(f(s,a) &= f'(\phi(s),a) \\ g(s,a) &= g'(\phi(s),a) \end{aligned}\right\} \text{for all } a \in I.$$

If ϕ is further a one-one and onto function, then M is said to be **isomorphic** to M'.

Finite state machines are used in compilers where they usually perform the task of a scanner. The machine in such a case identifies variable names, operators, constants, etc. A machine which performs this scanning task is called an **acceptor**.

9.2 FINITE STATE AUTOMATA

A finite state automaton is a special kind of finite state machine.

Definition 9.7

A **finite state automaton** $M=\{I, S, O, s_0, f, g\}$ is a finite state machine in which finite set O of output symbols is $\{0, 1\}$ and where the current state determines the last output.

Those states for which the last output was 1 are called **accepting states**.

EXAMPLE 9.6

Let

$$I=\{a, b\},\ S=\{s_0, s_1, s_2\},\ s_0 \text{ is initial state,}$$
$$O=\{0, 1\}$$

and

$$f(s_0, a)=s_1,\quad f(s_0, b)=s_0,$$
$$f(s_1, a)=s_2,\quad f(s_1, b)=s_0,$$
$$f(s_2, a)=s_2,\quad f(s_2, b)=s_0,$$
$$g(s_0, a)=1,\quad g(s_0, b)=0,$$
$$g(s_1, a)=1,\quad g(s_1, b)=0,$$
$$g(s_2, a)=1,\quad g(s_2, b)=0.$$

Draw transition table and transition diagram for this finite state automaton.

Solution.

The transition table of this automaton is

	f		g	
S \ I	a	b	a	b
s_0	s_1	s_0	1	0
s_1	s_2	s_0	1	0
s_2	s_2	s_0	1	0

We note that,

If we are in state s_0, then $g(s_2, b)=0$ and so the last output is 0. If we are in state s_1, then $g(s_0, a)=1$ and so the last output is 1. Similarly, if we are in the state s_2, then $g(s_2, a)=1$ and so the last output is 1. Thus M is a finite state automaton. **Since the last output was 1 when we are in the states s_1 and s_2, it follows that s_1 and s_2 are accepting states.** Therefore, the transition diagram is as shown in the Figure 9.5.

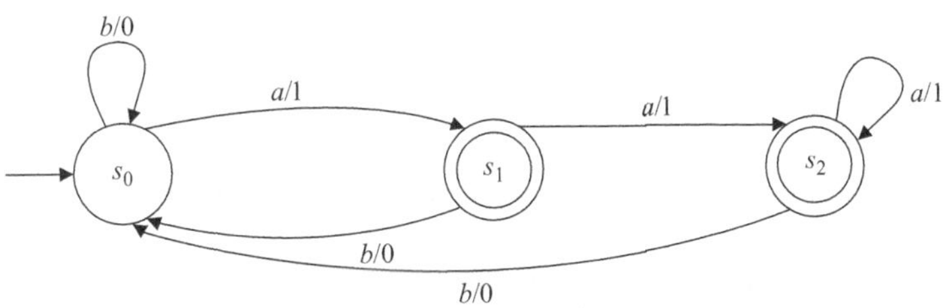

Figure 9.5

In the transition diagram of a finite state automaton, the **accepting states are shown in double circles** and the output symbols are omitted. So the output symbols 0 and 1 should also be omitted form the above diagram.

Thus, we observe that

(i) A finite state machine defined by a transition diagram is a finite state automaton if $O=\{0, 1\}$ and if in all states s, the incoming edges to s have the same output label.
(ii) Incoming edges in an accepting state has output 1.
(iii) Incoming edges in non-accepting state has output 0.

So, an alternate definition, without output, of a finite state automaton is the following:

Definition 9.8

A **finite state automaton** (*FSA*) or simply an **automaton** M or **finite state acceptor** consists of

1. A finite set I, called the input alphabet of input symbols
2. A finite set S of states
3. A subset A of S of accepting states
4. An initial state s_0 in S
5. A next state function f from $S \times I$ into S

Such an automaton is denoted by $M=(I, S, A, s_0, f)$. Thus, finite automaton does not have an output alphabet; instead it has a set of acceptance state. The plural of automaton is **automata.**

EXAMPLE 9.7

Find the transition diagram of the finite state automaton $M=(I, S, A, s_0, f)$, where

$I=\{0, 1\}$, $S=\{s_0, s_1, s_2\}$, $A=\{s_2\}$, s_0 is initial state, and the transition function f is given by the table

$$f(s_0, 0)=s_1, \quad f(s_0, 1)=s_0,$$
$$f(s_1, 0)=s_2, \quad f(s_1, 1)=s_0,$$
$$f(s_2, 0)=s_2, \quad f(s_2, 1)=s_0.$$

Also, redraw the transition diagram of this *FSA* as a transition diagram of a finite state machine.

Solution.

The transition table for the given finite state automaton is

	f	
$S \backslash I$	0	1
s_0	s_1	s_0
s_1	s_2	s_0
s_2	s_2	s_0

The transition diagram of this finite state automaton is drawn in the Figure 9.6.

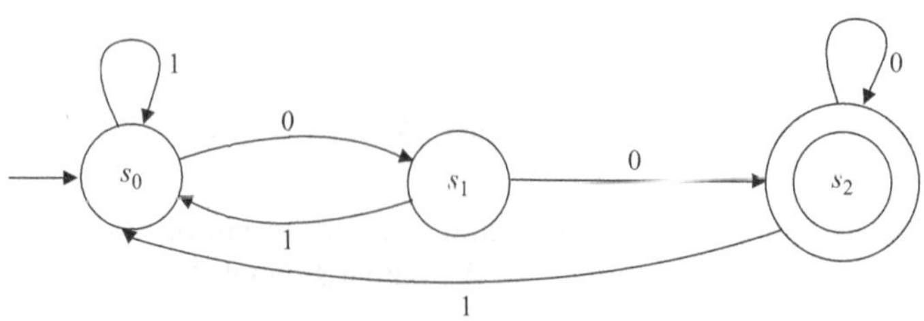

Figure 9.6

Now a finite state automaton is a finite state machine with outputs 0 and 1. The incoming edges in an accepting state has output 1 and incoming edges in a non-accepting state has output 0. Thus, the given finite state automaton is a finite state machine whose transition diagram is given in the Figure 9.7.

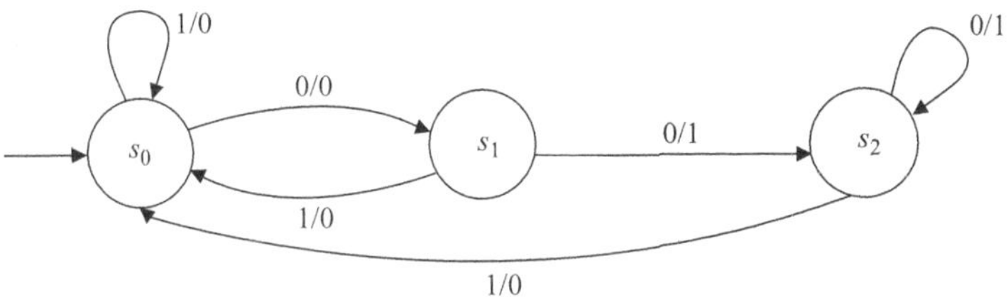

Figure 9.7

EXAMPLE 9.8

Let

$I=\{a, b\}$, $S=\{s_0, s_1, s_2\}$, $A=\{s_2\}$, $s_0 \in S$, the initial state and f is given by the table

	f	
I \ S	a	b
s_0	s_0	s_1
s_1	s_0	s_2
s_2	s_0	s_2

The transition diagram of a finite state automaton is usually drawn with accepting states in double circles. Thus the transition diagram for the example is shown in the Figure 9.8.

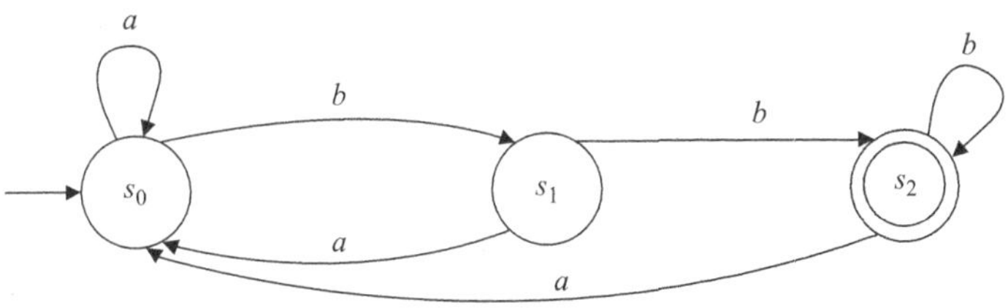

Figure 9.8

EXAMPLE 9.9

Let

$I=\{a, b\}$, input symbols,

$S=\{s_0, s_1, s_2\}$, internal states,

$A=\{s_0, s_1\}$, yes states (accepting states),

s_0=initial state,

Next state function $f: S\times I \rightarrow S$ defined by

$$f(s_0, a)=s_0, f(s_1, a)=s_0, f(s_2,a)=s_2,$$
$$f(s_0, b)=s_1, f(s_1, b)=s_2, f(s_2,b)=s_2.$$

Then, $M=(I, S, A, s_0, f)$ is a finite state automaton. Its transition table is

I \ S	f: a	f: b
s_0	s_0	s_1
s_1	s_0	s_2
s_2	s_2	s_2

and the transition diagram is shown in the Figure 9.9.

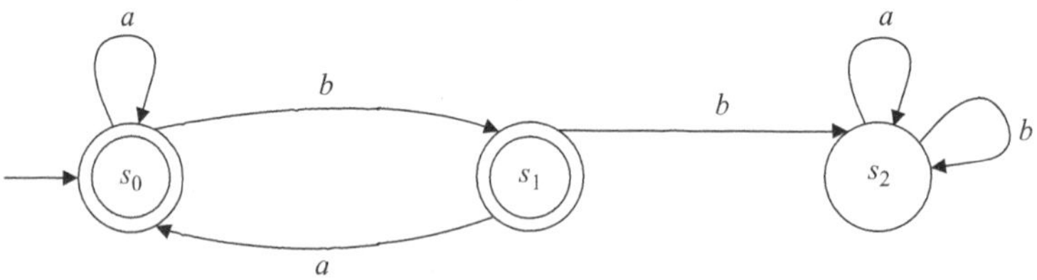

Figure 9.9

If a string is input to a finite state automaton, we will end at either an accepting or a non-accepting state. The status of this final state determines whether the string is accepted by the finite state automaton.

Definition 9.9

Let $M=(I, S, A, f, s_0)$ be a finite state automaton. Let $x_1 \ldots x_n$ be a string over I. If there exist states $s_0, s_1, \ldots, s_n$ such that $f(s_{i-1}, x_i)=s_i$ for $i=1, 2, \ldots, n$ and $s_i \in A$, then we say that the string $x_1 \ldots x_n$ is **accepted by** M.

We call the directed path $P\,(s_0, \ldots, s_n)$ the path representing $x_1, \ldots, x_n$ in M.

Thus M **accepts** $x_1 \ldots x_n$ **if and only if path** P **ends at an accepting state.**

EXAMPLE 9.10

Design a finite state automaton that accepts those strings over $\{0, 1\}$ such that the number of zeros is divisible by 3.

Solution.

Let $M=(I, S, A, s_0, f)$ be the required finite state automaton, where $I=\{0, 1\}$.

Let s_0 be the initial state of M. Thus s_0 is the state of M after zero number of 0 have been input. Since zero is divisible by 3, s_0 must be an accepting state. Now let s_1 be the state of M after one 0 has been input and s_2 be the state, where two 0's have been input. We observe that

(i) From the state s_0, three 0's are needed to reach a new total divisible by 3.
(ii) From the state s_1, two 0's are needed to reach a new total divisible by 3.
(iii) From the state s_2, one 0 is needed to reach a new total divisible by 3.

Further, after three 0's are input, three more 0's are needed to get a new total divisible by 3. In general, if $k \geq 0$ be integers and $(3k)$ 0's are input, then 3 more 0's are needed to get a new total divisible by 3.

Thus, we will have a transition diagram of M to be of the type shown in the Figure 9.10.

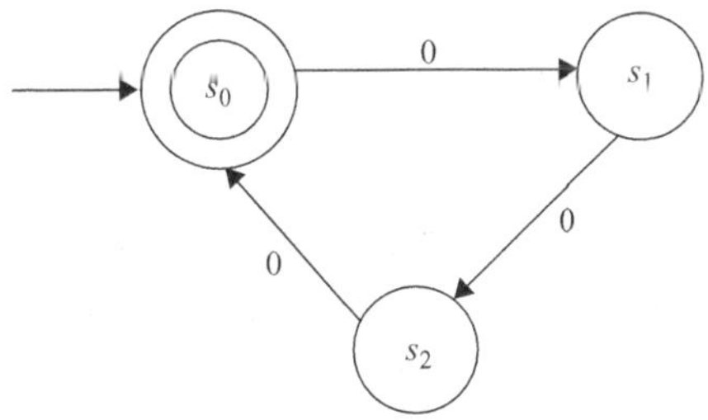

Figure 9.10

Further, if M is in any state and 1 is input, the total number of 0's in the input string remains unchanged. Thus, there is loop at each state labelled 1. Hence the transition diagram of the finite state automaton is as shown in the Figure 9.11.

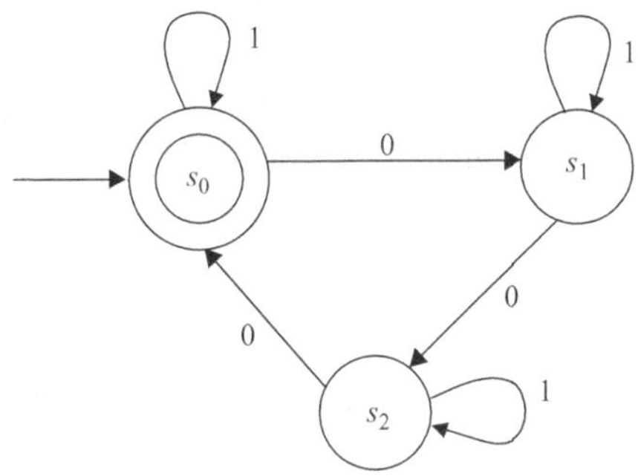

Figure 9.11

We thus have
$I=\{0, 1\}$ $S=\{s_0, s_1, s_2\}$, $A=\{s_0\}$, s_0 as the initial state and next-state function f defined by $f(s_0, 0)=s_1$, $f(s_0, 1)=s_0, f(s_1, 0)=s_2, f(s_1, 1)=s_1, f(s_2, 0)=s_0, f(s_2, 1)=s_2$.

EXAMPLE 9.11

Design a finite state automaton that accepts the set of all strings over $\{0, 1\}$ that contains an even number of 1^s.

Solution.

Let $M=(I, S, A, s_0, f)$ be the required finite state automaton, where $I=\{0, 1\}$. Let s_0 be the initial state of M. Thus, s_0 is the state of M after zero number of 1's have been input. Since 0 is even, s_0 must be an accepting state. Now let s_1 be the state of M after one 1 has been input. We observe that

(i) From the state s_0, two 1's are required to reach a new even number of 1's.
(ii) From the state s_1, one 1 is needed to reach a new even number.

Further, after two 1's are input, two more 1's are required to get a new even number of 1's. In general, if k is non-negative integer and $(2k)$ 1's are input, then 2 more 1's are needed to get even number of 1's.

Therefore, the transition diagram of M should be of the type drawn in the Figure 9.12.

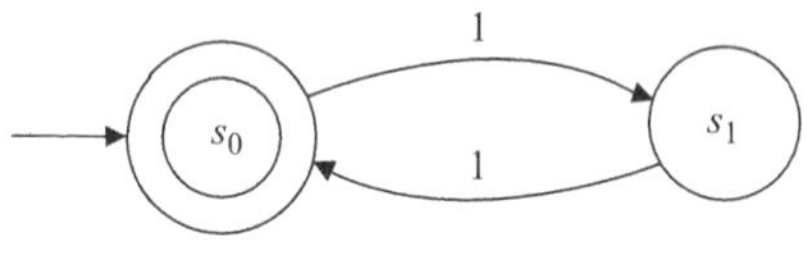

Figure 9.12

Further, if M is in any state and 0 is input, then the total number of 1's in the input string remains unchanged. Thus we have loop at each state labelled 0. Hence the transition diagram of the finite state automaton is as shown in the Figure 9.13.

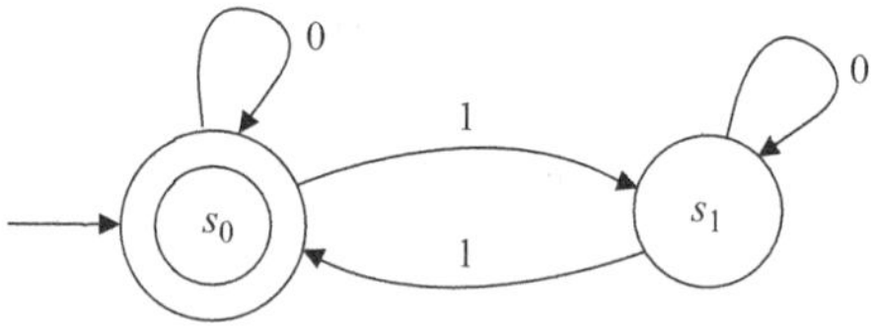

Figure 9.13

Thus, we have

$$I=\{0, 1\}, \quad S=\{s_0, s_1\}, \quad A=\{s_0\},$$

s_0 as the initial state and the next state function f defined by

$$f(s_0, 0)=s_0, \quad f(s_0, 1)=s_1,$$
$$f(s_1, 0)=s_1, \quad f(s_1, 1)=s_0.$$

EXAMPLE 9.12

Construct a finite state automaton that accepts those strings over $\{0, 1\}$ for which the last two input symbols are 1.

Solution.

Let $M=(I, S, A, s_0, f)$ be the required finite state automaton. We are given that $I=\{0, 1\}$. As per our hypothesis, acceptance of a string by M depends on the values of two consecutive inputs. Therefore, M requires three states s_0, s_1 and s_2, where s_0 is initial state, s_1 is the state in which last input character was 1 and s_2 is the state in which last two input characters were 1's.

Suppose now that 0 is input in the state s_0, this contributes nothing to have a string of two consecutive 1's. Therefore, the finite state automaton M will remain in the state s_0. When 1 is input in the state s_0, then M goes to the state s_1 to get input character of the string as 1.

If 0 is input in the state s_1, then M should go to s_0 otherwise two 1's will not be together in the input. If 1 is input to s_1, then M goes to s_2 to yield two consecutive 1's.

If 0 is input in the state s_2, then M goes to s_0 to await the input of more 1's. If 1 is input in the state s_2, then to get final two symbols of the input string as 1, M should stay in the state s_2.

Hence, the transition diagram of the finite state automaton is as shown in the Figure 9.14.

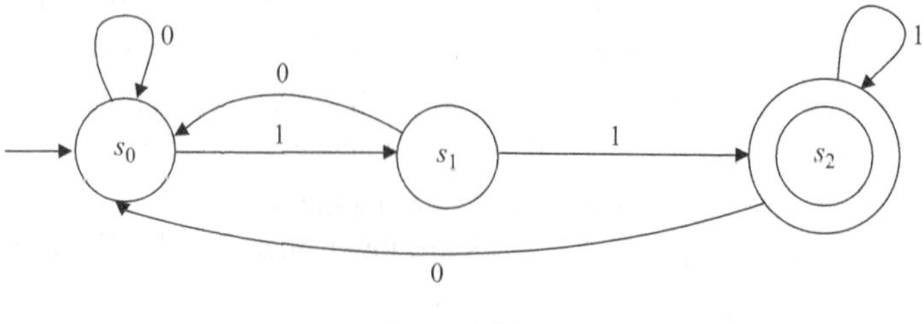

Figure 9.14

Thus,

$$I=\{0, 1\}, \quad S=\{s_0, s_1\}, \quad A=\{s_2\}$$

and the next state function f is defined by

$$\begin{aligned} f(s_0, 0)&=s_0, & f(s_0, 1)&=s_1, \\ f(s_1, 0)&=s_0, & f(s_1, 1)&=s_2, \\ f(s_2, 0)&=s_0, & f(s_2, 1)&=s_2. \end{aligned}$$

EXAMPLE 9.13

Design a finite state automaton which accepts the set of all strings of 0's and 1's and containing exactly three 1's.

Solution.

Let $M=(I, S, A, s_0, f)$ be the required finite state automaton. We have $I=\{0, 1\}$. The automaton M must have at least four distinct states:

s_0: initial state
s_1: state when the input string contains exactly one 1
s_2: state when the input string contains exactly two 1's
s_3: state when the input string contains exactly three 1's

Thus a partial transition diagram of M is shown in the Figure 9.15.

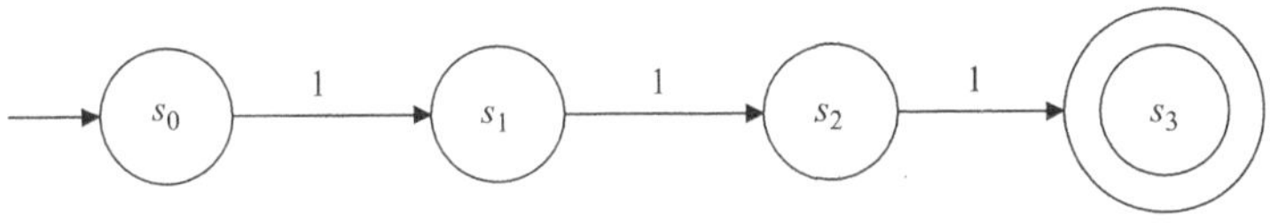

Figure 9.15

If M is in the state s_0 and 0 is input, then there will be no transition. When 1 is input and M is in s_0, then M goes to s_1.

If M is in state s_1 and 0 is input, then there shall be no transition because the input string has so far only one 1. But if 1 is input, then M shall go to s_2 and there will be two 1's in the input string.

If M is in state s_2 and 0 is input, then there shall be no transition because the input string so far has only two 1's. If 1 is input, then M goes to s_3 and the input string will have three 1's.

Now, if M is in state s_3 and 0 in input, then the input string still has three 1's and so M stays in s_3. On the other hand, if 1 is input, then M must move to some other state since M does not accept string with more than three 1's. So, we require one more state s_4 from which there is no return to s_3.

Hence the transition diagram of M is as shown in the Figure 9.16.

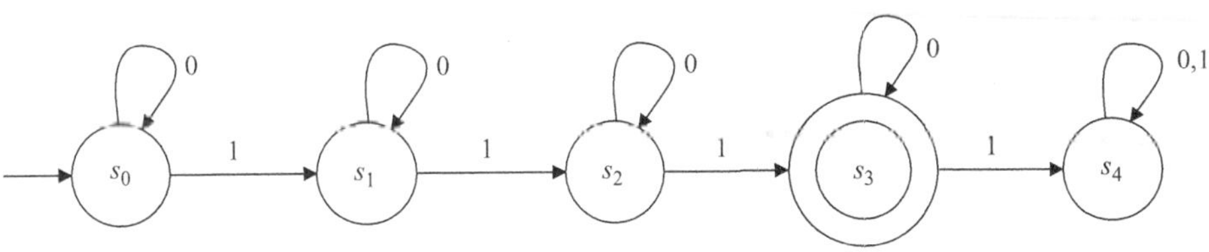

Figure 9.16

Thus,

$$I=\{0, 1\}, \quad S=\{s_0, s_1, s_2, s_3, s_4\}, \quad A=\{s_3\}$$

and the transition function f is defined by

$$\begin{aligned} f(s_0, 0)&=s_0, & f(s_0, 1)&=s_1, \\ f(s_1, 0)&=s_1, & f(s_1, 1)&=s_2, \\ f(s_2, 0)&=s_2, & f(s_2, 1)&=s_3, \\ f(s_3, 0)&=s_3, & f(s_3, 1)&=s_4, \\ f(s_4, 0)&=s_4, & f(s_4, 1)&=s_4. \end{aligned}$$

EXAMPLE 9.14

Design a finite state automaton with input set $I=\{a, b\}$ that accepts of set of all strings that start with *ab* or *ba*.

Solution.

The readers may verify that the required finite state automaton has the transition diagram shown in the Figure 9.17.

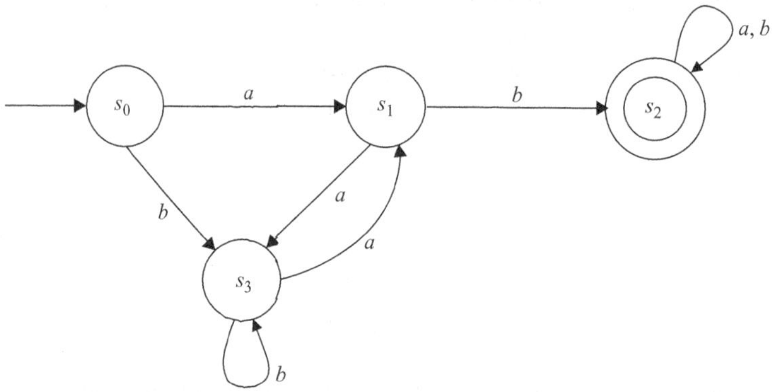

Figure 9.17

EXAMPLE 9.15

Design a finite state automaton that accepts precisely those strings over $\{a, b\}$ that contain no *a*'s.

Solution.

We want to have two states:

A: an *a* was found,
NA: No *a*'s were found.

The state *NA* is the initial state and the only accepting state (see Figure 9.18).

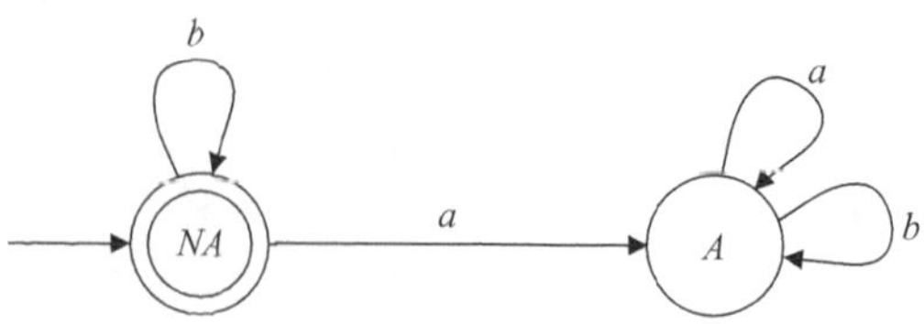

Figure 9.18

If f is next-state function, then

$$f(NA, a)=A, \quad f(NA, b)=NA,$$
$$f(A, a)=A, \quad f(A, b)=A.$$

EXAMPLE 9.16

Design a finite state automaton that accepts precisely those strings over $\{a, b\}$ that contains an odd number of a's.

Solution.

There shall be two states:

E: An even number of a's was found,
O: An odd number of a's was found.

The initial state is E and the accepting state is O (see Figure 9.19).

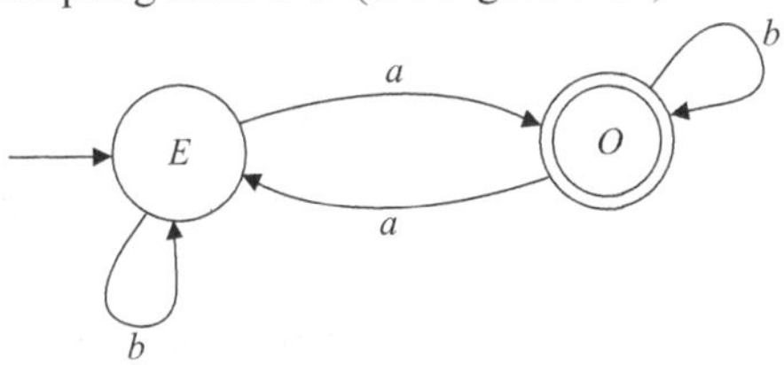

Figure 9.19

If f is next-state function, then we have

$$f(E, a)=O, \quad f(E, b)=E,$$
$$f(O, a)=E, \quad f(O, b)=O.$$

EXAMPLE 9.17

Let $M=\{I, S, A, s_0, f\}$ be a finite state automaton with

$$I=\{0, 1, 2, 3, 4, 5, 6, 7, 8, 9\},$$
$$S=\{s_0, s_1, s_2\},$$
$$A=\{s_0\},$$
$$a \in \{0, 3, 6, 9\}, b \in \{1, 4, 7\}, c \in \{2, 5, 8\}.$$

Next-state function f defined by

$$f(s_0, a)=s_0, \quad f(s_0, b)=s_1, \quad f(s_0, c)=s_2,$$
$$f(s_1, a)=s_1, \quad f(s_1, b)=s_2, \quad f(s_1, c)=s_0,$$
$$f(s_2, a)=s_2, \quad f(s_2, b)=s_0, \quad f(s_2, c)=s_1.$$

Draw transition table and transition diagram for this FSA. Does this automaton accept 258 and 142?

Solution.

The transition table for FSA is

	f		
S \ I	a	b	c
s_0	s_0	s_1	s_2
s_1	s_1	s_2	s_0
s_2	s_2	s_0	s_1

The transition diagram for this FSA is shown in the Figure 9.20.

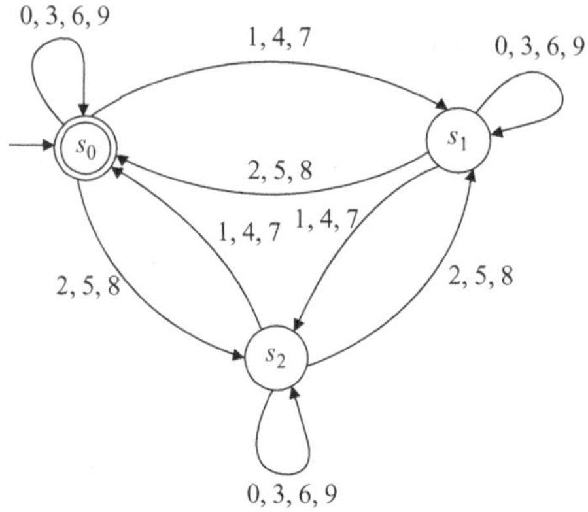

Figure 9.20

Here $A=\{s_0\}$ is the initial state and also is an acceptor. Further, we note that

$$\begin{aligned} f(s_0, 258) &= f(f(s_0, 25), 8) \\ &= f(f(f(s_0, 2), 5), 8) \\ &= f(f(s_2, 5), 8) \\ &= f(s_1, 8) = s_0 \in A. \end{aligned}$$

Thus, the string 258 determines the path

$$s_0 \xrightarrow{2} s_2 \xrightarrow{5} s_1 \xrightarrow{8} s_0 \in A.$$

Hence 258 is accepted by the given finite state automaton.
On the other hand,

$$\begin{aligned} f(s_0, 142) &= f(f(s_0, 14), 2) \\ &= f(f(f(s_0, 1), 4), 2) \\ &= f(f(s_1, 4), 2) \\ &= f(s_2, 2) = s_1 \notin A. \end{aligned}$$

Thus, the string 142 determines the path

$$s_0 \xrightarrow{1} s_1 \xrightarrow{4} s_2 \xrightarrow{2} s_1 \notin A.$$

Hence 142 is not accepted by the given finite state automaton.

EXAMPLE 9.18

Determine whether the string *abbaa* accepted by the finite state automaton having transition diagram given in the Figure 9.21.

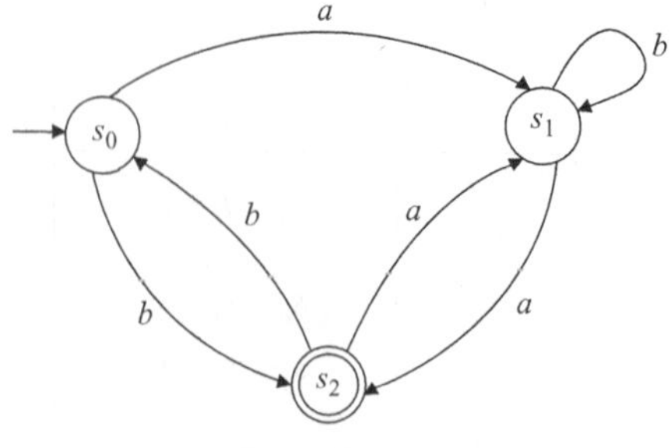

Figure 9.21

Solution.

Here s_0 is the initial state and s_2 the accepting state. The input set $I=\{a, b\}$. The set of states is $S=\{s_0, s_1, s_2\}$, the set of accepting states is $A=\{s_2\}$ and the transition table is

	f	
$S \backslash I$	a	b
s_0	s_1	s_2
s_1	s_2	s_1
s_2	s_1	s_0

We have

$$\begin{aligned} f(s_0, abbaa) &= f(f(s_0, abba), a) \\ &= f(f(f(s_0, abb), a), a) \\ &= f(f(f(f(s_0, ab), b), a), a) \\ &= f[(f(f(f(f(s_0, a), b), b), a), a)] \\ &= f[f(f(f(s_1, b), b), a), a)] \\ &= f[f(f(s_1, b), a), a)] \\ &= f[f(s_1,a), a)] \\ &= f(s_2, a) = s_1 \notin A \end{aligned}$$

Thus the word $w=abbaa$ determines the path

$$s_0 \xrightarrow{a} s_1 \xrightarrow{b} s_1 \xrightarrow{b} s_1 \xrightarrow{a} s_2 \xrightarrow{a} s_1.$$

The final state is not in A. Hence $w=abbaa$ is **not accepted** by the finite state automaton $M=\{I, S, A, s_0, f\}$.

EXAMPLE 9.19

Let $I=\{a, b\}$. Construct an automaton M such that $L(M)$ consists of those words which begin with a and end with b.

Solution.

Let s_0 be the initial state. If we define f as

$$f(s_0, a)=s_1, \quad f(s_1,b)=s_2 \quad \text{(accepting state)}.$$

Then, we have three states as shown in the Figure 9.22.

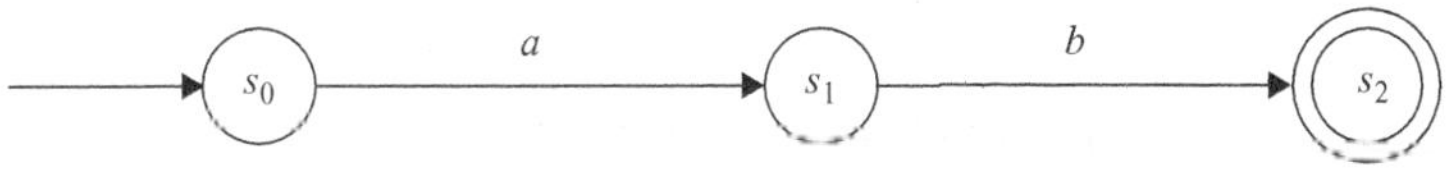

Figure 9.22

Now we cannot take $f(s_0, b)=s_0$ or $f(s_0, b)=s_1$, because then b will be the starting letter. So, we have to take $f(s_0, b)=s_3$. We cannot take $f(s_3, a)=s_0, s_1, s_2$ because in that case the string would end in a.

We cannot have $f(s_2, a)=s_2$, because then a will be the last letter. Thus the automaton M will be as shown in the Figure 9.23.

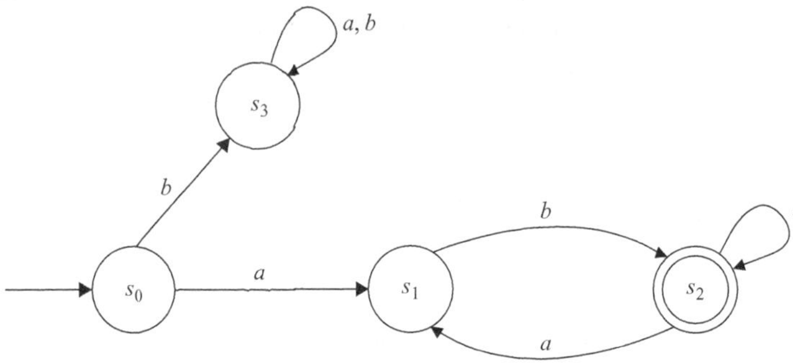

Figure 9.23

In this automaton, any string shall begin with a and end in b.

9.3 NON-DETERMINISTIC FINITE STATE AUTOMATON

Definition 9.10

A non-deterministic finite state automaton is a 5-tuple $M=(I, S, A, s_0, f)$ consisting of

1. A finite set I of input symbols
2. A finite set S of states
3. A subset A of S of accepting states
4. An initial state $s_0 \in S$
5. A next state function f from $S \times I$ into $P(S)$

Thus, in a non-deterministic finite state automaton, the next state function leads us to a set of states, whereas in a finite state automaton, the next state function takes us to a uniquely defined state.

EXAMPLE 9.20

Find the transition diagram for the NDFSA

$$M=(I, S, A, s_0, f),$$

where

$$I=\{a, b\}, \quad S=\{s_0, s_1, s_2\}, \quad A=\{s_0\}$$

and the next state function f is given by the table given below:

	f	
$S \backslash I$	a	b
s_0	ϕ	$\{s_1, s_2\}$
s_1	$\{s_2\}$	$\{s_0, s_1\}$
s_2	$\{s_0\}$	ϕ

Solution.

Here the initial state is s_0 and the accepting state is also s_0. The transition diagram of this non-deterministic finite state automaton is shown in the Figure 9.24.

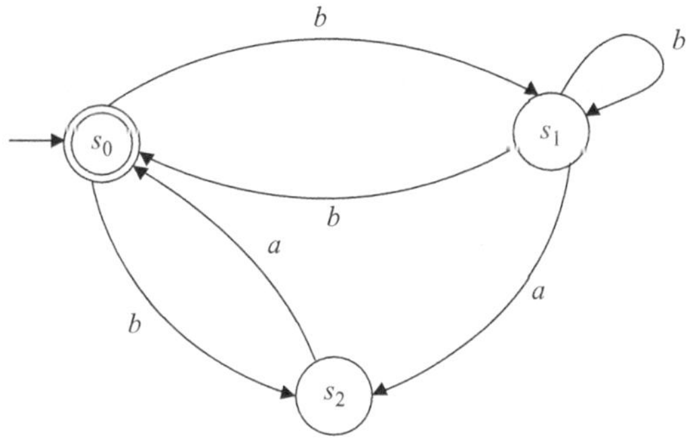

Figure 9.24

EXAMPLE 9.21

Find the transition diagram for NDFSA

$$M=(I, S, A, s_0, f),$$

where

$$I=\{0, 1\}, \quad S=\{s_0, s_1, s_2, s_3\}, \quad A=\{s_2, s_3\}$$

and the next state function f is given by the following table:

S \ I	f: 0	f: 1
s_0	$\{s_0, s_1\}$	$\{s_3\}$
s_1	$\{s_0\}$	$\{s_1, s_3\}$
s_2	ϕ	$\{s_0, s_2\}$
s_3	$\{s_1, s_2, s_3\}$	$\{s_1\}$

Solution.

Here the initial state is s_0 and the accepting states are s_2 and s_3. The transition diagram of this NDFSA is shown in the Figure 9.25.

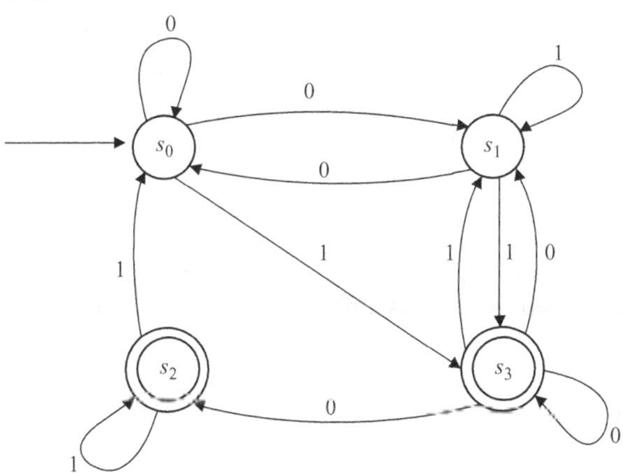

Figure 9.25

EXAMPLE 9.22

Describe the non-deterministic finite state automaton represented by the transition diagram given in the Figure 9.26.

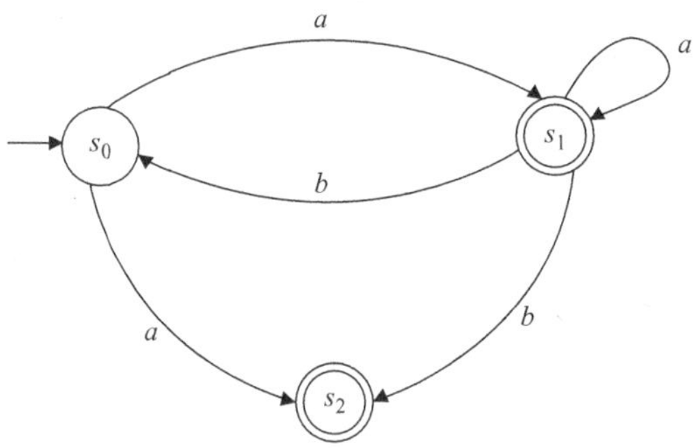

Figure 9.26

Solution.

Let $M=(I, S, A, s_0, f)$ be the given NDFSA. Then, we observe that

$$I=\{a, b\}, \quad S=\{s_0, s_1, s_2\}, \quad A=\{s_1, s_2\}$$

and s_0 is the initial state and the next state function f is given by the following transition table:

	f	
$S \backslash I$	a	b
s_0	$\{s_1, s_2\}$	ϕ
s_1	$\{s_1\}$	$\{s_0, s_2\}$
s_2	ϕ	ϕ

that is,

$$f(s_0, a)=\{s_1, s_2\}, \quad f(s_0, b)=\phi \text{ and so on.}$$

A string w is accepted by a non-deterministic finite state automaton M if there is some path representing w in the transition diagram of M **beginning at the initial state and ending in an accepting state.**

Definition 9.11

Let $M=(I, S, A, s_0, f)$ be a non-deterministic finite state automaton. The null string is **accepted** by M if and only if $s_0 \in A$. If $w=a_1 a_2 \ldots a_n$ is a non-null string over I and there exist states $s_0, s_1, \ldots, s_n$ such that

1. s_0 is the initial state
2. $s_i=f(s_{i-1}, a_i)$, $i=1, 2, \ldots, n$
3. $s_n \in A$

then we say that w is accepted by M.

We denote by $AC(M)$, the set of strings accepted by M and say that M accept $AC(M)$.

Definition 9.12

Two non-deterministic finite state automata M and M' are said to be **equivalent** if

$$AC(M)=AC(M').$$

EXAMPLE 9.23

Let $M=(I, S, A, s_0, f)$ be a NDFSA with $I=\{0, 1\}$, $S=\{s_0, s_1, s_2, s_3, s_4\}$, $A=\{s_2, s_4\}$, s_0 as the initial state and the next-state function defined by the transition table given below:

$S \backslash I$	f: 0	f: 1
s_0	$\{s_0, s_3\}$	$\{s_0, s_1\}$
s_1	ϕ	$\{s_2\}$
s_2	$\{s_2\}$	$\{s_2\}$
s_3	$\{s_4\}$	ϕ
s_4	$\{s_4\}$	$\{s_4\}$

Determine whether M accepts the words (i) $w=010$ and (ii) $w=01001$.

Solution.

(i) The word $w=010$ determines the path

$$s_0 \xrightarrow{0} \{s_0, s_3\} \xrightarrow{1} f(s_0, 1) \cup f(s_3, 1)=\{s_0, s_1\} \cup \phi=\{s_0, s_1\}$$
$$\xrightarrow{0} f(s_0, 0) \cup f(s_1, 0)=\{s_0, s_3\} \cup \phi=\{s_0, s_3\}.$$

But $A \cap \{s_0, s_3\}=\{s_2, s_4\} \cap \{s_0, s_3\}=\phi$. Hence the word $w=010$ is not acceptable to the given non-deterministic finite state automaton.

(ii) We have seen above that

$$s_0 \xrightarrow{0} \{s_0, s_3\} \xrightarrow{1} \{s_0, s_1\} \xrightarrow{0} \{s_0, s_3\}.$$

Therefore the word $w=01001$ determines the path

$$s_0 \xrightarrow{0} \{s_0, s_3\} \xrightarrow{1} \{s_0, s_1\} \xrightarrow{0} \{s_0, s_3\} \xrightarrow{0} f(s_0,0) \cup f(s_3, 0)$$
$$= \{s_0, s_3\} \cup \{s_4\}=\{s_0, s_3, s_4\} \xrightarrow{1} f(s_0, 1) \cup f(s_3, 1\} \cup f(s_4, 1\}$$
$$= \{s_0, s_1\} \cup \phi \cup \{s_4\}=\{s_0, s_1, s_4\}$$

so that

$$A \cap \{s_0, s_1, s_4\}=\{s_2, s_4\} \cap \{s_0, s_1, s_4\}=\{s_4\} \neq \phi.$$

Hence the string 01001 is acceptable to the given NDFSA.

9.4 EQUIVALENCE OF DFSA AND NDFSA

We have seen that in the definition of finite state automaton, the next state function is from $S\times I$ into S, whereas in the definition of NDFSA, the next state function is from $S\times I$ into $P(S)$. Thus, **every DFSA is an NDFSA,** that is, the class of languages accepted by NDFSA includes the languages accepted by DFSA. However, these are the only languages accepted by NDFSA. In other words, **for every NDFSA, we can construct an equivalent DFSA.** In this direction, we have the following:

Theorem 9.4

Let L be a set accepted by a non-deterministic finite automaton. Then there exists a deterministic finite automaton that accepts L.

Proof. Let $M=(I, S, A, s_0, f)$ be an NDFSA accepting L. Define a DFSA,

$$M'=(I, S', A', s_0', f')$$

as follows:

The states of M' are all the subsets of the set of all states of M, that is, $S'=2^S$. Also $s_0'=\{s_0\}$ and A' is the set of all states in S' containing a final state of M, that is, $A'=\{s \in S': s \cap A \neq \phi\}$. Further, for $s \in S'$ and $a \in I$, let

$$f'(s,a) = \bigcup_{\sigma \in s} f(\sigma, a).$$

To prove that M' accepts the same language as M, it is sufficient to show that for any string $x \in I^*$ (the set of strings formed by I),

$$f'^*(s_0', x)=f^*(s_0, x). \tag{1}$$

We shall prove (1) by using induction on the length of the input string x.
If $x=\wedge$, then

$$\begin{aligned} f'^*(s_0', x) &= f'^*(s_0', \wedge) \\ &= s_0' \text{ (by definition of } f'^*) \\ &= \{s_0\} \text{ by the definition of } s_0' \\ &= f^*(s_0, \wedge) \text{ (by the definition of } f^*) \\ &= f^*(s_0, x). \end{aligned}$$

Thus (1) holds for $|x|=0$ (i.e., for $x=\wedge$).
The induction hypothesis is that x is a string satisfying

$$f'^*(s_0', x)=f^*(s_0, x)$$

and we want to show that

$$f'^*(s_0', xa)=f^*(s_0, xa) \quad \text{for} \quad a \in I.$$

To show it, we have

$$\begin{aligned} f'^*(s_0', xa) &= f'(f'^*(s_0', x), a) \text{ (by the definition of } f'^*) \\ &= f'(f^*(s_0, x), a) \text{ (by induction hypothesis)} \\ &= \bigcup_{\sigma \in f^*(s_0, x)} f(\sigma, a) \text{ (by the definition of } f') \\ &= f^*(s_0, x\,a) \text{ (by the definition of } f^*). \end{aligned}$$

We know that a string x is accepted by M' if $f'^*(s_0', x) \in A'$ that is, if $f^*(s_0, x) \in A'$ and using the definition of A', it follows that this is true if and only if

$$f^*(s_0, x) \cap A \neq \phi,$$

that is, if $f^*(s_0, x) \in A$, that is, if x is accepted by M. Thus x is accepted by M' if and only if x is accepted by M. This completes the proof of the theorem.

EXAMPLE 9.24

Construct deterministic finite state automaton equivalent to the following non-deterministic finite state automaton:

$$M=(\{0, 1\}, \{s_0, s_1\}, s_0, \{s_1\}, f),$$

where f is given by the table

$S \backslash I$	f: 0	f: 1
s_0	$\{s_0, s_1\}$	$\{s_1\}$
s_1	ϕ	$\{s_0, s_1\}$

Solution.

Let

$$M'=\{\{0, 1\}, \{\phi, \{s_0\}, \{s_1\}, \{s_0, s_1\}\}, s_0'=\{s_0\}, A', f'\}$$

be the DFSA, where

$$A'=\{s \in \{\phi, \{s_0\}, \{s_1\}, \{s_0, s_1\}: s \cap \{s_1\} \neq \phi\}$$
$$= \{s_1\}, \{s_0, s_1\} \textbf{ (accepting states)}$$

and

$$f'(s, a)= \bigcup_{\sigma \in s} f(\sigma, a) \text{ for } s \in \{\phi, \{s_0\}, \{s_1\}, \{s_0, s_1\}\}.$$

We have
$\{s_0\}$ as the initial state
The finite set of states is $\{\phi, \{s_0\}, \{s_1\}, \{s_0, s_1\}\}$,
The finite set of inputs is $\{0, 1\}$,
The accepting states are $[s_1]$ and $[s_0, s_1]$.
Now

$$f'(\phi, 0)=\phi \quad \text{and} \quad f'(\phi, 1)=\phi,$$
$$f'([s_0], 0)=f(s_0, 0)=[s_0, s_1],$$
$$f'([s_0], 1)=f(s_0, 1)=[s_1],$$
$$f'([s_1], 0)=f(s_1, 0)=\phi,$$
$$f'([s_1], 1)=f(s_1, 1)=[s_0, s_1],$$
$$f'([s_0, s_1], 0)=f(s_0, 0) \cup f(s_1, 0)= \{s_0, s_1\} \cup \phi$$
$$= [s_0, s_1] \cup \phi=[s_0, s_1],$$
$$f'(\{s_0, s_1\}, 1)=f(s_0, 1) \cup f(s_1, 1)$$
$$= \{s_1\} \cup \{s_0, s_1\}=[s_0, s_1].$$

Hence the next state function and the transition diagram for DFSA are as given below (Figure 9.27).

S \ I	f' 0	f' 1
ϕ	ϕ	ϕ
$[s_0]$	$[s_0, s_1]$	$[s_1]$
$[s_1]$	ϕ	$[s_0, s_1]$
$[s_0, s_1]$	$[s_0, s_1]$	$[s_0, s_1]$

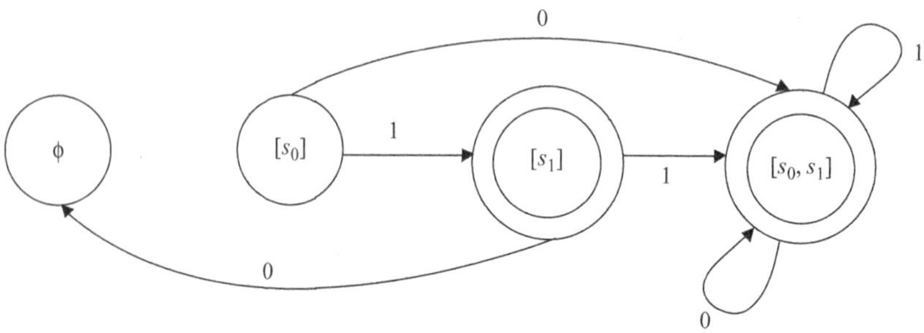

Figure 9.27

It may be mentioned here that a state which is never entered may be deleted from the transition diagram. In view of this, the above transition diagram becomes as shown in the Figure 9.28.

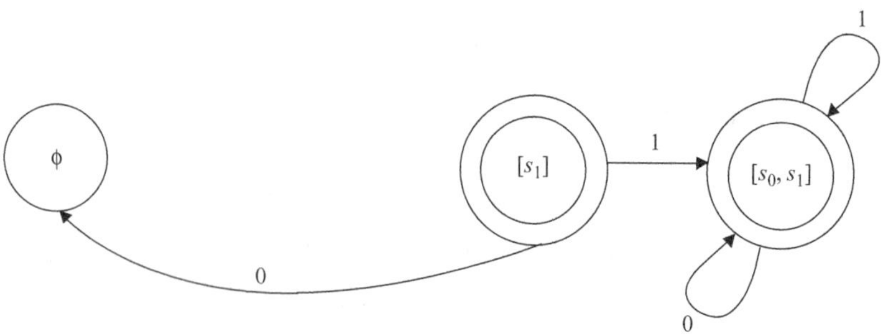

Figure 9.28

Thus, we note that if NDFSA has n states, then DFSA will have 2^n states.

EXAMPLE 9.25

Find a deterministic acceptor equivalent to NDFSA

$$M=(\{0, 1\}, \{s_0, s_1, s_2, s_3\}, s_0, \{s_3\}, f),$$

where f is given by the following table:

	f	
I / S	0	1
s_0	$\{s_0\}$	$\{s_0, s_1\}$
s_1	$\{s_2\}$	$\{s_2\}$
s_2	$\{s_3\}$	$\{s_3\}$
s_3	ϕ	ϕ

Solution.

The transition diagram of the given NDFSA is shown in the Figure 9.29.

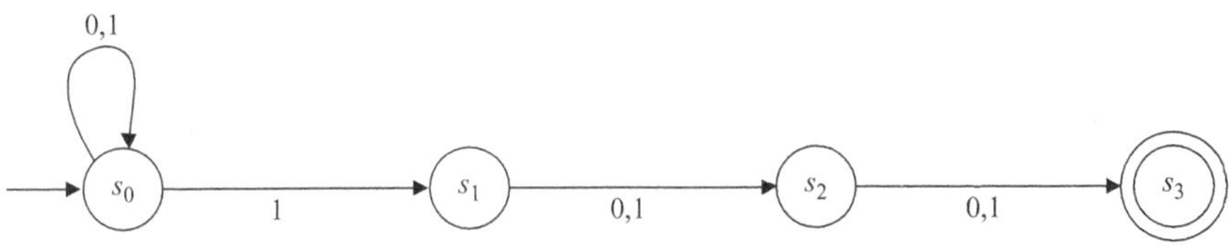

Figure 9.29

Let
$M'(\{0, 1\}, \{\phi, [s_0], [s_1], [s_2], [s_3], [s_0, s_1], \ldots, [s_0, s_1, s_2, s_3]\}, s_0' = [s_0], A', f')$ be the equivalent DFSA, where the set of accepting states (A') is equal to

$$\{s \in \{\phi, [s_0], \ldots, [s_0, s_1, s_2, s_3]\}: s \cap \{s_3\} \neq \phi\}$$
$$= \{s_3\}, \{s_0, s_3\}, \{s_1, s_3\}, \{s_2, s_3\}, \{s_0, s_1, s_2, s_3\}, \{s_0\ s_1\ s_3\}\ \{s_0\ s_2\ s_3\}, \{s_1, s_2, s_3\}, \{s_0, s_1, s_2, s_3\}.$$

Further,

$$f'(\phi, 0) = \phi \quad \text{and} \quad f'(\phi, 1) = \phi,$$

$$f'([s_0], 0) = f(s_0, 0) = [s_0], f'([s_0], 1) = [s_0, s_1],$$

$$f'([s_1], 0) = f(s_1, 0) = [s_2], f'([s_1], 1) = f(s_1, 1) = [s_2],$$

____

____

$$f'([s_0, s_1, s_2, s_3], 0) = f(s_0, 0) \cup f(s_1, 0) \cup f(s_2, 0) \cup f(s_3, 0)$$
$$= \{s_0\} \cup \{s_2\} \cup \{s_3\} \cup \phi = [s_0, s_2, s_3],$$

$$f'([s_0, s_1, s_2, s_3], 1) = [s_0, s_1, s_2, s_3].$$

Hence the next state function f' for DFSA is given by the following table:

$S \backslash I$	f' 0	1
ϕ	ϕ	ϕ
$[s_0]$	$[s_0]$	$[s_0, s_1]$
$[s_2]$	$[s_3]$	$[s_3]$
$[s_3]$	ϕ	ϕ
$[s_0, s_1]$	$[s_0, s_2]$	$[s_0, s_1, s_2]$
$[s_0, s_2]$	$[s_0, s_3]$	$[s_0, s_1, s_3]$
$[s_0, s_3]$	$[s_0]$	$[s_0, s_1]$
$[s_2, s_3]$	$[s_3]$	$[s_3]$
$[s_0, s_1, s_2, s_3]$	$[s_0, s_2, s_3]$	$[s_0, s_1, s_2, s_3]$
$[s_0, s_1, s_2]$	$[s_0, s_2, s_3]$	$[s_0, s_1, s_2, s_3]$
$[s_0, s_2, s_3]$	$[s_0, s_3]$	$[s_0, s_1, s_3]$
$[s_0, s_1, s_3]$	$[s_0, s_2]$	$[s_0, s_1, s_2]$

The states $[s_1]$, $[s_1, s_2]$, $[s_1, s_3]$, $[s_1, s_2, s_3]$ are never entered and so omitting these states, the transition diagram of the deterministic finite state automaton is as shown in the Figure 9.30.

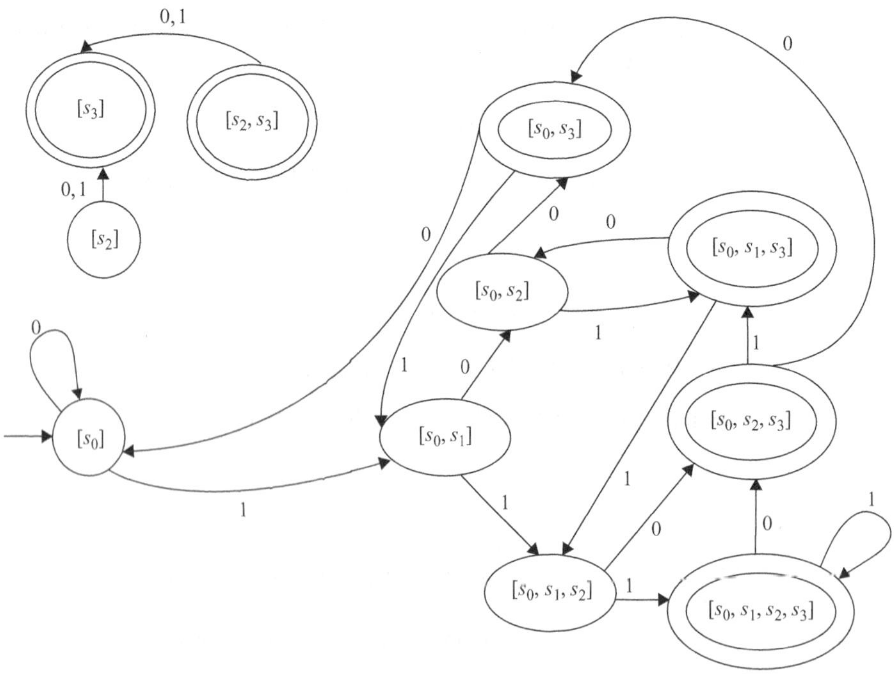

Figure 9.30

Now the states $[s_2]$, $[s_3]$ and $[s_2, s_3]$ are not entered and so may be omitted. Thus the transition diagram of the BFSA becomes as shown in the Figure 9.31.

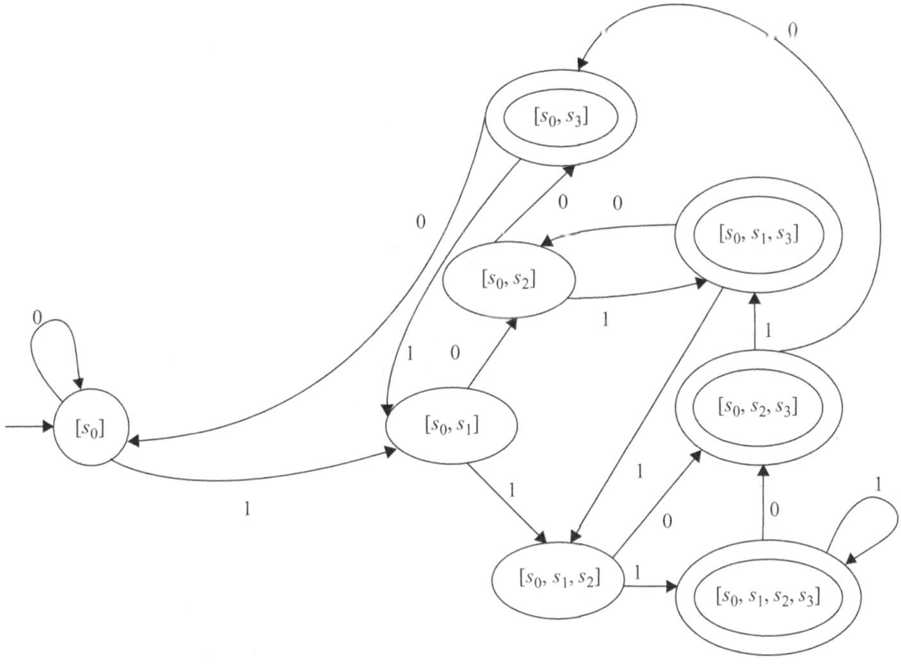

Figure 9.31

EXAMPLE 9.26

Draw transition diagram of the NDFSA

$$M' = (\{a, b\}, \{s_0, s_1, s_2\}, \{s_0\}, s_0, f),$$

where f is given by

$S \backslash I$	f: a	b
s_0	ϕ	$\{s_1, s_2\}$
s_1	$\{s_2\}$	$\{s_0, s_1\}$
s_2	$\{s_0\}$	ϕ

Also find equivalent DFSA.

Solution.

Here
Initial stage is s_0,
Set of accepting state is $\{s_0\}$,
Finite set of states is $\{s_0, s_1, s_2\}$,
Finite set of inputs is $\{a, b\}$.

Hence the transition diagram is shown in the Figure 9.32.

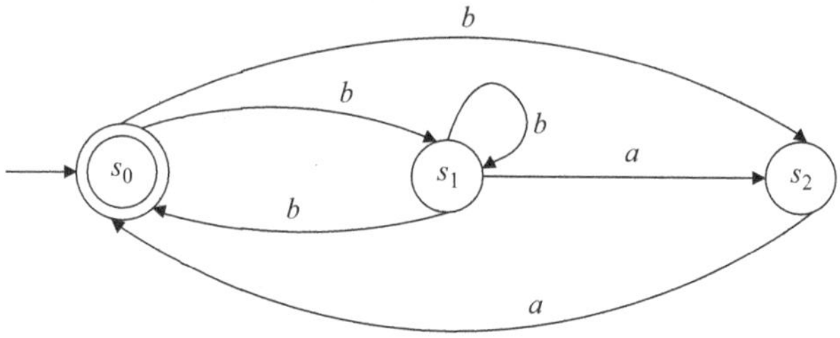

Figure 9.32

Let

$$M'=(\{a, b\}, \{\phi, \{s_0\}, \{s_1\}, \{s_2\}, \{s_0, s_1\}, \{s_0, s_2\}, \{s_1, s_2\} \{s_0, s_1, s_2\}\} s_0', A', f')$$

be the equivalent DFSA, where $s_0'=[s_0]$ and set of accepting states is

$$A'=\{s \in \{\phi, \{s_0\}, \ldots, \{s_0, s_1, s_2\}: s \cap \{s_0\} \neq \phi\}$$
$$= [s_0], [s_0, s_1], [s_0, s_2], [s_0, s_1, s_2].$$

Further,

$$f'(\phi, a)=\phi, \quad f'(\phi, b)=\phi,$$
$$f'([s_0], a)=f(s_0, a)=\phi, \quad f'([s_0],b)=f(s_0, b)=[s_1\ s_2],$$
$$f'([s_1], a)=f(s_1, a)=[s_2], \quad f'([s_1], b)=f(s_1, b)=[s_0\ s_1],$$
$$f'([s_2], a)=f(s_2, a)=[s_0], \quad f'([s_2], b)=f(s_2, b)=\phi,$$
$$f'([s_0\ s_1], a)=f(s_0, a) \cup f(s_1, a)=[s_2], \quad f'([s_0\ s_1], b)=[s_0\ s_1\ s_2],$$
$$f'([s_0\ s_2], a)=f(s_0, a) \cup f(s_2, a)=[s_0],$$
$$f'([s_0\ s_2], b)=f(s_0, b) \cup f(s_2, b)=[s_1\ s_2],$$
$$f'([s_1\ s_2], a)=f(s_1, a) \cup f(s_2, a)=[s_0\ s_2],$$
$$f'([s_1\ s_2], b)=f(s_1, b) \cup f(s_2, b)=[s_0\ s_1],$$
$$f'([s_0\ s_1\ s_2], a)=f(s_0, a) \cup f(s_1, a) \cup f(s_2, a)=[s_0\ s_2],$$
$$f'([s_0\ s_1\ s_2], b)=f(s_0, b) \cup f(s_1, b) \cup f(s_2, b)=[s_0\ s_1\ s_2].$$

Thus, the transition table of DFSA is as shown below:

	f'	
$S \backslash I$	a	b
ϕ	ϕ	ϕ
$[s_0]$	ϕ	$[s_1\ s_2]$
$[s_1]$	$[s_2]$	$[s_0\ s_1]$
$[s_2]$	$[s_0]$	ϕ
$[s_0\ s_1]$	$[s_2]$	$[s_0\ s_1\ s_2]$
$[s_0\ s_2]$	$[s_0]$	$[s_1\ s_2]$
$[s_1\ s_2]$	$[s_0, s_2]$	$[s_0, s_1]$
$[s_0\ s_1\ s_2]$	$[s_0\ s_2]$	$[s_0\ s_1\ s_2]$

The state $[s_1]$ is never entered and so may be omitted from the transition diagram. Thus the transition diagram of the DFSA is as given in the Figure 9.33.

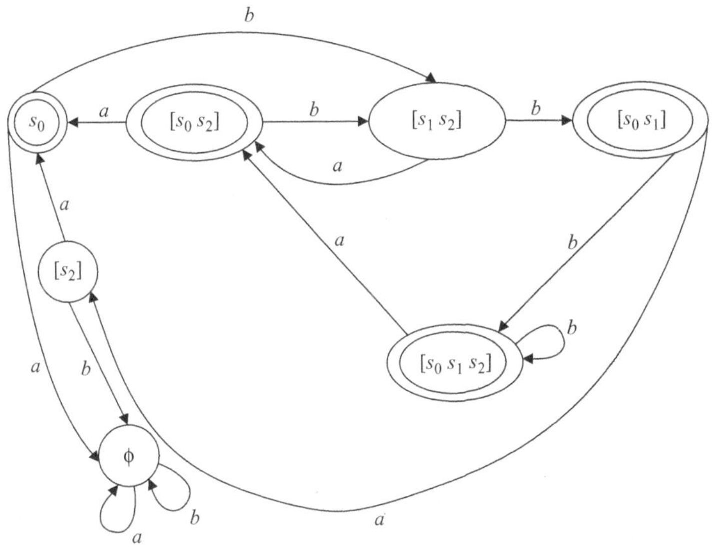

Figure 9.33

9.5 MOORE MACHINE AND MEALY MACHINE

We have seen that in case of finite automaton, the output is limited to a binary signal "accept" or "don't accept". However, some models in which the output is chosen from some other alphabet have also been considered. There are two different approaches.

1. If the output function depends only on the present state and is independent of the current input, the model is called a **Moore Machine.**
2. If the output function is a function of transition (a function of present state and the present input) the model is called a **Mealy Machine.**

9.5.1 Moore Machine

A **Moore machine** is a six-tuple

$$(I, S, O, s_0, f, g),$$

where

1. I is a finite set of input symbols
2. S is a finite set of internal states
3. O is a finite set of output symbols
4. s_0 is the initial state
5. f is the transition (next-state) function from $S \times I$ into S
6. g is the output function mapping S into O

The output in response to input $a_1\ a_2\ a_3 \ldots a_n$, $n \geq 0$ is $g(s_0)\ g(s_1) \ldots g(s_n)$, where $s_0, s_1, \ldots, s_n$ is the sequence of states such that

$$f(s_{i-1}, a_i) = s_i,\quad 1 \leq i \leq n.$$

Moore machine gives output $g(s_0)$ in response to input ε (empty string).

Obviously, **DFSA is a special case of Moore machine**, where the output alphabet is $\{0, 1\}$ and the state s is "accepting" if and only if $g(s)=1$.

EXAMPLE 9.27

Let

$$M=(I, S, O, s_0, f, g)$$

be a Moore machine, where

$$I=\{0, 1\}, \quad S=\{s_0, s_1, s_2, s_3\},$$
$$O=\{0, 1\}, \quad s_0 \text{ is initial state,}$$

f is transition function such that

$$f(s_0, 0)=s_3, \quad f(s_0, 1)=s_1,$$
$$f(s_1, 0)=s_1, \quad f(s_1, 1)=s_2,$$
$$f(s_2, 0)=s_2, \quad f(s_2, 1)=s_3,$$
$$f(s_3, 0)=s_3, \quad f(s_3, 1)=s_0$$

and g is the output function such that

$$g(s_0)=0, \quad g(s_1)=1, \quad g(s_2)=0, \quad g(s_3)=0.$$

Determine the transition table for M and the output string for the input string 0111.

Solution.

The transition table for this Moore machine is given below:

	f		g
S \ I	0	1	
s_0	s_3	s_1	0
s_1	s_1	s_2	1
s_2	s_2	s_3	0
s_3	s_3	s_0	0

The input string is 0111. We note that
For empty string ε, the output is $g(s_0)=0$

$$f(s_0, 0)=s_3 \text{ and } g(s_3)=0,$$
$$f(s_3, 1)=s_0 \text{ and } g(s_0)=0,$$
$$f(s_0, 1)=s_1 \text{ and } g(s_1)=1,$$
$$f(s_1, 1)=s_2 \text{ and } g(s_2)=0.$$

Thus the output string is 00010.

9.5.2 Mealy Machine

A **Mealy machine** M is a six-tuple (I, S, O, s_0, f, g), where

1. I is a finite set of input symbols
2. S is a finite set of internal states

3. O is a finite set of output symbols
4. s_0 is the initial state
5. f is the transition (next-state) function from $S \times I$ into S
6. g is the output function mapping $S \times I$ into O

The output given by M in response to input $a_1\, a_2 \ldots a_n$ is $g(s_0, a_1)\, g(s_1, a_2)\, g(s_2, a_3) \ldots g(s_{n-1}, a_n)$, where $s_0, s_1, \ldots, s_n$ is the sequence of states such that $f(s_{i-1}, a_i) = s_i$, $1 \le i \le n$.

Note that the output sequence in case of Mealy machine has length n, whereas the length of output sequence in case of Moore machine is $n+1$. Further, Mealy machine gives output ε for the input string ε.

EXAMPLE 9.28

Let

$$M = (I, S, O, s_0, f, g)$$

be a Mealy machine, where

$$I = \{0, 1\}, \qquad S = \{s_0, s_1, s_2\}, \qquad O = \{y, n\},$$

s_0 is the initial state,
f is next-state function such that

$$\begin{aligned} f(s_0, 0) &= s_1, & f(s_0, 1) &= s_2, \\ f(s_1, 0) &= s_1, & f(s_1, 1) &= s_2, \\ f(s_2, 0) &= s_1, & f(s_2, 1) &= s_2 \end{aligned}$$

and the output function is defined by

$$\begin{aligned} g(s_0, 0) &= n, & g(s_0, 1) &= n, \\ g(s_1, 0) &= y, & g(s_1, 1) &= n, \\ g(s_2, 0) &= n, & g(s_2, 1) &= y. \end{aligned}$$

Determine the transition table and transition diagram for M. Find the output string corresponding to the input string 01100.

Solution.

The transition table for the given Mealy machine is shown below:

	f		g	
S \ I	0	1	0	1
s_0	s_1	s_2	n	n
s_1	s_1	s_2	y	n
s_2	s_1	s_2	n	y

The transition diagram for the Mealy machine is as shown in the Figure 9.34.

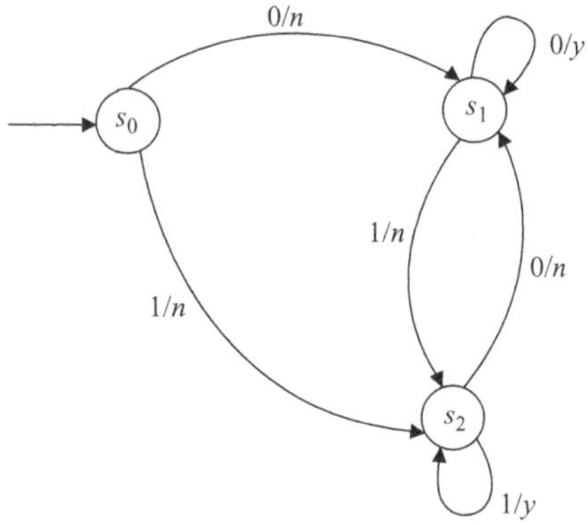

Figure 9.34

The input string is 01100. We have

$$\begin{aligned} f(s_0, 0)&=s_1, & g(s_0, 0)&=n, \\ f(s_1, 1)&=s_2, & g(s_1, 1)&=n, \\ f(s_2, 1)&=s_2, & g(s_2, 1)&=y, \\ f(s_2, 0)&=s_1, & g(s_2, 0)&=n, \\ f(s_1, 0)&=s_1, & g(s_1, 0)&=y. \end{aligned}$$

Hence, the output string is *nnyny*.

9.5.3 Equivalence of Moore and Mealy Machines

We know that the output string length in case of Mealy machine is one less than the output string length in case of Moore machine.

Neglecting the response of a Moore machine to input ε, we say that Moore machine M and Mealy machine M' are **equivalent** if for all input string v

$$k\,\Delta_{M'}(v)=\Delta_M(v),$$

where $\Delta_{M'}(v)$ and $\Delta_M(v)$ are output produced by M' and M on input v and k is output of M for its initial state.

Theorem 9.5

Let $M_1=(I, S, O, s_0, f, g)$ be a Moore machine. Then there is a Mealy machine $M_2=(I, S, O, s_0, f, g')$ which is equivalent to M_1.

Proof. Define $g' : S\times I \to O$ by

$$g'(s, a)=g(f(s, a)), \quad \text{for all} \quad s\in S \quad \text{and} \quad a\in I.$$

Then M_1 and M_2 enter the same sequence of states on the same input and with each transition M_2 emits the output that M_1 associates with the state entered.

EXAMPLE 9.29

Let the transition table of a Moore machine $M_1=(\{0, 1\}, \{s_0, s_1, s_2, s_3\}, \{0, 1\}, s_0, f, g)$ be as given below:

	f		g
$S \backslash I$	0	1	
$\rightarrow s_0$	s_3	s_1	0
s_1	s_1	s_2	1
s_2	s_2	s_3	0
s_3	s_3	s_0	0

Construct a Mealy machine M_2 equivalent to M_1.

Solution.

Let $M_2=(\{0, 1\}, \{s_0, s_1, s_2, s_3\}, \{0, 1\}, f, g', s_0)$ be the equivalent Mealy machine, where $g'(s, a)=g(f(s, a)), s \in S, a \in I$. Thus,

$$g'(s_0, 0)=g(f(s_0, 0))=g(s_3)=0,$$
$$g'(s_0, 1)=g(f(s_0, 1))=g(s_1)=1,$$
$$g'(s_1, 0)=g(f(s_1, 0))=g(s_1)=1,$$
$$g'(s_1, 1)=g(f(s_1, 1)=g(s_2)=0,$$
$$g'(s_2, 0)=g(f(s_2, 0)=g(s_2)=0,$$
$$g'(s_2, 1)=g(f(s_2, 1)=g(s_3)=0,$$
$$g'(s_3, 0)=g(f(s_3, 0)=g(s_3)=0,$$
$$g'(s_3, 1)=g(f(s_3, 1)=g(s_0)=0.$$

Thus the transition table for Mealy machine is as shown below:

	f		g'	
$S \backslash I$	0	1	0	1
$\rightarrow s_0$	s_3	s_1	0	1
s_1	s_1	s_2	1	0
s_2	s_2	s_3	0	0
s_3	s_3	s_0	0	0

Theorem 9.6

Let $M_1=(I, S, O, s_0, f, g)$ be a Mealy machine. Then there is a Moore machine $M_2=(\mathrm{I}, S', O, s_0', f', g')$ which is equivalent to M_1.

Proof. Let b_0 be arbitrary member of finite set O of output symbols. Set

$$M_2=(I, S\times O, O, [s_0, b_0], f', g').$$

Thus the states of M_2 consists of pairs $[q, b]$, where $q \in S, b \in O$.

Define f' by

$$f'([q, b], a)=[f(q, a), g(q, a)]$$

and g' by

$$g'([q, b])=b.$$

The component b in a state $[q, b]$ is the output made by M_1 on some transition into state q. Only the first component of M_2's states determine the moves made by the machine M_2. Induction on n shows that if M_1 enters states $q_0, q_1, \ldots, q_n$ on input $a_1\, a_2 \ldots a_n$ and emits outputs $b_1, b_2, \ldots, b_n$, then M_2 enters states $[q_0, b_0], [q_1, b_1], \ldots [q_n, b_n]$ and emits outputs $b_0, b_1, b_2, \ldots, b_n$.

EXAMPLE 9.30

Let M_1 be a Mealy machine whose transition table is given below:

	f		g	
I \ S	0	1	0	1
s_0	s_3	s_1	0	1
s_1	s_1	s_2	1	0
s_2	s_2	s_3	0	0
s_3	s_3	s_0	0	0

Find equivalent Moore machine M_2.

Solution.

The states of M_2 are

$$[s_0, 0], [s_0, 1], [s_1, 0], [s_1, 1], [s_2, 0], [s_2, 1], [s_3, 0], [s_3, 1].$$

We select $b_0=0$ making $[s_0, 0]$ as start state for M_2.
The transitions and outputs of M_2 are as follows:

$$f'([s_0, 0], 0)=[f(s_0, 0), g(s_0, 0)]=[s_3, 0]; \quad g'([s_0, 0])=0,$$

$$f'([s_0, 0], 1)=[f(s_0, 1), g(s_0, 1)]=[s_1, 1]; \quad g'([s_0, 0])=0,$$

$$f'([s_0, 1], 0)=[f(s_0, 0), g(s_0, 0)]=[s_3, 0]; \quad g'([s_0, 1])=1,$$

$$f'([s_0, 1], 1)=[f(s_0, 1), g(s_0, 1)]=[s_1, 1]; \quad g'([s_0, 1])=1,$$

$$f'([s_1, 0], 0)=[f(s_1, 0), g(s_1, 0)]=[s_1, 1]; \quad g'([s_1, 0])=0,$$

$$f'([s_1, 0], 1)=[f(s_1, 1), g(s_1, 1)]=[s_2, 0]; \quad g'([s_1, 0])=0,$$

$$f'([s_1, 1], 0)=[f(s_1, 0), g(s_1, 0)]=[s_1, 1]; \quad g'([s_1, 1])=1,$$

$$f'([s_1, 1], 1)=[f(s_1, 1), g(s_1, 1)]=[s_2, 0]; \quad g'([s_1, 1])=1,$$

$$f'([s_2, 0], 0)=[f(s_2, 0), g(s_2, 0)]=[s_2, 0]; \quad g'([s_2, 0])=0,$$

$$f'([s_2, 0], 1)=[f(s_2, 1), g(s_2, 1)]=[s_3, 0]; \quad g'([s_2, 0])=0,$$

$$f'([s_2, 1], 0)=[f(s_2, 0), g(s_2, 0)]=[s_2, 0]; \quad g'([s_2, 1])=1,$$

$$f'([s_2, 1], 1)=[f(s_2, 1), g(s_2, 1)]=[s_3, 0]; \quad g'([s_2, 1])=1,$$
$$f'([s_3, 0], 0)=[f(s_3, 0), g(s_3, 0)]=[s_3, 0]; \quad g'([s_3, 0])=0,$$
$$f'([s_3, 0], 1)=[f(s_3, 1), g(s_3, 1)]=[s_0, 0]; \quad g'([s_3, 0])=0,$$
$$f'([s_3, 1], 0)=[f(s_3, 0), g(s_3, 0)]=[s_3, 0]; \quad g'([s_3, 1])=1,$$
$$f'([s_3, 1], 1)=[f(s_3, 1), g(s_3, 1)]=[s_0, 0]; \quad g'([s_3, 1])=1.$$

Thus the transition table and transition diagram of Moore machine M_2 which is equivalent to given Mealy machine M_1 are given in the table and Figure 9.35.

	f'		g'
$S \backslash I$	0	1	
$\rightarrow [s_0, 0]$	$[s_3, 0]$	$[s_1, 1]$	0
$*[s_0, 1]$	$[s_3, 0]$	$[s_1, 1]$	1*
$*[s_1, 0]$	$[s_1, 1]$	$[s_2, 0]$	0*
$[s_1, 1]$	$[s_1, 1]$	$[s_2, 0]$	1
$[s_2, 0]$	$[s_2, 0]$	$[s_3, 0]$	0
$*[s_2, 1]$	$[s_2, 0]$	$[s_3, 0]$	1*
$[s_3, 0]$	$[s_3, 0]$	$[s_0, 0]$	0
$*[s_3, 1]$	$[s_3, 0]$	$[s_0, 0]$	1*

and

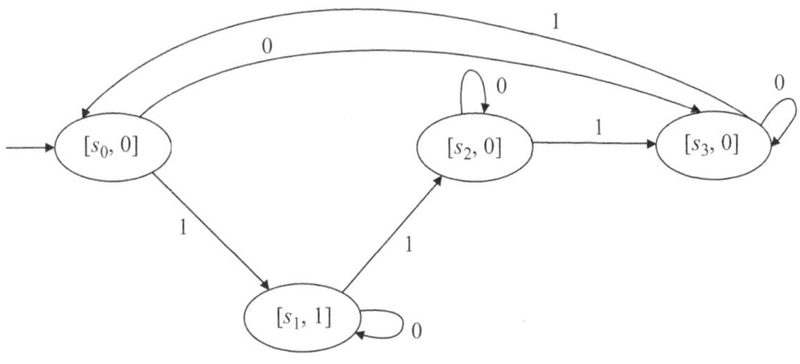

Figure 9.35

The states $[s_0, 1]$, $[s_1, 0]$ $[s_2, 1]$, $[s_3, 1]$ can never be entered and so have been removed from the diagram.

Leaving aside the outputs corresponding to the removed states which have been **marked by** * in the transition table, the outputs are 0, 1, 0, 0.

EXAMPLE 9.31

Find an equivalent Moore machine for the Mealy machine shown in the Figure 9.36.

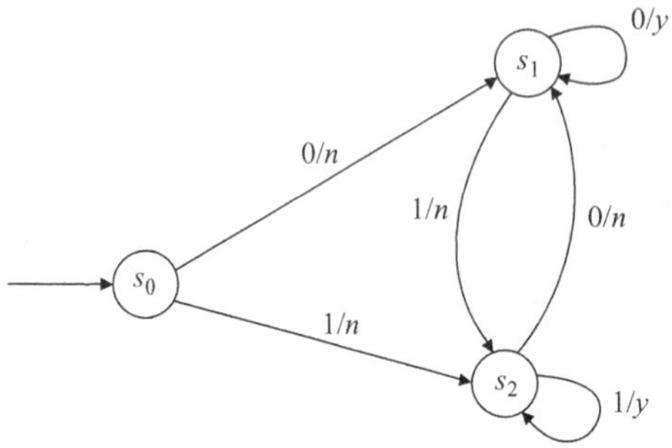

Figure 9.36

Find the output string corresponding to the input string 01100.

Solution.

The states of the Moore machine are

$$[s_0, n], [s_0, y], [s_1, n], [s_1, y], [s_2, n], [s_2, y].$$

We select $b_0 = n$ making $[s_0, n]$ as the initial state for the Moore machine. The transition and outputs of the required Moore machine are:

$$f'([s_0,n],0) = [f(s_0,0), g(s_0,0)] = [s_1,n]; \quad g'([s_0,n]) = n,$$

$$f'([s_0,n],1) = [f(s_0,1), g(s_0,1)] = [s_2,n]; \quad g'([s_0,n]) = n,$$

$$f'([s_0,y],0) = [f(s_0,0), g(s_0,0)] = [s_1,n]; \quad g'([s_0,y]) = y,$$

$$f'([s_0,y],1) = [f(s_0,1), g(s_0,1)] = [s_2,n]; \quad g'([s_0,y]) = y,$$

$$f'([s_1,n],0) = [f(s_0,0), g(s_1,0)] = [s_1,y]; \quad g'([s_1,n]) = n,$$

$$f'([s_1,n],1) = [f(s_1,1), g(s_1,1)] = [s_2,n]; \quad g'([s_1,n]) = n,$$

$$f'([s_1,y],0) = [f(s_1,0), g(s_1,0)] = [s_1,y]; \quad g'([s_1,y]) = y,$$

$$f'([s_1,y],1) = [f(s_1,1), g(s_1,1)] = [s_2,n]; \quad g'([s_1,y]) = y,$$

$$f'([s_2,n],0) = [f(s_2,0), g(s_2,0)] = [s_1,n]; \quad g'([s_2,n]) = n,$$

$$f'([s_2,n],1) = [f(s_2,1), g(s_2,1)] = [s_2,y]; \quad g'([s_2,n]) = n,$$

$$f'([s_2,y],0) = [f(s_2,0), g(s_2,0)] = [s_1,n]; \quad g'([s_2,y]) = y,$$

$$f'([s_2,y],1) = [f(s_2,1), g(s_2,1)] = [s_2,y]; \quad g'([s_2,y]) = y.$$

Thus the transition table of the Moore machine is given below:

	f'		g'
$S \backslash I$	0	1	
$\rightarrow [s_0, n]$	$[s_1, n]$	$[s_2, n]$	n
$[s_1, n]$	$[s_1, y]$	$[s_2, n]$	n
$[s_1, y]$	$[s_1, y]$	$[s_2, n]$	y
$[s_2, n]$	$[s_1, n]$	$[s_2, y]$	n
$[s_2, y]$	$[s_1, n]$	$[s_2, y]$	y

We note that the state $[s_0, y]$ never enters and therefore may be omitted from the transition diagram. Therefore the transition diagram of the Moore machine is as shown in the Figure 9.37.

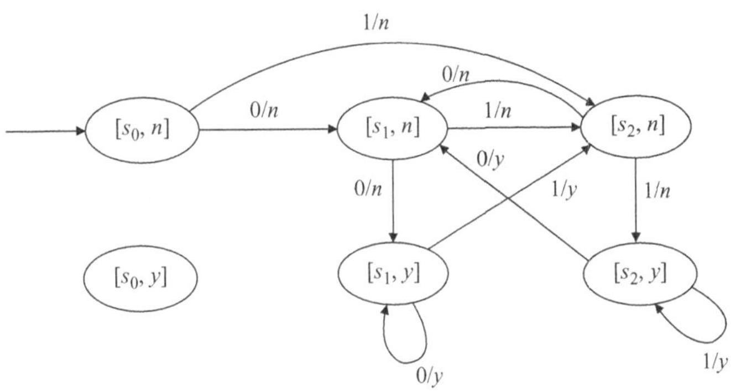

Figure 9.37

The output string corresponding to the input string 01100 is *nnnyn*.

9.6 GODEL NUMBERS

We know that any positive integer $n > 1$ can be expressed uniquely, except for order, as a product of prime numbers. Kurt Godel assigned a positive integer called Godel number to each sequence or word.

Definition 9.13

The **Godel number of the sequence** $s = \{n_1, n_2, \ldots, n_k\}$ of non-negative integers is the positive integers $c(s)$ defined by

$$c(s) = p_1^{n_1} p_2^{n_2}, \ldots, p_k^{n_k},$$

where the right-hand side is the prime decomposition of $c(s)$.

For example, consider the sequence $s = \{1, 0, 1, 2\}$. Then the Godel number of this sequence is

$$c(s) = 2^1 \cdot 3^0 \cdot 5^1 \cdot 7^2 = 490.$$

Definition 9.14

The **Godel number of a word** w on an alphabet $\{a_0, a_1, a_2, \ldots\}$ is the positive integer $c(w)$, where subscript of the ith letter of w is the exponent of p_i in the prime decomposition of $c(w)$.

For example, the word $w = a_3 a_1 a_3 a_1 a_2$ is coded by

$$c(w) = 2^3 \cdot 3^1 \cdot 5^3 \cdot 7^1 \cdot 11^2 = 2{,}541{,}000.$$

Obviously, the code of w is code for the sequence $\{3, 1, 3, 1, 2\}$ of the subscripts of the letters of w.

Further, the **Godel coding** is a one-to-one mapping $C : L \rightarrow N$ from any language L over a countable alphabet A into the set of natural numbers. It follows therefore that L is countable. Hence, any language L over countable alphabet is also countable.

9.7 TURING MACHINE

We know that a finite state machine could add, but other arithmetic operations like **multiplication** were beyond its range. A Turing machine is a powerful device for carrying out an algorithm. Such a machine can compute any partial recursive function. It is believed that any function that can be computed by some digital computer can be computed by some Turning machine. This assertion is known as "**Turing's hypothesis**" or **"Church's thesis"**. Church's thesis implies that a **Turing machine is the correct abstract model of a digital computer.** An algorithm can be defined in term of turning machine as "An algorithm is a Turing machine that, given an input string, eventually stops."

Like a finite state machine, a Turing machine is always in a particular state. The input string to a Turing machine is assumed to reside on a tape that is infinite in both directions. This tape is divided into cells, each of which can hold one character (symbol). A Turing machine scans one character at a time and after scanning a character, the machine either halts or does some, none or all of the following: alters the character, moves one position left or right, changes states. Thus, the input string can be changed. At any instant, the machine can be in exactly one of a finite number of internal states. The action taken by the machine is determined completely by the state of the machine and the character just scanned.

A Turing machine T accepts a string α if, when α is input to T, T halts in an accepting state. In fact, a language L is generated by a phrase structure grammar if and only if there is a Turing machine that accepts L.

A Turing machine T involves three disjoint non-empty sets:

1. A finite alphabet of tape symbols

$$A = \{a_1, a_2, \ldots, a_m\} \cup \{a_0\},$$

   where a_0 is the blank symbol, and $\{a_0\}$ is generally denoted by $\{B\}$.

2. A finite set of internal states

$$S = \{s_1, s_2, \ldots, s_n\} \cup \{s_0\} \cup \{s_H, s_Y, s_N\},$$

   where

   s_0 is the initial state,
   s_H is the Halting state,
   s_Y is the Yes (accepting) state,
   s_N is the No (non-accepting) state.

3. A direction set

$$d = \{L, R, N\},$$

   where

   L denotes "left"
   R denotes "right"
   N denotes "no movement" or "stay"

A finite sequence of elements from $A \cup S \cup d$ is called an **expression**. Thus, expression is a word whose letters come from A, S and d.

An expression in which the letters used are only from the tape set A is called a **tape expression.**

Intuitively, a Turing machine T can be represented as in the Figure 9.38.

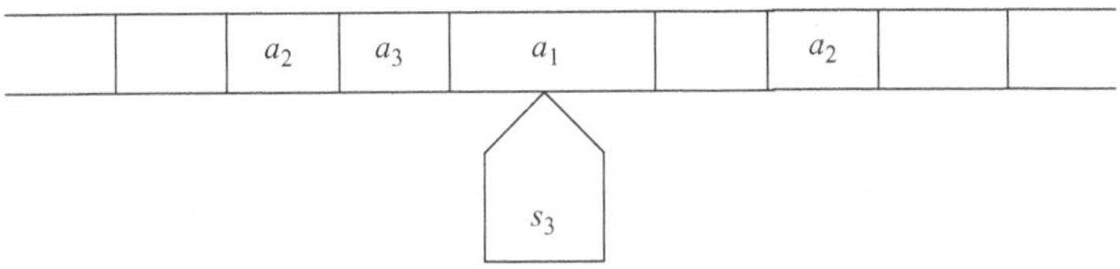

Figure 9.38

In this picture, a Turing machine is in state s_3 and is scanning the character a_1, where $a_2a_3a_1Ba_2$ is printed on the tape. Here B stands for blank symbol. This picture of the Turing machine can be represented by the expression

$$\alpha = a_2a_3s_3a_1Ba_2,$$

where we write the state s_3 before the tape symbol a_1 that is being scanned by the Turing machine.

Definition 9.15

An expression of the form $\alpha = Ps_ia_kQ$, where P and Q are tape expression (possibly empty), is called a **picture**.

Definition 9.16

If $\alpha = Ps_ia_kQ$ is a picture, we say that the Turing machine T is in state s_i scanning the character a_k and that the expression on the tape is Pa_kQ.

Definition 9.17

A **quintuple** q is a five letter expression of the form

$$q = \left(s_i, a_k, a_l, s_j, \begin{Bmatrix} L \\ R \\ N \end{Bmatrix} \right).$$

In this expression,

The first letter is a state symbol,
The second letter is a tape symbol,
The third letter is a tape symbol,
The fourth letter is a state symbol,
The last letter is a direction symbol L, R or N.

Now we are in a position to define a Turing machine.

Definition 9.18

A **Turing machine** T is a finite set of quintuples such that

(i) No two quintuples begin with the same first two letters
(ii) No quintuple begins with s_H, s_Y or s_N.

The condition (i) implies that T cannot do more than one thing at any given step, whereas, condition (ii) implies that T halts in s_H, s_Y or s_N.

Definition 9.19

Let S be the finite state set

$$S=\{s_1, s_2, \ldots, s_n\} \cup \{s_0\} \cup \{s_H, s_Y, s_N\}.$$

Then a Turing machine T is a **partial function** from $S-\{s_H, s_Y, s_N\}\times A$ into $A\times S\times d$, where partial function means that domain of T is a subset of $S-\{s_H, s_Y, s_N\}\times A$.

Definition 9.20

Let a, b, c be tape symbols and P and Q be tape expressions (possibly empty). Suppose α and β be pictures. We say that $\alpha\to\beta$ if any one of the following holds:

(i) $\alpha=Ps_iaQ$, $\beta=Ps_jbQ$ and T contains the quintuple $q=s_iabs_jN$.
(ii) $\alpha=Ps_iacQ$, $\beta=Pbs_jcQ$ and T contains the quintuple $q=s_iabs_jR$.
(iii) $\alpha=Pcs_iaQ$, $\beta=Ps_jcbQ$ and T contains the quintuple $q=s_iabs_jL$.
(iv) $\alpha=Ps_ia$, $\beta=Pbs_jB$ and T contains the quintuple $q=s_iabs_jR$.
(v) $\alpha=s_iaQ$, $\beta=s_jBbQ$ and T contains the quintuple $q=s_iabs_jL$.

We observe that

In (i), T replaces a on the tape by b and T does not move.
In (ii), T replaces a on the tape by b and T moves to the right.
In (iii), T replaces a on the tape by b and T moves to the left.
In (iv), T replaces a on the tape by b and T moves to the right. Further, since T is scanning the rightmost letter, it must add the blank symbol B on the right,
In (v), T replaces a on the tape by b and T moves to the left. Further, since T is scanning the leftmost letter, it must add the blank symbol B on the left.

Definition 9.21

A picture α is said to be **terminal** if there is no picture β such that $\alpha\to\beta$.

In particular, any picture α in one of the three halt states must be terminal since no quintuple begins with s_H, s_Y or s_N.

Definition 9.22

A **computation** of a Turing machine T is a sequence of pictures $\alpha_0, \alpha_1, \ldots, \alpha_n$ such that

$$\alpha_{l-1}\to\alpha_i,\ i=1, 2, \ldots, m$$

and α_m is a terminal picture.

Thus computation is a sequence.

$$\alpha_0\to\alpha_1\to\ldots\to\alpha_m,$$

which cannot further be extended since α_m is terminal.

EXAMPLE 9.32

Let T be a Turing machine. Determine the picture α corresponding to the following situation:

(i) T is in state s_2 and scanning the fourth letter of the tape expression $w=aabcab$,
(ii) T is in state s_3 and scanning the second letter of the tape expression $w=abca$.

Solution.

We know that the picture α is obtained by placing the state symbol before the tape symbol which is being scanned by the Turing machine. Thus, the pictures are

(i) $\alpha=aabs_2cab$,
(ii) $\alpha=as_3bca$.

EXAMPLE 9.33

Let $\alpha = aas_2ba$ be a picture. Determine β such that $\alpha \rightarrow \beta$ if the Turing machine T has the quintuple q, where

(i) $q = s_2bas_1L$,
(ii) $q = s_2bbs_3R$,
(iii) $q = s_2bas_2N$.

Solution.

In (i), T erases b and writes a, changes its states to as_1 and moves left. Hence $\beta = as_1aaa$.

In (ii), T does not change the scanned letter b, changes its state to s_3 and moves right. Thus $\beta = aab\ s_3a$.

In (iii), T erases b and writes a, does not change its state s_2 and does not move. Thus $\beta = aas_2aa$.

EXERCISES

1. Draw the transition diagram of the finite state machine (I, O, S, s_0, f, g), where $I = \{a, b\}$, $0 = \{0, 1\}$, $S = \{s_0, s_1, s_2, s_3\}$ and the transition table is

S \ I	f: a	f: b	g: a	g: b
s_0	s_1	s_2	0	0
s_1	s_0	s_2	1	0
s_2	s_3	s_0	0	1
s_3	s_1	s_3	0	0

2. The transition diagram of a finite state machine $(I, 0, S, s_0, f, g)$ is given in the Figure 9.39.

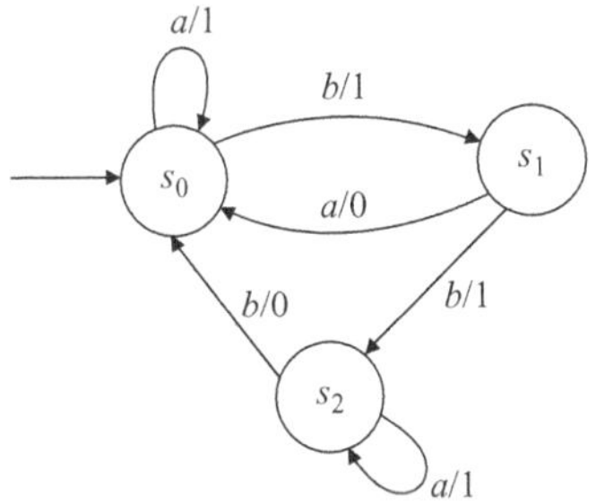

Figure 9.39

Determine I, O, S, s_0, and transition table for f and g.

3. Design a finite state automaton with input set $I = \{0, 1\}$ that accepts a set of all strings that start with 01.

4. Redraw the transition diagram of finite state automaton in Exercise 3 as a diagram of a finite state machine.

5. Design a finite state automaton with input symbols 0 and 1 that accept a set of all strings that contain even number of 0's and even number of 1's.

6. Does the finite state automaton in Exercise 5 accept the input string 110101?

7. Draw transition diagram of the non-deterministic finite state automaton (I, S, A, s_0, f), where

$I = \{0, 1\}$, $S = \{s_0, s_1, s_2, s_3, s_4\}$, $A = \{s_2, s_4\}$, s_0 is the initial state, and next state function f is defined by the table given below:

S \ I	f: 0	f: 1
s_0	$\{s_0, s_3\}$	$\{s_0, s_1\}$
s_1	ϕ	$\{s_2\}$
s_2	$\{s_2\}$	$\{s_2\}$
s_3	$\{s_4\}$	ϕ
s_4	$\{s_4\}$	$\{s_4\}$

8. Does the NDFSA in Exercise 7 accept the string 01001?

9. Let $M = (I, S, A, s_0, f)$ be a non-deterministic finite state automaton, where

$I = \{0, 1\}$, $S = \{s_0, s_1\}$, $A = \{s_1\}$, s_0: initial state and f is defined by

	f	
I \ S	0	1
s_0	$\{s_0, s_1)$	$\{s_1\}$
s_1	ϕ	$\{s_0, s_1\}$

Draw transition diagram of M and construct a deterministic finite automaton M' equivalent to M.

10. Draw the transition diagram of the non-deterministic finite state automaton whose transition table is given below:

	f	
I \ S	a	b
s_0	$\{s_0)$	$\{s_2\}$
s_1	$\{s_0, s_1)$	ϕ
s_2	$\{s_2)$	$\{s_0, s_1\}$

where $\{s_2\}$ is the accepting state.
Also find a finite state automaton equivalent to this non-deterministic finite state automaton.

11. Construct a Moore Machine over $\{0, 1\}$ that accepts those input sequences that contain the string 01 or the string 10 anywhere within them.
12. Design a Moore machine over the set $\{0, 1\}$ which print the residue modulo 3.
13. Construct a Mealy machine equivalent to Moore machine of Exercise 11.
14. Find an equivalent Moore machine for the Mealy machine shown in the Figure 9.40.

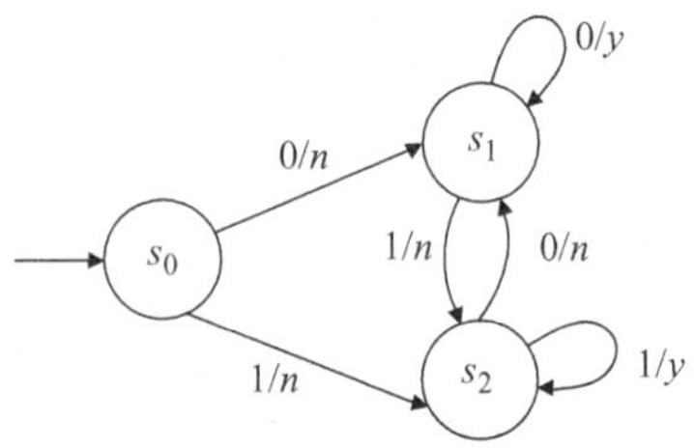

Figure 9.40

10 Languages and Grammars

The formal languages discussed in this chapter are used to model natural languages and to communicate with computers. These languages are completely determined by specified rules.

10.1 LANGUAGES AND REGULAR EXPRESSIONS

Definition 10.1

Let A be a finite set of symbols. A (formal) **language** L over A is a subset of A^*, the set of all strings over A.

For example, let $A=\{a, b\}$. Then the set L of all strings over A containing an odd number of a's is a language over A.

Similarly, $\{a, ab, ab^2, \ldots\}$ is a language over A. This consists of all words beginning with a and followed by zero or more b's.

Let L_1 and L_2 be languages over an alphabet A. Then the concatenation of L_1 and L_2, denoted by L_1L_2, is the language defined by

$$L_1L_2=\{uv: u \in L_1, v \in L_2\}.$$

Thus, L_1L_2 is the set of all words formed by the concatenation of a word from L_1 with a word from L_2. For example, let

$$L_1=\{a, b^3\}, \quad L_2=\{a^3, ab^2, b\}.$$

Then

$$L_1\ L_2=\{a^4, a^2b^2, ab, b^3\ a^3, b^3\ ab^2, b^4\}$$

is a language.

Since concatenation of words is associative, it follows that **concatenation of languages is associative.**

Definition 10.2

The **powers of a language** L are defined as

$$L^0=\{\varepsilon\}, \quad L^1=L, \quad L^2=LL, \ldots, L^{m+1}=L^mL, \quad m>1.$$

Definition 10.3

The unary operation L^* of a language L, called the **Kleene closure of** L, is defined as the infinite union

$$L^*=L^0 \cup L^1 \cup L^2 \cup \ldots=\bigcup_{k=0}^{\infty} L^k.$$

If we leave apart $L^0=\{\varepsilon\}$, then we write

$$L^{+}=L^{1}\cup L^{2}\cup\ldots=\bigcup_{k=1}^{\infty}L^{k}.$$

Definition 10.4

The regular expressions over an alphabet A and the sets they denote are defined recursively as follows:

1. The empty string ε is a regular expression and denotes the set $\{\varepsilon\}$.
2. ϕ or ( ) is a regular expression and denotes the empty set.
3. Each letter a in A is a regular expression and denotes the set $\{a\}$.
4. If r is a regular expression denoting the language R, then (r^*) is a regular expression and denotes the set R^*.
5. If r and s are regular expressions denoting the languages R and S, then $(r \vee s)$ or $(r+s)$ is a regular expression and denotes the set $R \cup S$.
6. If r and s are regular expressions denoting the languages (sets) R and S, then $(r\ s)$ is a regular expression and denotes the set $R\ S$.

Thus, a **regular expression r is a special kind of a string (word) which uses the letters of A and the five symbols**

( ), *, . , ∨, ε (or ∧).

For example,

(i) The regular expression $(0+1)^*$ denotes all the strings of 0's and 1's.
(ii) The regular expressions $(1+10)^*$ denotes all the strings of 0's and 1's and beginning with 1 and **not** having two consecutive 0's.
(iii) The regular expression $(0+1)^*\ 00\ (0+1)^*$ denotes all the strings of 0's and 1's with at least two consecutive 0's.
(iv) $(0+1)^*\ 0\ 1\ 1$ denotes all strings of 0's and 1's ending in 0 1 1.
(v) $0^*\ 1^*\ 2^*$ denotes all the strings with any number of 0's followed by any number of 1's followed by any number of 2's.

Definition 10.5

The **language** $L(r)$ **over** A defined by a regular expression r over A is as follows:

1. $L(\in)=\{\in\}$.
2. $L(\,(\,)\,)=\phi$, the empty set.
3. $L(a)=\{a\}$, where a is a letter in A.
4. $L(r^*)=(L(r))^*$, the Kleene closure of $L(r)$.
5. $L(r_1+r_2)=L(r_1)\cup L(r_2)$, the union of languages.
6. $L(r_1\ r_2)=L(r_1)\ L(r_2)$, the concatenation of the languages.

Definition 10.6

Let L be a language over A. Then L is said to be a **regular language** over A if there exists a regular expression r over A such that $L=L(r)$.

EXAMPLE 10.1

Let $A=\{a, b\}$. Then

(i) If $r=a^*$, then $L(r)$ consists of all powers of a.
(ii) if $r=a\ a^*$, then $L(r)$ consists of all positive powers of a, that is, all words in a excluding the empty word.

(iii) if $r=a+b^*$, then $L(r)$ consists of a or any word in b, that is, $L(r)=\{a, \varepsilon, b, b^2\ldots\}$.
(iv) if $r=(a+b)^*$, then $L(r)$ consists of all strings of a and b, i.e. all words (strings) over A.
(v) if $r=(a+b)^*\, a\, a$, then $L(r)$ denotes all strings of a and b ending in $a\, a$, i.e. $L(r)$ consists of the concatenation of any word in A with $a\, a$ (or a^2).

EXAMPLE 10.2

Let $L=\{a^m\, b^n: m, n>0\}$ be a language over $A=\{a, b\}$. Find a regular expression r such that $L=L(r)$.

Solution.

The given language L consists of strings beginning with one or more a's followed by one or more b's. Hence $r=a\, a^*\, b\, b^*$.

10.2 LANGUAGE DETERMINED BY A FINITE-STATE AUTOMATON

Let M be a finite state automaton with input set A. Then M defines a language over A, denoted by $L(M)$, as follows:

Let $u=a_1\, a_2\, \ldots\, a_n$ be a string on A. Then u determines a sequence of states

$$s_0 \to s_1 \to s_2 \to \ldots \to s_n,$$

where s_0 is the initial state and

$$f(s_{i-1}, a_i)=s_i \quad \text{for} \quad i\geq 1.$$

In other words, u determines the path

$$s_0 \xrightarrow{a_1} s_1 \xrightarrow{a_2} s_2, \ldots, \xrightarrow{a_b} s_n.$$

A finite state machine M is said to **accept** (recognize) the word u if the final state s_n belong to an accepting state in A (subset of internal states S).

The **language** $L(M)$ **of the finite state automaton** M is the collection of all words from the input set A which are accepted by M.

EXAMPLE 10.3

Determine the language $L(M)$ of the finite state automaton whose transition diagram is given in the Figure 10.1.

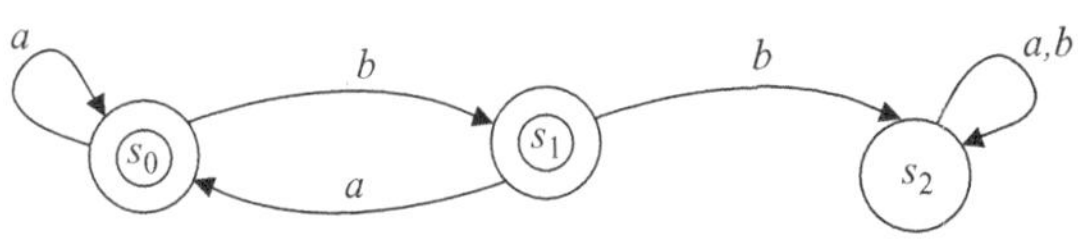

Figure 10.1

Solution.

Let $M=(I, S, A, s_0, f)$ be the finite state automaton. Then, we note that s_0 is the initial state, $S=\{s_0, s_1, s_2\}$ and

$$f(s_0, a)=s_0, \qquad f(s_0, b)=s_1,$$
$$f(s_1, a)=s_0, \qquad f(s_1, b)=s_2,$$
$$f(s_2, a)=s_2, \qquad f(s_2, b)=s_2.$$

Also, $A=\{s_0, s_1\}$ and $I=\{a, b\}$. We note that

(i) We can never leave s_2
(ii) The state s_2 is the only rejecting (non- accepting) state
(iii) A string in which there appears two successive b's is not accepted by M

Thus, $L(M)$ consists of all strings (words) from $I=\{a, b\}$ which do not have two successive b's.

EXAMPLE 10.4

Find the language accepted by the automaton M shown in the transition diagram (Figure 10.2):

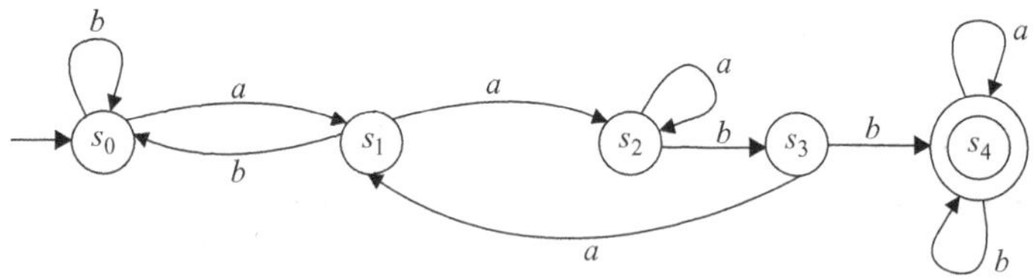

Figure 10.2

Solution.

Let $M=(I, S, A, s_0, f)$ be the FSA. Then, we have

$$I=\{a, b\}, \quad S=\{s_0, s_1, s_2, s_3, s_4\}, s_0 \text{ is the initial state, } A=\{s_4\}$$

and f is given by

$$f(s_0, a)=s_1,\ f(s_0, b)=s_0,\ f(s_1, a)=s_2,\ f(s_1, b)=s_0,\ f(s_2, a)=s_2,\ f(s_2, b)=b,$$
$$f(s_3, b)=s_4,\ f(s_3, a)=s_1,\ f(s_4, b)=s_4,\ f(s_4, a)=s_4.$$

We note that

$$s_0 \xrightarrow{a} s_1 \xrightarrow{a} s_2 \xrightarrow{b} s_3 \xrightarrow{b} s_4 \quad \text{(accepting state)},$$

$$s_0 \xrightarrow{b} s_0 \xrightarrow{a} s_1 \xrightarrow{a} s_2 \xrightarrow{b} s_3 \xrightarrow{b} s_4 \quad \text{(accepting state)},$$

$$s_0 \xrightarrow{b} s_0 \xrightarrow{a} s_1 \xrightarrow{a} s_2 \xrightarrow{b} s_3 \xrightarrow{b} s_4 \xrightarrow{a} s_4 \xrightarrow{b} s_4 \quad \text{(accepting state)}$$

Hence, $L(M)$ consists of all words which contain $a\,a\,b\,b$ as a subword.

10.3 GRAMMARS

Definition 10.7

A **phrase-structure Grammar** or simply a **Grammar** G consists of

1. A finite set N of **non-terminal symbols (or variables)**
2. A finite set T of **terminal symbols**, where $N \cap T=\phi$,
3. A finite subset P of $[(N \cup T)^* - T^*]\times(N \cup T)^*$, called **the set of productions**. Thus a production is an ordered pair (A, B), written as $A \rightarrow B$, where $A \in [(N \cup T)^* - T^*]$ must include at least one non-terminal symbol whereas $B \in (N \cup T)^*$ can consist of any combination of non-terminals and terminal symbols.
4. A **starting symbol** $\sigma \in N$.

A grammar G is denoted by $G(N, T, P, \sigma)$. Terminals will be denoted by lower case letters $a, b, c, \ldots$ whereas non-terminals will be denoted by $A, B, C, \ldots$

EXAMPLE 10.5

Let

$$N=\{\sigma, A\}, \quad T=\{a, b\},$$
$$P=\{\sigma\to b\,\sigma, \sigma\to b\,A, A\to a\,A, A\to b\},$$

where σ is the starting symbol. Then $G=(N, T, P, \sigma)$ is a grammar.

Since $\sigma\to b\,\sigma$, $\sigma\to b\,A$ and $A\to a\,A$, $A\to b$, we can also write the productions as

$$\sigma\to(b\,\sigma, b\,A), \quad A\to(a\,A, b).$$

Definition 10.8

Let $G=(N, T, P, \sigma)$ be a grammar and let $\alpha\to\beta$ be a production. If $x\,\alpha\,y\in(N\cup T)^*$, then $x\,\beta\,y$ is said to be **directly derivable** from $x\,\alpha\,y$ and we write $x\,\alpha\,y\Rightarrow x\,\beta\,y$.

Further, if $\alpha_i\in(N\cup T)^*$ for $i=1, 2, \ldots, n$, and α_{i+1} is directly derivable from α_i for $i=1, 2, \ldots, n-1$, we say that α_n **is derivable from** α_1 and write $\alpha_1\Rightarrow\alpha_n$.
We call $\alpha_1\Rightarrow\alpha_2\Rightarrow\alpha_3\Rightarrow\ldots\Rightarrow\alpha_n$, the **derivation of** α_n (from α_1).

By convention, any element of $(N\cup T)^*$ is derivable from itself.

Definition 10.9

The **language generated by a grammar** G, written as $L(G)$, consists of all strings over T derivable from the start symbol σ. Thus,

$$L(G)=\{v\in T^*: \sigma\Rightarrow\ldots\Rightarrow v\}.$$

EXAMPLE 10.6

Find the language $L(G)$ over $\{a, b\}$ generated by the grammar with productions

$$\sigma\to b\,\sigma, \quad \sigma\to a\,A, \quad A\to b\,A, \quad A\to b,$$

where σ is the starting symbol.

Solution.

We note that the string $bbab$ is derived from σ, written as $\sigma\Rightarrow b\,bab$ by the derivation

$$\sigma\Rightarrow b\sigma\Rightarrow bb\sigma\Rightarrow bbaA\Rightarrow bbab$$

Further, all possible derivation from σ are

$\sigma\Rightarrow b\sigma\Rightarrow bb\sigma\Rightarrow\ldots\Rightarrow b^n\,\sigma\Rightarrow b^n\,a\,A\Rightarrow b^n\,a\,b\,A\Rightarrow b^n\,abbA\Rightarrow\ldots$
$\Rightarrow b^n\,ab^{m-1}\,A\Rightarrow b^n\,ab^m$ for $n\geq 0$, $m\geq 1$. Hence

$$L(G)=\{b^n\,ab^m: n\geq 0, m\geq 1\}.$$

Definition 10.10

A **sentential form** is any derivative of the unique non-terminal symbol S.

The language $L(G)$ generated by the grammar G is the set of all sentential forms whose symbols are terminals.

EXAMPLE 10.7

Let $G=\{N, T, \sigma, P\}$ be a grammar, where $N=\{\sigma\}$, $T=\{a, b\}$, σ is starting symbol, and the production P are

$$P=\{\sigma\to a, \sigma\to\sigma\,a, \sigma\to b \text{ and } \sigma\to b\,\sigma\}.$$

Obtain sentential form and find the language generated by G.

Solution.

We note that

$$\begin{aligned}\sigma &\Rightarrow \sigma\, a\\ &\Rightarrow \sigma\, a\, a\\ &\Rightarrow b\, \sigma\, a\, a\\ &\Rightarrow b\, b\, \sigma\, a\, a\\ &\Rightarrow b\, b\, b\, a\, a.\end{aligned}$$

Thus $b^3a^2 = b\,b\,b\,a\,a$ is a sentential form. Hence the language generated by G is

$$L(G) = \{b^n\, a^m : n \geq 0,\ m \geq 0\}.$$

EXAMPLE 10.8

Find the language $L(G)$ generated by the grammar G with variables σ, A, B; $T = \{a, b\}$ and productions $P = \{\sigma \to a\,B,\ B \to b,\ B \to b\,A,\ A \to a\,B\}$.

Solution.

If we start from σ, we see that

$$\sigma \Rightarrow a\,B \Rightarrow a\,b \quad \text{(we reach at terminal symbol)}$$

or

$$\sigma \Rightarrow a\,B \Rightarrow a\,b\,A \Rightarrow a\,b\,a\,B \Rightarrow a\,b\,a\,b \quad \text{(we reach at terminal word)}$$

or

$$\sigma \Rightarrow a\,B \Rightarrow a\,b\,A \Rightarrow a\,b\,a\,B \Rightarrow a\,b\,a\,b\,A \Rightarrow a\,b\,a\,b\,a\,B \Rightarrow a\,b\,a\,b\,a\,b$$

and so on.
Thus, it follows that

$$L(G) = \{a\,b\,a\,b\,a\,b\,a\,b \ldots a\,b\} = \{(a\,b)^n,\ n \in \boldsymbol{N}\}.$$

Definition 10.11

Let λ be a null string. Then a grammar $G = (N, T, P, \sigma)$ is said to be **context-sensitive** (or **type-1**) **grammar** if every production is of the form

$$\alpha\, A\, \beta \to \alpha\, \delta\, \beta,$$

where $\alpha, \beta \in (N \cup T)^*$, $A \in N$, $\delta \in (N \cup T)^* - \{\lambda\}$.

Thus, type-1 grammar contains only productions of the form $\alpha \to \beta$, where $|\alpha| \leq |\beta|$.

Definition 10.12

A grammar $G = (N, T, P, \sigma)$ is said to be **context-free** (or **type-2**) grammar if the productions are of the form

$$A \to \delta,$$

where $A \in N$, $\delta \in (N \cup T)^*$.
Thus, in this case, we can replace A by δ regardless of where A appears.

Definition 10.13

A grammar $G=(N, T, P, \sigma)$ is said to be **regular** (or **type-3**) grammar if every production is of the form

$$A \to a \quad \text{or} \quad A \to a\,B \quad \text{or} \quad A \to \lambda,$$

where $A, B \in N$, $a \in T$.

Thus, in this case, we replace a non-terminal symbol by a terminal symbol, by a terminal symbol followed by a non-terminal symbol, or by the null string.

We further note that **a regular grammar is context-free grammar** and that **a context-free grammar with no productions of the form $A \to \lambda$ is a context-sensitive grammar.**

EXAMPLE 10.9

Name the type of the grammar G defined by $T=\{a, b, c\}$, $N=\{\sigma, A, B, C, D, E\}$, starting symbol σ and productions

$$\sigma \to a\,A\,B,\ \sigma \to a\,B,\ A \to a\,A\,C,\ A \to a\,C,\ B \to D\,c,\ D \to b,$$
$$C\,D \to C\,E,\ C\,E \to D\,E,\ D\,E \to D\,C,\ C\,c \to D\,c\,c.$$

Also find its language.

Solution.

The production $C\,E \to D\,E$ says that we can replace C by D if C is followed by E. The production $C\,c \to D\,c\,c$ says that we can replace C by $D\,c$ if C is followed by c. Thus the grammar is context-sensitive. We can derive $D\,C$ from $C\,D$ since $C\,D \Rightarrow C\,E \Rightarrow D\,E \Rightarrow D\,C$. We note that

$$\begin{aligned}\sigma &\Rightarrow a\,A\,B \Rightarrow a\,a\,A\,C\,B \Rightarrow a\,a\,a\,C\,C\,B \Rightarrow a\,a\,a\,C\,C\,D\,c\\ &\Rightarrow a\,a\,a\,C\,D\,C\,c \Rightarrow a\,a\,a\,D\,C\,C\,c \Rightarrow a\,a\,a\,D\,C\,D\,c\,c\\ &\Rightarrow a\,a\,a\,D\,D\,C\,c\,c \Rightarrow a\,a\,a\,D\,D\,D\,c\,c\,c \Rightarrow a\,a\,a\,b\,b\,b\,c\,c\,c.\end{aligned}$$

Thus $a^3\,b^3\,c^3$ is in $L(G)$. Proceeding in this way, we can show that

$$L(G)=\{a^n\,b^n\,c^n : n \in \mathbf{N}\}.$$

EXAMPLE 10.10

Determine whether the following grammar is context-sensitive, context-free, regular or none of these:

$$G=(N, T, \sigma, P),$$

where $N=\{\sigma, A\}$, $T=\{a, b\}$, starting symbol is σ and the productions are

$$\sigma \to b\,\sigma, \quad \sigma \to \sigma\,A, \quad A \to a\,\sigma, \quad A \to b\,A, \quad A \to a, \quad \sigma \to b.$$

Solution.

We note that

(i) $A \to \delta$, $A \in N$, $\delta \in (N \cup T)^*$. Hence the grammar is context-free grammar.

(ii) $A \to a$ or $A \to bA$, $A \in N$, $b \in T$. Hence the grammar is regular.

(iii) The grammar is also context-sensitive because

$$\alpha\,A\,\beta \to \alpha\,\delta\,\beta,$$

where $\alpha, \beta \in (N \cup T)^*$, $A \in N$, $\delta \in (N \cup T)^* - \{\lambda\}$.

EXAMPLE 10.11

Find a context-free grammar G which generates the language $L=\{a^n b^n: n>0\}$.

Solution.

Here

$$T=\{a, b\}.$$

If we consider the productions

$$\sigma \rightarrow ab, \quad \sigma \rightarrow a\sigma b,$$

then we note that

$$\sigma \Rightarrow a\,\sigma\, b \Rightarrow a\, ab\, b$$
$$\sigma \Rightarrow a\,\sigma\, b \Rightarrow a\, a\,\sigma\, b\, b \Rightarrow a\, a\, a\, b\, b\, b$$
$$\sigma \Rightarrow a\,\sigma\, b \Rightarrow a\, a\,\sigma\, b\, b \Rightarrow a\, a\, a\,\sigma\, b\, b\, b \Rightarrow a\, a\, a\, a\, b\, b\, b\, b$$

In general,

$$L(G)=\{a^n b^n, n>0\}.$$

Hence the grammar with production

$$\sigma \rightarrow ab, \quad \sigma \rightarrow a\,\sigma\, b$$

generates $L(G)$.

10.3.1 Backus–Naur Form (BNF)

This notation is used for describing the productions of a context-free grammar. In this form,

1. : :=is used instead of $\rightarrow$
2. Every non-terminal symbol is enclosed in brackets<>
3. All productions of the same non-terminal symbol are combined into one statement with all the right-hand side listed on the right of : :=, separated by vertical bars

For example,

$$A \rightarrow a\, A, \quad A \rightarrow a, \quad A \rightarrow B\, C$$

are combined into one statement

$$<A> ::= a <A> \mid a \mid <B> <C>$$

The bar "|" is read "or".

10.4 DERIVATION TREES OF CONTEXT-FREE GRAMMARS

Let G be a context-free grammar. An ordered rooted tree which represents any derivation of a word in $L(G)$ is called a **derivation Tree** or **parse tree**.

EXAMPLE 10.12

Consider the language

$$L=\{a^n b^n: n>0\}.$$

We have seen that context-free grammar which generates $L(G)$ is

$$N=\{\sigma\}, \quad T=\{a, b\}, \quad P=\{\sigma \rightarrow a\,\sigma\, b, \sigma \rightarrow a\, b\}.$$

The word $w = a\ a\ a\ b\ b\ b$ is derived as

$$\sigma \Rightarrow a\ \sigma\ b \Rightarrow a\ a\ \sigma\ b\ b \Rightarrow a\ a\ a\ b\ b\ b.$$

The Figure 10.3 will therefore be its derivation tree:

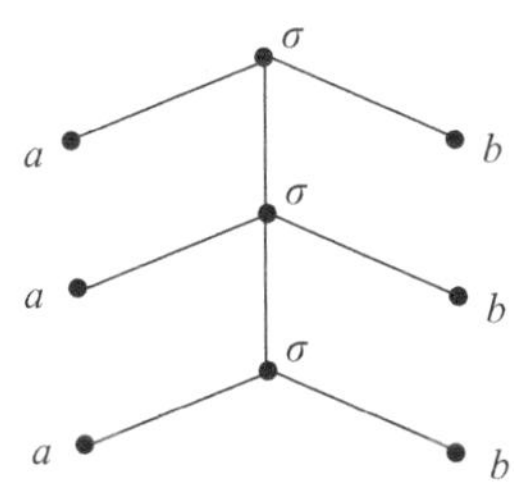

Figure 10.3

EXAMPLE 10.13

Find the derivation tree for the word $a\ a\ b\ a$ in $L(G)$, where G has the productions

$$\sigma \rightarrow a\,A, \quad A \rightarrow a\,B, \quad B \rightarrow b\,B, \quad B \rightarrow a.$$

Solution.

The word $a\ a\ b\ a$ is derived as

$$\sigma \Rightarrow a\,A \Rightarrow a(a\,B) \Rightarrow a\ a\ (b\,B) \Rightarrow a\ a\ b\ a$$

and therefore the derivation tree of $a\ a\ b\ a$ is as shown in the Figure 10.4.

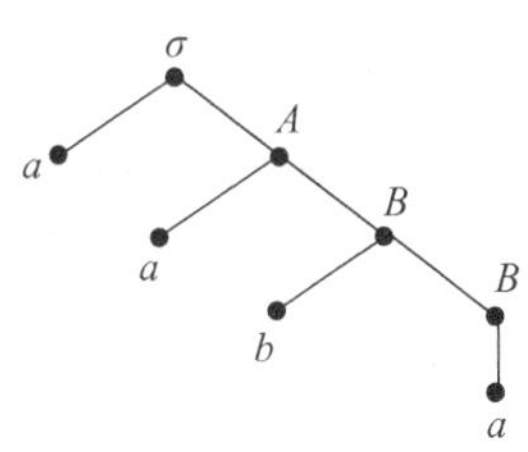

Figure 10.4

Definition 10.14

A language is said to be **context-sensitive** if there is a context-sensitive grammar G with $L = L(G)$.

Definition 10.15

A language is said to be **context-free** if there is a context-free grammar G with $L = L(G)$.

Definition 10.16

A language is said to be **regular** if there is a regular grammar G with $L = L(G)$.

EXAMPLE 10.14

Is the language $L=\{a^n b^n, n=1, 2, \ldots\}$ over $\{a, b\}$ context-free?

Solution.

Let G be grammar defined by $N=\{\sigma\}$, $T=\{a, b\}$, σ is staring symbol and production as $\sigma \to a\, \sigma\, b$, $\sigma \to a\, b$. Then derivation of σ are

$$\begin{aligned} \sigma &\Rightarrow a\, \sigma\, b \\ &\Rightarrow a\, a\, \sigma\, b\, b \\ &\text{\_\_\_\_\_\_\_\_\_\_} \\ &\text{\_\_\_\_\_\_\_\_\_\_} \\ &\Rightarrow a^{n-1}\, \sigma\, b^{n-1} \\ &\Rightarrow a^{n-1}\, a\, b\, b^{n-1} = a^n\, b^n. \end{aligned}$$

Thus the grammar G generates the language $L(G)$. Also the grammar G is context-free. Hence the language $L=\{a^n b^n, n=1, 2, \ldots\}$ is a context-free language.

We now show that **regular grammar and finite state automata are essentially the same**. After that we would be able to say that

"A language is a regular set (or just **regular) if it is accepted by some finite automaton."**

Theorem 10.1

Let M be a finite state automaton given as a transition diagram. Let σ be the initial state. Let T be the set of input symbols and let N be the set of states. Let P be the set of productions

$$s \to x\, s'$$

if there is an edge labelled x from the state s to the state s', and

$$s \to \lambda$$

if s is an accepting state. Let

$$G=(N, T, P, \sigma)$$

be the regular grammar. Then the set of strings accepted by M is equal to $L(G)$.

Proof. Let $AC(M)$ denote the set of strings accepted by M. We first show that $AC(M) \subseteq L(G)$. So, let $\alpha \in AC(M)$. If α is the null string, then σ is an accepting state. In this case G contains the production $\sigma \to \lambda$. The derivation $\sigma \Rightarrow \lambda$ shows that $\alpha \in L(G)$.

Now let $\alpha \in AC(M)$ and let α is not a null string. Then

$$\alpha = a_1\, a_2 \ldots a_n,\ a_i \in T.$$

Since α is accepted by M, there is a path

$$(\sigma, s_1, s_2, \ldots, s_n),$$

where s_n is an accepting state with edges successively labelled $a_1, \ldots, a_n$. It follows that G contains the productions

$$\begin{aligned} &\sigma \to a_1\, s_1, \\ &s_1 \to a_2\, s_2, \\ &\text{\_\_\_\_\_\_\_\_\_\_} \\ &\text{\_\_\_\_\_\_\_\_\_\_} \\ &s_{n-1} \to a_n\, s_n. \end{aligned}$$

Since s_n is an accepting state, G also contains the production $s_n \to \lambda$. The derivation

$$\begin{aligned}\sigma &\Rightarrow a_1\, s_1\\ &\Rightarrow a_1\, a_2\, s_2\\ &\Rightarrow a_1\, a_2\, a_3\, s_3\\ &\underline{\qquad\qquad}\\ &\underline{\qquad\qquad}\\ &\Rightarrow a_1\, a_2 \ldots a_n\, s_n\\ &\Rightarrow a_1\, a_2 \ldots a_n \quad (\because s_n \to \lambda) \qquad (1)\end{aligned}$$

shows that $\alpha = a_1\, a_2 \ldots a_n \in L(G)$. Hence $AC(M) \subseteq L(G)$.

It remains to show that $L(G) \subseteq AC(M)$. Suppose that $\alpha \in L(G)$. If α is the null string, then α must result from the derivation $\sigma \Rightarrow \lambda$. Thus the production $\sigma \to \lambda$ is in the grammar. Hence σ is an accepting state in M and so $\alpha \in AC(M)$.

Now let $\alpha \in L(G)$ be a non-null string. Then

$$\alpha = a_1\, a_2 \ldots a_n, \quad a_i \in T.$$

So there is a derivation of the form (1). If in the transition diagram, we begin at σ and trace the path

$$(\sigma, s_1, s_2, \ldots, s_n),$$

we can generate the string α. The last production used in (1) is $s_n \to \lambda$. Thus the last state reached is an accepting state. Therefore, α is accepted by M, that is, $L(G) \subseteq AC(M)$. Hence

$$L(G) = AC(M).$$

Thus, given a finite state automaton M, we can construct a regular grammar G such that the set of strings accepted by M is equal to $L(G)$.

EXAMPLE 10.15

Verify Theorem 10.1 for the finite-state automaton that accepts those strings over $\{a, b\}$ that contains odd number of a's.

Solution.

There are two states

E: An even number of a's was found

O: An odd number of a's was found.

The state E is the initial state and O is accepting state. The transition diagram is drawn in Figure 10.5.

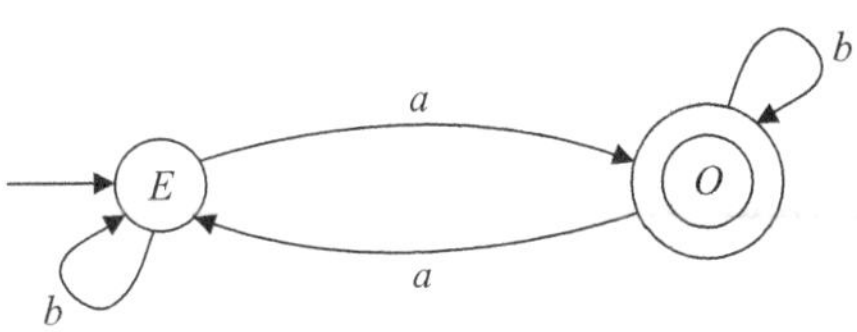

Figure 10.5

Let $T = \{a, b\}$, $N = \{E, O\}$, E is starting symbol. The productions are

$$E \to b\,E, \quad E \to a\,O, \quad O \to a\,E, \quad O \to b\,O.$$

Since O is accepting state, we have an additional production

$$O \to \lambda.$$

Then

$$E \Rightarrow b\,E \Rightarrow b\,a\,O \Rightarrow b\,a\,a\,E \Rightarrow b\,a\,a\,a\,O \Rightarrow b\,a\,a\,a.$$

Thus,

$$L(G) = \{b^n\,a^m : n > 0,\ m \text{ is odd}\},$$

which is same as the string accepted by the finite state automaton M.

EXAMPLE 10.16

Let $G(T, N, P, \sigma)$ be a regular grammar, where

$$T = \{a, b\},\ N = \{\sigma, A\},\ \sigma \text{ is starting symbol and}$$
$$P = \{\sigma \to b\,\sigma,\ \sigma \to a\,A,\ A \to b\,A,\ A \to b\}.$$

Does there exist finite state automaton corresponding to G?

Solution.

Let the input symbols be the terminal symbols and the states be the non-terminal symbols, where σ is the initial state.

For each production of the form

$$s \to x\,s',$$

draw an edge from state s to state s' and label it x. Thus the productions

$$\sigma \to b\,\sigma, \quad \sigma \to a\,A, \quad A \to b\,A$$

yield the graph shown in the Figure 10.6.

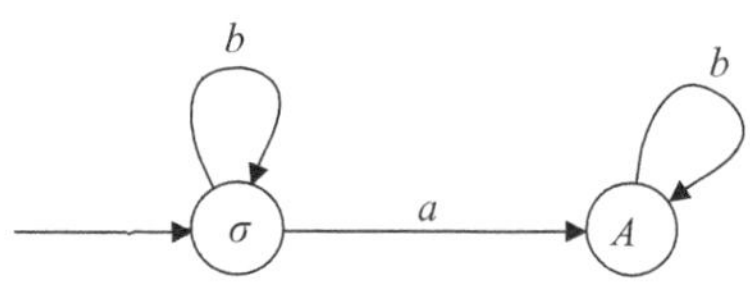

Figure 10.6

The last production $A \to b$ is equivalent to two productions

$$A \to bB \quad \text{and} \quad B \to \lambda,$$

where B is an additional and non-terminal symbol.

The productions

$$\sigma \to b\,\sigma, \quad \sigma \to a\,A, \quad A \to bA, \quad A \to bB$$

give us the graph shown in the Figure 10.7.

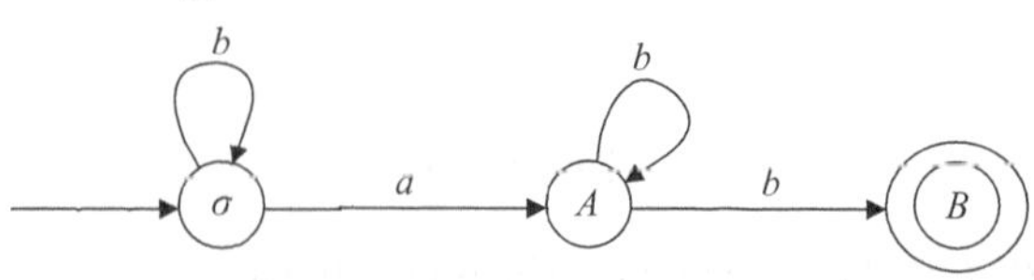

Figure 10.7

and the production $B \rightarrow \lambda$ indicates that B is an accepting state. We note that

(i) Vertex A has no outgoing edge labelled as a
(ii) Vertex B has no outgoing edge
(iii) A has two outgoing edges labelled as b

Thus, the above graph is not finite state automaton but a **non-deterministic finite state automaton** (I, S, A, σ, f), where $I = \{a, b\}$, $S = \{\sigma, A, B\}$, $A = \{B\}$, initial state σ and next state function f is defined by

	f	
$S \backslash I$	a	b
σ	$\{A\}$	$\{\sigma\}$
A	ϕ	$\{A, B\}$
B	ϕ	ϕ

We further notice that

(i) The string $b\ b\ a\ b\ b$ is in $L(G)$ since

$$\begin{aligned} \sigma &\Rightarrow b\,\sigma \\ &\Rightarrow b\,b\,\sigma \\ &\Rightarrow b\,b\,a\,A \\ &\Rightarrow b\,b\,a\,b\,A \\ &\Rightarrow b\,b\,a\,b\,b\,B \\ &\Rightarrow b\,b\,a\,b\,b. \end{aligned}$$

Also the string $b\ b\ a\ b\ b$ is accepted by the non-deterministic finite state automaton obtained above since the path

$$\sigma \xrightarrow{b} \sigma \xrightarrow{b} \sigma \xrightarrow{a} A \xrightarrow{b} A \xrightarrow{b} B,$$

which ends at state B (accepting state) represents the string $b\ b\ a\ b\ b$.

Theorem 10.2

Let $G(T, N, P, \sigma)$ be a regular grammar and let $I = T$, $S = N \cup \{F\}$, where $F \notin N \cup T$, σ as initial state, $A = \{F\} \cup \{s \colon s \rightarrow \lambda \in P\}$ and f be defined by

$$f(s, x) = \{s' \colon s \rightarrow x\, s' \in P\} \cup \{F \colon s \rightarrow x\, F \in P\}.$$

Then the non-deterministic finite state automaton $M = (I, S, A, \sigma, f)$ accept the strings $L(G)$.

(The proof is same as the proof for finite state automaton).

Also, we know that a non-deterministic finite state automaton can be converted into an equivalent finite state automaton.

Hence, we have the following result.

Theorem 10.3 (Kleene Theorem)

A language L is regular if and only if there exists a finite state automaton that accepts strings in L.

Theorem 10.4 (Pumping Lemma)

Let M be an automaton over A such that

(i) M has k states $s_0, s_1, \ldots, s_{k-1}$
(ii) M accepts a word v from A, where $|v|>k$.

Then

$$v=x\,y\,z,$$

where, for every m, $v_m=x\,y^m z$ is accepted by M.

Proof. Let $s_0, s_1, \ldots, s_{k-1}$ be the states of automaton M over A and let M accept a word $v=a_1\,a_2 \ldots a_n$ over A such that $n>k$. Let the sequence of states determined by the word v be

$$P=(s_0, s_1, \ldots, s_n).$$

Since $n>k$, two of the states in P must be equal. Suppose $s_i=s_j$, $i<j$. Letting

$$x=a_1\,a_2 \ldots a_i,\quad y=a_{i+1}\,a_{i+2} \ldots a_j,\quad z=a_{j+1}\,a_{j+2} \ldots a_n,$$

we see that $x\,y$ ends in $s_i=s_j$ and so $x\,y^2$, $x\,y^3$, $\ldots$, xy^m (for all m) also end in s_i. Thus for every m, $v_m=x\,y^m\,z$ ends in s_n, which is an accepting state.

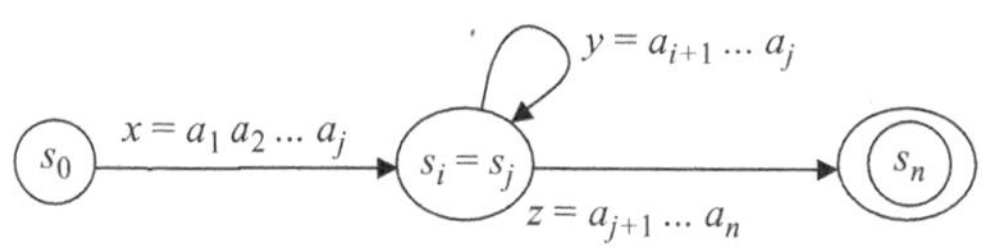

Figure 10.8 (Pumping Lemma)

EXAMPLE 10.17

Show that language $L=\{a^m\,b^m\colon m \text{ is positive}\}$ is not regular.

Solution.

Suppose on the contrary that L is regular. Then, by Kleene theorem, there exists a finite state automaton M which accepts $L=\{a^m\,b^m\colon m \text{ is positive}\}$. Suppose M has k states. Let $v=a^k\,b^k$ be a word. Then length of v is greater than k, the number of states in M. Therefore, by Pumping lemma,

$$v=x\,y\,z,\quad y\neq\lambda.$$

and $x\,y^2\,z$ is also accepted by M.

If y consists of only a's or only b's, then $v_2=x\,y^2\,z$ will not have same number of a's or b's. If y consists of both a's and b's, then v_2 will have a's following b's. In either case v_2 does not belong to L which is a contradiction. Thus L is not regular.

Definition 10.17

A context-free grammar G is called an **ambiguous grammar** if there is at least one string in $L(G)$ which has more than one derivation trees.

EXAMPLE 10.18

Show that grammar G with productions

$$S \to a\,S, \quad S \to S\,a, \quad S \to a$$

is ambiguous.

Solution.

We note that the string $a\ a\ a$ can be generated by four derivation trees shown in the Figure 10.9.

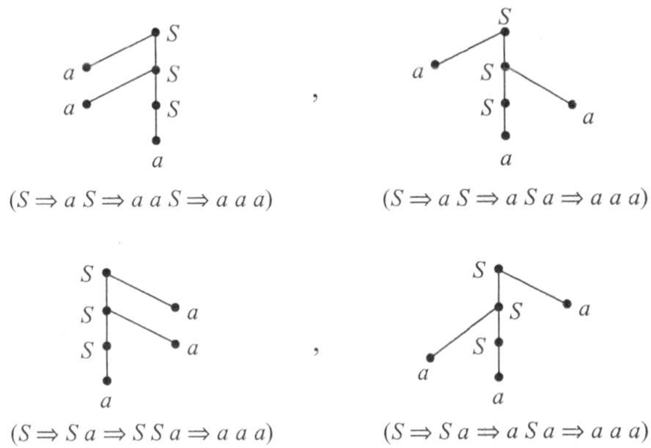

Figure 10.9

Hence G is ambiguous.

EXAMPLE 10.19

Show that the grammar $G=(\{S\}, \{a, +\}, S, P)$ with production $P=(S \to S+S, S \to a)$ is ambiguous.

Solution.

We note that word $a+a+a$ can be generated in two ways (shown in the Figure 10.10):

(i) $S \Rightarrow S+S$
$\Rightarrow S+S+S$
$\Rightarrow a+S+S \Rightarrow a+a+S$
$\Rightarrow a+a+a.$

(ii) $S \Rightarrow S+S$
$\Rightarrow S+S+S$
$\Rightarrow S+S+a$
$\Rightarrow a+S+a$
$\Rightarrow a+a+a.$

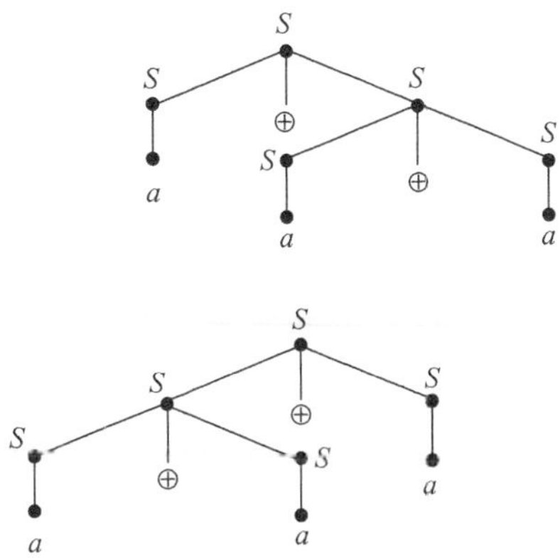

Figure 10.10

Thus, the word $a+a+a$ has two derivation trees. Hence G is ambiguous.

EXERCISES

1. Let $A=\{a, b\}$. Describe the language $L=\{ab^n: n>0\}$ and find a regular expression r such that $L=L(r)$.
2. If $r=(a \vee b \vee c)^*\, bbb$, describe the language $L(r)$.
3. What is represented by the following regular expression?

$$((a \backslash b).\, a)^*$$

4. The language $L(G)=\{a^n b^n c^n: n \geq 1\}$ is generated by the grammar $G=\{N, T, \sigma, P\}$, where $N=\{\sigma, A, B\}$, $T=\{a, b, c\}$, σ is the starting symbol and the productions P are

$\sigma \rightarrow a\,\sigma\, AB,\ \sigma \rightarrow a\, AB,\ BA \rightarrow AB,\ aA \rightarrow ab,$
$bA \rightarrow bb,\ bB \rightarrow bc,\ CB \rightarrow cc.$

Write the derivation for the string $a^2b^2c^2$.

5. Let $G=\{N, T, \sigma, P\}$, where $T=\{a, b\}$ and productions P are

$$\sigma \rightarrow a,\ \sigma \rightarrow \sigma a,\ \sigma \rightarrow b,\ \sigma \rightarrow b\sigma$$

Describe the set of strings generated by G.

6. Let $G=\{N, T, \sigma, P\}$, where $T=\{a, b\}$, $N=\{\sigma, c, D\}$ and productions P are

$\sigma \rightarrow a\, c\, D\, a,\ c \rightarrow b\, a\, c\, D\, b,\ D \rightarrow c\, a\, b,$
$ac \rightarrow b\, aa,\ b\, D\, b \rightarrow ab\, ab.$

Show that $ba\ a\ bb\ ab\ aaa\ bb\ a\ ba \in L(G)$.

7. A finite state automaton M is described by the Figure 10.11.

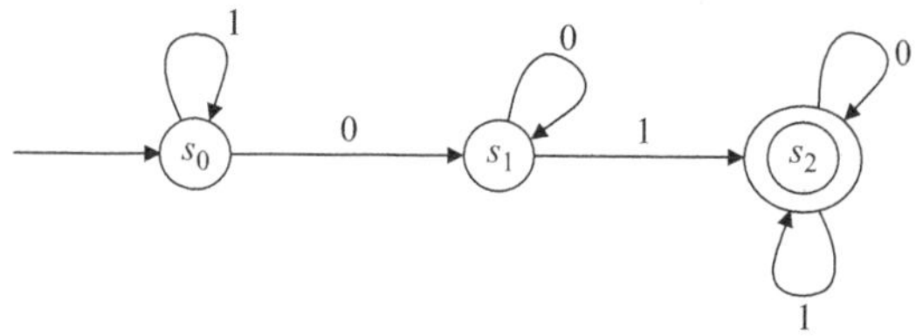

Figure 10.11

Find the corresponding grammar G and $L(G)$.

8. Let $G=\{N, T, \sigma, P)$ be a grammar, where $T=\{a, b\}$, $N=\{\sigma\}$, σ is the initial symbol and productions P are:

$$\sigma \rightarrow a\,\sigma\, b,\ \sigma \rightarrow ab.$$

Find the corresponding language $L(G)$.

Appendix

C PROGRAMMING FOR SOME ALGORITHMS

1. **To write a recursive function to print the factorial of a number and execute it to find the factorial of 6.**

```
#include <stdio.h>
#include <conio.h>
unsigned long fact(int);
void main()
{
      unsigned long f;
      int n;
      clrscr();
      printf("\nEnter a number: ");
      scanf("%d",&n);
      f=fact(n);
      printf("\nFactorial of %d is %ld",n,f);
      getch();
}
unsigned long fact(int n)
{
      unsigned long m;
      if(n==1)
            return(1);
      else
            m=n*fact(n-1);
      return(m);
}
```

The Execution of the program to find the factorial of 6 is yields:-

Enter a Number:
6
Factorial of 6 is 720

2. **To write a recursive function to print the Fibonacci series up to a given number and write the series up to 89.**

```
#include <stdio.h>
#include <conio.h>
unsigned long fib(int);
void main()
{
      int n,i;
      unsigned long f;
      clrscr();
      printf("\nEnter a number: ");
      scanf("%d",&n);
      printf("\nThe Fibonacci series upto %d Numbers
      is:\n",n);
      for(i=0;i<n;i++)
      {
            f=fib(i);
            printf("%lu ",f);
      }
      getch();
}
unsigned long fib(int n)
{
      unsigned long res;
      if(n==0)
            return(0);
      else
            if(n==1)
                  return(1);
            else
            {
                  res=fib(n-1)+fib(n-2);
                  return(res);
            }
}
```

The Execution of the program yields:-

Enter a number:
89
The Fibonacci series upto the number 89 is
0 1 1 2 3 5 8 13 21 34 55 89

3. **To write a program to implement the Tower of Hanoi.**

```
#include <stdio.h>
#include <conio.h>
Void transfer(int n,char from,char to,char temp);
void main()
{
      int n;
      clrscr();
      printf("Tower of Hanoi ");
      printf("\nEnter the Number of Disks");
      scanf("%d",&n);
```

```
    transfer(n,'O', 'D','T');
    getch();
}
transfer(int n,char from,char to,char temp)
{
/* Transfer of disks from one pole to another*/
/* n= number of disks
   From=origin
   To=destination
   Temp=temporary storage */
    if(n>0)
    {
        transfer(n-1,from,temp,to);
    /* move nth disk from origin to destination */
        printf("move disk %d from %c to
        %c\n",n,from,to);
    /* move n-1 disk from temporary to destination */
    transfer(n-1,temp,to,from);
    }
return;
}
```

Execution of program for n=3 yields:

Tower of Hanoi
Enter the Number of Disks 3
Move disk 1 from O to D
Move disk 2 from O to T
Move disk 1 from D to T
Move disk 3 from O to D
Move disk 1 from T to O
Move disk 2 from T to D
Move disk 1 from O to D

4. To write a program to find hamming distance between two binary codes.

```
\* Hamming distance between two binary codes*\
#include<stdio.h>
#include<conio.h>
void main()
{
int count=0,x,y,j,i=1,n,rem1,rem2,rem3[10],temp=0;
printf("enter the length of binary no.");
scanf("%d",&n);
printf("enter two binary numbers of length %d",n);
scanf("%d%d",&x,&y);
while(x!=0)
{
rem1=x%10;
x=x/10;
rem2=y%10;
y=y/10;
if(rem1==1 &&rem2==1)
{
rem3[i]=0+temp;
temp=1;
}
else
{
rem3[i]=rem1+rem2+temp;
temp=0;
}
i++;
}
if(temp==1)
{
rem3[i]=temp;
for(j=i;j>=1;j--)
{
if(rem3[j]==1)
count++;
printf("%d",rem3[j]);
}
printf("weight is %d",count);
}
else
{
for(j=i-1;j>=1;j--)
{
if(rem3[j]==1)
count++;
printf("%d",rem3[j]);
}
printf("weight is %d",count);
}
getch();
}
```

5. To write a program to find chromatic number of a given graph.

```
\*Chromatic number of a graph*\
#include<stdio.h>
#include<conio.h>
int mcolor(int);
int nextvalue(int);
int max=0,k=1,m=4,x[10],g[10][10],n,i,j;
void main()
{
clrscr();
printf("enter no of vertices");
scanf("%d",&n);
printf("enter adjacency matrix");
for(i=1;i<=n;i++)
for(j=1;j<=n;j++)
scanf("%d",&g[i][j]);
```

```
for(i=1;i<=n;i++)
x[i]=0;
mcolor(k);
for(i=1;i<=n;i++)
if(max<x[i])
max=x[i];
printf("\nchromatic no is %d",max);
}
mcolor(int k)
{
nextvalue(k);
if(x[k]==0)
{
printf("insufficient no of colors");
exit(1);
}
if(k==n)
for(j=1;j<=n;j++)
printf("%d",x[j]);
else
mcolor(k+1);
}
nextvalue(int k)
{
x[k]=(x[k]+1)%(m+1);
if(x[k]==0)
return 0;
for(j= 1;j<=n;j++)
{
if((g[k][j]!=0)&& (x[k]==x[j]))
break;
}
if(j==n+1)
return 0;
else
nextvalue(k);
}
```

6. To write a program to find path matrix by Warshall's algorithm.

```
/* Program to find path matrix by Warshall's algorithm
*/
#include<stdio.h>
#include<conio.h>
void main()
{
int i,j,k,n;
int adj[10][10],path[10][10];
clrscr();
printf("Enter number of vertices : ");
scanf("%d",&n);
printf("Enter  adjacency matrix :\n");
for(i=0;i<n;i++)
for(j=0;j<n;j++)
scanf("%d",&adj[i][j]);
printf("The adjacency matrix is :\n");
for(i=0;i<n;i++)
{
for(j=0;j<n;j++)
printf("%d",adj[i][j]);
printf("\n");
}
for(i=0;i<n;i++)
for(j=0;j<n;j++)
path[i][j]=adj[i][j];
for(k=0;k<n;k++)
{
for(i=0;i<n;i++)
{
for(j=0;j<n;j++)
path[i][j]=( path[i][j] || ( path[i][k] && path[k][j] ) );
}
}
printf("The Path matrix is :\n");
for(i=0;i<n;i++)
{
for(j=0;j<n;j++)
printf("%d",path[i][j]);
printf("\n");
}
getch();
}
```

7. To write a programme to find minimum spanning tree by Prim's algorithm.

```
/* Program to find minimum spanning tree by prim's
algorithm */
#include<stdio.h>
#include<conio.h>
#define T 1
#define F 0
void main()
{
int k,min,i,j,n,edge;
int graph[10][10],mst[10][10],output[10];
clrscr();
printf("Enter number of vertices : ");
scanf("%d",&n);
printf("Enter  adjacency matrix :\n");
for(i=1;i<=n;i++)
for(j=1;j<=n;j++)
scanf("%d",&graph[i][j]);
printf("The adjacency matrix is :\n");
for(i=1;i<=n;i++)
```

```
{
for(j=1;j<=n;j++)
printf("%d",graph[i][j]);
printf("\n");
}
for(i=1;i<=n;i++)
{
for(j=1;j<=n;j++)
if(graph[i][j]==0)
graph[i][j]=32767;
}
for(i=1;i<=n;i++)
{
for(j=1;j<=n;j++)
printf("%d",graph[i][j]);
printf("\n");
}
for(i=1;i<=n;i++)
output[i]=F;
for(i=1;i<=n;i++)
for(j=1;j<=n;j++)
mst[i][j]=0;
output[1]=T;
edge=1;
while(edge<n)
{
min=32767;
for(i=1;i<=n;i++)
{
if(output[i]==T)
for(j=1;j<=n;j++)
{
if(output[j]==F)
if(min>graph[i][j])
{
min=graph[i][j];
k=j;
}
}
mst[i][k]=1;
output[k]=T;
}
edge=edge+1;
}
printf("The MST is :\n");
for(i=1;i<=n;i++)
{
for(j=1;j<=n;j++)
printf("%d",mst[i][j]);
printf("\n");
}
getch();
}
```

8. To Write a Program to find minimal spanning tree of a given graph using Kruscal's Algorithm.

```
#include<stdio.h>
#include<conio.h>
void main()
{
int  sol[20],i1,j1,min=32766,a[10][10],i,j,n,k=1,count=0,
temp,flag1=0,flag2=0;
printf("enter no of vertices");
scanf("%d",&n);
printf("enter adjacency matrix");
for(i=1;i<=n;i++)
for(j=1;j<=n;j++)
scanf("%d",&a[i][j]);
for(i=1;i<=n;i++)
for(j=1;j<=n;j++)
if(a[i][j]==0)
a[i][j]=-1;
for(temp=1;temp<=2*n;temp++)
sol[temp]=0;
while(count<n-1)
{
min=32766;
flag1=flag2=0;
for(i=1;i<=n;i++)
for(j=1;j<=n;j++)
if(min>a[i][j] && a[i][j]!=-1)
{
min=a[i][j];i1=i;j1=j;
}
a[i1][j1]=a[j1][i1]=-1;
for(temp=1;temp<=2*n;temp++)
if(sol[temp]==i1)
flag1=1;
for(temp=1;temp<=2*n;temp++)
if(sol[temp]==j1)
flag2=1;
if(flag1==1&&flag2==1)
continue;
sol[k]=i1;
k++;
sol[k]=j1;
k++;
count++;
}
for(i=1;i<=k;i=i+2)
printf("%d%d\n",sol[i],sol[i+1]);
}
```

Answers to Exercises

CHAPTER 1

2. 5

3. 252

4. (i) No (ii) No (iii) Yes (iv) boundary of a unit circle and the relation C is called **Circle Relation**.

5. 2^{36}, 2^6

(*Hint:* Number of elements in $X \times X$ is 36 and therefore number of possible subsets is 2^{36}. There are 6 elements of the form (a, a) and so the number of reflexive relations is 2^6.)

7. The relation is not anti-symmetric

8. $R = \{(1, 1), (2, 2), (3, 3), (4, 4)\}$

9.

$$\begin{bmatrix} 0 & 1 & 0 & 0 & 0 \\ 0 & 0 & 1 & 0 & 0 \\ 0 & 0 & 0 & 1 & 0 \\ 0 & 0 & 0 & 0 & 1 \\ 0 & 0 & 0 & 0 & 0 \end{bmatrix}$$

10.

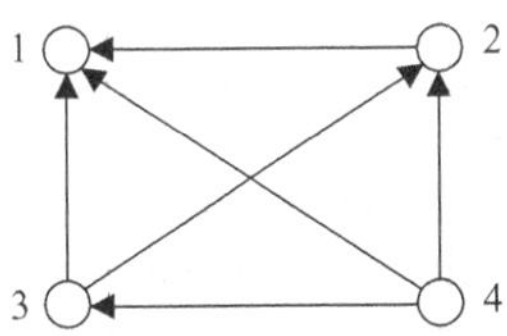

11.

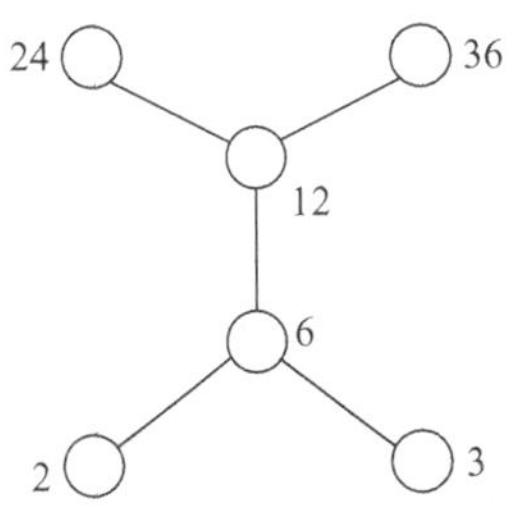

12. (i), since (ii) has a cycle of length 2

13. *Hint:* $g \circ f = I = f \circ g$

14. No, since f is not one-to-one

15. $R^\infty = \{(1, 1), (1, 2), (1, 3), (1, 4), (2, 1), (2, 2), (2, 3), (2, 4), (3, 4)\}$

16.

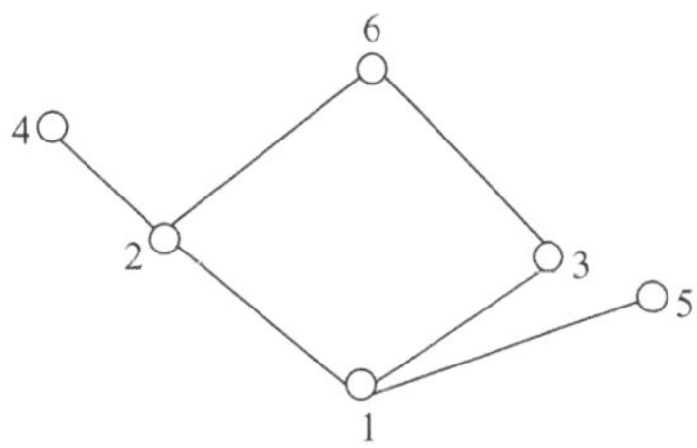

17. $R_1 \circ R_2 = \{(1, 1), (1, 2), (2, 1), (2, 2), (3, 1), (3, 2), (4, 2)$

$R_2 \circ R_1 = \{(1, 1), (1, 2), (3, 4), (4, 1), (4, 2)\}$

18. In the digraph of the relation for any vertices v and w, there is a directed edge from v to w

19.

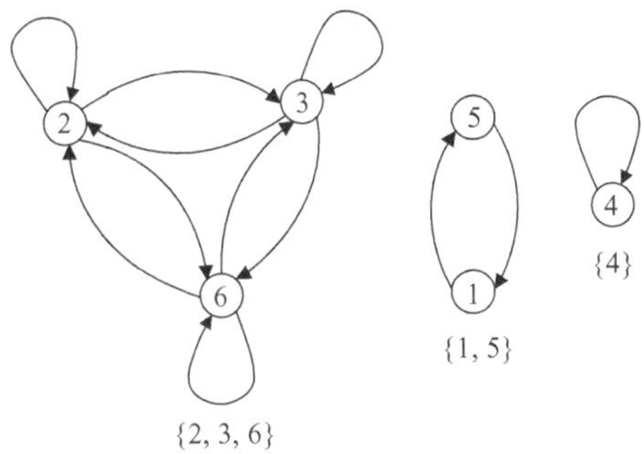

CHAPTER 2

3. 3

4. 27

(*Hint:* In a pack, there are 26 red cards. So, if we draw 27 cards, then certainly at least one of them shall be black).

5. 90

(*Hint:* Let $A = \{1, 2, \ldots, 9\}$, $B = \{10, 11, \ldots, 99\}$. All integers in A have no repeated digits. So there are nine such integers in A. In B, we have two digit numbers. The first digit can be selected in nine ways from 1, 2, …, 9, whereas the second digit can be any of the digit out of 0, 1, 2, …, 9 except the digit which has been taken as first digit. Hence the second digit can be selected in nine ways.

By multiplication rule B has $9 \times 9 = 81$ integers without repeated digits. Hence the required number of digits is $9 + 81 = 90$).

6. 333

9. $\frac{1}{18}$

10. $\frac{1}{5}$

12. $\frac{11}{18}$

13. $\frac{1}{3}$

(*Hint:* Use $P(A \cup B) = P(A) + P(B) - P(A \cap B)$ and $P(A \cap B) = P(A) \bullet P(B)$).

14. $\frac{21}{40}$

15. (i) $\frac{473}{729}$ (ii) $\frac{496}{729}$

16. $\frac{20}{3}$, $\sqrt{\frac{40}{9}}$

(*Hint:* $p = \frac{1}{3}, q = 1 - \frac{1}{3} = \frac{2}{3}, n = 20.$

Find n p and $\sqrt{npq}$)

CHAPTER 3

1. $S_n = 2^n$

2. $a_n = 3 + \frac{n(n+1)}{2}$

3. $a_n = 1 + 3n,\ n \geq 0$

4. $a_n = 2^n - n\, 2^{n-1}$

5. $\frac{1}{4}n^2 + \frac{13}{24}n + \frac{71}{288} + u(-3)^n + v(-2)^n$

6. $u(3)^n + v(2)^n + \frac{1}{2}$

7. $\left(2n + \frac{2}{3}\right)2^n$

9. 5, 9

10. $a_n = (-1)^n + (3)^n, n \geq 0$

12. $a_n = 3 \bullet 2^n + 7 \bullet 5^n$

CHAPTER 4

2. (i) If a figure is a square, then it is a rectangle
   (ii) If I live in Bangalore, then I am a citizen of India
   (iii) If Pakistanis win cricket match, then they have at least two left-handed batsmen

3. (i) Some rectangles are squares
   (ii) He takes his bath if and only if the water is not cold
   (iii) If it rains and they go for a walk

4. (a) The logical expression for the argument is

$$p \to q$$
$$p$$
$$\therefore q,$$

which is valid by Modus Ponens

   (b) In logical symbols, the premises and conclusion are

$$p \to q$$
$$\sim q$$
$$\therefore \sim p,$$

which is valid by Modus Tollen

5. **Converse:** If $|AB|^2 + |BC|^2 = |AC|^2$, then triangle ABC is a right triangle
   Inverse: If ABC is not a right triangle, then $|AB|^2 + |BC|^2 \neq |AC|^2$
   Contrapositive: If $|AB|^2 + |BC|^2 \neq |AC|^2$, then triangle ABC is not a right triangle

7. The argument is valid. The logical expression of the argument is

$$p \vee q$$
$$p \to r$$
$$q \to r$$
$$\therefore r$$

This is because $(p \vee q) \wedge (p \to r) \wedge (q \to r) \to r$ is a tautology

8. Let

   p: Julia live in Italy
   q: Julia speaks Italian
   r: Julia drives a car
   s: Julia travels by a train

Then the logical form of the given statement is: (i) $\sim p \to \sim q$ (ii) $\sim r$ (iii) $p \to s$ (iv) $\sim q \to r$

The following deductions can be made:

(i) $\sim q \rightarrow r$ by (iv) (ii) $\sim p \rightarrow \sim q$ by (i)

$\sim r$ by (ii) q by conclusion of (i)

$\therefore q$ $\therefore p$

(iii) $p \rightarrow s$ by (iii)

p by conclusion of (ii)

$\therefore s$

Hence Julia travels by a train.

9. We note that

(a) $q \rightarrow r$ and (b) $p \rightarrow q$

$\sim r$ $\sim q$ by (*a*)

$\therefore \sim q$ $\therefore \sim p$

Hence the argument is valid

10. (i) The graphs are not isomorphic

(ii) It did not rain

11. Let

$C(x)$: x is complete

$B(x)$: x is bipartite

$P(x)$: x is planar

Then the given statement is

$\forall x, [C(x) \wedge B(X) \rightarrow \sim P(X)]$

Using the result $\sim(p \rightarrow q) \equiv p \wedge \sim q$, the negation is

$\forall x, [C(x) \wedge B(x) \wedge P(x)]$

12. $\exists$ even integers m and n such that $24 = m + n$

CHAPTER 5

5. *Hint:* The composition table is

$\odot_{14}$	2	4	8
2	4	8	2
4	8	2	4
8	2	4	8

7. *Hint:* Suppose on the contrary, $Q = \langle q \rangle$, where $q = {}^{m}\!/_{n}$, $n \neq 0$ and m, n are integers. If $x \in Q$, then $x = qp$ for some integer p. Take $x = {}^{1}\!/_{2n}$, then ${}^{1}\!/_{2n} = qp = p\,{}^{m}\!/_{n}$ and so ${}^{1}\!/_{2} = mp$ (an integer) which is absurd. Hence Q is not cyclic under addition.

8. *Hint:* $U_9 = \{1, 2, 4, 5, 7, 8\}$. Thus, $O(U_9) = 6$. The positive integers less than 6 and prime to 6 are 1 and 5. Hence number of generators is two. The composition table is

	1	2	4	5	7	8
1	1	2	4	5	7	8
2	2	4	8	1	5	7
4	4	8	7	2	1	5
5	5	1	2	7	8	4
7	7	5	1	8	4	2
8	8	7	5	4	2	1

We have

$2^1 = 2, 2^2 = 2 * 2 = 4, 2^3 = 2 * 2^2 = 2 * 4 = 8$

$2^4 = 2 * 2^3 = 2 * 8 = 7, 2^5 = 2 * 2^4 = 2 * 7 = 5$

$2^6 = 2^3 * 2^3 = 8 * 8 = 1$

Hence 2 is a generator of U_9 and so $U_9 = \langle 2 \rangle$.

The other generator is $2^5 = 5$.

11. Yes (being groups of prime order)

12. *Hint:* $x \in A_4$, $a \in V$ must imply $x\ a\ x^{-1} \in V$.

13. *Hint:* Show that $x, y \in N(a) \Rightarrow xy \in N(a)$ and $x \in N(a) \Rightarrow x^{-1} \in N(a)$.

14. 14, $\overline{4}$, $\overline{2}$ and $\overline{2}$

Hint: The integers relatively prime to 6 are units of $\mathbf{Z}_6$. So units are $\overline{1}$, $\overline{5}$.

Find additive inverse and multiplicative inverses from the composition table.

16. *Hint:* Use ontoness of the mapping f and then use homomorphism.

18. *Hint:* $x, y, e \in R \Rightarrow y - e - yx \in R$. Then using $xy = e$ show $x(y + e - yx) = e$. Then uniqueness of y implies $y + e - yx = y$ or $yx = e$. Thus, $xy = yx = e$ and so x is invertible.

19. (a) 2 (b) 2 (c) 3

20. (a) 3 (b) 3 (c) 3

21. 2

23. *Hint:* We note that $e(001) = 00000$, $e(000) = 00000$. Thus two elements 000 and 001 are mapping on to the same element. Hence e is not one-one and so cannot be an encoding function.

24. 1

25. 1

26. $e_H(B^2) = \{00000, 01011, 10011, 11000\}$

27. $e(000) = 000000$, $e(001) = 001011$, $e(010) = 010101$

$e(011) = 011110$, $e(100) = 100110$, $e(101) = 101101$

$e(110) = 110011$, $e(111) = 111000$

corrected code is 001011 and decoded word is 001.

28. 101

CHAPTER 6

1. It is a lattice and the Hasse diagram is

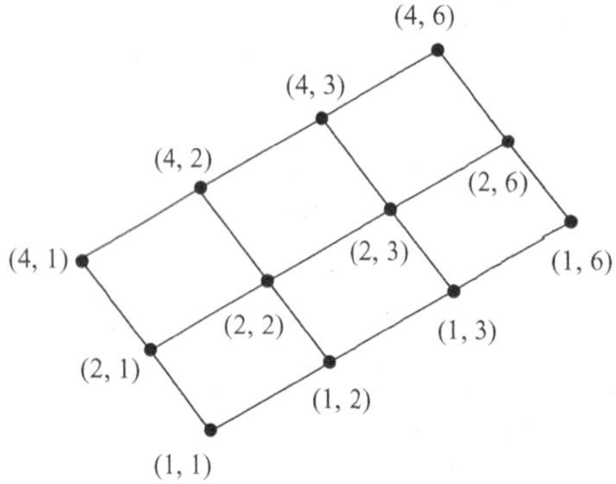

2. No, because glb (d, e) does not exist.
3. It is bounded, the bounds being 1 and 30. Further every element has complement. In fact, the complements are $1 \Leftrightarrow 30$, $2 \Leftrightarrow 15$, $3 \Leftrightarrow 10$, $5 \Leftrightarrow 6$.
4. (i) (0, 1, 0), (1, 1, 0), (1, 1, 1)
   (ii) (1, 0, 0), (0, 0, 1), (0, 1, 0)
   (iii) (1, 0, 0), (0, 1, 0), (0, 0, 1)
5. It is a lattice and the Hasse diagram is

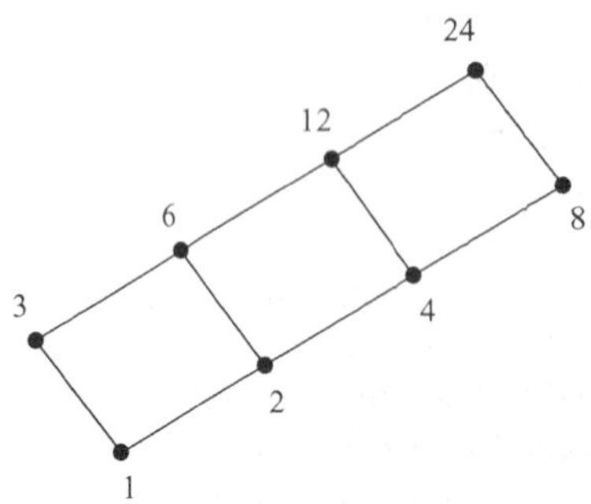

D_{24}

The atoms are 2 and 3. Join-irreducible elements are 2, 3, 8.

6.

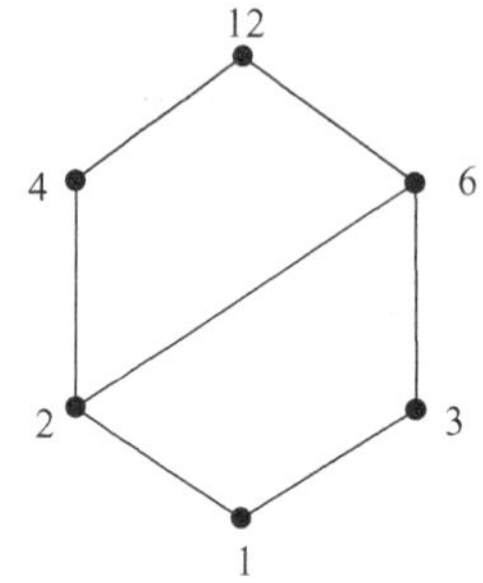

D_{12} (Distributive)

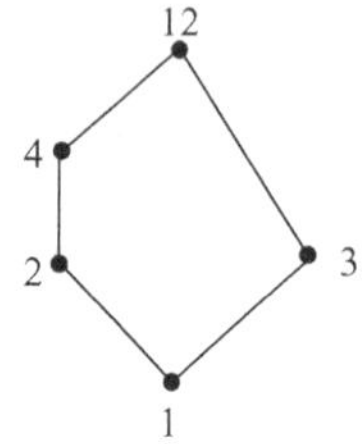

$(A, \leq)$ (Non-distributive)

In fact, for example,

$2 \wedge (3 \vee 6) = 2 \wedge 6 = 2$

and

$(2 \wedge 3) \vee (2 \wedge 6) =$
$1 \vee 2 = 2$

$2 \vee 3 = 12$
$4 \wedge (2 \vee 3) = 4 \wedge 12 = 4$
But $(4 \wedge 2) \vee (4 \wedge 3)$
$= 2 \vee 1 = 2$
Thus,
$4 \wedge (2 \vee 3) \neq (4 \wedge 2) \vee (4 \wedge 3)$

8. {1, 2, 3, 6}, {2, 4, 6, 12}

CHAPTER 7

1. No, since 60 cannot be expressed as product of distinct primes
2. No. In fact $D_{72} = \{1, 2, 3, 4, 6, 8, 9, 12, 18, 24, 36, 72\}$ and as such the number of elements in D_{72} is 12. But $12 \neq 2^n$, $n \geq 1$.
3. $D_{30} = \{1, 2, 3, 5, 6, 10, 15, 30\}$. The Hasse diagram for D_{30} is shown below.

The set of atoms of D_{30} is $A = \{2, 3, 5\}$ and we see that

$10 = 2 \vee 5$
$6 = 2 \vee 3$
$15 = 3 \vee 5$
$30 = 2 \vee 3 \vee 5$

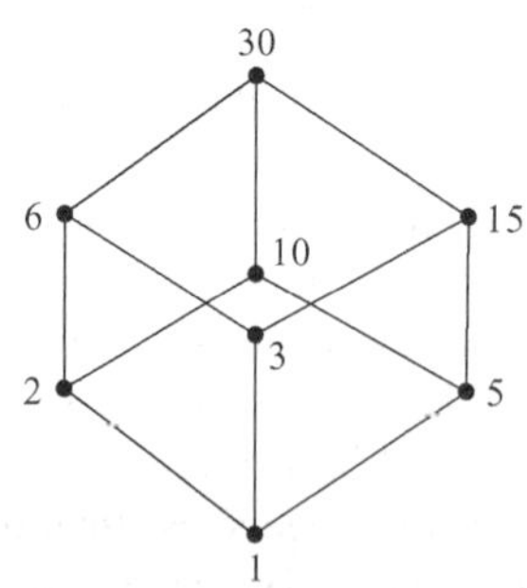

D_{30}

The Hasse diagram of $P(A)$ is:

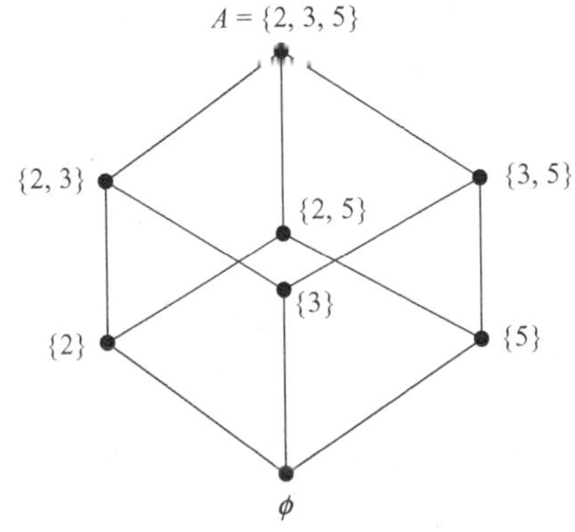

$P(A)$

Thus D_{30} and $P(A)$ are structurally the same.

4. We note that $S(f) = \{(1, 1, 1), (1, 1, 0), (1, 0, 0), (0, 1, 1), (0, 1, 0)\}$
Hence the Boolean expression for the function f is

$$E(x, y, z) = E_{(1,1,1)} \vee E_{(1,1,0)} \vee E_{(1,0,0)} \vee E_{(0,1,1)} \vee E_{(0,1,0)}$$

$$= xyz + xyz' + xy'z' + x'yz + x'yz'$$

which is already in disjunctive normal form.

5. We have $f(x_1, x_2, x_3) = (x_1 \wedge x_2') \vee (x_1 \wedge x_3)$
The special sequences for x_1, x_2, x_3 are
$x_1 = 00001111, \quad x_2 = 00110011,$
$z = 01010101, \quad x_2' = 11001100.$
Thus, $x_1 \wedge x'_2 = 00001100$ and $x_2 \wedge x_3 = 00000101$.
Hence the sum, which is nothing but $f(x_1, x_2, x_3)$, is
$f(x_1, x_2, x_3) = 00001101$
Thus, $T(L) = T(00001111, 00110011, 010101101) = 00001101$

6. $t = xy + yz'$

7. The logic circuit represent the Boolean expression $xy + x'y + x'y'$. The K-map for this expression is

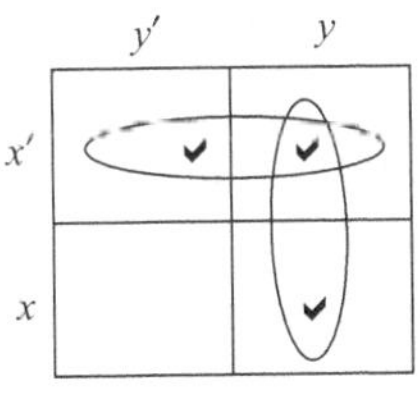

The prime implicants of the expression are x' and y. Thus the minimum sum is $x' + y$ and so the required logical minimal circuit is

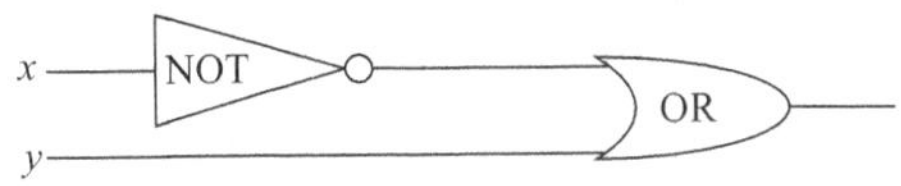

8. $E = xy' + xzt' + x'z't + y'z't$

9. $xyz + xy'z' + x'y'z'$

10.

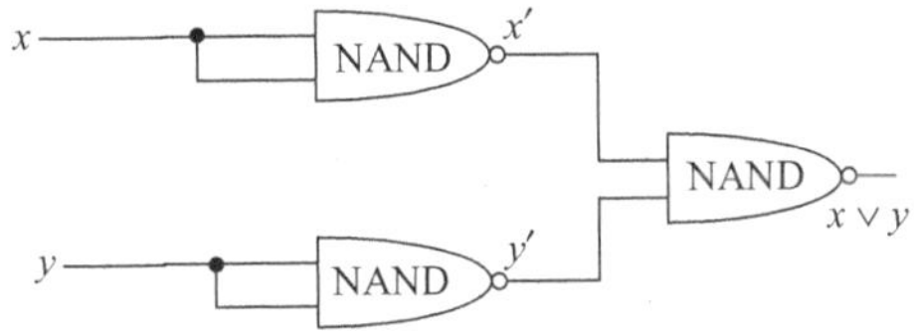

11.

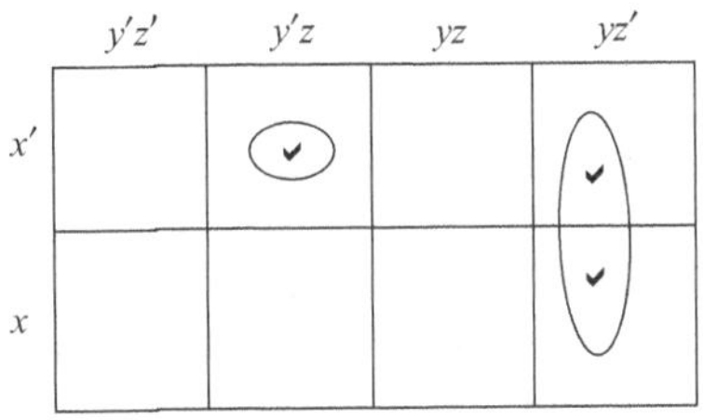

Minimal expression: $x'y'z + yz'$

CHAPTER 8

1. Yes. In fact if n = number of vertices $2n$ = number of edges, then using the formula for number of edges in a complete graph: $\frac{n(n-1)}{2}$, we have $2n = \frac{n(n-1)}{2}$ which yields $n = 5$. But in a complete graph K_5 the number of edges is 10. Thus the complete graph K_5, which is simple by definition, has five vertices and ten edges.

2. Yes, it is a bipartite graph, where the set of vertices V has been partitioned into two subsets, $V_1 = \{v_3, v_4\}$, $V_2 = \{v_1, v_2, v_5\}$. But it is not a complete bipartite graph.

3. The $m + n$ vertices having been divided into two subsets V_1 and V_2 containing m vertices and n vertices, respectively. Every verticex in V_1 is joined to all the n vertices in V_2. Thus the task of joining each vertex of V_1 to each vertex in V_2 can be performed in mn ways. Hence the number of edges in K_{mn} is mn.

4. Total degree of G is $1 + 4 + 3 + 7 + 3 + 2 = 20$ which should be twice the number of edges. Hence number of edges in G is 10.

5. Degree of each vertex in an n-cube in n. The number of vertices in an n cube in 2^n. Therefore total degree of n-cube in $n2^n$. Hence the number of edges is an n-cube is

$$\frac{n \cdot 2^n}{2} = n \cdot 2^{n-1}.$$

6. Complete graph is a connected graph. So it will have an Euler cycle if degree of each vertex is even.

7. Yes. The graph is connected and degree of each vertex is even. Hence the graph possesses an Euler circuit. For example, $v_1\, v_2\, v_5\, v_4\, v_5\, v_2\, v_3\, v_4\, v_1$ is an Euler cycle.

8. When both m and n are even. In such a case the degree of each vertex will be even.

9. Yes, because in such a case degree of each vertex of the graph of seven bridges problem shall be even.

11.

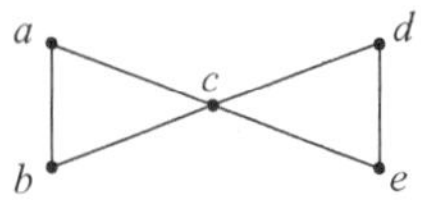

12.

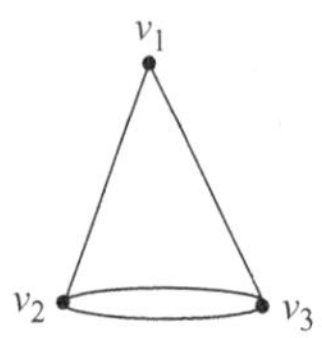

13.

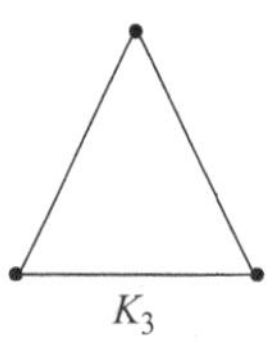

14. 6.7

15.

$$\begin{bmatrix} 0 & 1 & 1 & 0 & 0 \\ 1 & 0 & 0 & 1 & 1 \\ 1 & 0 & 0 & 1 & 1 \\ 0 & 1 & 1 & 0 & 1 \\ 0 & 1 & 1 & 1 & 0 \end{bmatrix}$$

16.

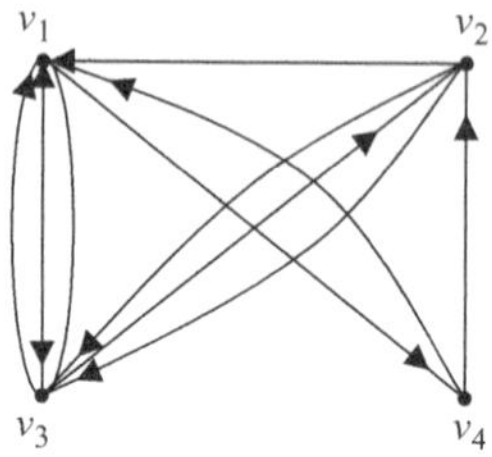

17. The graph G is connected because $B = A + A^2 + A^3 + A^4$ has non-zero entries off the main diagonal.

18. Vertex set of G_1 is $V(G_1) = \{a, b, c, d\}$ vertex set of G_2 is $V(G_2) = \{x, y, z, w\}$

Define $f: V(G_1) \to V(G_2)$ by

$f(a) = y, f(b) = w, f(c) = x, f(d) = z.$

Then f is bijective.

19. Graph G_2 is connected whereas G_1 is not connected. The connected components of G_1 are

20. The graphs do not have same number of vertices and so are not isomorphic.

21. The complement of the given graph is

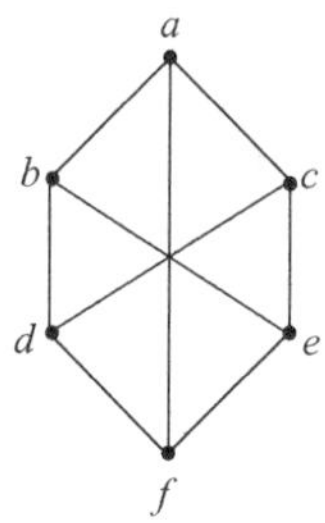

22. We note that

(i) $d(v_1, v_2) = 1, d(v_1, v_3) = 2, d(v_1, v_4) = 3$ and so $e(v_1) = 3$

(ii) $d(v_2, v_1) = 1, d(v_2, v_3) = 1, d(v_2, v_4) = 2$ and so $e(v_2) = 2$

(iii) $d(v_3, v_1) = 2, d(v_3, v_2) = 1, d(v_3, v_4) = 1$ and so $e(v_3) = 2$

(iv) $d(v_4, v_1) = 3, d(v_4, v_2) = 2, d(v_4, v_3) = 1$ and so $e(v_4) = 3$

Hence rad $(G) = \min \{3, 2, 2, 3\} = 2$. The centre is $\{v_2, v_3\}$.

24. Give one colour to A and D, give second colour to B, E and G, and third colour to C and F. Thus all the adjoining boundaries have different colours.

25. Three. We can paint v_1, v_2, v_3 with colours C_1, C_2, C_3. Then paint v_4 by C_1, v_5 by C_2, v_6 by C_3 and v_7 by C_1.

26. Number of edges in $K_{mn} = mn$. Total number of vertices in $K_{mn} = m+n$. If k_{mn} is a tree, then number of edges in K_{mn} should be $(m+n)-1$. Hence, to be a tree k_{mn} satisfies the condition

$$mn = (m+1)-1 \quad \text{or} \quad m+n = mn+1,$$

which is satisfied if either $m=1$ or $n=1$.

27. Suppose G is connected. The graph itself will be a tree or it can be made a tree by eliminating some edges. In either case it must have $n-1$ edges. But in the questions, number of edges is less than or equal to $n-2$. Hence G cannot be connected.

28.

$$\begin{bmatrix} \infty & 1 & 4 & 1 & 3 & 3 \\ 1 & \infty & 2 & \infty & 4 & \infty \\ 4 & 2 & \infty & 2 & \infty & \infty \\ 1 & \infty & 2 & \infty & 4 & 5 \\ 3 & 4 & \infty & 4 & \infty & 3 \\ 3 & \infty & \infty & 5 & 3 & \infty \end{bmatrix}$$

The diagonal is $\infty\ \infty\ \ldots\ \infty$ and the matrix is symmetric with respect to the diagonal.

29.

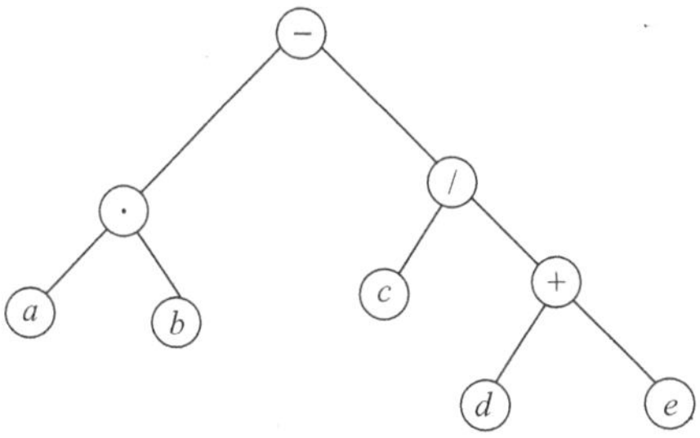

Prefix form: $-\cdot ab - c + de$
Postfix form: $ab \cdot de + c/-$

30.

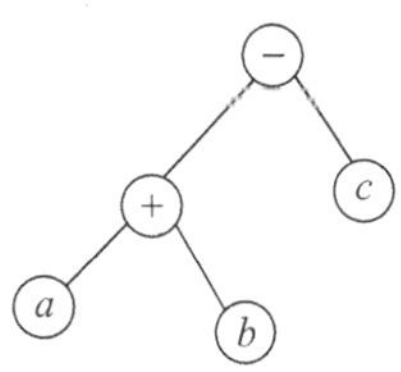

Prefix form: $-+abc$
Parenthesized form: $(a+b)-c$
Usual infix form: $a+b-c$

31. Using $\uparrow$ for exponentiation and $*$ for multiplication, the binary tree representing the algebraic expression is

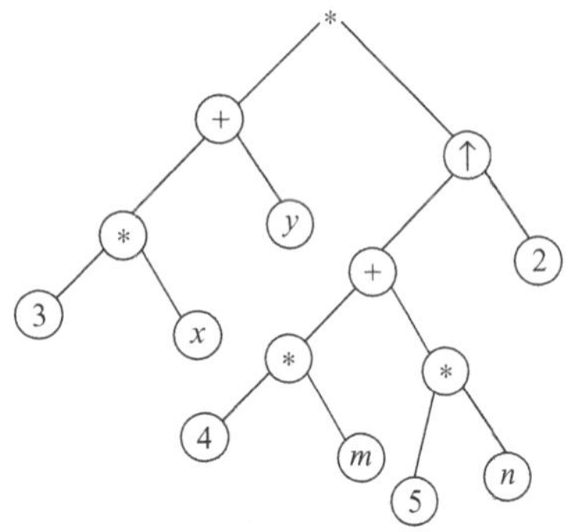

The prefix form is $* + * 3xy \uparrow + * 4m * 5n$

$$\begin{aligned} 232* \uparrow + 233 - 24 &= * \uparrow (2+3)\ 3\ (2-4) \\ &= * (2+3)^3 (2-4) \\ &= (2+3)^3 * (2-4) \\ &= 5^3 * (-2) \\ &= -250 \end{aligned}$$

33. $\{c, e\}$; $\{\{a, b\}, \{a, d\}\}$; $\{\{a, d\}, \{b, c\}\}$; $\{\{e, f\}, \{e, g\}\}$ $\{h, g\}\}$; $\{g, i\}$; $\{\{e, h\}, \{h, g\}\}$. Try for other cut edges.

34.

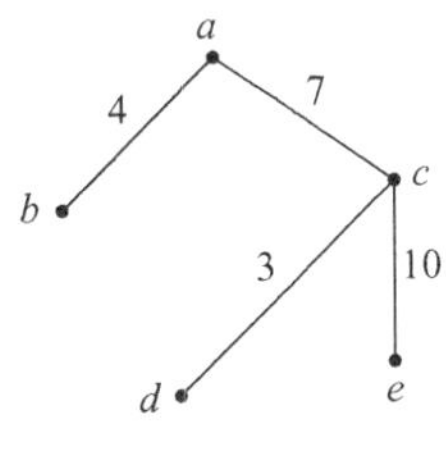

35. First remove $\{a, d\}$, then $\{d, c\}$, $\{a, e\}$, $\{b, e\}$. We cannot remove $\{c, e\}$ now because the graph will be disconnected by doing so. But we can delete $\{b, d\}$. Now delete any of $\{a, c\}$ or $\{b, c\}$. We will get two minimal trees.

36. Each spanning tree must have $n-1 = 4-1 = 3$ edges. There are eight such trees.

37. Value (F) for max flow$=7$, minimum cut (k) is shown in Figure 8.248.

38. Max flow$=6$.

CHAPTER 9

1.

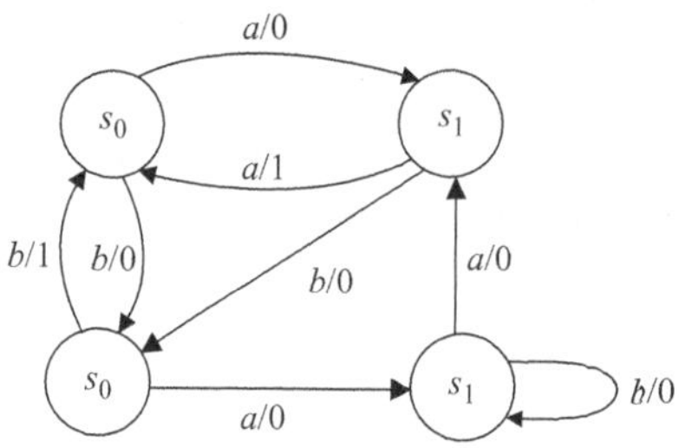

2. $I = \{a, b\}$, $O = \{0, 1\}$, $S = \{s_0, s_1, s_2\}$, initial state: s_0, the transition table is

	f		g	
S \ I	a	b	a	b
s_0	s_0	s_1	0	1
s_1	s_0	s_2	0	1
s_2	s_2	s_0	1	0

3.

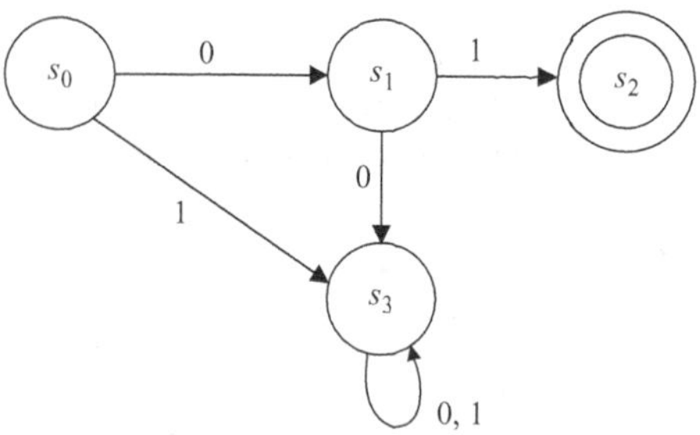

4.

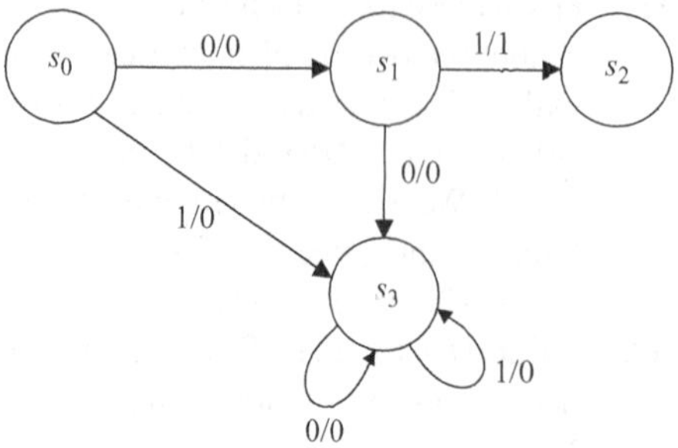

5.

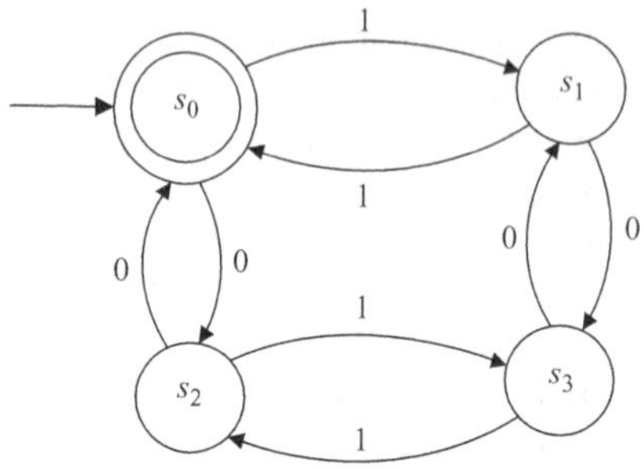

6. Yes. In fact

$$s_0 \xrightarrow{1} s_1 \xrightarrow{1} s_0 \xrightarrow{0} s_2 \xrightarrow{1}$$
$$s_3 \xrightarrow{0} s_1 \xrightarrow{1} s_0 \text{ (accepting state)}$$

7.

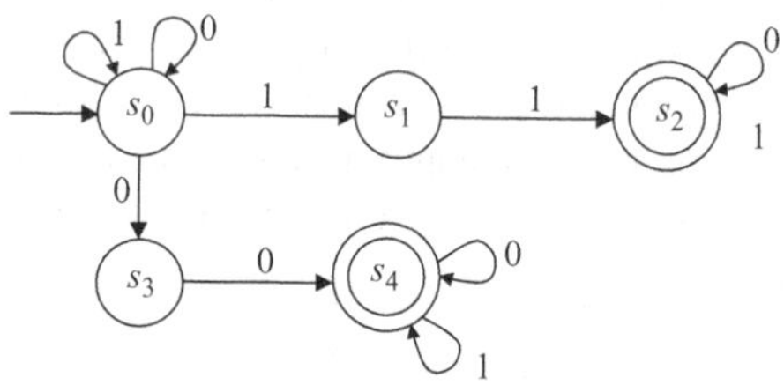

8. Yes. In fact,

$$f(s_0, 0) = \{s_0, s_3\}$$
$$\begin{aligned} f(s_0, 01) &= f(f(s_0, 0), 1) = f(\{s_0, s_3\}, 1) \\ &= f(s_0, 1) \cup f(s_3, 1) \\ &= \{s_0, s_1\} \end{aligned}$$

$$\begin{aligned} f(s_0, 010) &= f(f(s_0, 01), 0) = f(\{s_0, s_1\}, 0) \\ &= f(s_0, 0) \cup f(s_1, 0) \\ &= \{s_0, s_3\} \cup \phi \\ &= \{s_0, s_3\} \end{aligned}$$

$$\begin{aligned} f(s_0, 0100) &= f(f(s_0, 010), 0) = f(\{s_0, s_3\}, 0) \\ &= f(s_0, 0) \cup f(s_3, 0) \\ &= \{s_0, s_3\} \cup \{s_4\} \\ &= \{s_0, s_3, s_4\} \end{aligned}$$

$$\begin{aligned} f(s_0, 01001) &= f(f(0100, 1) = f(\{s_0, s_3, s_4\}, 1) \\ &= f(s_0, 1) \cup f(s_3, 1) \cup f(s_4, 1) \\ &= \{s_0, s_1\} \cup \phi \cup \{s_4\} \\ &= \{s_0, s_1, s_4\} \end{aligned}$$

and $\{s_0, s_1, s_4\} \cap \{s_2, s_4\} \neq \phi$

9.

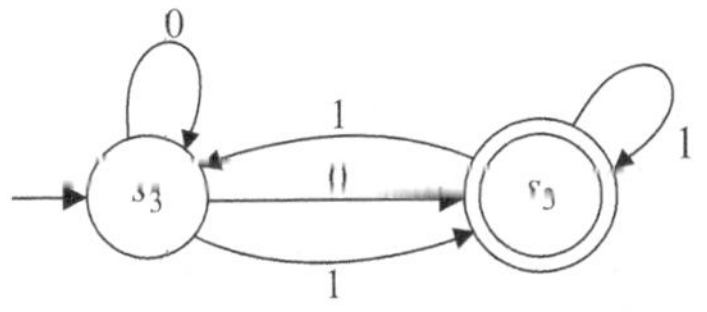

M

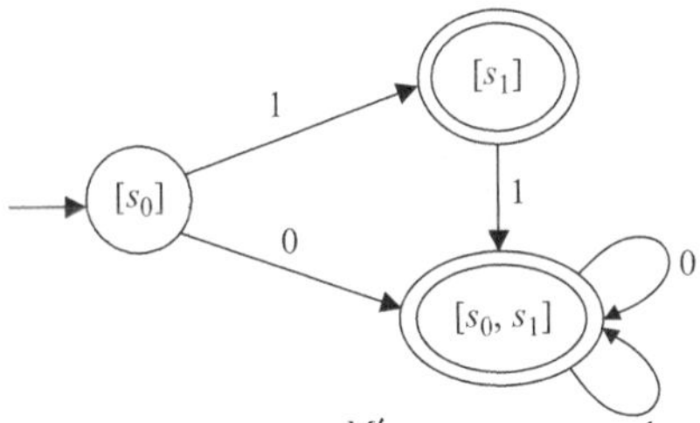

M'

10. Let M be the non-deterministic finite state automation and M' be the equivalent deterministic finite state automaton. Then the transition diagrams are:

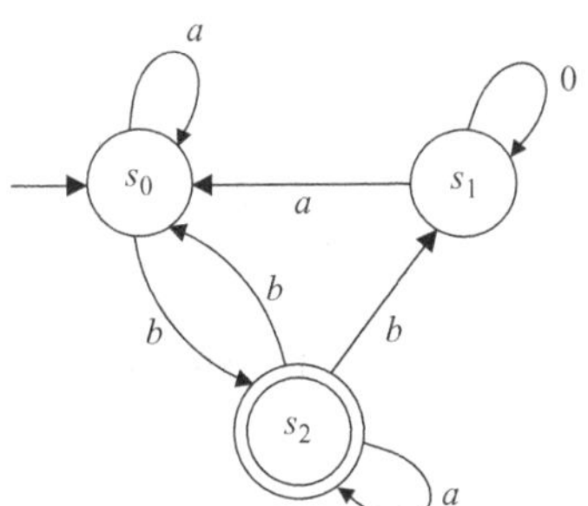

M

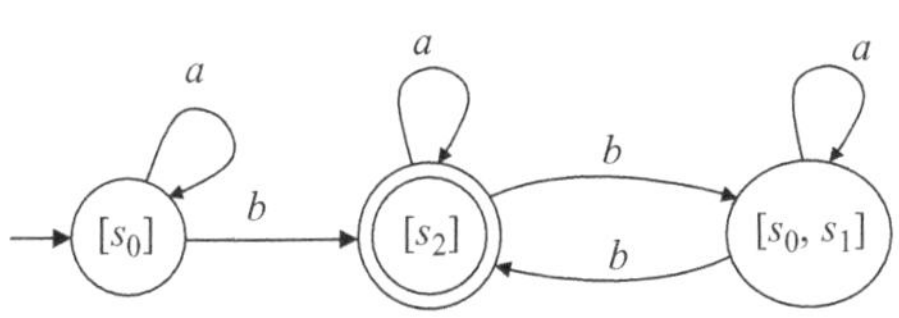

M'

(The states $[s_1]$, $[s_0, s_2]$, $[s_1, s_2]$ and $[s_0, s_1, s_2]$ are never entered and hence have been omitted from the transition diagram of M')

11. The required Moore machine (finite state automaton) is

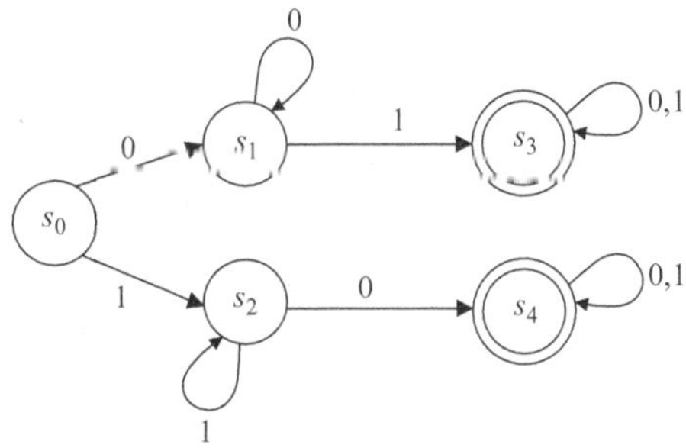

12. The output is residue modulo 3. Thus $O = \{0, 1, 2\}$. In Moore machine the output depends on the present state of the machine and not on the input. The state s_j is entered if and only if the residue is j.

Thus the Moore machine is

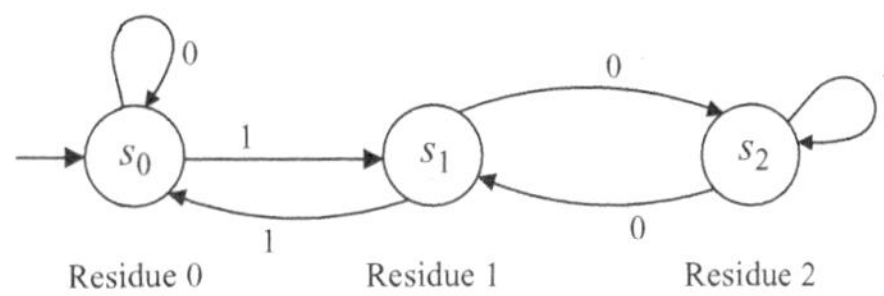

13. Let $M_1 = (I, S, O, s_0, f, g)$ be the Moore machine of Exercise 11. Its transition table is

	f		g
S \ I	0	1	
s_0	s_1	s_2	0
s_1	s_1	s_3	0
s_2	s_4	s_2	0
s_3	s_3	s_3	1
s_4	s_4	s_4	1

Let $M_2 = \{I, S, O, f, g')$ be equivalent Mealy machine, where

$$g'(r, a) = g(f(s, a)),\ s \in S,\ a \in \mathrm{I}$$

Using this we get

$$g'(s_0, 0) = g(f(s_0, 0) = g(s_1) = 0,$$
$$g'(s_0, 1) = g(f(s_0, 1)) = g(s_2) = 0$$
$$g'(s_1, 0) = g(f(s_1, 0)) = g(s_1) = 0$$
$$g'(s_1, 1) = g(f(s_1, 1)) = g(s_3) = 1$$
$$g'(s_2, 0) = g(f(s_2, 0)) = g(s_4) = 1$$

$$g'(s_2, 1) = g(f(s_2, 1) = g(s_2) = 0$$
$$g'(s_3, 0) = g(f(s_3, 0)) = g(s_3) = 1$$
$$g'(s_3, 1) = g(f(s_3, 1)) = g(s_3) = 1$$
$$g'(s_4, 0) = g(f(s_4, 0)) = g(s_4) = 1$$
$$g'(s_4, 1) = g(f(s_4, 1)) = g(s_4) = 1$$

Hence the equivalent Mealy machine is

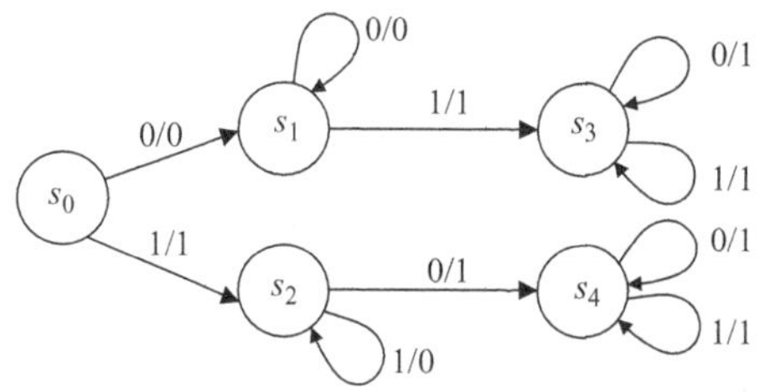

14.

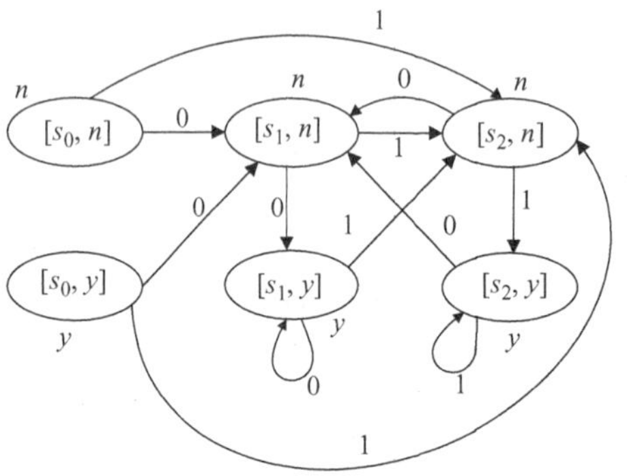

The state $[s_0, y]$ is never entered and therefore may be omitted.

The outputs are n, n, y, n, y.

CHAPTER 10

1. The language L consists of words w = *abbb*..., that is, the words beginning with *a* and followed by one or more *b*. Thus the regular expression r is given by $r = abb^*$.

2. $L(r)$ consists of concatenation of any word in $\{a, b, c\}$ with *bbb*. Thus $L(r)$ consists of all strings ending in b^3.

3. The given regular expression represent the infinite expression set $\{\ldots, aa, ba, aaaa, baba, baaa, \ldots\}$

4.
$$\sigma \Rightarrow a\,AB$$
$$\Rightarrow aa\,AB\,AB$$
$$\Rightarrow aa\,AA\,BB$$
$$\Rightarrow aab\,ABB$$
$$\Rightarrow aabbBB$$
$$\Rightarrow aab\,bcB$$
$$\Rightarrow aa\,bb\,cc$$

7. $G(N, T, s, P)$, where $N = \{s_0, s_1, s_2\}$, $T = \{0, 1\}$, s_0 is starting symbol. The productions are $s_0 \to 1\,s_0$, $s_0 \to 0\,s_1$, $s_1 \to 0\,s_1$, $s_1 \to 1\,s_2$, $s_2 \to 0\,s_2$, $s_2 \to 1\,s_2$, $s_2 \to \lambda$ (since s_2 is accepting state).

Further,
$$s_0 \Rightarrow 1\,s_0 \Rightarrow 10\,s_1 \Rightarrow 1\,00\,s_1 \Rightarrow 1001\,s_2$$
$$\Rightarrow 10010\,s_2$$
$$\Rightarrow 100101\,s_2$$
$$\Rightarrow 100101$$

Thus $L(G)$ consists of strings of the form 100101 which have same number of 0's and 1's. The strings are same which are accepted by the given finite state automaton.

8. $\sigma \Rightarrow a\,\sigma\,b \Rightarrow aa\,bb$

or

$$\sigma \Rightarrow a\,\sigma\,b \Rightarrow aa\,\sigma\,bb \Rightarrow aaa\,bbb$$

Thus,
$$L(G) = \{a^n b^n, n > 0\}.$$

Index

S

T

U

V

W